DIESEL FUEL SYSTEMS

DIESEL FUEL SYSTEMS

Robert N. Brady

Department Head,
Diesel Mechanic/Technician Program
Vancouver Vocational Institute
Vancouver, British Columbia

RESTON PUBLISHING COMPANY, INC.
A Prentice-Hall Company
Reston, Virginia

Library of Congress Cataloging in Publication Data

Brady, Robert N
Diesel fuel systems.

Includes index.
1. Diesel motor—Fuel systems. 2. Diesel motor—
Maintenance and repair. I. Title.
TJ797.B7 621.43′7 80-24440
ISBN 0-8359-1293-0
ISBN 0-8359-1292-2 (pbk.)

Reston Publishing Company, Inc.
A Prentice-Hall Company
Reston, Virginia 22090

10 9 8 7 6 5 4 3 2 1

Printed in the United States of America

I would like to dedicate this, my first book, to my wife, Linda, and my two daughters, Alanna and Alicia, for all those free hundreds of hours that I spent wrapped up in writing this book.

I would also like to thank all of the mechanics whom I was fortunate enough to serve under during my apprenticeship many years ago. They instilled in me confidence, ability, and, most importantly, pride in workmanship, which is indeed the mark of a true craftsman. My humble thanks also to all those students, instructors, mechanics, and industry personnel who encouraged my efforts to produce this book.

CONTENTS

PREFACE

This book has been written for all those associated with the maintenance and repair of all types of diesel engines.

There are few fields today in which the diesel engine does not play a major part, either directly or indirectly. Recent commitment by a number of large worldwide automotive manufacturers is further testimony to the popularity of diesel power.

At this time, the diesel engine is being used in over 4500 applications worldwide, including logging, mining, industrial, marine, on- and off-highway trucking, oil field exploration, electrical power generation, farming, pipeline, public transportation, railways, road building, and too many more to name here. It has been suggested that if all diesel power were to grind to a halt, the everyday world as we know it would experience industrial chaos.

The diesel industry today offers a unique job experience for those dedicated to being skilled craftsmen. World travel, diversified work experience, and future promotion are available to those with the desire and enthusiasm to pursue a career in this field. With the ever increasing number of both engines and equipment, the present shortage of skilled diesel mechanics will become even more severe throughout the 1980's.

There are many books and manufacturers' service manuals available as reference material to both the mechanic and apprentice. The author, however, has written this book from the viewpoint of operation, tune-up, and troubleshooting in the hope that the information contained herein will prove to be a theoretical reference, as well as a useful tool in the everyday practical tune-up and troubleshooting of all of the major diesel engines presently in use.

Having been in the diesel business almost 22 years, 14 of these instructing, this book is a culmination of the requests of many people, both students and mechanics, for a comprehensive book on diesel fuel systems. I am sure that it is not possible, or intended, to duplicate the excellent manufacturers' own service manuals; however, suggestions have been received and constructive criticism duly noted from literally hundreds of personnel associated with the diesel industry.

To everyone, individually or as a group, and those companies who gave so freely of their time and efforts to assist me with current information, I remain indebted. Such personnel continue to keep the diesel industry as exciting and challenging as it is.

Robert N. Brady

chapter 1

History and Types of Fuel-Injection Systems

1-1

History of Fuel Injection

Through the years, many companies and individuals have worked hard to develop and improve fuel-injection systems capable of withstanding the requirements placed upon them. However, let us step back in time and consider some of the problems that faced Rudolf Diesel. In 1892 a patent was issued to him whereby he proposed an engine in which air would be compressed to such an extent that the resulting temperature would exceed by far the ignition temperature of the fuel. He calculated and anticipated that the fuel injection from top dead center (TDC) on would take place so gradually that combustion due to the descending piston and the expansion of the gas would occur without material pressure or temperature rise. Further expansion of the gas would continue after fuel injection ceased.

Diesel obtained the financial backing of Baron von Krupp and the giant Machinenfabrik Augsburg Nurnberg Company, and set out to build an engine that would burn coal dust, then a useless by-product, mountains of which were piled up in the Ruhr valley. The actual commercial development of self-ignition of fuel in an internal combustion engine was carried out by Dr. Lauster and the engineers of the M.A.N. Company in Germany in collaboration with Diesel from 1893 to 1898. The first experimental engine that was built in 1893 used coal dust and high-pressure air to blast this fuel into the combustion chamber. This engine is said to have exploded, and further experiments with coal dust as fuel were unsuccessful. A successful compression ignition engine, however, was developed; it used oil as fuel, and a number of manufacturers were licensed to build similar engines.

The original oil burning engine experiments used mechanical injection; however, the results obtained were unsatisfactory, owing for the most part to the crude injection equipment with large dead volume.

Diesel again resorted to using air blast injection, which served the dual purpose of providing atomization of the fuel and turbulence of the mixture. His tests with air injection proved so successful that this became the accepted method of injection for many years. Failure of his first engines were due to his attempt to compress air to a pressure of 1500 psi and the lack of any provision for cylinder cooling. Diesel attempted to utilize so much of the heat of combustion by a long power stroke that no further cooling would be required.

His third engine, built in 1895, was a success. It was a four-stroke cycle with a compression pressure of 450 psi, which is comparable to many present-day engines. It was water cooled, and fuel was injected by a blast of high-pressure air. It developed 24% brake thermal efficiency, which was actually 35% indicated, and was a great improvement over all previous engines. Much of the subsequent improvement and progress in diesel engine development has been largely dependent upon improvements in fuel-injection technology.

Robert Bosch was most probably the one individual who contributed most to the success of the Diesel engine as we know it today by producing the first mass-produced fuel-injection pumps for use as early as 1927. Robert Bosch spent some of his early years outside Germany, specifically with Edison in the United States and then with Siemens in Great Britain. He subsequently designed and produced a magneto and established the Bosch Magneto Company in 1906. This company started production in a factory in Springfield, Massachusetts, in 1909–1910, and by 1914 the U.S. output was higher than that in Germany. By 1914 there was also a branch in Japan.

On December 28, 1922, Bosch decided to embark upon the development of a fuel-injection system for diesel engines. There had been up to that time some diesel activity at Bosch; however, now there was commitment to intensive research and design. By 1923, a dozen pumps were tested, and by the end of 1924 a pump that could meet the requirements was ready. In 1925 Bosch signed an agreement with a developer, Franz Lang, to use a diesel engine system including its fuel-injection equipment developed by Lang. This particular pump did not meet the particular needs; therefore, further development led to the design of an acceptable unit by 1927, and in August of that year, 1000 pumps had been produced; by March 1934, 100,000 pumps had left the factory.

As early as 1927, however, a diesel engine was installed in a passenger car, which was a 2.1-liter Acro engine; it operated until 1929 and accumulated 35,000 kilometers. Then in 1932 a small truck diesel was put into a car for testing.

Bosch applied for a patent for the field of diesel injection equipment in 1926 and along the way acquired licenses or protective rights to the developments of Acro, REF-Apparatebau (established by the L'Orange company), Deutz, Schnurle, and Atlas Diesel, which all contributed to injection development in some part. Since then, over 1000 patents have been granted for fuel-injection designs.

With the increasing demand for fuel-injection equipment, Bosch in Germany was unable to keep up with the demand; therefore, other agreements were reached with companies that would subsequently produce injection equipment. In France, the Société des Ateliers de Construction Lavalette SA, which Bosch had already been a part owner of since 1928, was chosen as a partner. A manufacturing license was signed on October 10, 1931.

In England, a subsidiary of Joseph Lucas Ltd. had been producing injection pumps under license from REF-Apparatebau in limited quantity. Negotiations between Lucas and Bosch led in 1931 to a contract whereby Bosch would be a 49% participant in a Lucas subsidiary, C. A. Vandervell (bearings). One of the stipulations was that this company take on the name of CAV-Bosch, which would be supported by Bosch in the creation of production facilities for diesel equipment. This was dated from October 21, 1931.

In the United States, production facilities were established in Springfield, Massachusetts, under the name American Bosch. This contract was signed on January 1, 1934.

On August 10, 1938, a licensing agreement was signed with a group of Japanese, and the Diesel Kiki Company was founded, with which Bosch is still associated.

Today, licenses have been granted to manufacturing companies in Argentina, Australia, Brazil, India, Japan, Rumania, Spain, Turkey, and the United States; therefore, there are very few manufacturers of diesel fuel-injection equipment today who do not use some form of Bosch designs. When diesel injection equipment is talked about in any circle, the name Robert Bosch stands at the forefront of any discussion.

1-2

Current Types of Fuel Systems

Fuel systems now in use fall into one of the categories illustrated in Figures 1-1 through 1-5. In the system in Figure 1-1, each engine cylinder is served by an individual *jerk pump* that supplies fuel under high pressure to an injection nozzle. All these pumps are timed to the engine and are interconnected to a common speed control (throttle/governor). These units are found on large slow-speed high cylinder power output stationary and marine engines, although some small medium-rpm engines also use this setup.

Jerk Pump System. This system is used extensively by many of the current producers of fuel-injection pumps. It is shown in simplified form in Figure 1-2.

The pumping plunger used in this system

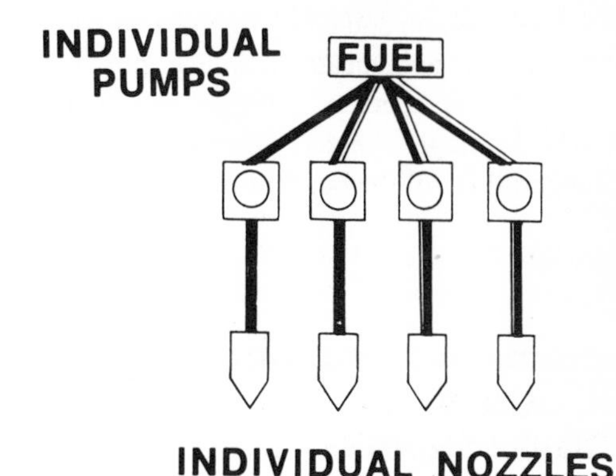

Figure 1-1
Typical Slow-Speed Engine Fuel System.
(Courtesy of Deere & Company)

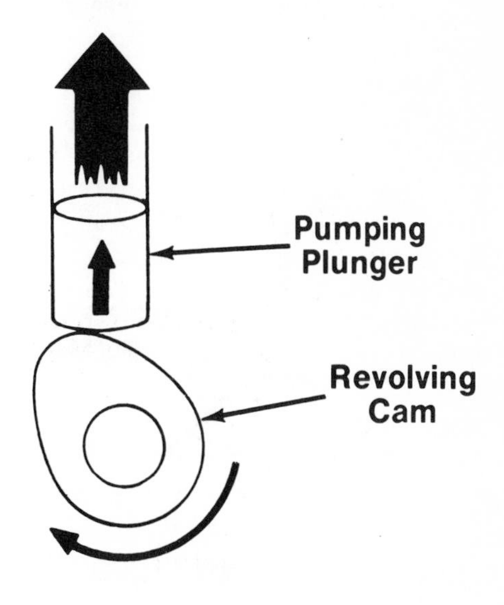

Figure 1-2
Basic Operation of a "Jerk Pump System."
(Courtesy of Deere & Company)

can be moved up and down by the action of the rotating cam underneath. It is also possible to rotate the plunger by throttle and governor action through linkage connected to the plunger; this will be covered in Chapters 2 and 4.

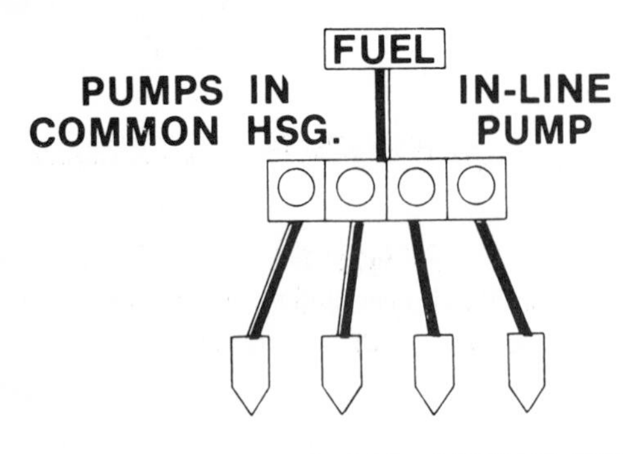

Figure 1-3
Medium- and High-Speed Diesel Injection Pump. Layout.
(Courtesy of Deere & Company)

The cam shape used with these systems basically controls the following:

1. The lift of the plunger and quantity along with the helix cut onto the plunger.
2. The basic timing of the plunger.
3. Rate of delivery.

The system in Figure 1-3 still operates on the jerk pump concept; however, all the pumps are contained within one common housing. Fuel is supplied to the injection pump housing at low pressure by a lift-type transfer pump, which is itself usually driven from the injection pump.

This type of system is more common to medium- and high-speed high power/displacement engines. Timing the pump to the engine automatically times all the other pumps within the housing.

The type of pump in Figure 1-4 uses the same basic principle of operation as that of an ignition distributor used on a gasoline engine. Again, as with the previous type, fuel is delivered to the injection pump at pressures as high as 130 to 200 psi (896.35 to 1379 kPa) by a transfer pump built into and driven from the

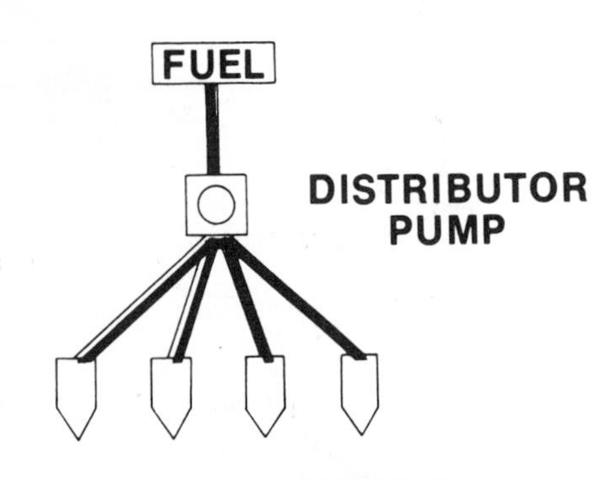

Figure 1-4
The Distributor Injection Pump.
(Courtesy of Deere & Company)

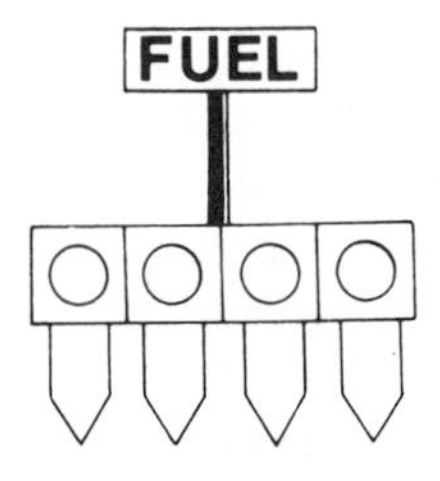

Figure 1-5
Unit Fuel Injector Concept.
(Courtesy of Deere & Company)

injection pump itself. A hydraulic head assembled to the nondrive end of the injection pump contains as many fuel line fittings as there are injection nozzles.

In the *unit injector* type of system in Figure 1-5, a small engine-driven transfer pump supplies fuel to an inlet manifold cast internally within the cylinder head or header (common rail) at between 60 and 70 psi (413.7 and 482.65 kPa). The unit injector then times, atomizes, meters, and pressurizes the fuel prior to combustion of the injected fuel. Each injector is timed when assembled and then adjusted in place with a timing pin and for rack travel.

In general, then, we will find one of these types of systems or a derivative thereof. Figure 1-6 shows the three major types of fuel systems being used today and the manufacturer or engine user.

Multiple Pump System

Fuel is drawn from the tank first and usually passes through a primary filter or strainer before entering the fuel transfer pump. Fuel under pressure at between 10 and 35 psi (68.95 and 241.325 kPa), depending on the manufacturer, leaving the pump is forced through a secondary filter and then on into the injection pump housing, where all individual pump units are subjected to this fuel oil. Many of the newer pumps do not use any lube oil in their base to lubricate the rotating camshaft, but have the internal area of the injection pump filled with fuel oil at all times. Older pumps and some current ones have a separate lube oil compartment for the injection pump camshaft. Fuel at each individual jerk pump within the common housing is timed, metered, and pressurized, and then delivered through a high-pressure steel-backed line to each injector nozzle in firing order sequence at pressures ranging anywhere from a low of 100 atm (atmospheres) (1470 psi) to as high as 300 atm (4410 psi). Some nozzles have a leak-off connection line on them to allow some fuel that is used for internal lubrication of the nozzle to return to the fuel tank.

The *multiple pump system* is presently commonly used by the following manufacturers of injection equipment and end users (engines).

1. American Bosch (AMBAC Industries, Inc.). Typical engine users of this type of pump are Mack, International Harvester, Leyland, and Onan.
2. Robert Bosch; typical engine users can be found in Chapter 4.
3. Diesel Kiki Co. Ltd. and Nippondenso Co. Ltd. in Japan, who build under license from Robert Bosch.
4. CAV Ltd. and their licensees, such as Roto Diesel in France, Condiesel and Simsa in Spain, Lucas do Brazil, Inyec Diesel in Mexico, and Nihon-CAV in Japan. The list of engine manufacturers using this fuel-injection equipment is also long and varied, with some of the more common ones being Perkins, Rolls-Royce, Leyland, Ford, Lister, Paxman, Petters, and Ruston.
5. L'Orange of Germany.
6. SIGMA of France.
7. Caterpillar Tractor Company, which uses its own system and also a sleeve metering fuel system (see Chapter 7).

Pressure Time System and Distributor Pump

The *pressure time* fuel system shown in Figure 1-6 is a variation of a distributor pump designed and built by the Cummins Engine Company for use on their own line of diesel engines. The fuel-injection pump has its own built-in fuel transfer pump (gear type), with one fuel line feeding the injectors through a *common rail* type of setup; refer to Chapter 9 for further information.

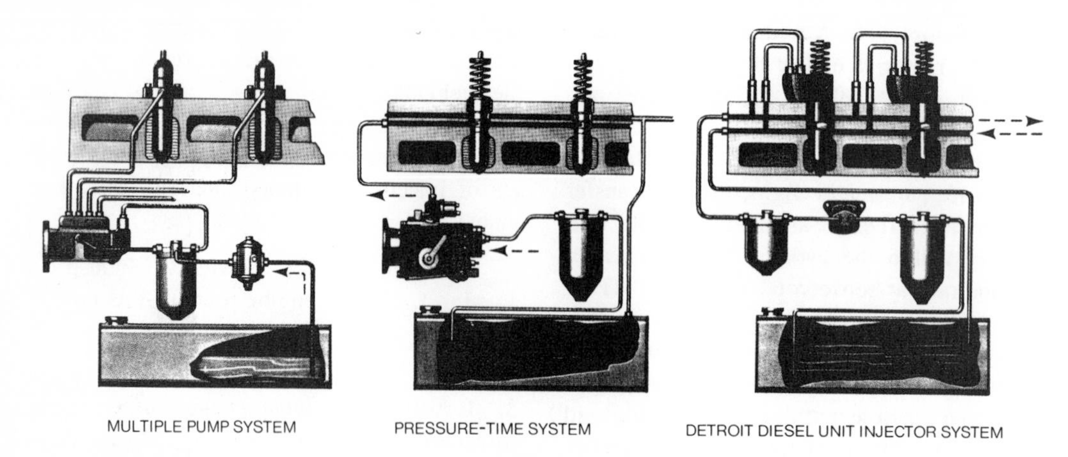

Figure 1-6
Types of Fuel Systems.
(Courtesy of Detroit Diesel Allison, Division of GMC)

The distributor-type pump, as with the Cummins pump, draws fuel from the tank through a primary filter, some to a lift-type supply pump, through a secondary filter and on into a vane-style transfer pump mounted in the end plate of the injection pump (opposite the drive end), where it is delivered to a charging passage inside the injection pump at pressures of around 130 psi (896.36 kPa) maximum. The fuel is then timed, metered, and sent under pressure to the individual injection nozzles from the hydraulic head area of the injection pump at pressures of from 150 atm (2205 psi) to as high as 260 atm (3822 psi) as a mean average. Each injection nozzle is directly connected to the pump by a high-pressure line. A fuel return or leak-back line is generally used with this system also.

Distributor pumps are presently manufactured by American Bosch (AMBAC) Industries, Inc., Robert Bosch (Germany) and its licensees, CAV Ltd. (England) and its licensees, Roosa Master Stanadyne/Hartford Division (USA), and SIGMA (France). Such pumps are found on a wide range of diesel engines, such as British Leyland, Perkins, Peugeot, Volkswagen, Neuss, Volvo-Penta, Oldsmobile-GMC, John Deere, and International Harvester.

Unit Injector Fuel System

Fuel is drawn from the tank by an engine-driven pump (transfer) through a primary filter and sent out of the pump under pressure of between 60 and 70 psi (413.7 and 482.65 kPa) to a secondary filter, where it then enters an inlet manifold cast within the cylinder head (common rail) or header line, flows through individual fuel jumper lines, through a small stainless-steel wire filter within the injector inlet, and down into the unit injector itself. The injector now times, meters, pressurizes, and atomizes the fuel as it enters the combustion chamber. As much as 50% of the fuel delivered to the injector is used for cooling and lubrication purposes, and is routed out of the injector to a return manifold cast within the cylinder head (or line) and returned to the fuel tank. This system is commonly known as a recirculatory type of fuel system because of the large volume of returned fuel. It is, however, very simple in design, lightweight, compact, and easy to service.

It is presently used by Detroit Diesel Allison (GMC) in all of their engines, both 2 and 4 stroke, Bendix USA, and L'Orange of Germany. The Electro-Motive Division (EMD) of General Motors also uses the unit injector system.

Since Detroit Diesel and Electro-Motive engines are sold extensively worldwide and serviced through a large network of distributor/dealers, they are found in almost every application and type of equipment. The Bendix unit injectors are found in larger slow-speed industrial and marine engines and the L'Orange injectors in European applications.

Injection Nozzles

We will not delve into the details of injectors at this time; however, they are of two basic types, an inward-opening and an outward-opening style. The inward-opening nozzle is often referred to as a *closed nozzle* owing to the fact that it opens away from the combustion chamber during fuel injection and closes toward the combustion chamber when injection is complete.

The outward-opening nozzle is referred to as an *open nozzle* since it moves toward the combustion chamber during injection and away from it when injection is complete. However, either term is correct. Figure 1-7 shows these two typical types.

In addition to the type, nozzles can be of a style known as a *multihole* or *pintle*. Basically,

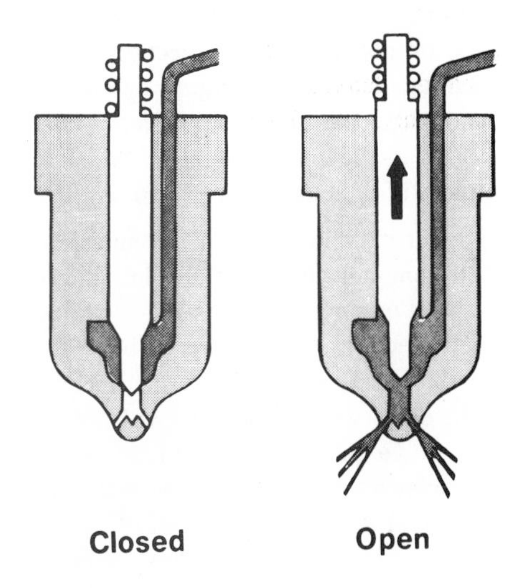

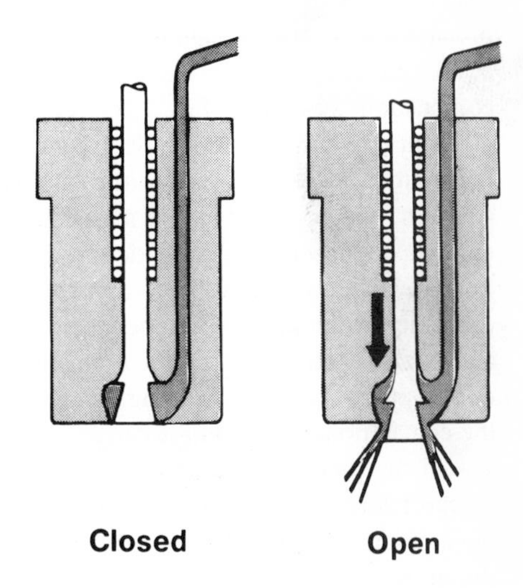

OUTWARD-OPENING NOZZLE

Figure 1-7
Types of Nozzles.
(Courtesy of Deere & Company)

the difference is that the former has more than one hole, and as many as eight, for example, drilled in the injector spray tip; this is generally an inward-closing nozzle and is found usually on open chamber or *direct injection* engines. The pintle nozzle employs one hole only, with the exception of the *pintaux* described later under Bosch fuel systems, which is found on precombustion chamber engines.

Fuel-Injection Pumps

You have probably heard the statement that "the fuel-injection system is the actual heart of the diesel engine." When you consider that indeed a high-speed diesel could not be developed until an adequate fuel-injection system was designed and produced, and that even Rudolf Diesel ran into problems basically associated with lack of a good injection system, then this statement takes on a much broader and stronger meaning.

From our previous discussion related to combustion systems, you will recollect that efficient combustion is dependent upon the fuel being injected at the proper time and rate. In addition, the injection pressure must be high enough for adequate atomization and penetration. Involved in this is the compressibility and dynamics of the fuel column between the pump and nozzle, plus the mechanical characteristics of the pump, discharge tubing, and nozzle of the conventional jerk pump system.

In the preceding section, various methods of mechanical injection and metering control were described. Many of these have since been discarded; others have been improved. There have been many important developments in pumps, nozzles, and unit injectors for high-speed diesel engines.

Prior to delving into the specifics of individual injection pumps, let us consider what the actual demands and functions of a good injection system are.

1-3

Functions of a Fuel-Injection System

The requirements of a fuel-injection system can be summarized as follows:

1. In order to receive equal power from all cylinders, the amount of fuel injected must remain constant from cycle to cycle, and obviously from cylinder to cylinder. A smooth-running engine is dependent on even fuel distribution to each cylinder throughout the speed range; otherwise, cylinder balance will be upset and some cylinders will be working harder than others. Overloading and overheating would result. This function is commonly referred to as *metering*.
2. As engine load and speed vary owing to application and operating conditions, the point of actual injection for a given load and speed will vary with this condition. Therefore, the injection system has to adjust the timing or point of injection to the fluctuating demands of engine operation. In summary, the injection system must *inject fuel* at the correct point in the cycle regardless of the engine *speed* and *load*.
3. In (direct-injection) open combustion chamber engines and especially in modern high-speed engines, a slow start or ending of the injection period affects both the initial and final portions of the injected fuel. In other words, the fuel will not be broken down or atomized into as fine a fuel droplet as it would be, for example, with a rapid start and rapid cutoff at the end of the injection period. Thus *injection must begin and end very quickly*.
4. Since fuel is compressible, there is a time lag between the actual beginning of delivery by the pump and the actual beginning of discharge from the nozzle; also the rate of delivery from the pump is not identical with the rate of discharge from the nozzle.

Therefore, by controlling the rate at which fuel is injected the performance of many engines can be improved. One of the most important characteristics is the spray duration, particularly at full load, since it directly affects engine power, fuel consumption, and exhaust smoke. In some engines a small amount of fuel is injected 8 to 10 degrees ahead of the main injection charge so that it is already burning when the main injection process occurs. This produces smoother combustion and a relatively slow rate of pressure rise in the cylinder. The type of nozzle used can to some extent control the actual rate of injection. In summation, the injection system must inject fuel at a rate necessary to control combustion and the rate of pressure rise *during combustion*.

5. Good combustion is related to the degree of fuel atomization; therefore, the type of combustion chamber and engine speed affect these requirements. The type and size of nozzle plus the injection pressure will control the degree of atomization. The injection system must then *atomize the fuel charge* as required by the particular type of combustion chamber in use.

6. The volumetric efficiency of an engine generally decreases with an increase in speed, because of increasing resistance to air flow and inertia of air in the actual intake system. Therefore, the *power* developed by the engine and the completeness of *combustion* are really dependent on air flow and the uniformity of fuel distribution throughout the air charge within the combustion chamber. In direct-injection or open-chamber-type engines, the fuel must penetrate into the air mass in all directions within the combustion chamber. In smaller high-speed engines, adequate penetration is also required; however, it is very undesirable to allow the fuel spray to strike either the piston crown or cylinder wall. Burning of the crown from direct flame impingement or cylinder wall wash, causing lube oil dilution, crankcase oil dilution, possible piston to liner scuffing, and eventual seizure, could result. Incomplete combustion would be a further result, creating carbon deposits and ring sticking. In summation, then, the injection system must *distribute fuel evenly throughout the air mass in the combustion chamber*.

1-4

High-Pressure In-Line Fuel-Injection Pumps

The *in-line* style of jerk pump shown in Figure 1-8 is used extensively by the major manufacturers of fuel-injection equipment worldwide. In addition to Robert Bosch, Lucas CAV, American Bosch, Caterpillar, and licensees of these companies use this basic concept, although with some variation in general design features and available options. Although this type of injection pump is used extensively, the rotary or distributor-type pump is becoming increasingly popular on smaller high-speed automotive-type engines. The major difference on Caterpillar's type of jerk pump is that it is a nonadjustable unit (see Chapter 7).

Plunger Operation

In Figure 1-9(a), with the plunger at the bottom of its stroke, both ports are uncovered by the top of the plunger, and the pump reservoir is full of fuel that will enter through both ports. The fuel delivery valve is closed at this time.

In Figure 1-9(b), as the plunger is starting its upward travel (by rotation of the injection pump camshaft), fuel will spill back out both ports into the reservoir until they are covered by the top land area of the plunger. This is termed *port closing* and is the basic start to injection. Note that the delivery valve is still on its seat.

In Figure 1-9(c), the fuel pressure will continue to rise until it is high enough to lift or force the discharge valve off its seat, therefore allowing the displaced fuel to pass through the fuel line to the nozzle and on into the cylinder, with the delivery valve open. Injection will continue until the lower helical land uncovers the control port. This port is uncovered slightly ahead of the actual ending of the upward-moving plunger, which will displace the remaining fuel back to the reservoir.

In Figure 1-9(d), the displaced fuel is allowed to escape down the relief area of the plunger and out the control port to the reservoir. The discharge valve closes, and the plunger completes its stroke and is positively returned to the next intake stroke [Figure 1-9(a)] by the plunger return spring.

At the end of injection when the control port is uncovered, the high pressure in the pump chamber flashes back into the reservoir; therefore, to prevent eventual erosion of the pump housing, this pressure is deflected by the hardened end of the barrel locating screw. Two

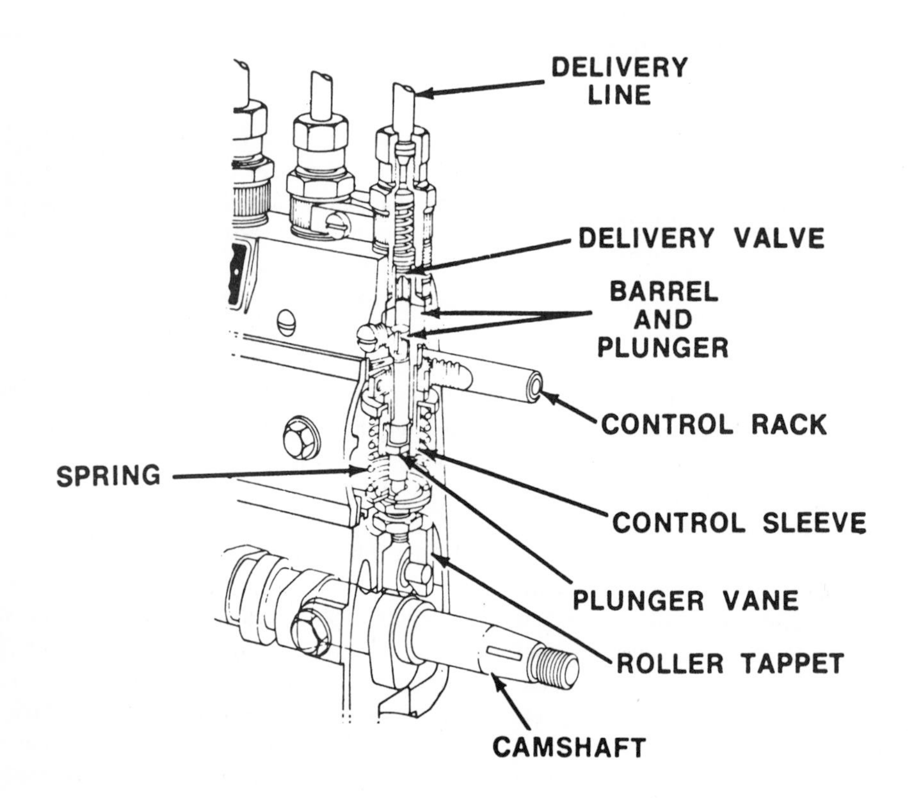

Figure 1-8
Typical In-Line High-Pressure Fuel-Injection Jerk-Type Pump.
(Courtesy of Robert Bosch, GmbH)

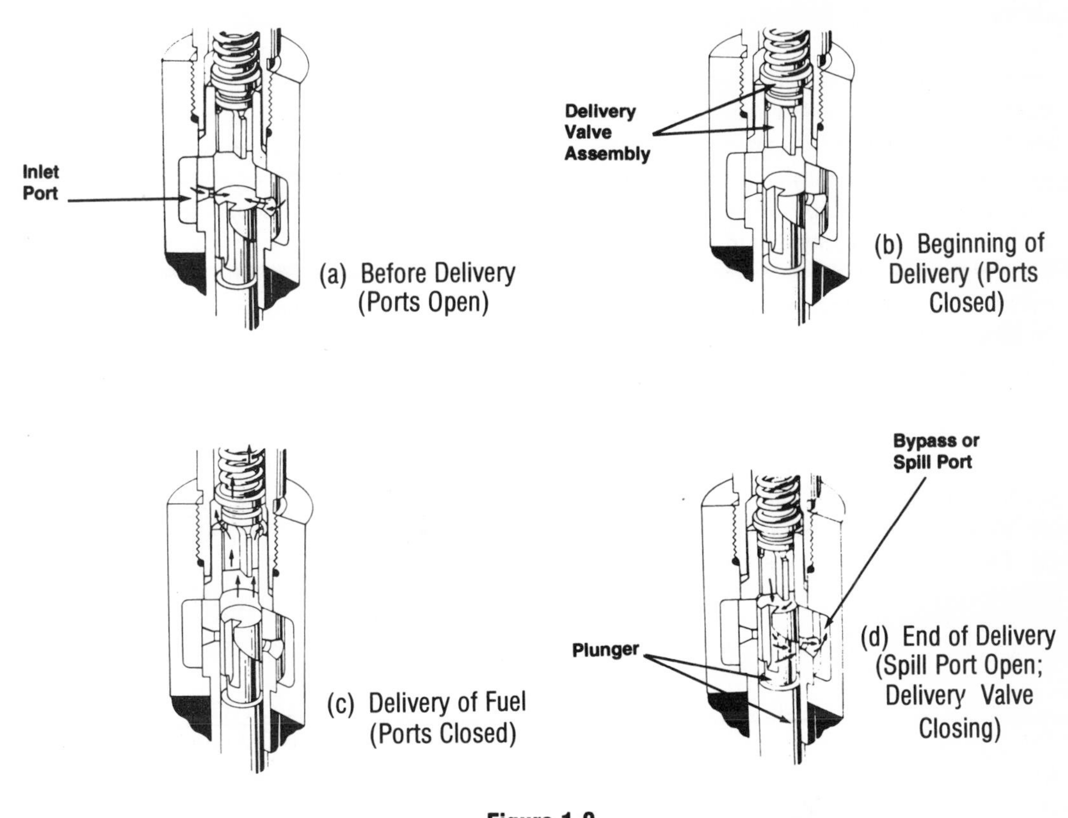

Figure 1-9
Plunger Operation: By providing a helical groove or land on the plunger and arranging to rotate it, the effective plunger stroke can be varied to control the quantity of fuel delivered per stroke. (Courtesy of Robert Bosch, GmbH)

ports will cut deflection pressures in half, while the conical port will disperse the spill deflection.

Metering Principle

The amount or volume of the fuel charge is regulated by rotating the plunger in the barrel as shown in Figure 1-10 to effectively alter the relationship of the control port and the control helix on the plunger. This is done by means of a rack and a control collar or control sleeve.

The *rack* is basically a rod with teeth on one side, which is supported and operates in bores in the housing. The rack is in turn connected to a governor. The geared segment or control collar is clamped to the top of the control sleeve with teeth that engage the rack. The control sleeve is a loose fit over the barrel and is slotted at the bottom to engage the wings on the plunger so that as the rack is moved it will cause rotation of the collar, sleeve, and plunger.

The operation of Robert Bosch in-line pumps is basically the same as that of CAV and American Bosch in-line pumps; however, let us quickly review the pumping plunger's operation and excess fuel device so that we thoroughly understand the principle.

The plunger within the barrel is moved up and down by the action of the rotating camshaft within the injection pump housing; it can also be rotated by the movement of the fuel control rack connected to the throttle and governor linkage. Anytime that the stop control is moved to the engine shutdown position, the plunger is rotated as shown in Figure 1-11, whereby the vertical slot machined in the plunger will always be in alignment with the supply or control port. Therefore, regardless of the plunger's vertical position within the barrel, fuel pressure can never exceed that deliv-

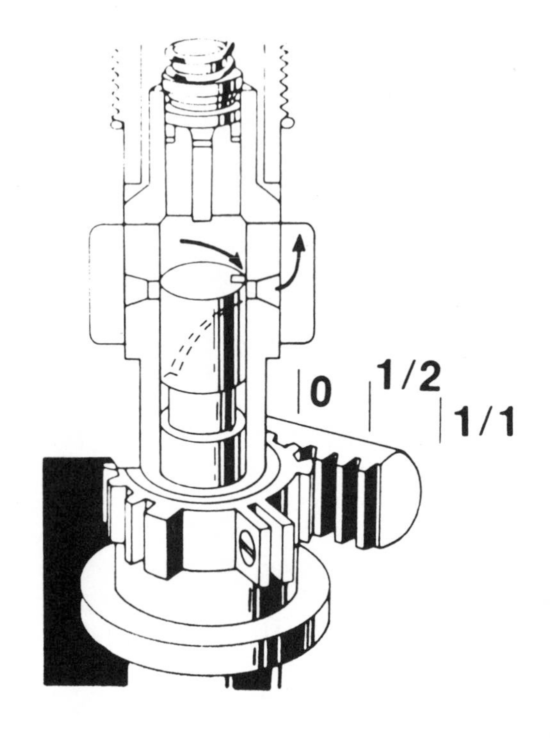

Figure 1-11
No Fuel Delivery.
(Courtesy of Robert Bosch, GmbH)

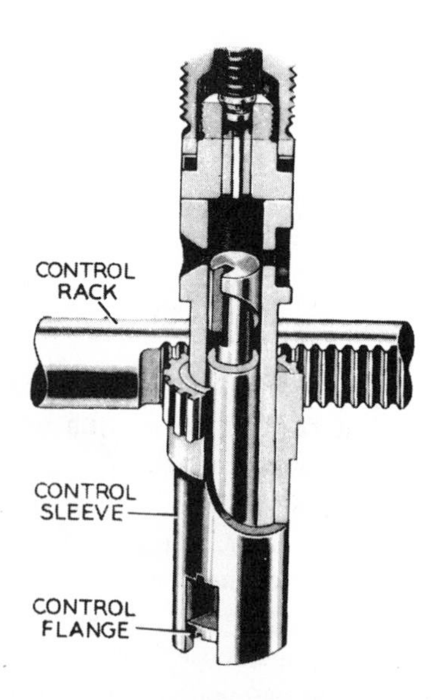

Figure 1-10
Regulating Fuel Charge.
(Courtesy of AMBAC Industries)

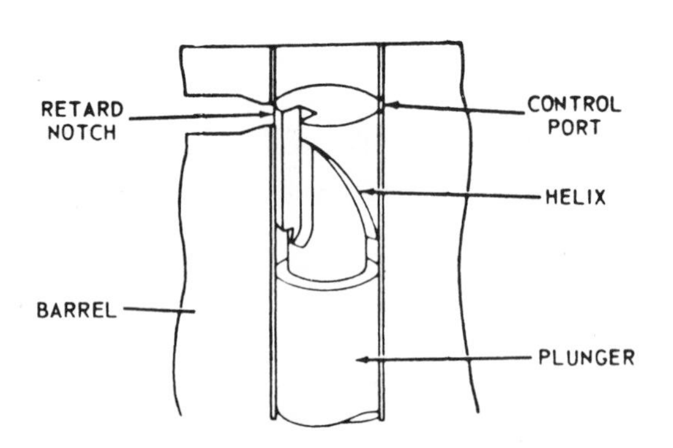

Figure 1-12
Excess Fuel Delivery and Retard Notch.
(Courtesy of Robert Bosch, GmbH)

ered by the fuel-transfer pump. This pressure will never be able to overcome the force of the delivery valve spring, so no fuel can be sent to the injectors.

Figure 1-12 shows the location of the *excess fuel delivery* and *retard* notch somewhat enlarged. Excess fuel is possible only during starting, since while the engine is stopped the speed control lever is moved to the *slow idle* position, thereby moving the fuel rack to place the plunger in such a position that excess fuel can be delivered. The instant the engine starts, however, the governor will move the fuel rack to a position corresponding to the position of the throttle lever. The retard notch, also in alignment with the control port, delays port closing and therefore retards timing during starting.

During any partial fuel delivery situation, the amount of fuel supplied to the injector will be in proportion to the *effective stroke* of the plunger, which simply means that the instant the supply port is covered by the upward-moving plunger, fuel will start to flow to the injector. This will continue as long as the control port is covered; however, as soon as the upward-moving plunger helix uncovers this port, fuel pressure to the injector is lost and injection ceases. Therefore, we only effectively deliver fuel to the injector as long as the control port is covered; this is shown in Figure 1-13(a) for any partial throttle position. This will vary in proportion to the throttle and rack position from idle to maximum fuel.

When the operator or driver moves the throttle to its maximum limit of travel, the effective stroke of the plunger, due to the rotation of the plunger helix, will allow greater fuel delivery because of the longer period that the control port is closed during the upward movement of the plunger by the pump camshaft. This is shown in Figure 1-13(b).

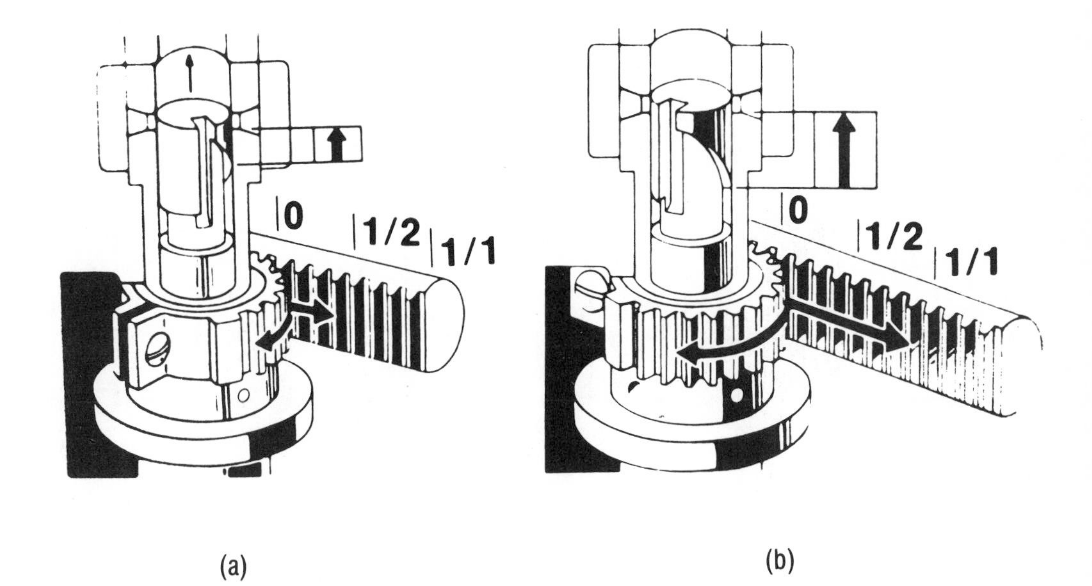

Figure 1-13
Fuel Delivery: (a) Partial; (b) Maximum.
(Courtesy of Robert Bosch, GmbH)

1-5

Helix Shapes and Delivery Valve

Plungers are manufactured with metering lands having lower or upper helixes (see Figure 1-14) or both to give constant port closing with a variable ending, variable port closing with a constant ending, or both a variable beginning and ending. With ported pumps, good control of injection characteristics is possible due to the minimum fuel volume that is under compression. However, a disadvantage of conventional port control pumps is the rising delivery characteristics as speed increases. This is caused by the fuel throttling process through the ports resulting in less fuel being bypassed before port closing and after port opening as the speed of the pump increases.

When the plunger is rotated so that the vertical slot on the plunger is in line with the control port (locating screw side), all the fuel will be bypassed; therefore, there will be no injection. With the rack in the full-fuel position, the plunger is able to complete almost its entire stroke before the helix will uncover the control port. Remember, as the plunger is rotated it will uncover the port earlier or later in the stroke (see Figure 1-15).

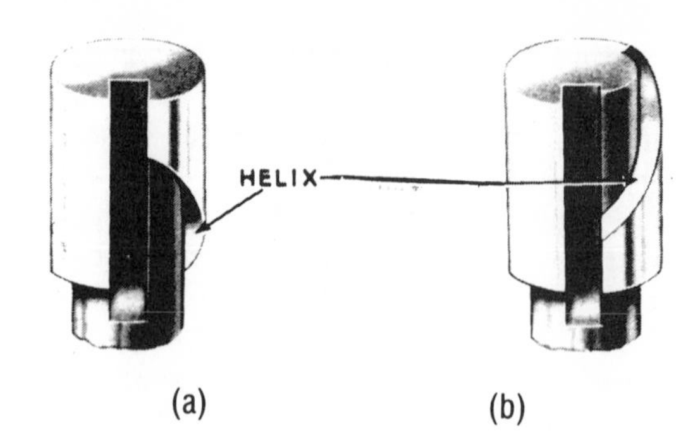

Figure 1-14
Plunger Types: (a) Lower Helix Plunger; (b) Upper Helix Plunger. (Courtesy of American Bosch—AMBAC Industries)

The standard basic type in use by American Bosch is the lower-right-hand helix where the start of injection is constant with regard to

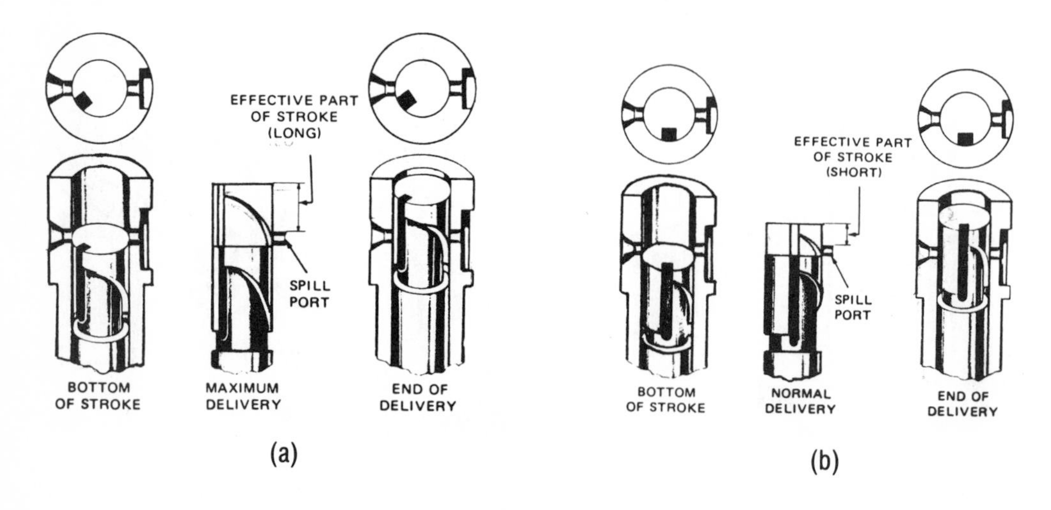

Figure 1-15
Plunger in Position for Fuel Delivery: (a) Maximum; (b) Normal. (Courtesy of American Bosch—AMBAC Industries)

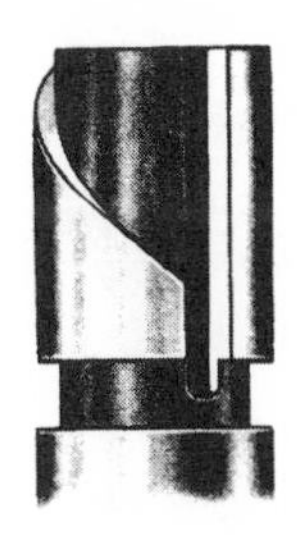

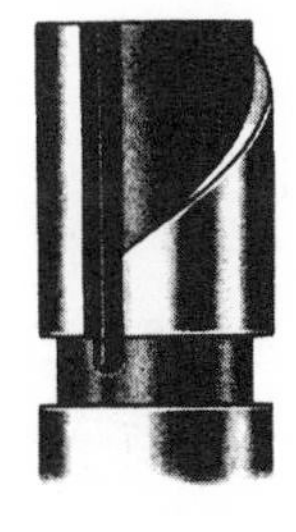

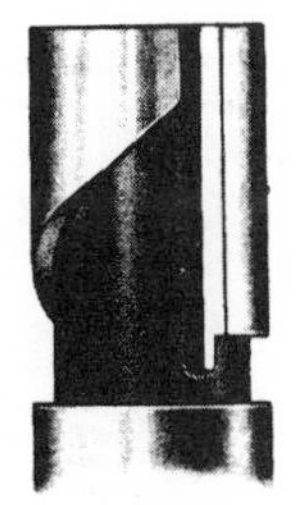

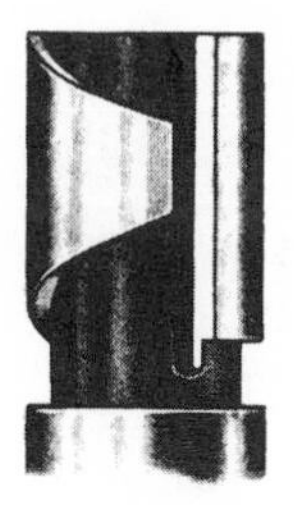

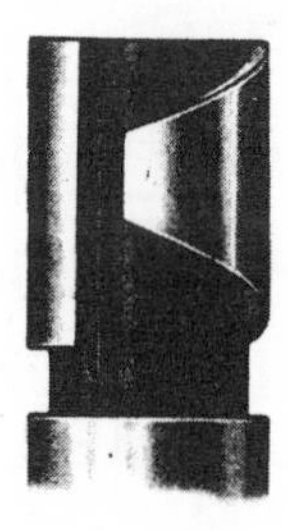

(a) (b) (c)

Figure 1-16
Helix Designs.
(Courtesy of American Bosch—AMBAC Industries)

timing; however, the ending is variable. In some applications it is advantageous to advance timing as the fuel rate is increased. This is achieved by the use of an upper helix, which gives a variable beginning and a constant ending. The helix may be cut on the left- or the right-hand side of the plunger. It does not alter the injection characteristic except that the rack must be moved in opposite directions to increase or decrease fuel. There are other special adaptions, such as a short, shallow helix on top to give a slight retarding effect to the injection timing on engines that operate in the idle range for extended periods, and a

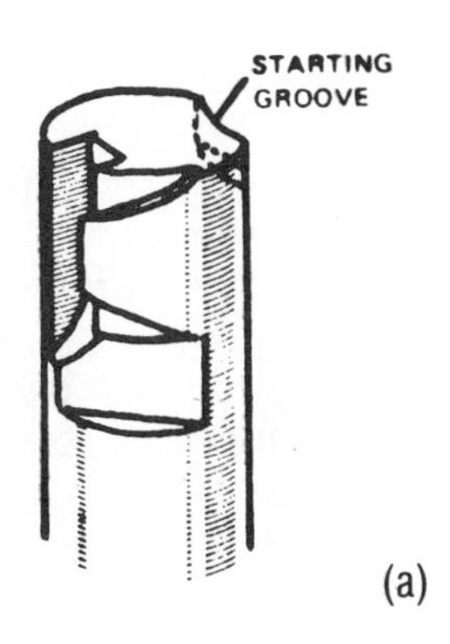

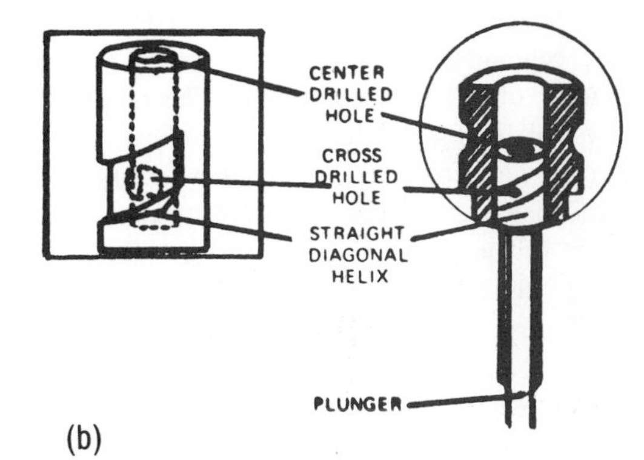

(a) (b)

Figure 1-17
(a) To assist the engine when starting by supplying more fuel and retarding delivery, a small starting groove is cut above the helix;
(b) a constant beginning and variable ending lower helix design with a center and cross-drilled hole, and also a diagonally machined groove.
(Courtesy of Robert Bosch, GmbH, and CAV Ltd.)

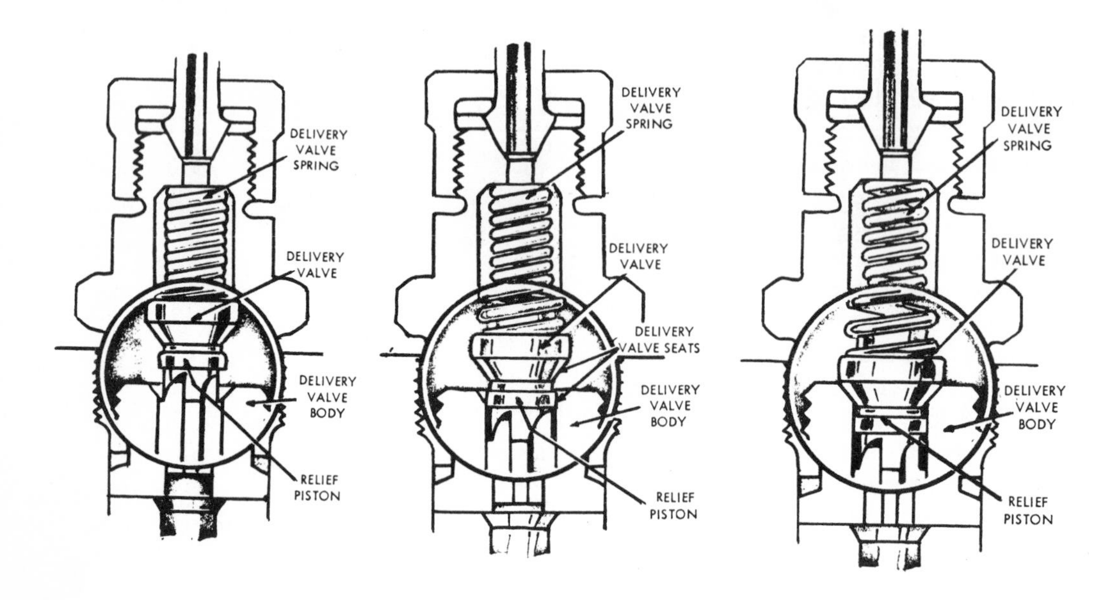

Figure 1-18
Delivery Valve Cycle.
(Courtesy of American Bosch—AMBAC Industries)

double helix used by some manufacturers to provide rapid response with minimum rack movement.

Figure 1-16(b) shows a lower helix design; the beginning of delivery is constant and the ending of delivery is variable. The reason for the helix being on opposite sides is that the one on the left would be employed when the governor is on the left or when the fuel rack is in front of the plunger. Figure 1-16(a) shows an upper helix design; the delivery has a variable beginning but a constant ending. Figure 1-16(c) shows plungers with both upper and lower helixes; both the beginning and ending of delivery are variable. Other features of a helix are shown in Figure 1-17.

Delivery Valve

The delivery valve, Figure 1-18, or what is sometimes referred to as a discharge valve, is specially designed to assist in providing a clean, positive end to injection. Below the valve face is a collar that is a precision fit in the valve bore. When pressure is created in the pump above the plunger by the closing of the ports, the valve must be raised far enough off its seat for the collar to clear the bore.

At the end of injection when pressure in the pump chamber is relieved by the opening of the control port, the valve drops down on its seat assisted by spring pressure. A volume of fuel equal to the displacement volume of the valve is added to the line and nozzle, reducing this pressure and allowing the nozzle valve to snap shut without the cushioning effect of pressure retained in the line and nozzle, such as with the closing of an ordinary valve. This is commonly called *line retraction,* which lessens the possibility of secondary injection or after-dribble at the spray nozzle. It is accomplished by an antidribble collar (accurately fitted relief or displacement piston) located at the upper end of the valve stem just below the seat.

chapter 2
American Bosch Fuel Systems

Over 40 years ago, American Bosch began assembling fuel-injection pumps at Springfield, Massachusetts, and their leadership in fuel-injection technology has been a significant factor in diesel power industries since that time. Figure 2-1 shows standard American Bosch pumps. Service facilities and parts are provided through a network of more than 400 authorized service agencies in the United States and Canada, as well as 110 service agencies in 59 other countries.

American Bosch leadership in fuel-injection technology has been a significant factor in diesel power industries since the first injection pumps were assembled at Springfield, Massachusetts, over 40 years ago. In 1950 their introduction of the first commercially practical single-plunger distributor series (PS) pump opened up new areas of opportunity for diesel power, especially in the farm tractor field. Improvements and refinements to this basic design over the intervening years have resulted in even more compact, simplified distributor pumps for application on modern high-speed automotive diesels, farm, and construction equipment.

American Bosch APE pumps are of the accepted standard design used on multicylinder engines the world over. They are readily adaptable to high-speed diesels and are of in-line construction with as many pumping elements as there are fuel discharge outlets. The lower pump compartment contains the camshaft and serves as a reservoir for the lubricating oil. It also contains the end plates with the camshaft roller bearings and oil seals, adjustable tappet assemblies, and closing plugs. Located in the middle section are the plunger and barrel assemblies, control sleeves with adjustable gear segments, plunger return springs, spring seats, and the fuel control rack. The upper portion of the housing contains the fuel sump, delivery valve assemblies, and union nuts for connection to the injection tubing.

Control of fuel quantity per stroke is accomplished by the longitudinal movement of

Figure 2-1
Pumps Manufactured by American Bosch. (Courtesy of American Bosch—AMBAC Industries)

the fuel control rack, which is connected to each plunger through the gear segment and control sleeve. The control rack projects through the end of the pump and is linked to a mechanical centrifugal governor, which is normally mounted on the end of the pump housing. An SPA self-regulating plunger-type fuel supply pump, driven from the pump camshaft, is mounted on the pump housing.

Newest in the line of APE series pumps, the APE 8VG injection pump of V8 configuration is a major redesign of the proven APE 8VBB pump, which has been in service since 1968. The VG design reflects a wide range of improvements drawn from the performance and service experience of current 8VBB equipment. The 8VG pump incorporates improved seals and gaskets to provide leak-free operation, a strengthened camshaft for higher injection pressures, redesigned pumping elements for improved engine performance, and a redesigned governor with added travel for longer service life and precision engine operation to meet today's exhaust emissions requirements.

An optional feature available for APE 6BB, APE 8VBB, and APE 8VG pumps is a dual-speed governor for highway diesel truck applications to control the top speed within national highway limits. This governor incorporates an air piston design operated by air from the trucks own air brake tank through the transmission air splitter valve. It restricts the engine speed between 1800 and 2000 in fifth gear or highway cruising speed, depending on the particular model and make of transmission; there is no reduction in normal engine speed in the lower-range transmission gears.

2-1

Single-Plunger Pumps

Many engines, both large and small, use one individual-type jerk pump per engine cylinder, as shown in Figure 2-2. These pumps of the single-plunger type are flange mounted directly over the engine camshaft, which has a fuel cam for each pump. The tappet assemblies are also part of the engine. Since each pump can

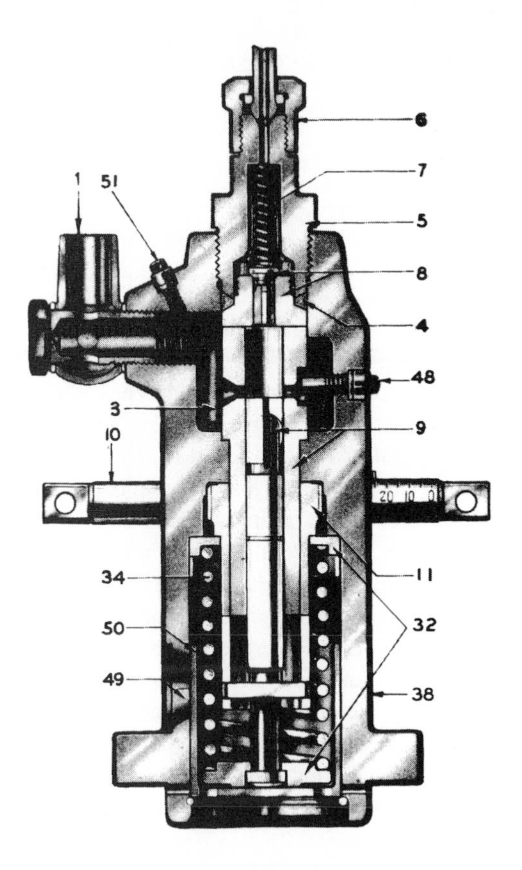

Figure 2-2
Cross Section of Type APF Fuel-Injection Pump. (Courtesy of American Bosch—AMBAC Industries)

Index:
1—Fuel inlet banjo
3—Fuel passage
48—Barrel locating screw
49—Timing inspection window
51—Fuel/air bleed screw
38—Housing
50—Plunger guide cup
9—Plunger and barrel assembly
11—Toothed control sleeve
34—Plunger return spring
32—Spring seats
10—Control rack
8—Delivery valve assembly
4—Gasket
7—Delivery valve spring
5—Delivery valve holder
6—Union nut
(The letters PF in the pump code name stand for an individual-type pump.)

be located adjacent to each engine cylinder, this has the advantage of using short fuel lines. Almost all these pumps today are of the port control type.

The pump shown in Figure 2-2 is produced in several sizes with plunger diameters of 5 to 35 mm and is suitable for cam lifts of 7 to 40 mm. The housing is made of rugged cast iron with an integral mounting flange, a guide cup for transmitting the motion of the tappet to the plunger, a closely fitted assembly of ported barrel and helical grooved plunger, a slotted control sleeve engaging lugs near the bottom of the plunger and meshing with the control rack, a plunger return spring and spring seats, a delivery valve assembly, a delivery valve spring, and a delivery valve holder. Some of these pumps are suitable for injection pressures as high as 15,000 psi. The barrel locating screws serve the additional function of absorbing the impact of the spilled high-pressure fuel.

Individual Jerk Pumps

Individual jerk pumps are found extensively on the larger diesel engines used in stationary and marine applications of slow- to medium-speed rpm ranges. Timing of the individual pumps to the engine is done in the same basic way regardless of the make; however, the following sequence is of a general nature and is not all-encompassing.

1. Refer to the flywheel markings on the engine to establish the base circle of the

camshaft for the particular pump being installed.

2. Place the pump unit onto its mounting base, and bolt it down.
3. Check plunger movement through the inspection window, as shown in Figure 2-3 (a), (b), and (c). With the proper flywheel timing mark aligned with the stationary pointer on the engine, the timing line on the pump plunger and inspection window should be as shown in Figure 2-3(c).
4. If the pump timing lines do not appear as in Figure 2-3(c), double-check to ensure that the engine flywheel marks correspond to the pump cylinder.
5. To correct the timing, some pumps employ a tappet adjusting screw to effectively raise or lower the plunger; however, some units require the use of selective shims under the pump base to correct this condition. Once adjusted, with the pump at the bottom and top of its stroke, the timing line on the plunger should stay in view, as shown in Figure 2-3(a) and (b), respectively.

2-2

APE Multiple-Plunger Pump

Now that we have studied the basic fuel flow path for this type of system, let us consider the actual injection pump itself (see Figure 2-4). All these pumps are of the ported type, with helix metering plungers available in various helix configurations and diameters from 5 to 13 mm. Camshafts are also available with several different cam shapes. The pump assembly shown is made up of an aluminum housing dividing into the camshaft compartment, plunger and spring compartment, and header section. The lower compartment contains the pump camshaft and also serves as a reservoir for lubricating oil. Later model APE pumps are lubricated by engine oil pressure. Pumps lubricated this way do not use a breather cap. APE pumps that are pressure

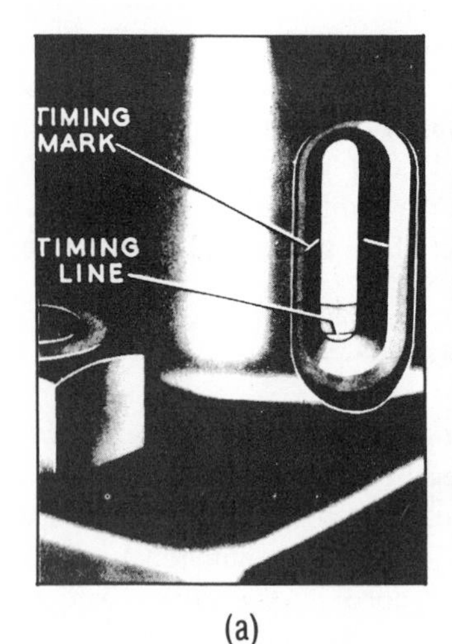

(a)

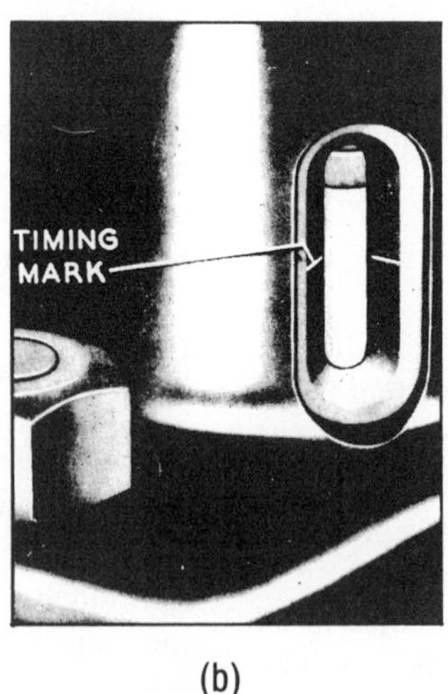

(b)

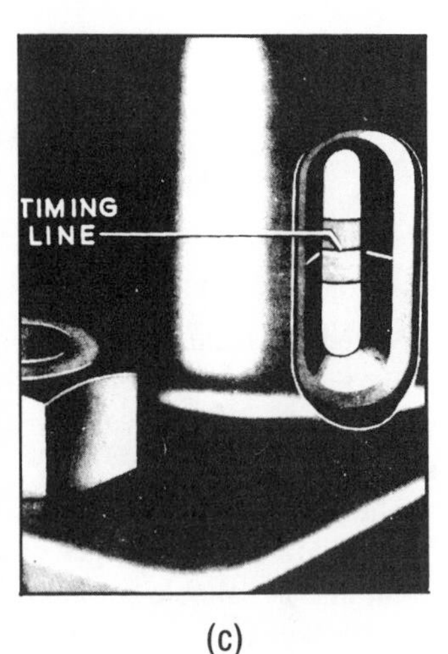

(c)

Figure 2-3
Plunger Movement.
(Courtesy of American Bosch—AMBAC Industries)

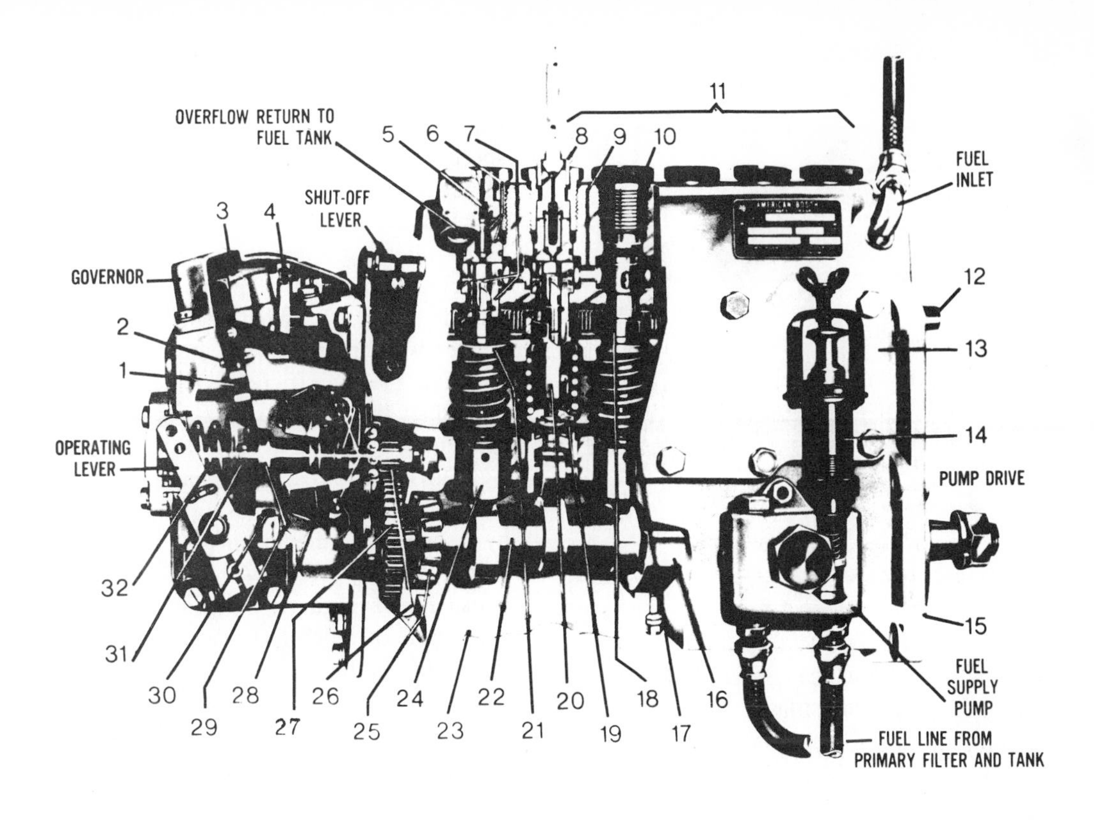

Figure 2-4
APE 6BB Sectionalized Pump.
(Courtesy of American Bosch—AMBAC Industries)

Index

1—Fulcrum lever assembly
2—Droop screw
3—Torque control cam
4—Adjustable stop plate
5—Delivery valve spring and holder
6—Delivery valve assembly
7—Plunger and barrel assembly (turned 90° from standard for illustration purposes)
8—Tubing union nut
9—Fuel sump
10—Retaining nut
11—Fuel discharge outlets (6)
12—Control rack
13—Inspection cover
14—Hand primer (optional)
15—Lubricating oil outlet
16—Camshaft center bearing
17—Plunger spring
18—Control sleeve gear segment
19—Lower spring seat
20—Control sleeve
21—Upper spring seat
22—Camshaft
23—Closing plug
24—Tappet assembly
25—Camshaft bearing
26—Driven gear, governor
27—Drive gear and friction clutch assembly
28—Flyweight assembly
29—Sleeve assembly
30—Speed adjusting screw
31—Fulcrum lever pivot pin
32—Inner/outer governor springs

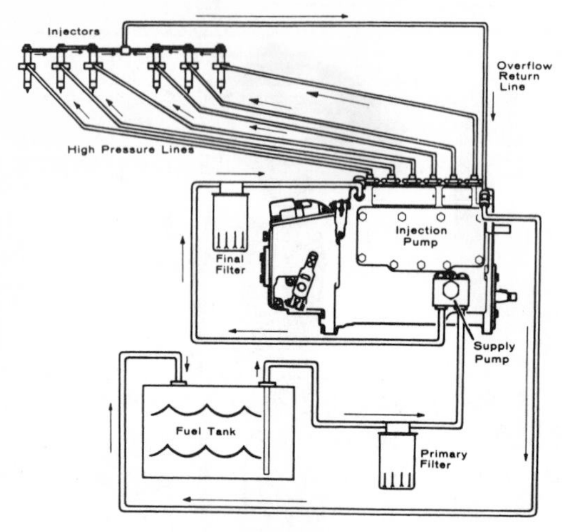

Figure 2-5
Typical Fuel Circuit.
(Courtesy of American Bosch—AMBAC Industries)

lubricated receive oil intermittently from the bore of the no. 6 tappet as it moves up and down, thereby permitting a quantity of lube oil to flow into the pump camshaft compartment with each rotation of the camshaft. Lube oil drains back to the engine crankcase through a drain hole at the drive end of the pump.

The end plates that support the camshaft contain ball bearings (bearing races separately mounted) and oil seals. Between the two lower compartments a partition contains the bores for the tappet or lifter assemblies. Timing adjustment for each plunger is provided by tappet screws on earlier model APE pumps; or by installing a thicker tappet spacer to decrease plunger lift to port closing, or to increase it, on newer model pumps. In addition to containing the plunger-barrel assemblies and return springs, the control sleeves with adjustable gear segments meshing with the control rack are also located in the middle section. The upper part of the housing contains the fuel supply sump, retracting delivery valves and springs, and discharge outlets. These pumps may be driven at half engine speed or engine speed for 4-cycle and 2-cycle engines, respectively.

Both oil level and drain plugs are provided with oil level dipsticks supplied on some models. Below each pump unit is a base plug threaded into the casing sealed with either a copper washer or beveled joint. On some of the later models, friction-type cups are used instead of the threaded plugs, with each plug containing a felt pad to act as a cam wiper.

The barrel of each pumping unit fits into a bore of the aluminum housing and is supported by a shoulder in the housing. It is located by a barrel locating screw, which threads into the housing and engages an oval recess in the upper part of the barrel. The locater screw is sealed against fuel leakage by a copper or fiber washer. The barrel is held in place also by the fuel discharge valve housing and the fuel stud, which threads into the top of the aluminum injection pump housing. The joint between the top of the barrel is a lapped or precision fit to the fuel discharge valve housing; however, there is also a copper or copper and fiber washer used between the fuel stud and discharge valve housing.

There are two ports in the upper part of the barrel that are open to the fuel reservoir, and they are diametrically opposite, with one port being open to the locater screw recess, and the other port having a conical outer end. Although both ports serve as intake ports, the one that is located beside the locater screw recess serves as the actual control port.

Control of fuel quantity per stroke is accomplished by the longitudinal movement of the fuel control rack, which is connected to each plunger through the gear segment and control sleeve. The control rack projects through the end of the pump and is linked to a mechanical centrifugal governor, which is normally mounted on the end of the pump housing.

A typical fuel circuit is shown in Figure 2-5.

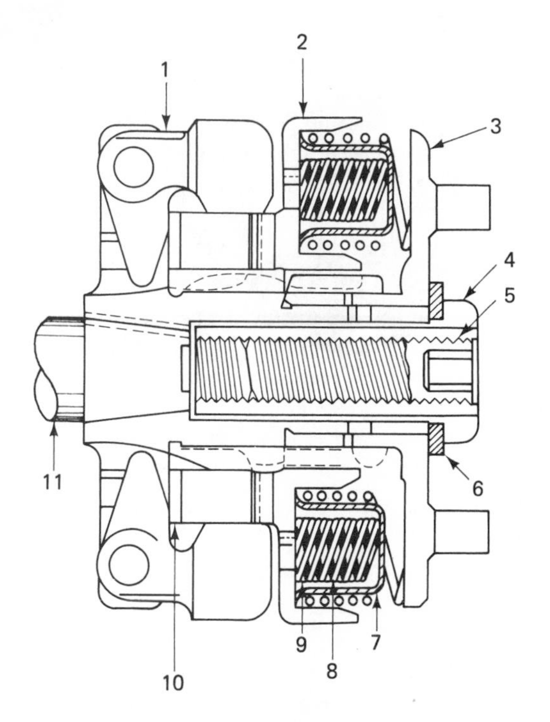

Figure 2-6
Synchrovance Unit.
(Courtesy of American Bosch—AMBAC Industries)

Index:
1—Synchrovance weight
2—Collar
3—Coupling
4—Camshaft nut
5—Lockscrew
6—Washer
7—Spring spacer
8—Inner spring
9—Outer spring
10—Thrust washer

Optional Synchrovance Timing Device for APE Pumps

An optional timing advance mechanism is available on some APE-type injection pumps (used by Mack and others). The synchrovance unit is enclosed by a housing mounted on the front of the injection pump (see Figure 2-6). The drive end of the housing corresponds with the mounting flange of the engine. The synchrovance unit is nonadjustable and provides the desired timing advance characteristic for the engine throughout its speed range to ensure and obtain optimum performance. The synchrovance assembly is used by Mack on naturally aspirated engines only.

As the engine turns the flyweight assembly (see Figure 2-7), centrifugal force tends to throw it outward. However, the preloading of the coil springs prevents this action until a preselected rpm is reached, at which point the force of the springs and the weights is in a state of balance. As engine speed increases centrifugal force continues to throw the flyweights outward, and the weight arms force the collar toward the coupling.

As a direct result of this movement, the injection pump camshaft is rotated slightly in advance of the coupling owing to the internal helical splines in the collar and the corresponding external splines on the spider and coupling. Therefore, the timing of the injection pump with relation to the TDC position of the engine piston is advanced within a given range. Maximum advance is 12 degrees.

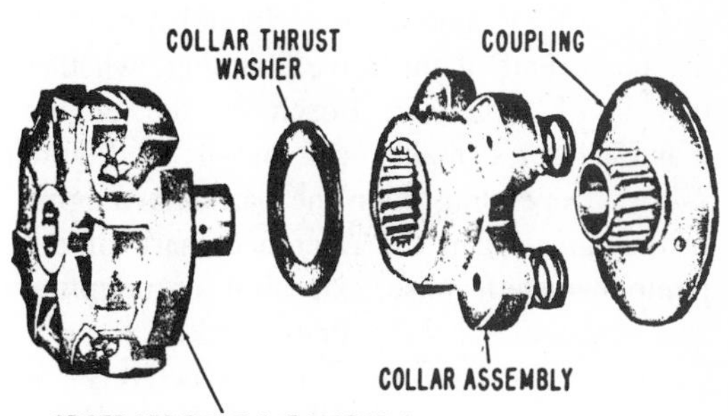

Figure 2-7
Exploded View of Injection Pump Advance Mechanism.
(Courtesy of American Bosch—AMBAC Industries)

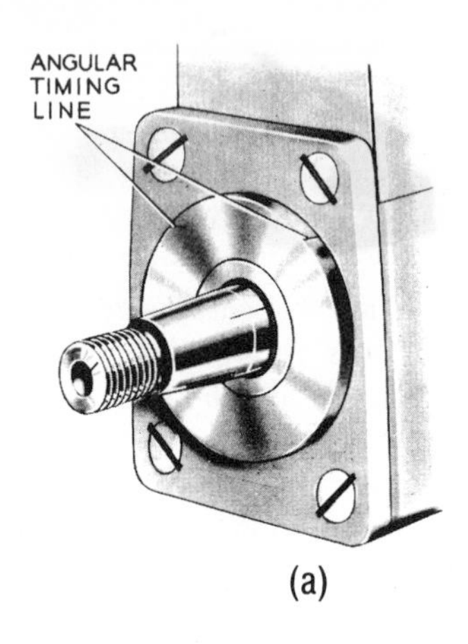

(a)

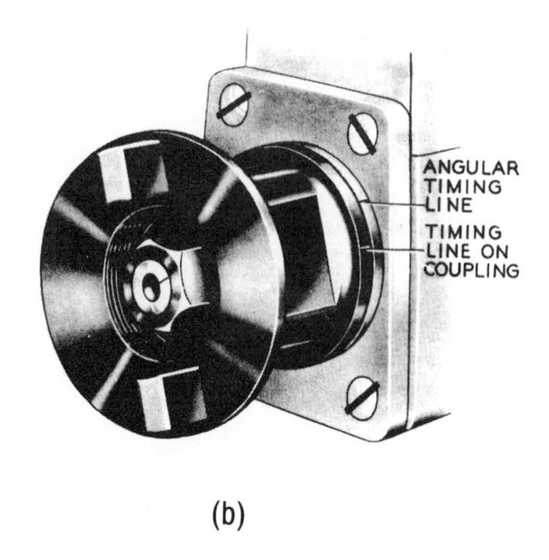

(b)

Figure 2-8
Timing Marks on (a) Camshaft Nose and (b) Pump Housing.

Timing of the Injection Pump to the Engine

Timing reference markings in line form are provided on all APE fuel-injection pumps to assist in proper installation and performance of the engine. These markings are scribed on both the camshaft nose extension and the pump housing end plates, respectively. See Figure 2-8(a).

The camshaft marking is parallel or longitudinal to it and is properly transferred to the pump half of the drive coupling, whether this is of American Bosch or the engine manufacturer's make, as shown in Figure 2-8(b). Depending upon the particular engine installation, when the reference mark of the pump camshaft nose extension or coupling half registers with the lines marked R or L (right and left-hand rotation, respectively), it indicates that the plunger in pump cylinder 1 has just reached its timed position, either port closing or port opening, depending upon the particular engine installation.

Since each particular engine manufacturer produces a workshop or overhaul manual for their engine, it is always wise to spend some time in reviewing the procedure outlined therein prior to attempting to install and time the injection pump on the engine. Failure to become familiar with every aspect of this procedure can prove to be costly and often disastrous. Therefore, the following timing procedure, although common to the APE type of fuel-injection pump, is general in nature and is not intended to supersede or replace the engine manufacturer's own sequence for fuel pump installation and timing.

Step 1 is always to turn the crankshaft of the engine over until the piston of cylinder 1 is approaching TDC (top dead center) on the compression stroke. This can be done by removing the valve rocker cover and checking the position of the intake and exhaust valves (4 cycle only). There should be clearance between both rocker arm assemblies and their valve stems; otherwise, if they are both under tension you are actually on the exhaust stroke and are half a revolution out on the crankshaft's rotation. If so, turn the engine around one more turn.

Now check the timing marks on the flywheel, which should indicate a mark with reference to #1 INJ stamped on the flywheel. This mark should coincide with a stationary mark or pointer on the flywheel housing or inspection cover plate. Always turn the engine in the same direction as its normal running direction.

Caution: Remember, if you are standing at the back of the engine, its rotation will be reversed [clockwise (cw) from the front, counter-clockwise (ccw) from the rear].

In addition, as you turn the engine over in its proper direction, also remember that the *injection* mark will appear before the cylinder TDC mark. (Opposite to a polar valve timing diagram whereby your eye is moving to follow the sequence of events; on the engine as it rotates your eye is stationary and the engine flywheel is now rotating.) If the TDC mark appears before the *injection* mark for the particular cylinder that you are trying to time, then you are turning the engine in the wrong direction. (Check the respective engine service manual for the correct specifications.)

With the necessary marks aligned, the injection pump can now be mounted on the engine, and the pump and engine coupling halves arranged in such a manner that the timing marks on the pump half of the coupling and the respective marking on the pump end plate are in line with one another.

Caution: When mounting two couplings together, it is extremely important to ensure that they are in fact properly aligned. Obviously, several factors come into play here; however, for the purpose of the injection pump to engine drive coupling arrangement, it is necessary to ensure that when the couplings come together you have one height and two parallel faces. In other words, with the use of a short straight edge and a spirit level, check that both are in fact level. In addition bring both couplings together, and with the aid of feeler gauges check if the faces are parallel by alternately inserting the feeler gauges at the 12, 3, 6, and 9 o'clock positions. It may be necessary to shim or space the pump to achieve this condition.

Any misalignment will cause severe stresses in both the injection pump camshaft and the engine drive shaft, which in turn can cause damage to one or possibly both. Although the flexible coupling through which the pump is driven provides some protection from slight vibrations due to minor misalignment, the coupling cannot compensate for poor installation procedures.

Drive couplings, such as that shown in Figure 2-9, are used principally on APE base-mounted pumps, which are driven from the end of an engine accessory drive shaft. A

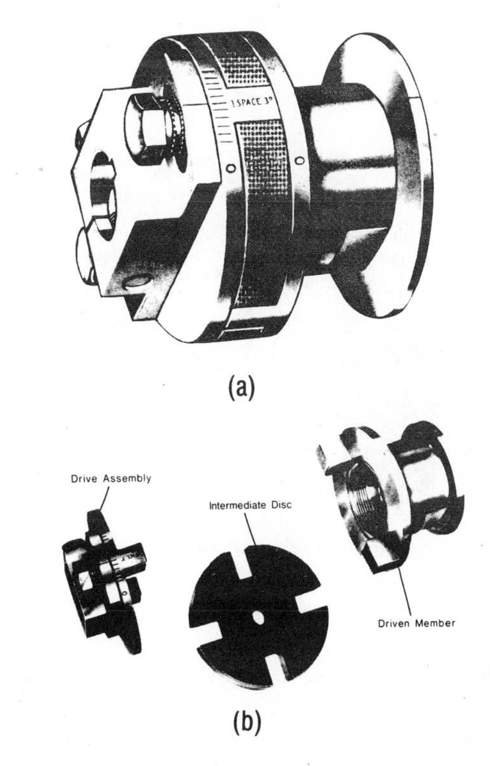

Figure 2-9
(a) Adjustable Drive Coupling;
(b) Adjustable Drive Member Components.
(Courtesy of American Bosch—AMBAC Industries)

vernier adjustment can be employed, which provides a means of *advancing* or *retarding* timing for precise mating to the engine. These couplings are available in 75- and 90-mm diameters with driving and driven members of various bores and tapers to suit.

2-3

Model 100 Distributor Pump

The Model 100 pump is the latest in the line of American Bosch single-plunger pumps and is similar in basic makeup to the PSB and PSJ pumps. Figure 2-10(a) and (b) shows two typical Model 100 pumps.

Figure 2-10(a) shows a Model 100 pump as currently fitted to an International Harvester DT 466 engine. The pump is equipped with a Bowden wire shutoff device and a *puff limiter* to reduce smoke during transient power changes. It is also available with a failsafe electric shutoff. Figure 2-10(b) shows a right-hand-mounted Model 100 for the British Leyland 500 series engine of medium horsepower for truck and bus engines.

Figure 2-11 gives the nomenclature and location of the various parts of the Model 100. Table 2-1 lists basic specifications.

By November 1977, over 200,000 Model 100 fuel-injection pumps had been delivered since 1970 for installation on International Harvester farm tractors and trucks.

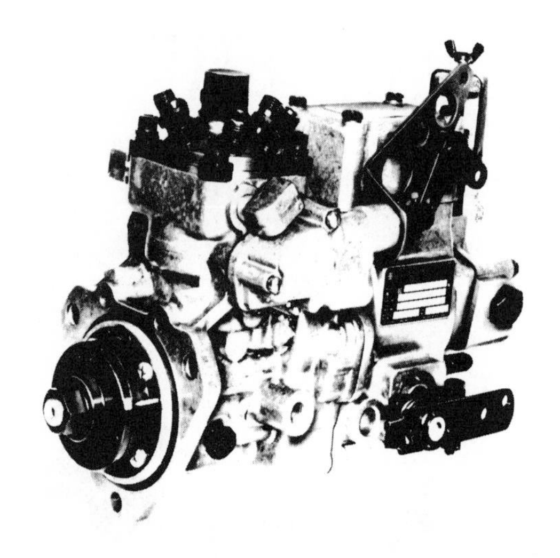

(a)

(b)

Figure 2-10
Typical Model 100 Pumps.
(Courtesy of American Bosch—AMBAC Industries)

Table 2-1
MODEL 100 SPECIFICATION DATA

Plunger diameter (mm)	9		10	
Number of cylinders	4–6	8	4–6	8
Maximum fuel delivery (mm³/stroke)				
Using a 45-mm³ relief volume delivery valve at 2400 rpm	85	75	115	105
Injection pressure: design maximum (psi)	10,000	10,000	10,000	9,000
Maximum rated speed (rpm)	3,300	3,000	3,200	2,900
Maximum effective stroke at (mm)	2.25		2.25	
Nominal 1.0 mm lift to port closing				
Cam lift (mm)	3.5		3.5	

PSJ, PSM, and Model 100

PSJ, PSM, and Model 100 pumps are of the same basic design; the major differences are explained in the following paragraphs. Figures 2-12, 2-13, and 2-14 show the lube oil flow within the injection pump body, a basic operational diagram, and an exploded view of the hydraulic head, respectively.

Model 100 PS Pump

The latest model of American Bosch single-plunger distributor PS pump is the Model 100, which is a flange-mounted single-plunger design of the constant stroke, distributor plunger, sleeve control type. It consists of the following major components, which are enclosed in an aluminum diecast housing along with the camshaft and drive arrangements.

Hydraulic Head Assembly. This includes the head block, plunger, metering sleeve, plunger face gear, delivery valve assembly, plunger return spring, spring seat, plunger button, and fuel discharge outlets suitable for Ermeto or swaged tubing connections.

All-Speed Mechanical Governor. This controls fuel delivery as a function of engine speed and load.

Intravance Unit. This is an internal timing device that provides a single and fully automatic hydraulically actuated means of varying the injection timing through the use of engine lube oil pressure; it is controlled by a servopiston and weight assembly. Functioning independently of the regular pump governing system, the intravance does not alter the pump space envelope (see Figure 2-15). The basic function of the Intravance unit is to sense engine speed and set the beginning of injection at the optimum position with respect to crank angle for best engine operation. The constant phase Model 100 intravance can provide up to 20 degrees advance, including retard for starting when required.

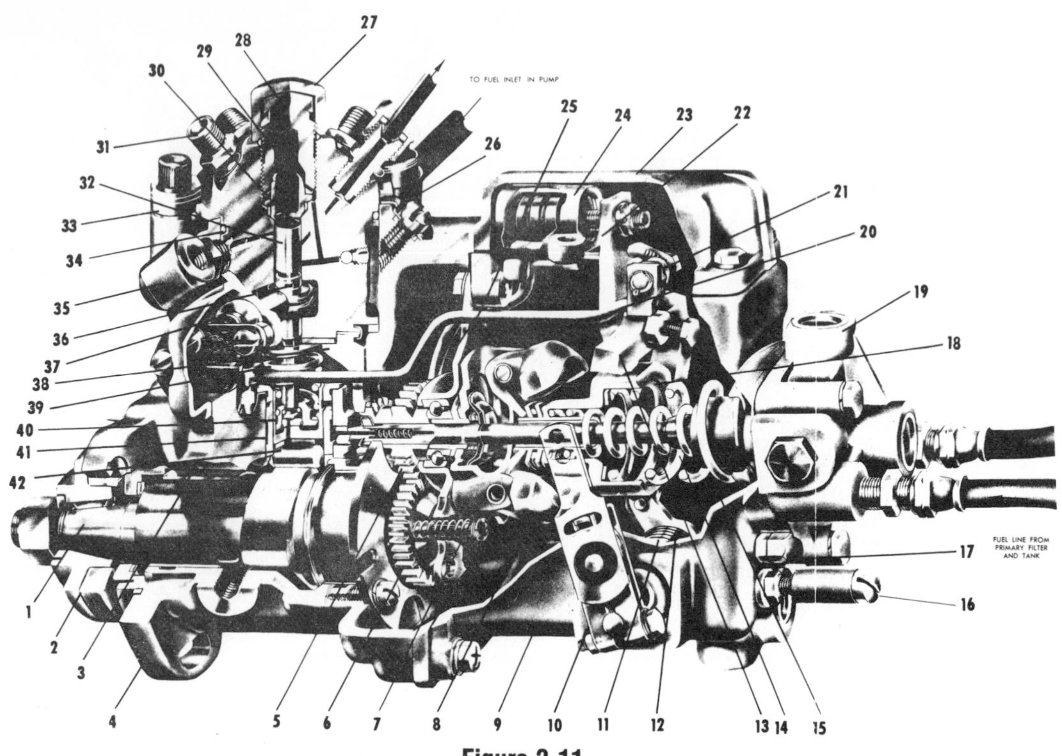

Figure 2-11
Single-Plunger Distributor Model 100 Fuel-Injection Pump.
(Courtesy of American Bosch—AMBAC Industries)

Index:

1—Camshaft
2—Drive plate
3—Camshaft bearing
4—Pump mounting flange
5—Governor and hydraulic drive gear
6—Camshaft gear
7—Governor weight spider
8—Governor weights
9—Governor housing
10—Operating lever
11—Operating shaft spring
12—Fulcrum lever bracket
13—Fulcrum lever
14—Inner governor spring
15—Outer governor spring
16—Low idle screw (spring loaded)
17—High idle screw
18—Governor sliding sleeve
19—Fuel supply pump
20—Control rod
21—Torque cam
22—Stop plate
23—Governor top cover
24—Excess fuel starting device
25—Ball bearing support plate
26—Head indexing plate
27—Delivery valve cap nut and gasket
28—Delivery valve holder
29—Delivery valve spring
30—Delivery valve and spring guide
31—Fuel discharge outlet
32—Hydraulic plunger
33—Hydraulic head clamping screw and holder
34—Hydraulic head assembly
35—Overflow valve
36—Fuel metering sleeve
37—Control unit assembly
38—Face gear
39—Plunger return spring
40—Plunger button and spring seat
41—Tappet guide
42—Tappet roller

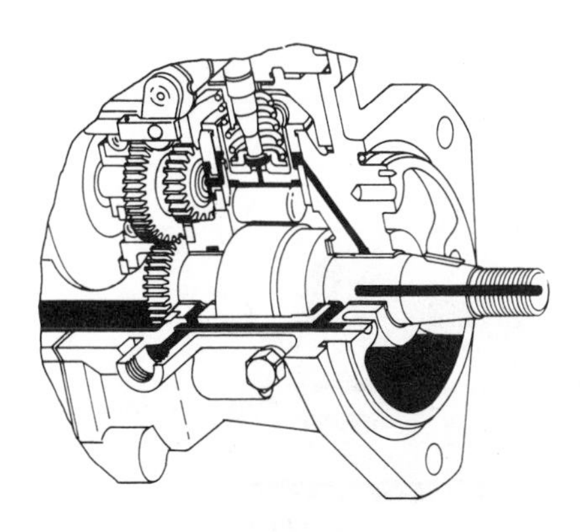

Figure 2-12
Lube-Oil Flow Within Pump Body.
(Courtesy of American Bosch—AMBAC Industries)

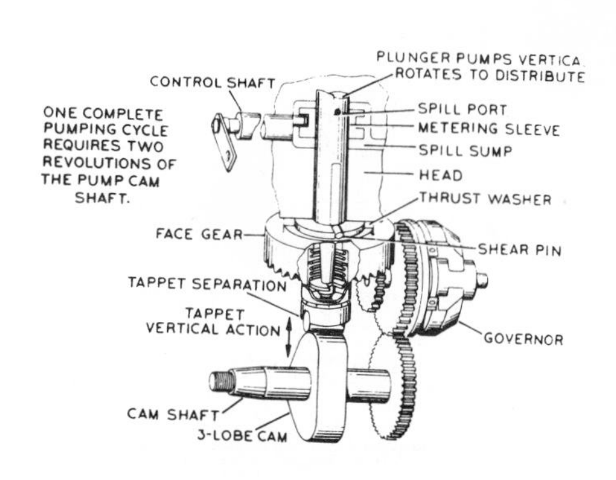

Figure 2-13
Operational Diagram.
(Courtesy of American Bosch—AMBAC Industries)

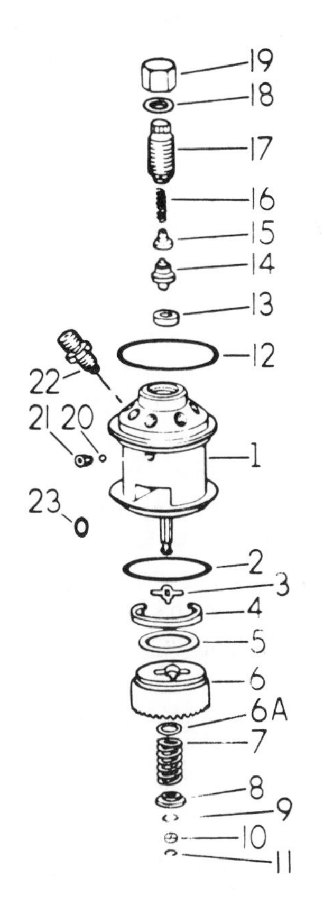

Index:

1—Hydraulic head
2—O ring
3—Oldham drive
4—Retaining ring
5—Thrust washer
6—Face gear
6A—Washer
7—Plunger spring
8—Spring seat lower
9—Spring seat retainers
10—Plunger button
11—Spring ring
12—O ring
13—Spacer, delivery valve
14—Delivery valve
15—Spring guide
16—Spring, delivery valve
17—Holder, delivery valve
18—Gasket
19—Cap nut
20—Sealing ball
21—Set screw
22—Discharge fitting
23—O ring, control unit

Figure 2-14
Exploded View of Hydraulic Head.
(Courtesy of American Bosch—AMBAC Industries)

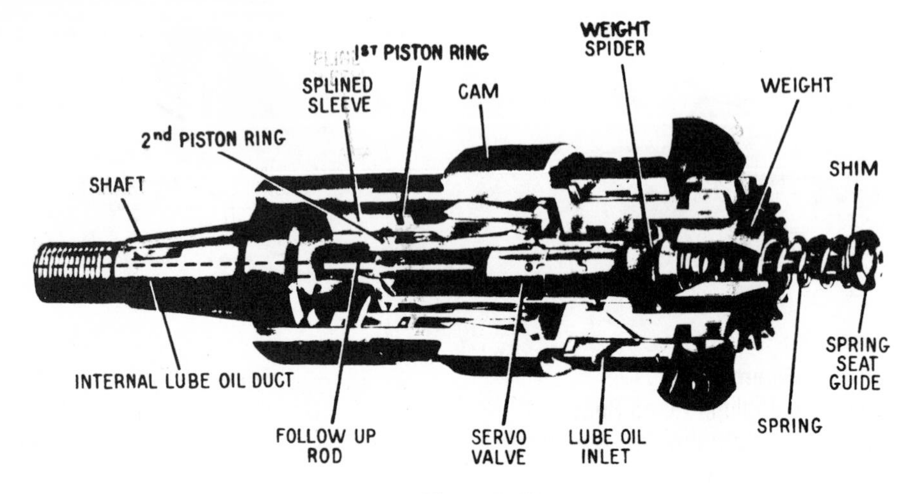

Figure 2-15
Intravance Unit.
(Courtesy of American Bosch—AMBAC Industries)

Intravance Internal Timing Device: Operation (PSJ and Model 100)

An exploded view of an *intravance assembly* is shown in Figure 2-16.

Although the present maximum mechanical advance of this unit is 20 degrees, the particular advance will vary between different makes of engines. However, the sequence of operations remains the same. For example, the maximum mechanical advance of the intravance unit used with the PSJ pump on Mack engines is 9.5 degrees.

The camshaft timing assembly of the intravance unit is what actually determines the amount of advance. Numbers stamped or etched on the threaded end of the camshaft assembly as follows indicate this:

Current Numbers (Model 100 Pumps)

1. An 83 identifies a CT 8583A with an advance limit of 13 to 16 degrees at full-load governed speed, with the advance starting at between 500 and 800 rpm.
2. An 87 identifies a CT 8587A and 87C identifies a CT 8587AC. Both have an advance of 13 to 16 degrees at full-load governed speed; however, if spring SP 8520-18 (double silver-stripe identification) is used, advance starts at 500 to 800 rpm, whereas with spring SP 8520-23 (brown-stripe spring), the start of advance is between 200 and 600 rpm with a 3 to $4\frac{1}{2}$ degree advance at 1000 rpm.
3. A 90 identifies a CT 8590A and 90C identifies a CT 8590AC. Both have a 13 to 16 degree advance at full-load governed speed, with the start of advance at 200 to 600 rpm and a 3 to $4\frac{1}{2}$ degree advance at 1000 rpm.
4. A 91 identifies a CT 8591A with a 13 to 16 degree advance at full-load governed speed, with a preadvance of approximately 6 degrees that occurs before low idle rpm.
5. A 92 identifies a CT 8592A with a 4 to 6 degree maximum advance at full-load governed speed, with a preadvance of ap-

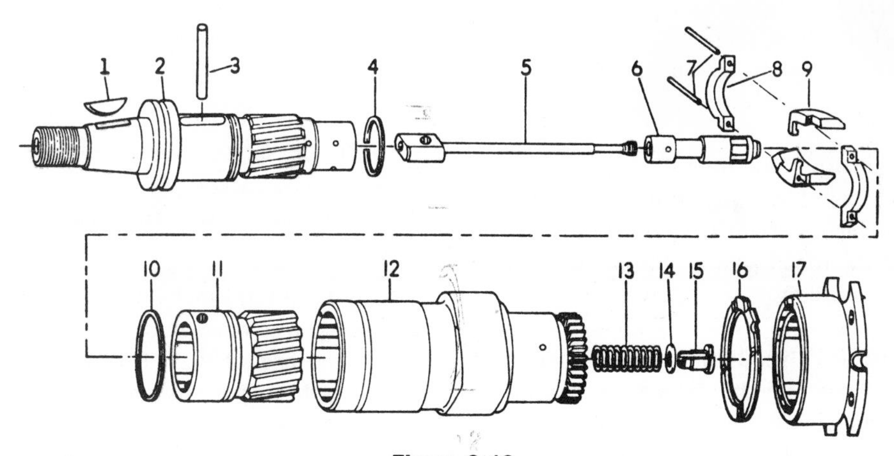

Figure 2-16
Exploded View of Intravance Assembly.
(Courtesy of American Bosch—AMBAC Industries)

Index:

1—Woodruff key
2—Control valve shaft assembly
3—Pin, sleeve and rod
4—Ring seal shaft
5—Follow-up rod
6—Control valve
7—Weight pin
8—Weight spider
9—Weight (2)
10—Ring, seal sleeve
11—Splined sleeve
12—Cam
13—Advance spring
14—Washer (available in different thicknesses)
15—Spring seat guide
16—Thrust washer (available in different thicknesses)
17—Bearing plate assembly

proximately 4 degrees that occurs before low idle speed.

6. A 108 identifies a CT 85108A with an advance of 13 to 16 degrees at 2300 rpm. Advance starts at between 200 and 700 rpm, with 10 to 12 degrees of advance by 1750 to 1850 rpm.

From the foregoing, you can see how the actual timing advance will vary between different engines to suit their particular needs.

From Figure 2-16 you will note that the cam mates with the sleeve by means of the helical splines machined into its inside diameter. The helix can be either of left- or right-hand cut for use on cw and ccw rotation engines. A machined flat on the hub of the cam permits electronic checking of the operation of the intravance by use of a magnetic pickup with adapter, special amplifier, and strobe light.

To grasp fully and thoroughly understand just how the intravance unit operates, let us consider one by one the function of the components that make up the assembly. Refer to Figure 2-16 for consideration of the parts breakdown.

Camshaft: Item 12. Machined into the large OD of this shaft is a groove with drilled diagonal oil inlet holes leading to an annulus in the small ID of the shaft. The cam mates with the sleeve by means of internal and external splines.

Follow-up Rod: Item 5. A pin holds the splined sleeve and follow-up rod together. Lube oil bleeds by to the flat portion of the rod, and the round area of the rod has a close fit to the camshaft.

Control Valve: Item 6. This is the most important part of the intravance assembly, since it contains the weight fingers, and provides a lube oil path that opens or closes (meters oil flow) at the rear edge of the large annulus through its axial movement, therefore directly affecting the timing advance of injection. Engine oil under pressure from idle speed to 1000 rpm is directed to oil holes in the rear bearing and on into a circular groove in the shaft, where three oil holes direct the oil into a reservoir area in the inside diameter of the shaft.

As long as the engine is running at less than *cut-in* speed, as shown in Figure 2-17, the oil will be trapped here and cannot flow past the metering edge of the control valve. At this speed range, the built-in intravance weights are unable to overcome the force of the follow-up rod spring; therefore, the control valve traps the pressurized lube oil within the recess located in the inside diameter of the shaft.

As engine speed increases, the centrifugal force of the rotating weights will be able to move the control valve to the left just slightly, as shown in Figure 2-18. This allows the previously trapped lube oil to enter the circular groove of the control valve or (annulus), where the oil passes through two holes in the shaft to the pressure side of the splined sleeve in which a relief area is machined, causing the oil to force the splined sleeve to the right, carrying the cross pin and rod with it. These items will move within the confines of the slot and the cam until maximum advance for that cam setup and rpm is attained. Because the splines are *helical* in design, the rotation of the cam will create a rotation to the assembly in an advance direction.

Note: If the splines were straight cut, there could be *no advance*. Since they are helical, any movement forward must cause rotational movement also.

As the follow-up rod moves to the right, the spring will be placed under compression against the end of the control valve, thereby forcing both the weights and valve back into a position that throttles and then stops the flow of high-pressure oil to the sleeve. If the engine speed was to decrease now, the centrifugal force of the weights would reduce and the

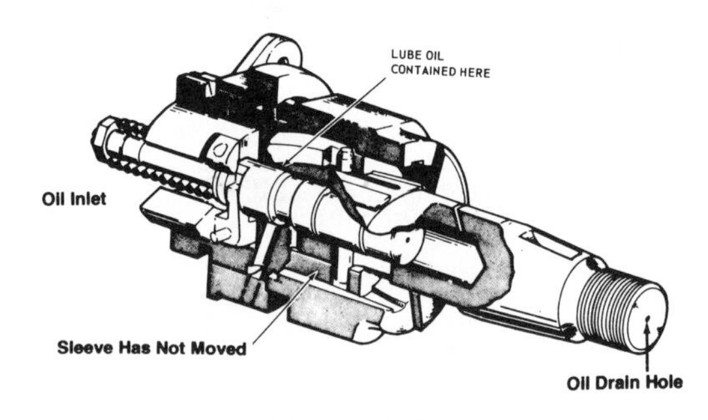

Figure 2-17
Oil Position Before Cut-in Speed.
(Courtesy of American Bosch—AMBAC Industries)

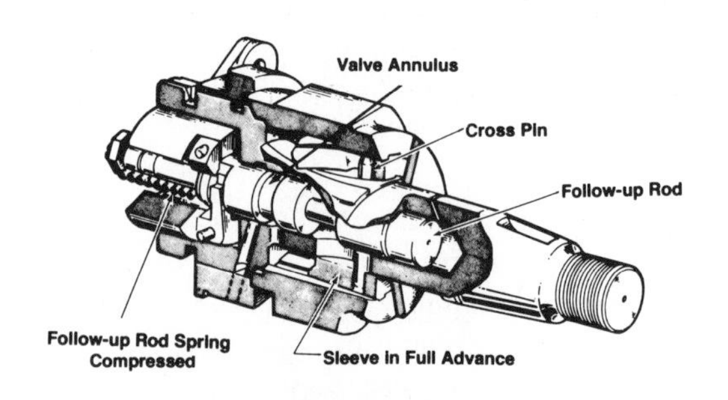

Figure 2-18
Oil Position at Cut-in Speed.
(Courtesy of American Bosch—AMBAC Industries)

metering edges of the control valve would stop the flow of lube oil, causing the cam to move back and in effect retarding the timing owing to the spring pressure on the follow-up rod.

The intravance operation is completely independent of the governor function within the pump.

2-4

Model 100 Injection Pump

Figure 2-19 shows a sectional view of the Model 100 injection pump with the location of components clearly shown. Injection pump overhaul should, as always, be attempted only by personnel experienced in the intracacies of fuel-injection equipment, with the special test equipment and tools readily available; otherwise, leave it alone.

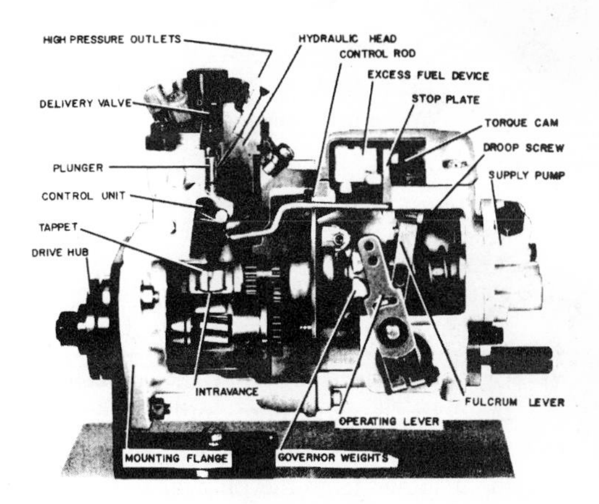

Figure 2-19
Sectional View of Model 100 Injection Pump (Courtesy of American Bosch—AMBAC Industries)

Internal Timing and Testing

With the pump on the engine, timing can be checked by turning the engine over manually until the flywheel or damper pulley timing mark is aligned with the stationary mark on either the flywheel housing or front engine cover. With the engine on its compression stroke for cylinder 1, check to see that the timing marks on both the pump and engine are correct. The line mark on the pump face gear should be aligned with the raised line mark in the pump housing timing window, as shown in Figure 2-20.

If the pump has been removed from the engine prior to disassembly or sending it to your local fuel-injection dealer for repair, the internal pump timing can be checked by manually turning the pump camshaft until the line mark on the face gear aligns with the raised line mark of the pump housing window, as shown in Figure 2-20. If the camshaft keyway is exactly at the 12 o'clock position, the pump is correctly timed.

If the Model 100 injection pump is to be overhauled, obtain a copy of Form 4000 from American Bosch directly or from your local fuel-injection dealer.

Sometimes you may simply be installing a new or the original hydraulic head assembly as shown in Figure 2-13; if so, the pump timing must be checked. Proceed as follows:

1. Refer to Figure 2-13; remove the delivery valve cap nut (19), gasket (18), delivery

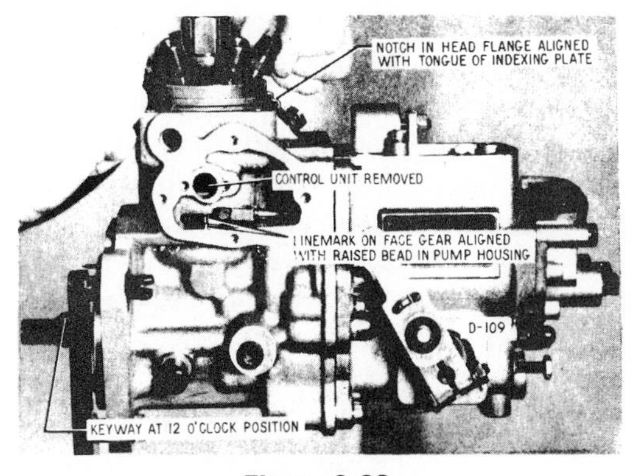

Figure 2-20
Checking Internal Timing.
(Courtesy of American Bosch—AMBAC Industries)

valve holder (17), spring (16), spring guide (15), and delivery valve assembly (14) along with its spacer (13).

2. Remove the control unit cover, as shown in Figure 2-20.
3. Align the pump timing marks as stated earlier, such as when checking timing with the pump removed from the engine.
4. Carefully insert a depth micrometer into the delivery valve bore until it makes contact with the internal single plunger, check the distance from the plunger to the top of the hydraulic head, and record it.
5. To ensure that the reading has in fact been taken at the base circle of the pump camshaft, reinsert the delivery valve holder (17) finger tight only.
6. Make up from an old fuel line a port closing adapter, which will be necessary to determine actual port closing (see Figure 2-21).
7. Assemble the port closing adapter as shown in Figure 2-21 to the no. 1 hydraulic head discharge fitting.
8. As Figure 2-21 also shows, to determine which fitting is in fact no. 1, face the control unit cover; on counterclockwise-rotating pumps, the no. 1 will be left of center; on clockwise-rotating pumps, the no. 1 will be to the right of center. If you are in doubt as to the pump rotation, look at the pump housing just above the control unit cover; however, head rotation and location of the no. 1 outlet may be stamped on the front of the pump housing (this is found on some particular applications).
9. Arrange a fuel tank with a gravity feed (overhead setup) approximately 16 to 20 in. (0.405 to 0.505 m) (40.64 to 50.8 cm) higher than the fuel inlet to the pump housing. Be sure to use either clean, filtered diesel fuel oil no. 2 (Bacharach no. 67-4038 or Bacharach test oil no. 67-4035).

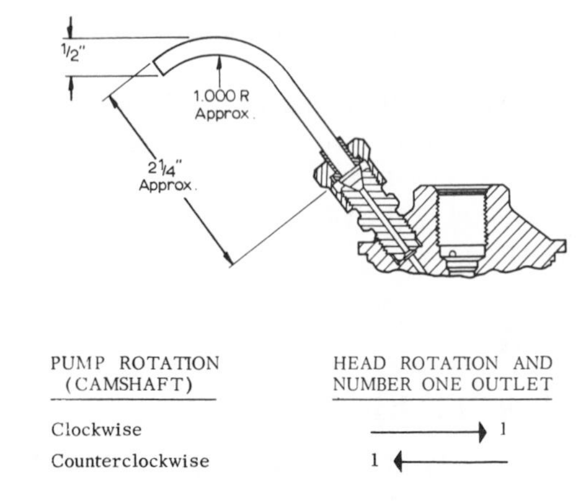

PUMP ROTATION (CAMSHAFT)	HEAD ROTATION AND NUMBER ONE OUTLET
Clockwise	⟶ 1
Counterclockwise	1 ⟵

Figure 2-21
Port Closing Adapter.
(Courtesy of American Bosch—AMBAC Industries)

10. Ensure also that the pump and test oil are at room temperature.
11. With the pump throttle operating lever in the low idle position, open the supply valve to the pump from the overhead tank, and allow fuel to flow through the connecting fuel line and pump to ensure that the pump is in fact filled with oil and free of any air lock.
12. Place the operating lever in the normal full-fuel position, and manually rotate the pump camshaft (normal direction) until the fuel flow from the closing adapter line of no. 1 is one drop per every 5 to 10 seconds, which indicates *port closing*.
13. Lock the pump camshaft in this position, shut off the fuel supply line, and again insert the depth micrometer as in step 4.
14. The difference between the reading obtained now and in step 4 is the actual single-plunger lift to port closing, which *must* be within the specified limits for that pump as stated in the applicable pump parts list.

15. If plunger lift is too much, install a *thicker* plunger button to decrease the lift; if too little, install a *thinner* button to increase it (button is item 10 in Figure 2-13).
16. If the specified plunger lift to port closing is still not within suggested tolerances, replace the tappet and recheck the timing, or replace the hydraulic head assembly and recheck.
17. If necessary after a timing adjustment, remove the old port closing mark and scribe a new timing mark on the drive hub.

If desired, a port closing tool can be manufactured as shown in Figure 2-22. Use this tool as follows:

1. Install it into the hydraulic head carefully until it comes in contact with the plunger; then zero in a dial indicator in contact with the pin or port closing tool.
2. After finding the port closing position as described earlier, read the face of the dial gauge, which will indicate the actual lift to port closing.

Pressure Testing the Injection Pump for Leakage

After any work on an injection pump has been completed, check it as follows for any signs of fuel leakage:

1. Refer to Figure 2-23 and connect a hand-operated injector nozzle test stand to the fuel inlet housing area of the pump.
2. Manually raise the fuel pressure as registered on the test pump gauge until a pressure of 400 psi (2758 kPa) is attained. Should the gauge hand drop rapidly, a fuel leak is evident.
3. Check all areas of the pump for signs of leakage, and replace any gaskets or O rings as necessary; however, a slight sweat at the control unit is acceptable, as long as no drops actually form.
4. A rapid gauge needle drop, with no signs of fuel leakage, is usually caused by a damaged lower hydraulic head gasket; therefore, after replacing the gasket, repeat the check.
5. If the condition explained in step 4 continues to exist, it well may be a damaged or worn plunger or a crack in the pump housing.

Pump Testing

Refer to Figures 2-24 and 2-25, which show the pump connected up to a test stand. Since the pump is pressure lubricated, a minimum supply of 2 gal/min (7.57 liters) at 40 psi (275.8 kPa) is recommended. Since the only way that the injection pump can be properly

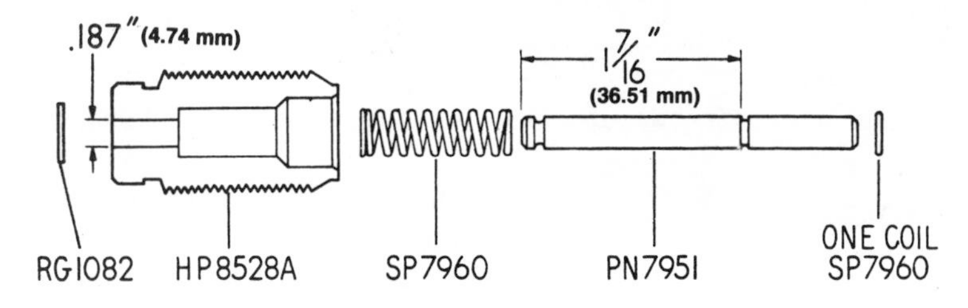

Figure 2-22
Port Closing Tool.
(Courtesy of American Bosch—AMBAC Industries)

Figure 2-23
Pressure Testing the Injection Pump.
(Courtesy of American Bosch—AMBAC Industries)

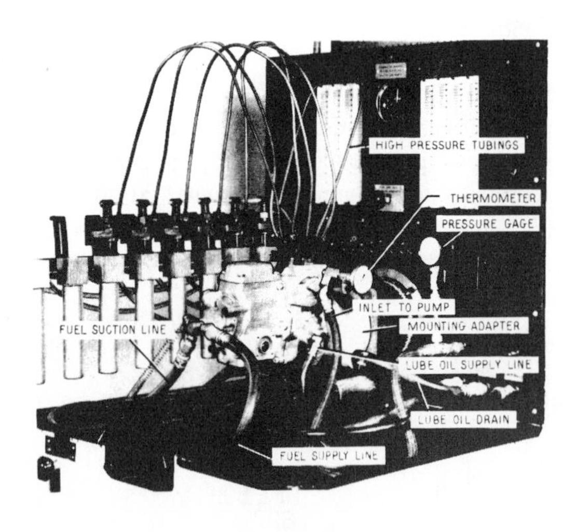

Figure 2-24
Pump Connected to Test Stand.
(Courtesy of American Bosch—AMBAC Industries)

tested is on a test stand, those personnel that have access to such a machine are most probably familiar with the actual setup and running of the machine. Since a variety of special tools and equipment is required, refer to American Bosch Service Instruction Booklet Form No. 4000, which deals in detail with the Model 100 single-plunger pump and nozzle holder assemblies.

Gear-Type Fuel Supply Pump

The fuel supply pump is of positive displacement gear type and is attached to the governor housing. An optional hand primer can be mounted directly into the supply pump to provide positive pump priming by forcing fuel into the supply tubing that feeds the injection pump. Figure 2-26 indicates the component

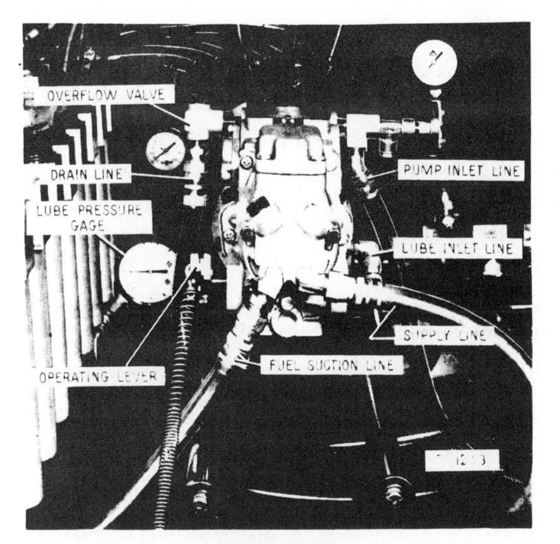

Figure 2-25
Pump Connected to Test Stand.
(Courtesy of American Bosch—AMBAC Industries)

parts of the typical supply pump used with the Model 100 injection pump.

Operation (Supply Pump). Referring to Figure 2-27, we see that rotation of the pump gears by an extension of the governor drive shaft causes fuel oil to be drawn from the fuel tank, where it will naturally flow through the inlet valve and duct to the suction side of the drive and idler gears. Fuel oil trapped between the gear teeth cavities and the pump housing is placed under pressure where it will exit at the pump outlet.

Figure 2-28 shows a pump with a relief valve installed in the supply pump cover. An internal passage allows fuel oil under pressure to be applied to the seat of the bypass relief valve; therefore, when pump delivery pressure exceeds the spring tension of the valve, the valve will be forced open, allowing excess fuel under pressure to be vented or bypassed back to the suction side of the pump. An arrow inscribed on the end of the pump housing or

Index:
1—Relief valve screw
2—Relief valve spring
3—Relief valve
4—Pump body
5—Pumping gears
6—Cover
7—Plugs
8—O ring
9—Hand primer

Figure 2-26
Components of Typical Supply Pump.
(Courtesy of American Bosch—AMBAC Industries)

Figure 2-27
Rotation of Pump Gears.
(Courtesy of American Bosch—AMBAC Industries)

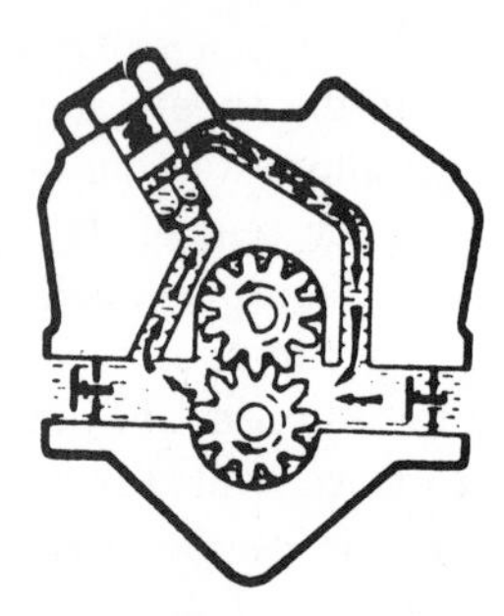

Figure 2-28
Pump with Relief Valve in Supply Pump Cover.
(Courtesy of American Bosch—AMBAC Industries)

the letters IN and OUT inscribed on the fuel connection bosses indicate the direction of fuel flow.

2-5

Hand Priming Pumps and Overflow Valves

By adding an optional hand priming pump to any American Bosch fuel system, the low-pressure portion or side of the fuel-injection system can be filled or primed manually to assist in venting or expelling air from the system prior to starting after changing fuel filters or if one or more fuel lines have been disconnected.

These pumps are plunger-type, positive displacement fuel supply units and are normally mounted directly to the supply pump itself. American Bosch, however, offers a remote-mounted unit that is adaptable to all types of fuel-injection pumps.

Figure 2-29 depicts a typical pump-mounted hand priming unit. It includes three main sections, which are as follows:

1. Item 4, which is the threaded barrel, acts as a fuel reservoir and both a mounting base and housing for the plunger.
2. Item 5, which is the plunger and its piston, sealing gaskets, and knob handle.
3. Item 1, which is the bracket assembly and wing nut that holds the plunger and piston at the bottom of the barrel when the unit is not in use. This prevents fuel from entering the reservoir of the barrel as the supply pump is operating. (On some other styles of primers, the plunger is held in place with the top section of the plunger stem, which acts as both a knob and retaining nut that fastens to a threaded extension of the barrel.)

Operation. Pulling up on the plunger creates a vacuum or suction effect that opens the check valve on the suction side of the fuel supply pump, allowing fuel to enter the barrel reservoir, as shown in Figure 2-30, whereas in Figure 2-31 on the downward stroke the plunger piston forces the fuel out of the reservoir, closing the check valve on the pump's suction side and opening the check valve on the outlet or discharge side. The pressurized fuel will be directed to the injection pump by the supply line. A buildup of pressure or increased resistance to the hand primer plunger movement indicates that the injection pump has been primed sufficiently, and starting of the engine can be attempted.

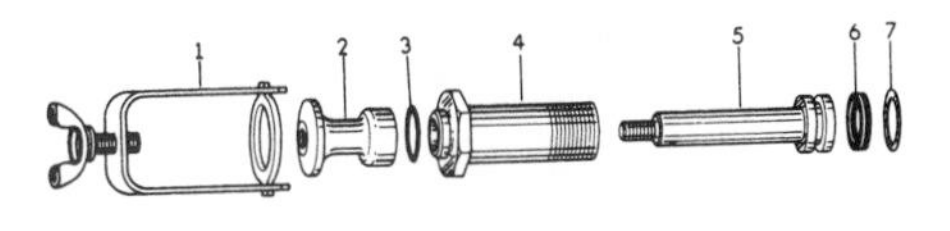

Figure 2-29
Typical Pump-Mounted Hand Priming Unit. (Courtesy of American Bosch—AMBAC Industries)

Index:
1—Bracket assembly
2—Knob
3—Sealing gasket
4—Barrel
5—Plunger
6—Plunger gasket
7—Sealing gasket

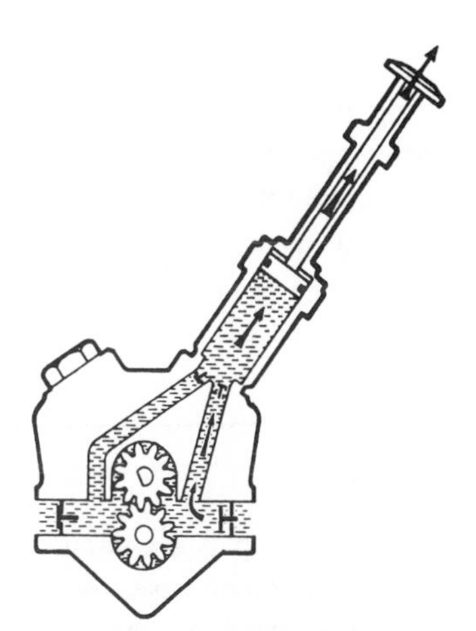

Figure 2-30
Operation of Hand Primer: Upward Stroke. (Courtesy of American Bosch—AMBAC Industries)

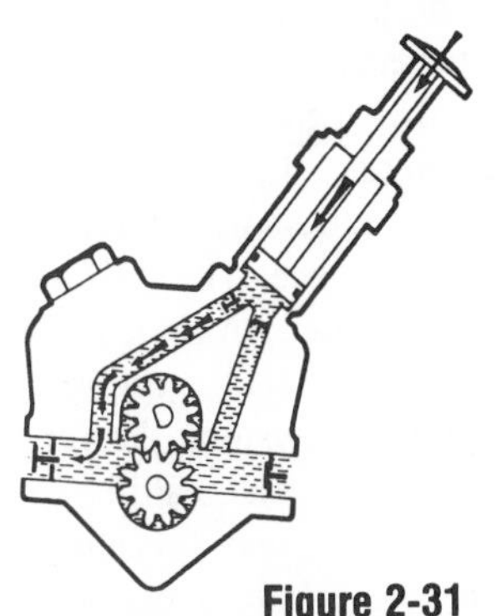

Figure 2-31
Operation of Hand Primer: Downward Stroke. (Courtesy of American Bosch—AMBAC Industries)

A typical remote-mounted hand primer is shown in Figure 2-32. If an external hand primer such as this is used, it can be employed with a special type of SGD fuel supply pump that incorporates a two-way regulating valve, as shown in Figure 2-33. When the hand primer is used on the suction side of the line, fuel will be forced through the top portion of the valve, thereby overcoming the ball check valve spring pressure that in normal operation would keep the internal passage closed. This setup therefore allows fuel oil to flow through the valve to the supply pump outlet and on to the injection pump.

Overflow Valves

The use of an overflow valve plumbed into any American Bosch fuel system is designed to control the maximum pressure that can be developed in the injection pump fuel sump. This will ensure adequate fuel for proper injection; also, excess fuel under low pressure will be able to flow back to the fuel tank. A typical overflow valve is shown in Figure 2-34.

In addition to the parts shown in Figure 2-34, the valve body has two tapped outlets, one of which accommodates the fuel tank return line, with the other set up to permit nozzle leak off back to the fuel tank. Location of the overflow valve on in-line injection pumps was shown earlier; it is located at the end of the pump opposite the fuel inlet. However, on the single-plunger type of fuel-injection pump such

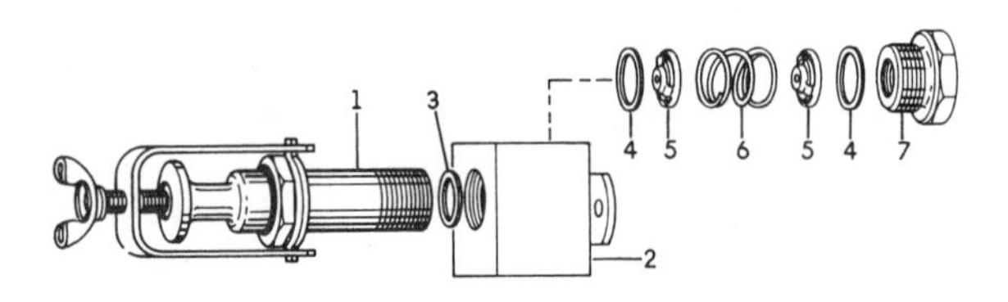

Index:
1—Hand primer assembly
2—Valve body
3—Sealing gasket
4—Check valve gasket
5—Check valves
6—Check valve spring
7—Inlet fitting

Figure 2-32
Typical Remote-Mounted Hand Primer. (Courtesy of American Bosch—AMBAC Industries)

as shown in Figure 2-35, the overflow valve is threaded into the hydraulic head. Also shown is a remote-mounted hand priming pump.

The two main functions of the overflow valve are as follows:

1. To maintain adequate fuel pressure in the sump of the injection pump under all conditions.
2. To relieve any excess fuel pressure buildup and trapped air back to the fuel tank.

In addition to these functions, the bypassing action of the overflow valve also tends to cool the injection pump by carrying away heat from the pump during the bypass stage. Overflow valves have idle regulating pressure limits of from 10 to 50 psi (68.95 to 344.75 kPa), with the average weight of an assembly being 5 ounces.

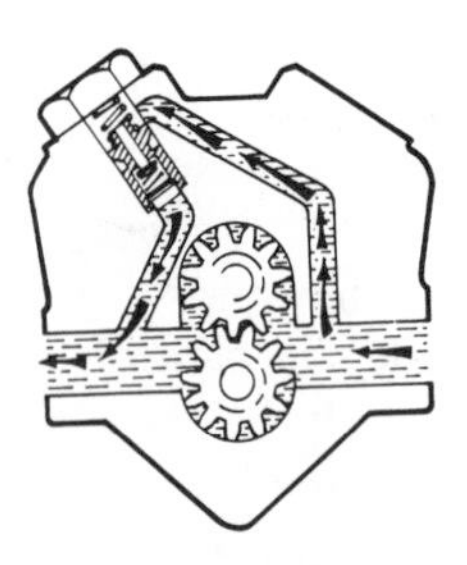

Figure 2-33
SGD Fuel Supply Pump. (Courtesy of American Bosch—AMBAC Industries)

2-6

PSU Model Pump

This highly compact pump is similar in many respects to the head assembly used with other PS series units. It is used on many ONAN diesel generator sets.

The PSU fuel-injection pump (shown in Figure 2-36) has no separate drive mechanism but is actuated by a cam and gear on the engine camshaft. This substantially reduces the cost of the pump and eliminates the need for the intermediate pump drive gear, shaft, coupling, and cover on the engine. The engine camshaft drive setup assures extremely accurate fuel-injection pump performance. The PSU uses the well-known and patented American Bosch hydraulic head and the fuel pumping, metering, and distributing system. Fewer parts and hydraulic head interchangeability permit quick, easy, and economical servicing. The PSU will handle the whole range of commercial fuels, enabling the engine manufacturer to develop a power plant with multi-fuel capabilities.

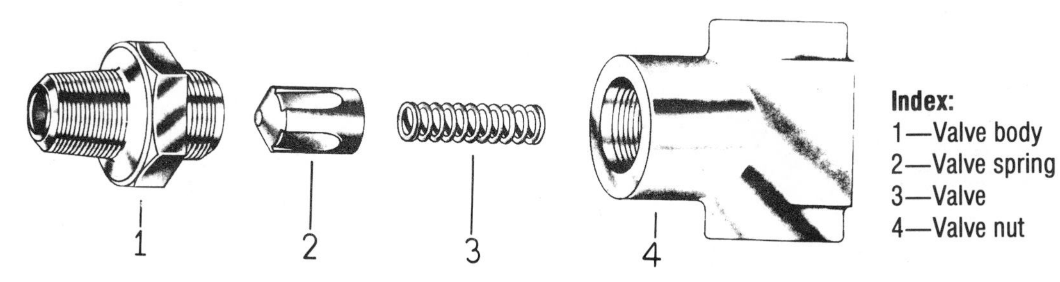

Figure 2-34
Overflow Valve.
(Courtesy of American Bosch—AMBAC Industries)

Available in 2-, 3-, 4-, and 6-cylinder configurations, the pump is ideally suited to small and medium high-speed diesels with displacements up to 110 in.^{3} per cylinder. Related American Bosch components are available to complete the fuel-injection system. The pump is so compact that it measures only 4 in.^{2}. (65.54 cm^3).

The PSU will operate in a cw or ccw direction. It is driven at camshaft speed for 4-cycle engines and at crankshaft speed for 2-cycle engines. Speed limitation is 3500 rpm

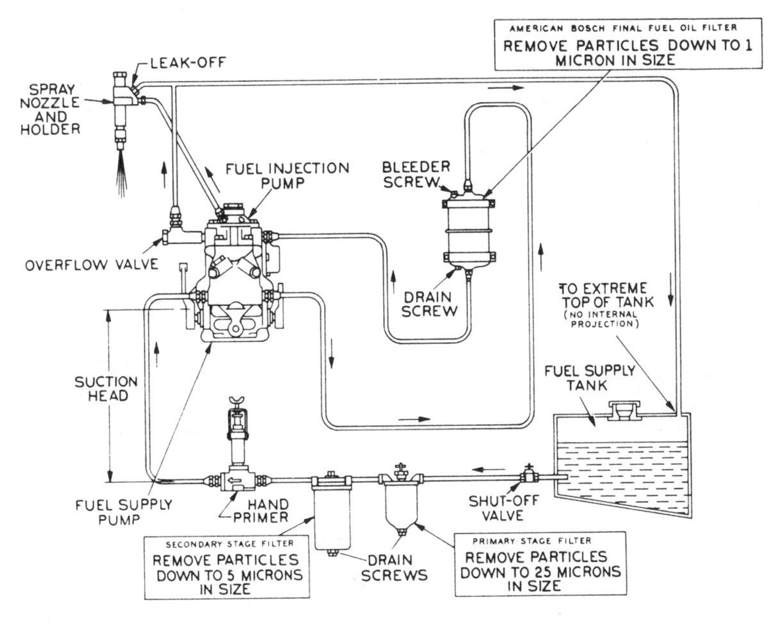

Figure 2-35
Overflow Valve Threaded into Hydraulic Head.
(Courtesy of American Bosch—AMBAC Industries)

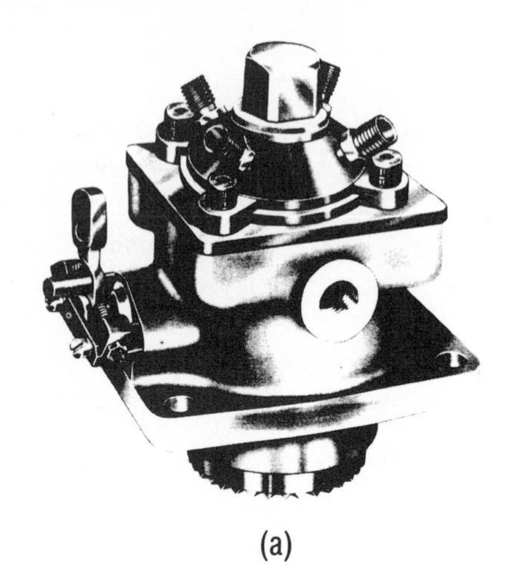

(a)

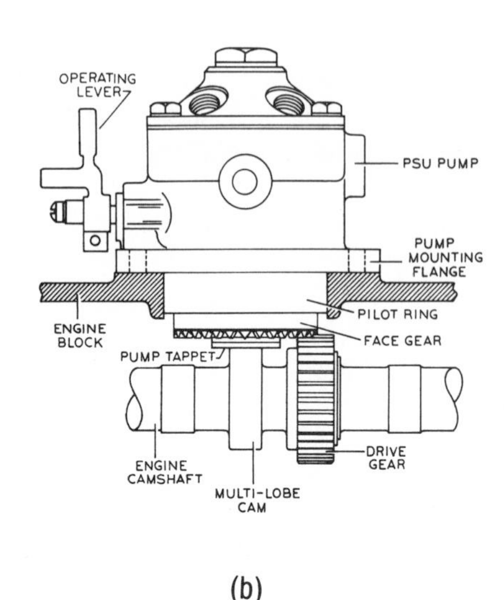

(b)

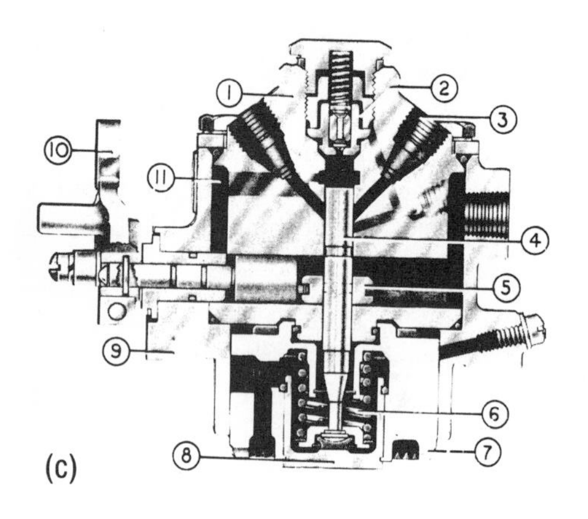

(c)

Figure 2-36
(a) PSU Fuel-Injection Pump; (b) Method of Mounting and Driving PSU Injection Pump; (c) Cross Section of PSU Injection Pump. (Courtesy of American Bosch—AMBAC Industries)

Index
1 —Nitrided steel hydraulic head
2 —Delivery valve assembly
3 —Fuel outlets
4 —Lapped and fitted through-hardened steel plunger
5 —Fuel metering control sleeve
6 —Plunger tappet return spring
7 —Face gear
8 —Tappet assembly
9 —Pump mounting flange
10 —Pump operating lever
11 —Fuel sump

for a 4-cycle engine. The pump is equally adaptable to open, precombustion, and energy-cell type combustion chambers. All lap-fitted components receive lubrication from the fuel oil, while the remaining internal parts are splash or pressure lubricated from the engine.

2-7

Injection Nozzles

The purpose of an injection nozzle is to deliver, in a metered quantity from the fuel-injection pump, a supply of fuel that will enter the cylinder combustion chamber at a high enough pressure to penetrate the high pressure of compressed air toward the end of the compression stroke. The fuel must enter the combustion chamber in a very fine atomized spray pattern so as to produce ignition with a minimum of delay, thereby producing optimum combustion and a smooth-running economical engine. Regardless of the make of fuel-injection nozzle, they must be manufactured to exacting tolerances. The nozzles produced by American Bosch are of two basic types: (1) the pintle nozzle, and (2) the hole nozzle. Figure 2-37 shows these two basic nozzle types. These nozzles are known as *closed differential hydraulically operated types*. They have an internal spring arrangement such as is shown in Figure 2-43, which normally holds the tapered-style needle valve against a lapped seat in the nozzle tip. Fuel under high pressure from the injection pump delivered to the nozzle tip acts upon the tapered needle valve face, causing the valve to lift away from the combustion chamber, thereby allowing the fuel to enter the cylinder or precombustion chamber, as the case may be. The spring within the nozzle holder will rapidly seat the needle valve back onto its seat after the injection period—therefore the term *closed-type nozzle*.

The *hole-type* nozzles are found on direct-injection combustion chamber engines, whereas the *pintle-type* nozzle is used on engines having precombustion, divided, air-cell or energy-cell type combustion chambers.

Since injection nozzles are manufactured to extremely close tolerances, the nozzle needle valve and its body are matched to one another; therefore, they must be replaced as a unit, that is, as a matched and mated assembly, if damaged.

These two types of nozzles can be found in a wide variety of styles, since the nozzle holder is designed to meet many applications and makes of engines.

Hole-Type Nozzles

Hole-type nozzles are available in ADA, ADB, and ADL models, with the difference being that the number, arrangement, and size of the spray holes are dependent upon the spray pattern that the particular engine combustion chamber requires, along with the physical size of the nozzle. The conical tip at the base of the nozzle valve stem is ground to a slightly different angle with respect to the valve seat, which results in *line contact seating,* thereby

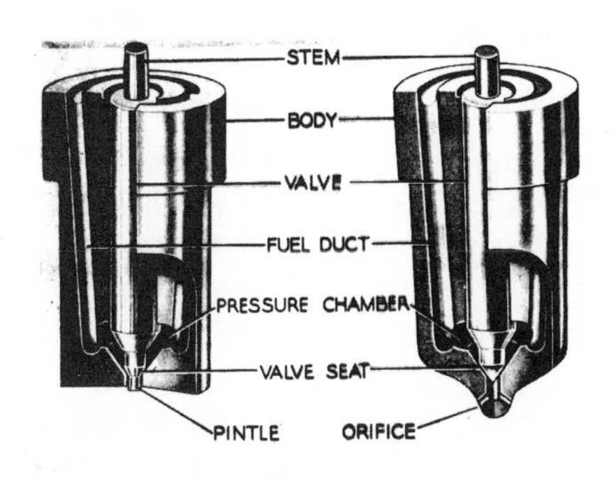

Figure 2-37
Two Basic Nozzle Types: (a) Pintle Type; (b) Hole Type.
(Courtesy of American Bosch—AMBAC Industries)

preventing leakage that could cause an increase in fuel consumption, unburned fuel and thus smoke emanating from the exhaust stack, and carbon buildup around the nozzle tip, creating futher problems.

A model ADC nozzle tip, which is normally equipped with one or two spray holes (orifices), is usually found on farm tractors and agricultural equipment. Figure 2-38 shows this unit. If you compare this nozzle with the hole type shown in Figure 2-37, you will notice that the spray hole chamber and nozzle tip (or button) are separate from the nozzle body, allowing for removal or replacement of the button separately.

The hole-type nozzle can have either a single hole with an angle of 4 to 15 degrees (turbulence chamber engines) or from 3 to 18 holes with orifice diameters ranging from 0.006 to 0.033 in. (0.076 to 0.838 mm), depending on engine size.

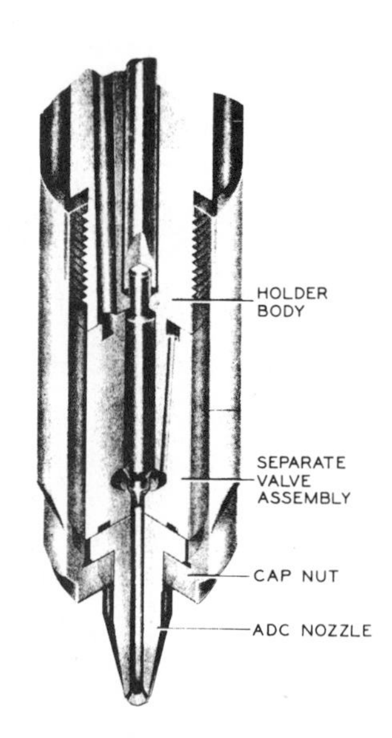

Figure 2-38
Model ADC Nozzle Tip.
(Courtesy of American Bosch—AMBAC Industries)

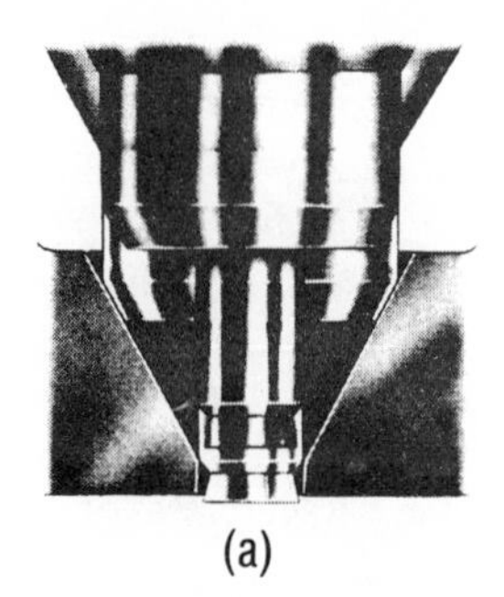

(a)

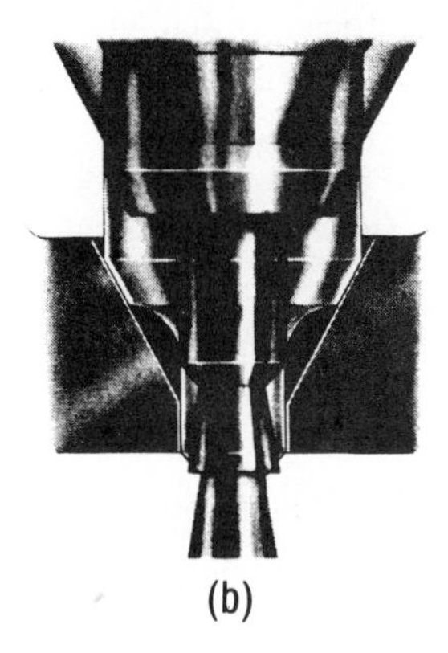

(b)

Figure 2-39
(a) Standard and (b) Throttling Pintle Nozzles.
(Courtesy of American Bosch—AMBAC Industries)

Pintle-Type Nozzles

The pintle nozzle derives its name from the fact that the valve tip terminates in a pin or *pintle* shape that extends past the end of the valve seat and actually protrudes through a close-fitting hole in the base of the nozzle body. The actual shape or profile of the pintle determines the spray angle and the pattern desired for the particular engine that it is to be used on. The spray pattern with this type of nozzle is therefore *cone shaped* (spray patterns are discussed later).

American Bosch pintle nozzles ADE and ADF are found on precombustion chamber and energy-cell engines.

Throttling-Type Nozzles (Pintle) Model ADN

The throttling nozzle is found on small-bore high-speed engines. Figure 2-39 shows the basic difference between the standard and throttling pintle nozzles; the pintle and hole in the base of the throttling pintle are longer, which produces a throttling or delaying action to the fuel, thereby allowing only a small amount of fuel to enter the combustion chamber at the beginning of injection. However, as the valve continues to lift (by fuel pressure), the rate of fuel flowing through into the combustion chamber is progressively increased.

The nozzle shown in Figure 2-40 is commonly known as a *pintaux pintle nozzle,* which is similar to the standard pintle type, but with an additional small plain orifice located so as to direct a finely atomized spray of fuel into the hottest part of the turbulence chamber during the starting period to assist cold starting of the engine. During engine operation, when the fuel pressure delivered to the nozzle is higher, the nozzle valve will be withdrawn from the pintle hole, allowing the major portion of the fuel to enter the combustion chamber through the main orifice.

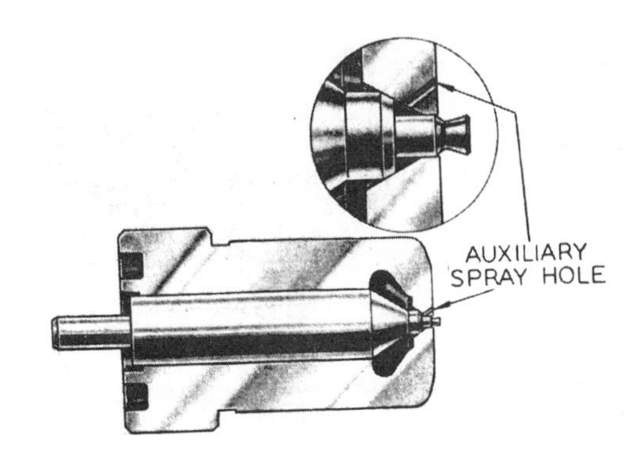

Figure 2-40
Pintaux Pintle Nozzle.
(Courtesy of American Bosch—AMBAC Industries)

Nozzle Holders

The actual shape of the nozzle holder will vary somewhat between different engines owing to manufacturers' design requirements; therefore, American Bosch nozzle holders are available in three basic types, as follows:

1. *Clamp type:* the holder body is held in place by clamping yokes (types AKB, AKF, AKK, AKL, AKN).
2. *Flange type:* have holes for mounting studs or bolts (types AKB, AKF, AKP).
3. *Screw type:* the nozzle holder is actually screwed into the cylinder head (types AKC, AKS).

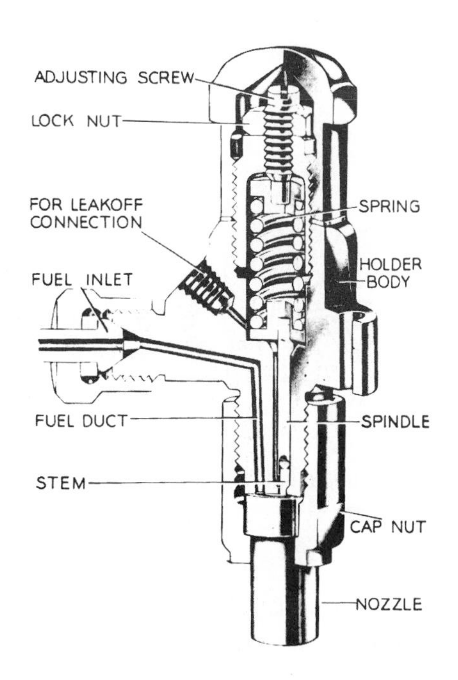

Figure 2-41
Type AKB Nozzle Holder.
(Courtesy of American Bosch—AMBAC Industries)

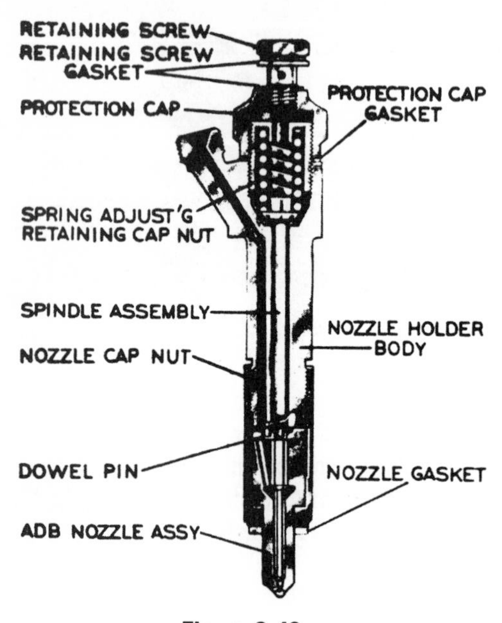

Figure 2-42
Type AKF/ADB Nozzle Holder.
(Courtesy of American Bosch—AMBAC Industries)

Type AKB, AKC, AKF, AKL Nozzle Holders. Figures 2-41 and 2-42 show two of these nozzle holders. The nozzle holder consists of a steel body, spindle, pressure adjusting spring, pressure adjusting screw or shims, nozzle retaining nut (cap nut), fuel inlet and fuel leak off, and a protection cap. The small amount of fuel that leaks internally past the spindle for lubrication purposes returns through the leak-off line. Removal of the protection cap allows easy access to the pressure adjusting screw or shims in order to adjust the spring pressure and therefore the opening pressure of the nozzle itself.

Type AKK and AKN Nozzle Holders. Figure 2-43 shows this type of holder, which differs from the previous units in that neither type AKK nor AKN is equipped with a spindle; also, the pressure adjusting spring and shims along with the upper and lower spring seats are all located in a valve stop spacer.

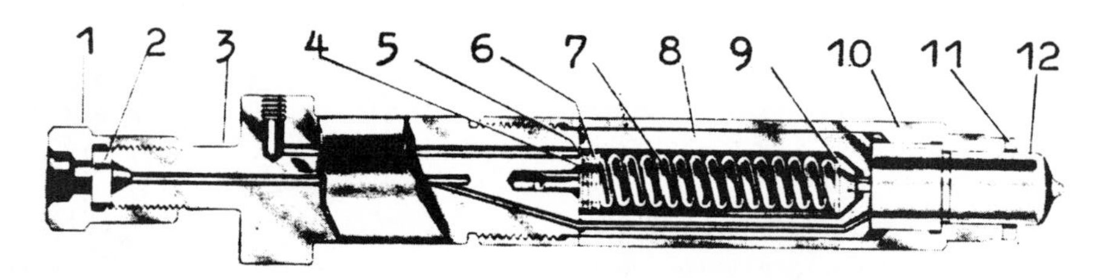

Index:

1—Inlet nipple nut
2—Washer
3—Body, holder
4—Spacer
5—Pressure adjusting shim
6—Upper spring seat
7—Pressure adjusting spring
8—Valve stop spacer
9—Lower spring seat
10—Nut, nozzle cap
11—Gasket
12—Nozzle

Figure 2-43
Type AKK/AKN Nozzle Holder.
(Courtesy of American Bosch—AMBAC Industries)

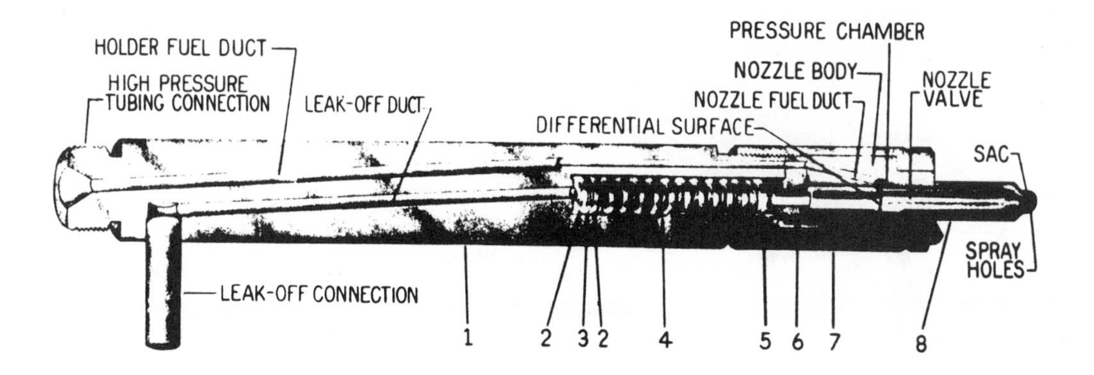

Index:
1—Body
2—Spring adjusting spacers
3—Spacers
4—Pressure adjusting spring
5—Lower spring seat
6—Nozzle spacer
7—Nozzle cap nut
8—Nozzle

Figure 2-44
AKN 120M 6547 Nozzle Holder.
(Courtesy of American Bosch—AMBAC Industries)

Type AKN M and SM Nozzle Holders. The AKN is a fairly recent addition to the current line of nozzle holders. Its configuration is similar to that of the AKK or AKN except for size. Figure 2-44 shows such a nozzle holder assembly. The M and SM nozzle holder assemblies are used to retain a nozzle in a cylinder head.

The nozzle spacer (6) contains two locating pins that position the nozzle (8) radially to ensure proper spray pattern distribution within the combustion chamber. Fuel entering the high-pressure tubing connection from the injection pump flows down through the nozzle fuel duct to the nozzle valve, where this fuel pressure, acting upon the valve, will raise the valve against spring pressure away from the combustion chamber, thereby allowing fuel injection. Any fuel leaking into the spring compartment will lubricate these parts and t[illegible] return to the fuel tank through the leak-off li[illegible]

When the plunger within the injection pump reaches the end of its effective stroke, the sudden drop in the fuel line pressure causes both the nozzle valve spring and delivery valve spring within the injection pump to immediately reseat the valves at both ends of the fuel line. This closes the nozzle and the delivery valve at the pump.

Type AKP and AKS Nozzle Holders. Both of these nozzle holders contain no actual working parts but serve simply as a body extension for the nozzle itself to hold it in the cylinder head and to transfer the high-pressure fuel from the injection pump and line. A model ADE pintle-type nozzle is used with these holders as shown in Figures 2-45 through 2-47. The pintle valve's maximum movement with these units is controlled and limited by the distance between the bottom of the valve retainer and the shoulder stop in the nozzle body. The particular spring that is installed controls the opening pressure of the nozzle.

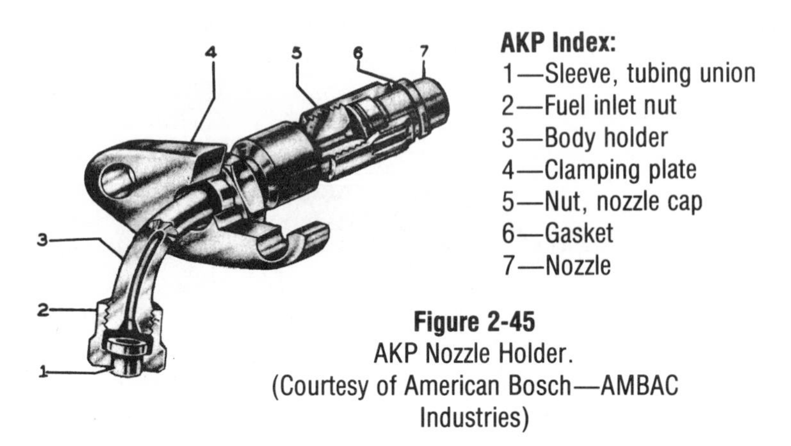

AKP Index:
1—Sleeve, tubing union
2—Fuel inlet nut
3—Body holder
4—Clamping plate
5—Nut, nozzle cap
6—Gasket
7—Nozzle

Figure 2-45
AKP Nozzle Holder.
(Courtesy of American Bosch—AMBAC Industries)

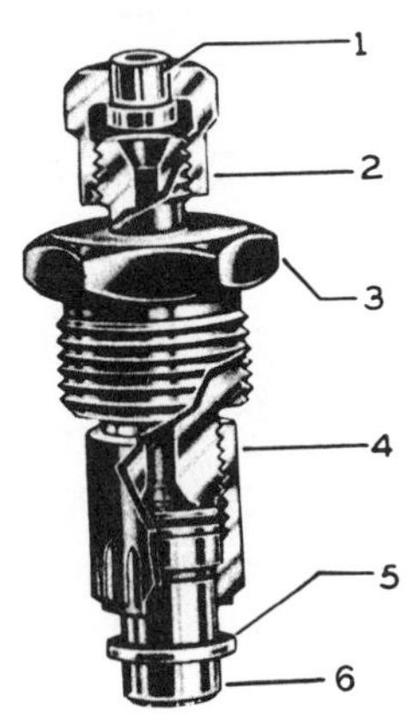

AKS Index:
1—Sleeve, tubing union
2—Fuel inlet nut
3—Body holder
4—Nut, nozzle cap
5—Gasket
6—Nozzle

Figure 2-46
AKS Nozzle Holder.
(Courtesy of American Bosch—AMBAC Industries)

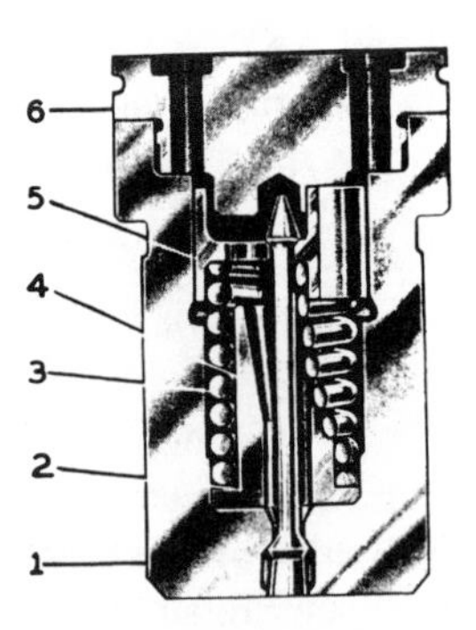

ADE Index:
1—Body nozzle
2—Pintle valve
3—Pressure adjusting spring
4—Spacer
5—Retainer valve
6—Cap, nozzle body

Figure 2-47
ADE Pintle Nozzle.
(Courtesy of American Bosch—AMBAC Industries)

Nozzle-Type Designations

Holder Assembly

Example: AKN 137M 6567 A1

AKN: Type of nozzle holder

137: Length of holder in millimeters from the clamping surface to the face of the cap nut

M: Diameter of holder shank is 17 mm; if SM, 21 mm; both with ADB M size nozzles

6567: Specification number; identifies special features

A: Edition letter; succeeding letters identify minor changes

1: Code number; denotes required nozzle opening pressure and/or nozzle

Nozzle

Example: ADB 155M 169-7

ADB: Denotes a hole (orifice)-type nozzle with a long shank

155: Included angle of spray cone

M: Miniature nozzle

169: Variation number; identifies special features

7: Identifies valve and body material

General Nozzle Maintenance

The servicing and maintenance of *any* fuel-injection system must be done with care, and must be carried out in an area of extreme cleanliness at all times. Failure to follow this major consideration will only lead to problems with repaired components. Too often when a lack-of-power complaint exists, the mechanic or serviceman immediately suspects either fuel-injection pump or nozzle problems, when often this is not the case. Once other possibilities have been checked and corrected, then the nozzles can be checked.

A quick check prior to removing the nozzles is to run the engine at an idle speed and loosen the nozzle fuel line about one half-turn. There should be a reduction in rpm and a postive change to the sound of the engine when this is done, since you are cutting out any fuel being delivered to that cylinder. If there is no change to the engine when this is done, check the operation of the nozzle after it has been removed. Do this check with each nozzle to establish if any appear to be misfiring.

If all nozzles check out satisfactorily, then injection pump timing should be checked (assuming that the condition of the fuel system up to the injection pump is satisfactory). If pump timing is correct, the pump should be removed for service. Remember to have checked and corrected all other possibilities first, especially such things as the items listed in the general troubleshooting charts and Chapter 12 of this book.

Some engine manufacturers of on-highway truck engines suggest that between 50,000 and 75,000 miles (80,450 and 120,675 km) it is advisable to check nozzles for proper opening pressure and spray pattern.

Removal

1. Always clean the area around the nozzles and fuel lines with solvent or a quick steam clean.
2. Remove leak-off lines, taking care not to lose the small copper gaskets.
3. Remove the high-pressure delivery line from the nozzle and immediately install protective shipping caps over both the fuel line and nozzle.
4. Remove nuts from the nozzle hold-down clamp studs (unless it is a screwed in type holder) and carefully remove the nozzle from its bore. If the nozzle is tight, insert a small heel bar under the hold-down flanges close to the nozzle body. Then

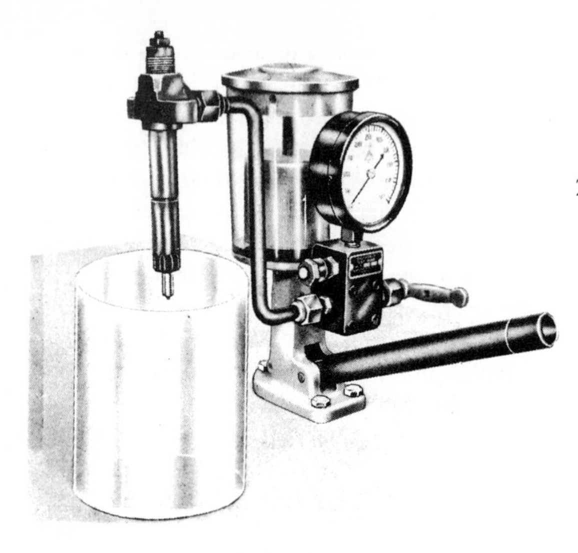

Figure 2-48
Nozzle Holder Assembly Attached to Pop Test Stand.
(Courtesy of American Bosch—AMBAC Industries)

again, if the nozzle is especially tight, a small amount of penetrating oil between the sleeve and the holder may help.

5. Install plastic-type shipping caps into the nozzle holder bores of the cylinder head if the nozzles are being left out for more than a few minutes.

Testing Nozzles

The test sequence for all nozzles follows the same basic pattern; therefore, the notes herein can be used to check out American Bosch, Robert Bosch, Simms/CAV, and similar-type nozzles. For explanation purposes, reference will be made to testing of typical Bosch 21/17 mm nozzles currently being used on a variety of engines worldwide; examples are Mack and Scania/Vabis.

Procedure

1. Using a small *brass* bristle brush, soak the tip of the nozzle in clean diesel fuel and carefully remove any carbon from it. *Never* use a wire brush or wheel for this purpose, as damage to the tip and spray holes is likely.
2. Attach the nozzle holder assembly to a suitable test stand as shown in Figure 2-48. Close the pressure gauge valve to protect the gauge and rapidly actuate the test stand lever at about 25 to 30 strokes/min to expel any air from the nozzle and also to settle the pressure regulating spring and nozzle loading column. Figure 2-48 shows one type of injector *pop tester,* which can be used for checking a variety of different nozzles. This is a fairly common piece of equipment in many maintenance shops.

Caution: Since fuel under pressure from the nozzle can penetrate the skin and destroy tissue, and possibly enter the bloodstream and cause blood poisoning, *keep your hands away from the nozzle spray.*

3. Open up the pressure gauge valve one half-turn and push down on the operating lever slowly, which will raise the fuel pressure. Take careful note on the gauge as to when the nozzle opens and compare this reading with the manufacturer's specifications.
4. If the nozzle requires adjusting to meet the specified opening pressure, depending on the particular type of nozzle being tested, add shims or adjust the opening pressure adjusting screw to obtain this reading.
5. American Bosch nozzles that require shims to affect the nozzle opening pressure, for example, are the AKC-S, AKB-S, AKK, to name a few, whereas the AKB-R and AKB-S have a pressure regulating spring adjusting screw.
6. On 17- and 21-mm nozzles the shim pack is located on top of the pressure adjusting spring. Adding one 0.002 in. (0.0508 mm)

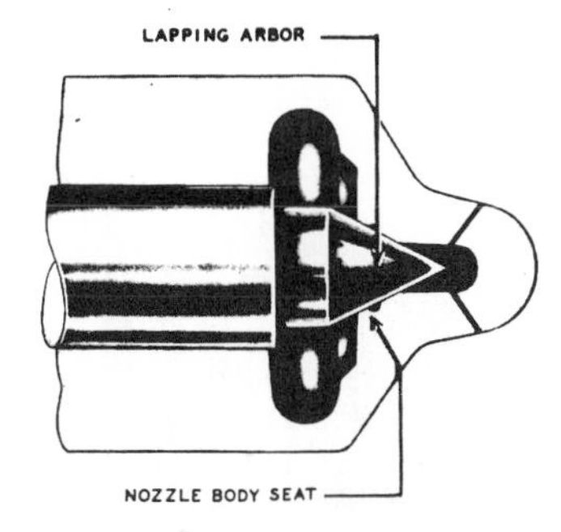

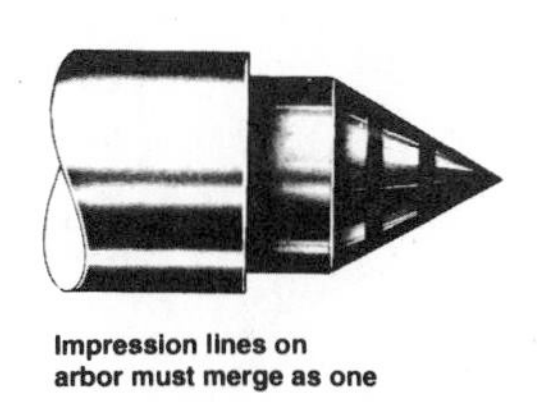

Figure 2-53
Use of an Arbor to Refurbish Nozzle Body Seat. (Courtesy of American Bosch—AMBAC Industries)

destroy the line contact (create different angles on valve and seat).

Any parts that show unusual wear patterns, pitting, scratching, or nicks should be lapped where possible or replaced.

Although nozzle valves and seats are capable of being cleaned with tallow, the valve is harder than the seat in the body; therefore, most wear will occur on the body seat. The body seat can be refinished by using a lapping arbor (dummy valve) with a small amount of very fine lapping compound such as BM 10007 or, for the initial lapping, coarse lapping compound such as TSE 7723 can be used followed by the fine.

Special equipment is required to grind the arbor to either 59 degrees on current nozzles or 89 degrees on earlier units. These angles are the included seat angle of the nozzle body; therefore, the grinding machine for the arbor must be set at half this angle, that is, 29.5 or 44.5 degrees.

Figure 2-53 shows a lapping arbor as it would contact the seat and also the impression of two light lines with a space between which indicates that the arbor angle is correct.

It will be necessary to periodically clean the body with solvent and to blow it clean with compressed air to allow a clean surface for lapping. As the arbor wears, it must be reground or another properly ground arbor can be used to speed up the process.

The seat should be lapped with compound and the arbor until the two lines on the arbor seat shown in Figure 2-53 merge and become one. Once the valve and nozzle seat have been thoroughly cleaned, the valve should be free in the nozzle and capable of dropping by its own weight back into the nozzle after lifting it out partway.

Important: Once the nozzle seat has been lapped, you must measure the valve lift, which is the distance from the valve stop shoulder to the high-pressure face of the body.

Figure 2-54 shows how to check the valve lift dimension. Variations in valve lift are caused by the following:

1. When lapping, the amount of material removed from the body seat.
2. When using a valve grinding refinisher, the amount of material removed from the valve seat.

If excessive lift is evident, a new nozzle assembly (valve and seat body that are matched units) must be used; otherwise, nozzle pounding will occur, giving short life and poor performance.

The nozzle lift can be done in two ways; one method is shown in Figure 2-54, where the nozzle can be held in its body as noted, with a small straightedge held across the lapped surface of the body. *Zero* the dial indicator over the end of the needle valve, and with a pair of tweezers gently lift the valve until it contacts the straightedge and note the reading. A simpler and easier way is to use a dial gauge, such

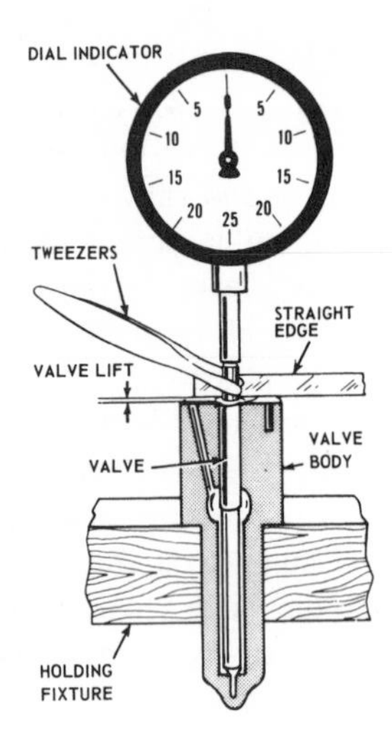

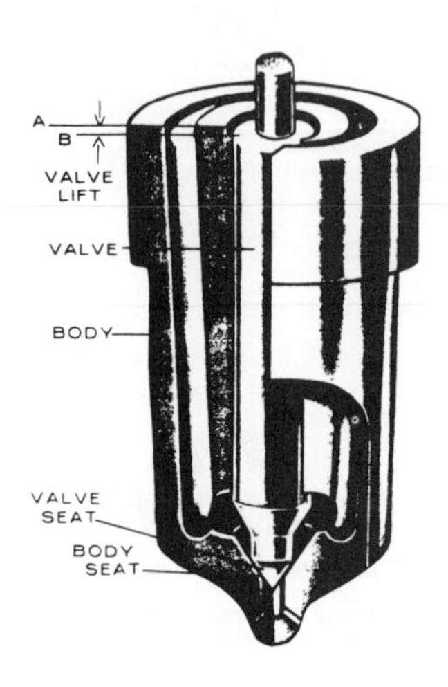

Figure 2-54
Measuring the Valve Lift.
(Courtesy of American Bosch—AMBAC Industries)

as a Kent-Moore J 9642-02 used on Detroit Diesel engine injectors.

Reassembly

1. Dip all component parts in clean test oil or diesel fuel, and lightly clamp the nozzle holder body into a soft-jawed vise or holding fixture.
2. Refer to Figure 2-55, which is a typical chart used for reassembling nozzle assemblies.

Note: If a new pressure adjusting spring is being used on reassembly, always adjust the nozzle opening pressure 10% higher than the listed specification to allow for settling of the spring after a short time.

Current 17- and 21-mm nozzles used on 6- and 8-cylinder Mack truck engines have a screw adjustment opening pressure.

Installation: Nozzle to Engine

1. Remove the protective shipping caps from the nozzle bore in the cylinder head. If not done earlier, remove the copper nozzle tip gasket from the bore area.
2. Clean the bore with reamer J-23303 and wire brush AC-12 or equivalent; then crank the engine over to blow out any loose material from the nozzle bore. You may have to use a lint-free rag to wipe the gasket surface clean.
3. A light coat of *never-seize* applied to the shank area of the nozzle holder assembly will generally prevent binding and freezing of the nozzle in its bore, making removal easier next time around.

4. Install new copper gaskets with a small amount of grease, which will allow them to stick to the nozzle during installation into the bore. Also install new dust gaskets.
5. Tighten the nuts on the 17- and 21-mm nozzles to 14 to 17 lb-ft (18.98 to 23.04 N·m) by pulling them down evenly.
6. Install the high-pressure fuel lines on the nozzles and torque to specification. Install the leak-off lines and copper gaskets.

Note: High-pressure lines will have to be left loose until the fuel system has been bled of all air. See section dealing with bleeding the fuel system.

Testing Pintle-Type Nozzles

The sequence of events for pintle-type nozzles is the same as that for hole-type nozzles already described. The fuel spray characteristic is, however, considerably different and should

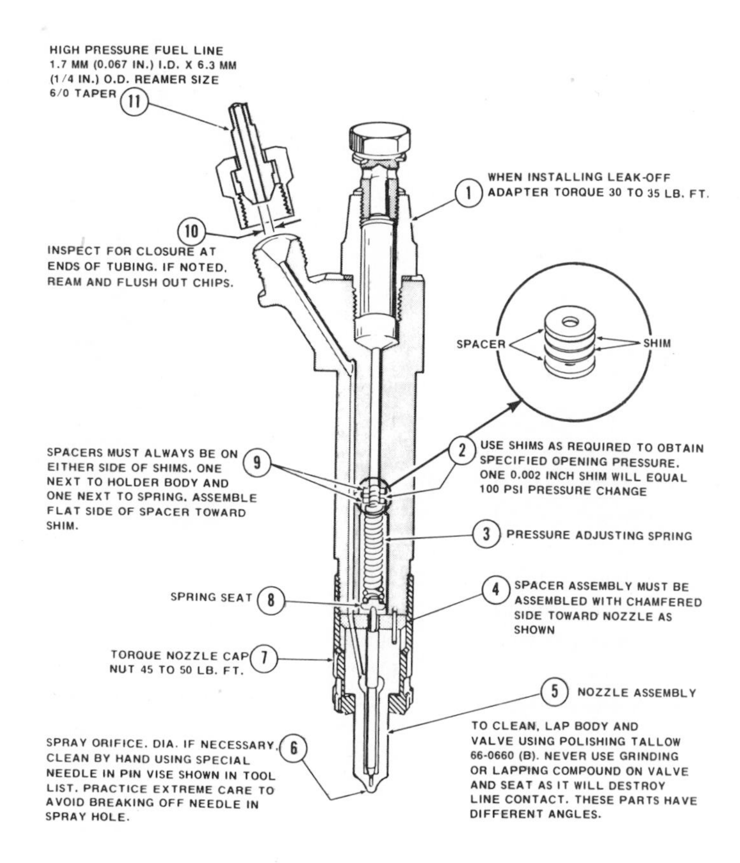

Figure 2-55
Reassembling Nozzle Assemblies. (Courtesy of American Bosch—AMBAC Industries)

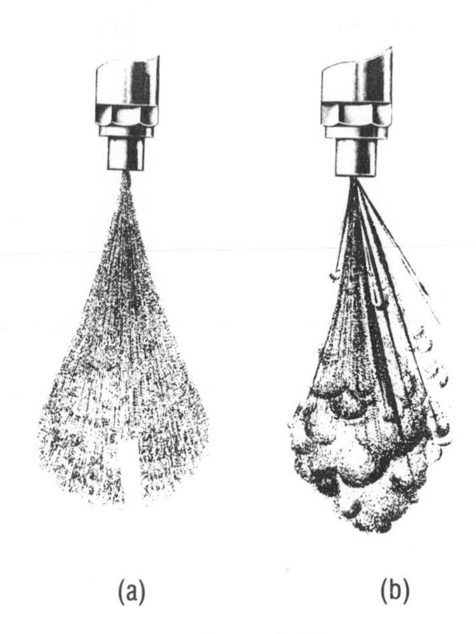

Figure 2-56
Fuel Spray Patterns—Standard Pintle Nozzle Assembly.
(Courtesy of American Bosch—AMBAC Industries)

appear as shown in Figures 2-56 and 2-57.

The standard pintle nozzle is depicted in Figure 2-56. The fuel spray pattern should be as shown in Figure 2-56(a). If the spray pattern appears as in Figure 2-56(b), the nozzle must be replaced.

Throttling or pintaux-type nozzles, because of their internal construction, require much shorter, quicker strokes of the pump handle when checking the spray pattern. Chatter also may or may not occur with this nozzle; therefore, it is not in poor condition because it does not chatter.

Figure 2-57(a) is the kind of spray pattern that should exist on a throttling nozzle, whereas Figure 2-57(b) is the kind of spray pattern that should exist when a pintle throttling nozzle with an auxiliary hole is used. These nozzles are identified by the letter P after the part number stamped on the nozzle body.

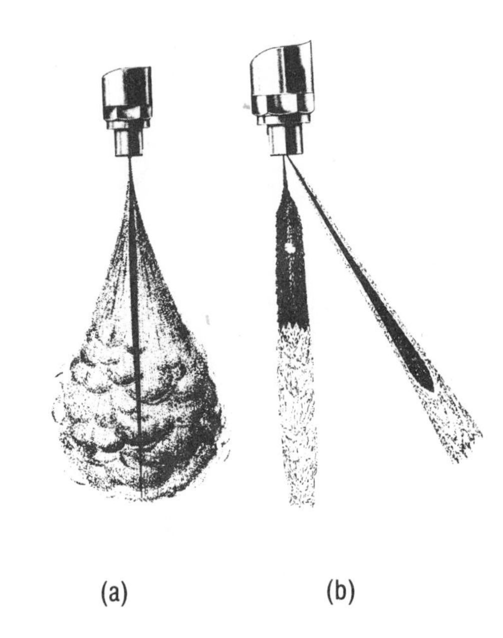

Figure 2-57
Standard Pintaux Nozzle Spray.
(Courtesy of American Bosch—AMBAC Industries)

chapter 3

Mack Engines Fuel System

3-1

Fuel-Injection Pump Installation and Timing to Engine

One of the main users of both American Bosch and Robert Bosch fuel-injection pumps is Mack and Mack/Scania. The following instructions relate to American Bosch APE and Robert Bosch PE types of injection pumps for 6- or 8-cylinder Mack diesel engines.

Since Mack engines use either American or Robert Bosch fuel-injection pumps, much of the applicable information for Mack systems can be found in Chapters 2 and 4.

Latest additions to the 6-cylinder American Bosch in-line injection pump include a vented overflow valve from the housing, which does away with the necessity of an air bleed hose from the secondary fuel filter. Therefore, on newer style pumps, the top of this filter has no attachment for this line, or it has been plugged off.

Another improvement to this pump is that a one-piece snubber valve is now used instead of the former three-piece unit. This is located above the delivery valve at the high-pressure fuel line outlet to the nozzle and is designed to reduce cavitation erosion by dampening out high-pressure fuel line flow.

V8 ENGINES

1. Rotate the engine over by hand in its normal direction of rotation, which is cw from the front or ccw from the rear. Continue to turn the engine over until piston 1 is at TDC on compression (cylinder 1 is on your left when facing the engine from the front). TDC is reached when the timing indicator and the timing marks on the auxiliary drive shaft gear align with the timing mark on the auxiliary drive idler. As shown in Figure 3-1, the pin hole location will be at the 5 o'clock position and the coupling tangs are at 12 degrees 36 minutes off the horizontal.

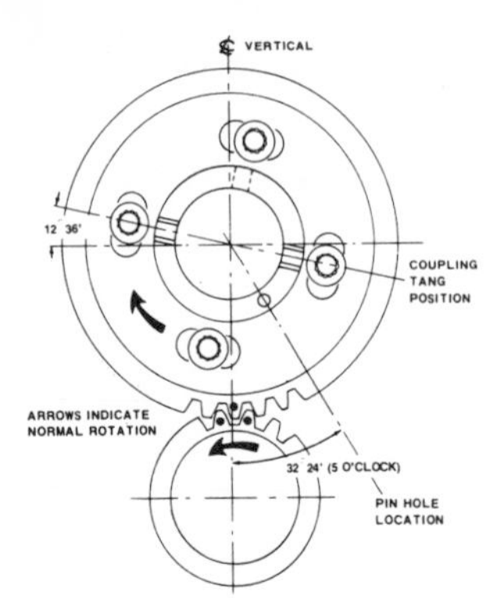

Figure 3-1
Position of Drive Coupling Tangs in Relationship to Timing Marks with Piston 1 at TDC. (Courtesy of Mack Trucks, Inc.)

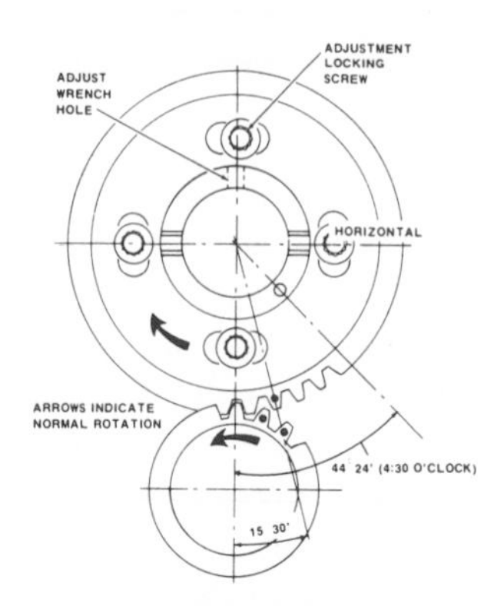

Figure 3-2
Approximate Position of Drive Coupling Tangs in Relationship to Timing Marks Prior to Fuel-Injection Pump Installation. (Courtesy of Mack Trucks, Inc.)

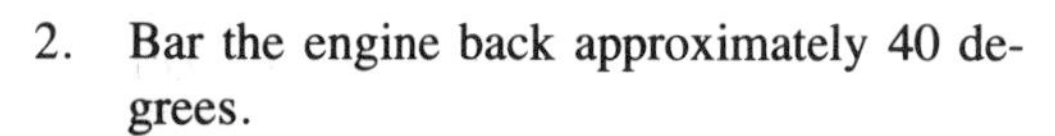

2. Bar the engine back approximately 40 degrees.
3. Bar the engine over again in its normal direction of rotation until the timing indicator reads the number of degrees BTDC as specified on the fuel-injection pump nameplate. Referring to Figure 3-2, it will be noted that in this position the tangs on the drive coupling will be approximately horizontal, and the drive coupling pin hole will be approximately at the 4:30 o'clock (1630) position. The gear marks will be approximately 15 degrees 30 minutes to the right.

Figure 3-3
Injection Pump and Ring and Pin Assembly Prior to Installation. (Courtesy of Mack Trucks, Inc.)

4. Install the pin and ring assembly in position on the injection pump drive coupling as shown in Figure 3-3.
5. Install the injection pump.
6. Recheck the pump for port closing in relationship to the specified timing degrees BTDC.
7. If it is necessary to make any adjustments, remove the auxiliary housing inspection cover and loosen the four adjustment locking screws on the fuel pump coupling.
8. Move the coupling to obtain port closure at the degrees desired BTDC as specified for that pump assembly on the engine.

Six-Cylinder Engines

The procedure for the 6-cylinder engine is similar to the V8 and is as follows:

1. Manually bar the engine over in its normal direction of rotation until cylinder 1 is on compression and the damper or flywheel timing mark indicates the correct number of degrees BTDC as recommended on the valve rocker cover escutcheon plate. The injection pump drive shaft flange lugs should now be in a horizontal plane and indexing pin hole at the 4 o'clock (1600) position.
2. Grease the front (noncounterbored) face of the coupling and mount on the pump drive shaft flange and center the coupling ring laterally.
3. Remove the pump adapter inspection hole cover.
4. Rotate the pump coupling driven flange until the lugs are in a vertical plane and the indexing hole is at the 7 o'clock (1900) position.
5. Mount the pump assembly on the engine. As the pump moves into position, the indexing pin can be observed through the adapter inspection hole.
6. Install the adapter to the cylinder block cap screws and tighten lightly.
7. Install and tighten the 3/8-in. (9.52-mm) cap screw connecting the top and bottom halves of the support bracket. Tighten the lower bracket mounting screws. This will align both halves of the bracket. Then loosen the $\frac{3}{8}$-in. connecting cap screw.
8. Torque the adapter to the cylinder block cap screws and tighten the 3/8-in. connecting cap screw.
9. Check the coupling ring through the adapter inspection hole for approximately $\frac{3}{32}$ in. (2.38 mm) end float.
10. Again bar the engine over in its normal direction of rotation until cylinder 1 is on its compression stroke and the flywheel or vibration damper timing mark is just coming up to the number of degrees BTDC as recommended on the valve cover escutcheon plate.

Timing

Important: Before timing the engine, verify the setting on the fuel-injection pump nameplate or EPA (Environmental Protection Agency) engine emission plate.

The following procedures are to be used when setting the static injection pump timing (port closure) using either the portable high-pressure PC (port closure) stand or the low air pressure PC checking method.

Recommended Tools

1. Portable high-pressure port-closing timer (Bacharach P/N 72-7010) or high-pressure hand supply pump (Robert Bosch P/N 1 687 222 039)
2. Timing plug gauge (J24345-1 for American Bosch injection pump, J24345-2 for Robert Bosch injection pump).

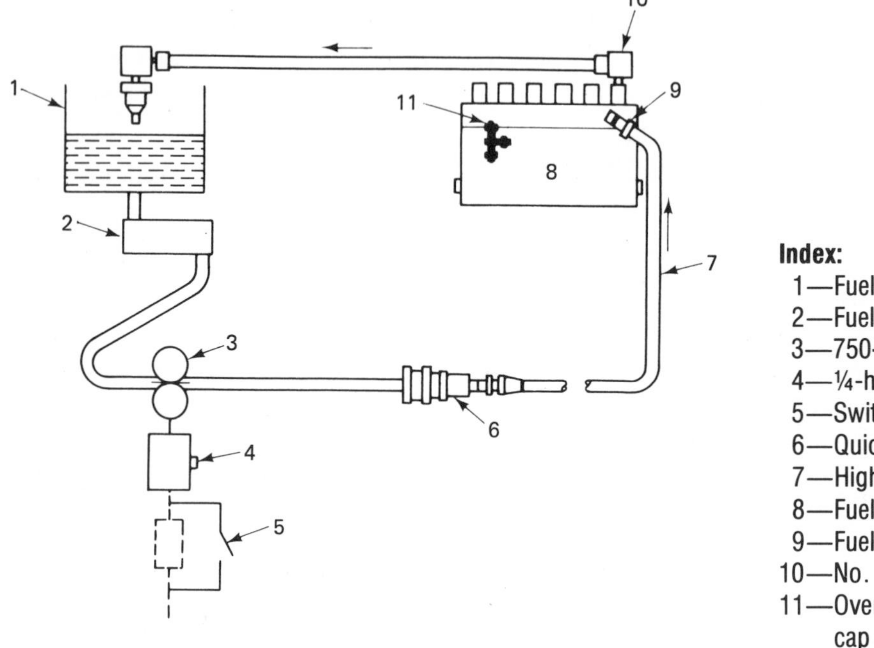

Index:
1—Fuel reservoir
2—Fuel filter
3—750-psi pump
4—¼-hp motor
5—Switch
6—Quick disconnect
7—High-pressure hose
8—Fuel-injection pump
9—Fuel inlet gallery
10—No. 1 delivery valve
11—Overflow relief valve cap

Figure 3-4
High-Pressure Port Closing Fuel System Diagram.
(Courtesy of Mack Trucks, Inc.)

Method 1: Using High-Pressure PC Stand. Refer to Figure 3-4 and connect up the injection pump as shown.

1. Cap or connect the injection lines on all delivery valve units except delivery valve 1 outlet on both the 6- and 8-cylinder engine fuel-injection pumps.
2. Remove all return fuel feed lines at the overflow relief valve union fitting and cap the valve port connections.
3. Connect the high-pressure line from the portable PC stand to the fuel pump inlet of the pump gallery.
4. Connect the return line from the cylinder 1 delivery valve holder to the portable PC stand.
5. Loosen the engine to fuel pump drive coupling cap screws and move the injection pump drive ccw while viewing the engine from the front on 6-cylinder engines, or cw while viewing the engine from the front on V8 engines, until the pump drive is at the end of the adjusting slots; snug up the cap screws.
6. Bar the engine over clockwise from the front until cylinder 1 is coming up on its compression stroke and stop at the recommended static port closing degrees BTDC for the particular engine involved. The number of degrees for the BTDC position is visible on either the vibration damper or flywheel.
7. Remove the control rack cap plug and

insert the correct timing plug gauge (J24345-1 for American Bosch injection pumps and J24345-2 for Robert Bosch injection pumps).

Note: Injection pumps not featuring retard start do not require the use of the timing plug gauge. Injection pumps fitted with the Mack puff limiter and/or torque limiter cylinder do not require timing plug gauges when setting the static injection pump timing (port closure); however, 80 to 120 psi (551.6 to 827.4 kPA) air pressure must be applied and held to the system air cylinder.

8. Secure the stop lever in the running position.
9. Activate the throttle lever several times and secure it in the full-load position.
10. Introduce fuel pressure to the pump gallery.
11. Turn the injection pump drive cw when viewing the engine from the front on 6-cylinder engines and ccw when viewing the engine from the front on V8 engines until the fuel flow from the delivery valve 1 outlet changes from a solid stream to the formation of drops. This is known as port closing.
12. Lock the drive gear cap screws securely, being very careful not to disturb the relative position of the gear of the drive shaft and thus the port closing setting.
13. Turn the engine crankshaft ccw, opposite engine rotation, a minimum of one half-turn, followed by cw rotation to check port closing at the correct position as indicated on the vibration damper or flywheel.

Note: If correct timing is not obtained during this check, deactivate the port closing stand, bleed residual pressure from it, and repeat steps 5 through 12.

14. Torque the drive gear cap screws to 35 lb-ft (47.45N·m) on 6-cylinder engines and 26 lb-ft (35.25 N·m) on V8 engines.
15. Remove the timing plug gauge and replace the control rack cap plug.
16. Remove the caps from the overflow relief valve union port connections and reconnect all fuel return lines.

Caution: When pressure timing the Robert Bosch fuel-injection pump as applied on the Mack/Scania engine, the damper cylinder must be removed from the pump to prevent cylinder damage.

Note: The high-pressure timing procedure is the recommended method for American Bosch APE and Robert Bosch PE injection pumps. The low-pressure timing procedure (method 2) should only be used in emergency cases where high-pressure timing is not available.

Method 2: Using Low Air Pressure

1. Remove the delivery valve 1 holder from the injection pump and take out the delivery valve and spring and, on Robert Bosch fuel-injection pumps, the spring shim.
2. Install a suitable air line onto the IN fitting of the pump gallery.

Caution: Ensure that the air line is equipped with a separator and pressure regulator. Moisture-laden air can cause serious damage to injection pump parts.

3. Attach a locally fabricated fixture to the delivery valve holder similar to that shown in Figure 3-5.
4. Secure the stop lever in the running position.
5. Activate the throttle lever several times and secure it in the full-load position.
6. Remove the control rack cap plug and insert the correct timing plug gauge (J24345-1 for American Bosch, J24345-2 for Robert Bosch).

Note: As in method 1, injection pumps not featuring retard start do not require the use of

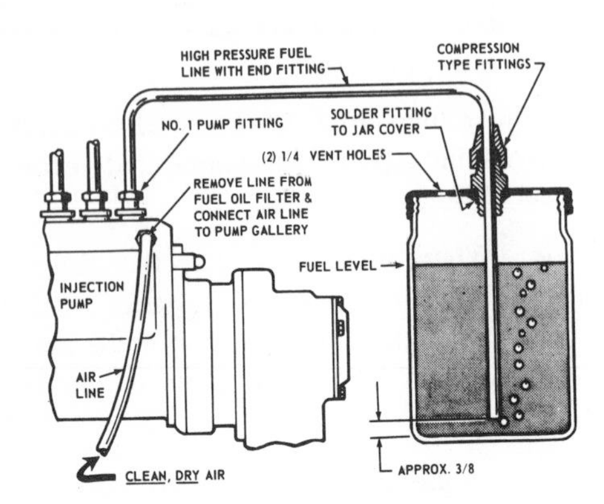

Figure 3-5
Fuel-Injection Pump Air Flow Checking Method. (Courtesy of Mack Trucks, Inc.)

the timing plug gauge. Injection pumps fitted with the Mack puff limiter and/or torque limiter cylinder do not require timing plug gauges when setting the static injection pump timing (port closure); however, 80 to 120 psi (551.6 to 827.4 kPA) air pressure must be applied and held to the system air cylinder.

7. Turn on the air supply and just crack the regulator so that a steady flow of air bubbles is seen in the fixture jar without excessive turbulence.
8. Rotate the crankshaft slowly in its normal direction of rotation. Observe the flow of air bubbles in the fixture jar, and the instant the bubbles stop, discontinue rotating the crankshaft.
9. Check the position of the flywheel or vibration damper timing indicator. If properly timed the indicator must register the recommended number of degrees of BTDC stamped on the valve rocker cover escutcheon plate.
10. If the timing checks out, repeat steps 8 and 9 to ensure accuracy.
11. If the timing does not check out, bar the engine over in its normal direction of rotation until cylinder 1 is on the compression stroke and the timing mark indicates the correct number of degrees BTDC as recommended on the valve rocker cover escutcheon plate.
12. Loosen the engine to fuel pump drive coupling cap screws and move the injection pump drive ccw on 6-cylinder engines when looking at the engine from the front, and cw when viewing the engine from the front on V8 engines, until the pump drive is at the end of its adjusting slots; tighten up the cap screws.
13. Slowly turn the injection pump drive cw when viewing the engine from the front on 6-cylinder engines, and ccw when viewing the engine from the front on V8 engines, while checking air bubbles in the fixture jar. Stop turning as soon as the air bubbles cease. This is known as port closing.
14. Turn the engine crankshaft ccw, opposite engine rotation, a minimum of one half-turn, followed by cw rotation in the direction of normal engine rotation to check port closing at the desired timing on the vibration damper or flywheel.
15. Tighten the two opposite cap screws to secure the hub, and rotate the crankshaft and recheck the timing.
16. When the timing is correct, tighten the two remaining cap screws to 35 lb-ft (47.45 N·m) on 6-cylinder engines and to 26 lb-ft (35.25 N·m) on V8 engines.
17. Remove the timing plug gauge and replace the control rack plug.
18. At the completion of a satisfactory timing check, proceed as follows:

American Bosch Pumps:

Replace the delivery valve and spring.

Torque the delivery valve holder nut as follows:

Cadmium coated: 60 to 65 lb-ft (81.34 to 88.12 N·m)

Black oxide: 85 to 90 lb-ft (115.24 to 122 N·m)

Robert Bosch Pumps:

Replace the delivery valve, spring shim, and spring.

Reinstall the delivery valve holder with a new O ring and copper gasket.

Torque the delivery valve holder to 50 to 55 lb-ft (68 to 4.56 N·m).

3-2

High and Low Idle Pump Adjustment

The following adjustments should only be done by a local Mack or authorized dealer if the engine is still under warranty.

Low Idle Adjustment

On both the American and Robert Bosch fuel-injection pumps, the low idle adjusting screw is located by the pump throttle lever. With the engine at operating temperature, loosen the idle screw locknut and adjust to the desired rpm.

High Idle (Maximum) Engine Speed

If a lack of top end power exists, check the maximum engine rpm as follows:

American Bosch Pumps

1. The high idle screw is located under the two-piece protective cover by the throttle lever at the pump. Break the seal to get at it, and remove the throttle lever temporarily from the pump shaft.
2. With the throttle lever removed, remove the cover, break the seal, and loosen the high idle screw locknut.
3. Reinstall the throttle lever, and run the engine to its maximum rpm. Adjust the screw to obtain the recommended top speed; lock the screw and reseal.

Robert Bosch Pumps

Repeat steps 1 through 3 for American Bosch pumps; however, if the rpm is low after high idle adjustment, then the injection pump requires recalibration and should be removed from the vehicle. If the maximum rpm is too high, readjust and reseal.

High Idle Adjustment for Maxi-Miser Units (Maxidyne Engines with American Bosch Pumps)

Apply the vehicle's spring parking brakes, and place the transmission in neutral.

1. Disconnect the accelerator linkage springs, and loosen and push to the side the air line from the rear of the governor.
2. Slide the loosened air line to the maxi-miser cylinder off to the side. Also remove the air inlet fitting from the dual-speed governor, and loosen the air inlet cap.
3. Start and warm up the engine; obtain an accurate tachometer. Place the throttle in the high idle position and lock it there.
4. Using an Allen wrench, turn the adjusting screw in the end of the maxi-miser air cylinder at the governor either cw to increase the rpm or ccw to decrease the rpm.
5. Tighten the air inlet cap at the end of the maxi-miser air cylinder; install the air inlet fittings and connect up the air line to the dual-speed governor.
6. Reconnect all linkage and install new lead seals as required.

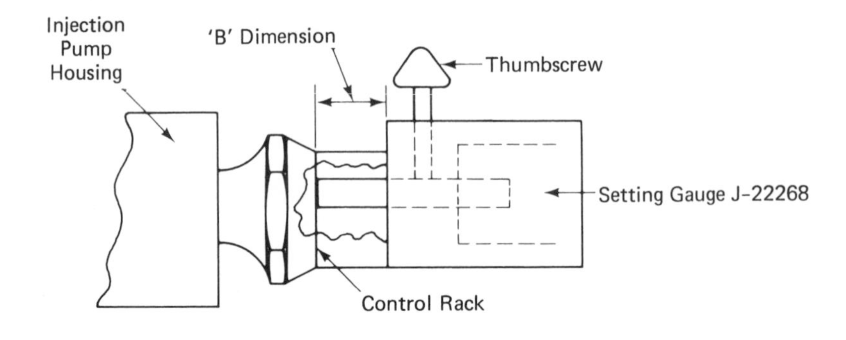

Figure 3-6
Checking the B Dimension.

3-3

Robert Bosch Pump Static Rack Setting

The *static rack setting* or *B* dimension is stamped on the governor housing of each pump just below the removable cover. To check this dimension, proceed as follows:

1. Remove the plug from the end of the control rack cap (end opposite the governor).
2. With the engine stopped, manually move the throttle lever to the full-fuel position and hold it there. Move the stop lever to the stop position and back to run again several times to index the internal cam nose with the cam plate. If the cam nose does not index on the cam plate, check the pump for a problem here.
3. With the cam nose and plate indexed, obtain rack setting gauge J-22268, and hold the gauge flush against the end of the control rack cap; push in the pin until it contacts the end of the rack (see Figure 3-6).
4. Lock the gauge internal pin in position while it is against the fuel rack; then remove the gauge and check the distance from the outer edge of the gauge to the internal pin with a depth micrometer. This is the B dimension.
5. If the measured B dimension is different from that stamped on the pump governor housing by more than 0.005 in. (0.127 mm), the setting is incorrect and the injection pump will have to be recalibrated on a test stand.

3-4

American Bosch Pump Static Rack Setting

On the American Bosch pumps, the same basic dimension that we just discussed for Robert Bosch pumps is known as the K dimension. It is also stamped on the housing. This dimension is the horizontal distance between the upper face of the stop plate and the inner vertical surface of the governor housing.

1. Cut the wire seal, remove the governor top cover, and move the shutoff lever to the stop position.
2. Loosen the thumbscrew on tool TSE-

79100, and pull the knurled pin all the way outward (see Figure 3-7). Make sure that the dowel pins face toward the fuel pump when the bracket tool (TSE-79100) is attached on top of the pump with the governor cover removed, and push the pin in until it contacts the cam nose stop plate. Make sure that the tool is pressed against the inner governor front housing when installed. Also make certain when the pin is pushed in that it only makes contact with the stop plate.

3. Tighten the thumbscrew and remove the tool from the pump. Lay a depth micrometer on the tool so that its flat base bears against the dowel pins. Run the micrometer in until its extension lightly contacts the round head of the pin, and note the reading.

4. If the dimension is incorrect, the pump requires recalibrating.

Maxi-Miser (Maxidyne Engines with American Bosch Pumps Only)

On many highway truck engines, a dual-speed variable-speed governor is used, which does the following. In the first four transmission gears, air pressure from the truck air system is directed to a piston within the governor housing, which forces a plunger forward allowing the stop bracket to move away from the throttle fulcrum lever bracket; this permits full rated rpm to be reached in all these gear ranges.

However, when the operator or driver shifts into fifth gear, the shift rail triggers open an exhaust valve, which vents the air from the governor maxi-miser, and a spring returns the stop bracket of the plunger to a position whereby the throttle fulcrum lever bracket will now be limited in its travel within the internal governor housing. Therefore, maximum rpm of the engine in top gear is less than in all other ranges, which reduces fuel consumption and lowers the maximum speed of the engine and vehicle. If air pressure is lost from the

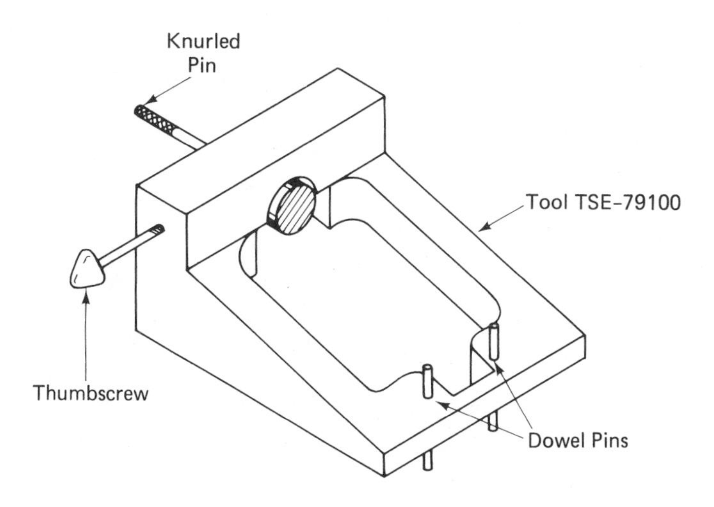

Figure 3-7
Tool TSE-79100, Used to Check the K Dimension.

maxi-miser cylinder at any time, then maximum rpm would also be available in top gear.

3-5

Puff Limiter Operation and Checks

The Mack puff limiter is an external pneumatic device for controlling the acceleration smoke puff common to turbocharged engines due to turbocharger lag. Figure 3-8 shows the puff limiter arrangement. It is designed to act directly on the fuel control rack of the injection pump and is independent of the governor action. The puff limiter consists basically of a reversing relay and a normally retracted air cylinder. The reversing relay provides an output pressure signal to the fuel rack control air cylinder that decreases in direct proportion to the increase in engine intake manifold turbocharger air boost pressure. During low manifold air pressure conditions, the puff limiter air cylinder controls the maximum fuel to the engine.

The air pressure to the system is provided by the vehicle's main air tanks (brakes), and will be 100 + 20 psi (689.5 kPa + 138 kPa); however, the reversing relay output pressure will be regulated to 28 to 29 psi (193 to 200 kPa) at zero intake manifold air pressure. The controlling or sensing pressure is zero to maximum inlet manifold air pressure.

Operation

During vehicle operation, the reversing relay will receive spring brake air system pressure required to activate the air cylinder. When parking the truck, always apply the spring brakes. You should also ensure that the spring brakes are applied prior to starting the engine, which will assure that maximum starting fuel will be delivered by the injection pump.

The port marked with an S on the relay is connected to the engine inlet manifold, and it is this pressure that controls the relay's output. The output port is in turn connected to the air cylinder on the fuel-injection pump.

The pressure coming out of the relay will be inversely proportional to the signal pressure from zero to 28 psi (193 kPa), or when the truck is moving the maximum pressure will be controlled by the inlet manifold pressure. Therefore, the reversing relay is adjusted so that at 100 psi (689.5 kPa) inlet pressure the reversing relay will put out 28 psi (193 kPa) when zero signal pressure is applied.

However, an increase in the signal pressure (from the inlet manifold) from zero to 5 psi (34.5 kPa) will raise the relay output pressure to 23 to 24 psi (159 to 165.5 kPA). With an additional signal pressure increase, the relay output pressure will decrease a like amount. Therefore, when a signal pressure (turbocharger boost) of 28 psi (193 kPa) is attained, the relay output pressure will be zero psi.

Normally, when the truck is operating, full rack travel will be possible as the signal pressure reaches 12 psi (83 kPa), with the relay output pressure reduced to 16 to 17 psi (110 to 111 kPa), since the puff limiter air cylinder piston shaft will retract enough to allow this condition.

The diameter of the PL air cylinder is $\frac{3}{4}$ in. (19.05 mm) and is normally a retracted air cylinder. As long as the turbocharger air boost pressure is at its high end, maximum fuel rack travel is possible for greatest fuel delivery, such as at full-load and lugdown situations. However, remember that the reduction in turbocharger boost pressure allows the air cylinder piston to move to the extended fuel position, thereby limiting rack travel.

Adjustment

The PL cylinder is *shim* adjustable, with its dimension stamped on the governor pump

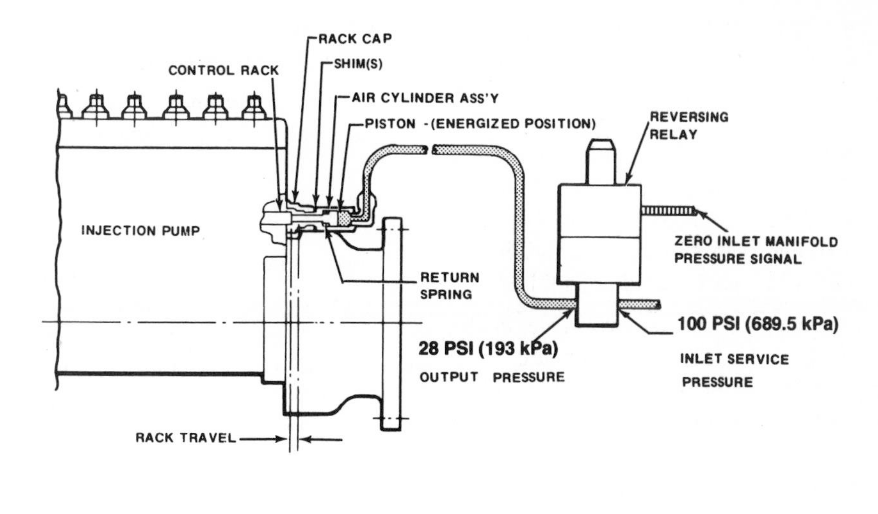

(a)

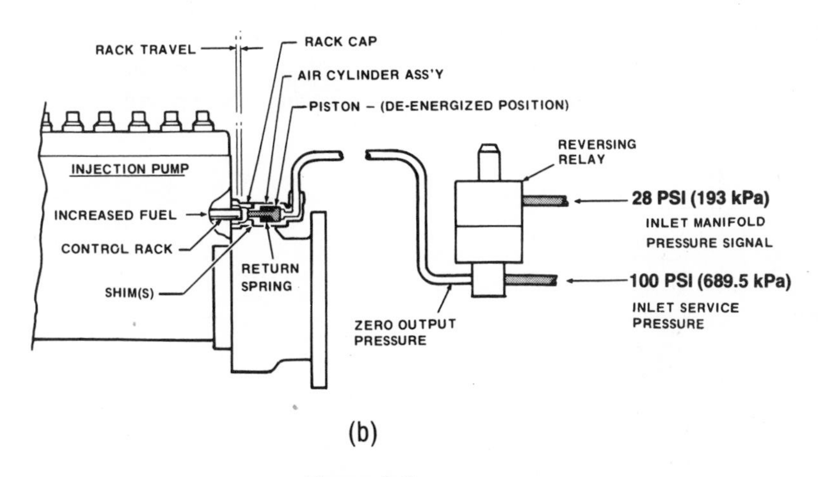

(b)

Figure 3-8
Puff Limiter: (a) Puff limiting position; (b) puff limiter normal operating position. (Courtesy of Mack Trucks, Inc.)

housing or puff limiter cylinder. The puff limiter extension (PLE) number is stamped on the injection pump or governor housing. To adjust, proceed as follows:

1. Remove both the PL and plunger, and apply 60 psi (413 kPa) shop air to the cylinder; measure the air cylinder extension from the plunger to the machined shoulder of the cylinder at the top end of the threads.
2. The dimension obtained in step 1 is known as the Y dimension. Therefore, subtract

the PLE number stamped on the injection pump from the Y dimension to obtain the difference, if any. Any difference would be the size of the shim pack required to obtain the correct setting. For example, if the Y dimension was 1.077 in. (27.35 mm) and the PLE number was given as 1.037 in. (26.33 mm), a 0.040-in. (1.016 mm) shim pack would be required.

3. Since the shims are available in several thicknesses, use a micrometer to ensure that the shim pack is within 0.005 in. (0.127 mm) of the specification number. Shims are available in thicknesses of 0.010 in. (0.254 mm), 0.015 in. (0.38 mm), 0.020 in. (0.508 mm), 0.025 in. (0.635 mm), 0.050 in. (1.27 mm), and 0.100 in. (2.54 mm).
4. Place the shim pack on the threaded end of the cylinder, and screw it into the injection pump housing rack cap to within $\frac{1}{8}$ in. (3.175 mm) of bottoming.
5. Apply 242 Loctite to the threads, and tighten the cylinder (hand tight).
6. Connect the air line to the air cylinder from the reversing relay valve.

Torque Limiting Control

The puff limiter is also used with a torque limiting control arrangement found on all Mack truck chassis using the TRXL107 or TRXL1071 Maxitorque transmission and bogie (axle) loading exceeding 34,000 lb (15422.4 kg). Figure 3-9 shows the setup of the puff limiter and the torque control as applied to one of these trucks. The transmission control valve (governor) functions as an exhaust. Any time that the TRXL107 transmission is shifted into reverse, or the TRXL1071 into reverse or Lo-Lo, the transmission valve opens, exhausting any manifold air pressure in the control line. When this occurs, the reversing relay directs maximum

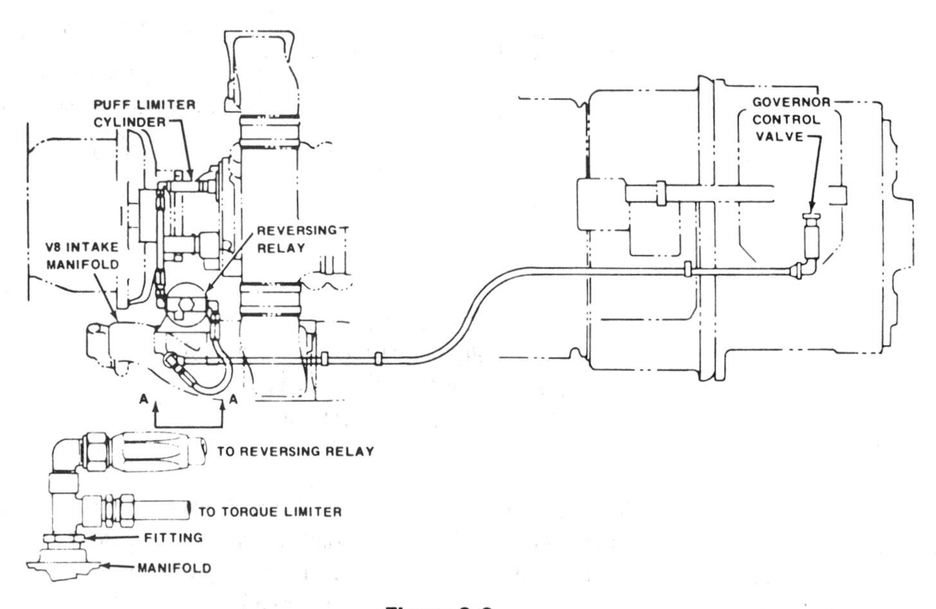

Figure 3-9
Puff Limiter Used with a Torque Limiting Control.
(Courtesy of Mack Trucks, Inc.)

Table 3-1
INJECTION PUMP IDLE SPEED AND FLOW CHART

Pump Style	*Engine Series*	*Pump RPM*	*Flow per 1,000 Strokes*
6 In-line	ENDT675/676	600	124 ± 4 cc**
V-8	ENDT865	800	108 ± 4 cc
V-8	ENDT866	800	100 ± 4 cc

**Flow specifications for Robert Bosch pumps 313GC597-P8 and P10, and 313GC5104-P2, applied on the ENDT676 series engine, is 116 ± 4 cc at 600 pump rpm.

output pressure to the air cylinder, which provides suitable torque limiting. The function of the special reducing orifice is necessary to meter the volume of air, which, if too great, could conceivably bleed over to the reversing relay, resulting in a condition similar to normal operation.

Additional Checks

There should be a fairly distinctive *click* sound from the PL air cylinder as the spring parking brakes are released. You may have to block the vehicle's wheels and have someone operate the spring brake control while you listen for this sound; or with your hand over the air cylinder, some vibration should be felt as the plunger contacts the injection pump rack.

With the signal line from the inlet manifold disconnected at the reversing valve relay port, install an adjustable air pressure regulator with a gauge into the port. Also disconnect the output pressure line at the reversing relay port connection, and install a 0 to 100 psi (689.5 kPa) gauge here. Run clean dry shop air to the pressure regulator, and with full system air in the truck air tanks, release the spring parking brakes, thereby directing full air pressure to the relay. Owing to the time delay within the relay, the gauge on the output side will momentarily indicate system pressure; however, the gauge should then record 28 to 29 psi (193 to 200 kPa) since the regulator for shop air will be set at zero, simulating zero inlet manifold air pressure. Adjust the regulator gradually to allow 28 to 29 psi (193 to 200 kPa) of shop air to feed into the relay. If the relay is operating correctly, the output gauge should read from 0 to 1 psi (0 to 6.895 kPa).

Repeated Low-Horsepower Complaint

If the PLE dimension is indistinguishable on the pump housing or a low-power complaint continues to exist, remove the pump from the engine, and mount it on a test stand. Proceed as follows:

1. With the pump set up and ready to be tested, apply a minimum of 60 psi (413.7 kPa) air pressure to the PL air cylinder.
2. Obtain the test specification sheet for your particular model of pump (includes *all* necessary information).
3. With the pump throttle lever at full load, run the pump on the test stand at the idle speed listed in Table 3-1.
4. If flow per 1000 strokes is not within specifications, proceed as follows.
5. Shut off test stand and release the air pressure from the PL air cylinder.
6. Remove the air cylinder and add shims if the flow is less than specifications, and remove shims if the flow is above specs. (*Note:* The addition of one 0.010 in.

(0.254 mm) shim will change the pump flow rate by 5 to 7 cc per 1000 strokes.)

7. Install the air cylinder, and repeat steps 1 to 4; if flow is still incorrect, repeat steps 5 and 6.
8. Shut off test stand, release air pressure from PL cylinder, and remove just the air cylinder in order to check the Y dimension with the shims in place. This dimension will in turn become the PLE dimension; therefore, stamp this on the injection pump housing.

CAV-Type Injection Pump

For detailed information on the internal workings of CAV pumps, refer to the CAV section of this book.

chapter 4 Robert Bosch Fuel Systems

4-1 Injection Pumps with Pneumatic Governors

More than any other company, the Robert Bosch Company of Stuttgart, Germany was responsible for the accelerated acceptance of the diesel engine by manufacturing precision fuel-injection equipment as early as 1922.

The American Bosch Company, Ambac Industries, Inc., was an offshoot of Robert Bosch, and many companies today manufacture precision fuel-injection equipment under license from Robert Bosch on a worldwide basis.

Table 4-1 lists those companies presently using Robert Bosch fuel-injection equipment, but does not take into account additional engine manufacturers using Robert Bosch equipment built under license. A good example of this is the Diesel Kiki Company Ltd. of Japan, who install Bosch pumps and nozzles on a wide variety of engines.

Injection pumps used by Robert Bosch are very similar in design and operation to those of American Bosch and CAV in-line units, with the injection nozzles also working on the same principle. One such example is shown in Figure 4-1, which is an in-line four multiplunger pump used on Mercedes-Benz cars (pump Model PES 4M).

Mercedes-Benz (Robert Bosch Injection Pump)

Since the operation of the injection pump is similar to that of American Bosch and CAV pumps, we will not delve into the operation; however, let us consider the basic adjustments that would be typical on the pneumatic governor used on the Mercedes-Benz cars. This pneumatic governor also functions along the same lines as the one used by American Bosch, that is, the type AEP/MZ described earlier.

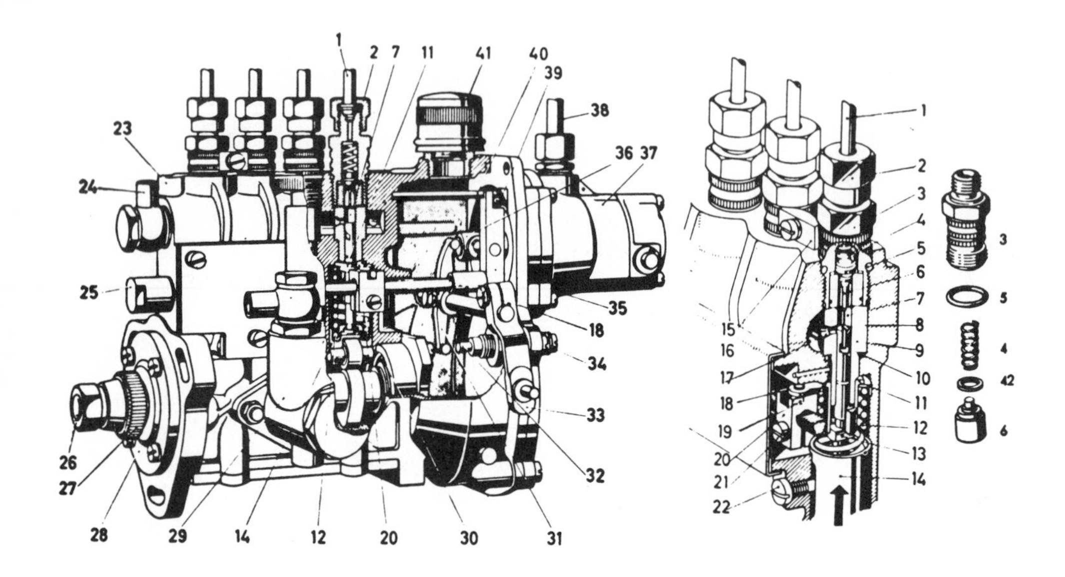

Figure 4-1
Typical Robert Bosch Injection Pump.
(Courtesy of Robert Bosch, GmbH)

Index:

1—Fuel line to injector
2—Cap nut
3—Pipe connection
4—Valve spring
5—Seal ring
6—Delivery valve and holder
7—Pressure area
8—Piston
9—Cylinder (pump element)
10—Seal
11—Control sleeve with guide lever
12—Tappet spring
13—Piston lug
14—Roller tappet
15—Clamping jaws
16—Suction area
17—Feed and return bore
18—Control rod
19—Bolt
20—Adjustable clamping piece
21—Clamp screw
22—Tappet guide screw
23—Pump housing
24—Fuel feed connection
25—Control rod guide sleeve stop
26—Camshaft (drive end)
27—Carrier
28—Bearing cap
29—Fuel transfer pump
30—Collar ball bearing
31—Double lever
32—Stop bolt for full load
33—Adjusting lever
34—Adjusting lever stop
35—Guide lever
36—Diaphragm bolt and compensating spring
37—Diaphragm block
38—Vacuum line
39—Diaphragm
40—Guide bolt
41—Air filter and oil filter hole
42—Seal ring

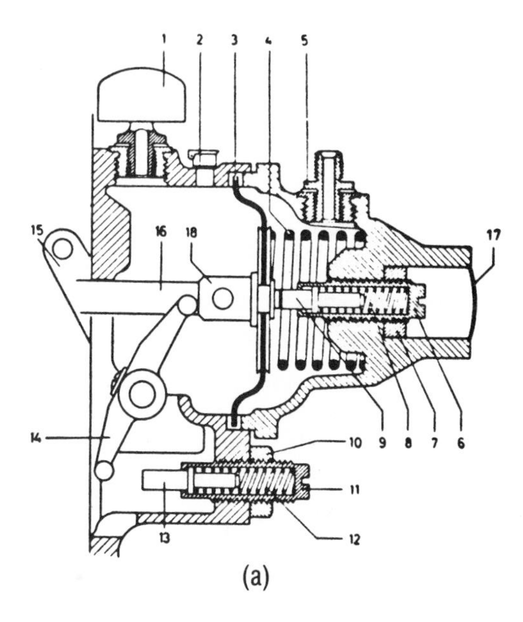

(a)

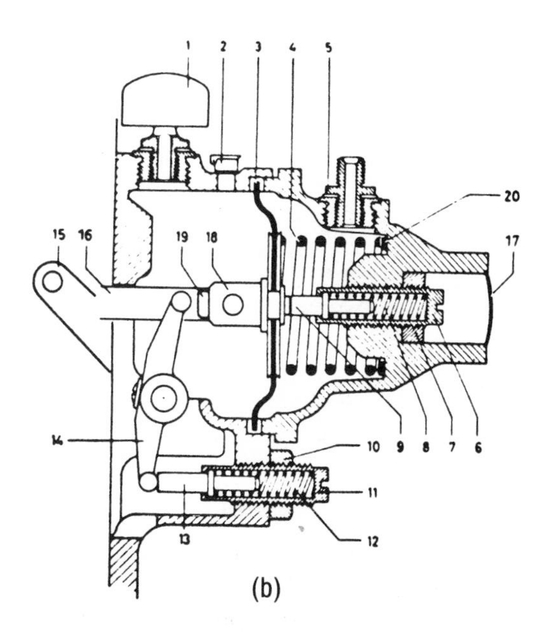

(b)

Figure 4-2
Cross-Sectional View of Pneumatic-Type Governor: (a) Governor in the stop position; (b) governor in the idling or partial load position.
(Courtesy of Daimler-Benz Aktiengesellschaft)

Index:

1—Air filter for atmospheric chamber
2—Cap oiler
3—Diaphragm
4—Control spring
5—Vacuum connection (hose from intake manifold)
6—Adjusting screw with booster spring
7—Locknut for adjusting screw
8—Booster or helper spring
9—Stop bolt for booster spring
10—Locknut
11—Adjusting screw with full-load stop
12—Spring
13—Stop bolt
14—Double lever (pivoting)
15—Adjusting lever
16—Control rod
17—Diaphragm housing cover end plate
18—Diaphragm bolt
19—Pressure bolt or adapting spring
20—Washer

The type of adjustments that can be undertaken without removing the injection pump and governor from the engine would be related to complaints such as the following:

1. Lack of horsepower (kilowatts) complaint.
2. Black smoke from the exhaust pipe.
3. Rough running.
4. Stalling and/or hunting (surging) during idling.
5. Increase in maximum governed engine rpm.

These adjustments when completed, should they fail to correct the problem, would require removal of the injection pump and governor assembly for further checks and tests on a fuel-injection pump test stand.

Figure 4-2 shows a cross-sectional view of the pneumatic-type governor used.

Table 4-1
ROBERT BOSCH APPLICATIONS

Customer	Pump Type	Governor Type
Domestic		
A. M. General (MAN)	P	RQ
Allis-Chalmers	P	RQV . . . K
Fiat-Allis	P	RQV . . . K
(Fiat)	A	RQV
J. I. Case	A	RSV
(David Brown)	A	RSV
(Unimog)	A	RSV
(Unimog)	M	RSV
(Unimog)	A	EP/M
(Scania)	P	RSV
John Deere	A	RSV
	P	RSV
International Harvester	A	RSV, RQV, (RQ)
	P	RSV, RQV, (RQ), RQV . . . K
	MW	RSV
(Neuss)	VA	
Mack	P	RQV . . . K
(Scania)	P	RQV (RQ)
Massey-Fersuson (Hanomog)	A	RSV
Murphy Diesel (MWM)	A	RSV
	P	RSUV
New Idea (GMC)	A	RSV
Thermo-King Mevosa (M-B)	A	RSV
Thansicold/Carrier Mevosa (M-B)	A	RSV
Waukesha	P	RSV
	Z	
	ZWM	
(Scania)	P	RSV
	P	RQV

Testing the Diaphragm Vacuum Housing for Leaks

Any leaks existing on the vacuum housing side of the pneumatic governor will cause poor governor reaction, since the force of the spring within the housing will be trying to push the fuel control rod toward an increased fuel position.

At the end of the injection pump opposite the governor, remove the protective cap from the pump fuel control rod (item 25 in Figure 4-1).

1. Disconnect the vacuum line or hose from the connection (5 in Figure 4-2).
2. Hold the adjusting lever (33 in Figure 4-1) in the *stop* position.

Table 4-1 (continued)

Customer	*Pump Type*	*Governor Type*
Imported		
Bomag (Hatz)	PFR	
DAF	A	RSV
Deutz (KHD)	A	RSV (RS)
	A	RQV (RQ)
	EP/VA	
Lamborghini	A	RSV
Lombardini	PFE,PFR	
Mercedes-Benz (car)	M	EP/MN
	MW	RW
Mercedes-Benz (bus)	A	RQV
Mercedes-Benz (truck)	A	RQV (RQ)
	A	RSV
	P	RQV
	P	EP/M
Mercedes-Benz (DO Brazil)	A	RQV
Peugeot (car)	EP/VA	
Peugeot (tractor)	PFR	
V. M. Motori	PFR	
	A	RSV
Volvo (truck)	P	RQV
	MW	RWV
Volvo-Penta (marine)	PFR	
	EP/VA	
	VE	
	P	RSV
	MW	RWV
Volkswagen (car)	VE	

Note: ( ) indicates imported engine.

3. Place a finger or thumb tightly over the end of the vacuum connecter (5 in Figure 4-2); then release the adjusting lever and watch the action of the fuel control rod at the end of the injection pump housing.
4. If there is "no leaks" internally at the diaphragm and the vacuum housing, the control rod will move out only a short distance by the action of the control spring (4 in Figure 4-2) and stay there. If, however, the control rod moves out all the way, which it may do slowly, there is a vacuum leak at either the diaphragm or housing cover.
5. Access to the diaphragm is accomplished by removing the four screws at the cover, then disengaging the diaphragm bolt (18 in Figure 4-2) at the control rod, and re-

moving the diaphragm. It is suggested that you replace the diaphragm if brittle or damaged, along with a new spring (control) that is matched to the diaphragm.

6. If the diaphragm appears to be satisfactory, check the cover and fittings for any possible leaks. Also make sure that the connections at both ends of the vacuum hose from the intake manifold are tight and leak free.

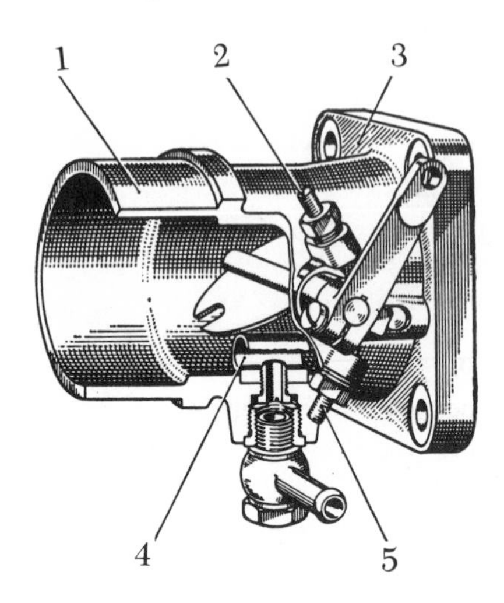

Figure 4-3
Adjusting Full-Load Stop Screw.
(Courtesy of Daimler-Benz Aktiengesellschaft)

Index:
1—Air intake (venturi)
2—Idling stop screw
3—Mounting flange
4—Auxiliary venturi pipe
5—Full-load stop screw

Checking Injection Pump Control Rod (Rack) Travel

Normally, once the injection pump has been set up on a test stand, further adjustment of the control rod should not be necessary. However, the full-load stop screw of the throttle butterfly and/or the adjusting screw of the feed volume full-load stop may require some alteration from time to time. Generally, these adjustments would be done only if the engine exhaust tends to smoke fairly heavily or if fuel consumption is more than normal.

This check is reasonably simple and involves the following:

1. Refer to current Robert Bosch or Mercedes-Benz service literature and establish what the control rod (rack) travel should be in millimeters.
2. Remove the protecting cap on the control rod (rack) at the end of the injection pump housing opposite the pneumatic governor (see item 25 in Figure 4-1).
3. If checking this in the vehicle, disconnect the start/stop cable at the adjusting lever (33 in Figure 4-1; 15 in Figure 4-2).
4. Check that the control rod (rack) moves freely with no tight spots.
5. Refer to Figure 4-1; hold the fuel adjusting (33) lever in the full-fuel position and measure the protrusion of the control rod (rack) from the end of the guide sleeve on the injection pump housing; compare to manufacturer's specifications.
6. Refer to Figure 4-2; push in the control rod (rack) 16 by turning the adjusting lever 15 so that the lower arm of the double lever 14 comes in contact with the housing, or the helper spring 8 is completely compressed to its stop position.

Note: If the control rod (rack) movement is greater than specified, the injection rate is too high. Tightening of the full-load stop causes a shorter amount of control rod travel and will *lower* the injection rate, whereas unscrewing or backing out the full-load stop will result in a longer control rod (rack) travel and consequently a *higher* rate of fuel injected.

Adjusting Maximum No-Load Engine Speed

1. If the maximum no-load speed with the throttle wide open is checked with an accurate tachometer, the full-load stop screw may have to be adjusted if the rpm is in excess of the published specification.

2. With the engine at normal operating temperature for this check, if an adjustment is necessary, refer to Figure 4-3 and adjust the full-load stop screw until the specified speed is attained.

3. On Mercedes-Benz engine models OM 636, make sure that the adjusting screw (item 6 on Figure 4-2) with the booster spring is adjusted properly.

4. If the throttle butterfly valve has reached its maximum open position by adjustment of the full-load stop screw, and the engine's maximum no-load speed is still lower than specified, it will be necessary to increase the initial tension of the control spring (item 4 in Figure 4-2) for OM 636 engines by installing shims and/or washers (item 20 in Figure 4-2). These washers or shims are available in the following thicknesses: (a) 1 mm thick, P/N 180 990 18 40; (b) 0.5 mm thick, P/N 180 990 17 40; (c) 0.2 mm thick P/N 180 990 16 40.

Installing a washer 1 mm thick will increase the maximum no-load speed of the engine by approximately 120 to 150 rpm. This variation is due to the individual characteristics of the control spring used.

Index:

1—Air filter
2—Guide rod
3—Guide lever
4—Adaptation spring
5—Diaphragm pin
6—Pressure pin for 4
7—Starting quantity stop
8—Control rod (rack)
9—Double lever
10—Diaphragm
11—Rubber buffer
12—Vacuum connection
13—Control spring
14—Spacer ring
15a—Engaging cam full-load position
15b—Engaging cam idling position
16—Automatic advance control lever
17—Stop pin
18—Additional spring
19—Sliding spring capsule
20—Full-load stop pin
21—Fixing nut
22—Spring
23—Full-load stop adjusting screw

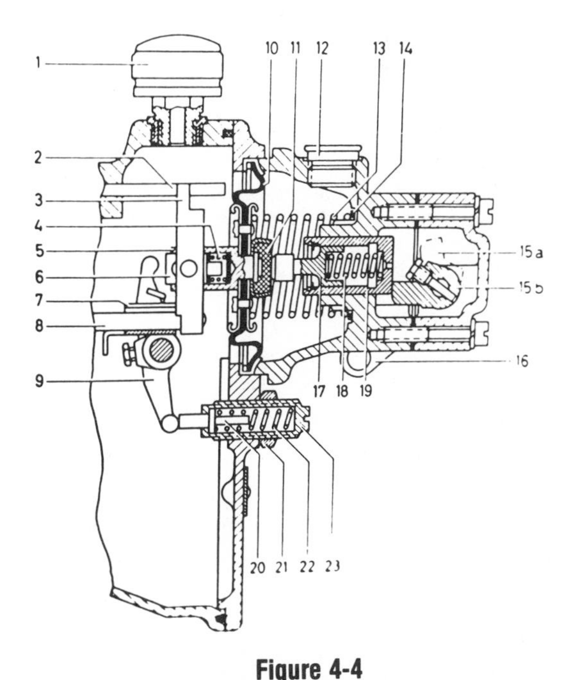

Figure 4-4
Governor Mechanism Used on OM 621 Engines. (Courtesy of Daimler-Benz, Aktiengesellschaft)

Figure 4-4 shows the governor mechanism used on an OM 621 engine, which varies slightly from that on the OM 636. In step 4 above, the same sequence would apply to the OM 621 engine, with the only exception being that the spacer rings (14) rather than washers are used to alter the control spring tension. They have the same part number as that for the washers.

After any adjustment to the maximum no-load speed, the vehicle should be road tested for acceleration and signs of black smoke emanating from the exhaust pipe. Should black smoke be visible, the maximum fuel discharge rate is too high. To correct this condition, proceed as follows:

1. Model OM 636 engines: refer to Figure 4-2 and turn in the adjusting screw (item 11) until there is no sign of smoke at the exhaust pipe.
2. Model OM 621 engines: refer to Figure 4-4 and turn in the adjusting screw (item 23) until the smoke disappears.

The smoke test should always be performed under *full-load* conditions, and a smoke tester/meter should be used such as a Robert Bosch Model EFAW 78, consisting of a suction pump with accessories EFAW 65 and a photoelectric evaluator EFAW 68. This can be hooked up during a road test, engine dynamometer, or chassis dynamometer test.

Adjusting Idling Speed

Prior to adjusting the idling speed, the engine must be at normal operating temperature (cooling water temperature at least 60°C or 140°F). This adjustment is done at the throttle linkage by the butterfly valve in the air intake, *not* at the injection pump. Proceed as follows:

1. On vehicles with a *bowden cable* idling control setup, make sure that you turn the knob on the instrument panel (dash) fully clockwise.
2. In this position, check that the throttle lever linkage is clear and not touching the adjustment ring of the Bowden cable. If it is, loosen the setting ring.
3. Refer to Figure 4-3 and adjust the idling screw until the engine speed is between 550 and 600 rpm on model OM 636 engines, and between 700 and 800 rpm on model OM 621 engines.
4. If the idle speed will not reduce to these speeds, proceed to check for an air leak at the following:
   a. Vacuum line from the intake manifold to the governor
   b. Diaphragm housing
   c. Diaphragm itself

   Any leaks existing in this system will cause the control rod (rack) to move toward an increased fuel position by the action of the control spring (4 in Figure 4-2).
5. On model OM 621 engines, a distance of approximately 1 mm must exist between the ball socket of the connecting throttle rod and the ball head on the angular lever at the cylinder head cover to ensure adequate throttle freeness. Adjust the length of the throttle connection rod if necessary.
6. On both OM 621 and OM 636 engines, after setting up the idling speed, ensure that the Bowden cable is correct.
   a. Turn the knob on the instrument panel (dash) fully clockwise.
   b. Clamp the setting ring on the cable so that a distance of between 0.1 and 0.2 mm exists between the setting ring and throttle lever or angular lever. Make sure that the cable is free to move without undue force being applied.

Adjustment to Correct Stalling or Hunting at Idle

Hunting, surging, or frequent stalling at an engine idle speed can be caused by vacuum variations in the line from the intake manifold to the diaphragm chamber of the pneumatic governor. This will cause small movements or oscillations of the diaphragm, which in turn, being connected to the fuel control rod (rack), will vary the amount of fuel being delivered to the cylinders and therefore cause the engine to hunt, surge, or stall.

To correct such a condition, the stop screw (6) must be adjusted as follows:

1. Bring the engine up to normal operating temperature.
2. Ensure that, with the knob on the instrument panel turned fully clockwise, the idling screw is against the stop at the throttle duct housing, or adjust if required.
3. Make sure that the engine is running at its proper idling rpm.
4. Refer to Figure 4-2 and screw in the adjusting screw (6) with Robert Bosch special wrench EFEP 95 with the engine running at an idle speed until all signs of hunting or surging disappear. Manually disturb the throttle; let the engine return to idle and make sure that it idles smoothly.

Note: On MZ-type governors (newer versions), remove the end plate (17 in Figure 4-2) for this adjustment. On MZ governor versions EP/MZ60A39 d and 51 d, the diaphragm housing cover must be removed to adjust screw (6).

5. After this adjustment, recheck the maximum no-load speed. If necessary readjust the full-load stop screw at the throttle duct. Also install a new end plate (17) in the governor housing. The governor should then be *lead sealed* to prevent tampering.

4-2

MW Model Fuel-Injection Pump

The model MW pump is the latest addition to the broad range of available Robert Bosch in-line style of injection pumps. It is presently being used extensively on International Harvester equipment, Mercedes-Benz cars, including the 5-cylinder 300D and turbocharged version, and on Volvo trucks and Volvo-Penta marine applications.

The basic operation of the pump follows the same sequence of events as other in-line Bosch pumps in that all pumping plungers are moved up and down by a camshaft contained within the base of the injection pump housing through roller-type lifters or followers. Rotary motion is by the accepted method of rack and gear, with the rack being controlled by a combination of operator or driver throttle position and by governor action.

With the plunger at BDC within its barrel, fuel delivered to the injection pump housing by its transfer pump is free to flow into the charging port in the conventional Bosch manner. The rotation of the pump camshaft will raise the plunger within its barrel until it closes the supply or charging port (this is commonly referred to as *lift to port closure*). Delivery to the injector nozzle can begin only after *port closure* since prior to this, the fuel will only be at transfer pump pressure, which is in itself too low to unseat the delivery valve within the top portion of the pump housing. The rotation of the injection pump camshaft will cause an increase of the trapped fuel within the barrel area, which will unseat the delivery valve, sending fuel at this high pressure to the injection nozzle and on into the combustion chamber.

The timing of this fuel to the injection nozzle depends on the point of port closure or amount of lift required by the camshaft to place the plunger into a position whereby the supply port at the lower end of the barrel is closed. To change the point of port closure on the model MW injection pump, the barrel in which the plunger operates must be raised or lowered, respectively, to retard or advance the start of injection or timing. This is easily done on the MW pump by moving the barrel up or down by the use of shims under its mounting flange, which rests on the top of the injection pump housing (see Figure 4-5). The shims used must all be of the same thickness for every barrel; otherwise, a variation in timing would result.

Once plunger 1 has been set at zero degrees, each remaining barrel and plunger would be set to the same lift from BDC to port closure in equally spaced pump camshaft degrees. For example, on a 4-cylinder 4-cycle engine pump, after setting cylinder 1, the other cylinders would be shimmed in firing order sequence 90 degrees apart. This adjustment is commonly called phasing or setting internal pump timing.

With the injection pump correctly timed, the following advantages will result:

1. Proper power developed
2. Good fuel economy
3. Ease of starting
4. Smoothness of operation
5. Proper exhaust smoke emissions

Remember that in any injection pump the quantity of fuel delivered is controlled by the plunger's *effective stroke*, which is the time between port closure and port opening. The start of the effective stroke is controlled by the shape of the helix cut on the plunger. The plunger rotation is controlled by injection pump rack movement coupled to the governor and throttle linkage.

When rack movement is checked by use of a dial gauge mounted on the end of the pump housing, we are in effect establishing the rotational movement of each and every plunger within its barrel, and thus the position of the plunger helix that controls the length of the effective stroke, since the plunger's up and down movement remains constant with the pump camshaft and roller lift. For any specified rack position, then, each plunger will produce a particular volume of fuel delivery.

To change the volume of fuel delivered by each plunger, we can simply rotate the barrel within the housing of the injection pump, which effectively varies the supply port position in relation to that of the plunger helix. This is accomplished by loosening the retaining nuts passing through the barrel flange at the top of the pump housing and gently tapping the flange to move the attached barrel to the front or rear of the injection pump housing, depending on whether the volume of fuel is to

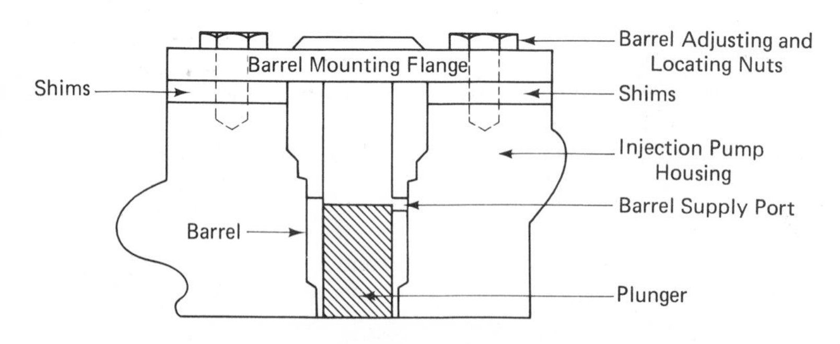

Figure 4-5
Use of Shims to Adjust the Point of Port Closure.

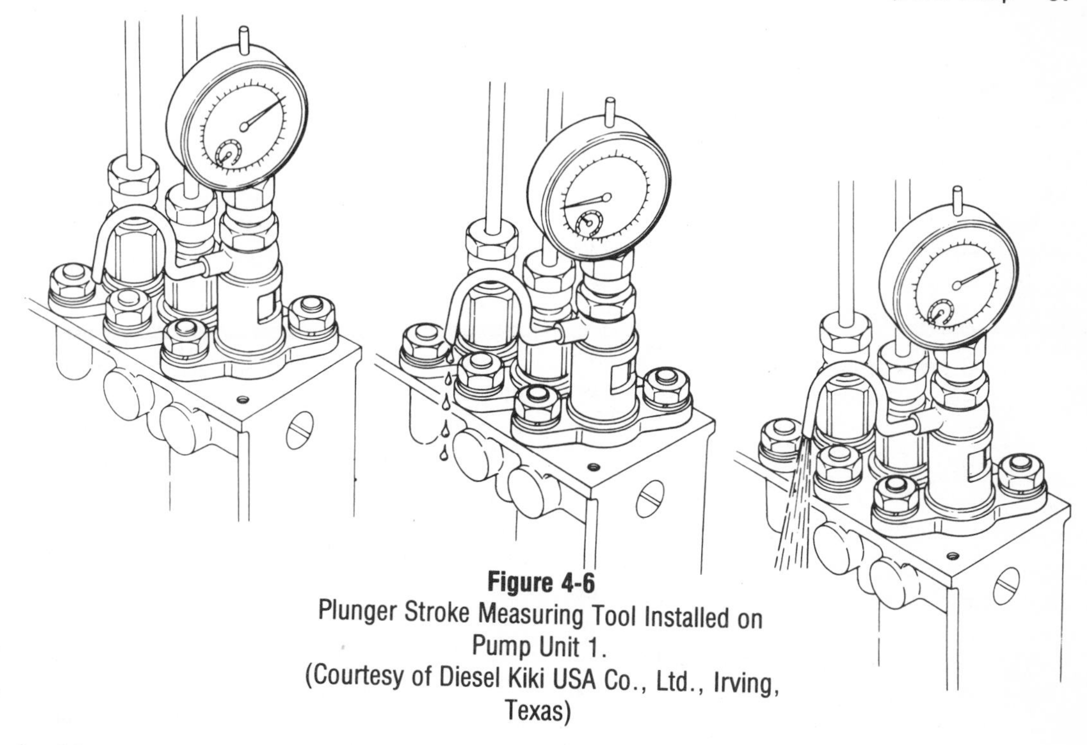

Figure 4-6
Plunger Stroke Measuring Tool Installed on Pump Unit 1.
(Courtesy of Diesel Kiki USA Co., Ltd., Irving, Texas)

be increased or decreased. This is how the MW injection pump is calibrated for equal delivery from each plunger.

In-Line Type Injection Pumps

Since the majority of Diesel Kiki injection pumps are manufactured under a technical license agreement with Robert Bosch Company, the operation and therefore the maintenance of pumps produced by both companies is the same. The newer models of Bosch-type pumps, such as the MW discussed earlier, and the PE (S)-P and PE (S)-PD..A, operate along the same lines; therefore, we shall limit this discussion to the basic adjustments required on the pump.

The pump must of course be attached to a suitable test stand at major service intervals in order to correctly carry out these adjustments. With the pump mounted on the test stand, proceed as follows:

Adjusting Injection Timing. This requires adjusting the plunger prestroke.

1. Each pump must be checked out in firing order sequence: therefore, begin by removing the delivery valve holder, spring, valve and gasket, and delivery valve stop from cylinder 1.
2. Refer to Figure 4-6; install the plunger stroke measuring tool to the flange sleeve by turning it as far as it will go (tool 5782-419 Diesel Kiki; not required on others which are simply spill timed by use of test stand degree wheel).
3. Place the fuel control rod in the full fuel position. Manually rotate the pump camshaft until, with the use of the installed dial gauge, bottom dead center for pump 1 is established. At this point, zero in the dial indicator needle.
4. Manually rotate the pump from its drive end until fuel from the attached nozzle pipe (Figure 4-1) just stops flowing. Make

sure that the pump is turned in the proper direction.

Note: The actual start of injection occurs when the plunger top just covers the fuel inlet hole in the barrel. By varying the thickness of the shims underneath the flange sleeve, the amount of camshaft lift necessary for the plunger head to completely close the inlet port in the barrel is increased or decreased respectively.

5. If the pump is equipped with reverse lead plungers, then the dial gauge would have to be zeroed with pump 1 in the TDC position. Then proceed to turn the pump camshaft manually in a reverse direction, and repeat the same shim adjustment as for a normal plunger-equipped pump.
6. If fuel delivery is early, add shims; if late, remove shims. Shim thicknesses are readily available from any Robert Bosch or Diesel Kiki dealer in graduated sizes to suit each pump model.

Adjusting Injection Interval. Once pump 1 has been adjusted, set the pointer on the pump test bench to the desired position for further measurement check.

1. Turn the pump camshaft manually in the normal direction, and establish that fuel stops flowing from each successive pump in firing order sequence, and also at the correct number of degrees.
   a. For 4-cylinder pumps, this is 90 degree intervals.
   b. For 6-cylinder pumps, this is 60 degree intervals.
   c. For 8-cylinder pumps, this is 45 degree intervals.
   d. For 10-cylinder pumps, this is 27 to 45 degree intervals; this would be given on the service data for the particular pump and engine combination.

   Allowable tolerance for all pumps is plus or minus (±) 0.5 degree.
2. When all pumping elements have been set, double-check the injection timing between the first and last cylinders (pumps). If timing is correct, the mark on the bearing cover should align with the drive coupling, spline gear, or timing advance device flyweight holder mark. If not, the timing mark should be erased and a new one scribed in place.

Adjusting Injection Quantity. To check that each pumping unit is delivering the same amount of fuel, the proper nozzle, nozzle holder, nozzle opening pressure, and transfer pump pressure must be as specified in the calibration test sheet for the particular pump model (obtainable at any local fuel injection repair shop). To change the amount of fuel delivered by these pumps simply requires turning the flange sleeve that rotates the plunger in the barrel. The flange sleeve is held in position by two nuts, which when loosened allow one to move the sleeve around the plunger up to a maximum of 10 degrees. The pump is then run on the test stand until each pumping unit has been adjusted to manufacturer's specifications.

Model PE-PD..A Plunger Block Assembly Replacement

Should a pumping assembly become damaged during engine operation in the field, the plunger block assembly can be replaced with a new one without adjusting the injection quantity, an operation that is normally done on the fuel pump test stand.

Removal

1. A variety of special tools is required for this purpose:
   a. Socket wrench 57914-050
   b. Delivery valve extractor 5792-004
   c. Stop bolt 57976-310
   d. Gauge 57990-520

e. Cam stroke measuring device 5782-424, plus a screwdriver, torque wrench, and wrench to loosen the fuel pipe nut at the delivery valve holder

2. Refer to Figure 4-7; remove the two screws that hold the top cover to the pump.
3. The fuel rack control rod must be placed into the mid-travel or centered position by removing the control rod cap and placing gauge 57990-520 against the control rod. Then lock the control rod in position by use of stop bolt 57976-310, as shown in Figure 4-7.
4. Since the flange sleeve is under plunger spring force, alternately loosen the retaining nuts; then using extractor 5792-004, pull the plunger block assembly from the injection pump housing.

Installation. Remove stop bolt 57976-310, and mount the normal control rod stop. With the replacement plunger block assembly mounted into position on the pump housing, refer to Figure 4-8 and line up the matching lines as shown. Remove the delivery valve assembly with tool 57914-050, and attach the cam lift measuring tool 5782-424 as shown in Figure 4-6.

By use of the transfer pump, prime the system and manually turn the injection pump camshaft until fuel stops flowing; read the dial gauge for lift. If necessary, change the shims until the reading is correct. Reassemble the delivery valve assembly, mount the top cover on the injection pump, attach all lines, and bleed the fuel system. Figure 4-9 indicates the fuel flow from the tank to all points in the system, and back to the tank (overflow fuel).

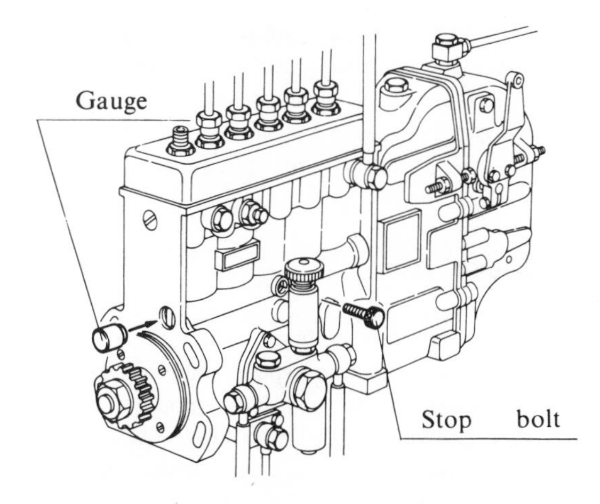

Figure 4-7
Placing the Fuel Rack into Its Centered Position. (Courtesy of Diesel Kiki USA Co., Ltd., Irving, Texas)

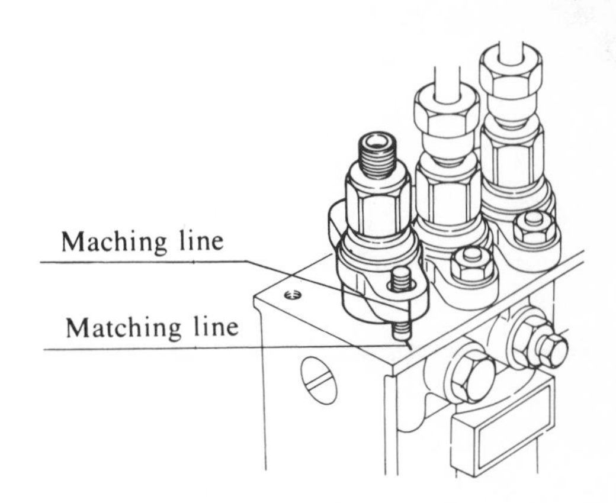

Figure 4-8
Aligning the Plunger Block Assembly to the Injection Pump Housing. (Courtesy of Diesel Kiki USA Co., Ltd., Irving, Texas)

4-3 In-Line Injection Pump Troubleshooting Guide

Figure 4-10 provides a detailed chart to aid in troubleshooting diesel fuel-injection systems for Robert Bosch equipment.

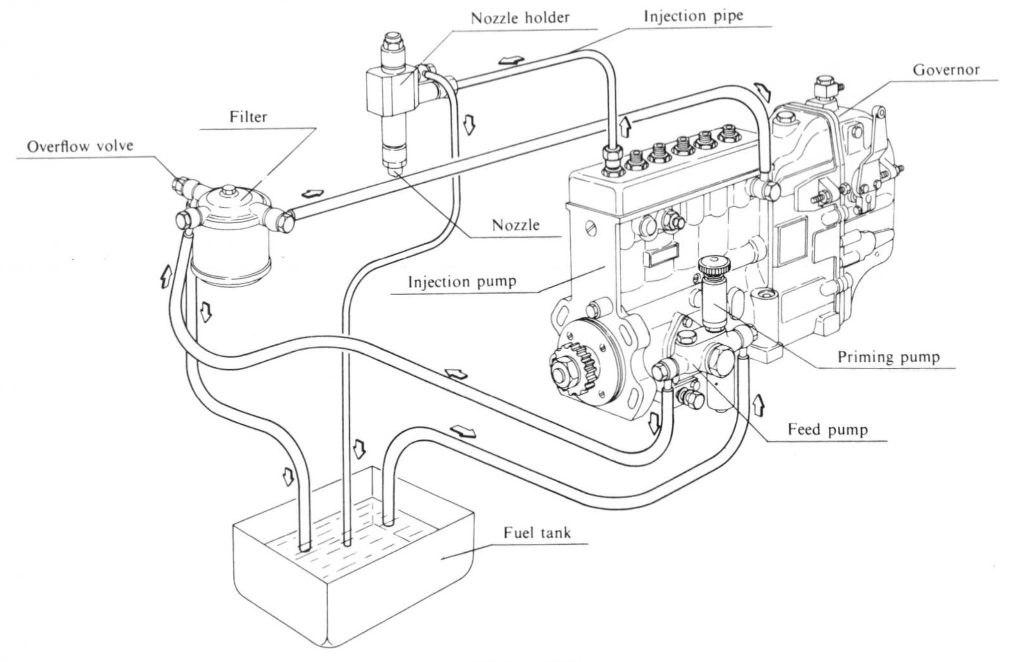

Figure 4-9
Typical Fuel System Flow.
(Courtesy Diesel Kiki USA Co., Ltd., Irving, Texas)

4-4

Robert Bosch Governors

Governor Operation

Although governor *terms* were discussed under general governors earlier, it would be helpful now if we were to again briefly review the basic operation of a governor mechanism and also to ensure that we understand the meaning of those terms used to express various reactions related to the governor.

Refer to Figure 4-11. In any governor used on diesel engines the basic principle is one of *weight* force against *spring* force. In other words, the tension of the governor spring assembly is always trying to *increase* the fuel delivered to the engine cylinders, and therefore the rpm, whereas the centrifugal force of the engine driven governor flyweights is trying to oppose and overcome the spring tension, and therefore decrease the amount of fuel injected.

Definition of Terms

1. *Initial speed change:* the immediate change in engine speed caused by a change in engine load. The amount of rpm increase or decrease is in direct proportion to the load change and is the main factor involved that initiates or causes the governor to react or respond.
2. *Sensitivity:* the amount of speed change that is necessary before the governor will sense or make a corrective change to the

It is assumed that the engine is in good working order and properly tuned, and that the electrical system has been checked and repaired if necessary.

Starting Problem	Engine surges at idle	Rough idle when engine is warm	Engine misses under load	Low Power	Excessive Fuel Consumption	Engine cannot be shut off	Poor performance **or** black smoke **or** low power	Fog-like exhaust in full-load range (white or blue)	Incorrect idle or maximum speed	Engine does not rev up	Injection pump runs hot	CAUSE	REMEDY
●	●		●	●				●		●		Tank empty **or** tank vent blocked	Fill tank/bleed system, check tank vent
●	●	●	●				●	●		●		Air in the fuel system	Bleed fuel system, eliminate air leaks
●						●						Shut off/start device defective	Repair or replace
●			●	●				●		●		Fuel filter blocked	Replace fuel filter
●			●	●			●	●				Injection lines blocked/restricted	Drill to nominal I.D. or replace
●			●	●						●		Fuel-supply lines blocked/restricted	Test all fuel supply lines—flush or replace
●			●	●	●							Loose connections, injection lines leak or broken	Tighten the connection, eliminate the leak
●			●					●		●		Paraffin deposit in fuel filter	Replace filter, use winter fuel
●			●	●	●		●	●		●		Pump-to-engine timing incorrect	Readjust timing
		●	●	●	●		●			●		Injection nozzle defective	Repair or replace
				●	●		●					Engine air filter blocked	Replace air filter element
●												Pre-heating system defective	Test the glow plugs, replace as necessary
●		●		●	●		●	●				Injection sequence does not correspond to firing order	Install fuel injection lines in the correct order
	●				●				●			Low idle misadjusted	Readjust idle stop screw
				●					●			Maximum speed misadjusted	Readjust maximum speed screw
			●	●							●	Overflow valve defective or blocked	Clean the orifice or replace fitting
		●										Delivery valve leakage	Replace delivery valve (max. of 1 on 4 cyl., 2 on 6 cyl.)
	●								●			Bumper spring misadjusted (RS . . . governors)	Readjust bumper spring
				●	●		●	●		●		Timing device defective	Repair or replace timing device
●		●		●			●			●		Low or uneven engine compression	Repair as necessary
	●			●	●	●	●		●	●		Governor misadjusted or defective	Readjust or repair
●	●	●	●	●	●	●	●	●				Fuel injection pump defective or cannot be adjusted	Remove pump and service

Figure 4-10
Troubleshooting Guide for Diesel Fuel-Injection System with Robert Bosch In-Line Fuel-Injection Pumps.
(Courtesy of Robert Bosch, GmbH)

fuel-injection mechanism. It is generally expressed as a percentage of the governed speed of the engine.

3. *Stability:* the ability of the governor to maintain a definite engine speed after a speed disturbance under either constant or varying load conditions with a minimum or false motions or overcorrections.
4. *Speed droop:* basically, the difference in engine speed (rpm) between the maximum governed *no-load* rpm and the maximum governed *full-load* rpm. The difference will vary depending on the type of governor used with the application. This rpm difference is expressed as a percentage of the no-load rpm.
5. *Hunting or surging:* a rhythmic variation of the engine speed that results from the governor response usually being out of phase with the feedback signal. The engine speed will continually fluctuate above and below the mean speed at which it is governed.

Figures 4-12 and 4-13 as illustrated describe the individual sequence of events involved by the governor action from no-load to full-load, and during varying degrees of throttle movement.

Boost Compensator Operation

The boost compensator device is only found on engines that are *turbocharged*, and allows for more fuel than the normal full-load injection amount, which is restricted by the use of the

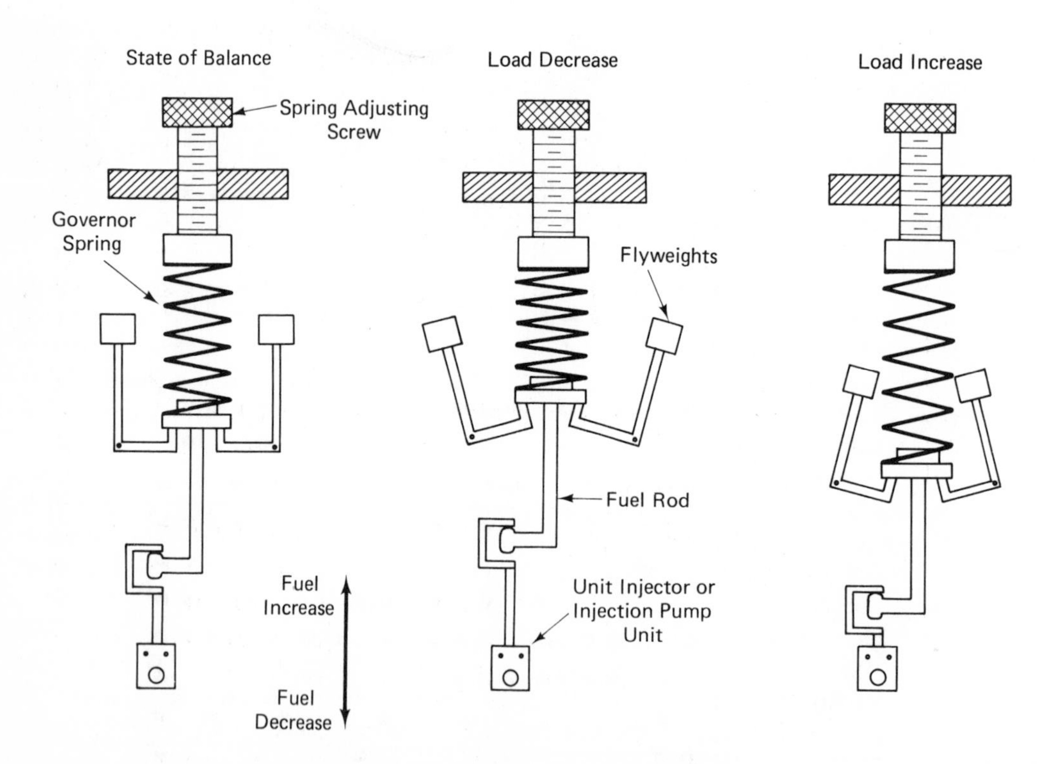

Figure 4-11
Basic Operation of a Governor Mechanism.

full-load stop bolt on standard RSV-type mechanical governors. Basically, the purpose of the boost compensator is not unlike an aneroid control on a Cummins engine or the hydraulic air–fuel ratio control used by Caterpillar on some of their engines. It ensures that the amount of injected fuel is in direct proportion to the quantity of air within the engine cylinder to sustain correct combustion of the fuel and therefore increase the horsepower of the engine.

The boost compensator unit is contained as shown in Figure 4-14 in the upper portion of the governor case. Very simply, it operates as follows: with the engine running, pressurized air from the cold end of the turbocharger passes through the connecting tube from the engine air inlet manifold to the boost compensator chamber. Inside this chamber is a diaphragm, which is connected to a pushrod, which is in turn coupled to the compensator lever. Movement of the diaphragm is opposed by a spring at its left-hand side; therefore, for any movement to take place at the compensator linkage, the air pressure on the right-hand side of the diaphragm must be higher than spring tension. As the engine rpm increases and the air pressure within the connecting tube becomes high enough to overcome the tension of the diaphragm spring, the diaphragm and pushrod will be pushed to the left-hand side.

This movement causes the compensator lever to pivot in a ccw direction around the fulcrum pin A shown in Figure 4-14. Similarly, the floating lever will also pivot around fulcrum pin D in a ccw direction, forcing the fuel control rack toward an increased fuel position. The boost compensator will therefore react to engine inlet manifold air pressure regardless of the action of the governor. If then the turbocharger boost air pressure reaches its maximum, the quantity of additional fuel injected will be equal to the stroke of the boost compensator in addition to the normal *full-load* injection amount that is determined by the governor *full-load stop bolt*.

RSVD Mechanical Governor

The characteristics of the RSVD governor are the same as the Robert Bosch/Diesel Kiki RSV type of governor, since it incorporates all the features contained in the RSV unit. The RSVD was originally designed for application to all Robert Bosch PE-A or PE-B size in-line-type fuel-injection pumps used on automative diesel engines. However, since it has variable-speed characteristics similar to those of the RSV unit, it can be applied to engines requiring variable-speed control or if necessary for both minimum–maximum and variable-speed governor control.

Referring to Figure 4-15, the flyweight carrier is mounted to the injection pump camshaft; therefore, pump rotation will cause the flyweights to transfer their motion axially to a sleeve and shifter mechanism via a ball-bearing thrust unit. A guide lever supports the shifter, which is suspended on the governor cover pin. As the engine is operated, the control lever shaft moves the lower end of the governor lever via the supporting lever, which is attached to the control lever shaft. Attached to a shaft located at the middle of the guide lever is the governor lever. Also at the lower end of the governor lever is a fixed pin, which acts as a pivot for the supporting lever at its lower end. At the top end of the governor lever is a link to which the fuel-injection pump rack is connected. Hooked to the top end of the guide lever is the starting spring, which is coupled at its opposite end to a spring eye on the governor housing. A pin passing through the governor cover also supports both the tension and guide levers. In addition, a swivel lever shaft is fitted into the governor cover bushing.

The main governor spring is hinged between both the tension and swivel levers, and also fitted to the swivel lever shaft (located in the governor cover) is the speed setting lever. Governor main spring tension is of a predetermined value specified by the engine manufacturer. At the lower end of the tension lever is

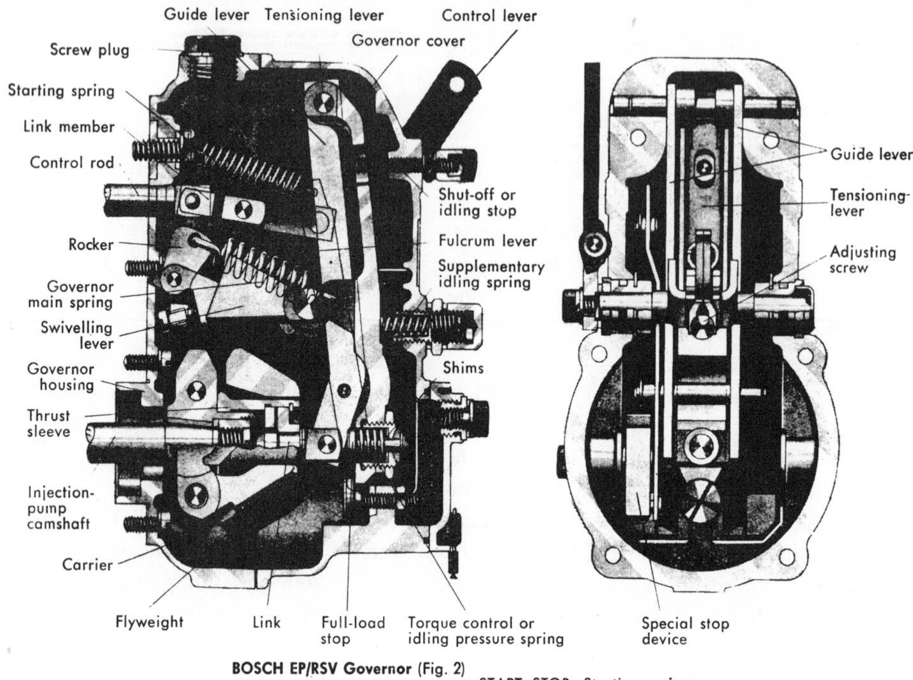

BOSCH EP/RSV Governor (Fig. 2)

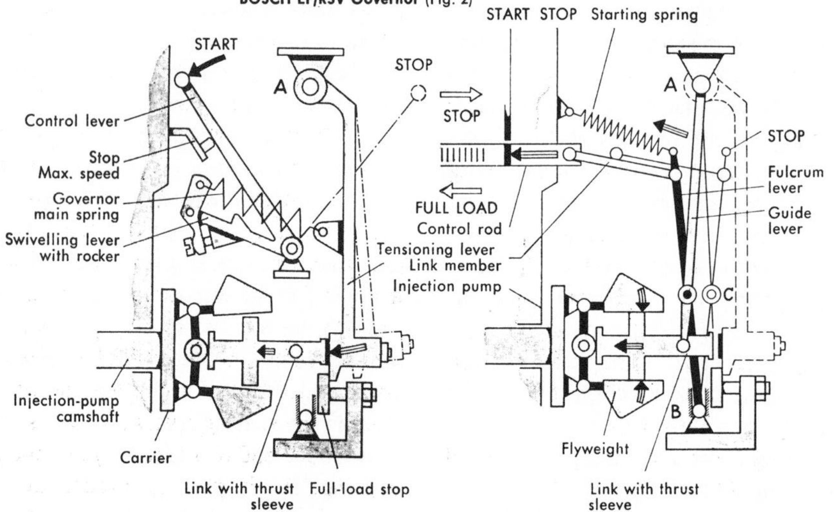

Movement I
With the control lever moved to START, the swivelling lever tensions the governor main spring. The tensioning lever is pulled against the full-load stop. The link connecting with the thrust-sleeve bearing and the thrust sleeve follow this movement. The control rod is in the full-load position.

Movement II
The swivelling lever projections have released the guide lever, causing the starting spring to pull the fulcrum lever and thus the control rod to the excess-fuel delivery position (START). The link connecting with the thrust-sleeve bearing and the thrust sleeve move to the left and the flyweights inwards.

Starting operation depicted in 2 phases (Fig. 5)

Figure 4-12
Method of Operation of Bosch EP/RSV Variable-Speed Governor. (Courtesy of Robert Bosch, GmbH)

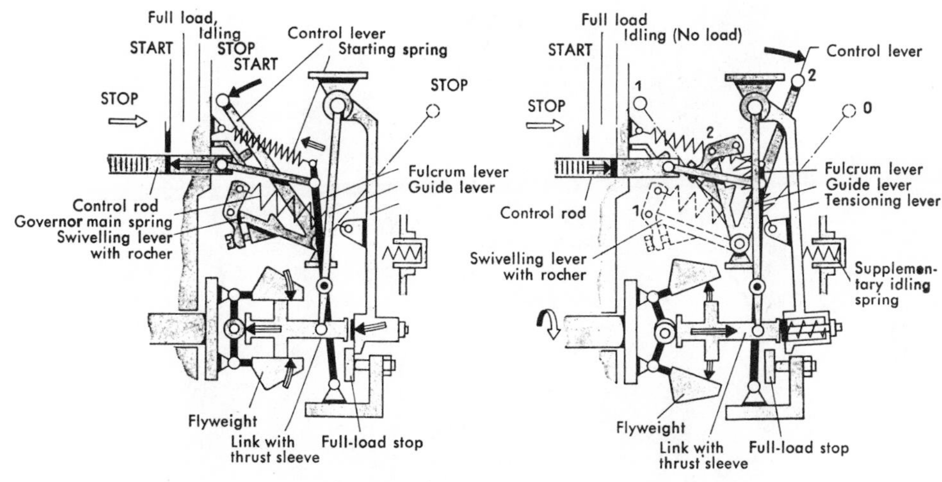

Starting operation of the BOSCH EP/RSV Governor (Fig. 6)

BOSCH EP/RSV Governor, Idling (Fig. 7)

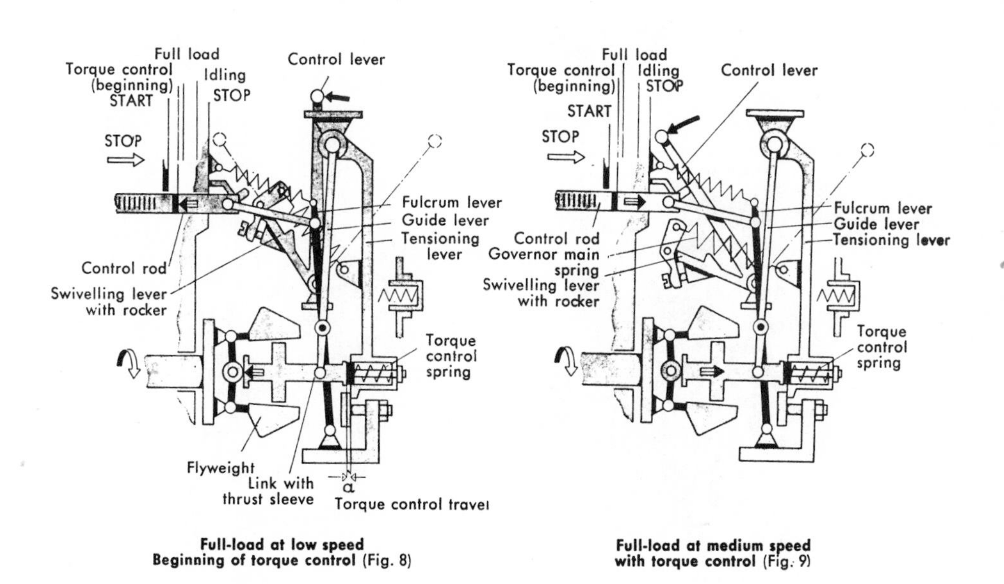

Full-load at low speed
Beginning of torque control (Fig. 8)

Full-load at medium speed
with torque control (Fig. 9)

Figure 4-12 (Continued)

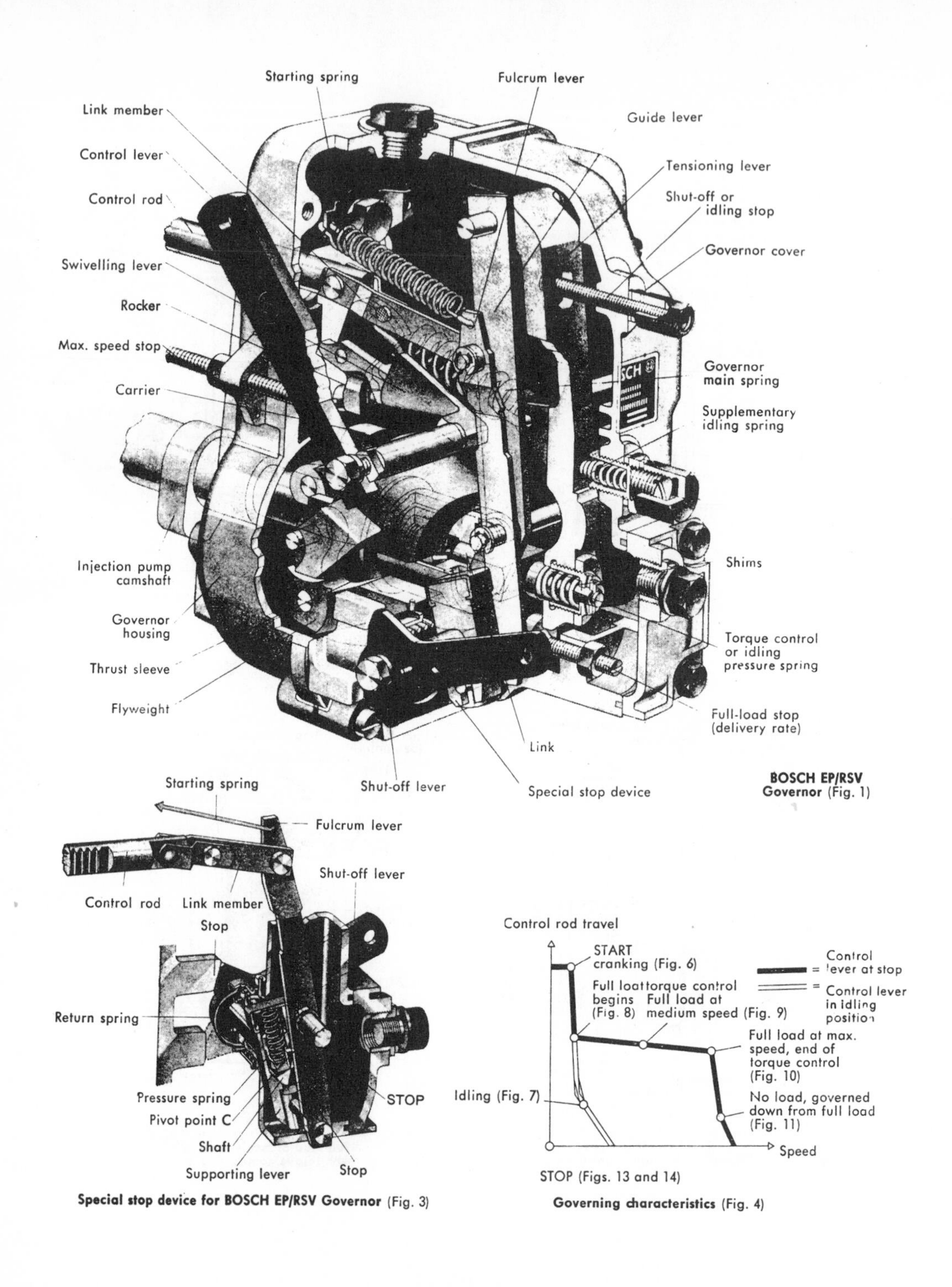
Starting spring
Fulcrum lever
Link member
Guide lever
Control lever
Tensioning lever
Control rod
Shut-off or idling stop
Governor cover
Swivelling lever
Rocker
Max. speed stop
Governor main spring
Carrier
Supplementary idling spring
Shims
Injection pump camshaft
Governor housing
Torque control or idling pressure spring
Thrust sleeve
Flyweight
Full-load stop (delivery rate)
Link
Shut-off lever
Special stop device
BOSCH EP/RSV Governor (Fig. 1)
Starting spring
Fulcrum lever
Shut-off lever
Control rod
Link member
Stop
Return spring
Pressure spring
Pivot point C
Shaft
STOP
Supporting lever
Stop
Special stop device for BOSCH EP/RSV Governor (Fig. 3)
Control rod travel
START cranking (Fig. 6)
Control lever at stop
Control lever in idling position
Full load torque control begins (Fig. 8)
Full load at medium speed (Fig. 9)
Full load at max. speed, end of torque control (Fig. 10)
Idling (Fig. 7)
No load, governed down from full load (Fig. 11)
Speed
STOP (Figs. 13 and 14)
Governing characteristics (Fig. 4)

Figure 4-13
Additional Governor Positions, EP/RSV Model.
(Courtesy of Robert Bosch, GmbH)

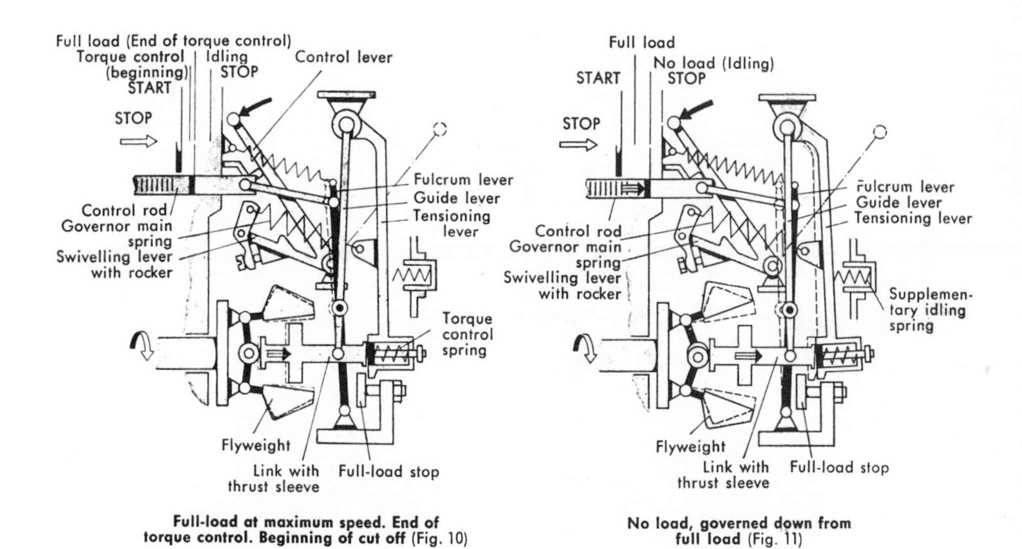

Full-load at maximum speed. End of torque control. Beginning of cut off (Fig. 10)

No load, governed down from full load (Fig. 11)

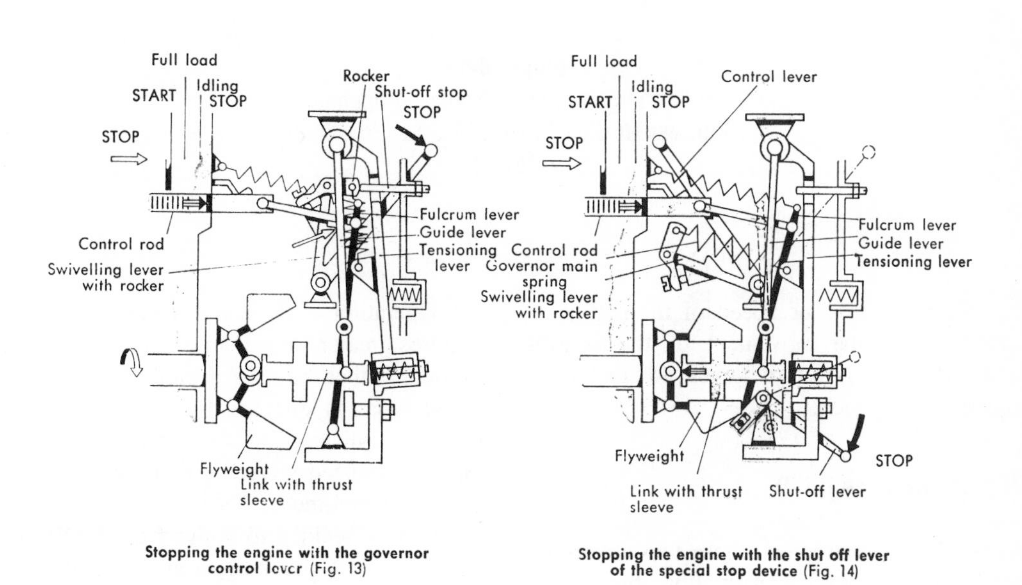

Stopping the engine with the governor control lever (Fig. 13)

Stopping the engine with the shut off lever of the special stop device (Fig. 14)

Figure 4-13 (Continued)

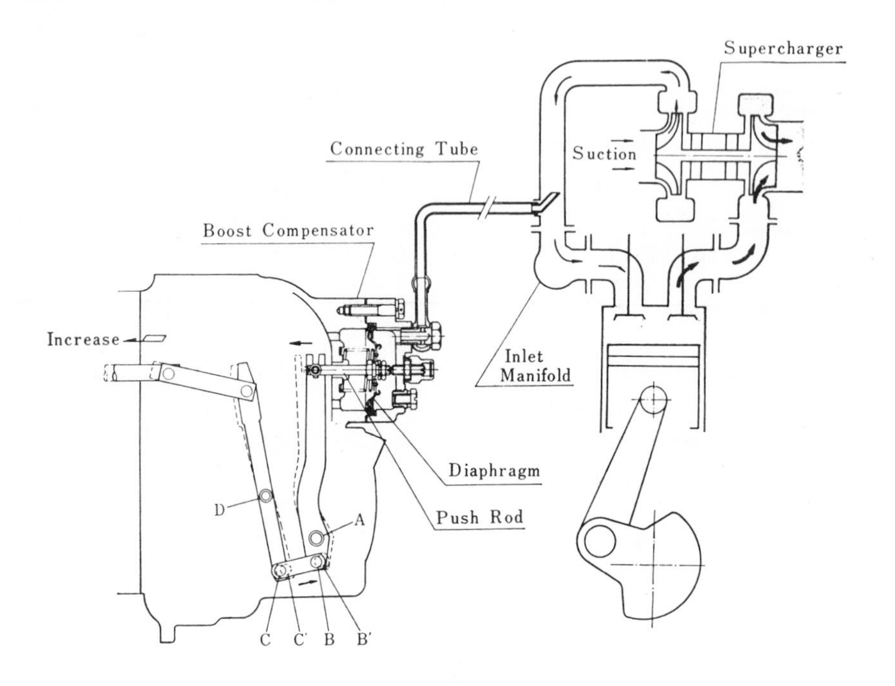

Figure 4-14
Turbocharged Engine, Boost Compensator.
(Courtesy of Diesel Kiki USA Co., Ltd., Irving, Texas)

the idling spring. The speed of the engine is controlled by the movement of the control lever via the governor lever to the fuel-injection pump rack.

RSV Compared to RSVD

As previously mentioned, these two governors are identical in basic construction, the exception being that the speed-control lever on the RSV is used as a speed setting lever on the RSVD. In addition, the governor lever on the RSVD is shaped in a crank-type fashion around the middle of the guide lever, which has the effect of placing the control lever face toward the front of the injection pump, as shown in Figure 4-15.

Operation of the control lever on the RSV requires greater force than on the RSVD owing to the fact that the RSV, being of the variable-speed type, is opposed by spring tension or reaction, whereas on the RSVD minimum–maximum speed governor there is no reaction force to overcome.

If it is decided to convert the RSVD to a variable-speed governor, a supplementary idling spring must be added; also the speed setting lever must be used as a control lever, with the other lever used as a stopping lever.

As far as performance of the RSV to the RSVD is concerned, the following holds true:

1. RSV: the regulated speed of the engine is set by manipulation of the control lever angle (remember that the control lever

does not move the fuel control rack directly on this unit).

2. RSVD: the control rack (fuel) is moved directly by manipulation of the control lever.
3. RSVD is a limiting speed type of governor; in other words, there is no governor control between the minimum and maximum speed ranges. Engine speed is controlled by movement of the control lever within these two ranges. This feature is of course a highly desirable quality on automotive applications since the operator will have more rapid response to throttle movement owing to the fact that he does not have to overcome control lever reaction force such as exists within the RSV unit.

RQV All-Speed Governor

The RQV governor is a widely used Robert Bosch unit that is designed to accomplish the following:

1. Maintain idle speed.
2. Control maximum speed.
3. Control speed throughout the operating range within the limits of its regulation.

This governor operates on the same basic setup as the RSV described earlier.

RQ Minimum–Maximum Governor

The RQ governor is used on highway truck applications and is therefore of the limiting speed type, since it will do the following:

1. Control idle speed.
2. Control maximum engine rpm.

Between idle and maximum speed there is no governor action, and the engine speed is controlled by the operator. It is found on highway trucks using Allison-type automatics (International Harvester is a good example).

RQV/RQ Combination Governor

Identified by Robert Bosch Co. as the RQV 300-800/1500 governor, this unit provides all-speed (RQV) governor control from 300 to 800 rpm (600 to 1600 engine rpm) and min–max (RQ) governor control from 800 to 1500 rpm (1600 to 3000 engine rpm). The primary function of the RQV/RQ governor is to provide an adequate range of all-speed-type governor control to handle the majority of PTO (power take-off) applications such as required on garbage and refuse trucks, tankers, and some concrete mixer truck applications, while also providing the degree of min–max type governor control necessary for Allison-type transmission applications and limiting maximum speed.

The all-speed portion of this governor control is limited to the speed range as noted above owing to the transmission's shifting characteristics. The remainder of the governor control (2000 to 3000 rpm) is exactly like that of the RQ governor in that it is *load sensitive*. In PTO applications with speed requirements below approximately 2000 rpm, the all-speed portion of RQV/RQ combination governor reacts similarly to the standard RQV all-speed governor.

RW Governor

The RW governor is being used by Robert Bosch on their model MW fuel-injection pump, discussed in detail in this chapter. This governor is found on such engines as Mercedes-Benz cars, with the RWV (variable speed) governor being found on Volvo trucks and Volvo-Penta marine applications.

The governor operates on the same principle as other Bosch units in that the centrifugal force created by a set of rotating flyweights is trying to reduce the position of the injection pump fuel rack, whereas the spring force within the governor is trying to increase the position of the fuel rack. As with any governor of this type, a *state of balance* can exist at an idle speed position or at the high end of the engine's rpm range.

Therefore, this governor controls the engine speed during starting, normal idling, high idle or no-load rpm, rated rpm or full-load speed, and breakaway, which is simply the speed range between full load and no-load rpm. At any speed above normal low idle, the operator or driver controls rack travel and therefore engine speed through the throttle linkage connected to the fuel control rack through the governor linkage.

Figure 4-16 shows the typical setup of the governor linkage and the actual component part stackup of the RW governor. Due to the fact that the RW governor contains linkage within the tensioning lever, Figure 4-12 depicts

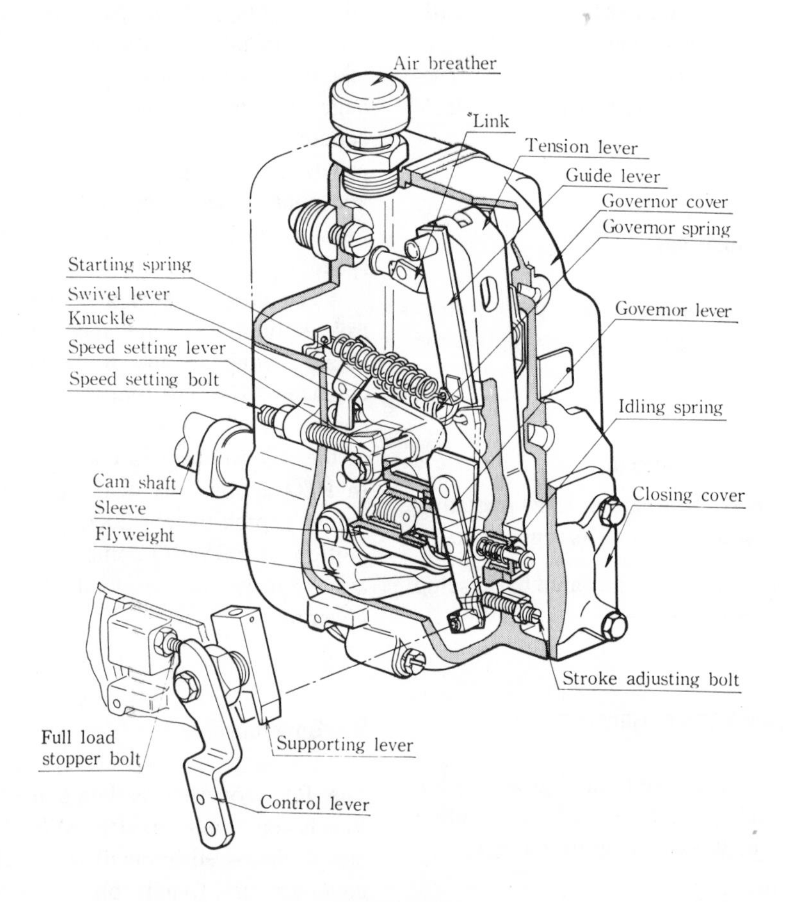

Figure 4-15
RSVD Mechanical Governor.
(Courtesy of Diesel Kiki USA Co., Ltd., Irving, Texas)

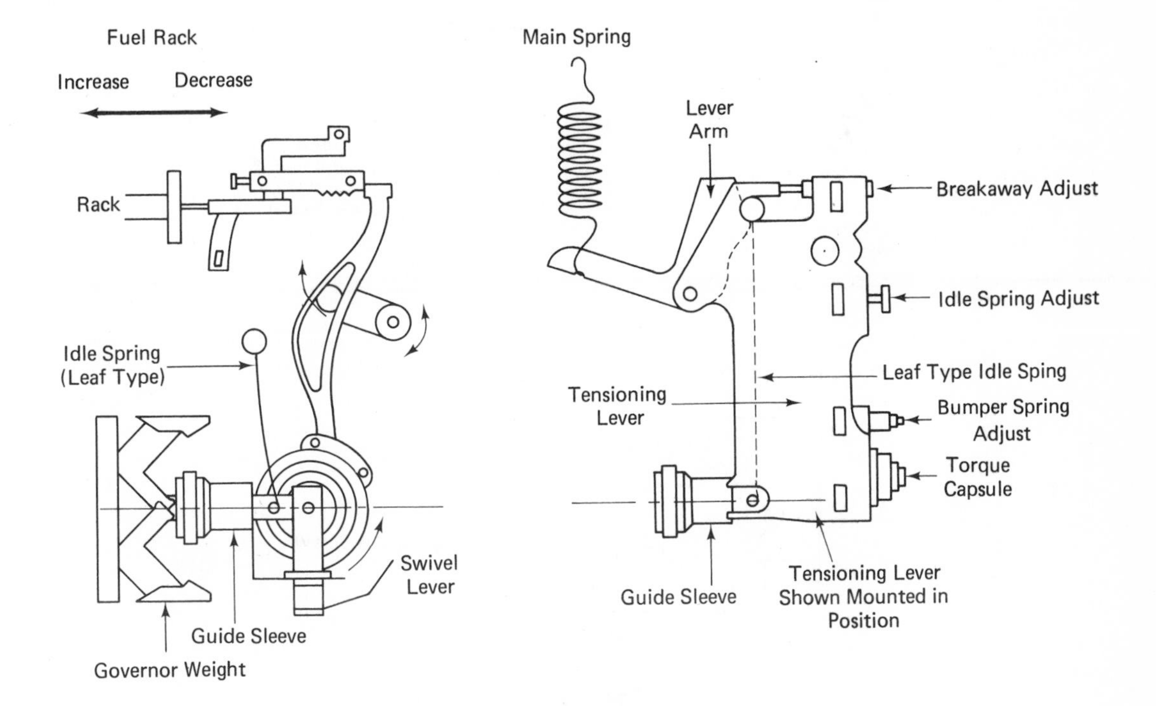

Figure 4-16
Typical Setup of RW Governor Linkage.

the governor linkage with the tensioning lever removed to show the parts hookup to the injection pump fuel control rack, with the right-hand view showing the tensioning lever in position with the various spring adjustments.

The initial force on the single leaf-type idle spring is established by the idle screw contacting up against it. In addition to the idle screw, a bumper spring adjustment is used whereby the idle regulation or stability is controlled by turning this bumper spring inward until it just touches the idle leaf spring with the engine running at its normal idle rpm.

At engine start-up, the fuel rack will therefore be in the maximum fuel position, and once the engine starts, the centrifugal force of the rotating governor flyweights will force the guide sleeve to the right, which will transfer this horizontal motion through the swivel lever and linkage to actually move the fuel rack to the right, also thereby changing the position of the rack toward less fuel. When the force of the rotating weights equals that of the idle spring, a state-of-balance condition will allow the engine to run at a steady idle speed.

During any part-load situation above idle speed, the driver controls the rack travel through the use of the throttle linkage and governor. During any full-load situation, the governor controls the rack travel, as we shall see later. At a full-load condition, a state-of-balance condition will exist; however, neither the weight nor spring force is controlling the rack, but the full-load stop screw butting up against the throttle lever controls its maximum travel.

The governor also contains a torque control device to avoid a smoking condition. At

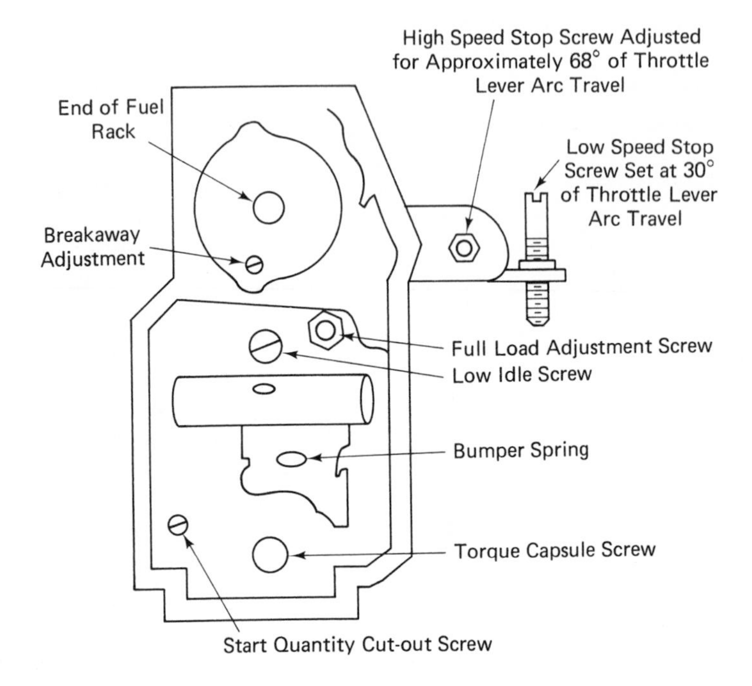

Figure 4-17
RW Governor Adjustment.

full-load the torque control allows the weights to pull the fuel rack to a decreased position, thereby reducing the full-load fuel limit.

As the rpm increases toward full-load, the weights oppose the internal torque capsule spring contained within the tensioning lever, which effectively reduces the full-load fuel delivery. The torque control capsule spring can expand, however, at lower engine speeds due to the reduced flyweight force and therefore push the fuel rack up to full load. With the engine under full power, the governor controls the rack for rated speed, at which time the force of the governor weights will cause the torque capsule to collapse. Turning of the collapsed torque capsule inward against the guide sleeve will increase the fuel delivery by increasing rack travel. This adjustment is commonly referred to as *torque backup*.

If at any time the engine were to rev past its rated rpm (called *breakaway*), the governor will pull the rack to a decreased fuel position, since at rated speed the weights will overcome all spring force in the tensioning lever. Just prior to breakaway the governor main spring continues to oppose the flyweight force working through the lever arm and tensioning lever.

Any adjustment of the breakaway screw inward causes the lever arm to stretch the governor main spring, which will therefore pull harder on the tensioning lever. This breakaway adjustment means higher rpm and a consequently higher weight force; therefore, after breakaway the main spring determines rack cutoff.

Figure 4-17 shows the position of the various governor adjustments when viewed from end on with the cover removed.

The preceding adjustments should be made only when the injection pump and governor are mounted on a suitable fuel pump test stand; otherwise, serious problems can result if attempted while on the vehicle.

Anytime that adjustments are to be made to the governor while it is mounted on the test stand, they must be done in the following sequence:

1. Low idle
2. Loading
3. Unloading
4. Bumper spring
5. Rated speed
6. Torque control
7. Full load
8. Breakaway
9. High idle
10. Fuel cutoff

Proceed as follows for the first four adjustments. Once the injection pump has been timed and phased, and all pumping elements have been adjusted to deliver the correct volume of fuel as per specifications, the governor can be adjusted. Check that the throttle control lever and fuel rack are free to move from the stop position to the full-load position.

Mounted onto the throttle lever on the side of the injection pump housing should be the degree protractor, which should be set to *zero* degrees when the throttle lever is at *zero*. Proceed to pretension the governor by screwing in the breakaway low idle screw, which will apply tension to the single leaf-type spring within the governor, tighten the start quantity cut-out screw, and finally unlock and tension the full-load adjustment. After tightening these adjustments, loosen both the bumper spring and screw in the torque capsule until it bottoms: then back it out seven complete turns. Screw in the low-speed stop and the high-speed stop, locking them in position with their locknuts.

With the low idle screw, adjust the pump while it is running on the stand to the specification for the particular model pump; then check the rack travel by mounting a dial indicator gauge onto a bracket connected to the pump housing with the gauge stem up against the rack. If the rack reading is incorrect, adjust the idle screw to obtain the specification. At this point, the pump's delivery should be checked at 1000 strokes and compared to the test specifications for that pump, with the throttle lever at the specified number of degrees and the pump at the specified rpm.

The pump can then be decreased or increased slightly in rpm by use of the test machine speed control handle, which will simulate both a loading and unloading situation. During this sequence, compare the rack travel on the dial gauge to the test specifications. If the reading is not according to specs, the basic guide sleeve dimension may need rechecking as per the test specification with the tool and dial gauge.

If when speeding up the test stand the governor does not react properly, you may have to back out the torque capsule even further to give the governor leaf spring room to govern.

When screwing in the bumper spring, do so until it just touches the leaf spring. This can be checked by watching the dial gauge up against the rack, which should move just a fraction of a millimeter above that established at the normal governor idle setting.

Proceed as follows for the final six adjustments. To set rated speed, let's take an example of 2200 rpm. Run the test stand at approximately 2180 rpm, and adjust the rack travel to specification. At this speed the torque capsule spring is collapsed; therefore, you are pressing against the guide sleeve to adjust the amount of torque control.

In the RW governor remember that the torque control spring will expand to push the rack forward as the engine or governor speed drops. With the test bench slowed to around 1600 rpm, the reduced governor flyweight force has been overcome by the torque capsule spring. You should see an increase in the rack travel from its normal rated speed setting; otherwise, you will have to replace the torque capsule. Take a measurement of the fuel delivery at this torque control point, and compare it without torque capsule action at full load. The full-load adjustment screw should be set to 1000 rpm at this time, which should produce a rack travel greater than was attained at 1600.

Adjust the full-load screw, lock it, and take a measurement of fuel delivery. After torque control, increase the rpm to 100 above rated speed (approximately 2300 rpm); the governor should pull the rack from rated speed. Adjust the breakaway screw until the rack moves about 1 mm to reduce fuel delivery; then check rack delivery at high-idle/no-load rpm to ensure that the rack is pulled back to specification (approximately 4 mm on the dial) and compare fuel delivery. Run the test stand at the pump's maximum rpm, and note that the governor returns the rack to within 1 mm of fuel cutoff to prevent any further overspeed with a wide-open throttle. Check the rack position as per the dial gauge; it is not necessary to check the fuel flow at this rpm.

With the stand running at 100 rpm to simulate starting, check that the rack movement is greater than at full load (21 mm approximately), and take a fuel delivery measurement at only 100 strokes, since fuel delivery at this speed is very slow. Make sure that the pump delivers its minimum amount as per specifications. Then check the altitude compensator with a vacuum gauge, at which time the rack travel should decrease by approximately 1 mm. If not, vary the shims or change the altitude compensator. With the throttle lever at 50 and 100 rpm below idle, fuel cutoff should occur. Remove the rack gauge and pin and install the rear cover; check the vacuum shutoff device by placing your finger over the tube connection with the plunger pushed in; it should hold. Install the shutoff device to the governor rear cover, and apply the specified vaccuum by the test stand pump to ensure that the rack will move to zero, thereby cutting off fuel.

4-5

Basic Injection Nozzles

The type of injection nozzle used by Robert Bosch and Company is similar to those used by American Bosch and CAV; therefore, refer to Chapter 2 for details regarding nozzle servicing.

As a means of general information, however, Figure 4-18 shows a nozzle and holder produced by Robert Bosch and used in Mercedes-Benz vehicles. This nozzle is used in a precombustion chamber; therefore, removal of this type of nozzle would be as follows:

1. Remove any accessories in your way (air cleaners and the like).
2. Loosen the fuel-injection nozzle line cap nut (item in Figure 4-18). Place plastic caps over the line and nozzle.
3. Loosen the hex nut (10) that secures the leak-off adapter (9) to the nozzle body.
4. If the nozzle starts to turn in the cylinder head when loosening nut 10, prevent it from turning with a 24-mm wrench.
5. Using a deep socket (24 mm) or box wrench, carefully unscrew the nozzle holder from the cylinder head. Remove the copper seal ring at the base of the bore once the nozzle has been withdrawn if it is not already stuck to the bottom of the nozzle. Place a plastic cap over the nozzle bore to keep out dirt.

4-6

Glow Plug Removal

Some model engines have the glow plugs screwed into the cylinder head at an angle; in other models they go in straight. This is due

Index:

1—Nozzle needle valve
2—Nozzle head
3—Nozzle holder insert
4—Pressure bolt
5—Cap nut
6—Pressure spring
7—Nozzle holder
8—Leak-off passage
9—Leak-off adapter
10—Hex nut
11—Fuel-injection line cap nut
12—Fuel inlet
13—Leak-off line return to tank
14—High-pressure fuel passage
15—Spring adjusting washers
16—Fuel inlet holes in nozzle holder insert
17—Fuel passage in nozzle head
18—Mounting threads
19—Fuel pressure chamber in nozzle head

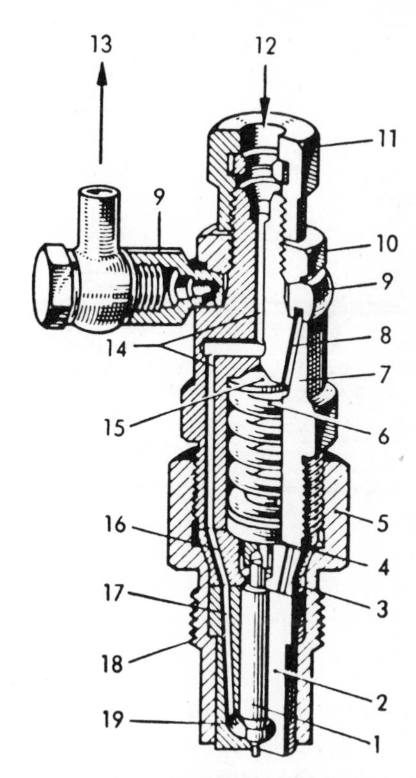

Figure 4-18
Injection Nozzle and Holder.
(Courtesy of Robert Bosch, GmbH)

simply to the difference in head casting between different model engines; however, the removal for both is the same. Proceed as follows with reference to Figure 4-19.

1. Unscrew the knurled nut that holds the electric cable to the top of the glow plug.
2. Using a deep 21-mm socket or box wrench, simply unscrew the glow plug.

4-7 Precombustion Chamber

Removal

Glow plugs must always be removed prior to pulling a precombustion chamber; otherwise, the heating wire of the glow plug that protrudes into the precombustion chamber will be sheared off.

A precombustion chamber should always be removed anytime that the nozzles are removed for checking purposes. Proceed as follows:

1. Refer to Figure 4-19; the threaded ring (12) must be removed by using a special tool (part no. 636 589 01 07 or P/N 636 589 02 07 for 9-mm-wide groove threaded ring). This tool is basically a keyed sleeve with two pegs extending from its base. It is circular in shape, with its size corresponding to the outside diameter of the threaded ring groove. Passing through the center of the sleeve is a spindle that is threaded at its base to screw into the inside threads of the threaded ring.

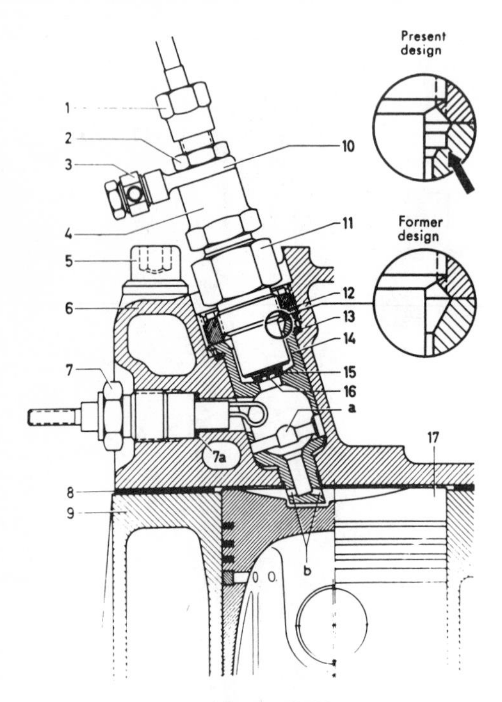

Figure 4-19
Glow Plug Removal.
(Courtesy of Daimler-Benz, Aktiengesellschaft)

Index:
1—Injection line nut
2—Hex nut
3—Leak-off line
4—Nozzle holder
5—Cylinder head bolt
6—Cylinder head
7—Glow plug
7a—Seal ring
8—Head gasket
9—Cylinder block
10—Banjo fitting
11—Nozzle nut
12—Threaded ring
13—Seal ring
14—Precombustion chamber
15—Seal
16—Nozzle needle
17—Piston
a—Ball pin
b—Outlet holes of precombustion chamber

2. With the spindle screwed into the threaded ring and the keyed sleeve properly located as to the groove of the threaded ring, tighten the spindle to the sleeve by tightening the large threaded nut on the spindle up against the sleeve.
3. By turning the keyed sleeve, the threaded ring can now be removed.
4. To extract the precombustion chamber, obtain special tool part no. 636 589 01 33, which is similar to the tool used for pulling the threaded ring. The center spindle of the tool is screwed into the precombustion chamber as far as possible. The puller itself that the spindle threads through must be rotated until a cutout groove is aligned with a small groove in the cylinder head in line with the glow plug. With this done, the precombustion chamber can now be removed by tightening up a hex nut that is threaded to the spindle and up against the puller sleeve.

Caution: While pulling the precombustion chamber, make certain that the puller sleeve itself does not turn; otherwise, the nose of the precombustion chamber can be sheared off.

Installation

1. Install a new sealing ring (13 in Figure 4-19) into the precombustion bore of the cylinder head. *Be sure* that the proper thickness ring is used, since these are available in different thicknesses, to properly establish the distance between the precombustion chamber and the cylinder head.
2. Install the precombustion chamber into its bore by reversing the removal procedure.
3. Screw in the threaded ring and torque to specification.
4. Install the glow plug.
5. Install the nozzle.

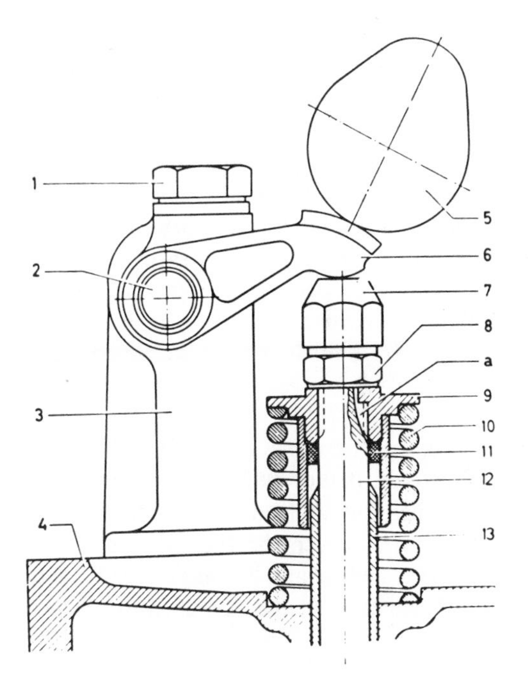

Index:

a—Valve stem groove
1—Necked down bolt
2—Rocker arm shaft
3—Rocker arm bracket
4—Cylinder head
5—Camshaft
6—Rocker arm
7—Cap nut
8—Counternut
9—Valve spring retainer with seal ring holder
10—Valve spring
11—Rubber seal ring
12—Valve stem
13—Valve guide

Figure 4-20
Valve Adjustment.
(Courtesy of Daimler-Benz, Aktiengesellschaft)

4-8

Valve Adjustment

Valve clearances should be adjusted when the engine is *cold* on engines using overhead camshafts. Proceed as follows:

1. Refer to Figure 4-20.
2. Prior to setting the valve clearance, double-check the cylinder head and rocker arm bracket *hold-down* bolts for the proper torque.
3. Bar the engine over until the base circle of the cam is in the position shown in Figure 4-20 and check the clearance. To adjust, loosen off the hex nut (8) and turn nut (7) as necessary.

4-9

Bleeding the Fuel System

Any time that the fuel filters are changed, the injection or fuel lift pump is serviced, the injection nozzles are removed and reinstalled, or the fuel lines have been removed, it will be necessary to *bleed* the fuel system of all entrapped air; otherwise, air in the fuel system can cause a heavy knocking sound from the engine cylinders, a reduction in engine performance, and hard starting characteristics.

1. On top of the main fuel filter is a bleeder screw (the one with the slotted screw and pin through it like a wing nut), which should be opened up between one and two turns.
2. Turn the control knob of the hand pump (fuel lift pump in Figure 4-21) counterclockwise until it is free from the body, and pump it up and down until fuel free from any air bubbles flows steadily out of the main filter bleeder screw; then close

the screw. *When bleeding the fuel system, place a suitable drain tray under the engine to retain dripping fuel.*

3. The location of the bleeder screw or screws will vary between different models of Robert Bosch injection pumps; however, there are two common locations:

   a. Two bleeder screws located on the injection pump usually with slotted heads, positioned on the same side of the housing as the fuel lift pump, but located high up on the housing, with one at the front and the other at the rear (one screw each opposite the no. 1 and no. 4 pump fuel lines).

   b. On those injection pumps with just one bleeder screw, it is located on the top of the housing at the rear of the pump just beside the fuel line to the nozzle.

   In either case, loosen the bleeder screw or screws and pump the handle of the lift pump until a good flow of fuel, free from any air bubbles, appears; then tighten up the screws. Also at this time, push down the handle of the lift pump and turn it clockwise until it is tight on the barrel of the housing.

4. Bleeding of the fuel lines is usually necessary only if they have been disconnected and therefore drained of fuel. The quickest way is to loosen the fuel line nut at the nozzle, and crank the engine over until fuel appears at the loosened line; then tighten it up.

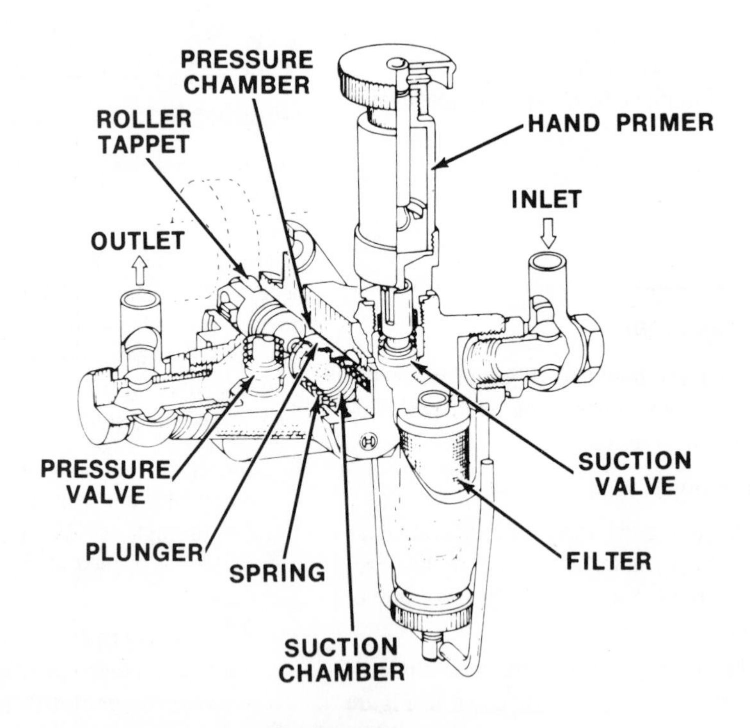

Figure 4-21
Fuel Transfer Pump.
(Courtesy of Robert Bosch, GmbH)

5. Once the engine fires and runs, you can quickly loosen each nozzle line one at a time to ensure that all air has in fact been removed; then tighten it again. Check all bleeder screws for signs of fuel leakage.

Fuel Lift (Transfer) Pump

Both the single- and double-acting lift pumps used by Robert Bosch are similar in design and operation to those used by American Bosch discussed earlier.

Automatic Timing Device (In-Line Injection Pumps)

The purpose of the automatic timing device used with in-line injection pumps is to advance the point of injection with a relative increase of the engine speed. Figure 4-22 shows the basic components of this unit.

Operation. The operation of this timing device is very similar to that used by Caterpillar diesel engines discussed in Chapter 7. Basically, the centrifugal force created by the weights is opposed by the tension of the springs. As the engine accelerates, the motion transfer of the weights causes relative movement of the drive plate to the injection pump, thereby advancing the point of initial injection to the engine.

4-10

Timing the Pump to the Engine

Anytime that a fuel-injection pump is to be reinstalled on an engine, consider the following:

1. Read over injection pump installation in Chapter 2, since the sequence of events for both these types of pumps is very similar.

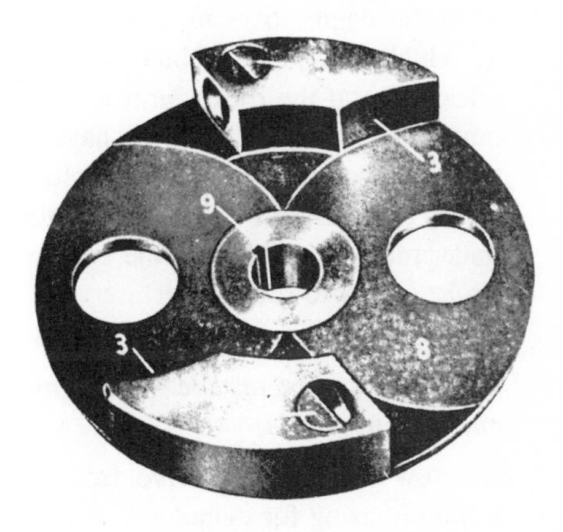

Figure 4-22
Basic Components of the Automatic Timing Device.
(Courtesy of Robert Bosch, GmbH)

Index:
1—Injection pump drive gear
2—Segment plate
3—Contact surfaces for the centrifugal weights on the segment plate and flange
4—Centrifugal weights
5—Spring seat bolts
6—Tension spring
7—Stop bolt in the tension spring for limiting adjustment
8—Segment flange
9—Woodruff key groove

2. Determine from manufacturer's specifications what the pump *spill timing* angle should be. In other words, how many crankshaft degrees BTDC (before top dead center) does fuel injection actually begin?
3. Timing marks will be scribed or stamped on either the flywheel or crankshaft pulley to assist during installation.
4. Generally, there is a *master spline* on the injection pump camshaft gear, or the pump has one tooth with a center punch mark on it, or the drive flange is marked in a similar manner.
5. Turn the engine over to TDC on cylinder 1, which should have both valves loose (clearance) at the rocker arm to camshaft; otherwise, the engine will have to be turned around one more revolution if either valve is still open. The engine cylinder must be at TDC on the compression stroke (ready to start power).
6. With the engine in this position, the injection pump can be installed by aligning the marks as in step 4.
7. Unscrew the delivery valve from the injection housing for cylinder 1 and take out the delivery valve and spring. Screw the delivery valve holder back into the housing and install a used fuel delivery line that can be used as an overflow pipe. If this is not available, leave the top of the delivery valve holder open.
8. Turn the engine over until the specified timing mark on either the flywheel or crankshaft pulley (initial injection timing) aligns with the stationary pointer. (Always turn the engine in its opposite direction of rotation first, approximately 45 to 60 degrees; then slowly turn it in its normal rotation toward the timing mark. This eliminates gear backlash in the drive train.)
9. The automatic timing device should be held in its idling position during this check by removing the inspection cover and temporarily holding it in this position by spring pressure.
10. As the engine approaches the timing mark for initial injection, the point of injection can be established by watching the overflow pipe or delivery valve holder, which will have fuel spill out of it. With the injection pump mounting flange bolts loose, manually grasp the pump and turn it back and forward until you just establish the point of fuel delivery at the holder or overflow line. Tighten up the injection pump mounting bolts.

Note: Turning the pump toward the engine will retard the timing, whereas turning it away from the engine will advance it on right-hand turning engines.

11. Reinstall the delivery valve and spring into the pump (clean).

4-11

Aneroid Control

On those engines using Robert Bosch injection pumps with a *turbocharged* engine, an aneroid control is used to prevent overfueling of the engine and hence black smoke during acceleration. This device controls the amount of fuel that can be injected until the exhaust gas driven turbocharger can overcome its initial speed lag and supply enough air boost to the engine cylinders. Such a device is used extensively by all 4-stroke-cycle engine manufacturers today to comply with Federal Environmental Protection Agency smoke emission standards.

The aneroid is mounted on top of the injection pump governor housing, with its linkage connected to the fuel control mechanism and a supply line running from the pressure side of the intake manifold (turbocharger outlet) to the top of the aneroid

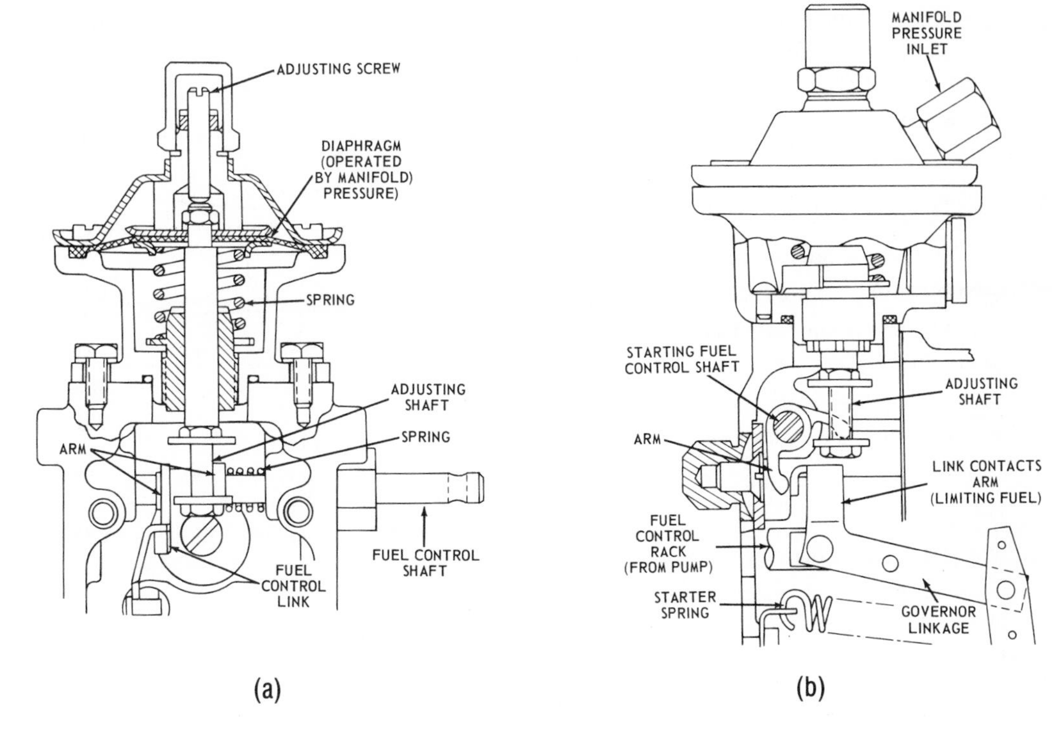

Figure 4-23
Aneroid Control Linkage.
(Courtesy of Robert Bosch, GmbH)

housing. Such a device is shown in Figure 4-23.

Figure 4-23(a) shows the position of the aneroid control linkage when the engine *stop* lever is actuated, which moves the aneroid fuel control link out of contact with the arm on the fuel-injection pump control rack. Figure 4-23 (b) shows the aneroid linkage position when the throttle control lever is moved to the *slow idle* position. This causes the starter spring to move the fuel control rack to the *excess fuel position*. During the cranking period *only* is excess fuel supplied to the engine, since the instant the engine starts we have the centrifugal force of the governor flyweights overcoming the starter spring tension, thereby moving the fuel control rack to a decreased fuel position. As this is occurring, the aneroid fuel control lever shaft spring will move the control link back into its original position. In Figure 4-23 (b), the fuel control rack arm will contact the aneroid fuel control link, thereby limiting the amount of fuel that can be injected to approximately half-throttle and preventing excessive black smoke upon initial starting. The same lever will control the rack position at any time that the engine is accelerated, preventing any further increase in fuel delivery until the turbocharger has also accelerated to supply enough air for complete combustion.

Mercedes-Benz Truck and Industrial Engines

Mercedes-Benz builds diesel engines in 4-, 5-, and 6-cylinder in-line configurations, along with 6, 8, 10, and 12 Vee types. All use Robert Bosch fuel-injection pumps, and other than the fact that some have a greater number of pumping elements than others, all operate on the same basic principle.

Models (Truck and Industrial Only)

OM 314, 4-cylinder rated at 31 to 85 hp (23 to 62 kW)

OM 352, 6-cylinder rated at 46 to 130 hp (34 to 95 kW)

OM 352A, 6-cylinder turbocharged rated at 168 hp (123 kW) maximum

OM 615, 4-cylinder rated at 22 to 60 hp (16 to 44 kW)

OM 616, 4-cylinder rated at 24 to 65 hp (18 to 48 kW)

OM 636, 4-cylinder rated at 19 to 43 hp (14 to 32 kW)

OM 401, V6 rated at 90 to 192 hp (66 to 136 kW)

OM 402, V8 rated at 121 to 256 hp (89 to 188 kW)

OM 403, V10 rated at 149 to 320 hp (110 to 235 kW)

OM 404, V12 rated at 195 to 400 hp (143 to 294 kW)

OM 404A, V12 turbocharged rated at 258 to 525 hp (190 to 386 kW)

Two of the more popular larger engines are the OM 402 and OM 403, which are V8 and V10 engines, respectively. All Vee-type engines are numbered with cylinder 1 located at the right front side of the engine, with cylinder numbering from front to rear. For example, on a V8 engine, 1–2–3–4 would be located on the right bank, with 5–6–7–8 being on the left bank. On a V10 engine, 1–2–3–4–5 would be on the right bank, with 6–7–8–9–10 being on the left bank. Left and right banks are established by standing at the flywheel end of the engine.

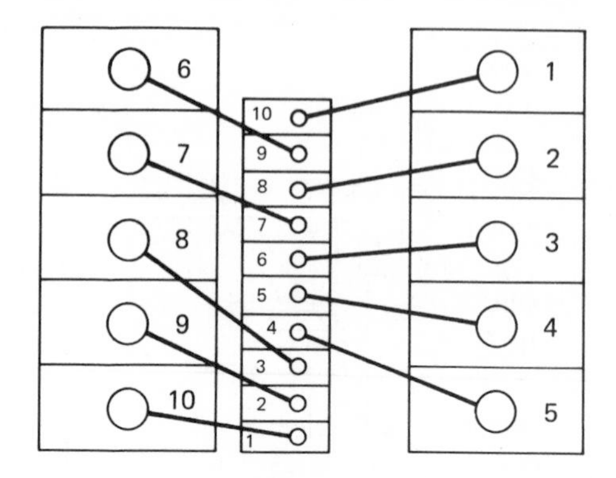

Figure 4-24
Injection Pump Lines from Pump to Engine Cylinders.
(Courtesy of Daimler-Benz, Aktiengesellschaft)

The V8 firing order is 1–5–7–2–6–3–4–8. The V10 firing order is 1–6–5–10–2–7–3–8–4–9.

Figure 4-24 shows the typical setup of the injection pump lines from the pump to the engine cylinders.

The V8 and V10 engine fuel-injection pumps are installed with piston 1 on its compression stroke and the engine 18° BTDC; then ensure that the injection pump to drive flange timing marks are aligned.

4-12

VE Model Injection Pump

The VE Model injection pump is presently being used on the Volkswagen Rabbit (Golf in Europe) diesel car, as well as by Volvo-Penta on their marine applications, and by Peugeot on their diesel cars. It will be found more and more in the immediate future with the ever-increasing diesel engine penetration of the passenger car and light truck market. Light marine (pleasure craft) and light industrial applications are also a growing market for this type of pump.

The pump takes its name from the German word *Verteiler,* which means *distributor*, although it is also referred to as a rotary pump. The E refers to the particular model of rotary injection pump produced by Robert Bosch Corporation. The pump is available in 2-, 3-, 4-, 5-, 6-, and 8-cylinder configurations to suit the applications previously mentioned.

The pump's operation is very similar in many respects to other distributor pumps on the market, but it does have some unique differences. If you are familiar with the operation of both a distributor pump and Caterpillar's sleeve metering fuel system, then you will quickly understand the pump operation of the VE model, since it incorporates a combination of these two systems.

Any distributor-type fuel-injection pump operates on the basic principle of rotation similar to that of a gasoline engine distributor rotor, and in fact utilizes a rotor (rotating plunger) for fuel distribution to the engine cylinders in firing order sequence. These types of pumps are much more compact than the more common in-line units; however, they also use approximately half as many component parts and usually weigh less than half as much as in-line pumps. Their popularity is steadily increasing on smaller high-speed diesel engines. Both the fuel transfer pump and governor mechanisms are contained within the injection pump housing.

Figure 4-25 shows the fuel delivery of a typical 4-cylinder VE injection pump. Figure 4-26 shows a cross-sectional view of the pump with all the major component parts identified.

Operation

Contained within the injection pump housing is a vane-type transfer pump, which is capable of lifting fuel from the fuel tank through a filter and into the main injection pump housing. The

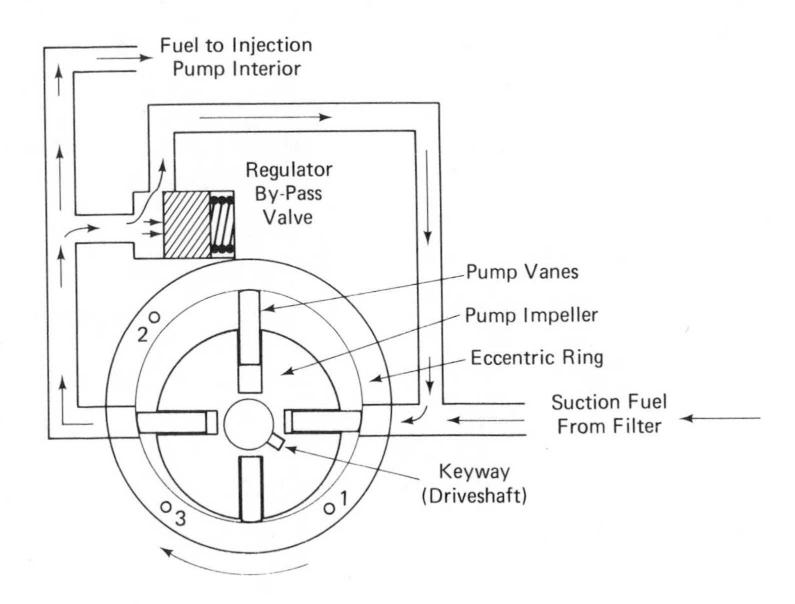

Figure 4-25
Four-Cylinder VE Injection Pump.

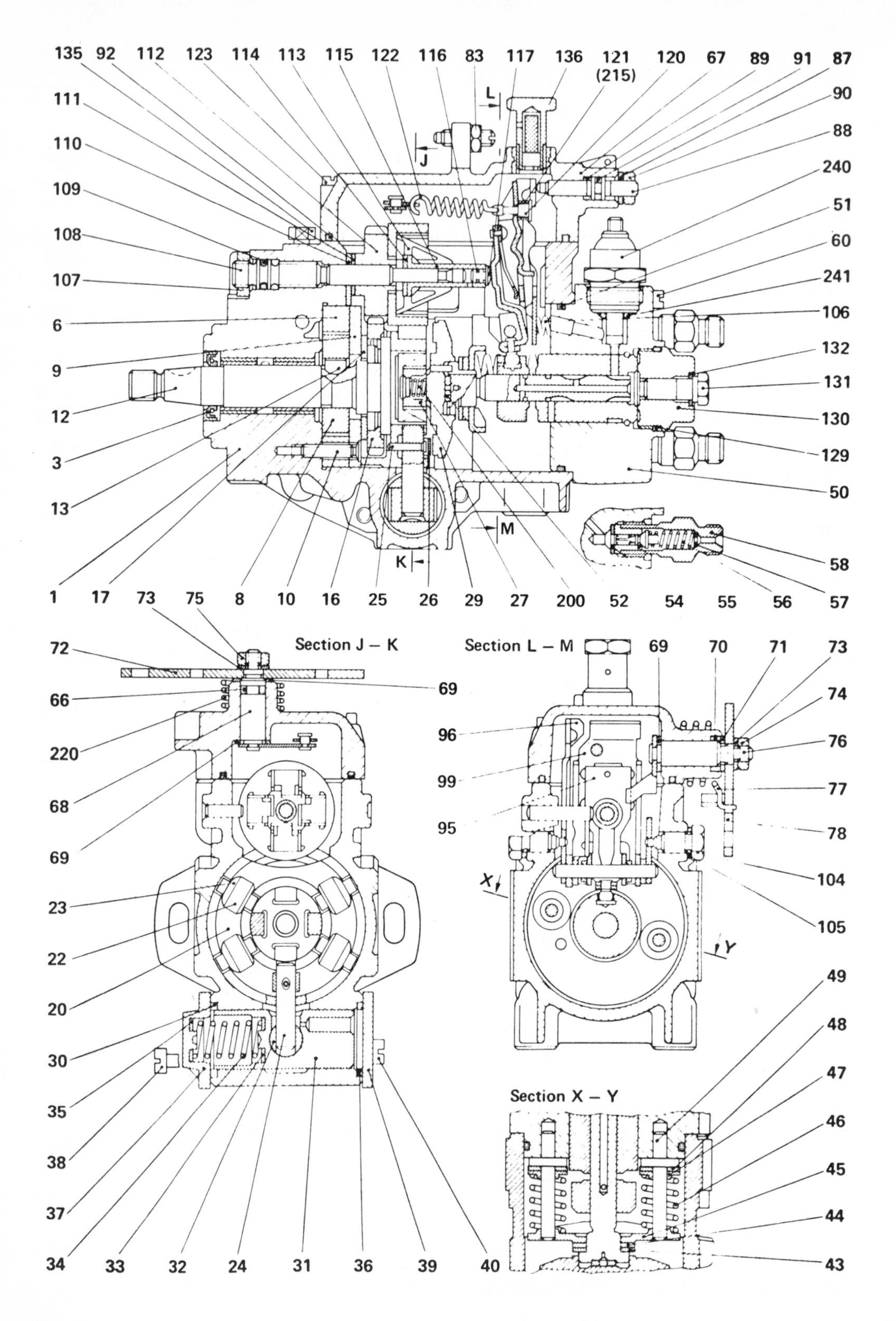
135 92 112 123 114 113 115 122 116 83 117 136 121 (215) 120 67 89 91 87
111
110
109
108
107
6
9
12
3
13
90
88
240
51
60
241
106
132
131
130
129
50
58
57
L
J
M
K
1 17 73 75 8 10 16 25 26 29 27 200 52 54 55 56
Section J – K
Section L – M
72
66
220
68
69
23
22
20
30
35
38
37
34 33 32 24 31 36 39 40
69
96
99
95
X
Y
69 70 71 73
74
76
77
78
104
105
Section X – Y
49
48
47
46
45
44
43

Figure 4-26
Cross-Sectional View of Figure 4-25.
(Courtesy of Robert Bosch, GmbH)

Index:

1—Pump housing
3—Radial lip-type oil seal
6—Eccentric ring
8—Impeller wheel
9—Support ring
10—Countersunk flat bolt
12—Drive shaft
13—Woodruff key
16—Gear wheel
17—Slotted disc
20—Cam roller ring
22—Rollers
23—Spring seat
24—Adjusting pin
25—Retaining pin
26—Retaining bracket
27—Cross-type disc
29—Cam plate
30—Seal ring
31—Timing device piston
32—Slider
33—Shim
34—Compression spring
35—Shims
36—Seal ring
37—Closing cover
38—Fillister head screw
39—Cover plate
40—Fillister head screw
43—Shim
44—Slotted disc
45—Spring seat
46—Compression springs
47—Spring seat
48—Spacers (shims)
49—Guide pins
50—Hydraulic head
51—Seal ring
52—Shim
54—Seal rings
55—Delivery valve assembly
56—Spring
57—Shims
58—Fitting
60—Fillister head screws
66—O ring
67—Housing cover
68—Setting shaft
69—Shim
70—Seal
71—Retainer
72—Control lever
73—Lock washer
74—Hex nut for shutoff lever
75—Hex nut
76—Lever shaft
77—Spring
78—Outer shutoff lever
83—Hex nut
87—Plain washer
88—Threaded pin
89—Spacer sleeve
90—Hex nut
91—O ring
92—Seal ring
95—Start lever
96—Correction lever
99—Tensioning lever
104—Slotted shoulder screw
105—Seal ring
106—Helical compression springs
107—Slotted nut
108—Governor shaft
109—O ring
110—Shim plate
111—Supporting plate
112—Governor flyweight assembly
113—Flyweights
114—Spacer ring
115—Sliding sleeve
116—Seal ring retainer
117—Blanking plug
120—Retaining pin
121—Helical compression spring
122—Extension spring
123—Fillister head screws
129—Seal ring
130—Screw plug
131—Bleeder screw
132—Seal ring
135—Control valve
136—Inlet union screw
200—Compression spring
215—Helical compression spring
220—Torsion spring
240—Solenoid stop valve
241—O ring

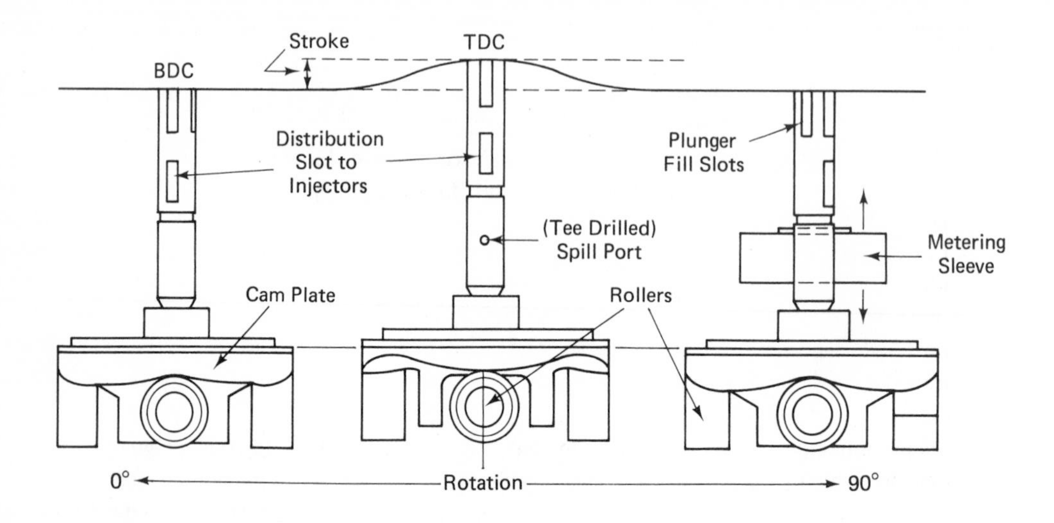

Figure 4-27
Connection Between Cam Rollers and Plunger.

transfer pump is capable of producing fuel at a delivery pressure of 100 psi (7 bar approximately) into the rotating pumping plunger. All internal parts of the injection pump are lubricated by this fuel under pressure; there is no separate oil reservoir. Maximum fuel pressure is controlled by an adjustable fuel pressure regulator screw. The distributor-type injection pump is driven at one-half engine speed on 4-stroke cycle engines, and is capable of delivering 2800 psi (approximately 200 bar) to the injection nozzles. An overflow line from the top of the injection pump housing allows excess fuel to return to the tank (carries heat away).

Anytime that you are working on this pump, take careful note of holes 1, 2, and 3. Hole 1 in the *eccentric ring* is farthest from its inner wall compared to hole 2. When looking at the eccentric ring, this hole must be in position 1 as shown in the diagram, for right-hand rotation injection pumps, and to the left for left-hand rotation pumps. Hole 3 should be on the governor side when the transfer pump is installed. Also, the pump vanes should always be fitted with the circular or crowned ends contacting the walls of the eccentric ring. Fuel from the vane-type transfer pump is delivered to the pumping plunger and then to the nozzles.

Let us study the action of the plunger before proceeding any further, since it is this unit that is responsible for the distribution of the high-pressure fuel within the system. Figure 4-27 shows the connection between the cam rollers and the plunger. Notice that the plunger is capable of two motions: (1) circular or rotational (driven from the drive shaft), and (2) reciprocating (back and forth by cam plate and roller action).

Reference to both Figures 4-26 and Figure 4-27 shows that the cam plate is designed with as many lobes or projections on it as there are engine cylinders. Unlike CAV and Roosa Master distributor pumps, the rollers are not actuated by an internal cam ring with lobes on it, but instead the cam ring is circular and attached to a circular cam plate. As the cam ring rotates with the drive shaft and plunger, the rollers, which are fixed, cause the cam lobe to lift every 90 degrees, for example, in a 4-cylinder engine. In other words, the rollers *do not* lift on the cam as in the conventional system; it is the cam ring, which is solidly attached to the rotating plunger, that actually lifts as each lobe comes into contact with each positioned roller spaced apart in relation to the number of engine cylinders.

With such a system, then, the plunger stroke will remain constant regardless of engine rpm. At the end of each injection stroke, two springs positively ensure a return of the cam ring to its former position (see Figure 4-27). Therefore, the back and forth motion of the individual pumping plunger is positive. Anytime that the roller is at the low point on the rotating cam ring lobe, then, the pumping plunger will be at a position commonly known as BDC (bottom dead center); and with the rotating cam ring lobe in contact with the roller, the pumping plunger will be at the TDC (top dead center) position.

Figure 4-28 shows the basic fuel flow in the system. Distribution of fuel to the injector nozzles is via plunger rotation, and metering (quantity) is controlled by the metering sleeve position, which varies the effective stroke of the plunger.

If we consider the plunger movement, that is, *stroke* and *rotation*, Figure 4-29 depicts the action in a 90 degree movement such as would be found on a 4-cylinder 4-cycle engine pump. Even though there is a period of *dwell* at the start and end of one 90 degree rotation (one cylinder firing), the plunger movement during this time continues.

The sequence of events shown in Figure 4-29 is as follows:

1. The fill slot of the rotating plunger is aligned with the fill port, which is receiving fuel at transfer pump pressure as high as 100 psi (7 bar approximately), one cylinder only.
2. The rotating plunger has reached the *port closing* position. The plunger rotates within a m-‵tering sleeve (see Figure 4-27

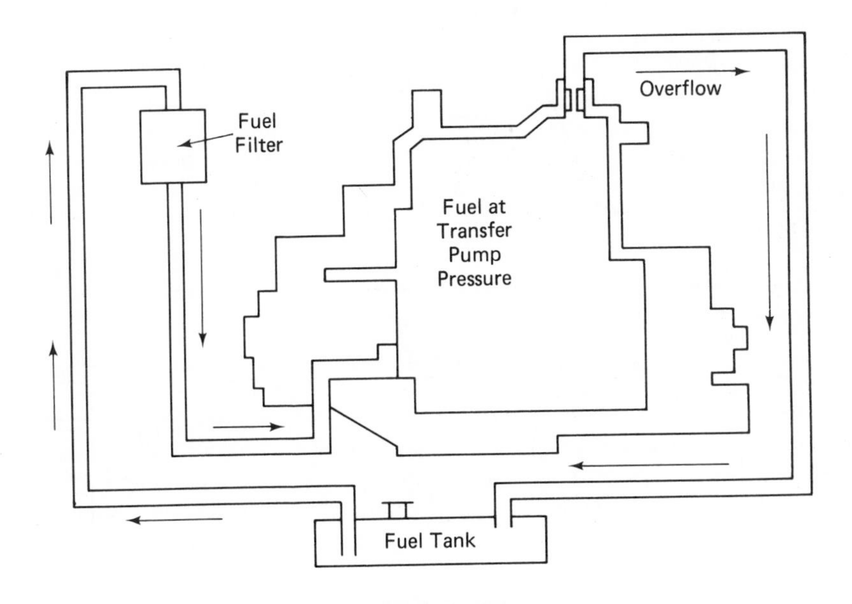

Figure 4-28
Basic Fuel Flow.

and Figure 4-30). The position of the metering sleeve is controlled by the operator or driver though linkage connected to and through the governor spring and flyweights. Because the plunger rotates as well as moving back and forth, the plunger must lift for port closure to occur; then delivery will commence. This portion of the operation is similar in manner to Caterpillar's sleeve metering fuel system, the difference being that the plunger in the Caterpillar system is not being turned continuously, and a rotating camshaft of the in-line type lifts the plunger, while the (metering sleeve) is controlled as in the VE pump by the operator or driver through linkage to the governor mechanism. Since the rotating plunger does stroke through the metering sleeve in the VE pump, this pump is classed as the *port closing* type. Therefore, even though the roller may be causing the cam–ring–plunger to lift, the position of the metering sleeve determines the amount of travel of the plunger or *prestroke*, so the actual effective stroke of the plunger is determined at all times by the sleeve position.

3. At the point of plunger lift (start of effective stroke), fuel delivery to the hydraulic head and injector line will begin in the engine firing order sequence.

4. The effective stroke is always less than the total plunger stroke. As the plunger moves through the metering sleeve it uncovers a *spill port,* opening the high-pressure circuit and allowing the remaining fuel to spill into the interior of the injection pump

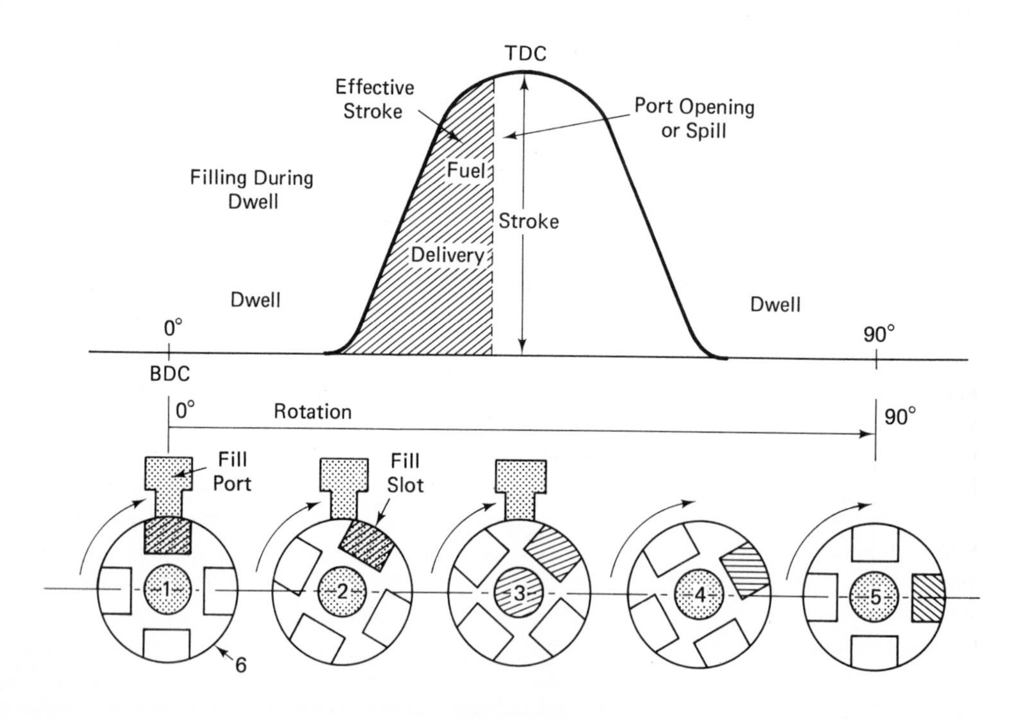

Figure 4-29
Plunger Movement in a 4-Cylinder, 4-Cycle Engine Pump.

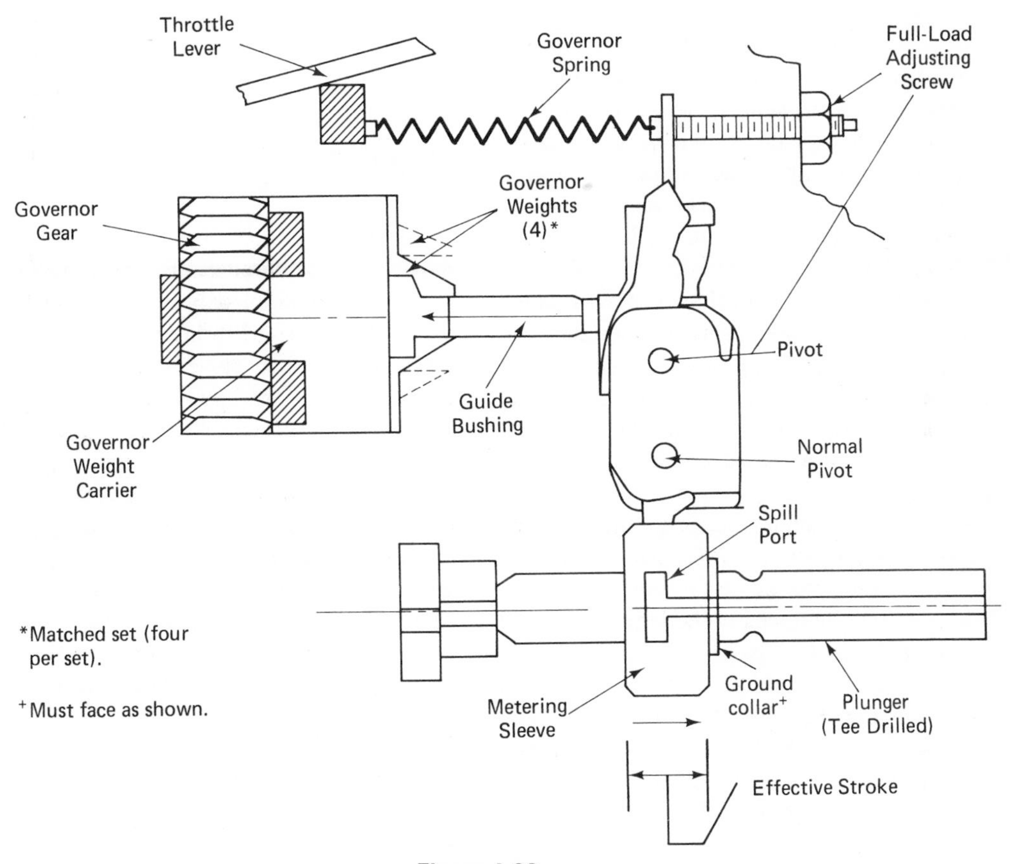

Figure 4-30
Governor Action.

housing. This then is *port opening* or spill, which ends the effective stroke of the plunger; however, the plunger stroke continues.

5. With the sudden decrease in fuel delivery pressure, the spring within the injector nozzle rapidly seats the needle valve, stopping injection and preventing after-dribble, unburnt fuel, and therefore engine exhaust smoke. At the same time the delivery valve for that nozzle located in the hydraulic head is snapped back on its seat by spring pressure.

In a 4-cylinder 4-stroke engine we would have four strokes within 360 degrees of pump plunger rotation, which is of course equal to 720 degrees of engine rotation. In summation, the volume of fuel delivered is controlled by the metering sleeve position, which alters the *effective stroke* time that ports are closed.

In Figure 4-29, item 6 is an annulus or circular slot located on the rotating plunger; this is the reason that the plunger must lift for port closure to occur. Only after the annulus lifts beyond the fill port do we have port closure. Port closing occurs only after a specified lift from BDC.

Governor Action

The metering sleeve is controlled by the action of the throttle and governor linkage; therefore, at low power demands, the spill port will uncover earlier in the stroke of the rotating plunger, and for high-power situations the spill port will uncover later, thereby lengthening the effective plunger stroke and delivering more fuel to the engine.

As with all mechanical governors, it is the force of the governor spring and weights that determines the metering sleeve position. The governor controls starting, low and high idle, rated speed, overspeed, and during any coasting situations that would occur on a vehicle application.

With the engine stopped, there is no flyweight force; therefore spring and throttle position cause the guide bushing shown in Figure 4-30 to be forced to the left, causing the governor lever to pivot, which moves the sleeve away from BDC and thereby gives maximum fuel delivery. Note that at start-up the plunger has its longest effective stroke since the metering sleeve has been moved to its farthest position from BDC, allowing even more fuel to be delivered than at full load.

As soon as the engine fires, the centrifugal force created by the revolving flyweights forces it to the right with a resultant transfer of motion through the governor lever, which will force the metering sleeve to the left, thereby shortening the effective stroke. The distance that the metering sleeve will move back is dependent on the throttle position, since once the weight force equals that of the governor spring a state of balance will occur.

However, under normal situations of start-up with the throttle at the idle position, the spring tension has been established by the adjustment screw. When the weight force balances the spring force, a state of balance will exist, holding the sleeve close to BDC and allowing a very short effective stroke before the spill port is uncovered.

At low idle, the governor controls and compensates for any temperature and load changes.

Speed Increase

When the operator steps on the throttle or moves a hand throttle, the governor linkage will upset the previously existing state of balance. The increased spring force will cause the metering sleeve to move away from the BDC position against the opposing weight force. By increasing the effective stroke of the plunger, more fuel is delivered and engine speed will increase. Any increase in engine rpm will also cause an increase in the centrifugal force created by the rotating governor flyweights, which are now working against the spring force. For any given throttle position, a state of balance can exist as long as the engine speed is such that the power being produced is capable of moving the vehicle or carrying the load placed upon it.

If the throttle is moved all the way to the high-speed stop screw, and at full load, the spring force is greater than the flyweight force, allowing the metering sleeve to be moved away from BDC so that the effective stroke is longer. With a load applied to the engine, full-load rated rpm will be attained. With no-load on the engine, and a higher rpm, the weight force will be stronger than the spring; therefore; the plunger's effective stroke is reduced, and this would be the high-idle position. If the operator or driver removes his or her foot from the throttle while coasting downhill, the governor gear is still being driven from the engine; therefore, the weights will move the sleeve toward BDC and injection pressures will never be developed at this position. If the vehicle speed reduces with the throttle still released, weight force will reduce and spring force will place the sleeve into a position to deliver fuel for normal idling as before.

Speed Decrease

The engine speed can be decreased either by an increase in load for a given throttle setting or by the throttle manually being moved to a decreased fuel position. If the engine is running at a fixed rpm, with an increased load applied, the engine rpm will tend to reduce. As this happens, the state of balance that previously existed between the weights and springs is upset in favor of the spring because of the reduction in engine rpm. The sleeve will be moved to increase the plunger's effective stroke without the operator having to move the throttle. If the engine has enough power to pick up the increased load, the engine will pick up speed and a state of balance will again exist.

However, if the engine is incapable of overcoming the additional load, the operator will be forced into downshifting, as in the case of a standard transmission in a car; if fitted with an automatic transmission, the downshift will occur automatically. If on a piece of industrial or marine equipment, the load would have to be reduced if the engine were incapable of maintaining a suitable rpm within its normal operating range.

Automatic Timing Advance

The automatic advance mechanism employs the same principle of operation as does CAV and Roosa Master distributor-type injection pumps. Fuel pressure from the transfer pump is delivered to a timing piston whose movement is opposed by spring pressure. At low engine speeds, the relatively low supply pump pressure has little to no effect on the timing piston travel. As engine speed increases, the rising fuel pressure will force the timing piston to overcome the resistance of the spring at its opposite end. At the center of the piston, as shown in Figure 4-31, is a connecting pin extending up into the roller ring. The movement of the piston transmits this motion through the pin, which in turn rotates the roller ring in the opposite direction to drive shaft rotation, thereby advancing the timing of the cam plate lift from BDC to begin the plunger stroke. The timing piston travel should not be toyed with, but should be checked while the injection pump is mounted on a test bench.

On some VE model pumps, a *cold start* device is used which advances the timing for the purpose of reducing start-up smoke due to lowered combustion temperatures and longer ignition delay at low ambient temperatures. This device is simply a small lever and cam that is manually operated by the operator or driver. This is also shown in Figure 4-31.

Shutoff Device

On most pumps the fuel shutoff is simply a small electrical solenoid screwed into the pump housing that closes the supply of fuel to the spill port and plunger when the ignition key is turned off. Some pumps use a straight mechanical shutoff, which is a lever that overrides the governor spring to move the metering sleeve to zero fuel delivery.

Prestroke Compared to Non-Prestroke Pumps

Some VE injection pumps use a plunger whereby all the fill ports are interconnected by an annulus or circular passage running around the circumference of the plunger as shown in Figure 4-32. With this type of plunger containing the annulus, the unit is known as a *prestroke* pump. With this type, the fill ports cannot close by plunger rotation alone. The plunger must lift for port closure to occur. Only after the annulus lifts beyond the fill port do we have port closure. The plunger must be adjusted for a specific lift from BDC for port closure to happen. With this type, fuel pressure buildup within the tee-drilled plunger takes a

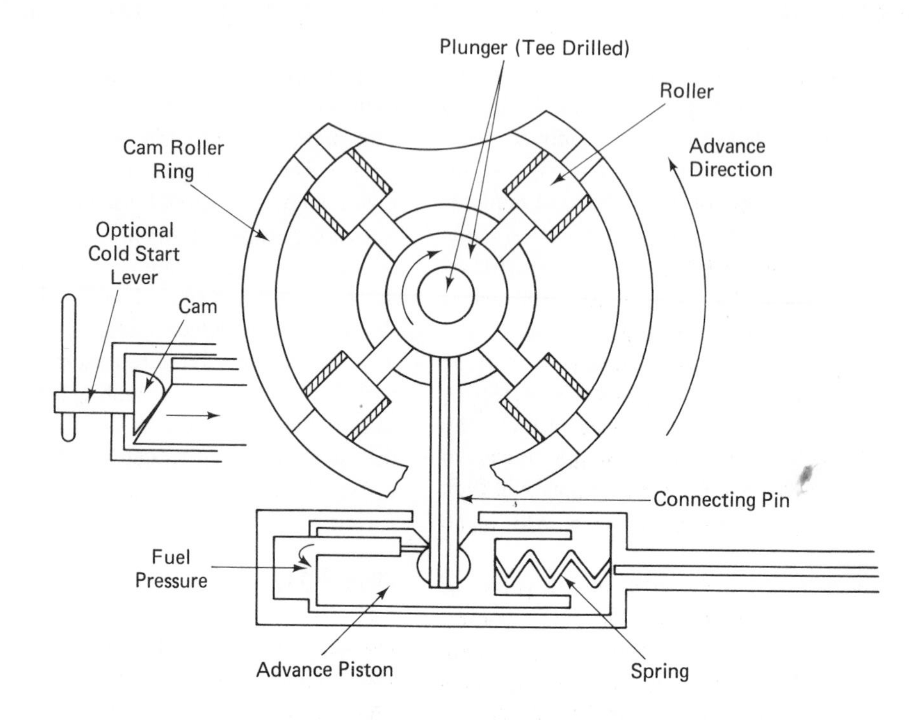

Figure 4-31
Automatic Timing Device.

few degrees longer than for the zero prestroke type, which does not have the annulus and wherein port closure occurs by plunger rotation alone: the plunger lifts from BDC after rotation from port closure.

Automatic Cold Start Device

Figure 4-33(a) and (b) show the relative positions of the injection pump linkage while the engine is cold and as it warms up.

Overhaul of the Injection Pump

As indicated continuously throughout this book, repair and major overhaul of any injection pump should only be undertaken by personnel trained in the diversified and intricate work of fuel-injection equipment. Since special tools and equipment are required, which are not always readily available to everyone, refer to the Robert Bosch publication 46, VDT-W-460/100 B, Edition 1, Repair of Distributor-type Fuel Injection Pump 04604-VE-F.

This is obtainable through your local Robert Bosch dealer, or from one of the Robert Bosch licensees mentioned in Chapter 1.

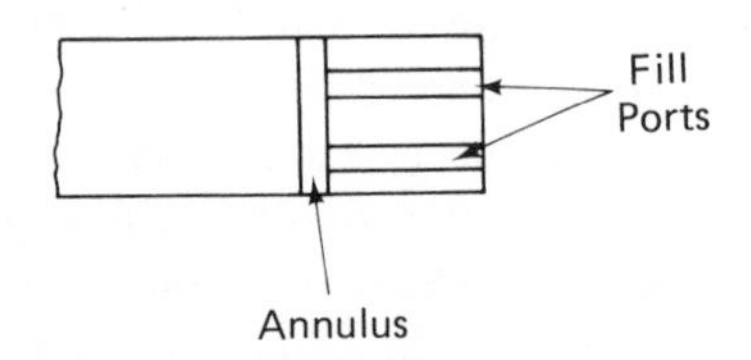

Figure 4-32
Annulus Interconnection.

ENGINE COLD

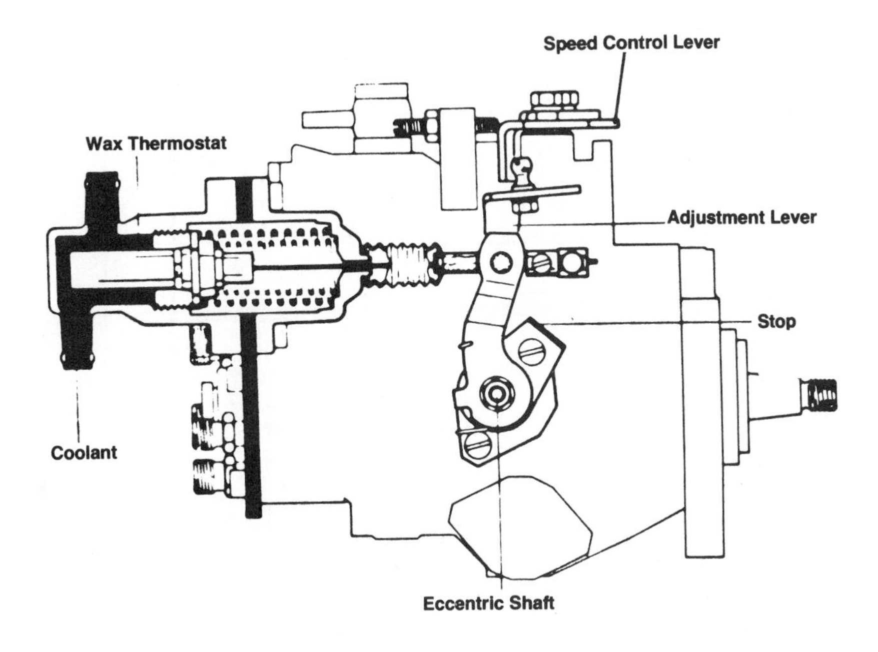

(a)

ENGINE WARM

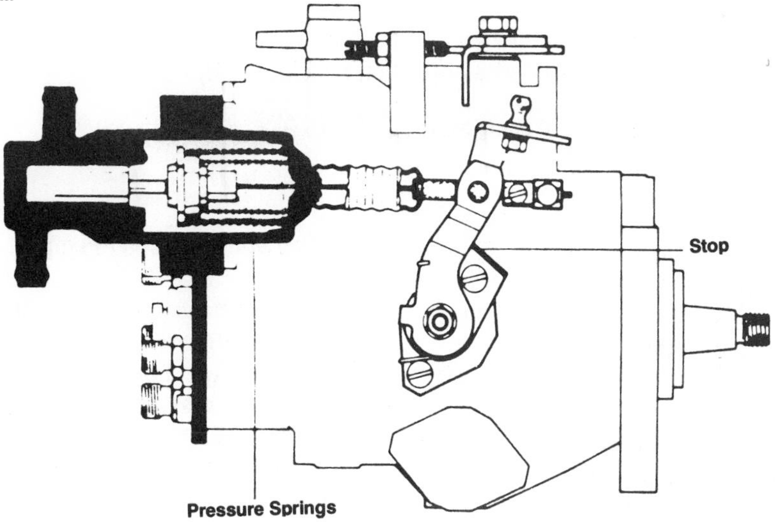

Stop

Pressure Springs

(b)

Figure 4-33
Injection Pump Linkage Positions.
(Courtesy of Robert Bosch, GmbH)

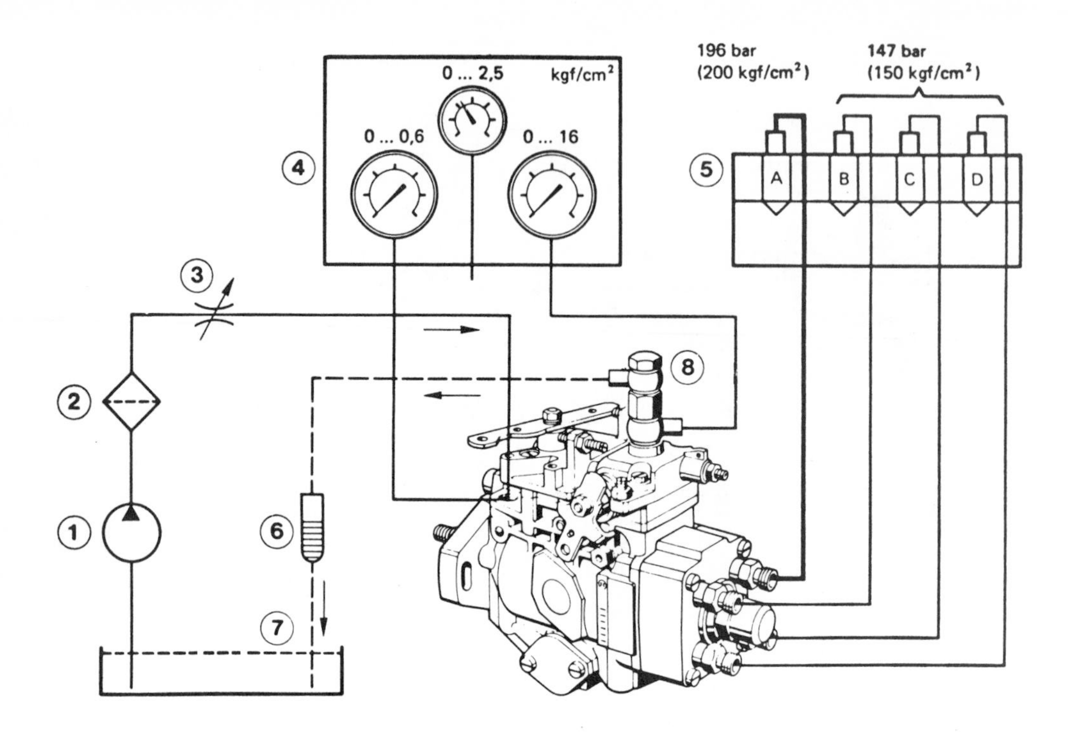

Figure 4-34
Testing the Injection Pump.
(Courtesy of Robert Bosch, GmbH)

Index:
1—Supply pump
2—Filter
3—Adjustable fuel pressure control inlet
4—Manometer: 0–0.6 kgf/cm² (feed pressure)
0–2.5 kgf/cm² (charging air pressure)
0–16 kgf/cm² (supply pump pressure)
5—Nozzle holder assemblies with nozzles set to 147 bar (2161 psi) or (150 kgf/cm²)
Nozzle holder assemblies with nozzles set to 196 bar (2881 psi) or (200 kgf/cm²)
Both of the above at outlet A for starting fuel delivery
6—Measuring beaker or glass for determining overflow (return) volume
7—Test stand oil reservoir
8—Overflow screw

Testing of the Injection Pump

Test instructions are given in detailed form in the Robert Bosch publication, *Testing of the Distributor-type Injection Pump 04604-VE-F, Number 46, VDT-W-460/300 B, Edition 1, or later*. However, for general information and to give you an appreciation of what is involved in the testing of the injection pump, the following sequence is general in nature. Figure 4-34 shows how a VE model injection pump would be connected up to a "test stand" for a normal sequence of testing and adjustment.

Prior to any test follow all the precautions given under pump testing in Chapters 6 and 8. Make sure that you have *all* the *special* tools required. Always follow the sequence of tests as specifically laid down by the manufacturer, since any deviation from this can create unnecessary problems.

Since the VE pump is used on a variety of applications, you must refer to the Robert Bosch test specification sheet determined by the injection pump nameplate, which lists the pump serial number and all other relevant data. Certain pumps use a higher test nozzle assembly as listed under (5) above.

With the injection pump assembly properly coupled and aligned to the test stand drive, proceed as follows:

Adjust the Governor Shaft. Refer to Figure 4-26, item 108.

1. Pumps without load-dependent port closing: screw the governor shaft inward with special adjustment tool KDEP 1082 until a distance of 1.5 mm (0.059 in.) measured with a vernier caliper exists between the pump housing and the end of the governor shaft.
2. Pumps with load-dependent port closing: as procedure 1; however, dimension should be 3 mm (0.118 in.) between the housing flange and end of the governor shaft.

In both instances, lock the governor shaft with the slotted round nut to the specified torque (item 107 in Figure 4-35).

Preheating Pump Test Oil. If the pump is equipped with an electrical shutoff solenoid, it must be activated to ''run position'' during testing with the correct voltage. Hold the throttle lever in the full-fuel position and tie it there for 10 min with the fuel control pressure adjusted as per the test specification sheet and the test stand running at the pump's rated speed. Make sure that fuel is being delivered from the pump during this time.

Preadjust Full-load Fuel Delivery. See item 90 in Figure 4–35.

1. Run the speed adjustment screw all the way out.
2. Lock the speed control lever hardup against the screw.
3. Run the pump at the speed given in the specification sheet, and turn the threaded pin (88 in Figure 4-36) to obtain the full-load fuel delivery.
4. A shorter threaded pin (88) may be required if the O ring (91) comes out of the port when adjusting to specifications.
5. On some pumps a longer pin (88) may be required.

Setting Full-Load Stop. Once the full-load fuel delivery has been adjusted as above, install a spacer sleeve (89) so that no more than one-third of a turn is possible on the threaded pin (88); then tighten the locknut (90 in Figure 4-35) to the torque shown.

Adjusting the Timing Device Travel and Supply Pump Pressure. The timing device is shown as items 30 through 40 in Figure 4-26. Hold the speed control lever wide open; bleed the measuring device which is installed into the timing device on the pump housing, and with the pump running at the speed given in the test specification sheet, note both the supply pump pressure and the travel of the timing piston on the installed test device.

Adjustment of Supply Pump Pressure. Refer to item 135 in Figure 4-35; contained within the control valve (135) is a vent plug whose installed distance will directly affect the supply pump pressure. Pressing the vent plug inward will both *increase* the transfer supply (vane) pump pressure and alter the timing device advancement (piston movement). Withdrawing, or moving the internal vent plug back, will create the opposite effect, that is, a reduction in the delivery pressure from the vane pump and a corresponding change to the degree of advancement of the timing piston and roller ring.

To remove the control valve (135), proceed as follows:

1. Use socket tool KDEP 1086 or equivalent to remove the valve.
2. Remove the bracing sleeve as shown in Figure 4-37.
3. With a special puller, remove the bracing sleeve and plunger along with the helical compression spring, and press the vent plug out.
4. Replace the compression spring and plunger.

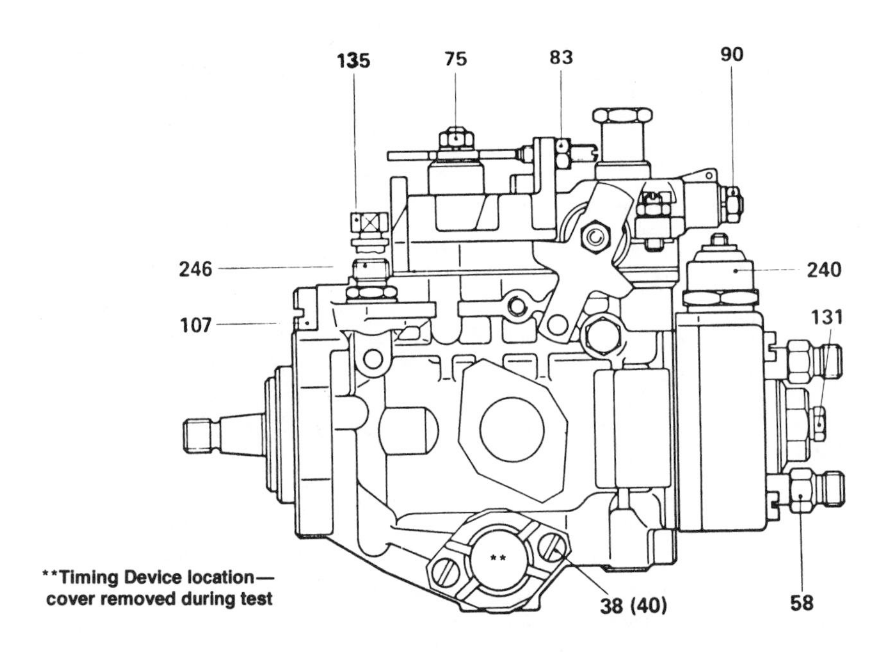

Figure 4-35
Torque Limits, VE Pumps.
(Courtesy of Robert Bosch, GmbH)

Index:

38 (40)—Cover screws (6 to 8 N·m, 4.5 to 6 lb-ft, or 0.6 to 0.8 kgf)
58—Fuel outlet fitting (35 to 45 N·m, 26 to 33 lb-ft, or 3.5 to 4.5 kgf)
75—Throttle lever shaft nut (5 to 10 N·m, 4 to 7.5 lb-ft, or 0.5 to 1 kgf)
83—Locknut (5 to 6 N·m, 0.5 to 0.6 kgf, or 3.5 to 4.5 lb-ft)
90—Full-load fuel adjusting screw (7 to 9 N·m, 5 to 6.5 lb-ft, or 0.7 to 0.9 kgf)
107—Governor shaft slotted nut (25 to 30 N·m, 2.5 to 3 kgf, or 18.5 to 22 lb-ft)
131—Fuel bleeder screw (8 to 10 N·m, 0.8 to 1 kgf, or 4.5 to 7.5 lb-ft)
135—Control valve (8 to 9 N·m, 0.8 to 0.9 kgf, or 4.5 to 6.5 lb-ft)
240—Electric solenoid (40 to 45 N·m, 4 to 4.5 kgf, or 29.5 to 33 lb-ft)
246—Fuel inlet fitting (20 to 25 N·m, 2 to 2.5 kgf, or 15 to 18.5 lb-ft)

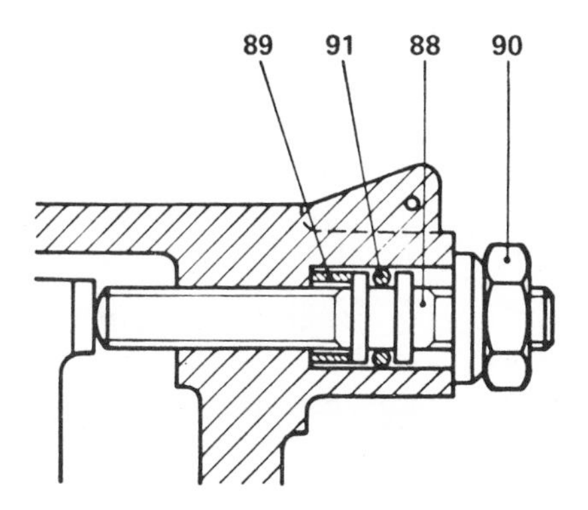

Figure 4-36
Enlarged View of Item 90 in Figure 4-35.
(Courtesy of Robert Bosch, GmbH)

5. Press a new bracing sleeve into the control valve as shown in Figure 4-37 until it is flush with the end of the valve.
6. Install the control valve (135) and torque to specifications. (Figure 4-35).

Adjust Timing-Device Travel. With the transfer pump pressure corrected as per specifications (test sheet), check that the travel of the timing piston is as specified with the use of the Robert Bosch timing gauge, which adapts and bolts in place on the timing cover (38 to 40 in Figure 4-35). The timing dimension can be changed by using shims and washers to alter the spring preload. Make sure that at least one disc is fitted at each side of the compression spring. The shims if used are inserted into the timing piston bore. Double-check the full-load fuel delivery as described previously.

Adjust Low-Idle Speed Regulation. Hold the throttle lever (item 72 in Figure 4-26) against the idle-speed adjustment screw, and with the pump running at the test speed quoted, adjust the specified idle fuel setting by use of the idle adjusting screw (82 in Figure 4-38).

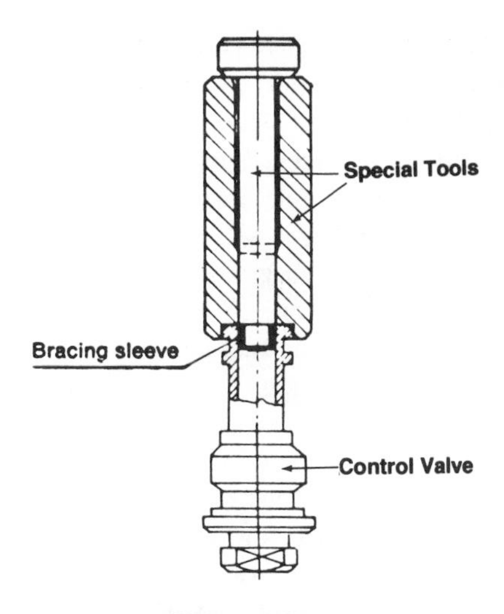

Figure 4-37
Removing the Bracing Sleeve.
(Courtesy of Robert Bosch, GmbH)

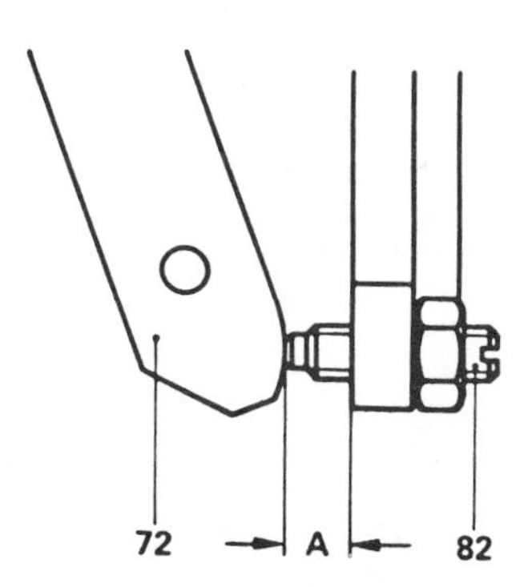

Figure 4-38
Adjusting Idle Fuel Setting.
(Courtesy of Robert Bosch, GmbH)

Referring to Figure 4-34, check and compare dimension A with the test sheet specification. If necessary, withdraw the throttle lever from its toothed (splined) control shaft, and reposition it until the correct distance A is obtained.

Some pump throttle levers have an O stamped on them, which indicates that they are offset by one-half a tooth compared to the nonmarked ones. Tighten the nut (75 in Figure 4-35) to specifications.

Measure Excess-Fuel Starting Dimension. With the nozzle assembly set to 196 bar (2281 psi), compare the fuel delivery with the specification test sheet. It is not necessary to have the throttle hard up against the idle stop screw. If the starting fuel delivery is incorrect, check the dimension MS shown in Figure 4-39, which is an enlargement of the same numbered items shown in Figure 4-26.

Attach spacer tool KDEP 1084 as shown in Figure 4-39 with the correction lever (96) hard up against the spacer, and push the tensioning lever (99) up against the stop pin. Measure the dimension MS and compare it with the test sheet specification. The plug (117) must be changed out by removing the complete governor flyweight assembly and the sliding sleeve.

Plug (117) comes in a variety of lengths, starting at 7.2 mm (0.2835 in.), and going up in additional lengths of 0.2-mm (0.008-in.) increments to a maximum length of 10.6 mm (0.417 in.) at the present time.

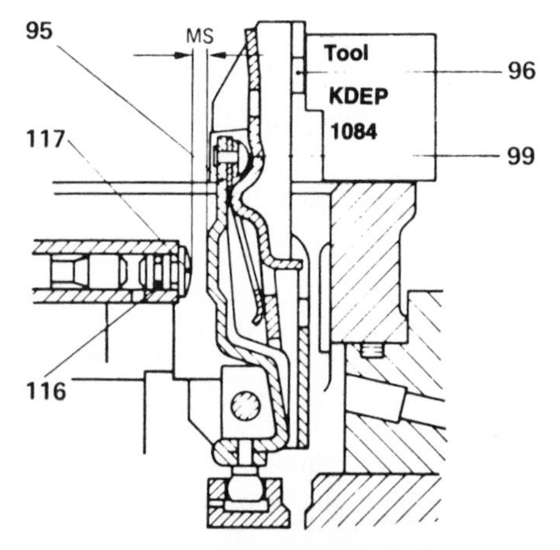

Figure 4-39
Checking the MS Dimension.
(Courtesy of Robert Bosch, GmbH)

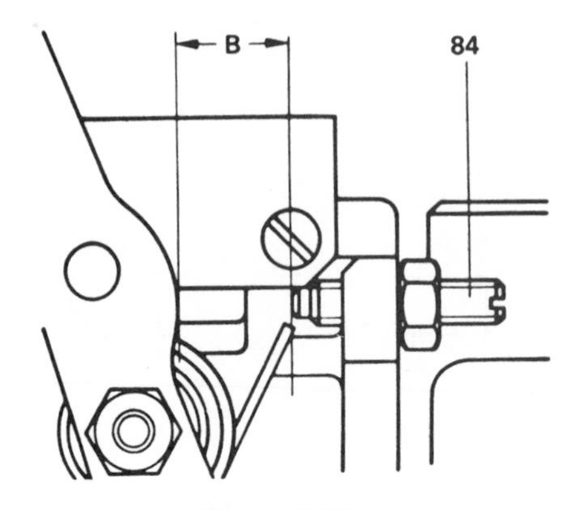

Figure 4-40
Adjusting Full-Load Speed Regulation.
(Courtesy of Robert Bosch, GmbH)

Adjust Full-Load Speed Regulation. Manually hold the throttle lever against the speed-adjusting screw; run the pump at the test specification speed, and turn the adjusting screw in or out to obtain the specified fuel delivery rate.

Refer to Figure 4-40 and measure dimension B, which is the distance between the idle and speed adjustment screws. Use a vernier caliper for this check. Ensure that the throttle lever is up against the idle-adjustment screw when taking the measurement. If the distance measured is incorrect, it is possible that the wrong governor spring may have been installed.

Adjust Load-Dependent Port Closing. Some VE model pumps operate with load-dependent port closing. An adjustment speed will be given under the test sheet for those pumps so equipped. Proceed as follows:

1. Adjust the governor shaft as described earlier.
2. Hold the speed control lever hard against the maximum speed screw.
3. Using special tool KDEP 1082, loosen the locknut on the governor shaft (item 107 in Figure 4-35).

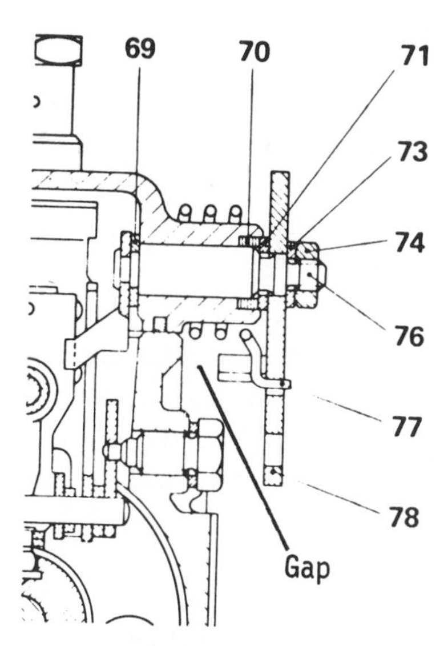

Figure 4-41
Adjusting the Zero Delivery Stop Linkage. (Courtesy of Robert Bosch, GmbH)

4. Set the test stand speed to that given in the specification sheet.
5. With an Allen head hex wrench, unscrew the governor shaft until the supply pump pressure just begins to decrease; then turn the shaft one-eighth turn back in again.
6. Using tool KDEP 1082, tighten up the adjustment.

Timing Device and Transfer Pump. Test the timing device travel and transfer pump supply pressure as explained earlier.

Measure the Fuel Overflow Quantity. With the throttle held in the wide-open position, run the test stand at the test speed given in the test sheet and measure the overflow or return fuel quantity.

Test Fuel Delivery and Breakaway Characteristic. Measure the volume of fuel delivered at the speeds stated in the test sheet.

Test Zero Delivery Stop

1. With the test stand running at the speed given in the test specification, zero fuel delivery must be attained with the specified voltage applied to the solenoid shutoff.
2. With the manual type of shutoff, place it hard against the stop position screw and check for zero fuel delivery as per specifications. Make up a spacer block of 27.7 mm (1.090 in.) and insert it between the inside of the pump housing cover and the internal shutoff lever linkage. Install the external shutoff lever (78) on the splined shaft so that the smallest gap exists between the shutoff lever (78) and the housing cover (see Figure 4-41). Tighten the shutoff lever with the hex nut (74).

Excess Starting Fuel. Measure the excess fuel for starting as described earlier.

Fuel Shutoff Solenoid. Check out the fuel shutoff solenoid to ensure that it energizes at the correct voltage.

Injection Pump Removal (from the Engine)

The injection pump is driven from a gear that is in itself driven by a toothed drive belt used to drive the overhead camshaft of the engine. Prior to removing the pump from the engine, clean the immediate area around the injection pump, its drive gear, and injector nozzles. Do not steam clean the pump with direct pressure and heat, since damage to the pump can occur. Tools required are Volkswagen no. 2065, camshaft locking bar, and Volkswagen no. 203b, gear puller.

Proceed as follows:

1. Bar the engine over until cylinder 1 is at TDC on the *compression stroke,* which can be determined by removing the plug (if

used) found in the engine flywheel (bell) housing, and lining up the no. 1 TDC flywheel mark with the corresponding line on the housing. (This is important.)

2. Using tool no. 2065, lock the camshaft against rotation and remove the toothed drive belt.
3. With the camshaft jammed against rotation, loosen the injection pump drive gear nut several turns.
4. Using gear puller no. 203b or equivalent, withdraw the gear until it butts up against the nut; then remove the puller and nut. Pull off the drive gear.
5. Slacken off all the injector fuel line nuts, remove all lines, and cover both the fuel outlets at the pump delivery holders and injector inlets with protective plastic shipping-type caps. Remove the fuel inlet and overflow line from the pump and plug off with caps.
6. Disconnect both the stop and accelerator cables (wire) or electric shutdown solenoid wire along with the cold start lever cable if used.
7. Remove the bolts from the injecion pump mounting flange while supporting the weight of the pump; then pull the pump from the engine.

Injection Pump Installation

1. On the top center of the pump mounting flange is a small boss area with a scribed line running through it parallel to the injection pump body. Also, on the engine mounting flange, which the drive shaft of the pump passes through, is another scribed line. When installing the pump, both of these marks must be aligned to one another.
2. With the engine still at no. 1 TDC, install the pump drive gear over its shaft keyway. Turn the gear until the timing mark on the gear and the engine mounting plate are aligned. Lock the pump gear in this position with Volkswagen pin no. 2064, and tighten up the nut on the drive gear to 4.5 mkg (33 lb-ft or 44.74 N·m) *only* after having secured the pump to the engine by tightening its flange bolts to 2.5 mkg (18 lb-ft or 24.4 N·m).
3. Individually remove the plastic shipping caps from the fuel lines and injectors; install and tighten the fuel pipes to the injectors to 2.5 mkg (18 lb-ft or 24.4 N·m).
4. Loosen the camshaft gear retaining bolt one half-turn and using a rubber or plastic hammer, knock the gear off its tapered shaft.
5. Double-check that the cylinder 1 TDC mark on the flywheel (view through flywheel housing hole) is positively aligned with the stationary reference mark.
6. Install the camshaft and pump drive belt (one belt); then remove the pin from the pump drive gear that was installed during step 2. The camshaft setting and locking bar should be in position at this time. By forcing the belt tensioner to its right, apply tension on the drive belt.
7. To check drive belt tension (extremely important), use Volkswagen tool no. 210 or any of the available reputable belt-tensioning gauges.
8. Tighten the camshaft bolt to 4.5 mkg (33 lb-ft or 44.74 N·m), and remove the camshaft setting and locking bar.
9. Manually bar the engine over at least two complete revolutions; then recheck the belt tension after striking the center of the belt (between the gears) with a rubber or plastic hammer. Always check the belt by placing the tension gauge midway between the gears.

10. Install the accelerator cable so that the ball end of the pump lever is pointing upward and with the mark in the elongated hole. The accelerator cable should be attached at the upper hole of the bracket.
11. With the accelerator pedal in the full throttle position, adjust the cable with the nuts so that the pump lever will contact the stop without undue strain.
12. Install the cold start lever if used, and with the lever at the pump in the zero position, pull the inner cable tight and secure the cable clamping screw.

Check Injection Pump Timing

Several methods can be used to ensure that the injection pump is properly timed to the engine. The timing can be checked with the engine in the application or on the shop floor. Proceed as follows:

1. When turning the engine over, always rotate it in its normal direction of rotation to ensure removal of all gear backlash as you approach the timing mark located on the flywheel.

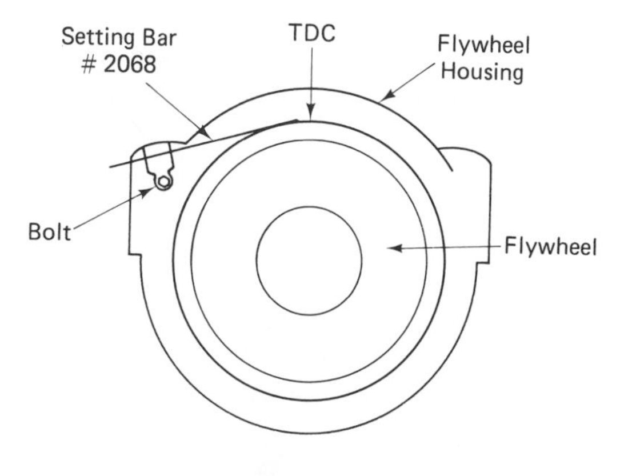

Figure 4-42
Use of Setting Bar 2068.

2. With the engine installed in its particular application, bar the engine over until the TDC mark on the flywheel is aligned with the boss marking on the flywheel (bell) housing. Also ensure that the scribed marks on both the injection pump and engine mounting plate are in alignment. If the engine is on the shop floor, a setting bar (Volkswagen P/N 2068) is required, which is located as shown in Figure 4-42. Rotate the engine so that the TDC mark on the flywheel is lined up with the edge of the setting bar, which can be moved in or out on its support bracket bolted to the housing.
3. Refer to Figures 4-26 and 4-35; take out the bleeder plug, item 131, from the center of the hydraulic head.
4. Install Volkswagen adapter 2066 mounted to a dial gauge in place of the bleeder plug. Preload the dial gauge until it reads 2.5 mm (0.097 in.).
5. Bar the engine over ccw until the dial gauge needle stops moving (do this carefully).
6. Zero the gauge now so that it has 1 mm (0.039 in.) of preload.
7. Slowly turn the engine cw (normal rotation) until the TDC mark is exactly aligned with the edge of the setting bar, and check that the reading on the dial gauge registers 0.83 mm (0.032 in.).
8. If the reading is incorrect, loosen the injection pump mounting bolts, grasp the pump housing firmly, and rotate it gently until the lift is 0.83 mm (0.032 in.). What you have just done is to adjust the actual plunger lift from the BDC position to the point of port closure by turning the cam ring away from or toward the rollers, depending on which way you turn the pump. In other words, you can effectively advance or retard the start of injection this way.

Caution: This check is extremely important to the end performance of the engine and its actual life between rebuilds. Field experience has shown that advancing the injection pump timing as little as 0.005 in. (0.127 mm) will help starting characteristics in cold weather environments; however, it greatly reduces piston life.

Be sure to repeat the checking procedure to ensure that the injection pump timing is in fact correct.

Valve Timing Check

Injection pump timing can also be checked in the following fashion:

1. Turn the engine over until piston 1 is at TDC by referencing the flywheel mark to the housing.
2. Hold the camshaft in this position by use of Volkswagen setting bar 2065, mentioned earlier under pump removal and installation.
3. With the engine in this position, check to see if timing pin 2064 will enter through the hole in the pump gear and engine mounting plate. It it does, the timing is correct; if it does not, proceed as follows:

    a. Turn the engine opposite its normal direction of rotation approximately one half-turn; then bring it back in its correct direction of rotation until piston 1 is at TDC as per the flywheel timing mark (both valves closed).

    b. Hold the camshaft in position with setting bar 2065.

    c. Loosen the camshaft gear bolt approximately one half-turn, and using a plastic or rubber hammer, knock the cam gear loose from its shaft.

    d. Turn the pump gear until the marks on the gear and the engine mounting plate are correctly aligned.

    e. Install pin 2064 through the pump gear and into the engine mounting plate.

    f. Proceed to adjust the camshaft belt via its tensioner and tighten up the cam gear bolt to specifications of 4.5 mkg (33 lb-ft or 44.74 N·m). Check the belt tension with a suitable belt-tensioning gauge between the two gears.

    g. Remove the camshaft setting bar and double check that the timing is still correct after also having removed the pin from the pump gear.

Bleeding the Fuel System

On any diesel fuel-injection system, air trapped within the pump or fuel lines must be vented to prevent hard starting and irregular operation. On the Volkswagen VE pump, simply loosen the small vent screw located on the top of the fuel filter housing and simultaneously manually operate the hand priming pump until a steady flow of fuel totally free of air bubbles flows from the filter; then tighten up the vent screw.

If the engine fails to start, loosen two injector line nuts at the cylinder head end and crank the engine over until fuel flows; then tighten the nuts.

Injectors

The injectors used operate in the same manner as other Robert Bosch injectors described in this book. Any servicing required can be found in this chapter.

Specified *popping pressures* for these nozzles (Volkswagen) is 1706 to 1849 psi (116 to 126 atm). This can be adjusted by unscrewing the upper portion of the injector from the lower portion and adding or removing shims on top of the injector needle valve spring. Shims are available in thicknesses ranging from 1.00 to 1.95 mm (0.039 to 0.077 in.) in 0.05-mm (0.019-in.) increments. Adding shims increases the popping pressure; removing shims reduces this pressure.

To check the injector for leakage after any repairs or adjustments have been made, with the injector in a pop-testing machine, operate the pump handle slowly and hold or maintain a pressure of about 110 atm (1617 psi) for about 10 seconds, during which time there should be no fuel leakage from the injector nozzle tip.

A quick and common method to determine if an injector is faulty is to run the engine slightly faster than its normal idle rpm, and then to loosen the fuel line nut at the injector, which should cause the engine speed to decrease. If the speed and sound of the engine remain the same, the injector is faulty.

Injectors can be removed with the use of tool US 2775 after removing the fuel lines. When replacing, always install new lower heat shields (washers) into the injector bore and tighten the injectors to 7.0 mkg (51 lb-ft) and the pipes to 2.5 mkg (18 lb-ft).

Idle and Maximum Speed Adjustment

A detailed sequence of injection pump adjustments was given under the section dealing with test stand run-in and testing. However, normal minor adjustment for both the engine idling and maximum speed can be done on the vehicle or in the application simply by loosening the locknut on both of these screws, which controls the arc of travel of the throttle lever on top of the injection pump housing.

On the Volkswagen car, Bosch dwell tachometer EFAW 166C along with adapter VW 1324 can be used to determine accurately these two speeds. A Sun model TDT-12 tachometer or equivalent magnetic tape pickup digital electronic tachometer is also suitable for this purpose.

The idle speed should be set between 850 and 950 rpm with the adjustment screw, and then locked and sealed. The maximum no-load rpm should be between 5400 and 5450, set by the adjustment screw, which should also be locked and sealed.

Note. Prior to either of these adjustments, it is imperative that the engine be at its normal operating temperature of 50 to 70°C (122 to 158°F) engine oil temperature.

4-13

Troubleshooting the VE Pump

Figure 4-43 provides a detailed guide to aid in troubleshooting diesel fuel-injection systems with Robert Bosch distributor injection pump model VE.

4-14

VA Model Pump

An earlier model of the Robert Bosch distributor pump, which is found on some International Harvester Company farm and light industrial equipment, Peugeot cars, Deutz (KHD) equipment, and Volvo-Penta marine applications, is the EP/VA . . ./H, commonly called the VA series distributor injection pump (Figure 4-44). It is very similar in operation to the current VE model discussed herein, the major difference being that the main plunger is shaped (stepped) in such a fashion that it supplies fuel under moderate pressure for the governing circuit.

The fuel quantity to the injectors is governed by the *control piston,* which reciprocates (back and forth) in a certain phase relationship with the main plunger, which itself rotates as well as reciprocating. The end of injection will occur earlier or later with respect to the travel

It is assumed that the engine is in good working order and properly tuned, and that the electrical system has been checked and repaired if necessary.

Starting Problem	Engine surges at idle	Rough idle when engine is warm	Engine misses under load	Low Power	Excessive Fuel Consumption	Engine cannot be shut off	Poor performance **or** black smoke **or** low power	Fog-like exhaust in full-load range (white or blue)	Incorrect idle or maximum speed	Engine does not rev up	Injection pump runs hot	CAUSE	REMEDY
●		●	●	●			●	●		●		Improper fuel (gasoline) in tank	Drain tank, flush system, fill with proper fuel
●	●		●	●				●		●		Tank empty **or** tank vent blocked	Fill tank/bleed system, check tank vent
●	●	●	●				●	●		●		Air in the fuel system	Bleed fuel system, eliminate air leaks
	●	●										Pump rear support bracket loose	Replace as necessary
●						●						Low voltage, no voltage **or** stop solenoid defective	Correct electrical faults/replace stop solenoid
●			●	●				●		●		Fuel filter blocked	Replace fuel filter
●			●	●			●	●				Injection lines blocked/restricted	Drill to nominal I.D. or replace
●			●	●				●		●		Fuel-supply lines blocked/restricted	Test all fuel supply lines – flush or replace
●			●	●								Loose connections, injection lines leak or broken	Tighten the connection, eliminate the leak
●			●					●		●		Paraffin deposit in fuel filter	Replace filter, use Diesel Fuel no. 1
●			●	●	●		●	●		●		Pump-to-engine timing incorrect	Readjust timing
		●	●	●	●		●					Injection nozzle defective	Repair or replace
				●	●		●					Engine air filter blocked	Replace air filter element
●												Pre-heating system defective	Test the glow plugs, replace as necessary
●		●		●	●		●					Injection sequence does not correspond to firing order	Install fuel injection lines in the correct order
					●				●			Low idle misadjusted	Readjust idle stop screw
				●					●			Maximum speed misadjusted	Readjust maximum speed screw
	●		●	●				●	●	●		Overflow fitting interchanged with inlet fitting	Install fittings in their proper positions
											●	Overflow blocked	Clean the orifice or replace fitting
●										●		Cold-start device not operating	Check bowden cable and lever movement
●		●		●			●			●		Low or uneven engine compression	Repair as necessary
●	●	●	●	●	●		●	●	●	●		Fuel injection pump defective or cannot be adjusted	Replace

Figure 4-43
Troubleshooting Guide for Diesel Fuel-Injection System with Robert Bosch Distributor Injection Pump VE in VW Rabbit Diesel.
(Courtesy of Robert Bosch, GmbH)

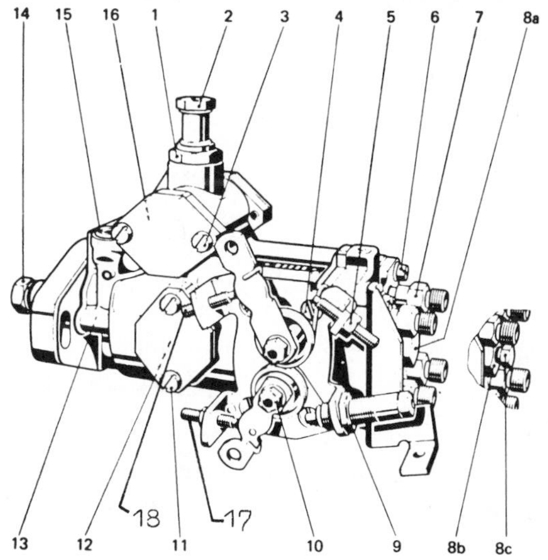

Index:

1—Overflow valve
2—Inlet union screw
3—Screw on cover plate for automatic timing device
4—Bolt for throttle and spill piston cover plate
5—Maximum speed stop screw
6—Slotted head screw
7—Injection nozzle delivery valve holder
8a/b—Screw plug with 8, 9, 10, 11, or 12 mm piston diameter
8c—Hex screw
9—Stop lever adjustment
10—Spill piston hex nut
11—Timing pointer cover screw
12—Timing pointer cover
13—Countersunk flat bolt for vane-type pump
14—Drive shaft nut
15—Fuel inlet
16—Pressure regulating valve
17—Full-load stop screw
18—Idling speed screw

Figure 4-44
VA Series Distributor Injection Pump.
(Courtesy of Robert Bosch, GmbH)

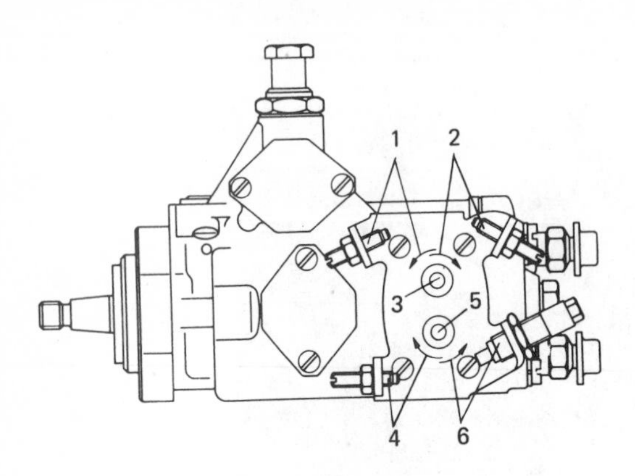

Figure 4-45
Throttle and Spill Piston Arrangements.
(Courtesy of Robert Bosch, GmbH)

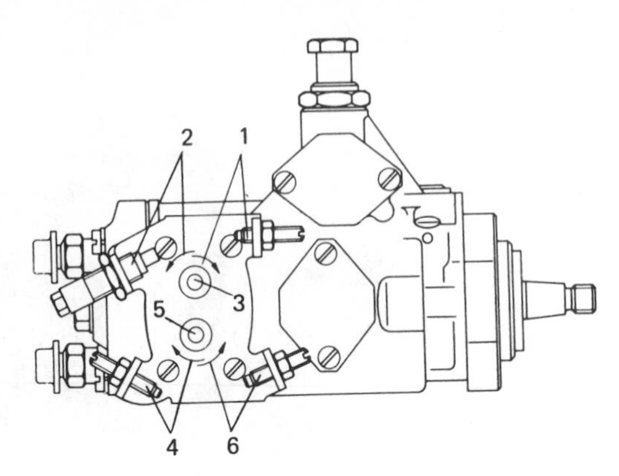

Figure 4-46
Throttle and Spill Piston Arrangements.
(Courtesy of Robert Bosch, GmbH)

Index:
1—Idling speed stop direction
2—Maximum speed stop direction
3—Throttle shaft
4—Full-load stop direction
5—Spill piston (delivery rate)
6—Stop direction

Index:
1—Full-load stop direction
2—Stop direction
3—Spill piston (delivery rate)
4—Maximum speed stop direction
5—Throttle shaft
6—Idling speed stop direction

of the main plunger, as engine load is changed with a given speed settting position.

Fuel delivery between full load and low idle is therefore achieved by this type of governing arrangement. The governor and automatic timing device are integral with the injection pump housing.

Figures 4-45 and 4-46 show the throttle and spill piston arrangements that are commonly found on the model VA pump.

With the injection pump installed on the test stand, the test sequence follows the same basic pattern as that for the VE model pump. The *notch* shown on the spill piston (5 in Figure 4-45) must point away from the throttle shaft in its initial position. With the pump running on the test stand at approximately idle rpm, refer to the test sheet, and adjust the throttle position until the correct quantity of fuel is delivered.

Note: Should the throttle be rotated one half-turn from this position, no delivery will occur.

Major overhaul instructions for this injection pump can be found in Robert Bosch booklet 46, VDT-WJP 161/4 B. Complete test instructions can be found in Robert Bosch booklet VDT-WPP 161/4 B.

chapter 5

Diesel Kiki Model VM Distributor Pump

If you are familiar with the Robert Bosch VA and VE type distributor pumps, you will readily understand the operation of the VM (Figure 5-1). The injection pump is driven from the engine on 4-cycle engines at one-half engine speed. Also driven from the engine is the diaphragm-type fuel lift pump, which draws fuel from the tank, which may first pass through a primary filter or strainer. If water in the fuel is a problem, a fuel water separator can be used. Fuel from the diaphragm feed pump is delivered to a secondary filter and then on into the vane-type transfer pump, which is driven from the injection pump drive shaft via a key and keyway.

Vane pump pressure can approach 100 psi (approximately 7 bars); it is controlled by the action of the main pressure regulator valve, which is adjustable. Fuel oil under pressure is directed through an external steel line to the fuel metering valve (14 in Figure 5-2) and to the piston of the automatic advance timing mechanism.

The cam disc, which is the same setup as the Bosch VE pump described earlier, has a number of cam lobe projections and rollers dependent on the number of engine cylinders. As with the VE pump, the rollers do not lift, but stay stationary; the cam ring, which is driven directly by the injection pump drive shaft as it rotates, will cause the plunger attached to it to lift as each cam lobe projection on the cam ring comes into contact with the individual rollers. Therefore, the plunger operates exactly as in the VE model pump, that is, with reciprocating and rotary motion (see Figure 4-23).

Unlike the VE pump, which uses a metering sleeve through which the rotating plunger travels (strokes back and forth), the VM pump uses a separate metering valve (14) whose position is controlled directly by throttle and governor action, as shown in Figure 5-2.

Fuel from the vane transfer pump is directed into and around the metering valve and its bore. This fuel can flow from the metering valve bore into the rotating plunger inlet port when the plunger and barrel inlet ports are in alignment.

The charging or filling sequence is as follows: as the plunger continues to rotate, inlet port closure occurs and, through pump timing, the plunger lifts as the cam ring lobe runs over the individual roller for that cylinder. Plunger lift opens the outlet port (distributor slot), allowing fuel at increasing pressure to be distributed to one cylinder at a time in engine firing-order sequence through the corresponding delivery valve in the hydraulic head. Remember that the plunger may have as many holes or ports drilled in it as there are cylinders, but it only has one outlet port; therefore, delivery can occur to only one cylinder at a time. Injection will cease the instant the cutoff groove of the barrel aligns with the plunger's circumferential groove, allowing the spilled fuel to enter the injection pump housing.

The VM model injection pump is not at all dissimilar in its operation to other Robert Bosch distributor-type pumps. With the ever-growing popularity of diesel engines, the import and export of a variety of makes of vehicles and equipment, the Diesel Kiki (Robert Bosch licensee) VM pump will be found more and more on exported Japanese equipment.

Figure 5-2 shows the basic flow of fuel within the injection pump and is followed by a more detailed explanation of its operation.

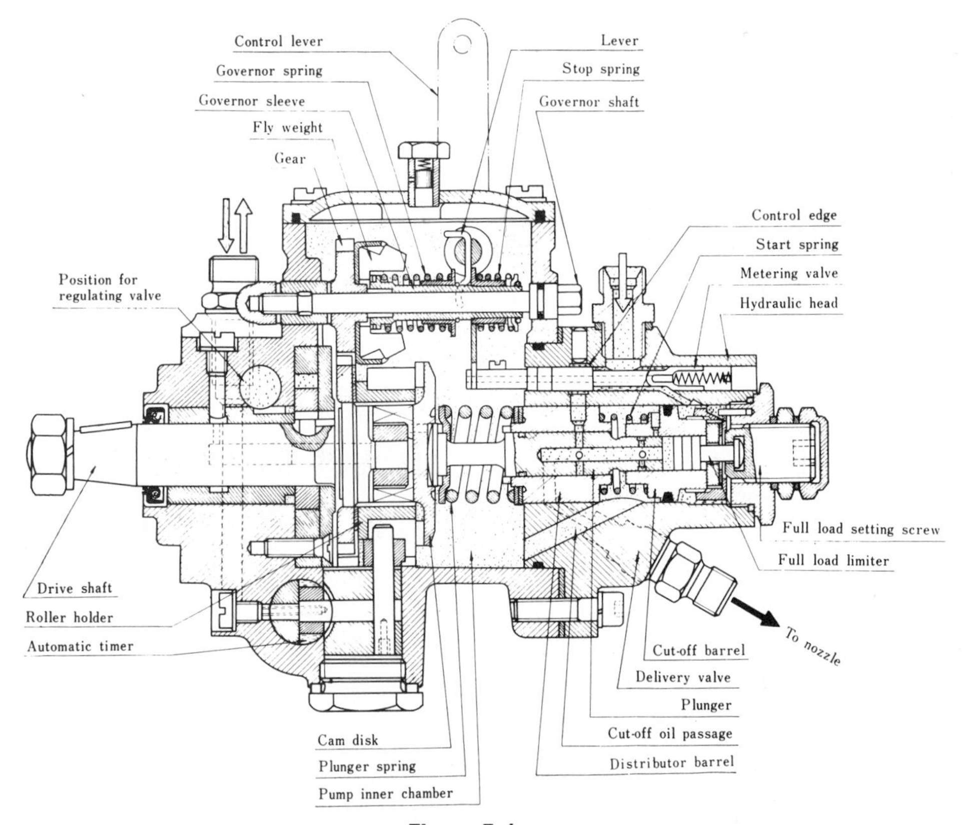

Figure 5-1
General Construction of Injection Pump.
(Courtesy of Diesel Kiki USA Co., Ltd., Irving, Texas)

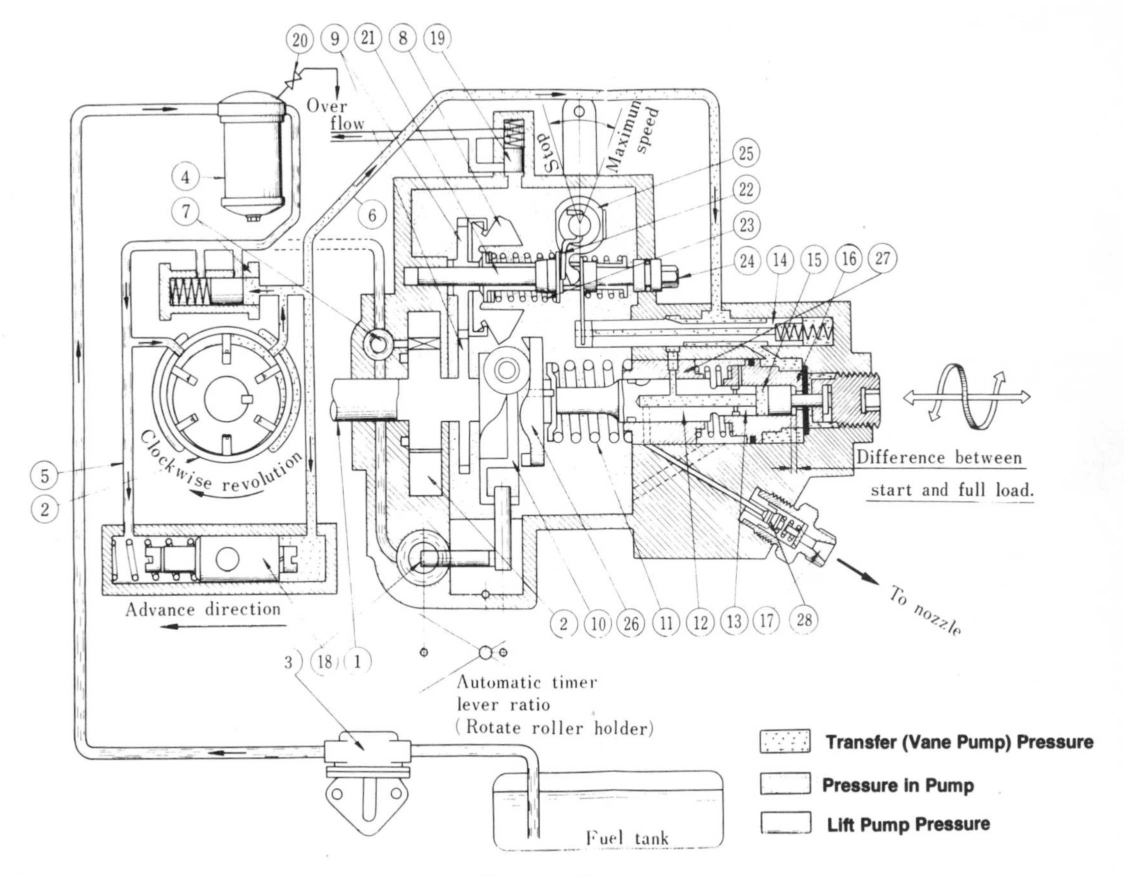

Figure 5-2
Basic Flow of Fuel Within the Injection Pump.
(Courtesy of Diesel Kiki USA Co., Ltd., Irving, Texas)

Index:

1—Injection pump drive shaft
2—Vane-type transfer pump
3—Diaphragm-type engine-driven lift (feed) pump
4—Fuel filter
5—Low-pressure fuel inlet to timing device from (3)
6—Fuel delivery pipe to injection pump housing
7—Vane pump pressure regulating (bypass) valve
8—Governor flyweights
9—Governor drive and driven gears
10—Plunger roller retainer cage
11—Plunger spring
12—Plunger
13—Plunger pressure passage
14—Fuel metering valve
15—Pumping chamber
16—Cutoff barrel
17—Delivery holder (fuel outlet to nozzle)
18—Automatic advance piston
19—Injection pump housing pressure relief valve (to tank)
20—Throttle
21—Governor sleeve
22—Governor lever
23—Governor spring
24—Governor shaft
25—Stop arm
26—Cam ring (actuated by rollers)
27—Plunger distributor barrel
28—Delivery valve

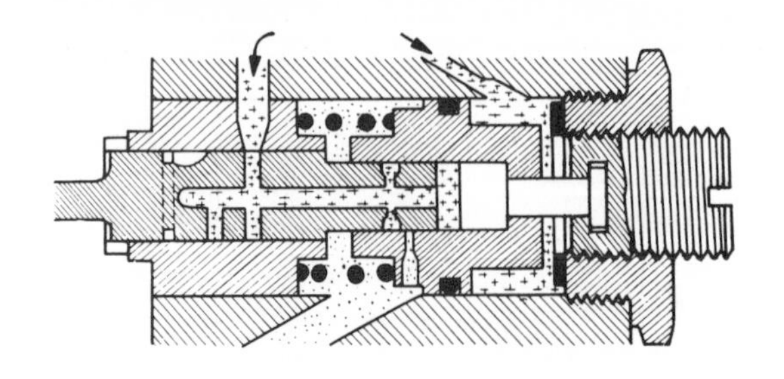

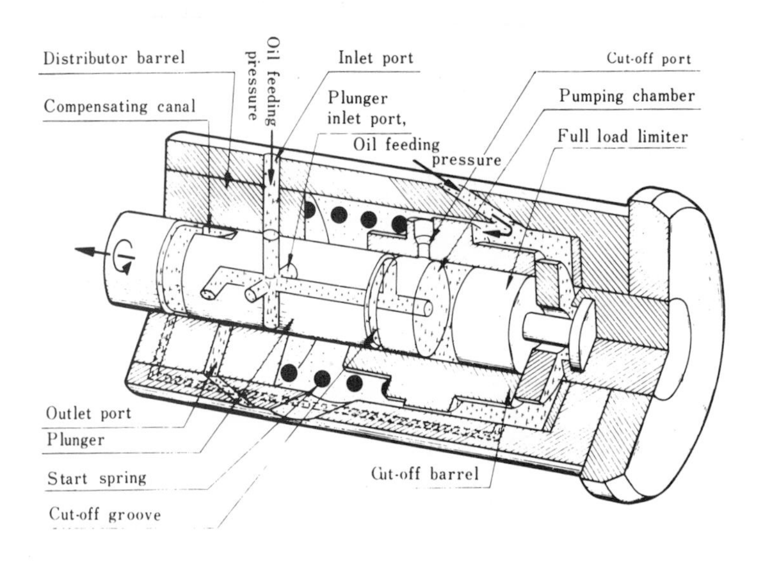

Figure 5-3
Port Filling or Charging Cycle.
(Courtesy of Diesel Kiki USA Co., Ltd., Irving, Texas)

Throttle position varies the opening of the inlet port between the metering valve and barrel and hence the volume of fuel that will be directed to the plunger fill ports. The position of the cutoff barrel spill port determines the maximum effective stroke of the plunger and, thus, the maximum amount of fuel that can be delivered.

Reference to Figure 5-3 shows that the larger diameter of the pumping plunger contains as many inlet ports as there are engine cylinders, while the smaller forward portion of the plunger contains the *cutoff groove,* which once in alignment with the barrel *cutoff port* will end injection.

Figure 5-4 shows the sequence of events once the fill port has rotated to the point that fuel supply to it has just stopped. As this is happening, simultaneously the rotating plunger cam plate or disc lobe is contacting the roller, which will start to pressurize the trapped fuel within the plunger. As the rotating plunger outlet port (one only) aligns with a delivery port in the distributor barrel, fuel at high pressure will unseat the delivery valve within its holder, sending fuel to the injection nozzle, which overcomes the valve spring seating force, thus opening the nozzle and spraying fuel into the combustion chamber.

Fuel will continue to be injected as long as there is adequate pressure to the nozzle. However, as the plunger strokes through the barrel, the cutoff groove around the outer circumference of the smaller plunger diameter will come into alignment with the spill port or cutoff port in the barrel. The instant this occurs fuel injection will cease, since the fuel is vented through the excess fuel passage drilled diagonally within the hydraulic head. This is shown in Figure 5-5.

To prevent the possibility of pressure surges, which would lead to irregular fuel pressures and possible unequal injection characteristics, a small semicircular groove, shown in both Figures 5-2 and 5-6, is machined out of the larger diameter of the plunger toward its base. Fuel oil at vane pump pressure is directed to this groove through the plunger groove. At the end of injection, this *compensating canal*, as it is commonly called, will come into alignment with the outlet port of the distributor barrel. The fuel from the delivery valve is then returned to a stable pressure, and normal injection for the next cylinder is assured.

Cold Start Device

Unlike the small cold start cam on the VE model pump, which is manually operated to advance the timing piston and so rotate the

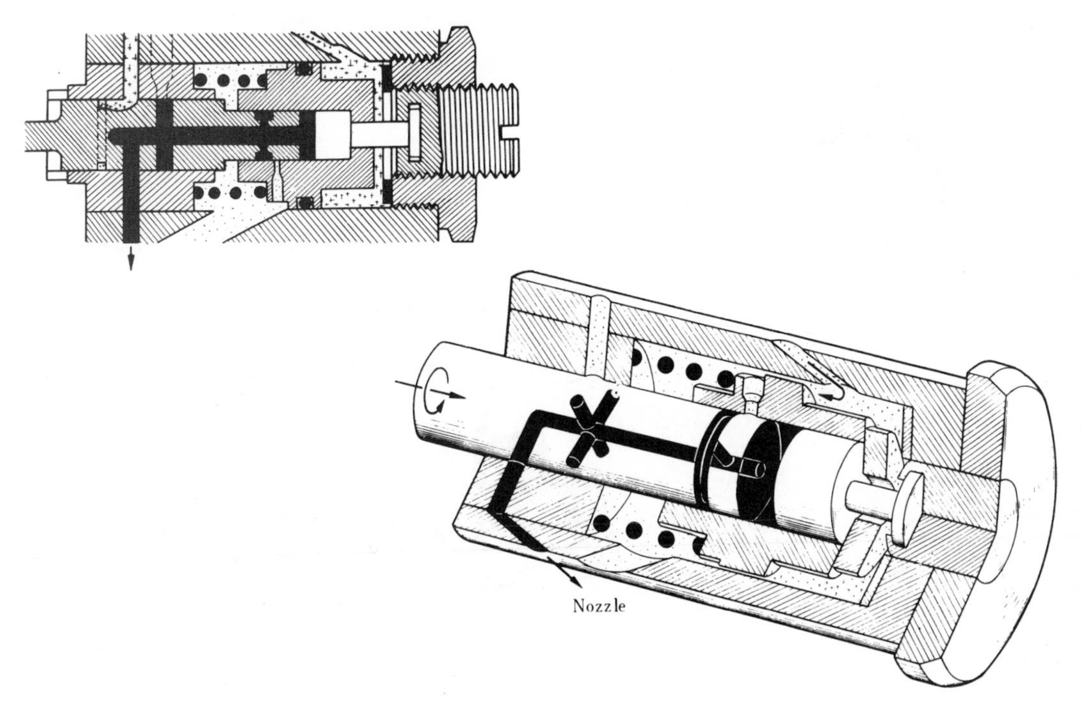

Figure 5-4
Injection Stroke of Plunger.
(Courtesy of Diesel Kiki USA Co., Ltd., Irving, Texas)

roller ring, the VM pump employs a starting spring that is sandwiched between the cutoff and distributor barrel. This spring can be adjusted by the use of shims, which has the net effect of forcing the cutoff barrel to the right when the engine is stopped, thereby retarding the finish of injection by allowing a longer effective stroke of the plunger, which will therefore increase the quantity of fuel delivered during starting to the engine cylinders.

Figure 5-7 shows the action of the cold start device. When the operator or driver cranks the engine over on the starter, fuel pressure from the vane-type transfer pump is increasing, which is not only being sent to the metering valve, but also to the area surrounding the cutoff barrel. When the engine starts, this fuel pressure will force the cutoff barrel back against spring pressure, which now effectively positions the cutoff barrel spill port in a position that will shorten the effective stroke of the plunger, thereby controlling the full-load injection quantity.

Automatic Advance Mechanism

The automatic advance device operates in the same manner as that described under the VE injection pump.

Governor Operation

The governor used with the VM injection pump is of the mechanical type. The governor spring force can be adjusted by shims internally and the travel of the throttle shaft lever

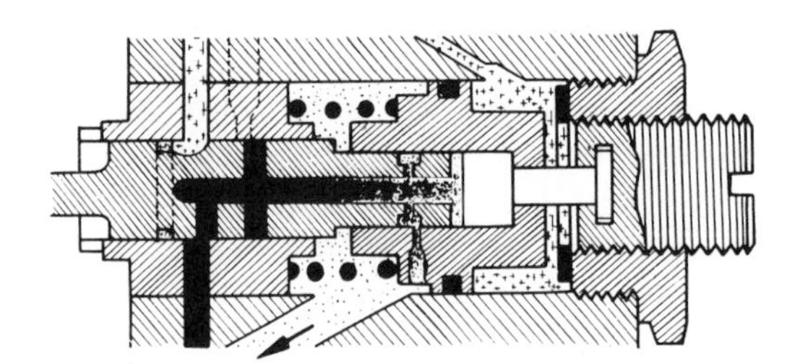

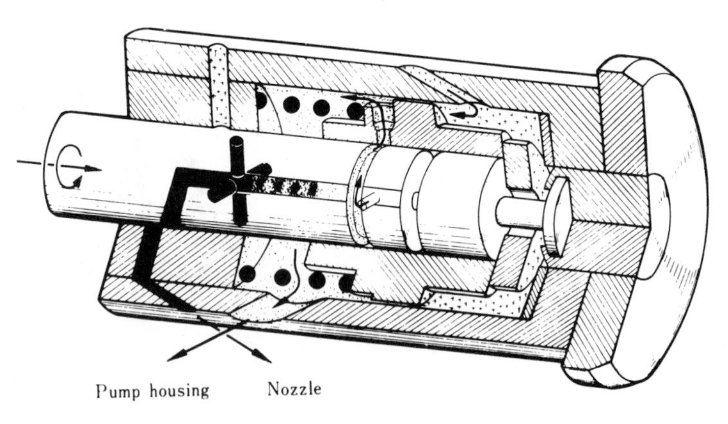

Figure 5-5
Conclusion of Injection.
(Courtesy of Diesel Kiki USA Co., Ltd., Irving, Texas)

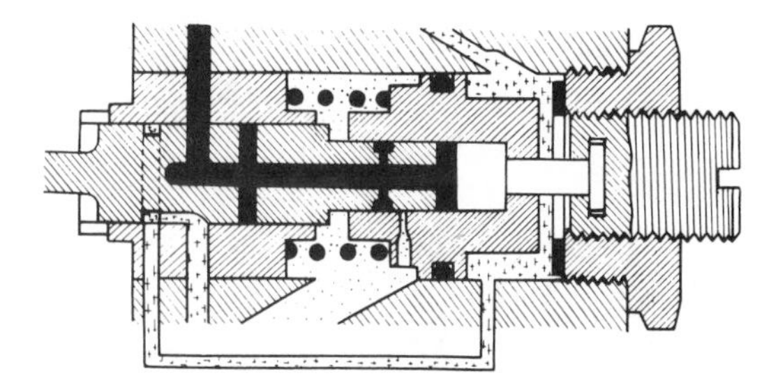

Figure 5-6
Compensating Groove or Canal.
(Courtesy of Diesel Kiki USA Co., Ltd., Irving, Texas)

through adjustment of the external screws. The governor operation is as follows:

Starting. Prior to starting the engine, the weights are in a stationary position, and the starting spring has placed the cutoff barrel into the retarded position as described earlier. The position of the throttle shaft will determine the actual position of the metering valve within its bore since this will manually be placed to a predetermined position by the adjustment of the idling screw stop bolt shown in Figure 5-8.

Idling. With the throttle control lever up against the idle stop bolt screw, the metering valve will be as shown in Figure 5-8. When the engine starts, fuel pressure forces the cutoff barrel back, and the centrifugal force of the rotating governor flyweights will balance out the force of the established governor spring. If there is a change in engine speed, the opening and closing of the plunger inlet port will vary in proportion to the speed change, thereby controlling the engine rpm by throttle position alone.

High-Speed Control. Normal speed regulation on this style of distributor pump can be controlled by throttle opening and therefore metering valve and inlet port opening time; however, at higher speeds, this function alone is not responsive enough to ensure accurate speed control under varying load conditions.

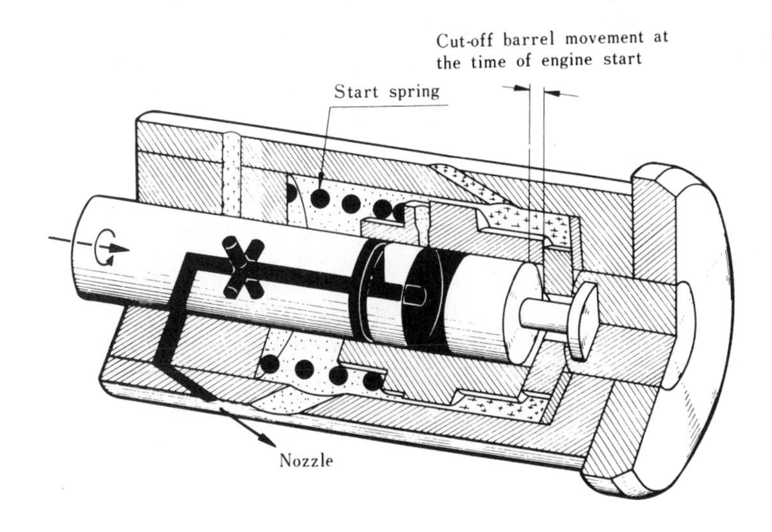

Figure 5-7
Cold Start Device.
(Courtesy of Diesel Kiki USA Co., Ltd., Irving, Texas)

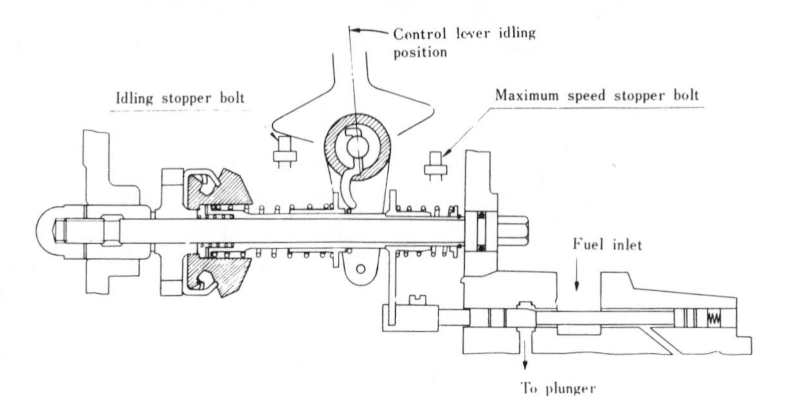

Figure 5-8
Idling Control.
(Courtesy of Diesel Kiki USA Co., Ltd., Irving, Texas)

As the throttle lever position is changed by the operator or driver accelerating the engine, the throttle lever will compress the governor spring, forcing the fuel metering valve to move to an increased delivery position. As engine speed increases, the centrifugal force of the revolving governor flyweights will oppose the spring force. When a state-of-balance condition exists between the weights and spring, the engine will run at a constant speed, as shown in Figure 5-9.

If the engine load increases, such as when climbing a hill in a car, the state-of-balance condition is upset in favor of the spring, which will increase the fuel to the engine. If the load is decreased, engine speed tends to increase; however, the greater force of the faster turning flyweight will move the metering valve to a decreased fuel position. In either situation, a corrected state of balance will occur as long as the engine is not overloaded or overspeed does not occur in too low a gear.

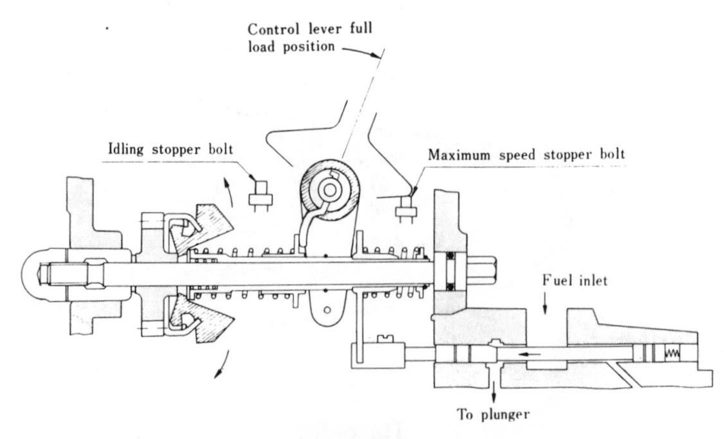

Figure 5-9
High-Speed Control.
(Courtesy of Diesel Kiki USA Co., Ltd., Irving, Texas)

chapter 6
Lucas CAV Fuel Systems

Lucas CAV is a subsidiary of Lucas Industries of Birmingham, England. The world headquarters for Lucas CAV Limited is in West London, England (PO Box 36, Warple Way).

CAV is presently the world's largest manufacturer of diesel fuel-injection equipment, with over one-third of total world production of diesel engines fitted with CAV equipment.

In Britain, the rotary pump product center is located at Medway, Kent, and comprises four factories in the area covering 85,000 m^2 of manufacturing area, and employing 3500 people. Production capacity at Medway is almost 1 million injection pumps per annum.

The in-line injection pumps are manufactured at Finchley in North London and employs 1800 people. The third product center in the United Kingdom is at Sudbury, Suffolk. This is the principal location for the manufacture of injectors and filters, with current production capacity for 8 million injectors and nozzles, and substantially more filters and filter elements. Sudbury employs 2500 people.

In France, a subsidiary company, CAV RotoDiesel, supplies two-thirds of the fuel injection systems for the French diesel industry. In Spain, associate company Condiesel supplies two-thirds of Spain's diesel fuel-injection systems.

In Brazil, subsidiary company CAV do Brazil makes rotary pumps, injectors, and filters, and supplies one in every three diesel engines made in Brazil with fuel injection equipment.

In Japan, subsidiary company Lucas CAV Kk produces rotary fuel-injection pumps for the Japanese market, along with injectors and filters to Japanese customers. In Mexico, associate company CAV InyecDiesel makes injectors and filters, producing more than half of the country's total requirement.

In the United States, Lucas CAV is based in South Carolina. The first phase of a major development program has been completed, and the company is producing rotary pumps for North American customers.

In Korea, Lucas CAV has established a partnership company, CAV Korea Limited, to manufacture injection nozzles for the country's rapidly expanding diesel engine industry.

In addition to these fuel-injection equipment operations in the United Kingdom and around the world, Lucas CAV has established licensee agreements with MEFIN in Romania, IPM in Yugoslavia, and WSK in Poland for the production of rotary pumps.

In the United Kingdom, three quarters of their output is either directly exported, or exported as part of a complete diesel engine from a British manufacturer.

After sales, service support for all CAV fuel-injection equipment is provided through the Lucas Service network, covering 4500 outlets worldwide in more than 130 countries.

Whenever fuel-injection equipment is discussed, Lucas CAV remains as one of the truly great leaders. Lucas CAV is the parent company of Lucas, Simms, and Bryce fuel-injection companies: therefore, fuel-injection equipment made by these companies or made under their names is similar in design and operation to those with the CAV name.

Lucas CAV manufactures injection pumps in both in-line and rotary configurations, with the Minimec, Majormec, and Maximec being the best known of the in-line pumps, and the legendary DPA and DP15 rounding out their distributor-type pumps. The fuel injectors manufactured by CAV operate on the same basic principles as those produced by both Robert and American Bosch companies.

Note: The DP15 distributor pump is no longer manufactured.

6-1

Minimec Injection Pump

The Minimec is an in-line-type injection pump available in 2-, 3-, 4-, 5-, 6-, and 8-cylinder versions, with outputs suitable for engines up to 1.5 liters (91.5 in.^{3}) per cylinder. This pump can be flange, cradle, or platform mounted for ease of application.

For example, the 8-cylinder pump is found on the Perkins V8 model engines, with (Dagenham) Ford using the 4- and 6-cylinder pump. Figure 6-1 shows a sectional view of such an injection pump. Figure 6-2 gives a parts breakdown. The Minimec pump is available with a boost control device for turbocharged engines. GP pneumatic-type governors or GMV mechanical governors can be used on this pump, depending on the particular engine application. They are also equipped with an automatic advance mechanism, and a diaphragm-type fuel feed pump is standard, driven from an eccentric on the injection pump camshaft. However, a piston-type fuel feed pump is available as an alternative. Injection pump lubrication is via the engine lube system, with both the feed and return flow being taken through drillings in the pump mounting flange.

Pumping Operation

Figure 6-3 shows an end cross section of the injection pump with the location of the respective components. The individual pumping plungers are of Simms standard design with a central spill passage located in the top of the plunger leading to a 45 degree spill groove on the side of the plunger. Figure 6-4 shows the unit. Machined onto the plunger stem to reduce fuel leakage and to prevent dilution of the lube oil in the pump camshaft area is a T-flat and a circular groove.

As with all injection pumps of this design, the plungers are spring loaded to keep them in contact with the camshaft via their roller tappets, which are fitted with a spacer for purposes of phasing adjustment. The plunger barrels are secured to the pump body by serrations, and each barrel has the common type of delivery valve holder screwed into the top of the pump housing.

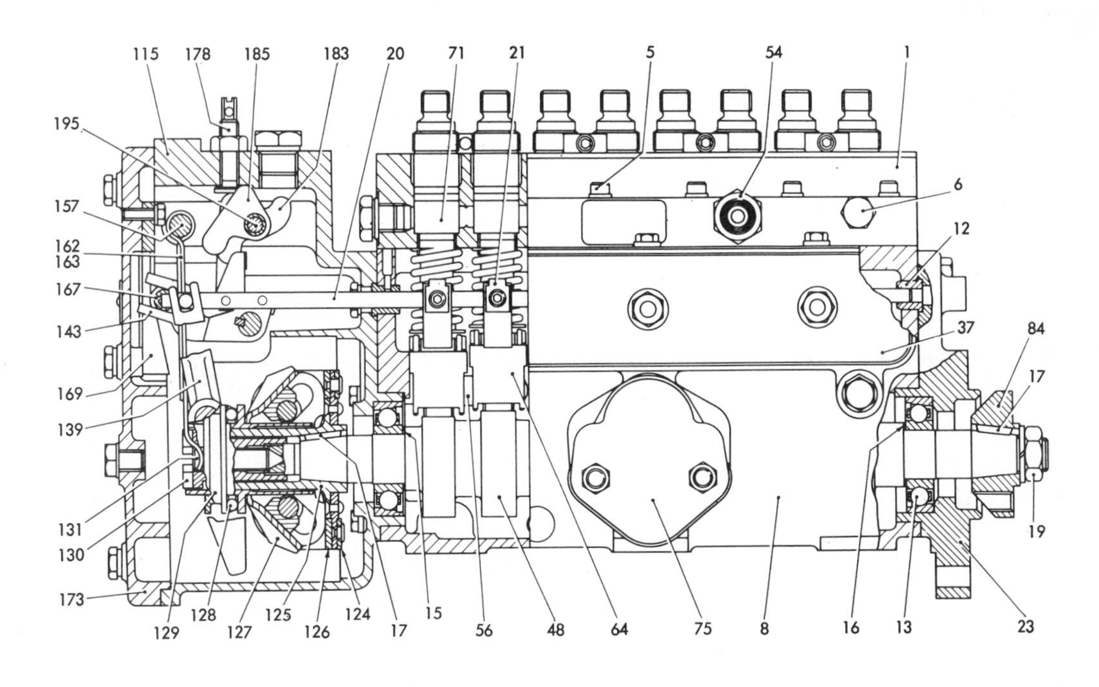

Figure 6-1
Simms 8-Cylinder Minimec Fuel-Injection Pump.
(Courtesy of Lucas CAV Ltd.)

Index:

1—Pump body
5—Pump body screw
6—Air vent screw
8—Cam box
12—Control rod bushing
13—Ball bearing
15—Shims
16—Shims
17—Woodruff key
19—Camshaft nut
20—Control rod
21—Control fork
23—Mounting flange
37—Inspection cover
48—Camshaft
54—Fuel inlet adaptor
56—Tappet locating T-piece
64—Tappet
71—Element
75—Blanking plate
84—Gear adaptor
115—Governor case
124—Cushion drive ring
125—Governor hub
126—Back plate
127—Weight carrier
128—Thrust bearing
129—Thrust pad
130—Hub nut
131—Lockscrew
139—Rocking lever
143—Roller control lever
157—Spring fulcrum pin
162—Primary governor spring
163—Secondary governor spring
167—Spring roller assembly
169—Ramp
173—Governor cover
178—Maximum fuel screw
183—Maximum stop lever
185—Stop shaft assembly
195—Excess shaft

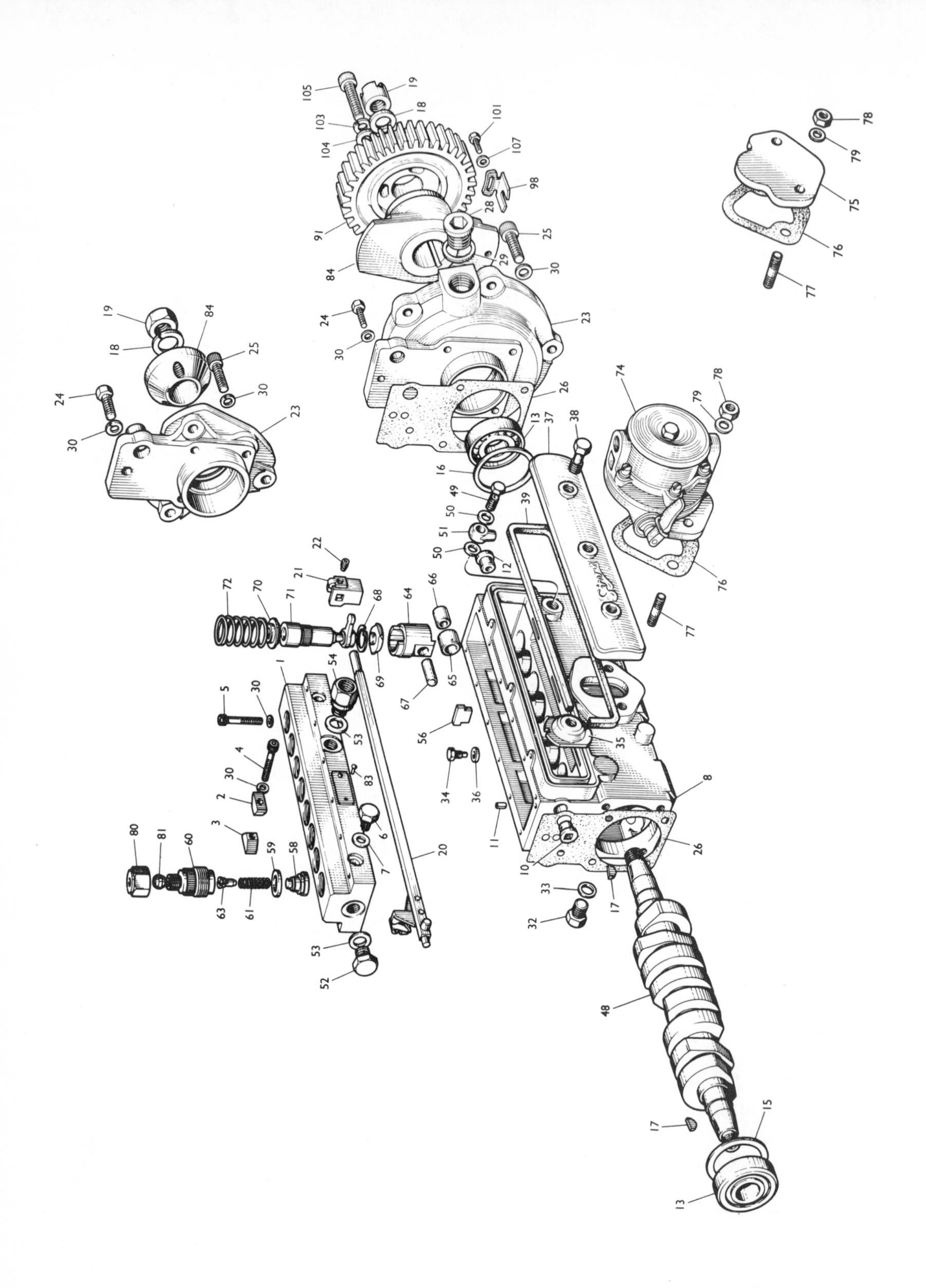

Figure 6-2
Parts Breakdown for Minimec SPE8M
Fuel-Injection Pump.
(Courtesy of Lucas CAV Ltd.)

Index:

1—Pump body
2/3—Delivery valve holder clamp
4—Screw, clamp
5—Screw, pump body to cam box
6—Air vent screw
7—Air vent screw washer
8—Cam box
10—Control rod bushing
11—Groverlok pin
12—Control rod bushing
13—Camshaft ball bearing
15—Shim, camshaft end float
17—Woodruff key
18—Spring washer
19—Slotted or locknut (varies with type of drive)
20—Control rod and maximum stop fork
21—Control fork
22—Control fork screw
23—Two types of mounting flange
24—Screw
25—Screw
26—Gasket
28—Timing inspection plug
29—Washer
30—Spring or sealing washer (varies with pumps)
32—Plug
33—Joint washer
34—Bridge plate locating screw
35—Bridge plate
36—Sealing washer
37—Inspection cover
38—Bolt
39—Sealing ring
48—Camshaft
49—Banjo union screw
50—Joint washer
51—Banjo
52—Sealing plug
53—Joint washer
54—Fuel inlet adapter
56—Tappet locating tee
58—Delivery valve and guide
59—Joint ring
60—Delivery valve holder
61—Delivery valve spring
63—Volume reducer
64—Tappet body
65—Tappet roller
66—Bushing
67—Pin
68—Bowed retainer
69—Tappet spacer (available in 3.9, 4, 4.1, 4.2, 4.3 mm)
70—Lower spring disc (available from 1B to 16B, starting at 0.6 mm in graduations of 0.1 mm up to 2.1 mm)
71—Barrel and plunger (9.0 mm) LH
72—Plunger return spring
74—Lift pump AC-VP series
75—Blanking plate
76—Gasket
77—Stud
78—Nut
79—Washer
80—High-pressure union nut
81—Olive
83—Drive rivet
84—Gear adaptor
91—Fuel pump drive gear, 36T
98—Timing tab
101—Screw
103—Spring washer
104—Plain washer
105—Bolt
107—Plain washer

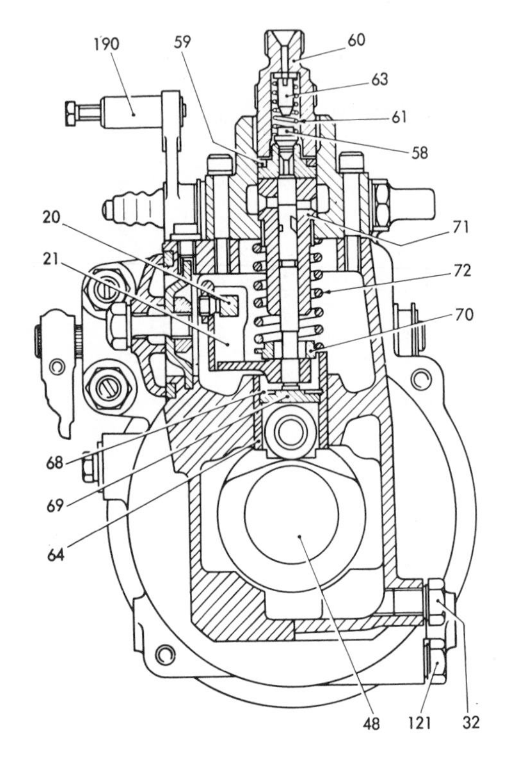

Index:

20—Control rod
21—Control fork
32—Cam box drain plug
48—Camshaft
58—Delivery valve and guide
59—Joint ring
60—Delivery valve holder
61—Delivery valve spring
63—Volume reducer
64—Tappet
68—Bowed retainer
69—Tappet spacer
70—Lower spring disc
71—Element
72—Plunger return spring
121—Governor case drain plug
190—Stop control lever

Figure 6-3
Injection Pump Cross Section.
(Courtesy of Lucas CAV Ltd.)

The sequence of events leading to fuel delivery is shown in Figure 6-4. The plunger element's up and down movement is affected by the camshaft rotation and tappet lift, whereas the rotational movement of the plunger within its barrel is via control forks clamped to the injection pump fuel control rod or rack and engaging the plunger arms. The control rod is in itself controlled by governor action.

The actual volume of fuel delivered by the plunger at any time is controlled by the radial position of the plunger spill groove in relation to the spill port of the barrel, that is, the effective stroke of the plunger, since if the spill groove uncovers the spill port earlier in the plunger's upstroke, then less fuel will be delivered. The longer the spill port remains covered, the greater the volume of fuel delivered. Placing the governor control to the shutdown position will cause rotation of the plunger to a point at which the spill groove and spill port will always be in an uncovered position regardless of the pump camshaft's rotation.

Governor Operation (Mechanical)

The operation of all mechanical governors work on the principle of spring force trying to always move the linkage to an *increased* fuel position, opposed by the centrifugal force of the rotating flyweights, which are trying to move the linkage to a *decreased* fuel position.

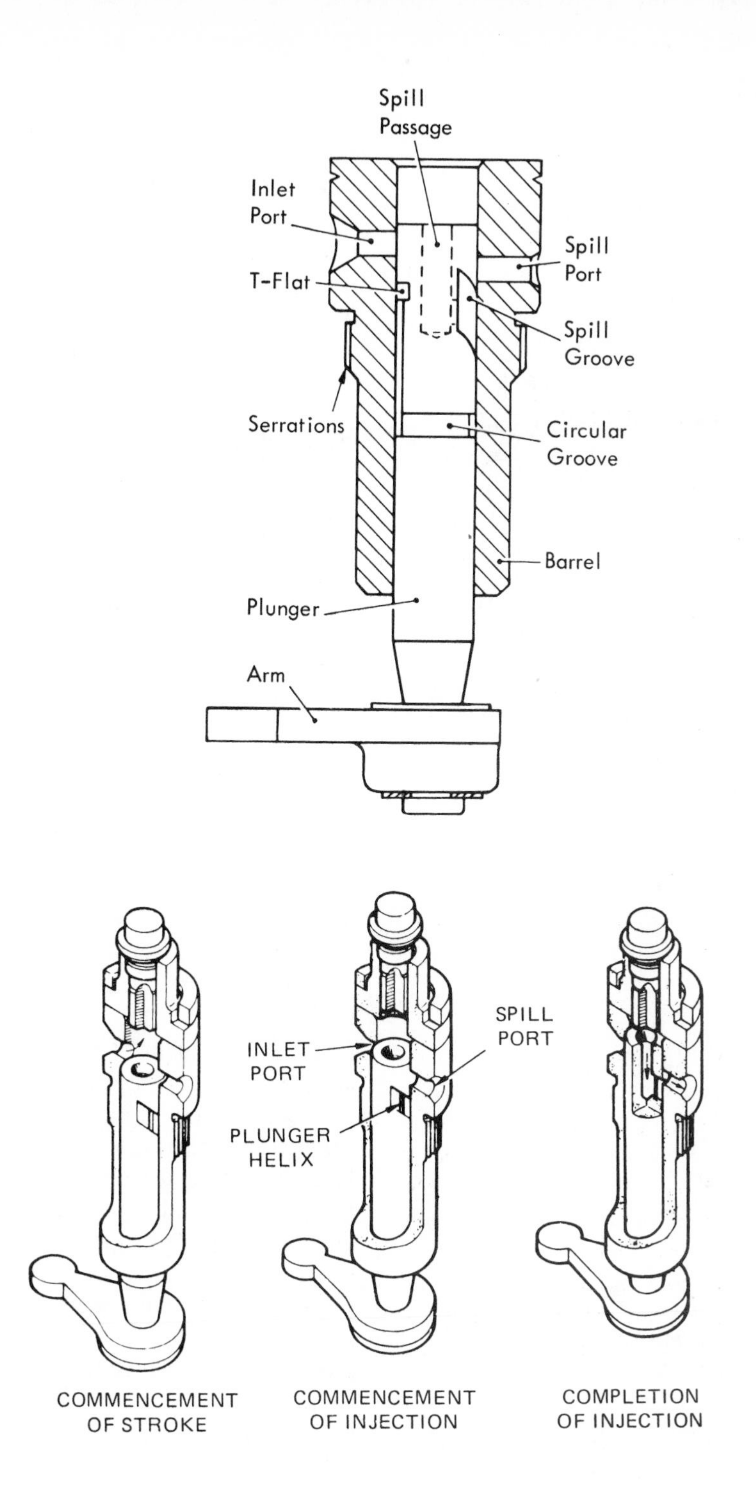

Figure 6-4
Pumping Cycle Events.
(Courtesy of Lucas CAV Ltd.)

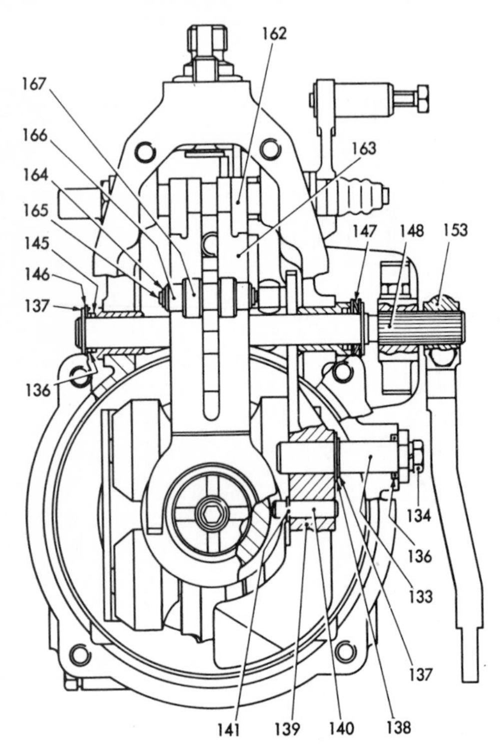

Figure 6-5
Governor Components.
(Courtesy of Lucas CAV Ltd.)

Index:
133—Pivot pin
134—Pivot pin screw
136—0 seal
137—E clip
138—Thrust washer
139—Rocking lever
140—Rocking lever pin
141—Circlip
145—Nu-lip ring
146—Washer
147—Bowed spring washer
148—Speed control shaft
153—Speed control lever
162—Primary governor spring
163—Secondary governor spring
164—Roller circlip
165—Roller pin
166—Ramp roller
167—Spring roller

For any given engine speed, a state of balance can exist between the weight and spring force as long as the engine rpm is capable of handling the load placed upon it.

The mechanical governor used with the 8-cylinder pump, for example on the Perkins V8 510 engine, is a model GEVM, which is of the *variable-speed* type in that it controls *all speeds* from idling to maximum rpm. If you are familiar with the Minimec injection pump leaf spring governor, you will readily associate with this unit, since the major difference is simply in the *cushion drive* of the governor itself.

Figure 6-5 in conjunction with Figure 6-1 shows the governor location and component parts. Drive to the governor assembly is through its hub (125 in Figure 6-1,) which is keyed to the pump camshaft. Attached to the hub is a riveted drive plate with three buttons and three slots into which the three buttons of the U-shaped back plate (126) locate. Fitted onto these six buttons is a $\frac{1}{16}$-in. (1.587-mm) thick flexible cushion drive ring. (On the Minimec pump, this back plate is bolted to the pump camshaft directly.)

Fitted on the weight carrier (127) are a thrust bearing (128), a grooved thrust pad (129), and two pairs of roller-type weights that bear against the back plate. The weights mount on a press-fit pin, which in turn supports two slippers that are free to slide along two inclined faces located at the center of the weight carrier, which is itself free to slide back and forth on the governor hub.

The rocking lever (139 in Figure 6-5) engages the control rod (rack) at its top end, with its bottom containing a pin (140) engaging a groove in the thrust pad. The rocking lever is so called because it pivots centrally on a pin (133) in the side of the governor case.

The governor spring is not of the usual coil type, but consists of two leaf springs (162 and 163 in Figure 6-1) supported on a fulcrum pin (157) bearing up against the thrust pad (129). Spring tension is established by the roller as-

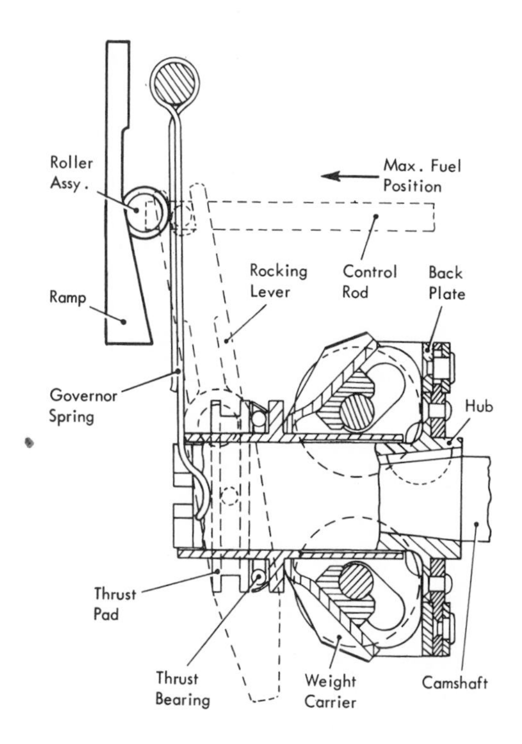

Figure 6-6
Relative Positions of Weights and Springs with Engine Stopped.
(Courtesy of Lucas CAV Ltd.)

sembly (167) located on the ramp (169) by the action of the speed control lever.

Figures 6-6 and 6-7 show the relative positions of both the weights and springs with the engine stopped and the engine running. Figure 6-8 provides a parts breakdown.

Engine Stopped

With the engine stopped, the force of the leaf springs pushes the injection pump control rod toward an increased fuel position, which assists starting of the engine.

Engine Starting and Running

When the engine is cranked over, the weights will start to move outward ever so slightly until the engine actually starts, at which time the centrifugal force of the weights causes the weight pins to apply a thrust through the slippers to the two inclined faces of the governor weight carrier. At this point, the engine speed has been established by the spring force, which has been set by the action of the "idling speed screw" shown in Figures 6-8 and 6-10.

When the force developed by the weights equals that of the springs (weight force is actually slightly greater), the weight carrier will be pushed away from the back plate by the thrust applied through the slippers, which, act-

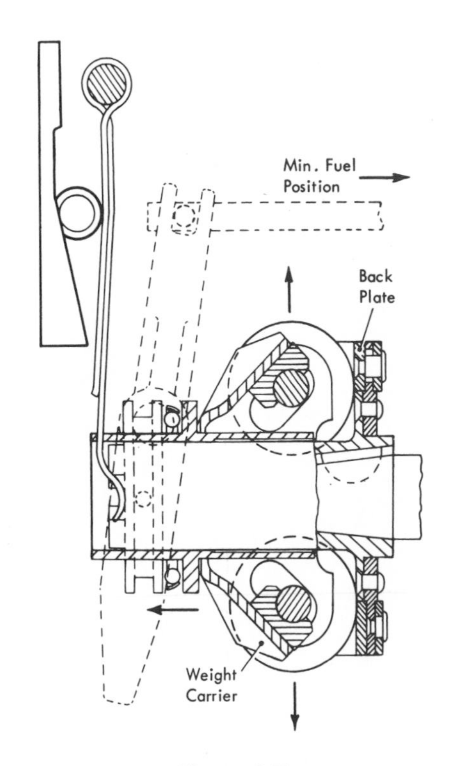

Figure 6-7
Relative Positions of Weights and Springs with Engine Running.
(Courtesy of Lucas CAV Ltd.)

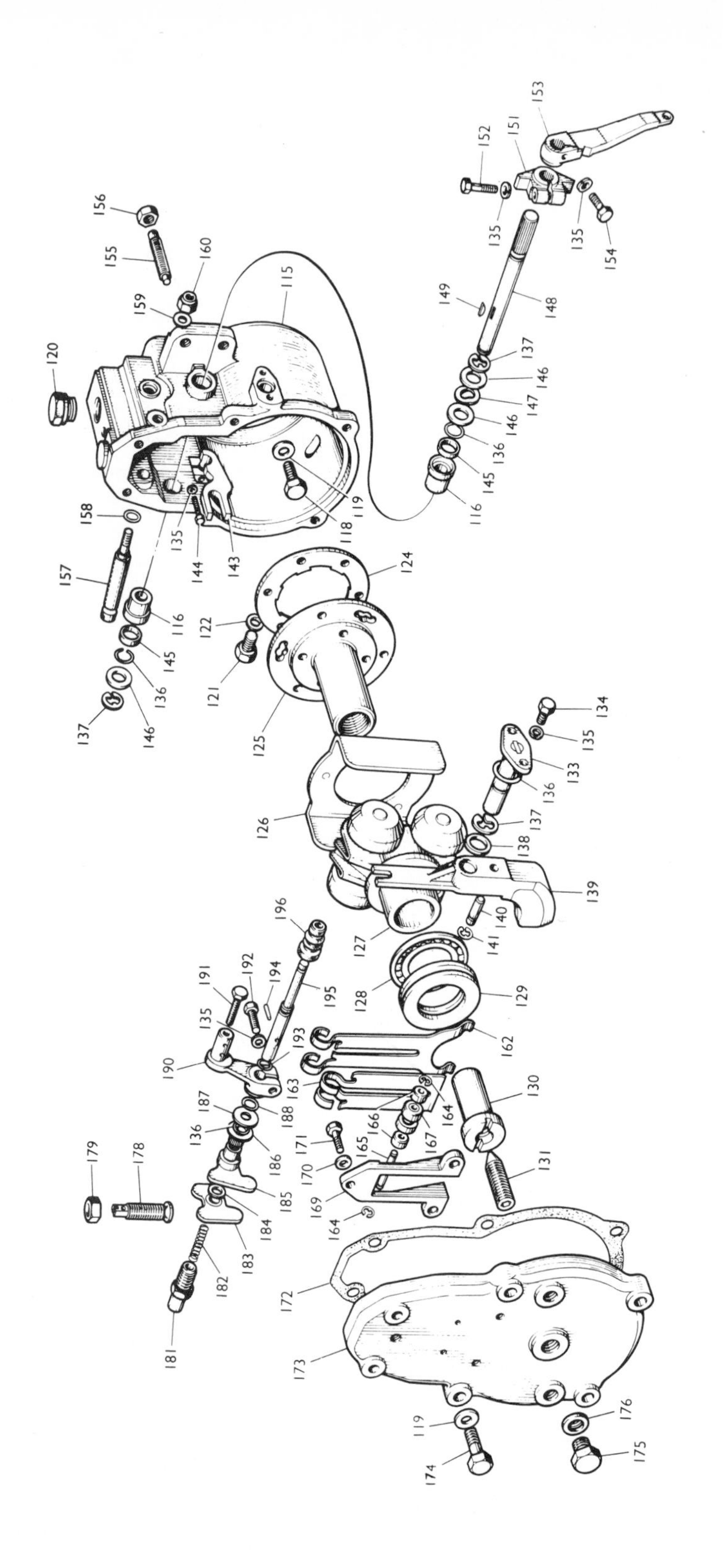
153
152
151
156
135
135
154
155
160
115
149
148
159
120
137
146
147
146
136
145
116
119
118
158
135
144
143
124
157
122
116
145
121
136
146
137
125
134
135
133
136
126
137
138
139
196
140
127
141
191
192
194
195
128
129
135
193
190
162
187
188
163
164
130
136
166
167
186
171
165
179
178
170
131
185
169
184
164
183
182
172
181
173
119
176
174
175

Figure 6-8
Parts List for GEVM Governor.
(Courtesy of Lucas CAV Ltd.)

Index:

115—Governor housing
116—Control shaft bushing
118—Screw
119—Spring washer
120—Oil filler plug
121—Plug
122—Joint washer
124—Cushion drive ring
125—Governor hub and cushion drive plate
126—Back plate
127—Governor weight
128—Thrust bearing
129—Thrust pad (use thickness required)
130—Nut
131—Lockscrew
133—Pivot pin and plate
134—Screw
135—Washer
136—O ring seal
137—E clip
138—Thrust washer
139—Rocking lever
140—Rocking lever pin
141—Rocking lever pin circlip
143—Roller control lever
144—Screw
145—Nu-lip ring
146—Washer
147—Bowed washer
148—Control shaft
149—Woodruff key
151—Stop quadrant
152—Screw
153—Control lever
154—Screw
155—Stop screw
156—Nut
157—Spring fulcrum pin
158—O ring seal
159—Joint washer
160—Nyloc nut
162—Primary governor spring
163—Secondary governor spring
164—Circlip, ramp roller
165—Ramp roller pin
166—Ramp roller
167—Spring roller and bushing
169—Ramp
170—Washer
171—Screw
172—Gasket
173—Governor cover
174—Screw
175—Oil level plug
176—Washer
178—Maximum fuel stop screw
179—Nut
181—Bearing, excess shaft
182—Return spring
183—Maximum stop lever
184—Sealing washer
185—Stop control shaft
186—Thrust washer
187—Washer
188—Spring ring
190—Stop control lever
191—Screw
192—Screw
193—O ring seal
194—Groverlok pin
195—Excess shaft
196—Shroud

ing on the pivoted rocking lever, will move the fuel control rod (rack) to a decreased fuel position. Its movement toward less fuel will stop when the centrifugal force developed by the rotating weights (now turning slower) is balanced by the force of the leaf springs, therefore establishing a steady idle speed.

Engine Being Accelerated

When the operator or driver steps on the throttle, this action is transferred to the speed control lever (153 in Figure 6-5), which, acting through the rocking lever (139), will force the control rod to an increased fuel position, also upsetting the state of balance that existed previously between the weights and springs. With the spring's force now greater due to throttle movement, as the engine speed increases, the weight force will also increase until for any given throttle setting the force of the weights will again balance out with that of the springs, thereby maintaining the engine speed in a fixed position.

Again, as long as the engine rpm is adequate for the load being applied, the speed will remain constant. The only things that will cause this to change are (1) the operator or driver manually moving the throttle and hence the speed control lever, or (2) a load increase or decrease applied or removed from the engine.

Engine Load Increase

If a vehicle running along the highway at a steady engine rpm encounters a hill, for the fixed throttle position that the operator or driver has his foot at, the engine would start to slow down. Whether the transmission would have to be downshifted (manual or automatic) would depend on the length of the incline and/or horsepower and torque characteristics of the engine as the rpm decreases.

In either case, if the engine speed starts to drop, the state of balance between the weights and springs will be upset in favor of the spring force, since the centrifugal force of the flyweights becomes less with a reduction in engine speed. The force of the springs will then push the fuel control rod (rack) to an increased fuel position in order to offset the increased load. If the engine has enough reserve power, then, without the operator having moved his foot on the throttle, the governor will automatically react to the change in load.

If, however, the engine does not have enough reserve power, the rpm would continue to drop, forcing the operator or driver to downshift the transmission; or in an automatic it would cause a downshift to prevent overloading of the engine. The governor reaction would then allow the engine speed to increase, and when the state of balance again exists, the engine will maintain a fixed rpm.

Engine Load Decrease

If the same vehicle running at a steady speed on a level road started into a downhill run, the operator or driver would have to control the speed of the vehicle by use of the brakes or a combination of brakes and downshifting. If this was not done, regardless of keeping the throttle at a fixed position or backing off to an idle position, engine overspeed due to the road wheels becoming the driving member is always a possibility.

When the engine rpm starts to increase, for a fixed throttle position, the centrifugal force of the flyweights will also increase, again upsetting the former state of balance in favor of the weights, which will push the control rod (rack) to a decreased fuel position, thereby preventing engine overspeed from too much fuel.

The bottom end of the rocking lever (139 in Figure 6-1) is weighted to prevent the possi-

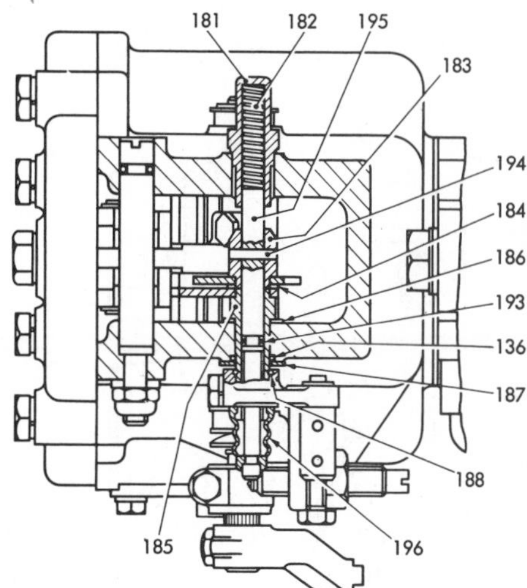

Index:
136—O seal
181—Excess shaft bearing
182—Return spring
183—Maximum stop lever
184—Sealing washers
185—Stop shaft assembly
186—Thrust washer
187—Washer
188—Spring ring
193—O seal
194—Groverlok pin
195—Excess shaft
196—Shroud

Figure 6-9
Excess Fuel Device.
(Courtesy of Lucas CAV Ltd.)

bility of engine stalling under very rapid deceleration due to extreme braking. This integral weight counterbalances the effect of momentum to prevent the control rod (rack) from moving to the minimum fuel position.

Excess Fuel Control Device

Figure 6-9 shows the excess fuel control device; its purpose is to facilitate starting of an engine in low ambient temperatures, in many instances without the use of (ether) starting fluid. During this time, the pump can deliver more fuel than its normal maximum. At any time that the excess shaft button is pushed in, it allows the maximum stop lever to be pushed clear of the maximum stop fork, which is riveted to the end of the control rod (rack), thereby letting the control rod be moved to an excess fuel position through the speed control lever movement. Once the engine starts, the control rod is acted upon by the governor, enabling the excess fuel shaft to be pulled back to its original position by its return spring.

Automatic Advance Unit

The injection pump drive, taken from the engine timing cover, is through a coupling attached to the automatic advance unit, which is fitted to some models of Minimec and Majormec pumps. It operates along the same basic lines as those used by Caterpillar, Robert Bosch, and American Bosch, described in Chapters 2, 4, and 7. A set of weights attached to a plate through a hub to the injection pump drive automatically advances the point of injection in relation to engine speed opposed by a set of weight springs.

6-2

Minimec External Governor Adjustments

Figure 6-10 shows the available external adjustments that are accessible on the governor housing of the fuel-injection pump.

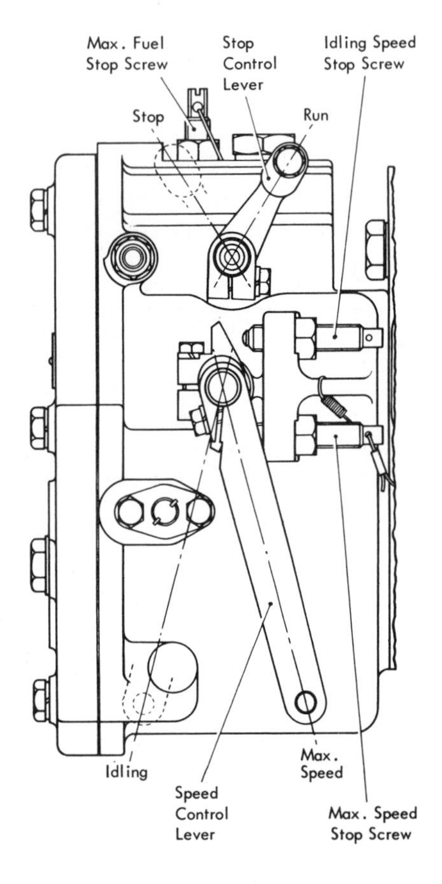

(a)

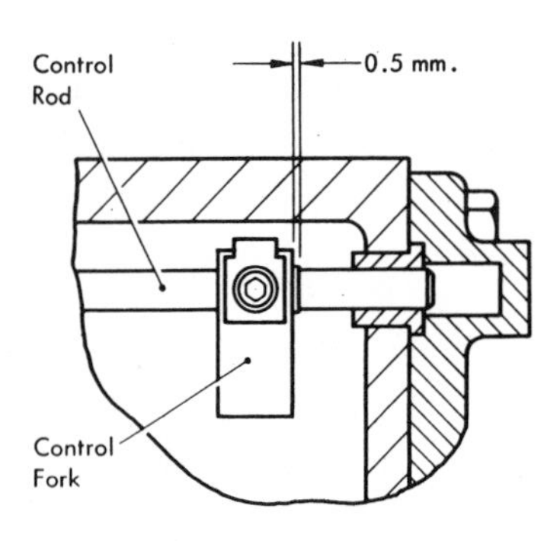

(b)

Figure 6-10
External Governor Adjustments. (Courtesy of Lucas CAV Ltd.)

6-3

Minimec Pump Disassembly and Assembly

Refer to Figure 6-1 for names of items to be taken apart. Pump teardown should only be done if you are familiar with the complete operation and servicing of such units and have both the expertise and available test apparatus, such as test stands, in order that the injection pump can be repaired according to acceptable industry standards.

Always work in a clean, well-organized area or, preferably, a fuel lab, and keep together all parts, such as pump plungers and barrels, delivery valves, and tappets, for that particular cylinder. Use plastic protective caps on delivery valve holders and the like.

Proceed as follows:

1. Drain the cam box of lube oil; then remove the pump inspection cover (37) and also the blanking plate (75).
2. Remove the retaining screws (5), which will allow you to separate the pump body (1) from the cambox (8).

Note: Carefully lay the pump on its back while doing this, and watch that the plungers, springs, and the like, do not all pop out.

3. As you remove the plungers and other parts, keep them together in individual sections of a suitable container.
4. From the pump body (1) remove the delivery valve holders (60), volume reducers (63), delivery valve springs (61), and the delivery valves (58).
5. You will most probably have to tap the bottom of each barrel with a plastic or soft-faced hammer, since they are held to the body by serrations. Also remove the delivery valve guides (58) and joint rings (59).

6. Lift out the tappets (64) and the tappet locating T-pieces (56).
7. Remove the drive gear adapter (84) with a suitable puller after removing the camshaft nut (19), and then the woodruff key.
8. Remove the governor cover (173) and roller assembly (167) from the control lever (143). Also remove pin (157) and springs (162 and 163).
9. Refer to Figure 6-5; remove E clip (137) and pivot pin and rocking lever (139).
10. Remove the weight carrier (127) from hub (125) after taking off nut (130). Also remove the thrust pad (129) and bearing (128).
11. Screw an extractor into the hub to pull it from the camshaft. Pull off the cushion drive ring (124), which will separate the back plate (126) from the hub.
12. Back off the control rod fork screws to allow removal of the control rod (20) and forks (21). Remove mounting flange (23) and gasket, and lift out the camshaft (48).

Examine all parts of the injection pump for wear or damage and replace as necessary. Fine vertical scratches on the plunger stem appear after a short time in service; therefore, do not replace the plunger unless the scoring is deep and/or the upper edge of the spill groove is ragged.

Pump Assembly

Putting the pump back together is simply the reverse of disassembly, starting with the camshaft and mounting flange. Make sure that the unserrated portion, or master-spline, on the barrel is aligned with the one in the pump body.

Pump Phasing

Pump phasing is the procedure required to ensure that the time interval (degrees) between each cylinder's injection period is correct or *in phase*. This check should be performed on an injection pump test stand; however, it can also be done with the pump mounted on a bench, along with a gravity feed fuel tank connected to a shut-off cock (valve) in the inlet fuel line to the pump body.

Phasing must be checked if any of the pumping assembly components have been changed, but it is strongly recommended that phasing be checked and adjusted any time that the pump has been taken apart.

Note: The settings for phasing and calibration for each particular injection pump must be obtained from a fuel setting data sheet, which is readily available from the manufacturer or your local fuel-injection dealer.

The phasing sequence for both the Minimec and Majormec pumps follows the same basic procedure, other than some change in specifications; therefore, whether the pump is of 2, 3, 4, 5, 6, or 8 cylinders, follow the same sequence for phasing.

Proceed as follows:

1. Refer to Figure 6-11 and assemble the barrels and delivery valves, followed by assembly of the delivery valve holders.
2. Refer to Figure 6-12; insert the plungers in their respective barrels complete with the lower spring disc, but without the spring at this time.
3. With the pump housing lying on its side, install the pump body (5 in Figure 6-17) with the plunger arms pointing downward onto the top of the main pump housing so that the plunger arms (6) locate into the control forks. Run in a retaining screw at each corner of the body to hold it in position for now.

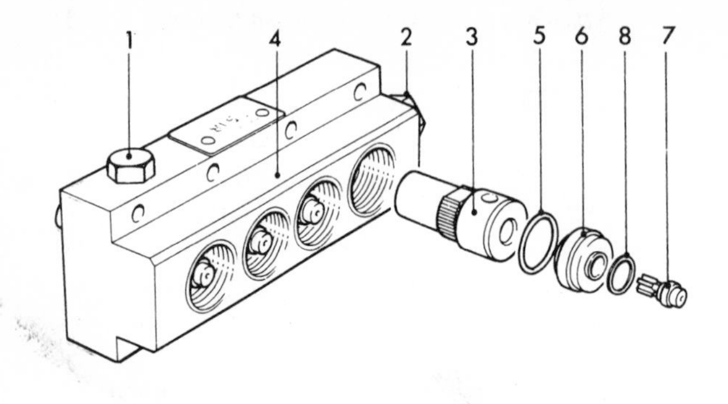

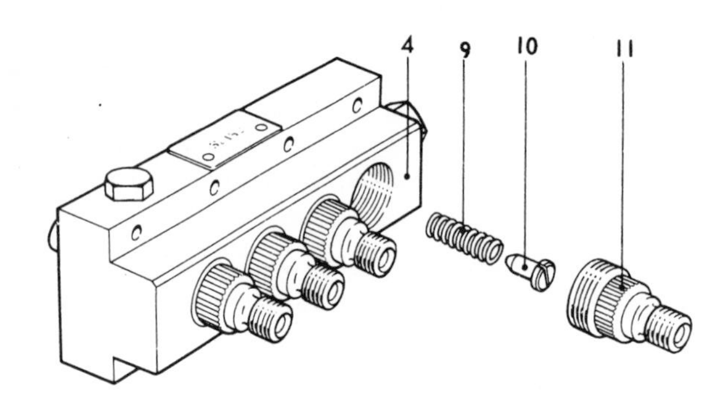

Index:
1—Air vent screw
2—Fuel inlet adapter
3—Barrel
4—Pump body
5—O seal ring
6—Delivery valve guide
7—Delivery valve
8—High-pressure seal washer
9—Spring
10—Volume reducer
11—Delivery valve holder

Figure 6-11
Assembly of Barrels, Delivery Valves, and Delivery Valve Holders.
(Courtesy of Lucas CAV Ltd.)

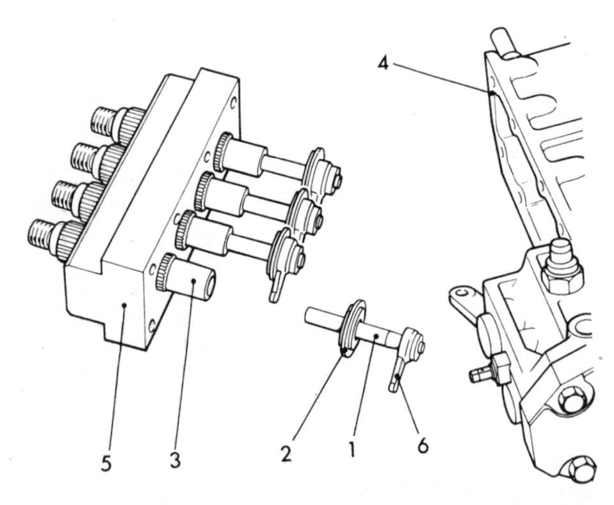

Index:
1—Pump plunger
2—Lower spring discs
3—Barrel
4—Pump housing
5—Pump body
6—Plunger arm

Figure 6-12
Assembly of Plungers and Barrels for Spill Timing.
(Courtesy of Lucas CAV Ltd.)

4. With the pump now mounted onto a test bench or work bench, install a degree wheel, such as shown in Chapter 2 or 7.
5. Connect a gravity feed or hand pump fuel feed supply to the fuel inlet adapter (item 2 in Figure 6-16).
6. Remove from pump 1 the delivery valve holder, volume reducer, delivery valve spring, and delivery valve. Assemble a plunger clearance gauge, such as tool no. 89558, and adapter for holding a dial indicator, such as tool no. 23764, in place of the delivery valve holder.
7. With a rack control rod gauge or equivalent, set up the control rod so that it is 10 mm (0.3937 in.) from the bushing (end with the square section).
8. With the inspection cover removed from the side of the injection pump housing, turn the pump camshaft over until pump 1 plunger is at the bottom of its stroke and *zero* in the dial gauge on top of the plunger.
9. Open up the fuel supply valve or cock to the gravity fuel supply or hand pump as the case may be.
10. Slowly rotate the camshaft in the pump's normal direction of rotation until the fuel just stops flowing past the stem of the dial gauge, which is the point at which the supply port to the plunger has just closed (start of injection).
11. Note the reading on the dial indicator when the fuel just stops flowing, since this is the actual travel distance of plunger 1 from the bottom of its stroke to the point of closing of its fuel inlet port. Compare this to the dimension given in the manufacturer's test data. For example, on the 8-cylinder Minimec pump used on the Perkins V8 510 engine, this distance should be between 2.9 and 3.1 mm or 0.114 and 0.118 in.
12. If the travel is correct, set the pointer of the degree plate to zero since pump 1 now becomes the reference point for all other pumping elements.
13. Reassemble the delivery valve parts back into pump 1 and torque the delivery valve holder after installing a new seal as recommended (usually between 35 and 40 lb-ft or 47.45 and 54.23 N·m).

Note: To change the travel of the plunger if it is incorrect, the tappet spacer (69 in Figure 6-3) can be replaced with one of a different thickness. These are readily available from numbers 1 to 12 with an increase in thickness from 1 to 12 of 0.1 mm (0.00393 in.). If the setting of pump 1 is changed, remember to reset the timing plate (degree wheel) to zero. Also ensure that the *convex* side of the bowed retainer is adjacent to the tappet.

14. Proceed to the next pumping element in the firing order sequence and repeat steps 6 through 13. The number of degrees between phasing of each pump element is dependent on the number of pumping units in the pump itself. Since these pumps are used on 4-stroke-cycle diesel engines, they are driven at half engine speed; therefore, when on a test stand they are driven at a 1:1 ratio, which means that we are dealing with only 360 degrees. Divide 360 by the number of pumping elements and you have the degrees between injections. For example, an 8-cylinder pump would have the phasing 45 degrees apart, a 6-cylinder pump would be 60 degrees apart, and so on. If then we were working on an 8-cylinder engine injection pump, after setting pump 1 to zero, we would proceed in firing order sequence to 45, 90, 135, 180, 225, 270, and 315 degrees. The allowable tolerance in degrees when phasing the pump is $\pm\frac{1}{2}$ degree. Once pumping element 1 has been set, never go back and

change its spacer to try and offset or correct the phase angle for some other pump.

15. With all the pumps now phased, check the vertical lift or float of the plunger arms. With the lower spring discs on the plungers, hold them firmly down onto the tappet body and move the plunger arm vertically. The clearance between the spring disc and plunger arm should be 0.05 to 0.2 mm or 0.002 to 0.008 in. Spring discs are available in sizes ranging from 02S to 16S in increments of 0.1 mm (0.004 in.).

Note: Change the spacers only after checking all pumps; a difference of 0.1 mm is equal to $\frac{1}{2}$ degree of camshaft rotation.

16. Remove the body from the pump housing to adjust the phase angles by fitting tappet spacers, as mentioned earlier.

Pressure Testing the Pump Body

If it is suspected that fuel leakage is occurring into the cam box area of the pump, prepare the pump body as in Figure 6-16; then proceed as follows:

1. Using a temporary plate, place the plungers about 10 mm from the top end of the barrel; don't worry about the position of the plunger arms.
2. Seal off the delivery valve outlets and attach a compressed air outlet line to the fuel inlet adapter with a regulator to maintain the air at a maximum pressure of 50 lb/in.^{2} (3.52 kg/cm^2).

Caution: Use only clean, dry, filtered air.

3. Submerge the body in clean test oil or diesel fuel and note any air bubbles. No leakage is permissible around the barrel seat area: if leakage occurs, the seats will require refacing with a tool such as an ST119. Rephase the pumps after this operation.

Calibration of the Injection Pump

Calibration can only be done on a suitable fuel-injection pump test stand; no other way is acceptable. Once the pump is mounted onto the test stand, all the necessary fuel supply and delivery lines must be connected. Fill the pump cam box with lube oil, as well as the governor case to the correct level. Ensure that pump 1 element control fork is 0.5 mm (0.020 in.) from the square section of the control rod as shown in Figure 6-10(b).

Start the machine and open up the fuel supply valve; then crack open the pump body air vent screws and run the pump for a few minutes to purge the system of air. Make sure that the test oil is at the proper temperature.

Refer to Figure 6-10(a); place the speed control lever in the maximum speed position, and run the pump on the test stand at the speed specified.

For example, certain models of 8-cylinder Minimec pumps used on Perkins V8 engines call for setting no. 1 fork 0.5 mm from the end of the squared section of the control rod [Figure 6-10(b)], then adjusting the maximum fuel stop screw [Figure 6-10(a)] to give between 12.2 and 12.6 cm^3 in 200 strokes with the pump running at 800 rpm for all pump elements. At an idle with the test stand running at 250 rpm and 200 shots, delivery should be 3 cm^3, with a maximum variation between pump elements of 0.5 cm^3.

The governed speed of the pump should be set up on the test bench by adjusting the idling speed stop screw [Figure 6-10(a)] to 250 rpm; the control rack should start to move at 1380 to 1420 rpm, and no fuel delivery should occur at 1560 to 1600 rpm.

All test stand adjustments should be done with the use of nozzle type N12S12 or equivalent. During testing, tighten the control fork screws to the proper torque.

6-4

Minimec Pump Installation

The injection pump is flange mounted on the Ford 4-242 and 6-330 engines to the rear of the timing gear housing, at the front right-hand side. It is driven from the governor end by a gear meshing with the engine camshaft gear. The pump assembly and gear can be removed simply by removing four bolts and one nut.

Pump Installation

1. Manually rotate the engine over until piston 1 is on its compression stroke, which will be indicated by the fact that both the intake and exhaust valve rocker arms will be free (have clearance) between the valve stem tip and themselves. If not, then the engine should be rotated another full turn (360 degrees).
2. Turn the engine over until the timing dot on the crankshaft pulley is aligned with the 22 degrees BTDC timing mark on the timing plate attached to the engine front timing cover. With the 22 degree mark aligned, look into the rear face area of the engine camshaft gear, where a timing mark should be visible at the injection pump aperture.
3. On the mounting flange of the injection pump is a large plug just to the side of a mounting flange slot. This plug should be removed, and timing tool C.9077 inserted.
4. Carefully rotate the gear until the spring-loaded plunger of the tool drops into the center punch mark in the rear face of the gear. Release the plunger, and turn the gear about $\frac{5}{8}$ in. (15.87 mm).
5. Place a new O ring seal around the front face of the injection pump.
6. Install the pump and its gear to the engine; then turn the pump body manually until the plunger of the tool engages the timing mark.
7. Bolt up the pump, and attach all fuel lines, injector lines, and control linkage.
8. Bleed the fuel system of all entrapped air by loosening the two bleed screws on top of the filter, and manually operate the lift pump priming lever until fuel free of air bubbles appears. If the priming lever seems to be inoperative, simply bar the engine over until the eccentric on the injection pump camshaft is against the pump priming handle. There will be a definite change to the feel now when operating the priming lever.
9. Once the filter has been bled, tighten the inlet and outlet screws on its cover, and open up the bleeder screws located on the injection pump body just above the inspection cover on the side. When all air has been expelled, tighten up the screw farthest from the inlet first, then the other. The engine should be ready to start; however, you may want to leave the injector nozzle nut loose during cranking until a steady flow of fuel appears there, and then tighten them up.

Pump Installation (Ford 4-220 and 6-330 with Exhauster)

Some models of Minimec injection pumps have an exhauster (for vacuum brakes) bolted to the rear of the pump housing by three bolts and driven by a splined extension on the injection pump camshaft. The pump is driven by the auxiliary drive shaft, and bolted to the engine block.

Installation follows the same pattern as for the previous Minimec description, other than the fact that the timing mark appears on the flywheel and should be set for 21 degrees BTDC with piston 1 coming up on its compression stroke.

Turn the injection pump manually until the timing mark on the pump coupling hub and pointer are aligned. Slide the coupling flange and fiber coupling rearward to engage with the pump coupling so that there is 0.005 to 0.015 in. (0.127 to 0.381 mm) of clearance between the fiber disc and coupling; then tighten the coupling flange clamp.

Make sure that with the flywheel timing marks in alignment the timing marks on the pump coupling coincide. Turn the engine opposite its normal direction of rotation at least a half-turn; then slowly bring it forward again (this removes all gear backlash). If the pump timing marks are not in alignment, loosen the pump mounting bolts, grasp the pump body firmly, and rotate the pump to bring the marks into alignment. Connect all fuel lines and the like, and bleed the system as described earlier.

Adjust the idle speed to 500 to 550 rpm by turning the idle stop screw on the governor housing back plate. The engine should be at its normal operating temperature. Hold the accelerator pedal to full fuel, and check that the maximum no-load speed is 2700 rpm. Adjust the clamp bolt on the pedal cross shaft lever to ensure that at this speed the pedal is $\frac{1}{4}$ in. from the floorboards.

The in-line injection pumps produced by CAV at present include the following:

1. The Minimec, which is used on engines up to 91.5 in.^{3} or 1.5 liters per cylinder.
2. The Majormec, which is a middle-range injection pump.
3. The newest addition, which is the Maximec, capable of being used on engines up to 100 bhp/cylinder (74.5 kW).

The Maximec pump is shown in Figure 6-13. The Maximec design follows the general pattern of other in-line type of current CAV injection pumps, wherein the lower portion

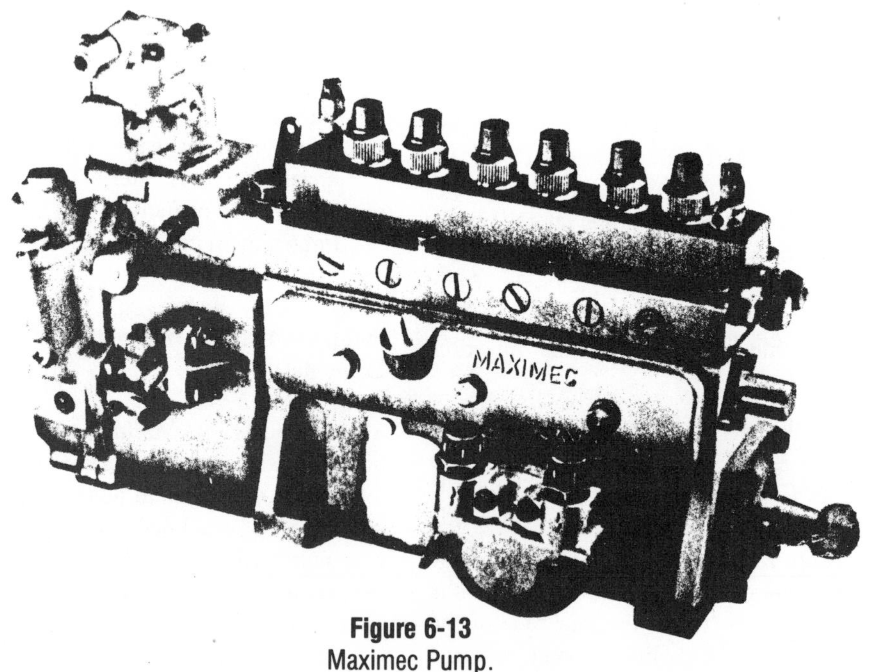

Figure 6-13
Maximec Pump.
(Courtesy of Lucas CAV Ltd.)

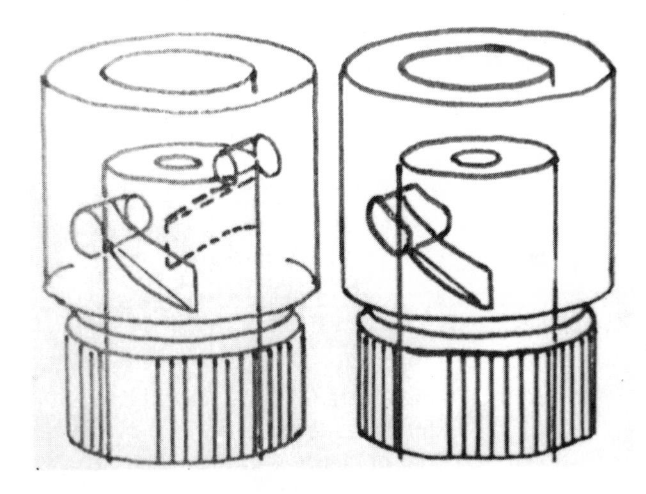

Figure 6-14
Twin Helix (left) and Elongated Port (right) Elements.
(Courtesy of Lucas CAV Ltd.)

(cam box) contains the camshaft, tappets, and control rod. The pumping head (pump body), complete with pumping elements and delivery valves and holders, is screwed to the cam box. The pump is available in 4-, 6-, 8-, or 12-cylinder configurations.

A unique feature of this pump is that the individual pumping elements can be ordered with either round or elongated ports, and in special circumstances a twin helix arrangement as shown in Figure 6-14 can be employed.

The use of elongated ports provides for a significant increase in plunger spill rates at the end of a pumping cycle. To achieve this condition, the port is elongated in the direction of the control helix, which allows for similar rates of port opening to that of a twin helix as well as performance characteristics.

Another unique feature of this pump's actual pumping element is a spiral groove that is cheaper to produce and operates better than the other common methods to allow fuel back leakage to return to the inlet fuel gallery (see Figure 6-15).

The cam box of the Maximec is considerably stronger than that of the Majormec, with

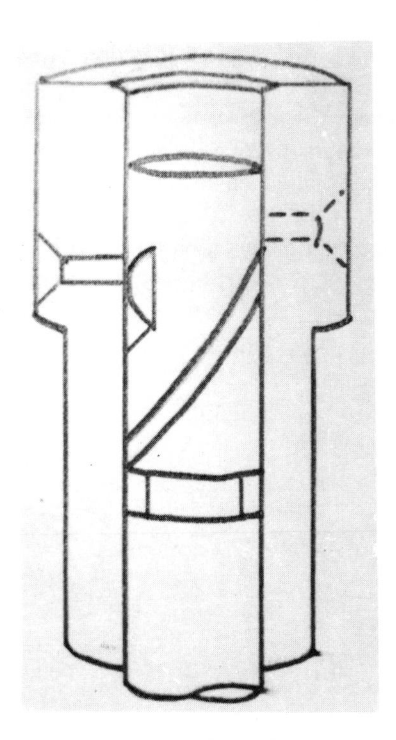

Figure 6-15
Spiral Groove Backleak Control.
(Courtesy of Lucas CAV Ltd.)

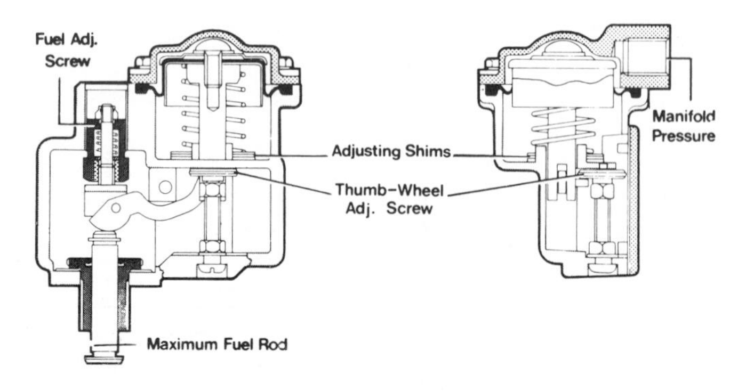

Figure 6-16
BC50 Boost Control Unit.
(Courtesy of Lucas CAV Ltd.)

the end result being a more rigid pump assembly with better sealing ability.

These features allow a performance of up to 400 mm³/stroke fueling, an injection rate of 57 mm³/camshaft degree, a top pump speed of 1350 rpm, and a peak injection pressure of 1000 bars.

Pump life expectancy is estimated to be 10,000 hours or 250,000 miles between major overhauls, based on maximum load and rpm for 10% of the time.

Injection pumps used on turbocharged engines use a boost control unit known as a BC50, which is suitable for use with the entire range of Lucas CAV in-line pumps. This unit is installed on the injection pump housing above the governor. Operation of the boost control unit is similar to other air–fuel units used by other engine manufacturers in that it allows low fuel delivery at low speeds during acceleration to prevent exhaust smoke due to low turbocharger flow rate under such conditions. The BC50 boost control unit can also be used to obtain several different horsepower

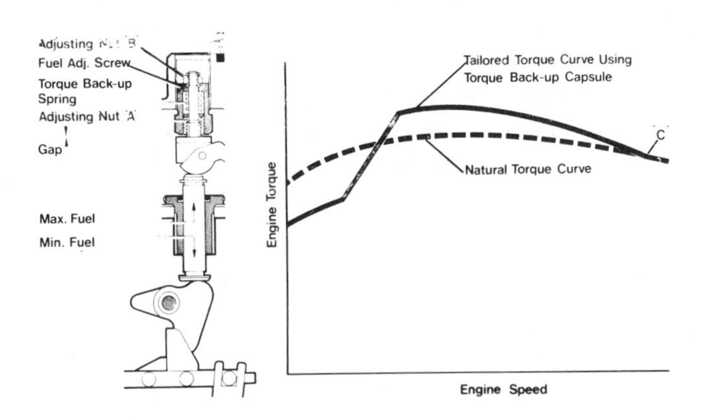

Figure 6-17
Torque Backup Device with Typical Curve It Produces.
(Courtesy of Lucas CAV Ltd.)

(kW) ratings from the same model of engine by assembling its components to produce the reverse of that normally done during low speeds, that is, to provide reduced fuel at the high end of the maximum speed range. Figure 6-16 shows the basic stackup of parts used with the BC50 boost control unit.

Additional *torque backup* can be obtained from these pump and engine combinations by using a *torque capsule*. This unit is shown in Figure 6-17. A decrease in engine rpm under load (maximum) allows the torque capsule spring to be overcome by the increasing governor spring force, thereby increasing fuel delivery and therefore engine torque in the mid-speed range, which gives the engine the ability of high torque rise when desired under these conditions.

Fuel transfer pumps used with the Maximec assembly are of the piston type operated from a roller follower on the injection pump camshaft. Fuel from the pump and oil within the cam box are sealed by a diaphragm each. Transfer pump capacity is 45 gal/h (203 liters/h) at 2 lb-ft/in.^{2} back pressure (13.8 kPa), reducing to 30 gal/h (135 liters/h at 15 lb-ft/in.^{2} back pressure).

6-5

Mack Truck Pump Models

Some models of 6-cylinder Mack truck engines use CAV in-line-type pumps. The pump operates in the same manner as other in-line models of CAV injection pumps other than the Minimec and Majormec, whose design and operation differ slightly. This pump operates along the same lines as the CAV type BPE and AA models described herein.

Figure 6-18 shows the type of fuel supply pump commonly used with these injection pumps. The supply pump is a combined plunger and diaphragm style mounted on the side of the injection pump housing in the same fashion as on other CAV units. An eccentric on the injection pump camshaft actuates the operating lever, which in turn causes the plunger and diaphragm to move. Plunger movement toward the injection pump creates a drop in pressure above the pump inlet valve, lifting it off its seat and drawing fuel into the pump. The return spring positively returns the plunger to its former position. Fuel is then sent under pressure to the fuel filter and injection pump gallery via the ball check valve. A hand priming lever on the side of the pump housing can be used at any time to prime the system.

Control Rack Travel Setting (Mack Trucks)

Prior to any checks on the injection pump, wash down the area around the injection pump and dry it off with compressed air. *Do not,* however, apply direct steam pressure to the injection pump housing, since damage to internal parts can occur, especially if the engine is running.

1. Carefully remove the inspection plate cover from the side of the injection pump after taking out the retaining screws.
2. Obtain tool J-21744, and insert the crimped tang end of the pointer gently between the split in the pump rack quadrant (rack segment gear) of plunger 2.

Caution: Never loosen the rack quadrant screw under any circumstances to insert the pointer. Otherwise, fuel delivery to that cylinder will be affected and recalibration of the pump will be required.

3. Refer to Figures 6-19 and 6-20; install the gauge bracket with its etched quadrant in position at the forward end of the pump. The center of the quadrant should be in line with the centerline of plunger 2. The locating buttons on the gauge will assist in correct gauge location; then fasten the gauge to the pump inspection cover studs with the aid of the two spacers and knurled nuts that come with the gauge kit.

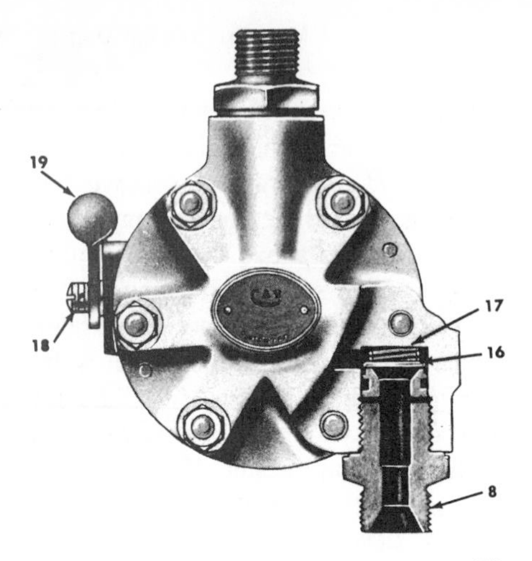

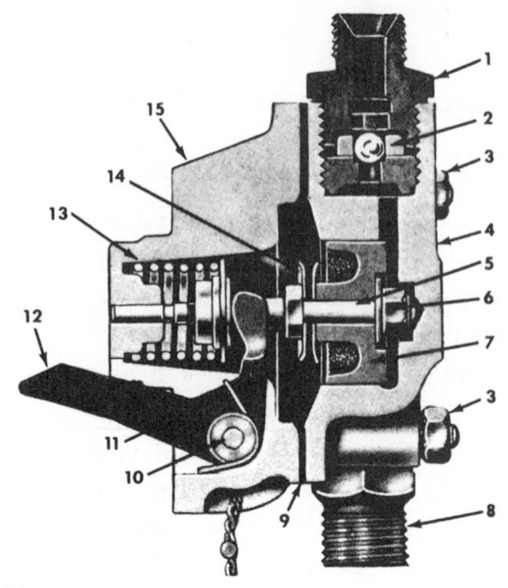

Figure 6-18
Typical Fuel Supply Pump.
(Courtesy of Lucas CAV Ltd.)

Index:

1—Outlet fitting
2—Ball check valve
3—Cover retaining nuts
4—Cover
5—Spindle
6—Plunger retaining nut
7—Plunger
8—Inlet fitting
9—Diaphragm
10—Fulcrum pin
11—Lever return spring
12—Operating lever
13—Spindle spring
14—Diaphragm support plate
15—Pump body
16—Inlet valve plate
17—Inlet valve spring
18—Priming lever retaining screw
19—Priming lever

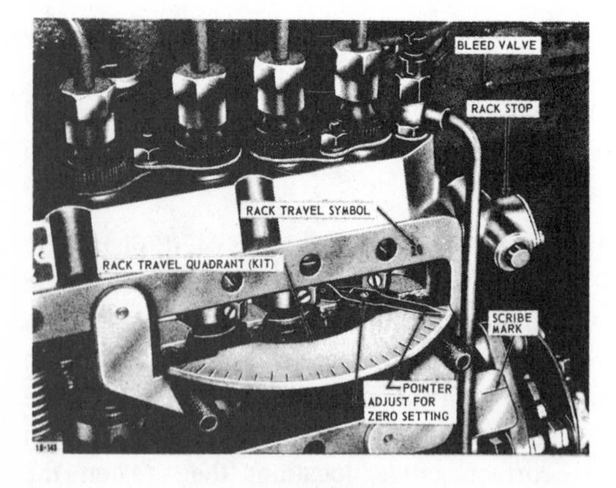

Figure 6-19
Gauge in Zero Position.
(Courtesy of Lucas CAV Ltd.)

Figure 6-20
Gauge in Maximum Position.
(Courtesy of Lucas CAV Ltd.)

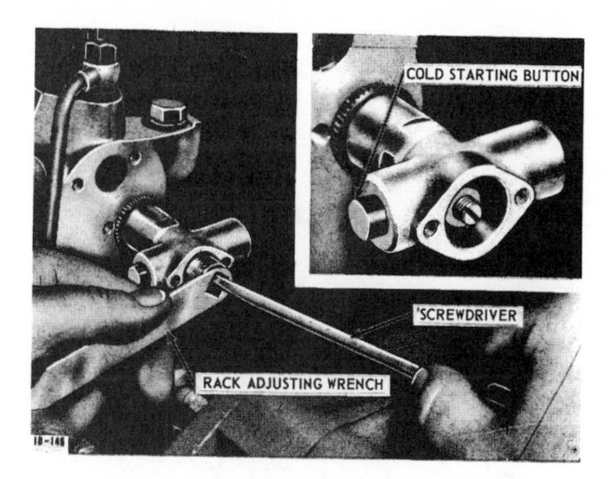

Figure 6-21
Adjusting the Fuel Control Rack.
(Courtesy of Lucas CAV Ltd.)

Note: The nuts have metric unit threads; do not substitute English/American threads.

4. Manually move the stop lever forward to place the rack in the fuel off position.
5. Gently line up the pointer tip to zero on the quadrant scale, which is the first line (long) on the right-hand side of the gauge.
6. Manually place the throttle or accelerator lever into the full-fuel position. Make sure that there is no bind as you go from one end of the range to the other. Repeat this several times, and carefully note the reading in *millimeters* of injection pump rack travel. (Each mark on the quadrant gauge represents 1 mm of rack travel.)
7. Refer to Figure 6-21; to adjust the fuel rack, a socket with a hole through it to allow adjustment of the rack screw with a screwdriver is required while the locknut has been loosened and held.
8. Stamped on the upper right-hand corner of the pump inspection cover pad is the injection pump rack travel specification, which means the following:
   a. Number 15: 11.5 mm (0.453 in.) rack travel
   b. Number 20: 12.0 mm (0.472 in.) rack travel
   c. Number 25: 12.5 mm (0.492 in.) rack travel

Note: If at any time the pump is rebuilt and recalibrated, new markings may have to be stamped on the pump. However, as always, only personnel experienced in the intricate details of fuel-injection pumps and with the proper test equipment available should undertake to rebuild and calibrate any injection pump.

To bleed the fuel system, open the vent screw on the filter and, using the supply pump hand primer, manually operate the pump to expel all air from the system; then open the injection pump fuel gallery bleed screw until fuel free of air bubbles appears. Close the filter screw first, and then the pump screw.

6-6

Model BPE Injection Pump

An older model of CAV injection pump, but one that is still very much in use, is the type BPE, which is a cam-operated, spring-return, plunger type that employs one pumping unit for each engine cylinder common to all in-line-style CAV pumps. It is available in single and multicylinder form, with the AA series available in multicylinder units only of 4 or 6. The BPE A series, which is the smaller of the series, also comes in 4- and 6-cylinder configurations. A B series is also used, which covers 2-, 3-, 4-, 5-, and 6-cylinder engines and is the largest pump within the BPE series.

Its principle of operation is much the same as other in-line CAV pumps. Figure 6-22 shows a typical BPE pump. Stamped on one end of the control rod (rack) at the top is the word *Stop* and an arrow that indicates the direction that the control rod should be moved to stop the engine.

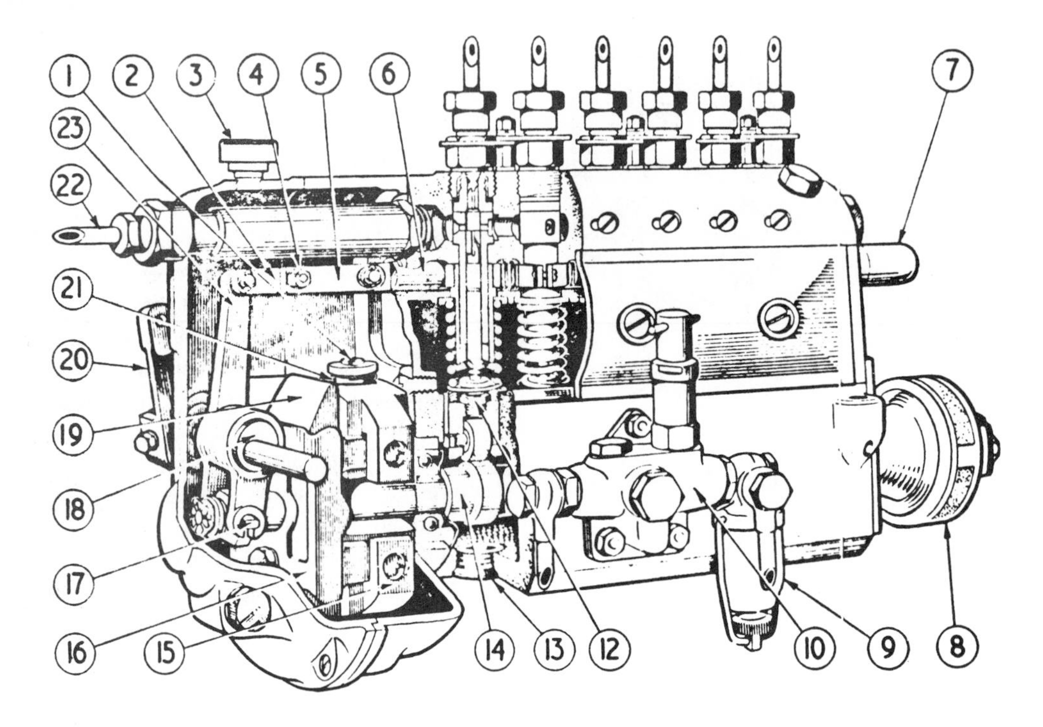

Figure 6-22
Typical BPE Pump.
(Courtesy of Lucas CAV Ltd.)

Index:
1—Adjusting nut
2—Outer link fork
3—Oil lubricator
4—Screw for link forks
5—Inner link fork
6—Control rod
7—Control rod stop
8—Drive coupling
9—Preliminary filter
10—Plunger-type feed pump
11—Oil dipstick
12—Tappet screw
13—Closing plug
14—Camshaft
15—Flyweights
16—Bell crank pin retaining cage
17—Coupling cross-head pin
18—Eccentric
19—Bell crank lever
20—Control lever
21—Governor spring
22—Fuel inlet connection
23—Floating lever

Governors used with this series of injection pump are (1) mechanical variable speed (control at all speeds), (2) limiting speed mechanical (idling and maximum speed control), (3) pneumatic, and (4) combined pneumatic and mechanical.

Drive

The pumps are provided with an extended camshaft with 20-mm (0.787-in.) taper cones at both ends so that they may be coupled to the engine at either end. This allows also for the installation of an injection advance device and governor. Drive from the engine is normally transmitted through a *close slot coupling,* as shown in Figure 6-23.

The injection pumps are supplied with their cams set in the same firing order as the engine cylinders, and since the cams are symmetrical in form, they can be driven in either direction without alteration other than resetting the couplings. It is possible to alter the firing order of a pump to suit a particular engine by crossing the delivery piping between the delivery connection of the injection pump and nozzle.

Mounting the Pump on the Engine

Prior to installing the injection pump onto the engine, make sure that the lines marked R and L for right- and left-hand rotation, respectively, on the half-coupling boss are in line with the mark on the pump body, relative to the direction in which the pump is to be driven.

1. Bar the engine over until the injection timing mark on the flywheel for cylinder 1 is aligned with the stationary pointer.
2. The pump can now be coupled to the engine drive through its close slot coupling.
3. The final adjustment for the pump can be done by means of the graduations provided on the coupling flange, with each division between the lines representing 3 degrees measured on the pump camshaft.
4. Slacken off the set screws provided on the coupling flange passing through the adjusting slots. Grasp the pump and turn it to obtain the correct amount of timing as stated for the particular engine; then tighten up the set screws.
5. Attach the fuel lines to the nozzles with particular reference to the firing order required.
6. Leave the nuts of the injection lines loose at the injectors for now.
7. With all other fuel lines connected, open the bleed screw on the secondary filter, and manually operate the transfer (feed) pump until a steady flow of fuel free of air bubbles emanates from the filter. Close the bleed screw.
8. Open both bleed screws on the injection pump body (see Figure 6-22), and continue to operate the transfer pump handle until a steady flow of fuel free of air bubbles flows from the injection pump housing. Tighten both bleeder screws.
9. Crank the engine over until fuel flows from each injection line and tighten them up in firing order sequence. The engine should now be ready to run (throttle wide open during cranking).
10. Once the engine fires, you may have to intermittently loosen each injection line one at a time in order to ensure that the system is free of air and that each injection nozzle is firing.

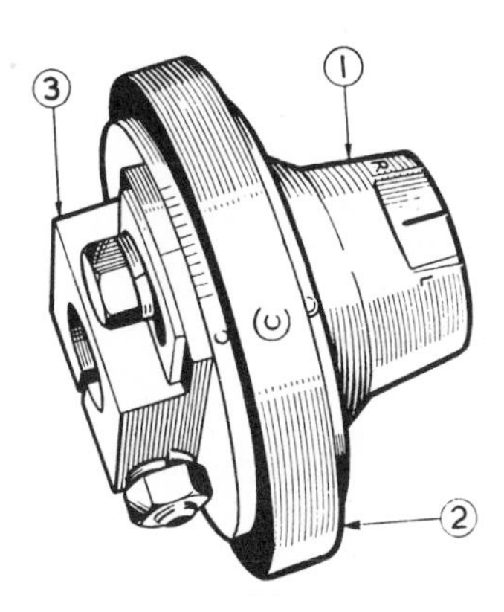

Figure 6-23
Close Slot Coupling.
(Courtesy of Lucas CAV Ltd.)

Index:
1—Pump side half-coupling
2—Center disc
3—Engine side half-coupling

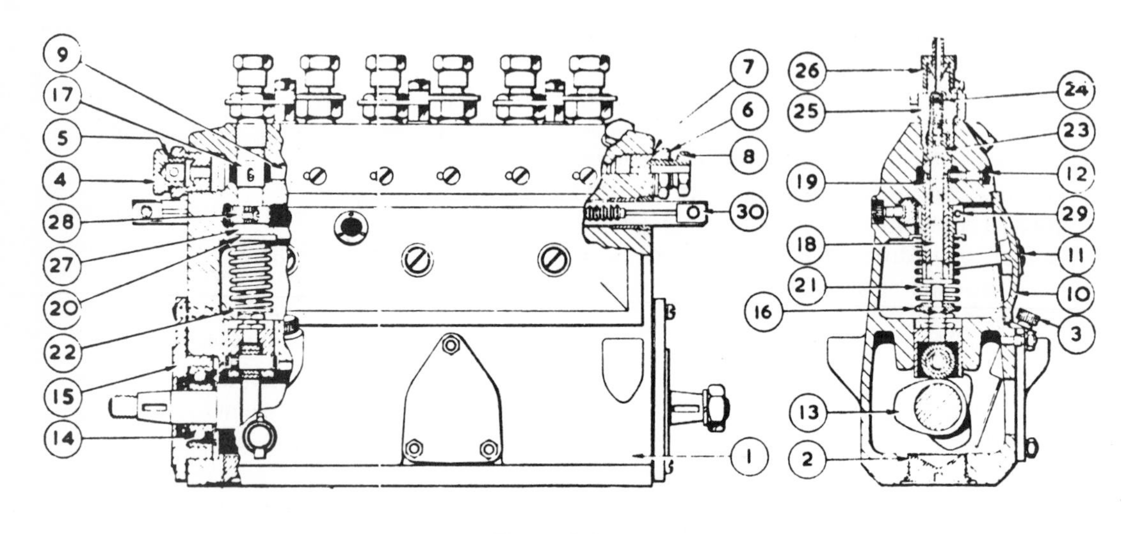

Figure 6-24
Internal Construction of Fuel-Injection Pump
Type BPE.
(Courtesy of Lucas CAV Ltd.)

Index:

1—Housing
2—Closing plug
3—Oil dipstick
4—Inlet closing plug
5—Joint for (4)
6—Inlet connection stud
7—Joint for (6)
8—Fuel inlet nipple nut
9—Suction chamber
10—Inspection cover plate
11—Screw with spring ring for (10)
12—Locking screw and joint
13—Camshaft
14—Ball bearing
15—Bearing end plate
16—Tappet assembly
17—Pump element (barrel and plunger)
18—Pump element plunger
19—Pump element barrel
20—Spring plate (upper)
21—Helical spring for (18)
22—Spring plate (lower)
23—Delivery valve and seating
24—Delivery valve spring
25—Delivery valve holder
26—Delivery nipple nut
27—Regulating sleeve
28—Regulating toothed quadrant
29—Clamp screw
30—Control rod

11. If the engine runs rough and stops, you must still have air in the system; therefore, bleed it again and check that there are no leaks on the suction side of the transfer (feed) pump.

Make sure that the oil level in the base of the pump housing is up to the correct level. This can be checked either by a small dipstick or level plug at the side of the housing. You may find that, by partly unscrewing the *inlet closing plug* shown in Figure 6-24 and cranking over the engine, bleeding of the fuel pump is more readily accomplished.

Injection Pump Overhaul

As with any injection pump, overhaul of the injection pump must be done in clean, dust-free, organized work areas, with a fuel lab being preferable. Second, no attempt should be made to rebuild a pump unless you are trained in the intricacies of fuel-injection equipment and have access to the necessary special tools and test stand equipment.

To check out a plunger spring and examine the plunger and the camshaft, first drain the oil from the pump. Then proceed as follows:

1. Refer to Figure 6-24 and remove the side plate inspection cover (10) from the pump housing.
2. Remove the bearing end plate (15) after rotating the camshaft (13), which places the tappet (16) at its TDC position; then insert the tappet holder (32) shown in Figure 6-25 under the head of the tappet adjusting screw.
3. Repeat step 2 for each pumping element. *Always* keep the parts from one pumping element plus the delivery valve parts together, since they are all matched components.
4. The end plates contain oil seals that should be removed and replaced at major overhaul. First, extract the outer race of the bearing, followed by the shim washer, and then the seal. Make sure when you replace the oil seal that the spring-loaded lip faces the bearing.
5. If the pump half-coupling is fitted at the opposite end to the bearing end plate, remove it from the cone of the camshaft with an extractor such as P/N ET 008, not with a hammer. Mark the shaft position so that upon reassembly the pump's firing order sequence will be correct.
6. Unscrew the closing plug (2 in Figures 6-24 and 6-25) with a tool such as ET 003

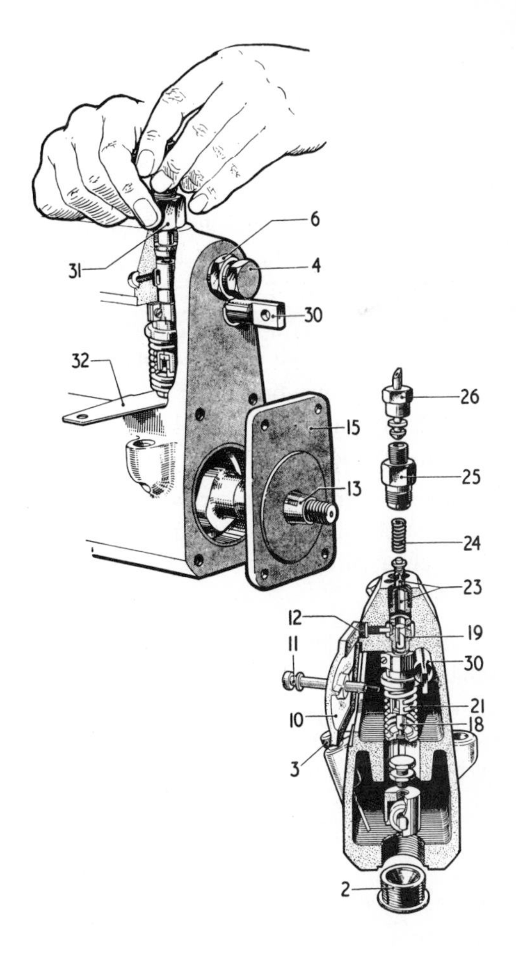

Figure 6-25
Dismantling and Reassembling Type BPE Fuel-Injection Pump.
(Courtesy of Lucas CAV Ltd.)

Index:

2—Closing plug
3—Oil dipstick
4—Inlet closing plug
6—Inlet connection stud
10—Inspection cover plate
11—Screw with spring ring for (10)
12—Locking pin and joint
13—Camshaft
15—Bearing end plate
18—Pump element plunger
19—Pump element barrel

for BPE-A pumps, ET 105 for BPE-B, or ET 815 for AA pumps, or a suitable equivalent.

7. With the plugs removed, push up the tappet (16) until you can withdraw the tappet holder; then remove the tappet assembly (16), lower spring plate (22), the plunger spring (21), and plunger (18).
8. Refer to Figure 6-25; unscrew the delivery valve holder (25), withdraw the spring peg (on AA pumps only), spring (24), and the delivery valve. Using tool 31, remove the valve seating and its joint.
9. Remove the locking screw (12), and push out the plunger barrel from the bottom with a soft punch. Repeat this for each barrel.

With the injection pump completely disassembled, inspect all component parts for any signs of wear, scoring, galling, scratching, overheating, damaged, off-square or weak springs, and any other visible wear patterns.

Pump Reassembly

Again, exercise extreme care and cleanliness during pump rebuild. Use clean, dry compressed air, and clean filtered test or fuel oil to lubricate all internal parts during assembly. Proceed as follows:

1. Refer to Figures 6-24 and 6-25 during reassembly.
2. Each barrel has a slot on its side that must be lined up with the hole for the locking screw (12) as it is inserted into the pump housing. Insert and tighten up the screw.
3. Locate the delivery valve seat and joint into the pump housing; then replace the delivery valve and its spring (24) into position. On AA pumps only, install the spring peg and then the delivery valve holder (25) in both cases, with its joint in position and screwed down tightly.
4. Install both the regulating quadrant (28) and regulating sleeve (27) along with the upper spring plate (20).
5. Insert the mated plunger with its spring (21) and lower spring plate (22) into the barrel (19) so that the lug on the lower plunger edge slides into the control sleeve slot.
6. Insert the mated tappet assembly (16), and push up against the spring so that you can install the tappet holder (32) between the tappet adjusting screw and the pump housing of each individual pumping unit as it is installed.
7. Carefully slide the camshaft into the pump housing so that the coupling or advance device is fitted to maintain the correct firing order for the engine.

Caution: You may recall that we mentioned earlier that the camshaft can be inserted from either end of the pump housing. When it was removed, we noted that it should be marked to assist in installation from the correct end. For example, a 4-cylinder camshaft with a firing order of 1–3–4–2 when reversed in the pump housing will have a firing order of 1–2–4–3.

8. Install the bearing end plate (15) and tighten up the holding screws.
9. Remove all the tappet holders (32) from under each tappet.
10. Coat the joint face of the closing plug (2) with one of the commercially available sealers and torque to specifications. (On AA pumps fit a new set of plugs.)
11. Fill the camshaft chamber with the recommended lube oil.
12. The pump must now be *phased* and *calibrated,* following the same basic procedure listed under that for the Minimec type of pump described earlier.

6-7

Rotary Fuel-Injection Pumps (DPA)

Over 20 years ago, CAV first produced their now legendary DPA (distributor pump assembly) model of fuel-injection pump. In 1975–1976 they released another pump of the same general design and operating principles, the DP15, which has now been discontinued in favor of the DPA.

The DPA pump is capable of being adapted to diesel engines with cylinder capacities up to 2 liters (122 in.^{3}) and is generally found on 2-, 3-, 4-, and 6-cylinder engines. It can be supplied with pumping plungers of 5, 6, 7, 8, 9, or 10 mm diameter with a normal maximum fuel output of 100 mm^3/stroke. These pumps do not require phasing or calibrating.

DPA Fuel-Injection Pump

The pump derives its name from the fact that its main shaft is driven, and runs through the center of the pump housing lengthwise. Fuel is in turn distributed from a single-cylinder opposed plunger control somewhat similar to a rotating distributor rotor in a gasoline engine. The pump can be hub mounted or gear driven since its shaft is very stiff to eliminate torsional oscillation and ensure constant accuracy of injection.

Figure 6-26 shows a cutaway view of a typical DPA fuel-injection pump with a mechanical governor. Figure 6-27 shows a DPA pump with a hydraulic governor.

All internal parts are lubricated by fuel oil under pressure from the delivery pump. The pump can be fitted with either a mechanical or hydraulic governor depending on the application, and a hydraulically operated automatic advance mechanism controls the start of injection in relation to engine speed.

The operation of the fuel distribution is similar to that found in Roosa Master distributor pumps in that a central rotating member forms the pumping and distributing rotor driven from the main drive shaft on which is mounted the governor assembly.

Fuel Flow

Mounted on the outer end of the pumping and distributing rotor is a sliding vane-type transfer pump that receives fuel under low pressure from a lift pump mounted and driven from the engine. This lift pump pressure enters the vane-type pump through the fitting on the injection pump end plate opposite its drive end and passes through a fine nylon gauze filter.

The vane-type pump has the capability of delivering more fuel than the injection pump will need; therefore, a pressure-regulating valve housed in the injection pump end plate allows excess fuel to be bypassed back to the suction side of the vane transfer pump. This valve is shown in Figure 6-28.

In addition to regulating fuel flow, the pressure-regulating valve also provides a means of bypassing fuel through the outlet of the transfer pump on into the injection pump for priming purposes. Referring to Figure 6-28, the regulating valve is round and contains a small *free piston* whose travel is controlled by two light springs. During priming of the injection pump, fuel at lift pump pressure enters the central port of the regulating valve sleeve and causes the free piston to move against the retaining spring pressure, thereby uncovering the priming port at the lower end of the sleeve, which connects by a passage in the end plate to the delivery side of the vane-type transfer pump, which leads to the injection pump itself.

Once the engine starts, we now have the vane-type transfer pump producing fuel under pressure, which enters the lower port of the regulating valve and causes the free piston to move up against the spring. As the engine is accelerated, fuel pressure increases, allowing

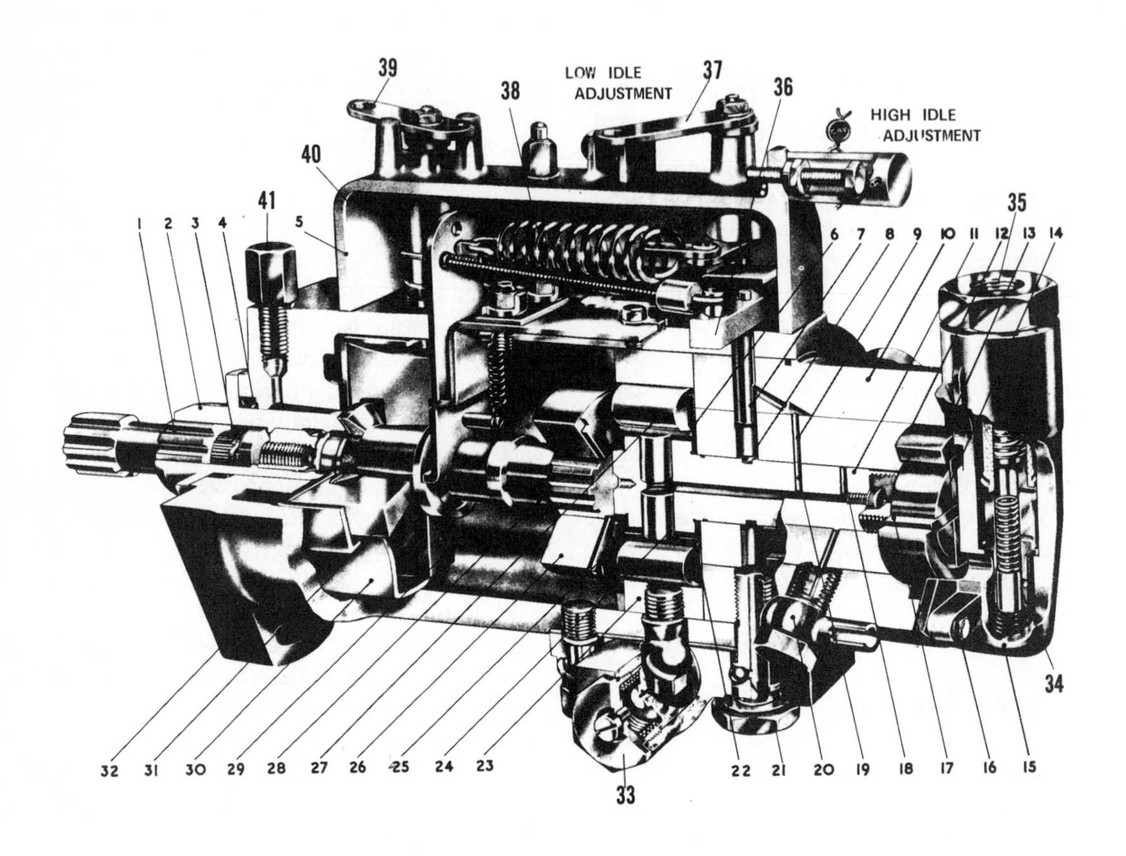

Figure 6-26
DPA Pump with Mechanical Governor.
(Courtesy of Lucas CAV Ltd.)

the free piston to progressively uncover the regulating port, thereby bypassing fuel from the outlet side of the vane-type pump. This action automatically controls the fuel requirements of the injection pump. Figure 6-29 shows the fuel pressure-regulating valve assembled into its bore and its action during *priming* and *regulating*.

DPA Distributor Pump Improvements

With the advent of diesel engine emissions legislation, Lucas CAV has now introduced an optional version of the legendary DPA injection pump, which uses four pumping plungers rather than the conventional two of the standard DPA pump. The four-plunger unit has the following inherent advantages:

1. A higher rate of injection.
2. Shorter injection periods.
3. Much more rapid termination of injection.
4. Faster rate of combustion.
5. Reduced smoke and gaseous emissions.

In addition to these advantages, because of the higher rate of injection, engine users can effectively retard the engine timing by up to 6

Index:

1—Pump drive shaft
2—Nose extension
3—Spring clip
4—Oil seal
5—Governor control cover
6—Bottom adjusting plate
7—Metering valve
8—Supply port to metering valve
9—Metering port (inlet)
10—Hydraulic head assembly
11—Distributor rotor
12—Hydraulic head sleeve
13—Internal (eccentric ring) transfer pump
14—Transfer pump
15—End plate and regulator valve housing
16—Retaining bolt
17—Transfer pump vane
18—Distributor port (outlet)
19—Rotor fuel inlet port
20—Banjo fitting fuel outlet to injector
21—Hydraulic head locking screw
22—Roller shoe
23—Internal cam ring
24—Pumping plunger
25—Top adjusting plate
26—Pumping and distributing rotor
27—Roller
28—Rotor drive shaft
29—Thrust sleeve
30—Weight carrier assembly
31—Governor weights
32—Pump housing mounting flange
33—Advance mechanism
34—Fuel pressure regulating valve assembly
35—Fuel inlet
36—Metering valve control lever
37—Throttle lever
38—Governor spring (main)
39—Shutoff lever
40—Governor spring link
41—Fuel back-leak connection

degrees without any increase in exhaust smoke levels. Also, a quieter engine results owing to the later start to injection, with a shorter ignition delay period. With the four-plunger design, cam contact stresses are within an acceptable level even at the peak injection pressures. Figure 6-30 shows a 6-cylinder version of the new four-plunger CAV DPA pump.

The four-plunger pump is available on engines ranging from 8 to 19 kW/cylinder (10 to 25 bhp/cylinder), which gives an application range of engines rated from 30 to 112 kW (40 to 150 bhp) in the premium performance and mildly turbocharged types. The first application for the four-plunger DPA is the Perkins turbocharged T6.354 for industrial and agricultural use, with automotive applications coming in the near future.

In addition to the Perkins contract, Volkswagen (Germany) has contracted with Lucas CAV to use the DPA injection pump on their VW model Golf diesel car sold in Europe, the equivalent of which is the diesel Rabbit in North America fitted with the Robert Bosch VE model injection pump. The DPA injection pump being supplied for the VW Golf car includes the following features:

1. A cold-start device to advance injection timing for quick starting in low ambient temperatures without blue smoke.
2. Automatic excess fuel to further ensure rapid starting under all cold conditions.
3. Ignition key to fuel solenoid fuel shutoff (stop and start).
4. Two-speed hydraulic governing for smooth accelerator reponse throughout the engine speed range.
5. Light-load advance to give optimum timing for minimum noise and smoke emissions.
6. A 5600 rpm maximum speed limit to avoid a sharp cutoff at the top of the engine performance range.
7. Use of new CAV *low spring* injectors.

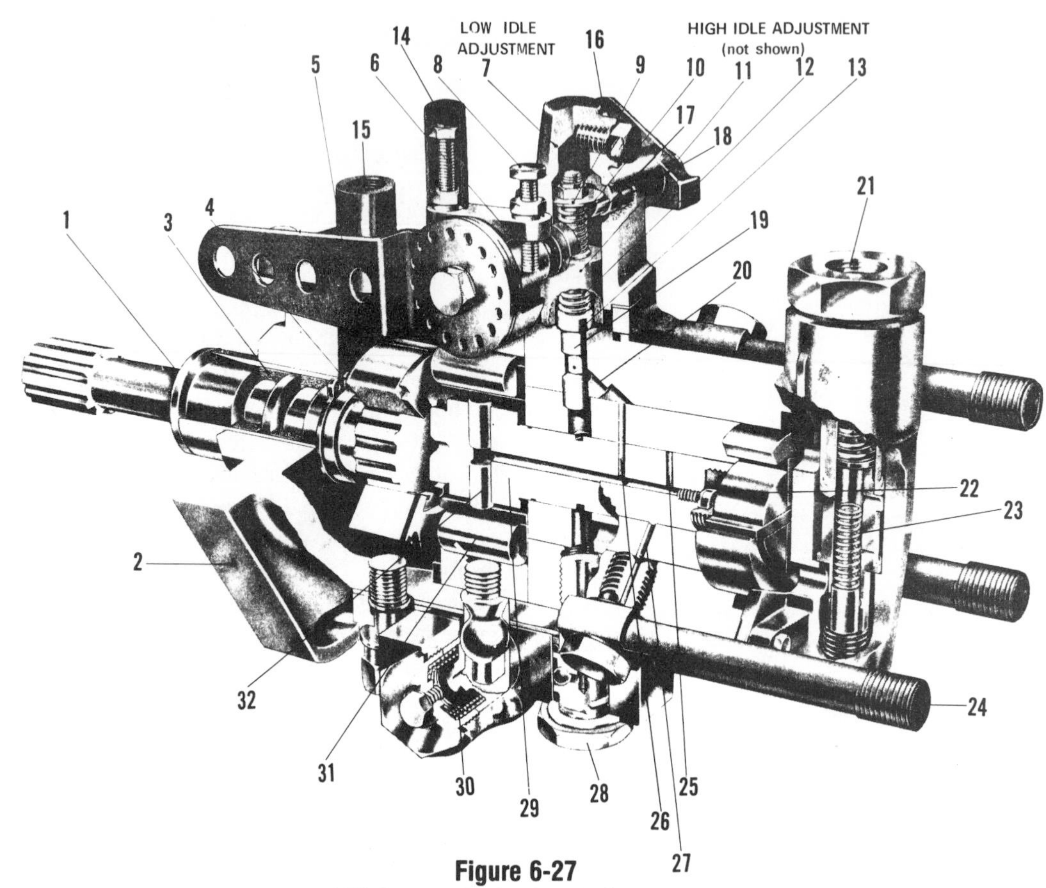

Figure 6-27
DPA Pump with Hydraulic Governor.
(Courtesy of Lucas CAV Ltd.)

Index:

1—Pump drive shaft
2—Pump mounting flange
3—Outer seal
4—Circlip
5—Throttle lever
6—Pinion shaft
7—Governor housing
8—Idling stop screw
9—Shutoff washer
10—Idling spring
11—Rack
12—Governor spring
13—Damping washers
14—Maximum speed stop screw
15—Back leakage mounted on inspection cover
16—Shutoff lever
17—Half-round cam end
18—Shutoff spindle
19—Metering valve
20—Metering port
21—Fuel inlet
22—Transfer pump
23—Pressure regulator assembly
24—To fuel injector
25—Distributor port (outlet)
26—Rotor fuel inlet port
27—Sleeve outlet port
28—Hydraulic head locking screw
29—Rotor
30—Advance mechanism
31—Roller
32—Plunger

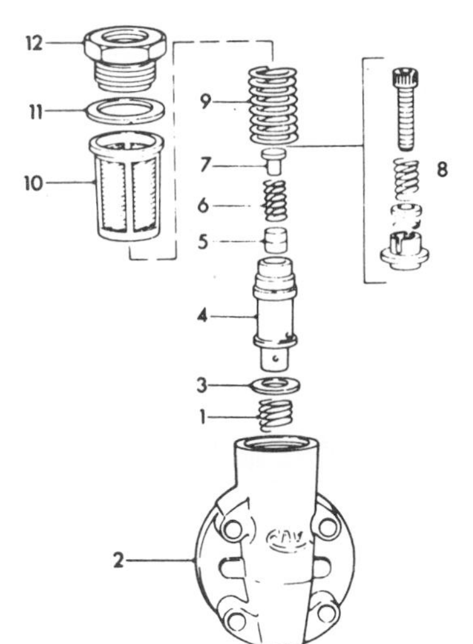

Index:

1—Piston retaining spring
2—End plate
3—Sealing washer
4—Regulating sleeve
5—Regulating piston
6—Regulating spring
7—Regulating plug
8—Transfer pressure adjuster
9—Sleeve retaining spring
10—Filter
11—Washer
12—Fuel inlet connection

Figure 6-28
Pressure-Regulating Valve.
(Courtesy of Lucas CAV Ltd.)

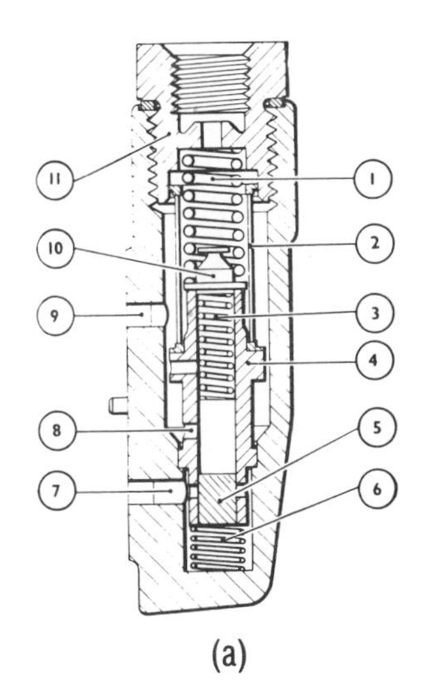

(a)

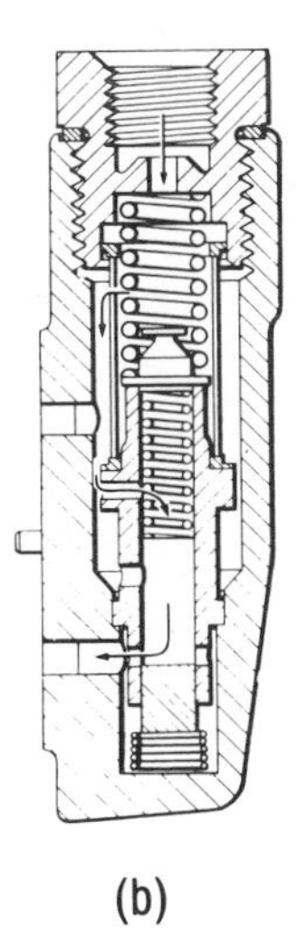

(b)

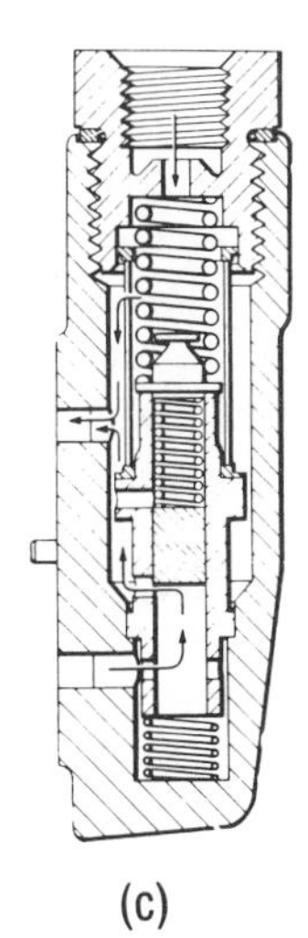

(c)

Figure 6-29
Action of the Pressure-Regulation Valve:
(a) At Rest; (b) Hand Priming; (c) Engine Running.
(Courtesy of Lucas CAV Ltd.)

Index:

1—Retaining spring
2—Nylon filter
3—Regulating spring
4—Valve sleeve
5—Piston
6—Priming spring
7—Fuel passage to transfer pump outlet
8—Regulating port
9—Fuel passage to transfer pump inlet
10—Spring guide
11—Fuel inlet connection

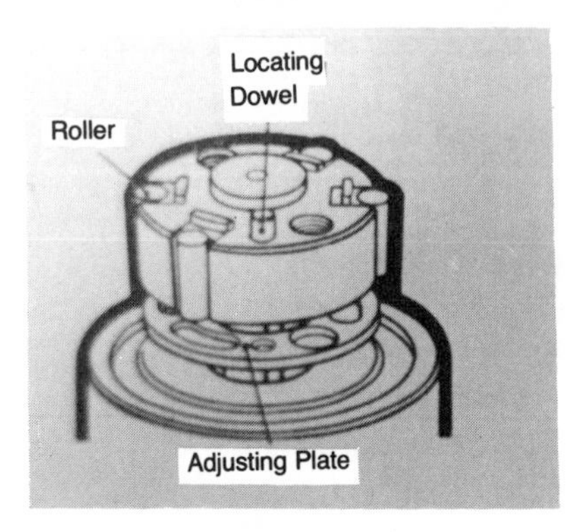

Roller and Shoe Assemblies

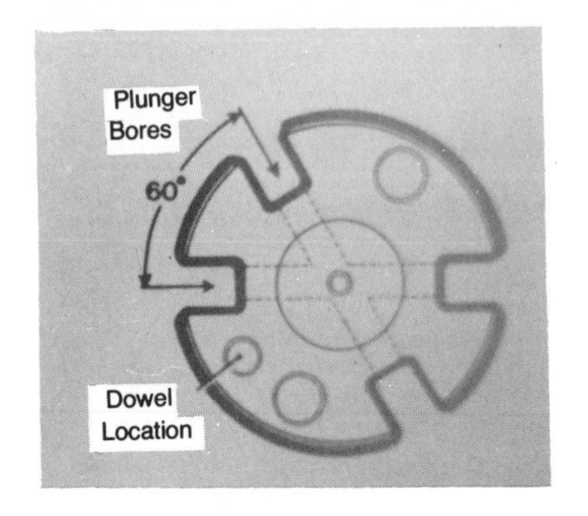

End View of Rotor

Figure 6-30
Six-Cylinder Version of the Four-Plunger DPA Pump. (Courtesy of Lucas CAV Ltd.)

Other diesel car manufacturers presently using Lucas CAV model DPA pumps and injectors include Peugeot and Citroen in France and Chrysler in Spain, supplied by CAV plants in those countries. The DPA pump on the Perkins engine is also used on the VW model LT van. The DPA injection pump is the world's best selling rotary fuel-injection pump, with more than 1 million a year being produced from Lucas CAV subsidiaries, associates, and licensees abroad.

DPA Excess Fuel Starting

The use of the excess fuel starting device cuts engine cranking time in half at minus 7°C in tests conducted on a typical 100-bhp (75-kW) automotive diesel. Both battery and starter life are prolonged, and a quicker run up to normal operating rpm is obtained, with elimination of initial hesitancy.

The excess fuel device is so arranged that an additional pair of plungers in the rotor head is brought into operation by an automatic valve when the engine is stopped. When the engine fires and reaches adequate rpm to continue running, the excess fuel starting valve automatically cuts off fuel supply to these two plungers, allowing the pump to operate as a standard DPA pump. The point of excess fuel cutoff can be adjusted for each particular engine, thereby offering good starting quality with minimum exhaust smoke. The CAV valve mechanism prevents reengagement of the excess fuel plungers with the engine running.

Figure 6-31 shows a typical flow diagram for a DPA pump using a mechanical governor. Let us study the action of the fuel under pressure once it leaves the vane-type pump and flows to the injection pump. The pumping and distributor rotor, which is driven from the drive on the engine, rotates within the stationary hydraulic head, which contains the ports leading to the individual injectors. The number of ports varies with the number of engine cylinders. Figure 6-32 shows the rotor during the charging cycle and delivery cycle. In Figure 6-32, fuel from the vane-type transfer pump passes through a passage in the hydraulic head to an annular groove surrounding the rotor and then to a metering valve (see Figure

6-26), which is controlled by the throttle position.

The flow of fuel into the rotor (volume) is controlled by the vane-type pump's pressure, which is dependent on the speed of the engine and hence throttle or governor position. Fuel flowing into the rotor [Figure 6-32(a)] comes from the inlet or metering port in the hydraulic head. These inlet ports are equally spaced around the rotor; therefore, as the rotor turns, these are aligned successively with the hydraulic head inlet port.

The distributor part of the rotor has a centrally drilled axial passage that connects the pumping space between the plungers with the inlet ports (the number depending on the number of engine cylinders) and single distributing port drilled radially in the rotor. As the rotor turns around, the single outlet port will successively distribute fuel to the outlet

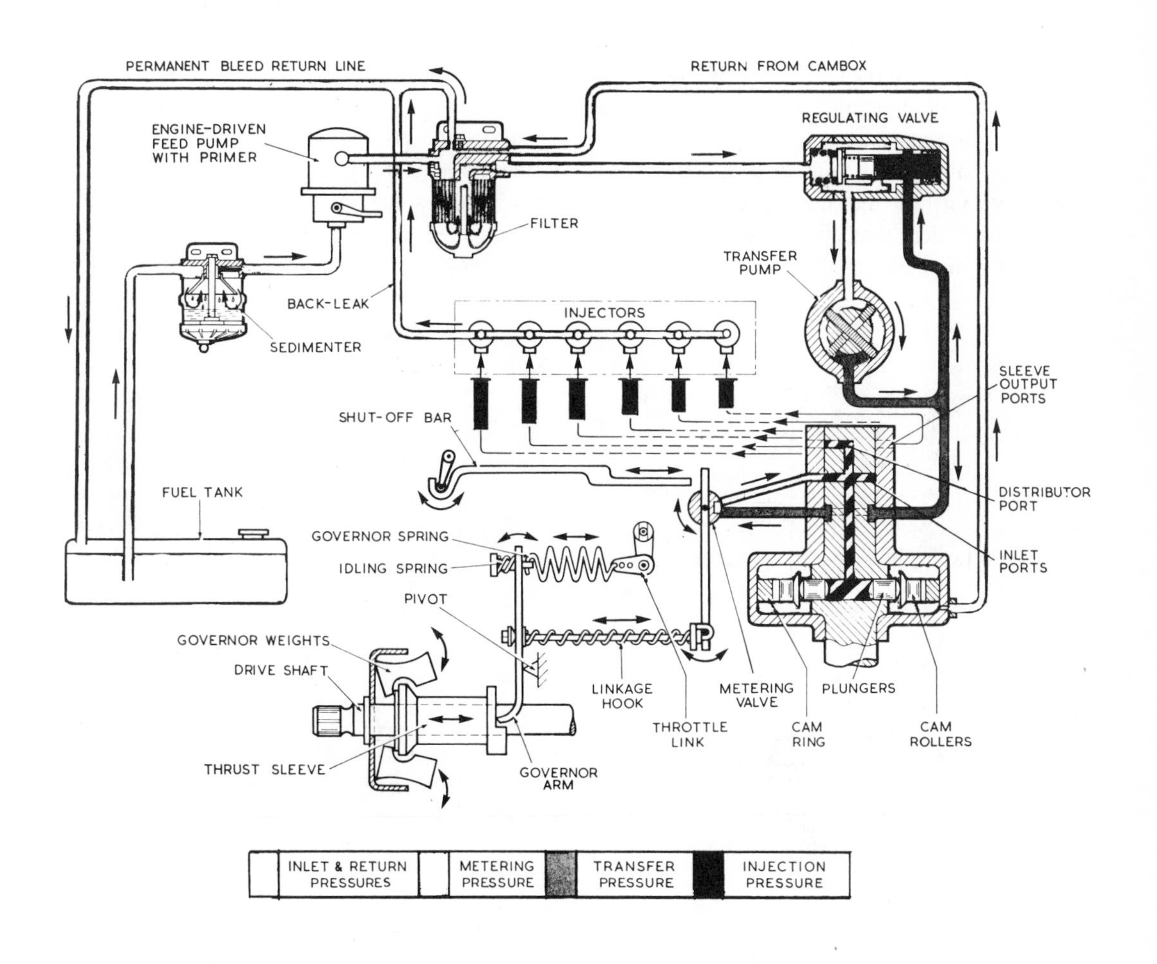

Figure 6-31
Flow Diagram for a DPA Pump Using a Mechanical Governor.
(Courtesy of Lucas CAV Ltd.)

ports of the hydraulic head and on to its respective injector.

The pumping section of the rotor has a cross-drilled bore that contains the twin opposed plungers, which are operated by means of a cam ring (internal) carried in the pump housing, through rollers and shoes that slide in the rotor. The internal cam ring has as many lobes as there are engine cylinders. For example, a 4-cylinder engine would have four internal lobes operating in diagonally opposite pairs.

The opposed plungers have no return springs and are moved outward by fuel pressure, the amount being controlled by throttle position, metering valve, and the time during which an inlet port in the rotor is exposed to the inlet port of the hydraulic head. As a result, the rollers that operate the plungers do not follow the contour of the internal cam ring entirely, but they will contact the cam lobes at points that will vary according to the amount of plunger displacement.

The maximum amount of fuel delivered to an injector is therefore controlled by limiting the maximum outward movement of the plungers. Refer to Figure 6-33, which shows an end-on view of the rotor with the cam rollers carried in shoes bearing against the plunger ends. These roller shoes sliding in slots in the rotor have projecting ears that engage eccentric slots in the top and bottom adjusting plates located against one another by two lugs in slots.

The top adjusting plate is clamped to the rotor by the drive plate; however, the adjusting plate is cut away at the locating screws to allow adjustment of the plates by rotation. In effect, then, the ears of the roller shoes coming into contact with the curved slot sides of the

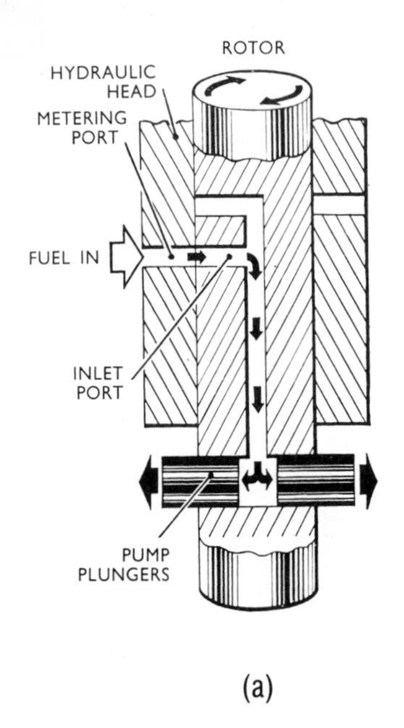

(a)

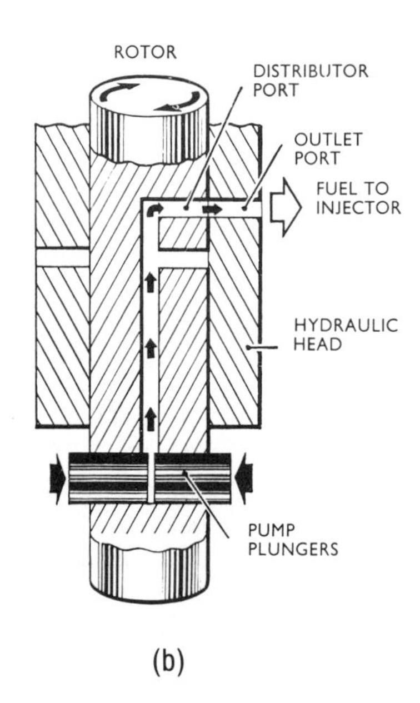

(b)

Figure 6-32
Rotor During Charging Cycle and Delivery Cycle:
(a) Inlet stroke; (b) injection stroke.
(Courtesy of Lucas CAV Ltd.)

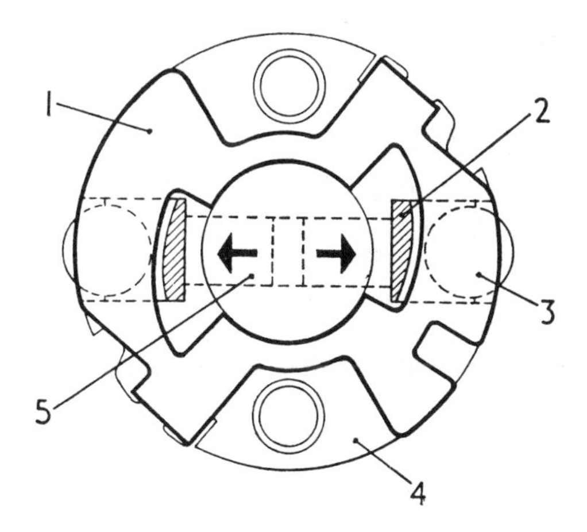

Figure 6-33
End View of Rotor.
(Courtesy of Lucas CAV Ltd.)

Index:
1—Top adjusting plate
2—Roller shoe ear
3—Roller
4—Pumping end of rotor
5—Pumping plunger

adjusting plates limit the maximum outward travel of the plungers. The slots are eccentric for adjustment purposes.

Control of DPA Pump Delivery

Figure 6-34 shows the method used to control injection pump delivery. The roller-to-roller dimension is checked with an outside micrometer; however, the stirrup pipe tool (item 1), part no. 7144-262A, must be connected to two of the hydraulic head high-pressure outlet ports along with a suitable relief valve (item 2), part no. 7144-155.

Connect an injector nozzle testing pump unit to the relief valve and raise the fuel pressure to the outlet ports until 30 atmospheres, or that specified in the pump test data sheet, is reached. By manually turning the pump rotor, the fuel pressure internally will force the pumping plungers and rollers to their maximum outward (full fuel) position. At this point, using the micrometer, measure the roller-to-roller dimension. To adjust the roller-to-roller dimension, the adjusting plates can be moved, then the drive plate screws torqued to specifications.

All necessary pump information can be found by consulting the proper test data sheet for your particular pump. This can be readily obtained from Lucas CAV or your local fuel-injection repair company, most of which are members of the Association of Diesel Specialists.

As indicated in earlier paragraphs, the maximum amount of fuel delivered can be regulated either by limiting the outward travel of the plungers or by limiting the stroke of a shuttle valve according to pump type, as shown in Figure 6-35.

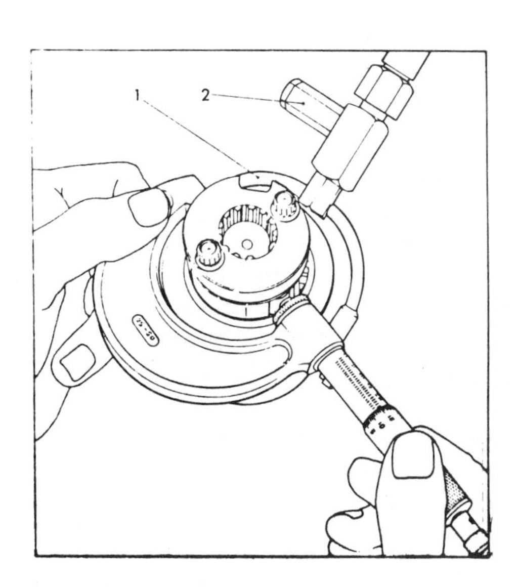

Figure 6-34
Measuring Roller-to-Roller Dimension.
(Courtesy of Lucas CAV Ltd.)

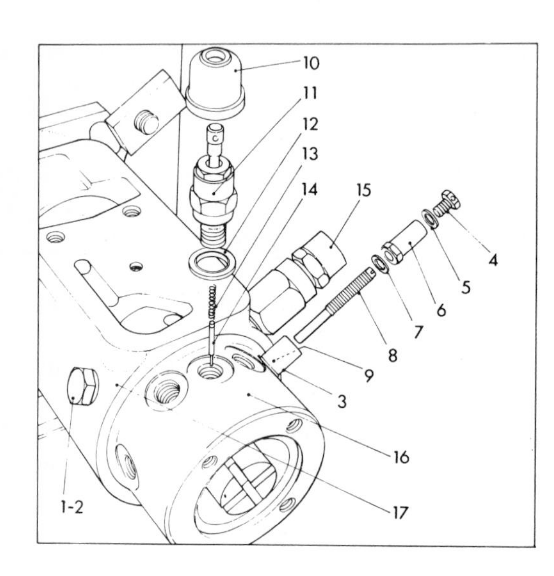

Index:

1 and 2—Hydraulic head locking screw and washer
3—Shuttle valve body
4—Shuttle plug screw
5—Screw plug washer
6—Shuttle tubular nut
7—Tubular nut washer
8—Shuttle stop screw
9—Shuttle valve piston
10—Cover (excess fuel)
11—Excess fuel selector assembly
12—Washer
13—Latch pin spring
14—Latch pin
15—Proportional pressurizing valve and washer
16—Hydraulic head and rotor assembly
17—Pump housing

Figure 6-35
Shuttle Valve Assembly Used on an Injection Pump Equipped with an Excess Fuel Device and External Maximum Fuel Adjustment. (Courtesy of Lucas CAV Ltd.)

Maximum Fuel Setting

Although the injection pump roller-to-roller dimension must be set as just discussed prior to assembling the pump, the pump can only be accurately checked for fuel delivery when mounted and run on a test stand. The specifications are listed on the fuel pump test data sheet available from Lucas CAV or a local dealer. A sample test data sheet is shown later. The maximum fuel setting on the DPA pump fitted with either the mechanical or hydraulic governor is basically the same, and would be done as follows:

1. The pump must be checked at a specified speed with both the throttle and shutoff controls fully open, and the maximum fuel delivery rate compared to specifications.
2. If the delivery rate is low, then the pump inspection cover, as shown in Figures 6-36 and 6-37, must be removed to gain access to the pump drive plate screws.
3. A special adjustment tool, part no. 7144-875, is available, which is used to engage the slot located in the periphery of the pump adjusting plate.
4. With the pump drive plate screws slackened off as mentioned in step 2, the end of the tool can be lightly tapped to adjust the internal pump plate. Although the direction of fuel increase and decrease will vary, depending on the type of adjusting plate fitted, the following holds true. Viewing the pump, from its drive end, if the top adjusting plate has a shallow slot 3 mm in depth (Figure 6-36), then turning

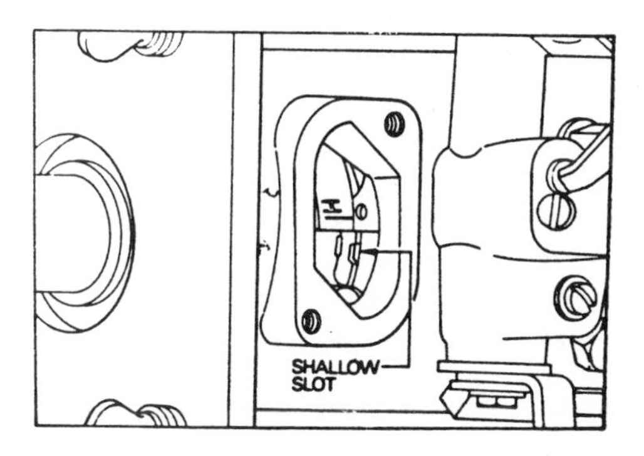

Figure 6-36
Top Adjusting Plate Shallow Slot (3 mm).
(Courtesy of Lucas CAV Ltd.)

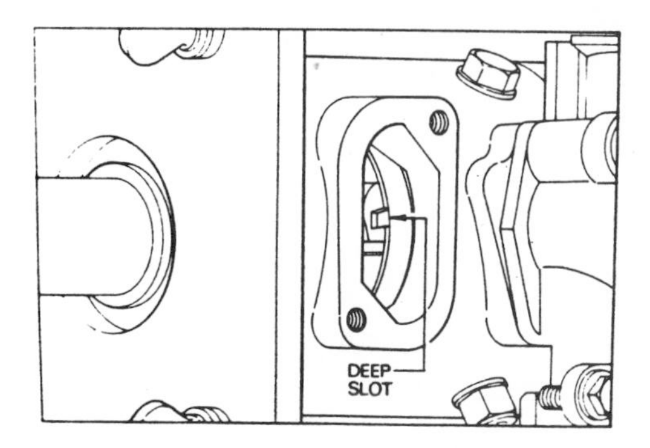

Figure 6-37
Top Adjusting Plate Deep Slot (5.5 mm).
(Courtesy of Lucas CAV Ltd.)

the plate in the counterclockwise direction will *increase* fuel delivery, whereas if the plate is turned in a clockwise direction, the fuel delivery rate will be *reduced*. If, however, the top adjusting plate has a deeper slot 5.5 mm deep (Figure 6-37), turning the top plate clockwise will increase fuel delivery, and counterclockwise will decrease it.

5. It is extremely important that the drive plate screws be torqued up evenly as listed in Table 6-1, using special adaptor tool, part no. 7144-482, spanner tool, P/N 7144-511A, and a torque wrench.
6. Install the pump inspection cover, prime the pump with fuel, and recheck the maximum fuel delivery rate. If necessary, repeat the previous procedure until the

Table 6-1
PUMP DRIVE PLATE SCREW TORQUES

	N·m	*kg/m*	*lb/in.*
Direct torque, plungers up to and including 7.5 mm diameter	18.1	1.85	160
Direct torque, plungers 8.00 mm diameter and above	28.4	2.90	250
Drive plate screw (A),[1] up to and including 7.5 mm diameter plungers	15.7	1.60	140
Drive plate screw (A),[1] 8 mm and above plunger diameter	24.3	2.48	215
Drive plate screw (B),[2] up to and including 7.5 mm plunger diameter	13.0	1.33	115
Drive plate screw (B),[2] 8.00 mm and above plunger diameter	20.6	2.10	180

[1](A): obtained with spanner 7144/511 and adapter 7144/482.

[2](B): obtained with spanner 7144/511A and adapter 7144/482.

Adapter and spanner *MUST* form a straight line when torquing up screws with a 127-mm (5.0-in.) adapter center to ring spanner.

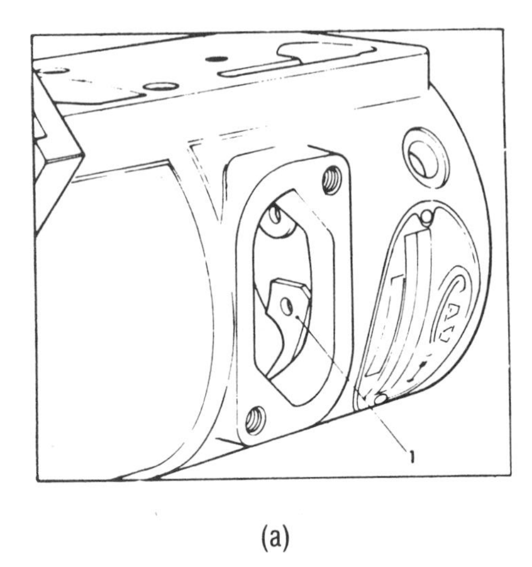

(a)

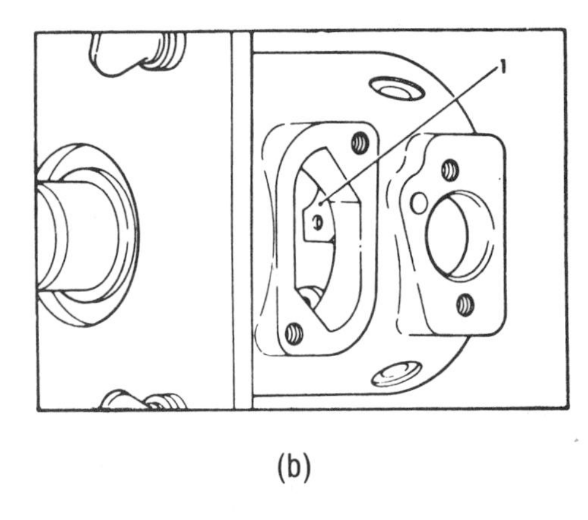

(b)

Figure 6-38
(a) DPA Pump Mechanical Governor Housing;
(b) DPA Pump Hydraulic Governor Housing.
(Courtesy of Lucas CAV Ltd.)

maximum fuel rate is as specified in the pump test data sheet.

7. On pumps with external maximum fuel adjustment, the shuttle stop screw is turned inward to reduce fuel delivery or outward to increase delivery. After adjustment, re-tighten shuttle tubular nut and replace shuttle plug screw.

Pump Timing

The correct timing in degrees for any DPA pump is given in the test data sheet. All pumps must be correctly timed. The sequence of events for DPA pumps with either a mechanical or hydraulic governor is basically the same. To check the timing, the inspection cover immediately above the pump advance unit must be removed.

When checking the timing, the pump should be removed from the test stand, and drained of fuel. You must then connect tool no. 7144-262A along with tool no. 7144-155, which was shown in Figure 6-34. This can be done by connecting the stirrup pipe tool to the fuel outlet specified on the test data sheet and to the diametrically opposite outlet. Bring the fuel pressure up to 30 atm by use of a nozzle testing pump unit, and proceed as follows:

1. Manually turn the pump drive shaft in its normal direction until resistance to further movement is felt, which is the *timing position*.
2. The timing ring is visible through the inspection cover.
3. Refer to Figure 6-38, and move the timing ring until the straight edge of the timing circlip, or the line scribed on the ring as is the case with the older-style clips, aligns with the mark on the drive plate as specified in the test plan sheet.
4. Circlips with two straight ears are only for spacing, and the circlip ends are positioned remote from the inspection hole aperture. Refit the inspection cover when timing is complete.
5. Pumps fitted with excess fuel devices have the timing marks machined on the drive plate and the cam ring, and are not adjustable. The pump is timed by removing the timing inspection cover and turning the rotor until the marks are in alignment.

A flange marking gauge tool, P/N 7244-

27, is available for use in remarking pump flanges at any time a change is required.

Automatic Light-Load Advance Mechanism

The light-load advance device automatically varies the start of injection in relation to the speed of the engine. Reference to Figure 6-39 will assist you in understanding its principle of operation. Rotation of the internal cam ring in which the rotor revolves affects the point at which the discharge port of the rotor will align with the hydraulic head outlet to the injector. Its action is similar to that of the advance mechanism used with an ignition distributor in a gasoline engine.

Screwed into the cam ring is a lever with a ball end on it, one side of which is acted upon by a spring-loaded piston, while the other side is subjected to fuel pressure or drain pressure, depending on the engine load and throttle position, from the metering valve, which feeds through a hollow locating bolt and port in the housing.

Operation: Full-Load. With the engine running under a situation of full load, the helical groove of the fuel metering valve will be aligned with the hydraulic head timing port, as shown in Figure 6-39(a); therefore, fuel at drain pressure is bled away from the advance piston, allowing the spring pressure to hold the internal cam ring in its fully retarded position.

Operation: Light-Load. With the engine running under light load, the fuel metering valve would be in a position whereby the flat at the lower end of the metering valve would now be aligned with the timing port to the piston. The transfer pump pressure would now flow to the advance piston, forcing it back against the spring and allowing the cam ring to move to its fully advanced position.

When starting the engine, due to the tension of the governor spring, the metering valve

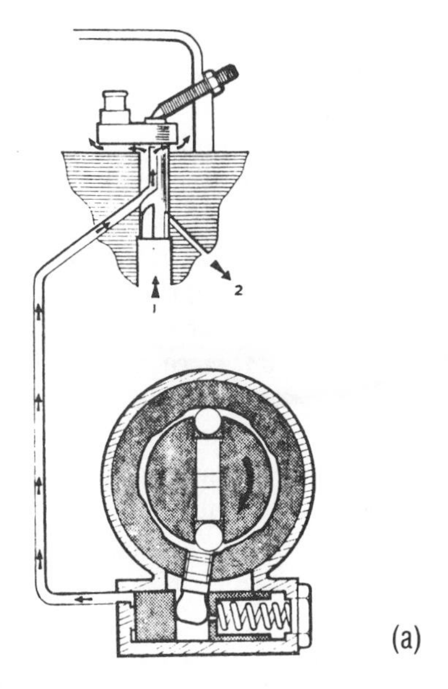

(a)

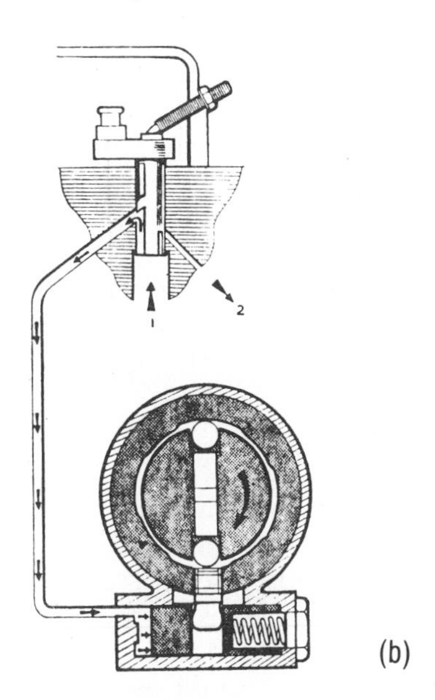

(b)

Figure 6-39
Automatic Light-Load Advance Mechanism: (a) Full Load; (b) Low Load. (Courtesy of Lucas CAV Ltd.)

Index:
1—Transfer pressure
2—To pumping element

would be in the full-load position with the advance piston subjected to fuel at drain pressure; therefore, the internal cam ring would remain in the fully retarded position.

A retraction curve machined on each lobe of the cam ring immediately after the peak of the cam allows the plungers to move slightly outward after the completion of an injection cycle. This is necessary in order to prevent secondary injection, since the distributor rotor port and the outlet port in the hydraulic head to the injector are still in partial alignment. This action allows rapid seating of the injector nozzle needle valve, thereby preventing *dribble,* which would lead to incomplete combustion and its associated problems.

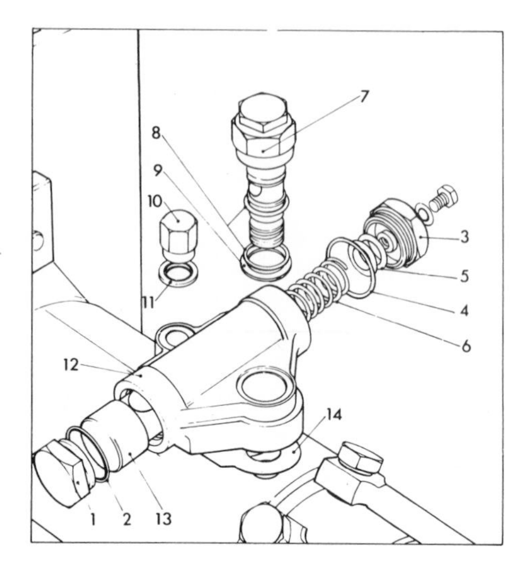

Figure 6-40
Removal of Start/Retard Device.
(Courtesy of Lucas CAV Ltd.)

Index:
1—Piston plug
2—Piston plug seal
3—Piston spring cap
4—Seal
5—Shim washers; altering the thickness of shims will change the pump advance
6—Outer piston spring
7—Head locating screw
8—Stud seals
9—Dowty washer
10—Cap nut
11—Washer
12—Autoadvance housing
13—Piston
14—Housing gasket

Automatic Two-Stage Start and Retard Device (Removal)

This device was discussed as the light-load advance mechanism earlier. It is located on the underside of the injection pump housing, and is easily removed for inspection as to its condition, by removing items 1, 3, 7, 10, and 12 in Figure 6-40 and items 1, 3, 11, 15 and 17 in Figure 6-41 on those pumps fitted with a damper assembly.

Mechanical Governor

The governor functions in the same manner as all mechanical units in that spring pressure is trying to increase the fuel delivered, while weight force opposing the spring is trying to reduce the fuel delivered.

Figure 6-42 shows the typical linkage hookup of such a mechanical governor found on DPA pumps.

Governor Operation. The weights (1) are held in the retainer (15), which is positioned between the injection pump drive hub and drive shaft (13) and thus rotates with the shaft as a unit. The weights will move outward under the influence of centrifugal force, the distance being related to engine rpm.

A thrust sleeve (14), which is hollow, is free to move lengthwise along the extension nose of the injection pump drive shaft under the influence of the governor weight force.

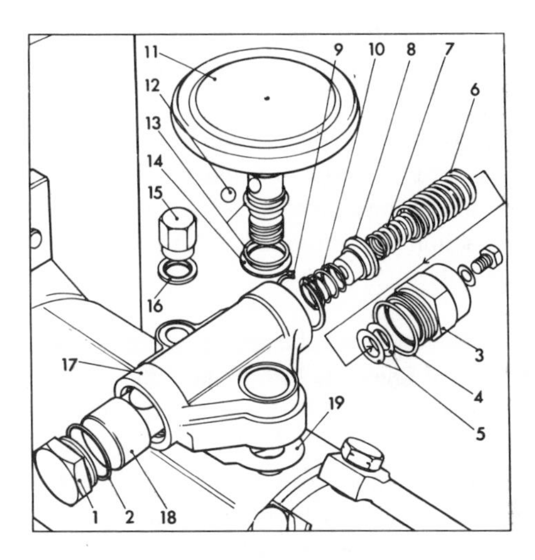

Figure 6-41
Removal of Start/Retard Device Fitted with a Damper Assembly.
(Courtesy of Lucas CAV Ltd.)

Index:
1—Piston plug
2—Piston plug seal
3—Piston spring cap
4—Seal
5—Shim washers
6—Outer piston spring
7—Inner piston spring
8—Spring plate
9—Housing circlip
10—Short piston spring
11—Damper assembly
12—Steel ball
13—Stud seals
14—Washer
15—Cap nut
16—Washer
17—Autoadvance housing
18—Piston
19—Housing gasket

Index:
1—Governor weight
2—Governor arm
3—Shutoff bar
4—Shutoff shaft
5—Idling spring
6—Governor spring
7—Throttle shaft
8—Linkage hook
9—Metering port
10—Metering valve
11—Timing port
12—Control bracket
13—Drive shaft
14—Thrust sleeve
15—Weight retainer

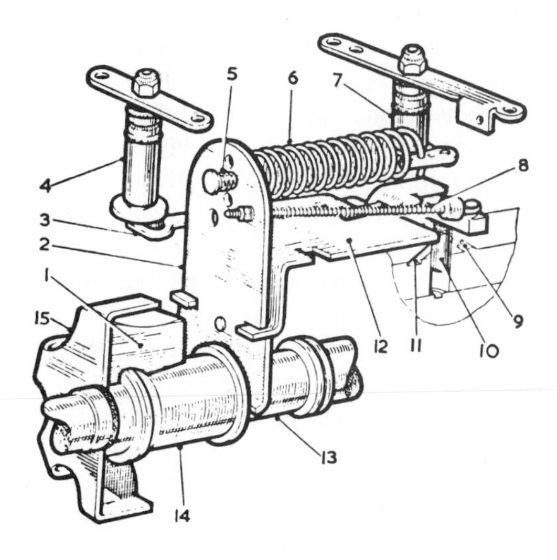

Figure 6-42
Typical Linkage Hookup of a Mechanical Governor.
(Courtesy of Lucas CAV Ltd.)

Any such movement would be transmitted by the governor arm (2) and the spring-loaded linkage hook (8) to cause rotation of the fuel metering valve (10). The governor arm (2) pivots around a fulcrum on the control bracket (12) and is held in contact with the thrust sleeve (14) by spring tension. The governor arm (2) and throttle arm (7) and shaft assembly are connected through the governor spring (6) and the idling spring (5) and its guide.

A shutoff bar (3) connected to an external shutdown control rotates the fuel metering valve to close off the metering port when it is desired to stop the engine.

Both throttle and governor action directly affect the rotation of the fuel metering valve and therefore the speed of the engine. Movement of the throttle arm (7) toward a fuel increase position causes the light idling spring (5) to be compressed as its guide is drawn through the governor arm (2). As the engine accelerates, the centrifugal force developed by the rotating flyweights will cause the thrust sleeve to move the governor arm (2) and therefore the metering valve (10) to a decreased fuel position.

However, as with all mechanical governors, within the governed range, for any fixed throttle position, the centrifugal force of the flyweights will be balanced by the force of the governor spring. Anytime that this occurs, the engine will run at a steady speed. This state of balance can occur at any speed range as long as the rpm of the engine is capable of carrying the load placed upon it.

This state of balance can be upset either by the operator or driver moving the throttle or by a change in load applied or removed from the engine. Therefore, an increase in load for a fixed throttle position will cause a reduction in speed, lowering the centrifugal force of the weights and upsetting the state of balance in favor of the spring, causing it to increase the fuel by rotation of the fuel metering valve. As the engine speed climbs again and a state of balance is reached, the engine will again run at a steady rpm. If the load were decreased, the rpm would increase, upsetting the state of balance in favor of the weights and causing a fuel decrease and engine rpm loss, until once again a corrected state of balance existed.

With the throttle at the idle position, governor action is controlled by the light idle spring only.

Mechanical Governor DPA Pump Linkage Adjustment

With the governor cover removed, set the link length as shown with the use of a vernier caliper. This dimension is listed on the correct test data sheet available from CAV or a local dealer, and is taken between the larger base diameters of the governor control cover stud (1) and the metering valve lever pin (2). See Figure 6-43. To adjust the linkage length, back off the lock nut (3), and turn nut (4). Hold light finger pressure against the control arm (6) to hold the metering valve in the fully open position when setting the desired length.

When installing the governor control spring to the governor control arm and throttle shaft link, ensure that it is inserted into the proper hole as shown above. This information is also given on the pump test data sheet.

Hydraulic Governor

The DPA pump fitted with a hydraulic governor operates on the same principle as the pump with a mechanical governor. However, with the hydraulic governor, there are no governor flyweights, elimination of the pump drive hub, and use of a sliding piston-type metering valve in place of the slotted semirotary type. The hydraulic governor is contained within a smaller housing than that of the mechanical, but it is also located on top of the pump housing. Since the basic injection pump components remain the same, we will only concern

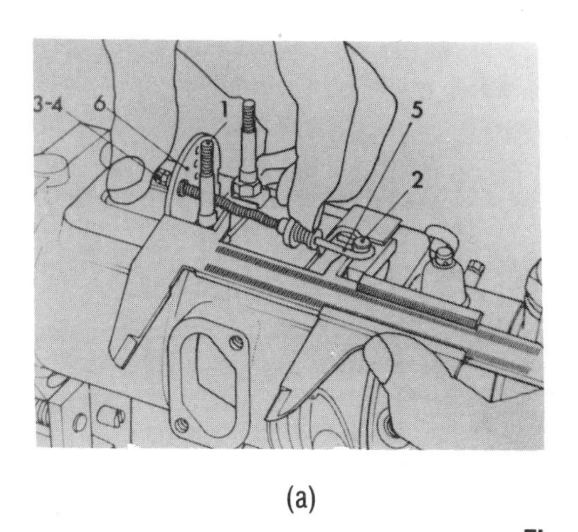
3-4
6
1
5
2

(a)

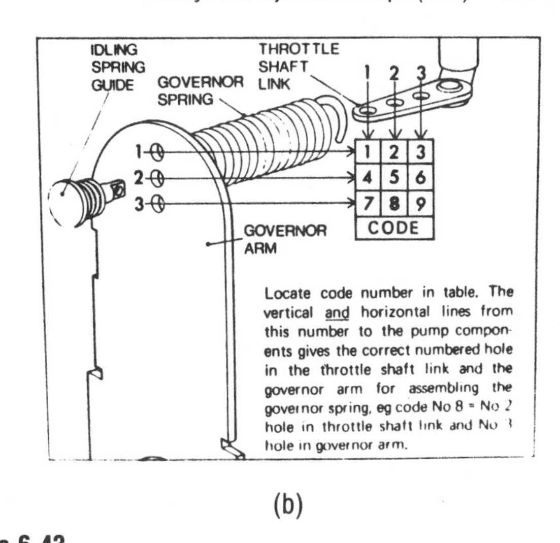
IDLING SPRING GUIDE
GOVERNOR SPRING
THROTTLE SHAFT LINK
1 2 3
1
2
3
1 2 3
4 5 6
7 8 9
CODE
GOVERNOR ARM
Locate code number in table. The vertical and horizontal lines from this number to the pump components gives the correct numbered hole in the throttle shaft link and the governor arm for assembling the governor spring, eg code No 8 = No 2 hole in throttle shaft link and No 3 hole in governor arm.

(b)

Figure 6-43
(a) Setting Governor Linkage; (b) Application of Setting Code.
(Courtesy of Lucas CAV Ltd.)

ourselves with the hydraulic governor mechanism itself.

Figure 6-44 shows the relative stackup of the governor components. Figure 6-45 shows an enlargement of the metering valve assembly, which along with Figure 6-44 will assist in explanation of the governor action.

The throttle lever (17 in Figure 6-44) connected to the metering valve pinion (2) engages the metering rack (6 in Figure 6-45) and therefore controls the movement of the metering valve assembly, which is free to move within a chamber of the hydraulic head, into which opens the diagonally drilled metering port. The damping washer (4) and floating washer (3) shown in Figure 6-45 act as a dashpot to dampen out any rapid movement of the metering valve either through throttle or governor action, thereby preventing the possibility of hunting or surging and poor stability.

The governor is operated by fuel at vane transfer pump pressure fed from the annular groove surrounding the pump rotor. The fuel then passes through the hollow metering valve via transverse holes to an annular space around the valve. Movement of the metering valve consequently varies the area of the metering port, which registers with the annulus around the valve. The port's effective area is the portion uncovered by the inner edge of the groove or annulus. The position of the throttle arm will vary the compressive force of the metering valve spring (5 in Figure 6-45).

Speed Increase. When the throttle is moved to an increased fuel position, the metering valve will be forced toward the maximum delivery position by the action of the spring (5). This allows the engine speed to increase; in so doing the vane-type transfer pump pressure also increases, thereby forcing the metering valve back against the spring pressure until a balance is reached, which can occur at any engine speed as long as the rpm is capable of carrying the load placed upon the engine.

Speed Decrease. Movement of the throttle toward the idling position allows the idling spring (7 in Figure 6-45) to be compressed. A state of balance will exist when the forces exerted by the idling spring and fuel pump transfer pressure equal that of the metering

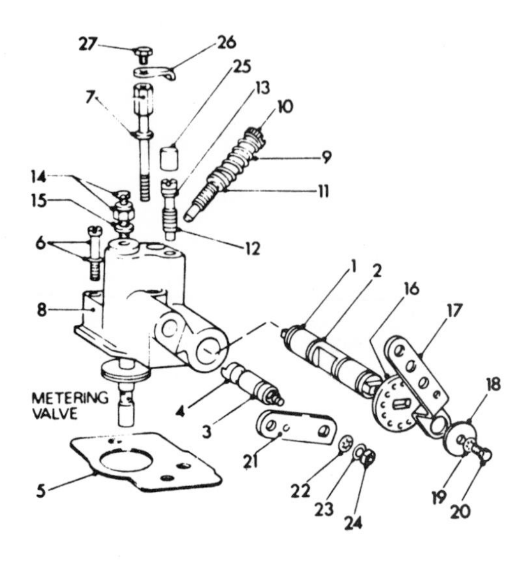

Figure 6-44
Stackup of Hydraulic Governor Components.
(Courtesy of Lucas CAV Ltd.)

Index:

1—0 ring seal
2—Metering valve pinion
3—0 seals
4—Shutoff shaft
5—Cover gasket
6—Short screw
7—Long screw
8—Governor housing
9—Idling spring
10—Idle adjusting screw
11—0 seal
12—0 seal
13—Maximum speed adjusting screw
14—Vent screw
15—Washer
16—Locking plate
17—Throttle arm
18—Washer
19—Lock washer
20—Lock screw
21—Shutoff lever
22—Washer
23—Plain washer
24—Nut
25—Rubber plug
26—Locking plate
27—Screw

valve spring (5). In a deceleration situation, the compression of the metering valve spring will become proportionately lower as the throttle approaches the idling stop screw, thus allowing the reducing transfer pump fuel pressure at low speeds to operate the metering valve and therefore perform the governing function throughout the idling range.

Shut-down Device. The shutoff shaft (4 in Figure 6-44) has a half-round end on it that contacts the underside of the shutoff washer (8 in Figure 6-45). When the shutoff lever (21 in Figure 6-44) is rotated, it lifts the metering valve to a position where the metering port is blanked off, thereby stopping the engine.

Reversible Hydraulic Governor

The reversible-type governor is identical in operation to the standard unit, with the only exception being that movement of the metering valve is obtained by using an eccentric formed onto the end of the throttle control shaft. Figures 6-46 and 6-47 show the basic component part layout.

Rotation of the throttle shaft (11) causes the eccentric to move the control sleeve of the metering valve against the spring (6), thus loading the metering valve. The shutdown is the same as for the other hydraulic governor.

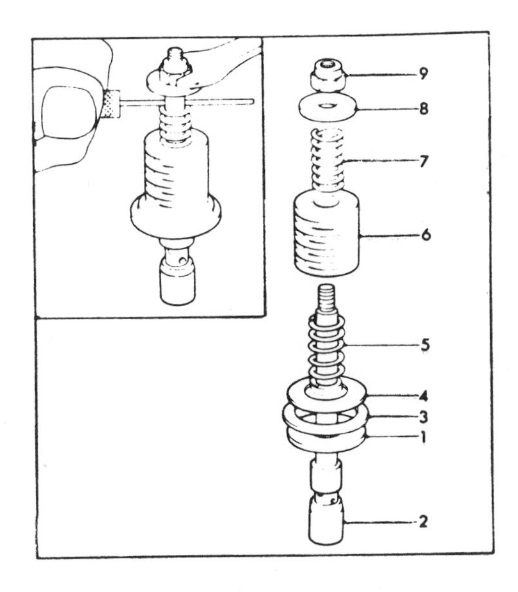

Figure 6-45
Enlargement of Metering Valve Assembly.
(Courtesy of Lucas CAV Ltd.)

Index:
1—Bottom seating washer
2—Metering valve stem
3—Floating washer
4—Top damper washer
5—Metering valve spring
6—Metering valve rack
7—Idling spring
8—Shutoff washer
9—Nut

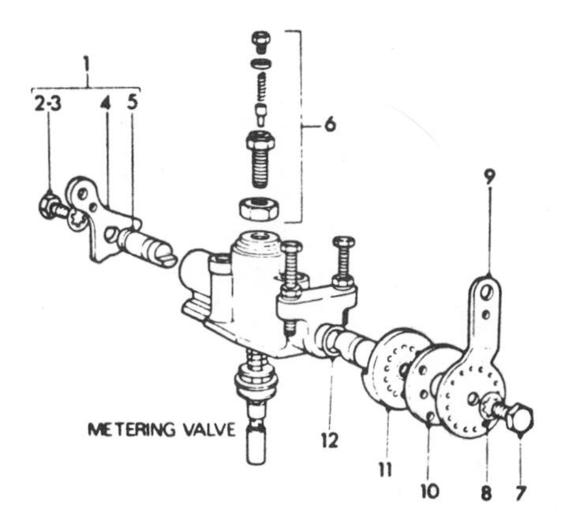

Figure 6-46
Component Parts Layout of Reversible Hydraulic Governor.
(Courtesy of Lucas CAV Ltd.)

Index:
1—Shutoff shaft assembly
2—Screw
3—Washer
4—Shutoff lever
5—Shutoff shaft and seal
6—Antistall and idling stop assembly
7—Lock screw
8—Washer (shakeproof)
9—Throttle lever
10—Vernier plate
11—Throttle shaft and plate assembly
12—O ring seal

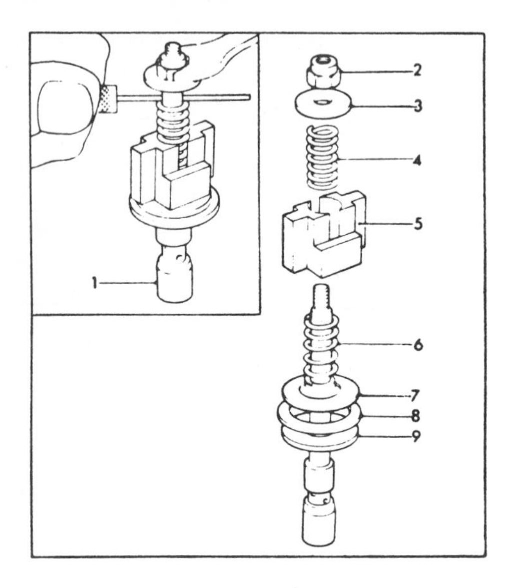

Figure 6-47
Metering Valve.
(Courtesy of Lucas CAV Ltd.)

Index:
1—Metering valve
2—Metering valve nut
3—Shutoff washer
4—Idling spring
5—Guided control sleeve
6—Metering valve spring
7—Top damper washer
8—Center floating washer
9—Bottom sealing washer

Perkins Diesel Engines

Perkins engines presently use either the DPA, DP15, or Simms/CAV fuel-injection pumps on their range of diesel engines. Older-model engines used in-line CAV pumps; however, on current in-line engines, the DPA style of rotary distributor pump is used, mounted either horizontally or vertically and driven from the front of the engine.

The V8-510 and 540 engines use the in-line Minimec pump along with the newer V8-640. A detailed description and operation of the DPA and Minimec pumps were given earlier in this chapter.

6-8

DPA Pump Installation

Top dead center (TDC) can be determined on these engines by either (1) looking through the inspection hole provided on the engine flywheel housing at the scribed marks located on the flywheel, or (2) by removing the front crankshaft nut washer and ensuring that the V groove on the front face of the crankshaft is at the top, and in alignment with the dot on the pulley (see Figure 6-48).

To check or reset the fuel-injection pump timing, a timing pin located in the lower half of the timing case can be unscrewed until it locates with the drilled hole in the rear face of the crankshaft pulley.

Checking DPA Type Pump Timing

1. Remove the side inspection cover from the pump housing shown under the fuel section of this book, which will expose the pump timing marks.
2. Turn the engine over manually until the scribed line in the pump inspection window marked H on the pump rotor coincides with the scribed line of the snap ring. (This letter can be different on other model engines.) See Figure 6-49.
3. Check that the flywheel timing mark is aligned at 29 degrees BTDC. This can also be checked by measuring that the distance from the TDC mark on the flywheel is 4.02 in. (102.108 mm) or the piston depth from TDC is 0.399 in. (10.134 mm).

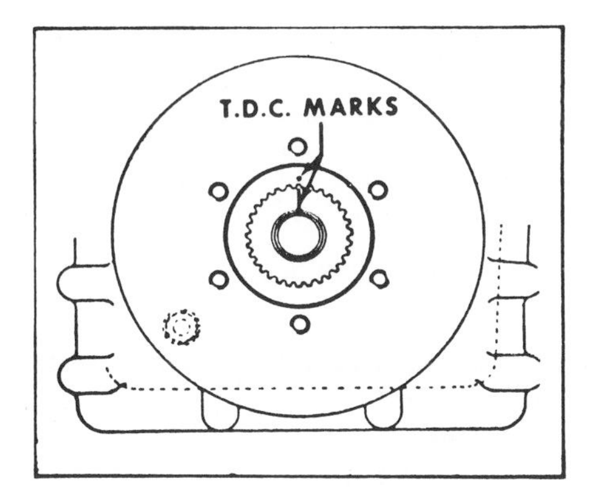

Figure 6-48
Determining TDC.
(Courtesy of Perkins Engines)

4. If the timing is incorrect, loosen the pump slotted flange nuts; manually grasp the pump housing, and turn the pump in the desired direction to change the timing.

Mechanical Knock Compared to Fuel Knock

Sometimes it is difficult to isolate a particular knock from the engine; to isolate it, run the engine at its maximum rpm and pull the stop control lever on the injection pump. If the knock disappears, it is due to a fuel problem; however, if it is still heard, it is a mechanical fault. When the fuel is cut off, the noise will fade in intensity, but will still be present.

Injection Pump Installation

A machined slot located in the hub of the worm gear on the injection pump drive shaft mates with a slot approximately $\frac{1}{8}$ in. wide (3.175 mm) in the adapter plate. With these slots aligned and piston 1 at TDC on the compression stroke, install the pump with the scribed line on its mounting flange in line with the scribed line on the pump adapter plate.

If, however, the machined slots are not in alignment, proceed as follows:

1. Place piston 1 at TDC on compression.
2. Remove the small auxiliary drive gear cover from the front cover timing case.
3. Remove the auxiliary drive gear by taking out the three retaining bolts.
4. Turn the auxiliary drive shaft until the machined slots are in alignment; then reinstall the auxiliary drive gear and bolt it into position, making sure that the drive shaft does not move during this operation.
5. Install the auxiliary drive gear cover back onto the timing case.

Engine Firing Orders

1. All 3-cylinder cw-rotation engines: 1–2–3.
2. All 4-cylinder cw-rotation engines: 1–3–4–2.
3. All 6-cylinder cw-rotation engines: 1–5–3–6–2–4.

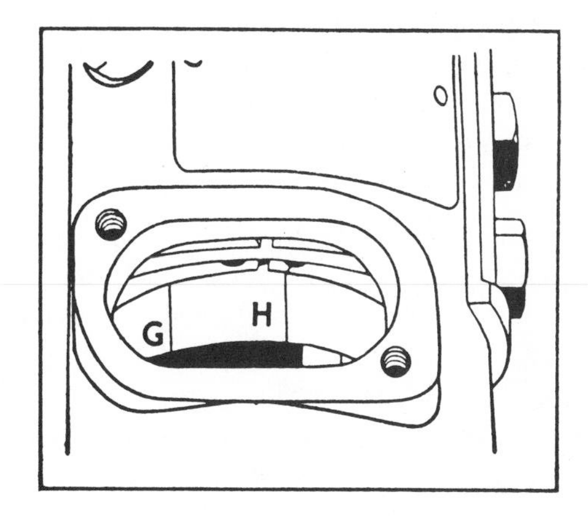

Figure 6-49
Checking DPA Pump Timing.
(Courtesy of Perkins Engines)

4. All V8 cw-rotation engines: 1–8–7–5–4–3–6–2.

DPA Injection Pump Installation: Ford 6-363 Engine

1. With the gear installed to the pump assembly, rotate the engine manually to align the flywheel timing mark with the V notch of the flywheel housing, which is 9 degrees BTDC.
2. Check that both valves on cylinder 1 are loose; otherwise, you are probably one turn out. Also ensure that the timing mark on the rear face of the camshaft gear is visible through the injection pump mounting flange opening on the engine.
3. Fit the pump to the engine flange after installing a new O seal ring on the mounting plate.
4. At the top center (front) of the injection pump mounting flange is located a timing locking bolt. If you are not sure if the pump is locked in the timed position, remove the side inspection cover on the pump body, and assemble the pump to the engine so that the timing marks between the pump cam ring and drive plate are in alignment. See Section 6-7 earlier in this chapter if in doubt about this aspect.
5. Secure the pump to the engine by tightening the bolts of the flange.
6. Remove the timing lock bolt from the top front center of the pump flange and its bracket. Install a sealing bolt with a copper washer to prevent fuel leakage when the engine is running.
7. Connect all fuel lines and linkage, bleed the pump, and start the engine.

No timing adjustment to this pump is necessary, since the preset time marks are machined on both the internal drive plate and cam ring of the pump. To ensure that the injection pump is timed properly prior to installation on the engine, proceed as follows:

1. Remove the side inspection cover from the pump body (see Section 6-7).
2. Turn the injection pump drive shaft around until the timing marks are aligned.
3. Remove the normally installed sealing bolt and copper washer from the pump mounting flange (top front center), and install the DPA timing lock bolt in its place torqued to 25 in.-lb. (2.82 N·m), which will lock the pump in its timed position.
4. If desired, a bracket can be installed under the timing bolt, which prevents pulling up the mounting bolts of the pump flange when installed; this ensures that the locking timing bolt attached to the pump is removed prior to final securing of the injection pump to the engine. Otherwise, if an attempt were made to start the engine with the timing pin locked in position, pump damage would result.
5. The system can be bled by opening the plugged connection on the filter head and operating the pump priming lever until fuel free of bubbles is apparent. Then close this screw and open the injection pump inlet connection until air-free fuel is present. Finally, open the bleed screw located on the governor housing closest to the engine block until fuel flows freely. You may also have to loosen the fuel line nut at the injectors until fuel flows from them in order to fully expel all air from the engine for satisfactory starting.

Injection Pump Overhaul

As with all fuel-injection pumps, they are precision manufactured equipment; therefore, unless you have the knowledge and expertise along with the necessary special tools and test equipment, *do not attempt* to overhaul or tinker with the pump. Send it to your local fuel-injection dealer.

The test procedure for any CAV injection pump can be found in specific test literature relating to the particular model of pump, along with the necessary test readings. The following procedure therefore is of a general nature, and gives an idea of the kinds of tests that are required during testing and overhauling of a pump. It is not meant to replace the manufacturer's service information, which is the best available.

These tests must be carried out under spotlessly clean surroundings, preferably in a fuel test lab, to prevent the possibility of problems associated with the entrance of dust or dirt.

6-9

Pump Testing

Some of the major factors to be considered prior to testing a pump are as follows:

1. During testing, only recommended fuel test oils should be used.
2. Carry out all tests in the sequence given by the manufacturer.
3. The pump *must* only be rotated in its normal direction of rotation; otherwise, severe damage can result.
4. Do not run the pump for long periods without having the fuel inlet valve to the pump open.
5. Do not at any time run the pump with either of the advance piston plugs loose or removed.
6. Oil leakage while testing the pump at the various adjusting screws is a normal occurrence. When finished, these will be corrected by the insertion of sealing caps.
7. Ensure that the fuel *test oil* temperature does not exceed that recommended by the manufacturer.

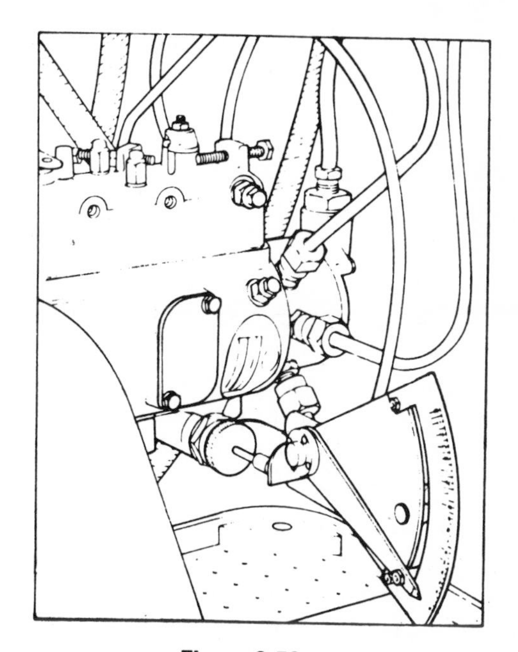

Figure 6-50
Mechanical Governor with Cam Advance Checking Tool Installed.
(Courtesy of Lucas CAV Ltd.)

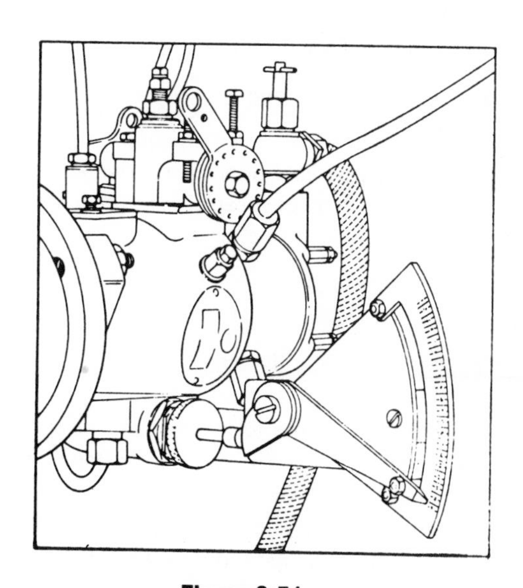

Figure 6-51
Hydraulic Governor with Cam Advance Checking Tool Installed.
(Courtesy of Lucas CAV Ltd.)

8. Do not run a cold pump on the test machine. Submerge the pump assembly in test oil for 20 minutes prior to mounting on the machine.
9. Ensure that the pump is mounted correctly and square to the drive, as well as being secure.
10. Make sure that you are familiar with the test machine controls before undertaking any tests. Do you know where the emergency control switch is in the electrical panel feeding the machine?
11. Make sure that you do not have any loose clothing on while testing the pump, especially around the drive shaft area.
12. Attach the pump specifications and test procedure to a suitable place on the machine so that you can read it during testing.
13. Do you have all the necessary tools and fittings handy?
14. Do you have a set of ear protectors or head phones available?
15. Do you have on a set of safety glasses?
16. Do you have enough wiping rags (lint free) available?
17. Is the immediate floor area around the test machine free of obstructions and spilled test oil?

Testing of Advance Devices

The checking of the DPA injection pump advance devices is similar on pumps with mechanical or hydraulic governors. Figures 6-50 and 6-51 show each pump type with the cam advance checking tool installed.

The special advance tool P/N 7244-50 (hydraulic governor) or P/N 7244-59 (mechanical governor) is inserted into the hole in the spring cap, after first unscrewing the small screw normally located in the spring cap. A feeler pin tool P/N 7244-70 used with both types of governors is inserted into the hole in the spring cap. The scale on the timing gauge generally covers 0 to 18 degrees. Zero the gauge by moving the scale relative to the pointer.

The pump should be reprimed with fuel after fitting the advance tool; then with the pump running on the test stand at 100 rpm, operate the throttle and press in and release the advance gauge pin several times. Reference to the test data sheet for the particular pump on test will indicate the advance at a given rpm.

The speed advance device on both the mechanical and hydraulic governor types is altered by changing the shims behind the piston spring and cap. On those units employing the combined load and speed advance, alter the shims beneath the inner and outer piston springs.

6-10

Fuel Transfer (Lift)

The common type of fuel lift pump used with CAV in-line and distributor-type injection pumps is a diaphragm-type unit manufactured by AC. It has a delivery pressure of between 2.75 and 4.25 psi (18.96 and 29.3 kPa). Such a pump is shown in Figure 6-52. This pump can be used to prime the fuel system by actuating the primer (18) on the side of the pump. The pump is normally engine driven from the camshaft. However, on in-line injection pumps, it can be mounted and driven from the injection pump camshaft.

6-11

Bryce–Berger Fuel Systems

Bryce–Berger is part of the British Lucas Company and has been involved in the manufacture of fuel-injection equipment for over

40 years. It presently produces jerk pumps in single-cylinder housing design with one exception, which is a two-cylinder unit. These pumps start with a plunger diameter of 8 mm and progressively increase to their largest unit with a plunger diameter of 50 mm (1.96 in.). To give you an idea of size, the smallest pump weighs in at 1.0 kg (2.2 lb), with a nominal stroke of 7 mm (0.275 in.) and a possible fuel delivery rate of 12 mm³/cam degree, whereas the largest pump weighs 218 kg (480 lb), with a nominal stroke of 50 mm and a fuel delivery rate of 3000 mm³/cam degree. These jerk pumps operate on the conventional constant stroke variable spill principle and are designed to meet the requirements of diesel engines with a power range of 1.5 to 2500 hp/cylinder (1.1 to 1865 kW).

Figure 6-53 shows one example of a model FFOAR jerk pump, which is the largest unit they make. Figure 6-54 shows a model FAOAN pump, which is one of their smallest, with a nominal stroke of 7.5 mm, plunger diameter of 9 mm, weight of 0.5 kg (1.1 lb), and a delivery capability of 16 mm³/cam degree.

These jerk pumps are mounted on the engine and require a tappet assembly, which is generally supplied by the engine manufacturer and driven from the engine camshaft, which has a fuel cam for each pump unit. The fuel control rack of several pumps can be interconnected and controlled by engine governor action. Installation and timing of each pump is similar to that of American Bosch APF pumps; therefore, refer to Chapter 2 for general information when installing.

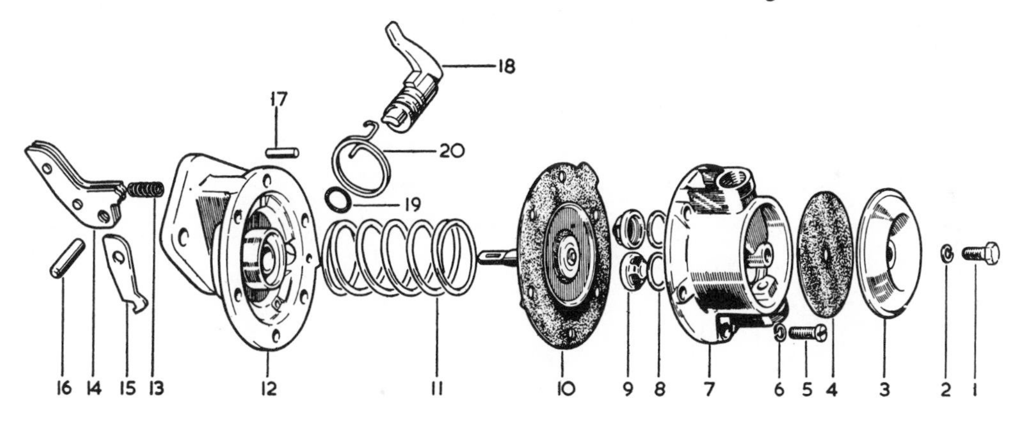

Figure 6-52
AC Fuel Lift Pump, Series VP.
(Courtesy of Lucas CAV Ltd.)

Index:

1—Pulsator cover screw
2—Pulsator cover washer
3—Pulsator cover
4—Pulsator diaphragm
5—Cover screw
6—Cover screw washer
7—Cover
8—Valve gasket
9—Valve assembly
10—Diaphragm assembly
11—Diaphragm spring
12—Body
13—Rocker arm spring
14—Rocker arm
15—Link
16—Rocker arm pin
17—Primer pin
18—Primer
19—Sealing ring
20—Primer spring

TYPICAL DPA PUMP TEST DATA SHEET

Pump type   3240530-3240539
Engine make   International Harvester BD154
Application   Tractor
(For test rig specification and general data, see explanatory notes)

Basic Pump Specification
Mechanical governor
Fixed phasing
Rotation (looking on drive end), clockwise
Governor link length 52.3 mm nominal ±1.0 mm
Governor control spring: no. 2 hole control arm and no. 1 hole throttle lever link
Roller-to-roller dimension: 50.25 mm

Test Procedure

Test No.	*Description*	*RPM (max.)*	*Requirements*
1	Priming	100	Fuel delivery at all injectors
2	Transfer pump vacuum	100	Note time to reach 16 in. (406 mm) Hg; max. time allowed is 60 seconds
3	Transfer pressure	100	18 lb/in.^{2} (1.2 kg/cm^2) min
4	Transfer pressure	900	41 to 64 lb/in.^{2} (2.9 to 4.5 kg/cm^2)
5	Back leakage	900	3 to 30 cm^3 for 100 shot time cycle
6	Max. fuel delivery	900[2]	Set to 9.4 ± 0.1 cm^3; spread between lines not to exceed 0.8 cm^3
7	Max. fuel delivery check	100[2]	Average delivery to be not less than average at test 6 minus 1.5 cm^3
8	Cutoff operation: shutoff lever closed (throttle open)	200	Average delivery not to exceed 1.0 cm^3
9	Throttle operation: throttle lever closed	200	Average delivery not to exceed 1.0 cm^3
10	Fuel delivery check	900	Record average delivery
11	Governor setting	1130	Set throttle by maximum speed adjustment screw to give maximum average delivery of 1.5 cm^3; no line to exceed 2.5 cm^3; lock stop screw
12	Fuel delivery check	900	With throttle set as at test 11, average delivery to be not less than average at test 10 minus 0.4 cm^3
13	Timing: Use outlet X (30 atm pressure); with indexing tool set to 22°, scribe line on housing flange		

[1]Check pump output at 200 strokes.

[2]Use 30 seconds glass draining time and allow fuel to settle for 15 seconds before taking reading.

Timing note: This pump must be timed using an indexing tool provided with a dowel that locates in a slot in the pump drive hub. A quill shaft must *not* be used when timing this pump owing to the possibility of timing error.

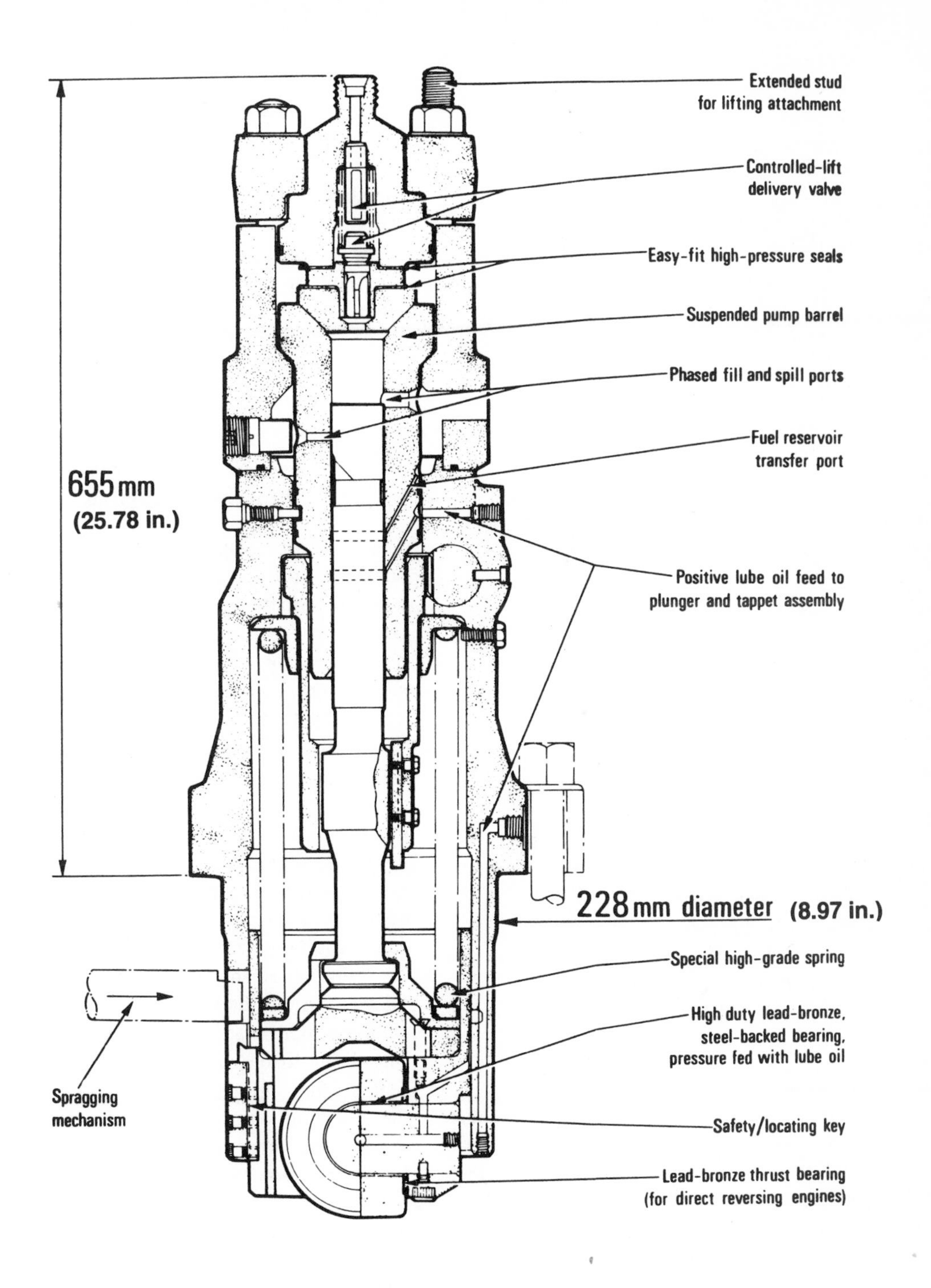

Figure 6-53
Bryce FFOAR Fuel-Injection Pump.
(Courtesy of Lucas CAV Ltd.)

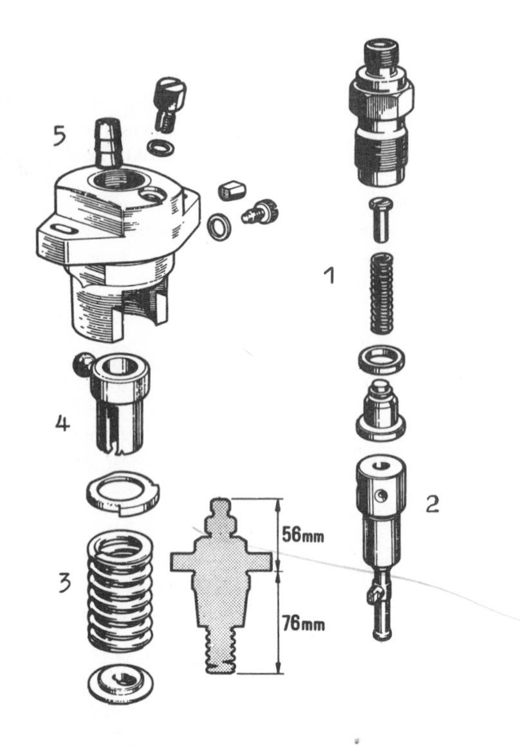

Figure 6-54
Bryce A-Size High-Flange FAOAN Pump. (Courtesy of Bryce-Berger, Lucas CAV Ltd.)

Index:
1—Delivery valve assembly
2—Plunger and barrel
3—Tappet spring assembly
4—Plunger control collar
5—Delivery valve body

Bryce pumps are offered with either *bucket* or *roller* tappet form, and employ separate lubrication to the pump plunger and other working parts, eliminating fuel leakage and ensuring smooth operation.

At the present time more than 50 diesel engine builders use Bryce–Berger fuel pumps and injectors as original equipment, from small engines to large stationary and marine applications. Service and parts dealers can be found in more than 130 countries of the world; therefore, they are well represented.

Injection Nozzles

Figure 6-55 shows the line of nozzles manufactured by Bryce–Berger. These nozzles range from a nominal shank diameter of 21.5 mm up to 65.0 mm, with many available as liquid-cooled units.

6-12

Lucas/CAV Injection Nozzles

The operation, design, and function of the fuel nozzles used by CAV are very similar to those used by American Bosch and Robert Bosch companies described in this book The servicing procedures for all these nozzles follow the same pattern, as do general maintenance and troubleshooting.

CAV and Simms injectors are basically of two main types, the multihole and the pintle. The type of spray is hard and of high penetrating power on the former, which is used on direct injection engines, whereas the pintle nozzle has a comparatively soft spray pattern of lower penetration and is found on pre-combustion-type engines.

Figure 6-56 shows a typical cross-sectional view of a multihole nozzle with the parts shown in both assembled and exploded views. The spring-loaded needle valve is forced off its seat by the high-pressure fuel delivered from the injection pump. This fuel, when forced through the spray holes, atomizes as it enters the combustion chamber. The fuel pressure required to lift the needle valve from its seat will vary between types of engines and injectors; however, it usually ranges from 100 to 200 atm (atmospheres) or 1470 to 2940 psi.

Spring pressure returns the valve to its seat upon completion of the injection cycle. A small amount of fuel will leak past the needle

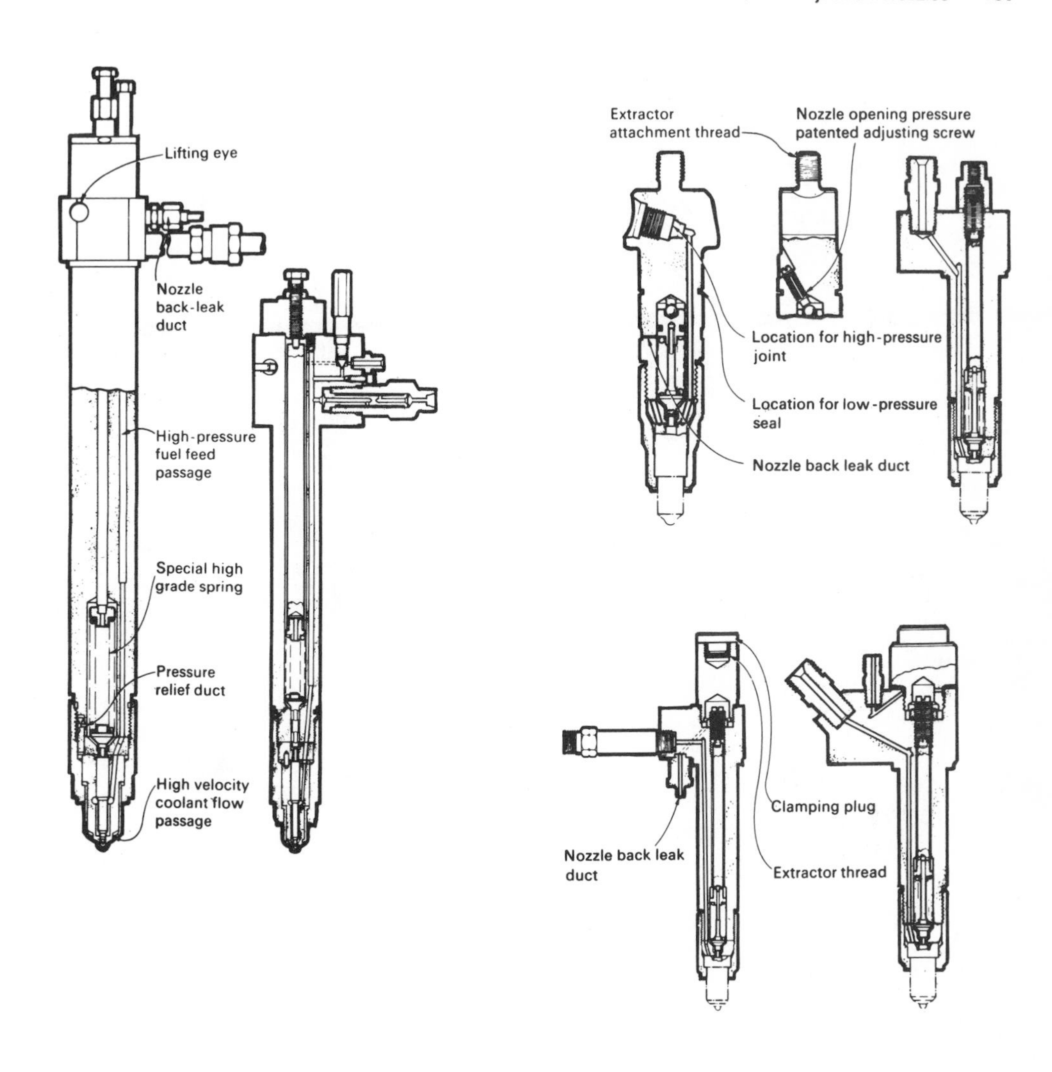

Figure 6-55
Bryce-Berger Nozzles.
(Courtesy of Bryce-Berger, Lucas CAV Ltd.)

valve stem for lubrication purposes and return via the leak-off connection on the nozzle back to the fuel tank.

The two basic types of multihole nozzles are (1) NL (long reach), and (2) NH (short reach). They are shown in Figure 6-57(a) and (b), respectively.

Adjustment of the nozzle release pressure is accomplished by the spring cap nut (6 in Figure 6-56), which varies the tension of the spring, thus holding the needle valve seated. Others, as shown in Figures 6-58 and 6-59, are adjusted by a spring-loading shim or an adjusting screw and locknut.

The lapped face of the nozzle contains an annular groove with either one or as many as

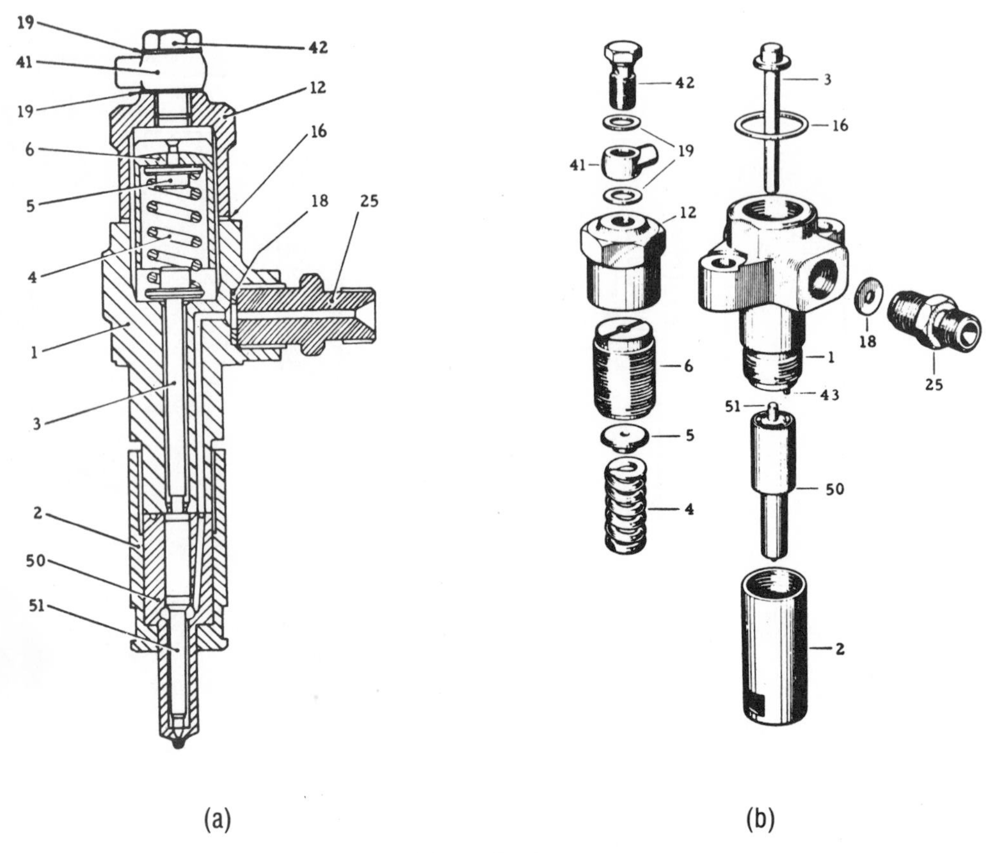

Figure 6-56
Cross-Sectional View of a Multihole Nozzle.
(Courtesy of Lucas CAV Ltd.)

Index:

1—Nozzle holder
2—Nozzle nut
3—Spindle
4—Spring
5—Upper spring plate
6—Spring cap nut
12—Cap nut
16,18,19—Joint washer
25—Inlet adaptor
41—Leak-off connection
42—Banjo screw
50—Nozzle
51—Needle valve

three supply holes leading down into the fuel gallery (Figure 6-57). There can be two, three, or four equally spaced holes in the tip leading from what is commonly referred to as the *sac*. Some nozzles are located to the holder by two dowels, while others have none. The needle valve and seat are a matched set and should never be intermixed with other parts. The lift of the needle can vary from 0.10 to 0.50 mm according to type.

Figure 6-58 is a cross-sectional view of an injector with a spring adjusting shim. Figure

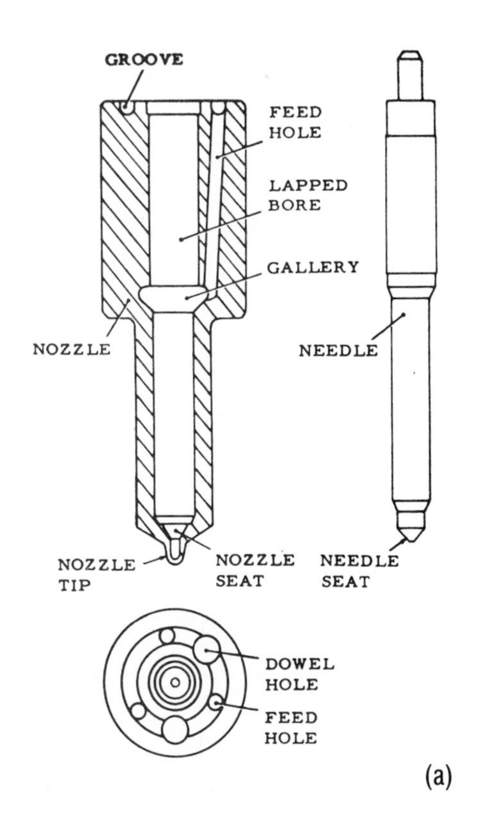

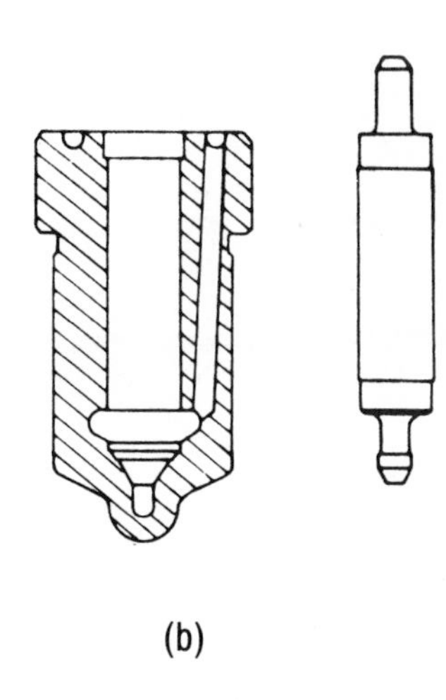

Figure 6-57
Basic Types of Multihole Nozzles: (a) NL (long reach); (b) NH (short reach). (Courtesy of Lucas CAV Ltd.)

6-59 is an injector with a spring adjusting screw.

Low Spring Injector

A recent addition to the Lucas CAV line of injectors is the model LR type (low spring range) designed for high rated output direct injection engines up to 100 bhp/cylinder (75 kW/cylinder). It is used along with the Maximec pump, for example, on both turbocharged and nonturbocharged engines, and the low spring injector has also been assigned for use in the VW Golf diesel car.

This injector is a further development of the well-established S injector, but the LR is capable of withstanding peak cylinder pressures in the region of 2000 psi before gas blow back, compared to 1500 psi for the earlier S-type injector.

The injector opening or popping pressure is controlled by shim adjustment of the spring. The LR barrel diameter is 21 mm, and the high-pressure inlet can be either the axial or right-angled version. In addition, the injector can be held in the cylinder head by either a clamp or gland nut arrangement with a radial locating peg.

The low spring range injectors implies that the spring is positioned virtually directly on top of the nozzle valve. This arrangement reduces the inertia of the moving parts, compared to that in the model S injector. Thus, faster valve closure is achieved at the end of the injection period.

This feature is beneficial in some applications to improve such factors as the exhaust emission characteristics. These low spring units are available with long stem hole-type and pintle-type nozzles.

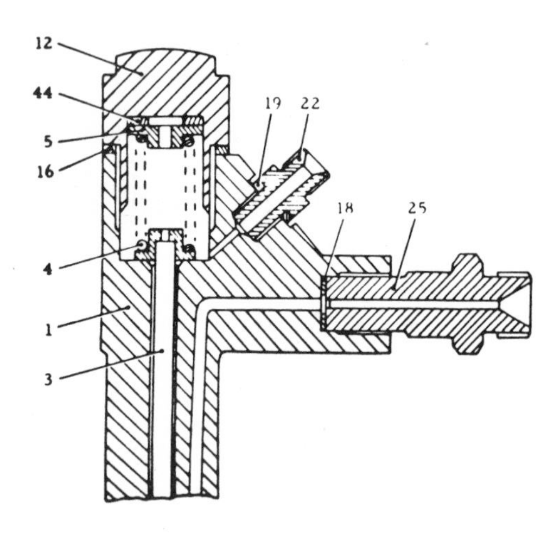

Figure 6-58
Injector with Spring Adjusting Shim.
(Courtesy of Lucas CAV Ltd.)

Index:
1—Nozzle holder
2—Nozzle nut
3—Spindle
4—Spring
5—Upper spring plate
6—Spring cap nut
8—Spring adjusting screw
9—Locknut
12—Cap nut
16 to 19—Joint washer
22—Leak-off adapter
25—Inlet adapter
27—Filter
28—Nipple
44—Spring adjusting shim
50—Nozzle
51—Needle

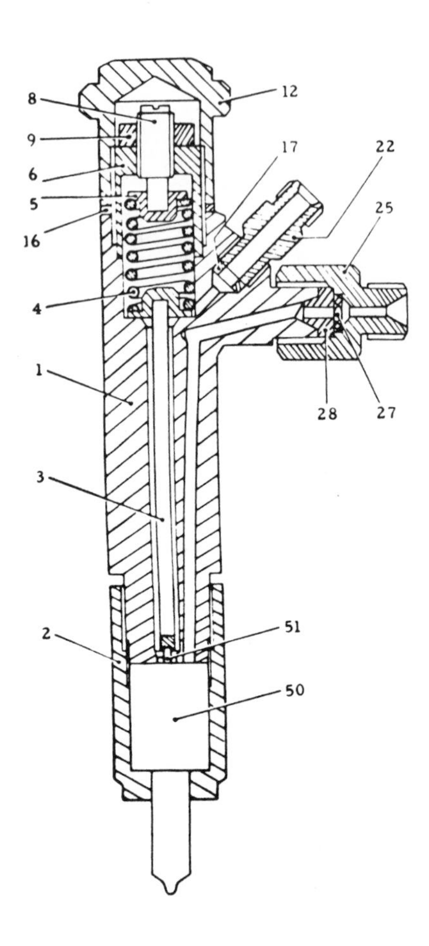

Figure 6-59
Injector with Spring Adjusting Screw.
(Courtesy of Lucas CAV Ltd.)

Nozzle Servicing

General nozzle servicing procedures for CAV injectors are very similar to those practiced for Robert Bosch and American Bosch units; therefore, the following is not intended to be all-encompassing, but will serve to give you an appreciation of the sequence involved when dealing with nozzle repair.

Figure 6-60 shows a typical nozzle cleaning kit that would be used for removing carbon from the exterior of the nozzle body and from the needle. Parts should be cleaned in benzene or one of the other widely used injector cleaning fluids readily available. Always use clean, dry filtered shop air to blow dry all parts and lint-free rags.

The pricker wires for the spray holes are available in a range of sizes to suit the particular tip; however, they are generally supplied in sizes of 0.25, 0.30, 0.35, and 0.40 mm diameter. It is suggested that you lightly stone or grind a 45 degree chamfer on the end of the pricker or cleaning wire, which facilitates cleaning of the tip.

Nozzle spray hole sizes are listed in CAV Bulletin G103; therefore, do not guess at the hole size. *Never* use a bench grinder wire wheel (brush) to clean the nozzle tip, since enlargement of the tip holes, plus scoring of the tip, can result.

Ultrasonic cleaning of the nozzle components will generally remove all traces of hard carbon deposits; however, if this system is unavailable, dissolve 2 ounces of caustic soda in 1 pint of water and add $\frac{1}{2}$ ounce of detergent. Measure carefully the amount of caustic soda, since more than a 15% concentration can corrode the bore and joint face on the nozzle, making them unserviceable. After cleaning, flush the parts in water; then immerse them in Shell Ensis 254 dewatering oil or equivalent. Blow dry with compressed air.

Reverse flushing of the nozzle can be carried out by connecting it to a *nozzle flushing holder* shown in Figure 6-61, then connecting it to a nozzle *pop testing* machine.

Examination of the nozzle and its associated parts should be done carefully with the use of a microscope and light. With no oil on the needle valve or nozzle bore, the needle should drop freely under its own weight into the seat area, and when turned upside down (carefully) should fall out. Examine all other related parts for any signs of damage or corrosion.

Injector part assembly should be done in spotless, dust-free surroundings. All parts should be assembled wet with clean fuel oil.

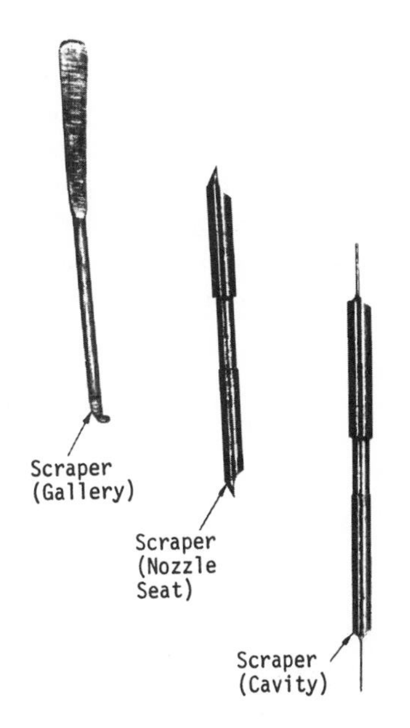

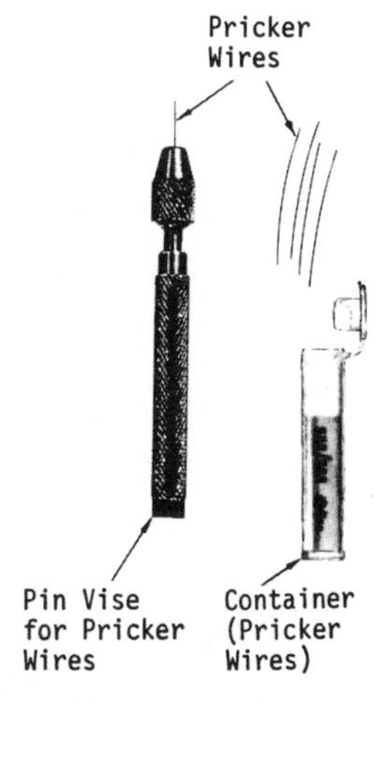

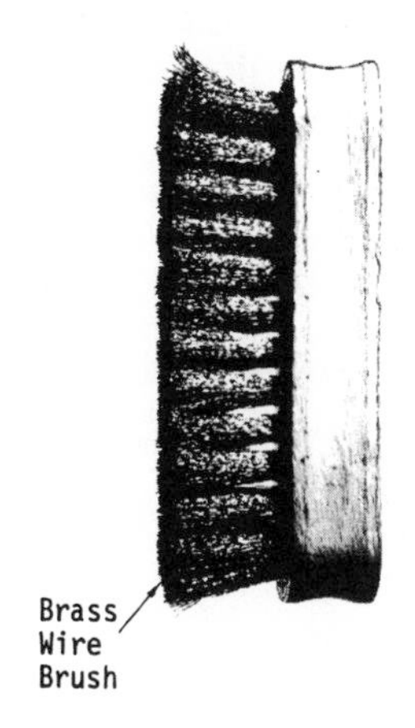

Figure 6-60
Typical Nozzle Cleaning Kit.
(Courtesy of Lucas CAV Ltd.)

Testing of the injectors should be done on a test pump at a rate of approximately 60 strokes/min. Check as follows:

Pressure Setting and Atomization

1. Nozzle opening pressures can be found in CAV Bulletin G103.
2. If Bulletin G103 is unavailable, contact your local fuel-injection dealer, who can give you the information you require.
3. If this cannot be done, the general rule of thumb is to set multihole nozzles to 150 atm or 2205 psi. Set pintle-type nozzles to 100 atm or 1470 psi.
4. If a new injector spring has been installed, set the release pressure to 5 ± 2.5 atm (73.5 ± 36.75 psi) higher than the specified setting, which will allow for initial spring seating.
5. The spray should be even and crisp from the tip when pop testing.

Back-Leakage Test

By timing the pressure drop on the test pump gauge, the fit of the needle valve to the nozzle can be established. General back-leakage criteria are as follows:

1. Below 160 atm: from 90 to 65 atm (1323 to 955.5 psi), should not be less than 10 seconds or more than 22 seconds.
2. At 160 atm and above: from 150 to 100 atm (2205 to 1470 psi), should not be less than 15 or more than 35 seconds.

These back-leakage tests are for typical new injectors. If the leakage test is less than 5 seconds, the nozzle and needle should be scrapped, but if not less than 5 seconds, it may be able to be reconditioned. Tight needle valves will obviously result in a high back-leakage time, whereas a badly worn one will have a low back-leakage time.

Seat Dryness Test

Pump the handle of the nozzle pop testing machine until the pressure is approximately 10 atm below the popping pressure, and check the tip area for signs of wetness. A very slight sweat is permissible; however, any wetness or dribble will cause incomplete combustion and exhaust smoke. Therefore, check for dirt between the needle and seat, scoring or pitting on the needle seat or nozzle, or internal corrosion.

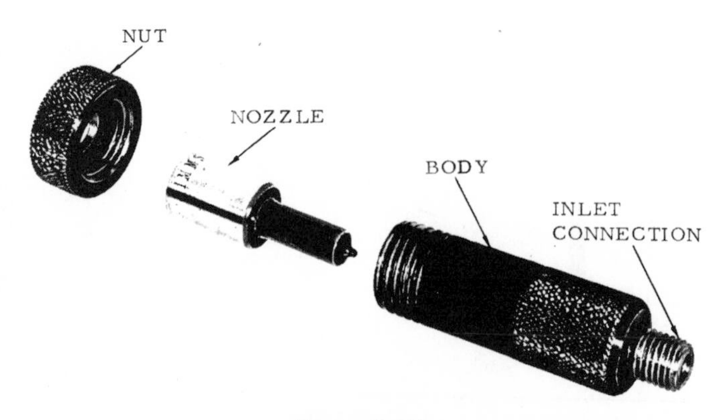

Figure 6-61
Nozzle Flushing Holder.
(Courtesy of Lucas CAV Ltd.)

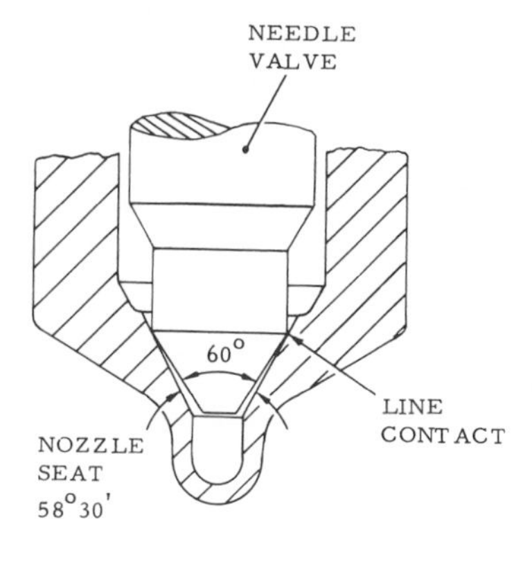

(a)

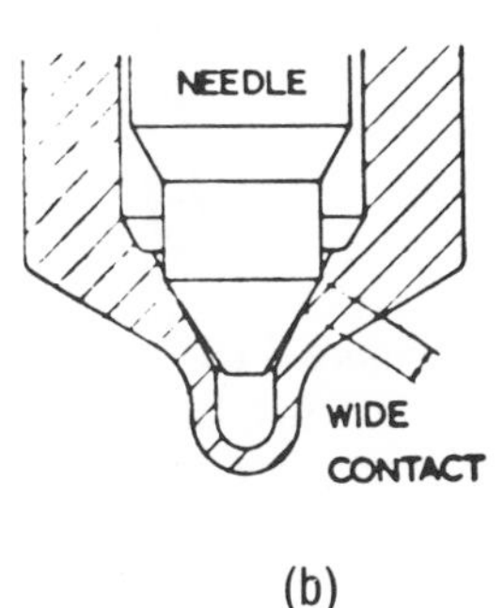

(b)

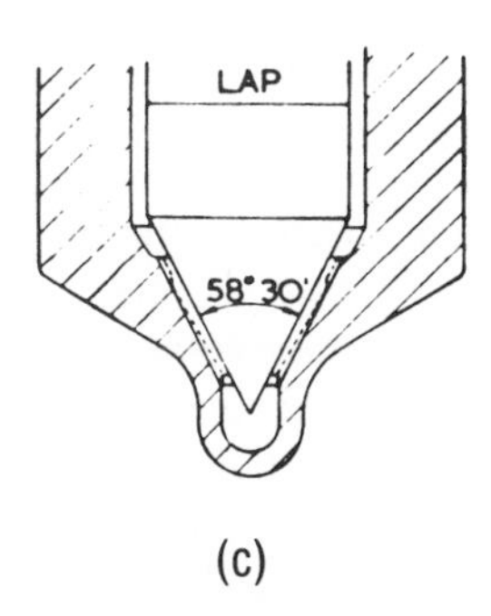

(c)

Figure 6-62
Angular Difference Between Nozzle Seat and Needle Valve.
(Courtesy of Lucas CAV Ltd.)

Nozzle Seat Angle

Figure 6-62(a) shows the angular difference between the nozzle seat and needle valve in multihole injectors. Figure 6-62(b) and (c) shows wide contact caused by wear and after lapping.

The angle of the nozzle seat is nominally 58 degrees 30 minutes; however, the angle of the needle seat is 60 degrees. This slight difference in angle provides a line contact between the needle and nozzle seat, and it is therefore imperative that this be maintained if regrinding is attempted. In pintle nozzles the nozzle seat is 59 degrees with the seat being 60 degrees.

Figures 6-63 through 6-66 show some of the special equipment needed to attempt the restoration of the needle to seat angle. If this is not available, replace the nozzle and seat if it shows any sign of damage. Remember, too, that with increasing labor charges today it may be cheaper to buy a new needle and seat. However, shops that are engaged in the rebuilding of pumps and nozzles on a large scale often recondition needles and seats as indicated. This operation is also described under American Bosch Nozzles.

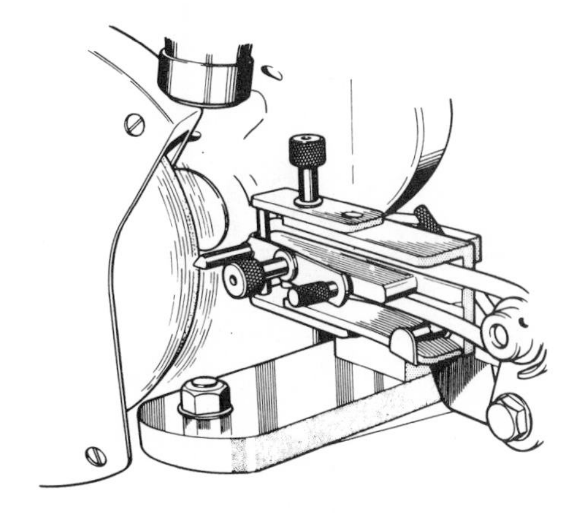

Figure 6-63
Grinding the Nozzle Seat Lap Tool.
(Courtesy of Lucas CAV Ltd.)

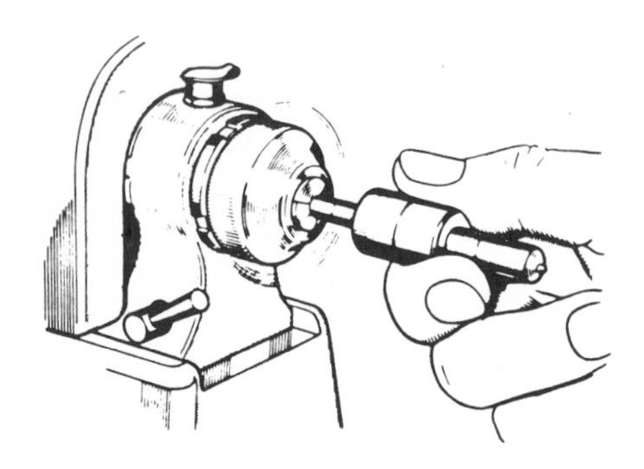

Figure 6-64
Lapping the Nozzle Seat.
(Courtesy of Lucas CAV Ltd.)

If the needle is lapped, make sure to check the needle valve lift, which can be altered. If the lift is greater than in CAV Bulletin G103, lap the joint face of the nozzle or nozzle holder.

The nozzle seat lapping tool is available in three different sizes of stem diameters, 6.00, 6.01, and 6.02 mm, to ensure a fine clearance fit in the bore and therefore proper concentricity of the nozzle seat to bore. Both the nozzle and needle seat must *never* be lapped together; otherwise, wide contact between them would occur, and line contact is what is desired. For this reason, the lapping tool must be reground as shown in Figure 6-63.

When reground, the lapping tool should be mounted into the collet of the machine shown in Figure 6-64; a small bead of fine lapping compound (such as chromic oxide) is applied to the lapping tool face; then the nozzle is held between the finger and thumb and moved back and forth quickly, as well as rotated occasionally, with the machine running. Do this for about 5-second intervals and for a maximum time of 30 seconds. Inspect the nozzle seat after washing in cleaning fluid; then carefully examine the seat condition through a microscope.

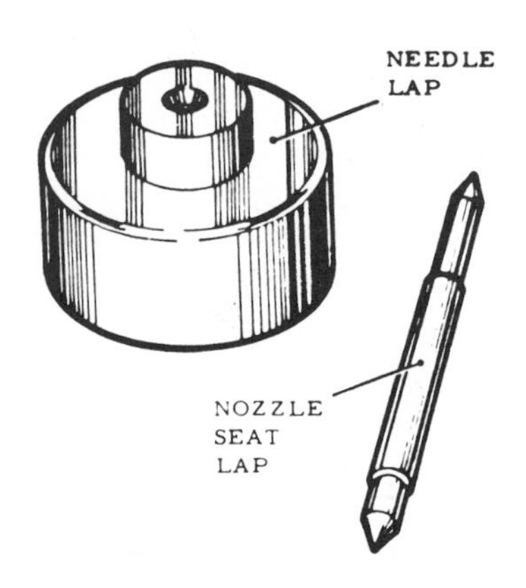

Figure 6-66
Needle and Nozzle Seat Lapping Tools.
(Courtesy of Lucas CAV Ltd.)

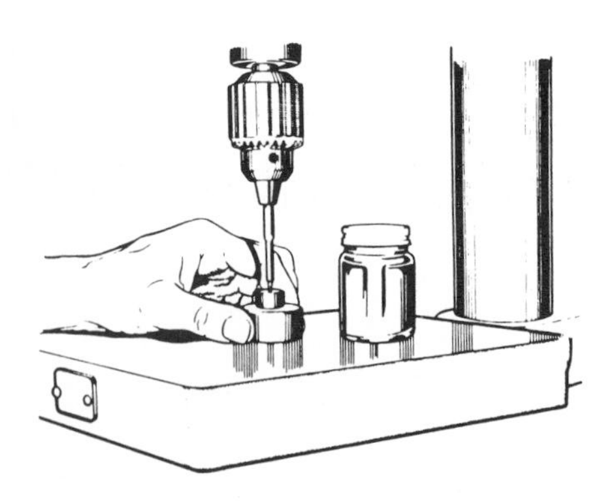

Figure 6-65
Lapping the Needle.
(Courtesy of Lucas CAV Ltd.)

Lapping of the needle is shown in Figure 6-65 with the chuck running at around 450 to 460 rpm. Follow the same procedure as for the seat; then inspect it.

chapter 7

Caterpillar Fuel Systems

One cannot consider heavy equipment without considering the name of Caterpillar, since it is a major force in the diesel equipment business. Although Caterpillar Tractor Company was formed on April 15, 1925, its birth was due to the merger of the two well-known U.S. West Coast firms, the Holt Manufacturing Company and the C. L. Best Gas Traction Company. Both of these companies were formed in 1869, and they both pioneered development of combined harvesters and large steam tractors between the years 1885 and 1900.

From the initial combine harvester sprung the development of a track-type traction engine, which was successfully tested for practical purposes in 1904 by Benjamin Holt near Stockton, California. The first production crawler was sold in 1906 for a price of $5,500, and by 1915 there were over 2000 Holt tractors working in 20 countries. Prior to 1906, steam power was used on the self-propelled agricultural machines; however, in 1906 gasoline engines were used by Holt with the Best Co. following three years later.

In 1913, C. L. Best developed the tracklayer machine, and Holt and Best as previously mentioned merged in 1925 to form Caterpillar Tractor Company. In the early 1920s, studies were underway to consider the feasibility of installing diesel engines into earthmoving machines. The only U.S. manufacturer of diesel engines that had an engine suitable for such equipment was the Atlas Engine Company of California; however, these engines were designed primarily for marine applications and were fairly bulky and heavy for actual use in earthmoving equipment.

Therefore, Caterpillar undertook a program that resulted in the introduction of a 4-stroke-cycle, water-cooled, precombustion chamber diesel engine in 1931. It was a 4-cylinder engine, with a bore of $6\frac{1}{8}$ in. (155.57 mm) and a stroke of $9\frac{1}{4}$ in. (234.95 mm) that produced 84 belt horsepower at 650 rpm. It was certainly a large displacement engine, both then and now, for a 4-cylinder unit: 1090.05 $in.^3$ or 17.87 liters.

Full-scale production of these engines began in 1932 since in 1931 five models of gasoline engines were available, but were sold on an as-is basis with no accessories or options offered. Many dealers of that day were converting these gasoline engines to run on natural gas for such applications as cotton gins.

The first industrial engine sold to an OEM (original equipment manufacturer) by Caterpillar in 1932 was to the Thew Shovel Company of Lorrain, Ohio, to power a $1\frac{1}{2}$ cubic yard (1.146 m^3) shovel. By 1933 three new diesel engines were introduced, the D6600, D8800, and D13000, which were 3-, 4-, and 6-cylinder engines, respectively, with a bore of $5\frac{1}{4}$ in. (133.35 mm) and a stroke of 8 in. (203.2 mm). This gave a displacement of 519.47 $in.^3$ (8.51 liters) for the 3 cylinder, 692.63 $in.^3$ (11.35 liters) for the 4 cylinder, and 1038.94 in. (17.03 liters) for the 6-cylinder engine. In 1935, these engines were redesigned, and were increased to a $5\frac{3}{4}$ in. (146.05 mm) bore. Also introduced at this time were the D4400 and D4600 engines, with a common bore and stroke of $4\frac{1}{4}$ by $5\frac{1}{2}$ in. (107.95 by 139.7 mm).

In the same year, the $5\frac{3}{4}$-in. (146.05-mm) bore line of engines was increased to include the 8-cylinder model D17000, which was not used in any earthmoving equipment, but was for industrial use only; it was built for 18 years before it was dropped in favor of newer models.

One problem with all early diesel engines, and not only Caterpillar's, was the common complaint of sticking piston rings; therefore, Caterpillar engineers along with the major oil companies arrived at a solution for this problem by compounding oils that were acceptable for diesel engine use. The first oil company to do this was the Standard Oil Company of California.

Injection systems were also a problem in early diesel engines, so Caterpillar designed and built their own system, one that required minimum maintenance and had no need to adjust or calibrate the pumping elements periodically.

In 1939, Caterpillar introduced their first diesel truck engine, the D468 model, followed by the D312 in 1940; however, when World War II broke out, both of these were discontinued and were not built again. It was not until the 1960s that Caterpillar actually entered the highway truck market with any seriousness, and the 1970s before they started any penetration of this market against such other well-known engine manufacturers as Cummins and Detroit Diesel in North America.

In 1942, Caterpillar built their last gasoline engine power plant, and by 1949 had introduced V8 and V12 configuration diesel engines of $5\frac{3}{4}$ in. (146.05 mm) bore, offered as both a naturally aspirated and roots blown model. In 1951, the models D337 and D326 appeared, followed in 1954 with modification and turbocharging of the D337. Caterpillar introduced their first aftercooled engine on their roots blown D375 marine engine in 1949 and for land vehicle use in 1957 on the model D337-powered DW21.

The engine division was first started in 1931 as the Special Sales group and was formed into the Engine Division of Caterpillar in 1953. Since 1972, Caterpillar Tractor Company has poured more than $1 billion into developing new diesels and into quadrupling its capacity to produce new high-speed high-horsepower engines capable of meeting the ever-increasing demand for such powerplants into the 1980s and beyond. They are predicting by 1981 that diesel engine sales could amount to $1.6 billion annually, or more than 20% of Caterpillar's total business.

Since Caterpillar was formed back in 1925, they have produced a long list of many firsts in the heavy equipment business, as well as adding to the successful completion of World War II. They presently have manufacturing plants in a variety of countries worldwide, with an extensive dealer network capable of responding to any need at any time desired by the customer.

As with most major engine manufacturers, Caterpillar have made many improvements to their available fuel-injection equipment over the years. They presently offer their diesel engines in both direct injection and pre-combustion chamber designs.

The three common types of fuel-injection pumps that are being used by Caterpillar at this time are (1) the flanged body style pump, (2) the compact body style pump, and (3) the sleeve metering style pump.

The basic operation of the flanged and compact bodies are somewhat similar; however, the sleeve metering system is of a design all its own. These three systems will be covered in the following pages, along with the direct injection nozzle and the pre-combustion chamber nozzle. Figure 7-1 shows a typical fuel system and its path of flow as employed by Caterpillar on a pre-combustion-type engine used on a highway truck. Direct injection engines are similar.

7-1

Flanged Body Pump System

The flanged body pump has been used by Caterpillar for many years, and is presently being used on their larger-model industrial and marine engine applications such as the D379, D398 V12, and D399 V16 engines. Other models also use this system to date. There is an individual pumping unit used for each engine cylinder with both the flanged body and compact style injection pumps.

Identification of the flanged body pump is easily made by the fact that each pumping unit is contained within a cast housing, which is located onto the main injection pump housing by dowels and then bolted into position. Figures 7-2 and 7-3 show one of the individual pumping units. If you are familiar with Bosch injection systems, you will find that these pumping units are not at all unlike the Bosch model APF units. Although the forged body system has the fuel-injection pumps exposed above the top of the main injection pump housing, contained within the main injection housing are the individual fuel-injection pumps for each cylinder along with the injection pump camshaft and fuel rack, which controls the position of the plungers within the barrel of the pumping units. Filtered fuel is delivered to a manifold in the injection pump housing, where it is then directed to each individual injection pump unit.

Each individual fuel pump housing contains a plunger with an attached gear segment that can be changed if damaged. As with other fuel-injection systems, the plunger is precision fitted and therefore mated and matched to the barrel contained within the housing; therefore, plungers and barrels should never be intermixed at any time.

The main injection pump housing is timed to the engine, and the gear segments of each pump unit are in turn timed to the main control rack. Therefore, the function of the individual pumping units is to meter and deliver fuel under pressure to the injection nozzles.

The action of the rotating camshaft within the main injection pump body causes vertical lift of the roller tappet or follower, as shown in Figure 7-2, which in turn transmits this motion to the pump plunger. It is this vertical motion that causes an increase in fuel pressure, which is sent through the fuel line to the injection nozzle and then into the engine cylinder or pre-combustion chamber as the case may be. Rotary motion of the plunger is accomplished by the back and forth movement of the fuel control rack, which is attached through linkage to the governor linkage and throttle. The plunger's rotary motion controls the quantity or volume of fuel delivered to the engine.

Let us study the action of the pump plunger as it completes a cycle of events lead-

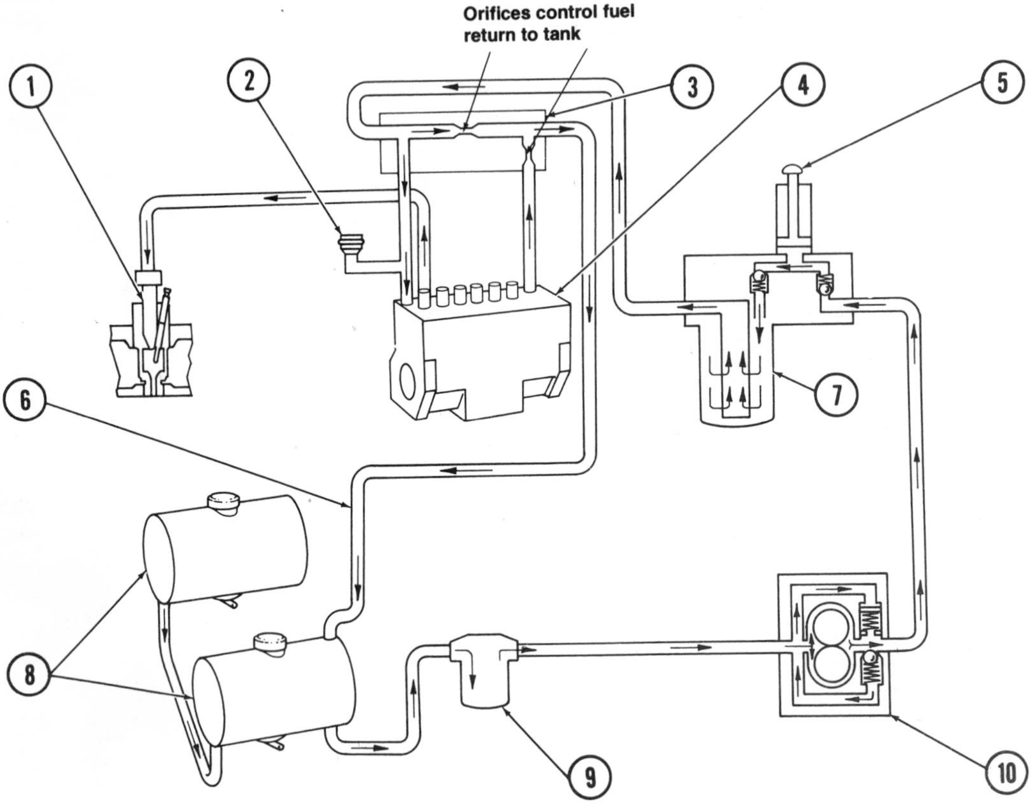

Orifices control fuel return to tank

Figure 7-1
Caterpillar Fuel System.
(Courtesy of Caterpillar Tractor Co.)

Index:

1—Injector (pre-combustion chamber engines); engines with direct injection use an adapter
2—Fuel pulsation damper
3—Antisiphon fuel block
4—Main injection pump housing
5—Hand priming pump (for priming or bleeding system)
6—Fuel return line
7—Secondary fuel filter
8—Dual fuel tanks
9—Primary filter or fuel and water separator
10—Fuel transfer pump

ing to fuel being injected into the cylinder. The sequence of events graphically illustrated in Figures 7-4 and 7-5 is as follows:

Step A: With the roller follower on the base circle of the injection pump cam lobe, fuel is free to flow into the area directly above the plunger at gear transfer pump pressure and down through the slot of the plunger into the recessed area of the plunger that contains the helical upper land and circumferential lower land.

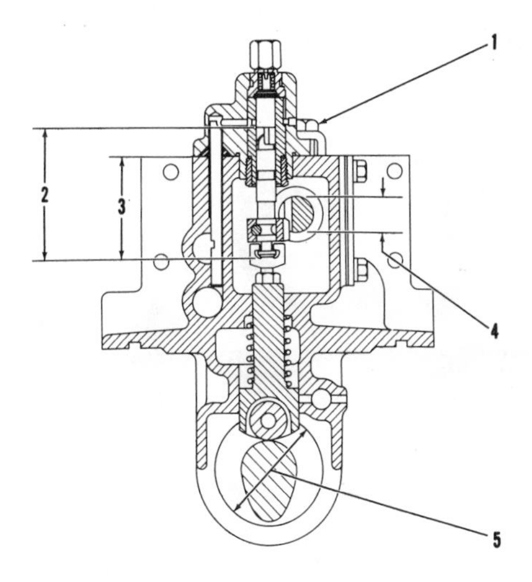

Figure 7-2
Flange Mounted Pump Assembly, Used on D379, D398, and D399 Engine Models. (Courtesy of Caterpillar Tractor Co.)

Index:
1—Clamp bolt
2—Length of pump plunger
3—Timing dimension
4—Fuel rack diameter
5—Camshaft

Step B: As the injection pump camshaft rotates, it pushes up the plunger within its barrel until the inlet port is covered or closed off. Trapped fuel above the plunger will be subjected to an ever-increasing pressure buildup, which when high enough unseats the spring-loaded check valve and enters the fuel injection line to the nozzle.

Step C: The upward moving plunger is displacing the fuel under pressure until it uncovers or opens the delivery port, allowing the remaining fuel to escape back through the port into the fuel manifold. Just before this happens, though, the delivery valve will be in the process of seating under spring pressure so that there is still fuel contained within the fuel-injection line, however at greatly reduced pressure. Therefore, injection ceases at this point.

Steps A through C illustrate the sequence of events during the upward or vertical motion of the plunger only; but you may recollect that we indicated that the plunger can also be rotated in a circular motion by the use of the rack and gear movement. Let us study then what happens during such rotary motion.

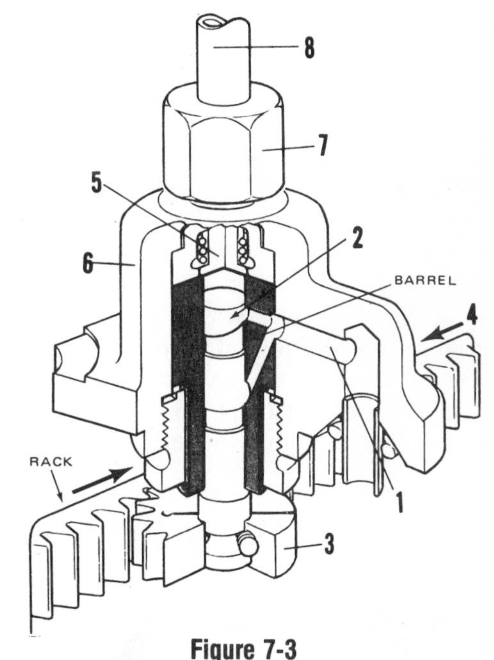

Figure 7-3
Flange Mounted Pump Arrangement. (Courtesy of Caterpillar Tractor Co.)

Index:
1—Fuel inlet port
2—Pump plunger
3—Gear segment
4—Fuel control rack
5—Delivery valve and spring
6—Fuel pump housing
7—Fuel-injection line nut
8—Fuel-injection delivery line

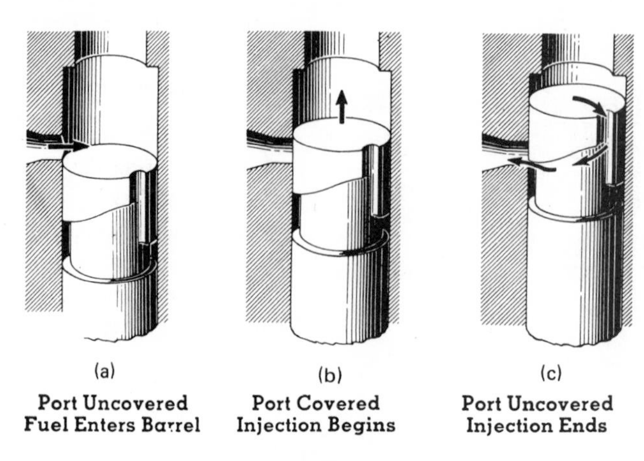

Figure 7-4
Action of the Pump Plunger.
(Courtesy of Caterpillar Tractor Co.)

As with all fuel-injection systems presently employed on diesel engines (except Caterpillar's sleeve metering system) that use vertical and rotary motion of individual pumping elements, in order to vary the quantity of fuel delivered per stroke of the plunger, a helix must be cut onto the recessed area of the plunger to allow control of how much fuel will be delivered at a given throttle position. By rotating the plunger within its barrel, the actual end of injection and opening of the supply port can be varied earlier or later in the upward motion of the plunger. This is therefore going to control the quantity of fuel that is metered during the plunger's upward movement. This is commonly referred to as the *effective stroke* of the plunger, since it is during the effective stroke that fuel injection is taking place. Any-

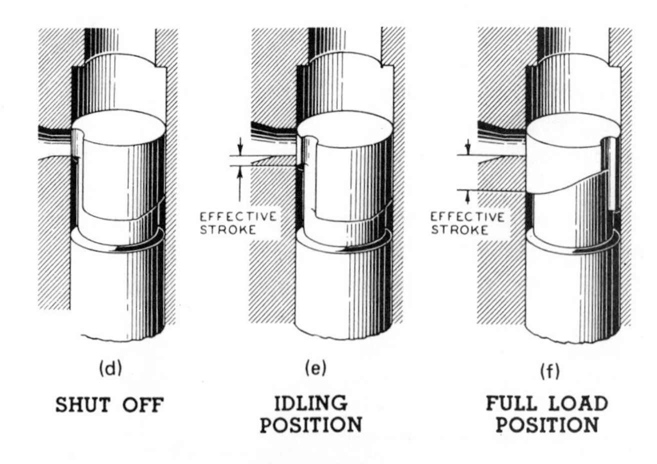

Figure 7-5
Action of the Pump Plunger.
(Courtesy of Caterpillar Tractor Co.)

time that the supply port is uncovered, there can be no fuel pressure above the plunger in excess of gear transfer pump pressure.

Step D: The plunger has been rotated by the control rack so that the slot in the upper end of the plunger is aligned with the supply port. Since the slot is connected with the recessed area of the plunger, fuel pressure at this position can never be more than gear transfer pump pressure; therefore, fuel injection is impossible. This then is the *shutoff* position.

Step E: As the engine is started, the plunger is rotated into a position whereby the narrow part of the plunger formed by the helix will in fact allow the supply port to be covered for a small period during the plunger's upward movement. The amount of fuel injected in this position will be proportional to the effective stroke of the plunger, in other words, as long as the supply port is covered.

Step F: If the control rack connected to the throttle is moved to the full-fuel position, the plunger will be rotated to its maximum position within its barrel, meaning that the supply port will be covered for a much longer time during the plunger's upward movement. A much greater quantity of fuel will therefore be injected during this time and at this position.

Injection Pump Removal

The injection pump assembly can be removed from the engine for repair, or individual pumping units can be taken off the main injection pump housing while it is still on the engine. Reasons for removing individual pumping units from the engine's main injection pump would be as follows:

1. Wear of injection pumping plungers and barrels affecting engine performance.
2. Lifter damage or spring damage, which would probably require inspection of the pump camshaft and therefore injection pump assembly removal if camshaft damage is found on inspection.
3. Fuel pump lifter settings and yoke inspection.
4. Fuel rack setting.
5. Checking pump gear tooth to rack timing.

If it is necessary to remove one or more pumping units from the injection pump, proceed as follows:

1. Clean the immediate area around the injection pump housing to prevent the entrance of any dirt.
2. Remove the fuel-injection line from the pump unit and cover both openings (pump and line) with plastic shipping caps, one a male type and the other a female.
3. Slacken off and remove both the hold-down bolts and clamp from the pumping unit; then lift the housing straight up in order to clear the main injection pump housing. See Figure 7-6.
4. Once the pump unit has cleared the injection pump housing dowel pins and plunger, place ferrule cap seals over both the fuel outlets and inlet.
5. Very carefully grasp the end of the plunger, which is protruding above the injection pump housing, sliding the lower end of the plunger out of the yoke that is attached to the lifter and withdrawing it. Place the plunger into its mating barrel immediately upon removal.
6. If more than one pump unit is being removed, make sure that you identify each pump unit as to its location on the main injection pump housing. Remove the side cover from the main pump housing.
7. If the main injection pump assembly has been removed from the engine, once the plungers have been removed, if the pump is to be stripped completely, then the fuel rack would be removed.
8. Continue to remove the three bolts that

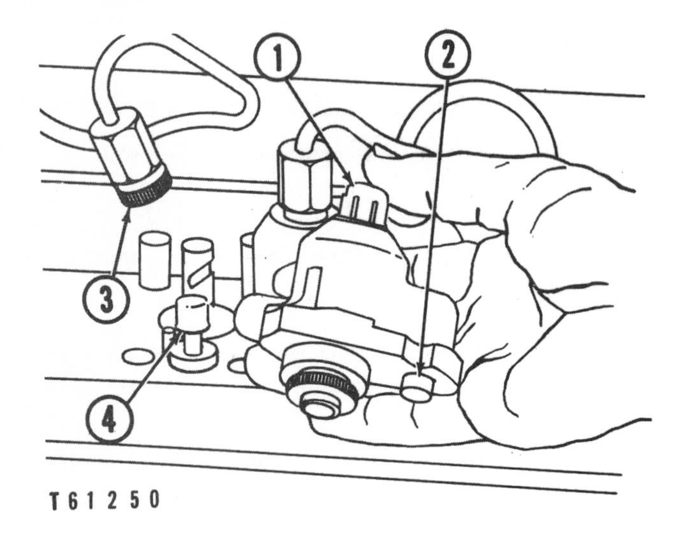

Index:
1—Female plastic shipping cover
2—Inlet plug
3—Fuel-injection line plastic shipping cap (male)
4—Ferrule cap seal

Figure 7-6
Removing a Flange Mounted Pump Assembly. (Courtesy of Caterpillar Tractor Co.)

hold the pump camshaft thrust plate in position. *(Note cam end markings).*

9. Hold the individual lifter yoke screws, loosen the locknuts, and remove them by screwing them out from the lifter. Identify each screw as to its location in the housing.
10. Install $\frac{3}{8}$-in. 24 NF by $3\frac{1}{4}$-in. long bolts and washers into the lifters with the bolts acting upon the injection pump housing so that when the bolts are tightened they will pull the lifters away from the camshaft.
11. Remove the injection pump camshaft from the housing with care. Proceed to remove the long bolts that were installed in step 10 from the lifters; then withdraw the lifters and springs from the bottom of the injection pump housing.
12. Remove both the control rack and camshaft bushings from the pump housing if worn beyond specifications.
13. On pump assembly, make sure that the oil holes in the camshaft bearings line up. Oil the lifters and install them into the housing *in the same position* from which they were removed, against their springs. Repeat step 10.
14. Oil the camshaft and *make sure* that you install it into the housing in the same position as when it came out (step 8).

Caution: Installing the camshaft in reverse will change the firing order. For example, on D398, D399, and D379 model engines, the camshaft must be installed with the arrow and the F mark at the front of the housing if the engine is of ccw rotation; whereas if the engine is of cw rotation, the camshaft must be installed with the arrow and F mark at the rear of the pump housing.

15. After the pump has been reassembled in the reverse procedure of teardown, make adjustments to the lifters prior to installing the plungers and pump housings as described under ''Fuel Injection Pump Lifter Setting (Off Engine)'' immediately following.

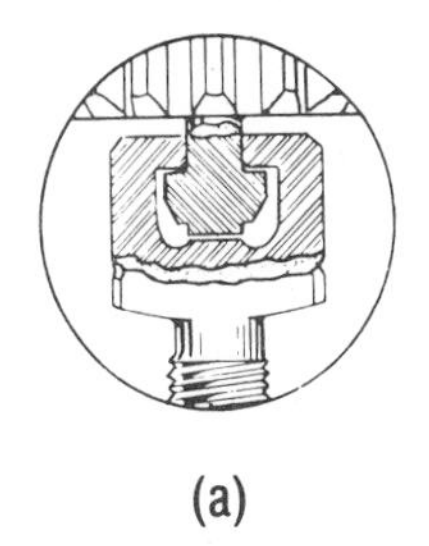

(a)

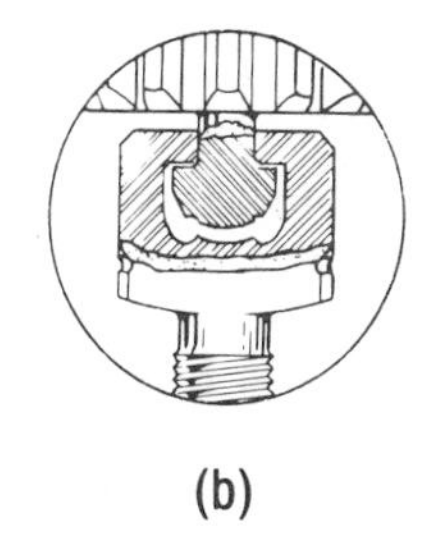

(b)

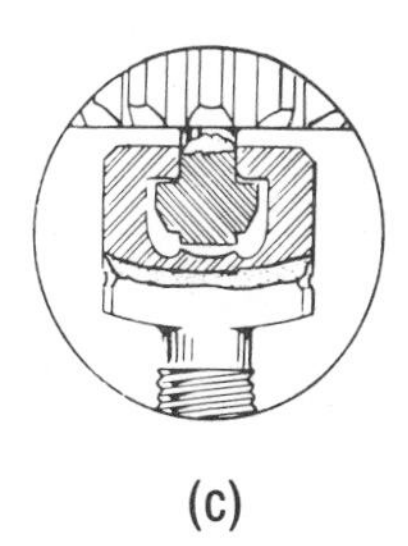

(c)

Figure 7-7
Plunger and Lifter Yoke.
(Courtesy of Caterpillar Tractor Co.)

Fuel Injection Pump Lifter Setting (Off Engine)

Prior to checking the individual pump lifter settings, it is advisable to check the condition of the lifter yokes for wear as shown in Figure 7-7. The condition of a new plunger and yoke would appear as shown in Figure 7-7(a), whereas the condition at (b) is one of obvious appreciable wear; Figure 7-7(c) shows that the installation of a new plunger alone does not correct the problem if the lifter yoke is itself worn. If a plunger appears to be in good condition, double-check that it is serviceable by measuring the length of it with a micrometer and comparing to specification for the particular engine that you are working on. If the plunger is worn, timing will be slow or retarded.

With the pump checked so far, you are ready to check the lifter setting. Special Caterpillar tools are required for this purpose; therefore, *do not attempt* this setting if you do not have the tools.

With the available tools at your disposal, proceed as follows: Refer to Figure 7-8; con-

Index:
1—Timing plate
2—Dummy shaft
6—Bolt
7—Timing pointer
10—Washer
11—Inner edge of pointer
12—Depth micrometer

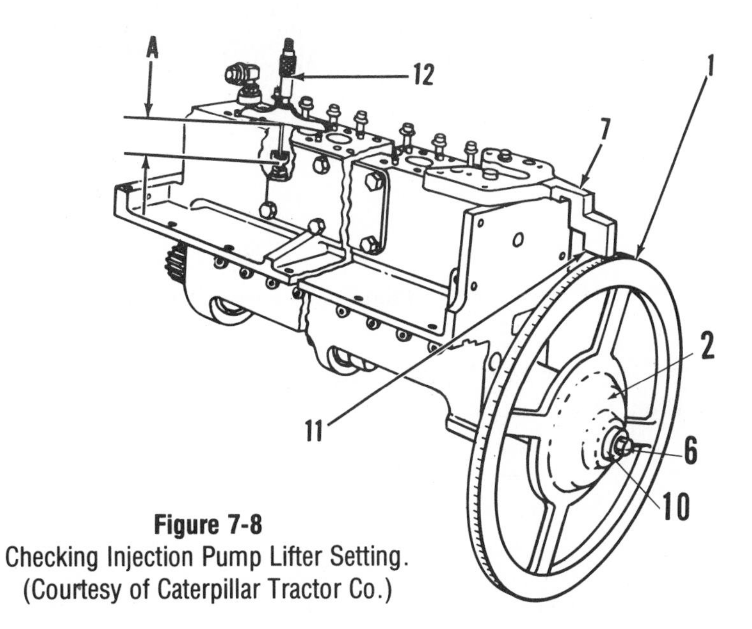

Figure 7-8
Checking Injection Pump Lifter Setting.
(Courtesy of Caterpillar Tractor Co.)

nect the timing plate or wheel (1) to the pump camshaft at the rear using dummy shaft (2), bolt (6), and washer (10). The dummy shaft is installed so that its key engages the camshaft slot, which allows the timing plate wheel to be attached and used for rotating the injection pump camshaft. The pointer (7) should then be placed over the dowel pin in the end of the housing and fastened in place.

Once the timing plate or wheel has been installed, refer to the timing chart for the particular lifter being set, and turn the plate in the direction of the pump's normal rotation, as indicated by the arrow on the end of the camshaft, until the correct setting in degrees aligns with the inside of pointer 7. Dimension A, which is also shown in Figure 7-2, should be 2.156 to 2.157 in. (54.76 to 54.79 mm) for all D379, D398, and D399 engines when the pump is off the engine.

Fuel Pump Lifter Settings (on the Engine)

Pump Timing Dimension (Piston at Top Dead Center)

D399 Engine: 2.076 to 2.080 in. (52.73 to 52.83 mm)

D398 and D379 Engines: 2.088 to 2.092 in. (53.04 to 53.14 mm)

It may be necessary to check and adjust the fuel pump lifter settings occasionally owing to such things as wear on lifters, pump plungers, and yokes, and even gear tooth wear. The procedure is simply to turn the engine crankshaft in its normal direction of rotation to the TDC position for the particular cylinder pump lifter that is being checked or set.

Be sure that if you turn the engine past TDC you back it up at least 60 degrees and

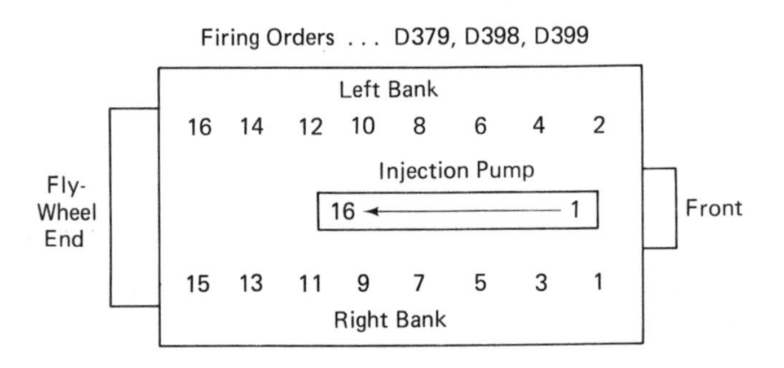

Figure 7-9
Firing Orders for D379, D398, and D399.

Counterclockwise Rotation (Taken from the Front of the Engine)

D399: 1–2–5–6–3–4–9–10–15–16–11–12–13–14–7–8

D398: 1–12–9–4–5–8–11–2–3–10–7–6

D379: 1–8–5–4–7–2–3–6

Clockwise Rotation

D399: 1–6–5–4–3–10–9–16–15–12–11–14–13–8–7–2

D398: 1–4–9–8–5–2–11–10–3–6–7–12

D379: 1–4–5–2–7–6–3–8

LIFTER SETTING IN DEGREES (OFF ENGINE)						
Lifter No. (numbered) in order from front to rear)	Timing Plate Degrees D379*		Timing Plate Degrees D398*		Timing Plate Degrees D399*	
	SAE Standard Rotation	SAE Opposite Rotation	SAE Standard Rotation	SAE Opposite Rotation	SAE Standard Rotation	SAE Opposite Rotation
1	0°	12°	0°	12°	179°	193°
2	210°	222°	210°	222°	209°	223°
3	270°	102°	240°	132°	269°	103°
4	120°	312°	90°	342°	299°	133°
5	90°	282°	120°	252°	224°	148°
6	300°	132°	330°	102°	254°	178°
7	180°	192°	300°	72°	134°	238°
8	30°	42°	150°	282°	164°	268°
9	–	–	60°	312°	314°	58°
10	–	–	270°	162°	344°	88°
11	–	–	180°	192°	44°	328°
12	–	–	30°	42°	74°	358°
13	–	–	–	–	89°	283°
14	–	–	–	–	119°	313°
15	–	–	–	–	359°	13°
16	–	–	–	–	29°	43°

*Timing plate must be turned in direction of arrow end of camshaft.

then bring it ahead again. This way you will be certain of removing all the backlash that exists in the gear train, therefore avoiding any possible error.

Continue the checks for each individual pump unit in the normal firing order of the engine by lining up the mark on the flywheel with the flywheel housing pointer.

Caution: Make sure that the machined surface of the injection pump housing is free of any nicks, scratches, or paint buildup; otherwise, the lifter setting will be greater than specified when you place the depth micrometer into position for the check.

Remember: Anytime you are turning the engine from the flywheel end you have to turn it opposite its normal rotation; otherwise, you will be turning it backward.

Installing Flanged Body Fuel Pumps

Prior to installation, ensure that the seals both at the large diameter of the pump unit and the fuel outlet ferrule are in good condition and in position. Proceed as follows:

1. Assuming that you have checked out the condition of all component parts, remove the plunger from its mating barrel and carefully wash both the plunger and barrel in clean, filtered diesel fuel; then install the plunger into its barrel.

2. Refer to Figure 7-10; rotate the pump plunger until the scribe mark on the pump gear tooth is aligned with the opposing scribe mark on the fuel control rack. Slide the plunger into position so that it engages the yoke of the lifter.

3. Very gently lower the pump housing over the plunger until it will slowly drop into place by its own weight, and guide it over the dowel pins. Install the pump clamp and hold-down bolts, and tighten them alternately on the inside first and then the outside to the specified torque, which is 32 ± 5 lb-ft (43.38 ± 6.77 N·m). Connect the fuel line nuts and torque them to 30 ± 5 lb-ft (40.67 ± 6.77 N·m).

Index:
5—Scribe mark on fuel rack
6—Mark on plunger gear tooth

Figure 7-10
Checking Timing of the Pump Plunger Gear to Fuel Control Rack.
(Courtesy of Caterpillar Tractor Co.)

Fuel Rack Adjustment (Using a Rack Setting Gauge)

The fuel rack can be set by the use of either a dial indicator or a rack setting gauge as shown in Figure 7-11. The fuel rack can be adjusted with the injection pump on or off the engine.

The rack setting gauge must be installed into the injection pump body (one pump unit removed) so that the timing marks on the gauge gear align with that on the rack, as shown in Figure 7-11. This can be double-checked quite easily by looking through the hole of the gauge. The rack setting gauge is constructed with an inner and outer calibrated scale. The longer lines represent 0.100 in. (2.54 mm), with the shorter ones being 0.050 in. (1.27 mm) on the outer circumference.

The inner dial face, which is directly connected to the gear, actually has two scales with each line equaling 0.005 in. (0.127 mm). The reason for the inclusion of two scales is so that the rack travel can be taken in both directions. The governor housing inspection cover must be removed in order to get at the necessary adjustments. Figure 7-12 shows the location of these adjustments.

With the rack setting gauge installed and aligned, loosen the locknut and turn the rack limiter adjusting screw cw so that the bell crank lever does not in any way limit the fuel rack movement. The speed limiter should be removed to allow complete and free rack travel.

To prevent deflection of the torque spring while adjusting the rack, you can place behind the torque spring and between the housing a spacer or block that is the same thickness as the torque spring spacer used with the particular model of engine. However, this is not usually necessary; but caution must be exercised when doing the rack adjustment to ensure that the torque spring is not deflected unduly.

Gradually move the fuel throttle to its full-fuel position, which causes the rack adjusting screw to come into contact with the torque spring. If no block or spacer is being used, just before the rack reaches full fuel pause and very gently come toward the point where the screw just touches the torque spring. You can see this and also feel a step up in effort; or place a feeler gauge between the screw and torque spring until it can be withdrawn with a slight drag (0.0015 in. feeler; 0.038 mm). Note the

(a)

(b)

Figure 7-11
(a) Flange Body Style Injection Pump Rack Setting Gauge; (b) Dial Face of Rack Setting Gauge.
(Courtesy of Caterpillar Tractor Co.)

dial gauge reading, and compare to specifications. Adjust the rack setting adjusting screw as necessary to obtain the correct rack travel.

Reading of the gauge is similar to reading a vernier caliper in that you note the position of the *zero* on the inner dial face, then find the lines that align between the inner and outer circumferences, and add this on to the reading obtained with the zero mark. For example, if the inner dial zero was aligned with the 0.400-in. (10.16-mm) mark on the outer circumference, the reading would be 0.400 in. However,

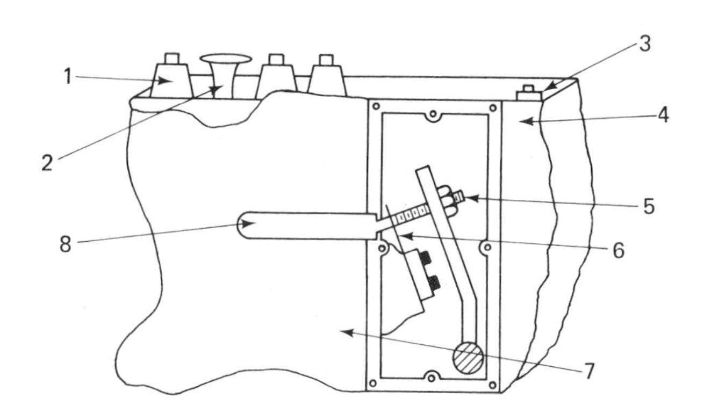

Index:
1—Pump unit body
2—Rack setting gauge
3—Rack limiter bell crank adjusting screw
4—Governor housing
5—Rack setting screw
6—Torque spring
7—Injection pump housing
8—Fuel pump rack

Figure 7-12
Location of Fuel Rack Adjustments.

if the zero was just past 0.400 in., look to the right of the zero and find the line that now comes into alignment with a line of the outer circumference. For example, if it was the third line to the right of the zero, this would be an additional 0.015 in. (0.381 mm) added on to the 0.400-in. reading, therefore giving a total reading of 0.415 in. (10.54 mm).

To adjust the rack limiter, loosen the locknut and turn the rack limiter screw until the dial gauge is as specified in the Caterpillar setting for your model engine.

7-2

Compact Body Pump System

In our previous discussion about forged body fuel-injection pumps, you will recollect that this system has its fuel-injection pumps exposed above the top surface of the main injection pump housing. The compact body style pump differs from the flanged body unit in that it has the injection pumps located completely inside the main injection pump housing. This then is where the terminology *compact body* comes from. Figure 7-13 shows a cross-sectional view of such a pump. The operation of both the flanged and compact body pumps is the same as far as fuel delivery is concerned. Refer to Figure 7-4 for the explanation of this sequence of events.

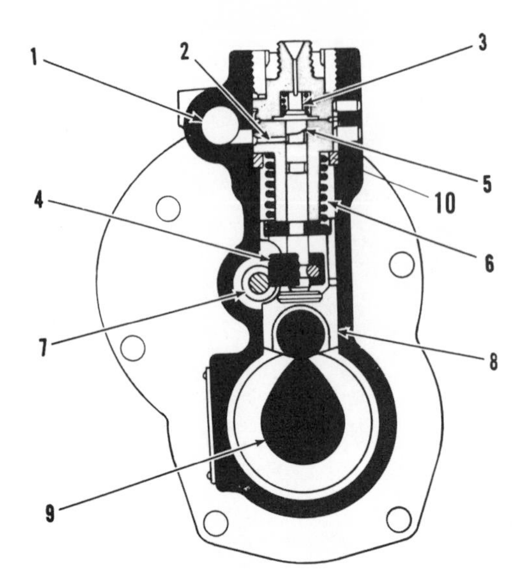

Figure 7-13
Compact Body Style Pump.
(Courtesy of Caterpillar Tractor Co.)

Index:
1—Fuel passage
2—Fuel inlet port
3—Check valve
4—Gear segment
5—Pump plunger
6—Spring
7—Fuel control rack
8—Lifter assembly
9—Camshaft
10—Phasing spacer

Centering the Fuel Rack

Prior to removing or installing individual injection pumping units from compact body pumps, it is necessary to position the fuel control rack of the main injection pump assembly in its center or mid-travel position. Failure to do this during individual pump removal can lead to damage or shearing of the locating dowel or pin in the bore area of the pump unit. During pump installation, failure to center the fuel rack can cause one or more injection pump units to be out of time, meaning that it will not deliver full fuel if it is installed a tooth off one way, and if installed the opposite way, can lead to excessive fuel delivery upon starting or possible jamming of the rack, leading to engine damage.

Since the compact body pump is used on a wide variety of Caterpillar engines, the actual centering of the rack will vary slightly among these different models. The following description covers these.

1. Refer to Figure 7-14. Some engines, such

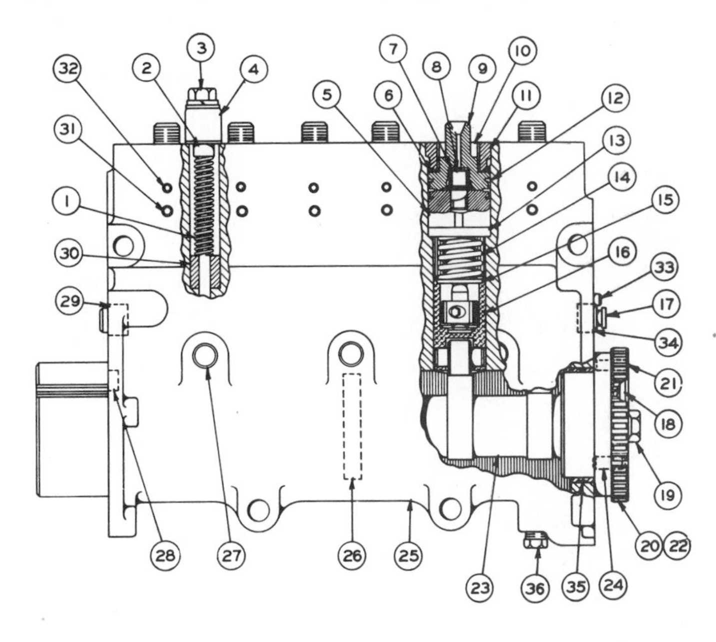

Index:

1—Spring
2—Pin
3—Bolt
4—Cover
5—Plunger and barrel assembly
6—Seal
7—Spring
8—Check valve
9—Bonnet
10—Washer
11—Bushing
12—Ring
13—Spacer
14—Spring
15—Washer
16—Lifter assembly
17—Rack
18—Plate
19—Bolt
20—Gear assembly
21—Dowel
22—Spring
23—Camshaft assembly
24—Dowel
25—Housing assembly
26—Bearing
27—Pin
28—Plug
29—Bearing
30—Bushing
31—Dowel
32—Dowel
33—Dowel
34—Bearing
35—Bearing
36—Plug

Figure 7-14
Fuel Pump Housing.
(Courtesy of Caterpillar Tractor Co.)

as the model D343 and the 1693 truck engine, employ a *rack centering pin*, which is located between the bore of the first two pump bores closest to the pump's drive end. To center the rack, bolt 3 must be removed along with cover 4, which allows access to the rack centering pin 2. Since the rack centering pin is spring loaded, it will pop out once the cover is removed; therefore, it must be depressed in order to hold the rack in the center position. Place cover 4 halfway over the rack centering pin with bolt 3 screwed in snug to hold it in place. When finished, remember to slacken off bolt 3, ensure that the rack centering pin returns to the disengaged position, and place cover 4 over the pin and secure it in place with bolt 3.

2. Refer to Figure 7-19, individual injection pump installation, for rack centering on engine models such as the D-330 engine.
3. To center the rack on the 3406 type engines, refer to Figure 7-15.

(a)

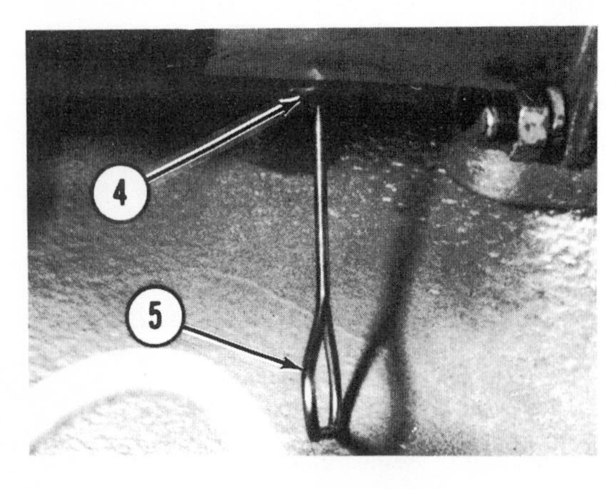

(b)

Index:
1—Stop
2—Spacer
3—Gaskets
4—9S8519 plug
5—9S8521 rod

Figure 7-15
Centering the Rack on 3406 Engines. (Courtesy of Caterpillar Tractor Co.)

Remove all three items shown in Figure 7-15(a) and install only the rack stop (1) back onto the main fuel pump housing from which it was removed (do not use any gaskets). On earlier engines equipped with a speed limiter, push the speed limiter plunger in as shown in Figure 7-15(b) with 9S8521 rod (5) so that the governor control lever can be moved to the fuel-on position, which will place the rack against the stop (1). The rod 5 is used in conjunction with plug 4 (9S8519), which screws into the hole in the governor housing to give a lock for the rod. Moving the governor control to the fuel-on position will place the fuel rack against the stop, thereby effectively placing the rack in its centered position, to allow removal or installation of individual injection pumps.

Speed Limiter

The function of a speed limiter is to prevent the engine speed from exceeding that of the normal engine rpm until adequate engine oil pressure has been attained after initial starting of the engine. It is generally accessible from the injection pump accessory drive housing or underside of the injection pump housing (Figure 7-16) and simply contains a small plunger that extends into the governor mechanism. With the engine stopped, the spring (2) forces the plunger (1) in front of the governor control lever linkage (internally), preventing increased fuel. With the engine running, oil pressure forces the plunger back, allowing the rack to be moved to more fuel.

The *injection pump camshaft* is fed by splash oil from the governor housing; therefore, rotation of the pump camshaft maintains a film of oil between it and the individual lifters. The oil drains through the accessory drive housing back to the engine crankcase. Drive to the injection pump camshaft is by an off-center tang on the end of the accessory drive shaft.

Injection Pump Removal

The injection pump assembly can be removed along with the governor as a unit from the engine, or the individual injection pumps can be removed separately from the injection pump assembly while it is still on the engine.

To remove the individual pumping units, center the rack first, and proceed as follows:

1. Clean the immediate area around the injection pump assembly.
2. Remove the injection line nut and place a plastic male shipping cap over the end of the line.
3. Remove the felt washer from under the retaining nut cap.
4. Install special wrench 1M6952 onto the fuel stud as shown in Figure 7-17(a) in order to remove the pump retaining bushing from the housing. Do *not* use a $\frac{1}{2}$-in.^{2} (12.7-mm)^2 drive socket extension in the end of the tool, which may cause damage.
5. Obtain knurled extractor 1M6954, thread it onto the fuel stud as shown in Figure 7-17(b), and lift the complete injection pump unit out of the injection pump housing.

With the injection pump assembly removed from the main pump housing, the seal can be removed first, followed by the retaining ring, which will allow you to separate both the barrel and bonnet assembly. Figure 7-18(a) shows the stackup of the component parts. Once this has been done, proceed to separate the spring, washer, and plunger from the barrel as shown in Figure 7-18(b).

Inspection of the component parts should be done with care and in clean, organized surroundings. Once the parts have been cleaned in clean diesel fuel, replace any worn or damaged parts, and remember that the

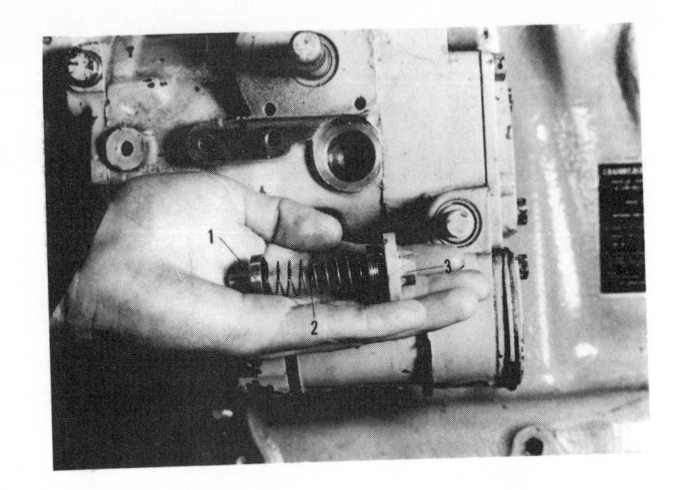

Index:
1—Plunger
2—Spring
3—Plug

Figure 7-16
Speed Limiter.
(Courtesy of Caterpillar Tractor Co.)

Figure 7-17
Injection Pump Removal.
(Courtesy of Caterpillar Tractor Co.)

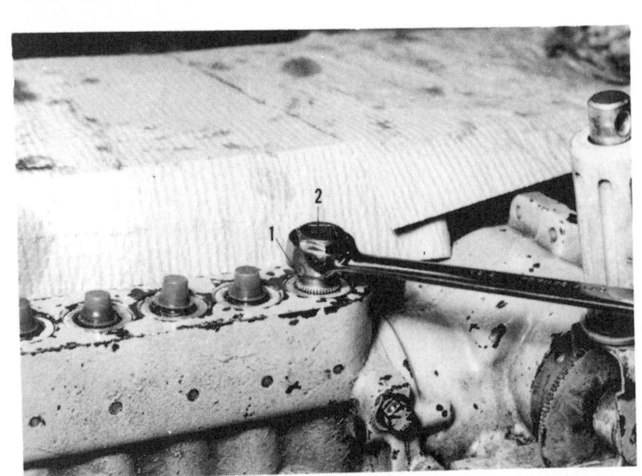

(a)

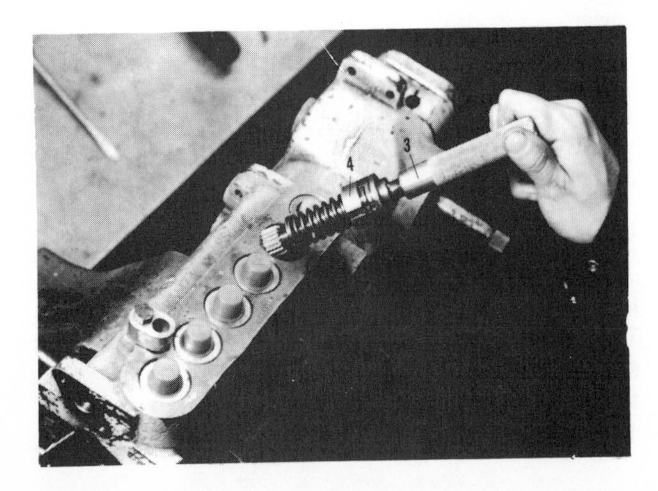

(b)

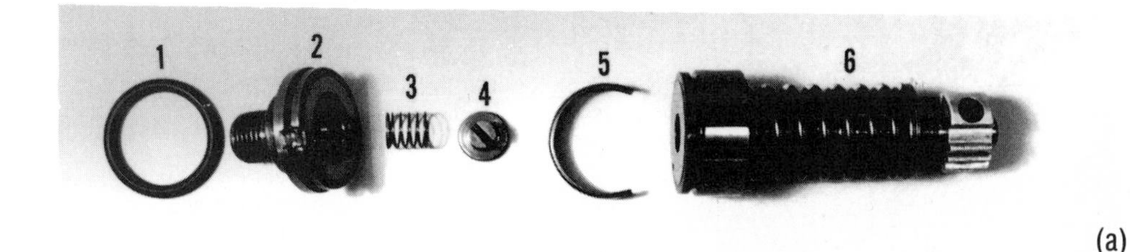

(a)

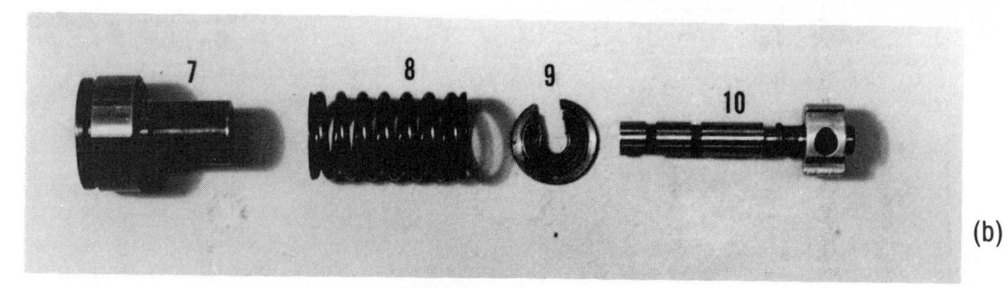

(b)

Figure 7-18
Component Parts of Injection Pump Assembly.
(Courtesy of Caterpillar Tractor Co.)

Index:
1—Seal
2—Bonnet
3—Spring
4—Check valve
5—Ring
6—Barrel assembly
7—Barrel
8—Spring
9—Washer
10—Plunger assembly

plunger and barrel are matched parts and cannot be intermixed.

Individual Injection Pump Installation

Unlike the forged body pump setup, one cannot see the mark on the rack or gear segment in order to line them up during installation; therefore, a rack setting gauge 8M530, which is a small-depth-type micrometer, must be used to position the fuel control rack of the injection pump in the mid-travel or centered position prior to installation of the individual injection pump units. To do this, proceed as follows:

1. Refer to Figure 7-19(a), and remove the small inspection cover on the end of the main injection pump housing.
2. Attach the gauge unit to the pump housing as shown with the two small bolts.
3. The rack should be positioned until the reading on the gauge is zero, which will ensure that the center notch on the rack will align with the center tooth of the gear segment of each injection pump plunger.
4. Refer to Figure 7-19(b), and ensure that prior to installation of the pumping unit the notches 2, 3, and 4 on the bonnet and barrel are in alignment with the slot on the plunger gear segment.
5. Extreme care must be exercised during this

(a)

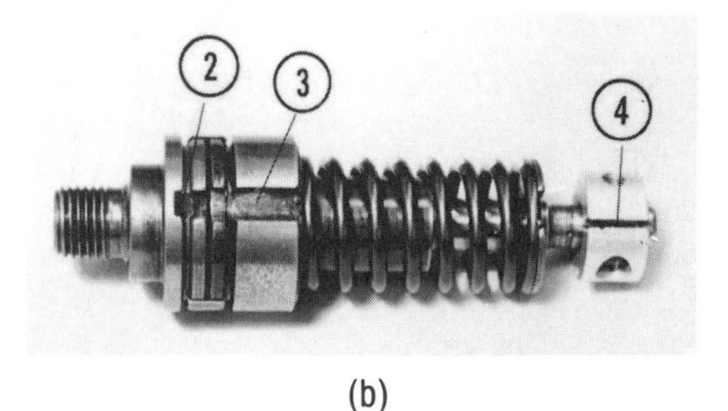

(b)

Figure 7-19
Rack Setting Gauge 8M530.
(Courtesy of Caterpillar Tractor Co.)

Index:
1—Rack Setting Gauge 8M530
2—Notch in bonnet
3—Notch in barrel
4—Slot in plunger gear segment

operation, because if the pump gear segment is misplaced in either direction, rack travel will be affected.

6. Thread tool 1M6954 as shown in Figure 7-17(b) onto the fuel stud of the injection pump unit and carefully insert the assembly into the bore of the pump housing in the reverse procedure of removal. Make sure that the notches of the bonnet and barrel engage the two locating dowels in the bore of the housing, which aligns the fuel inlet port with the fuel manifold outlet port.
7. Place a new rubber seal over the bonnet, install the retainer bushing, and using wrench 1M6952, shown in Figure 7-17(a), tighten the retaining bushing to specifications (150 ± 10 lb-ft ± 13.558N·m). Ensure that the bushing is flush with the top of the housing prior to final tightening.
8. To ensure that all the individual injection pump units are in fact aligned with the center tooth of the rack, measure the rack travel (one direction) with rack setting gauge 8M530 shown in Figure 7-19. Move the rack to its maximum fuel position and adjust the micrometer until it butts up against the end of the rack. For example, certain models of compact body pumps used with particular model engines specify a reading of plus 0.312 in. or 7.924 mm (reading obtained from rack center position only), which indicates that the plunger gear segment is in fact aligned properly with the rack.

Note: The rack collar must be removed in order to obtain this reading; therefore, remove the air–fuel ratio control to enable you to get at it. Failure to obtain the specified rack travel reading for your model engine as indicated means that you will have to pull the pump units again in order to reposition them correctly. If an injection pump unit is installed with a tooth out of alignment with the rack marking, it can cause possible engine overspeed and the resultant problems associated with such a condition. Check *rack travel* as you insert each pump unit.

9. Remove the plastic shipping caps from the fuel stud and the fuel line, install the felt washer, and connect the fuel-injection line.

Timing the Compact Body Fuel Pump (on the Engine)

This check involves establishing the distance from the pump lifter when installed in the injection pump housing to the top of a special gauge, which takes the place of the individual injection pump assembly. This check of pump timing should be done at any time that individual injection pumps are removed for service or replacement. It can readily indicate wear of the injection pump camshaft or pump lifter assembly. The checking procedure is as follows:

1. The engine must be placed at TDC on the compression stroke for the particular cylinder being timed by turning the engine in its normal direction of rotation.
2. With the injection pump unit removed from the pump bore, leave the lifter and button in place.
3. Refer to Figure 7-20, and with the phasing spacer in position in the housing, place the 2-in. (50.8-mm) spacer gauge 8S7167 on top of the phasing spacer.
4. The timing dimension setting for your particular engine must be obtained from available Caterpillar specifications. This distance is specified as a ±0.002 in. (0.0508 mm); otherwise, the phasing spacer must be changed. These spacers are available in graduated thicknesses of 0.004 in. (0.101 mm), from 0.170 in. (4.318 mm) to 0.198 in. (5.029 mm), in order to effectively alter the timing dimension on pre-combustion chamber type engines. On 3400 series engines with direct injection, spacer thickness starts at 0.195 in. (4.95 mm) and goes up in 0.004-in. (0.101-mm) graduations to 0.223 in. (5.664 mm).

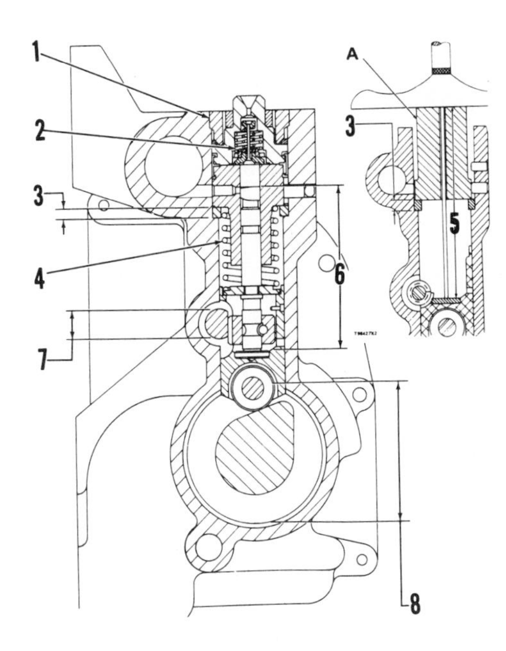

Figure 7-20
Setting the Timing Dimension.
(Courtesy of Caterpillar Tractor Co.)

Index:
1—Bushing
2—Spring
3—Phasing spacer
4—Spring
5—Timing dimension
6—Length of new pump plunger
7—Fuel control rack bore
8—Camshaft bore
A—Timing gauge 8S7167, 3406 engine

Injection Pump Timing Dimension Setting (off the Engine)

The sequence of events for this check is the same as described earlier for the flanged body injection pump. With the compact-body-type pump, however, it is advisable to check the injection pump lifter washer and pump plunger for proper contact. The following example gives the sequence of events required to do the timing check on a 3406 injection pump.

Index:
1—Depth micrometer
2—Stationary timing pointer
3—Timing plate wheel
4—Retaining bolt

Figure 7-21
Stationary Pointer Attached to Pump Housing.
(Courtesy of Caterpillar Tractor Co.)

1. Refer to Figure 7-17(a) and (b) and remove the individual injection pumps with the use of 8S4613 wrench and 8S2244 extractor.
2. Refer to Figure 7-21 and attach the stationary pointer 5P1768 to the pump housing by a short bolt.
3. Insert the short adapter shaft 5P1761 into the timing plate 1P7410, and install the plate onto the drive end of the injection pump camshaft.
4. Refer to Figure 7-15; remove plug 3A and insert the 8S2291 timing pin through the hole and into engagement with the slot in the camshaft.
5. Align the zero degree mark on the timing plate with the stationary pointer and tighten up the bolt (6). Remove the timing pin to allow rotation of the pump camshaft.
6. When checking the individual pump timing dimensions, do this in the normal firing order of the engine.
7. The charts shown in Figures 7-22 and 7-23 give the specifications for 3406 engines.
8. Gauge (spacer) 8S7167 similar to that shown in Figure 7-20, item (A), must be used along with the depth micrometer when checking the timing dimension of each individual pump bore.
9. Turn the timing plate counterclockwise when using until the degree setting for the particular pump lifter being checked is in alignment with the pointer attached to the housing.
10. Double-check all adjustments, especially if different-thickness phasing spacers have been used to correct the lifter settings.

Note: Refer to step 4 under Timing the Compact Body Fuel Pump on the Engine for available phasing spacer thicknesses.

Finding Top Dead Center (TDC)

Locating TDC on any Caterpillar engine can be done by reference to timing marks located on the engine's flywheel in alignment with a stationary pointer on the flywheel housing visible through an inspection cover. Normally, these marks are adequate for normal service procedures; however, if it is necessary to check the fuel-injection pump timing while on the engine, or to make a new TDC mark at any time, there is a recommended procedure that can be followed with the use of special tools and equipment.

Prior to covering this procedure, if you are simply double-checking that the injection pump assembly is properly timed to the engine, then on engines such as the D343, 1693 truck engine, and 3406 engines a timing pin can be installed through a locating hole of the injection pump body; the engine is then slowly rotated in its normal direction of rotation until the timing pin drops into and engages the slot of the fuel-injection pump camshaft. With the pin in position, a timing bolt should be able to be screwed into a locating hole in the engine flywheel from the flywheel housing side (same side as the injection pump). If this can be done, timing of the injection pump camshaft to the engine is correct.

Remember that the timing bolt will locate the top dead center position for either cylinder 1 or 6 piston; therefore, always check to ensure that both valves, intake and exhaust, are closed (rocker arms loose) on the cylinder for which you are checking.

The camshaft timing pin for D343 and 1693 engines is located on the injection pump housing at the drive end (side of pump). See Figure 7-24. The pin is screwed into a small

TIMING DIMENSION	FUEL INJECTION PUMP	ENGINE ARRANGEMENT
4.3939 ± .0020 in. (111.605 ± 0.051 mm) For 10° Timing	4N817 (with precombustion chambers)	9N2, 9N3, 9N4, 9N6, 9N8, 9N2091, 9N2092, 9N2280, 9N2290, 9N2291, 9N2292
4.3689 ± .0020 in. (110.970 ± 0.051 mm) For 28° Timing	6N1537 (with direct injection)	9N1, 9N5, 9N2090, 9N2169, 9N2170, 9N2171, 9L6270, 9L6677
4.3939 ± .0020 in. (111.605 ± 0.051 mm) For 29° Timing	4N2642 (with direct injection)	9N7, 9N9, 9L509

Figure 7-22
Specifications for 3406 Engines. (Courtesy of Caterpillar Tractor Co.)

LIFTER SETTING IN DEGREES (OFF ENGINE)			
Lifter Number (From Front To Rear)	Timing Plate Degrees		
	PC; 10°	DI; 28°	DI; 29°
1	355°	346°	345.5°
2	235°	226°	225.5°
3	115°	106°	105.5°
4	295°	286°	285.5°
5	55°	46°	45.5°
6	175°	166°	165.5°

Figure 7-23
Lifter Setting for 3406 Engines. (Courtesy of Caterpillar Tractor Co.)

Figure 7-24
Location of Camshaft Timing Pin. (Courtesy of Caterpillar Tractor Co.)

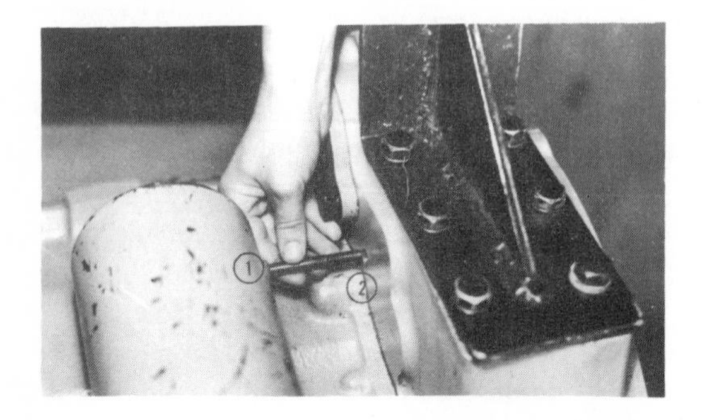

(a)

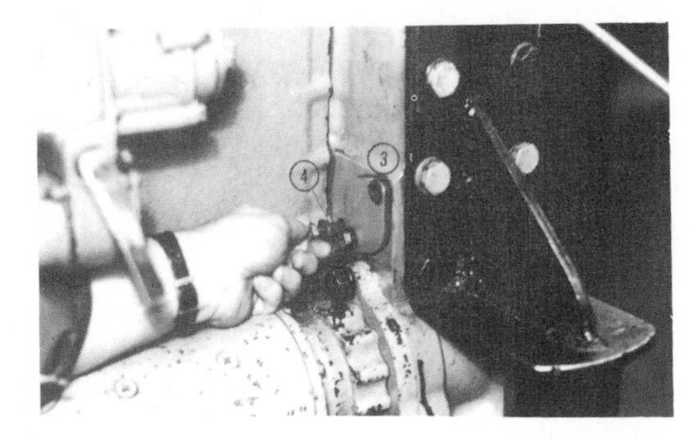

(b)

Figure 7-25
Location TDC on Piston 1: Timing Bolt Installed on (a) Starter Motor Side and (b) Oil Filter Side of Engine.
(Courtesy of Caterpillar Tractor Co.)

Index:
1—Timing bolt
2—Timing bolt location (left side)
3—Normal location of timing bolt when not in use
4—Timing bolt location (right side)

locating plate to retain it when not in use. The retaining plate is removed when the pin is used to allow the pin to engage with the pump camshaft slot. On the 3406 engine, the timing pin 8S2291 is required, which drops into the housing at the side after removal of a plug, as shown in Figure 7-14(a), item 3-A.

The flywheel timing bolt can be found on the front side of the flywheel housing screwed into the inspection plate immediately above the starting motor; it is the upper and outward bolt of the two. On 3406-type engines, the timing bolt can be installed into the left- or right-hand side of the flywheel housing and on into the flywheel in order to locate TDC on piston 1. See Figure 7-25(a) and (b) for this location.

Due to the fact that the timing bolt can be located on either the left or right side of the engine, the two holes in the flywheel of the 3406 engine are unequally spaced to prevent insertion of the timing bolt in the wrong hole. A similar setup is found on the D343 and 1693 engines. In either case, always turn the engine toward its normal rotation to eliminate gear backlash (do not turn it backward).

7-3

Checking Compact Pump Timing on the Engine

Although there are timing marks stamped on the flywheel that can be used to determine relative piston positions, and the 1P3500 timing light group can be used on a running engine to establish specific timing of the fuel-injection pump to the engine, wear in timing gears along with some possible slippage of the drive shaft coupling or automatic timing advance unit can cause some variation to this setting; therefore, the following can be used to check that the timing is, in fact, correct or in need of adjustment.

The 5P6524 tool group (Figure 7-26) contains all the tools necessary to accurately determine TDC on Caterpillar diesel engines. The 1P540 tool group consists of a fuel tank with a hand pump for field use, a pressure regulator that reduces shop air when used to a safe value of 10 to 15 psi (68.95 to 103.42 kPa), and necessary fuel pump tubes and assorted connecting hoses. For direct injection

3406 engines, a 5P3564 flow line assembly is available.

Figure 7-27 shows the typical setup of the timing indicator tools required to establish the relative piston position. The dial indicator reading can then be converted to a reading in degrees and compared to specifications.

Timing Procedure Check

1. From the available Caterpillar form SMHS7083-01 or the applicable parts list, select the necessary adapter, rod, and contact point.
2. Piston 1 at TDC (compression stroke) is the reference point for all timing checks.
3. When the fuel line from injector 1 has been removed, proceed to remove the injector on direct injection engines or the fuel-injection valve body and nozzle assembly from a pre-combustion chambered engine.
4. Install the correct adapter as stated in step 1. Exert *caution* here; tighten the adapter

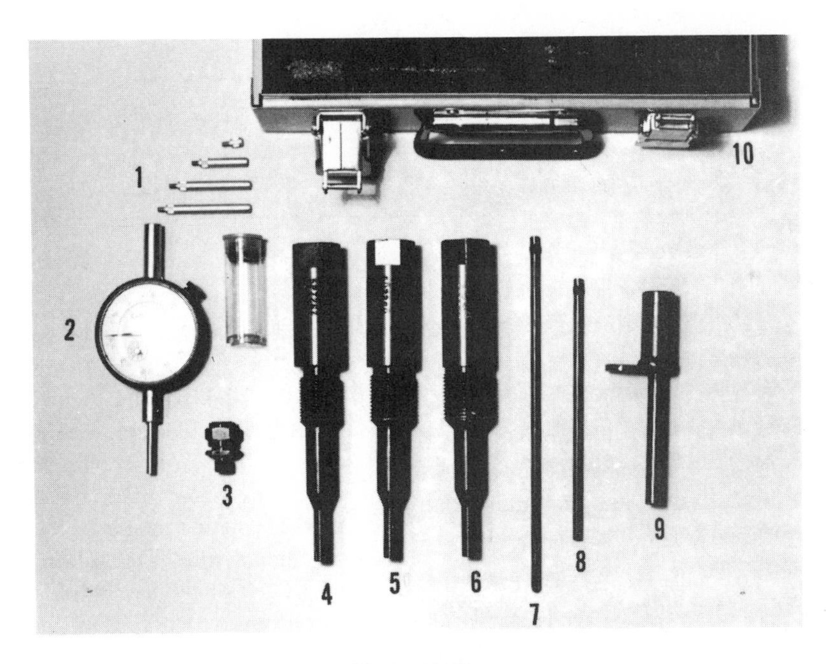

Figure 7-26
Parts List for 5P-6524 Engine Timing Indicator Group.
(Courtesy of Caterpillar Tractor Co.)

Index:

1—5P2393 contact point: 1.50 in. (38.10 mm)
3S3268 contact point: 0.25 in. (6.4 mm)
3S3269 contact point: 1.00 in. (25.4 mm)
3S3270 contact point: 1.75 in. (44.5 mm)
2—9S215 dial indicator
3—3P1565 collet
4—5P7267 adapter
5—5P7268 adapter
6—5P7265 adapter
7—3S3264 rod: 7.12 in. (180.9 mm)
8—8S2296 rod: 5.25 in. (133.4 mm)
9—5P7266 adapter
10—2P8290 box

Table 7-1

Engine	*Adapter*	*Rod*	*Contact Point*	*Dial Indicator and Collet*
D336	5P7268	3S3264	3S3268	9S215
1676				3P1565
1100	5P7266	8S2296	3S3268	9S215
3100				3P1565
3208				
Direct-injection	5P7268	8S2296	5P2393	9S215
3406				3P1565
Pre-combustion chamber 3406				
3408	5P7268	3S3264	3S3269	9S215
3412				3P1565
3204	5P7267	3S3264	3S3268	9S215
				3P1565
D353				
D379	5P7265	3S3264	3S3270	9S215
				3P1565
Others	5P7265	3S3264	3S3269	Same

only finger tight. Otherwise, damage may occur to the capsule seat.

5. Insert the correct extension rod (lightly oiled) in through the adapter; then install collet 3P1565. Install the dial indicator 9S215 along with the necessary contact point in the adapter.
6. With the timing indicator tools set up as shown in Figure 7-27, bar the engine over slowly until the reading on the dial indicator face is at its maximum. Carefully *zero in* the dial gauge.
7. Very gently continue to turn the engine in the same direction until a 0.020-in. (0.508-mm) reading registers on the dial face. Scribe or center punch a mark on either the flywheel, fan pulley, crank pulley, or vibration damper, but do not at any time place a punch mark on a viscous-type damper.
8. Slowly rotate the engine now in its opposite direction, past the zero reading, until a 0.020-in. (0.508-mm) reading registers on the dial face. At this point, scribe or center punch a second mark on the previously marked rotating member.

   Actual true TDC will be located midway between both of these marks; therefore, clearly mark TDC on the rotating member if it is different than the one already stamped there. Make sure to remove the old TDC mark if it is inaccurate.

Note: Be sure to *zero in* the dial indicator with the engine cylinder 1 now at true TDC.

Fuel-Injection Pump Timing Test (Procedure)

1. With TDC now established, the actual piston travel from the start of injection to

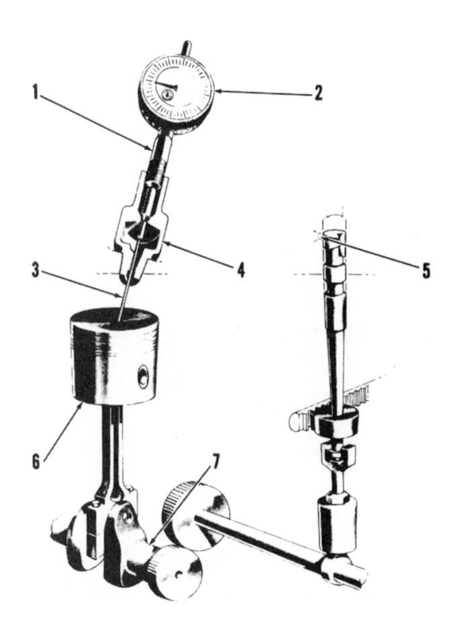

Figure 7-27
Typical Setup of Timing Indicator Tools. (Courtesy of Caterpillar Tractor Co.)

Index:
1—Adapter (varies with engine series)
2—Dial indicator
3—Rod
4—Pre-combustion chamber
5—Fuel inlet port
6—Piston
7—Engine crankshaft

TDC can be measured and converted to relative degrees of crankshaft rotation and compared to specifications to determine if the injection pump timing is correct.

2. With the dial indicator set up as before, bar the engine over in its reverse direction approximately 50 degrees, which will positively ensure removal of all gear backlash. Once this has been done, place the governor speed control lever in the full-fuel position.
3. Refer to Figure 7-28 and perform a fuel flow test through injection pump 1 to accurately establish the start of fuel injection (point of port closing, plunger to barrel inlet).
4. With the 1P540 flow checking tool group connected up as shown, open the fuel shutoff valve if the engine is so equipped, or if the engine uses a sleeve metering fuel system with a constant bleed valve, disconnect the fuel return line and place a cap on the constant bleed valve.

Note: On 3406 engines with direct injection, in place of fuel line (4) use 5P3564 flow line assembly, since it contains pins on each end that hold open the check valve in the fuel-injection pump bonnet. The end with the shorter pin attaches to the injection pump; however, if this fails to open the check valve, place the long pin end at the pump.

5. Disconnect the fuel line to the fuel filter and connect the hose (5) to the filter from the test tank, which should be filled with about 1 gallon (3.8 liters) of clean diesel fuel.
6. Pressurize the test tank to a maximum pressure of 15 psi (103.42 kPa) by use of the hand pump (7) on the top of the tank, or by use of a pressure-reducing regulator, apply shop air to this same value.

Caution: Under no circumstances apply more than 15 psi pressure.

7. Proceed to slowly bar the engine over in its normal direction of rotation (if turning it from the flywheel end, remember that you must turn it opposite to normal rotation) with a container positioned to catch the fuel coming out of injection pump 1 (line 4); slant line upward slightly. What you are actually doing here is shown in Figure 7-4(a) and (b).
8. As the plunger approaches the position shown in Figure 7-4(b), the flow of fuel

Index:

4—Fuel line from pump 1 to container
5—Fuel line (tank to filter)
6—Test fuel tank
7—Hand pump
8—Shop air supply (optional)
9—Air pressure regulator

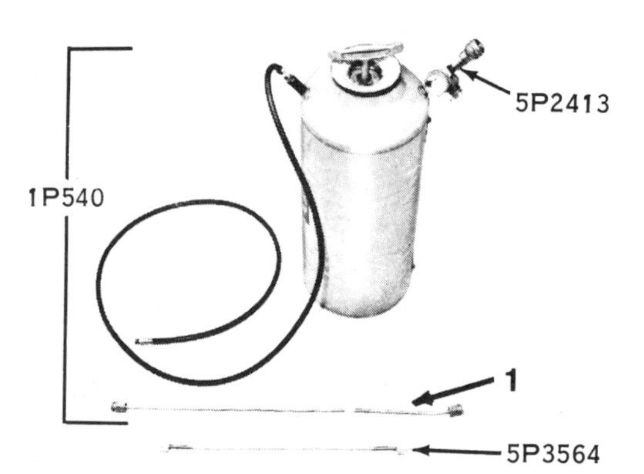

Index:

1P540—Flow checking tool group assembly
5P2413—Air regulator assembly (used to control 10 to 15 psi fuel pressure in fuel system during testing)
5P3564—Fuel flow line from pumping element 1: used only on 3406 DI (direct injection) engines
Item 1—Fuel flow line for use with other model engines

Figure 7-28
(a) Fuel Timing Check with IP-540 Flow Checking Tool Group; (b) IP540 Tool Components. (Courtesy of Caterpillar Tractor Co.)

from tube (4) will start to decrease to approximately 6 to 12 drops of fuel per minute on all engines, with the following exceptions:

a. On 5.4-in. (137.16-mm) bore engines with 3S1467 fuel pumps, it will be 12 to 30 drops per minute.

b. On sleeve metering fuel systems, it will be 12 to 18 drops per minute.

9. When the fuel rate is reduced to the point stated in paragraph 8, stop rotating the engine. Make reference to the chart shown in Figures 7-29 and 7-30 to determine the angle that corresponds to the dial indicator

reading obtained at this particular point, or cross-reference Caterpillar Special Instruction Bulletin No. 2MHS7083-01.

10. If the angle is not within specifications, reset the injection pump timing as per the following procedure (the allowable tolerance is ±1 degree):

Setting Pump Timing

Earlier sections have dealt with setting the injection pump timing dimension both on and off the engine; therefore, assuming that the internal adjustments such as the lifter yokes on flanged body pumps are correct, and the correct thickness of phasing spacer is installed on compact body pumps, the following is the typical procedure to correct pump *camshaft timing* if incorrect. See Figures 7-31 and 7-32.

3406 Engine

1. Refer to Figure 7-15, item 3A; remove the plug and install the 1P5550 fuel cam locking tool into position so that there is 6 to 12 drops of fuel per minute coming out of fuel line 1 into the container.
2. Refer to Figure 7-31 and slacken off bolt (2). Note that newer engines use four retaining bolts item (2) in place of one bolt used on earlier units. Strike the automatic timing advance unit (3) with a plastic or rubber hammer to loosen its hold on the end of the injection pump drive shaft. Ensure that it is in fact free to move.
3. From the front end of the engine, bar the crankshaft ccw approximately 90 degrees; then rotate it cw (clockwise) until the dial indicator attached to the cylinder adapter registers 0.600 in. (15.2 mm) and tighten up the bolt (2) only *hand tight*.
4. Slowly continue to turn the engine crankshaft clockwise until the dial gauge registers the correct specification shown in the chart in Figures 7-19 and 7-30.
5. Carefully torque the bolt (2) to 110 ± 10 lb-ft (149.13 ± 13.55 N·m) on earlier engines; on later engines with four retaining bolts, torque to 100 ± 5 lb-ft (135.58 ± 6.77 N·m).
6. Anytime that timing is altered, repeat the check from step 1 involving the use of the flow checking tool group.

D-343/1693 Engines

1. Refer to Figure 7-24. Insert the injection pump camshaft locating bolt into position and hold it there until 6 to 12 drops of fuel per minute, or 12 to 30 drops on 3S1467 fuel pumps, is attained from fuel-injection pump line 1.
2. Refer to Figure 7-32. After removing the bolts from the pump drive shaft inspection plate, slacken off the clamp bolt from the engine to the injection pump drive shaft as shown.
3. From the engine serial number located on the engine identification plate, establish from the applicable service manual the correct timing dimension for the particular engine arrangement and fuel-injection pump being used. Once this has been established, locate on the chart shown in Figure 7-29 the number of degrees that correspond to the desired reading on the dial indicator. Rotate the engine until the specification is attained; then torque up the clamp bolt.
4. Repeat the check from step 1 involving the use of the flow checking tool group.

7-4

Hydraulic Variable Timing Unit

Caterpillar diesel engines commonly use either a mechanical or hydraulic type of speed-sensing device to automatically advance the

	D311	D311H D320A	D315 D318	D330A D330B D333A D333B 1673	D326 D337	D326F D337F	D343 1693	641 650 651 657 660 666 D346 D348 D349	D339 D342 D8800 D13000	D353	D17000	D364 D375 D379 D386 D397 D398 D399	1674 D334	1673C D330C D333C 3304 3306	1676 D336 621 980	1140 1145 3145	1150 3150	1160 3160 3208	3406	3204	3408
3.0°									.007	.007				.006	.005				.006		.005
4.0°									.013	.013			.010	.010	.009				.010		.009
5.0°	.012	.013	.014	.014	.015	.016	.016	.016	.020	.020	.020	.019	.015	.016	.013	.010	.011	.013	.016	.008	.015
6.0°	.018	.018	.019	.020	.021	.024	.023	.022	.029	.030	.028	.028	.022	.023	.019	.014	.016	.018	.023		.021
7.0°	.024	.025	.026	.027	.029	.032	.031	.030	.040	.040	.038	.038	.029	.031	.026	.019	.022	.025	.032	.019	.029
7.5°							.036	.035	.046	.046			.034	.036	.030						
8.0°	.031	.032	.035	.036	.038	.042	.041	.040	.052	.052	.050	.049	.038	.041	.034	.025	.028	.032	.042		.038
8.5°							.046	.045	.059	.059			.043	.046	.039						
9.0°	.039	.041	.044	.045	.047	.053	.052	.050	.066	.066	.063	.062	.048	.052	.043	.032	.036	.041	.053	.034	.048
9.5°							.058	.056	.074	.074			.054	.058	.048						
10.0°	.048	.051	.053	.056	.059	.066	.064	.062	.082	.082	.078	.076	.060	.064	.053	.039	.044	.050	.065		.059
10.5°							.070	.068	.090	.090			.066	.070	.059						
11.0°	.059	.061	.065	.068	.072	.079	.077	.075	.099	.099	.094	.092	.072	.077	.064	.047	.053	.060	.078	.053	.071
11.5°							.084	.082	.108	.108			.079	.084	.070						
12.0°	.070	.073	.078	.080	.085	.094	.092	.089	.117	.118	.112	.110	.086	.092	.077	.056	.063	.072	.093		.084
12.5°							.099	.097	.127	.127			.093	.099	.083						
13.0°	.082	.085	.091	.094	.100	.111	.107	.104	.138	.138	.132	.129	.101	.107	.090	.066	.074	.084	.109	.076	.099
13.5°							.116	.113	.148	.149			.108	.116	.097						
14.0°	.095	.099	.106	.109	.115	.128	.124	.121	.159	.160	.153	.149	.117	.124	.104	.077	.086	.097	.127		.115
14.5°							.133	.130	.171	.171			.125	.133	.111						
15.0°	.108	.113	.121	.125	.133	.147	.143	.139	.183	.183	.175	.171	.134	.142	.119	.088	.098	.112	.145	.100	.132
15.5°							.152	.148	.195	.196			.143	.152	.127						
16.0°	.123	.129	.138	.143	.151	.167	.162	.158	.208	.208	.199	.194	.152	.162	.135	.100	.112	.127	.165		.150
16.5°							.172	.168	.220	.221			.161	.172	.144						
17.0°	.139	.145	.155	.161	.170	.188	.183	.178	.234	.235	.244	.219	.171	.182	.153	.112	.126	.143	.186	.131	.168
18.0°	.156	.162	.174	.180	.191	.215	.205	.199	.262	.263	.251	.246	.192	.204	.171	.126	.141	.160	.208	.148	.189
19.0°	.173	.180	.193	.200	.212	.234	.228	.221	.291	.292	.279	.273	.213	.227	.190	.140	.157	.178	.232	.166	.210
20.0°	.191	.199	.214	.221	.234	.259	.252	.245	.322	.323	.309	.302	.236	.259	.210	.155	.173	.197	.256	.184	.232
21.0°	.211	.220	.236	.243	.258	.286	.277	.270	.355	.356	.340	.333		.277	.231	.170	.191	.217	.282	.204	.256
22.0°	.231	.241	.258	.266	.282	.313	.303	.296	.389	.390	.372	.364		.303	.254	.187	.209	.238	.309	.225	.280
23.0°	.252	.263	.281	.290	.308	.341	.331	.322	.424	.425	.406	.398		.331	.277	.204	.228	.259	.337	.245	.305
24.0°	.274	.286	.306	.316	.334	.371	.360	.350	.461	.462	.442	.432		.360	.301	.221	.248	.282	.366	.267	.332
25.0°	.297	.309	.331	.342	.362	.402	.390	.379	.499	.501	.478	.478		.389	.325	.240	.268	.305	.396	.288	.359
26.0°																			.428		.388
27.0°																			.460		.417
28.0°																			.494		.448
29.0°																			.529		.479
30.0°																			.564		.512

Figure 7-29
Actual Piston Travel to TDC with Tool Arrangement Used in Figure 7-27 (in Inches). (Courtesy of Caterpillar Tractor Co.)

	D311	D311H D320A	D315 D318	D330A D330B D333A D333B 1673	D326 D337	D326F D337F	D343 1693	641 650 651 657 660 666 D346 D348 D349	D339 D342 D8800 D13000	D353	D17000	D364 D375 D379 D386 D397 D398 D399	1674 D334	1673C D330C D333C 3304 3306	1676 D336 621 980	1140 1145 3145	1150 3150	1160 3160 3208	3406	3204	3408
3.0°									0,18	0,18				0,15	0,13				0,15		0,13
4.0°									0,33	0,33			0,25	0,25	0,23				0,25		0,23
5.0°	0,30	0,33	0,36	0,36	0,38	0,41	0,41	0,41	0,51	0,51	0,51	0,48	0,38	0,41	0,33	0,25	0,28	0,33	0,41	0,20	0,38
6.0°	0,46	0,46	0,48	0,51	0,53	0,61	0,58	0,56	0,74	0,76	0,71	0,71	0,56	0,58	0,48	0,36	0,41	0,46	0,58		0,53
7.0°	0,61	0,64	0,66	0,69	0,74	0,81	0,79	0,76	1,02	1,02	0,97	0,97	0,74	0,79	0,66	0,48	0,56	0,64	0,81	0,48	0,74
7.5°							0,91	0,89	1,17	1,17			0,86	0,91	0,76						
8.0°	0,79	0,81	0,89	0,91	0,97	1,07	1,04	1,02	1,32	1,32	1,27	1,24	0,97	1,04	0,86	0,64	0,71	0,81	1,07		0,97
8.5°							1,16	1,14	1,50	1,50			1,09	1,16	0,99						
9.0°	0,99	1,04	1,12	1,14	1,19	1,35	1,32	1,27	1,68	1,68	1,60	1,57	1,22	1,32	1,09	0,81	0,91	1,04	1,35	0,86	1,22
9.5°							1,47	1,42	1,88	1,88			1,37	1,47	1,22						
10.0°	1,22	1,30	1,35	1,42	1,50	1,68	1,63	1,57	2,08	2,08	1,98	1,93	1,52	1,63	1,35	0,99	1,12	1,27	1,65		1,50
10.5°							1,77	1,73	2,29	2,29			1,68	1,78	1,50						
11.0°	1,50	1,55	1,65	1,73	1,83	2,01	1,96	1,91	2,51	2,51	2,39	2,34	1,83	1,96	1,63	1,19	1,35	1,52	1,98	1,35	1,88
11.5°							2,13	2,08	2,74	2,74			2,01	2,13	1,78						
12.0°	1,78	1,85	1,98	2,03	2,16	2,39	2,34	2,26	2,97	3,00	2,84	2,79	2,18	2,34	1,96	1,42	1,60	1,83	2,36		2,13
12.5°							2,51	2,46	3,23	3,25			2,36	2,51	2,11						
13.0°	2,08	2,16	2,31	2,39	2,53	2,82	2,72	2,64	3,51	3,51	3,35	3,28	2,57	2,71	2,29	1,68	1,88	2,13	2,77	1,93	2,51
13.5°							2,95	2,87	3,76	3,78			2,74	2,95	2,46						
14.0°	2,41	2,51	2,69	2,77	2,92	3,25	3,15	3,07	4,04	4,06	3,89	3,78	2,97	3,15	2,64	1,96	2,18	2,46	3,23		2,92
14.5°							3,37	3,30	4,34	4,34			3,18	3,38	2,82						
15.0°	2,74	2,87	3,07	3,18	3,38	3,73	3,63	3,53	4,65	4,65	4,45	4,34	3,40	3,61	3,02	2,24	2,49	2,84	3,68	2,54	3,35
15.5°							3,86	3,76	4,95	4,98			3,63	3,86	3,23						
16.0°	3,12	3,28	3,51	3,63	3,84	4,24	4,11	4,01	5,28	5,28	5,05	4,93	3,86	4,11	3,43	2,54	2,84	3,23	4,19		3,81
16.5°							4,37	4,27	5,59	5,61			4,09	4,37	3,56						
17.0°	3,53	3,68	3,94	4,09	4,32	4,78	4,65	4,52	5,94	5,97	5,69	5,56	4,34	4,62	3,89	2,84	3,20	3,63	4,72	3,33	4,27
18.0°	3,96	4,11	4,42	4,57	4,85	5,46	5,21	5,05	6,65	6,68	6,38	6,25	4,88	5,18	4,34	3,20	3,58	4,06	5,28	3,76	4,80
19.0°	4,39	4,57	4,90	5,08	5,38	5,94	5,79	5,61	7,39	7,42	7,09	6,93	5,41	5,77	4,83	3,56	3,99	4,52	5,89	4,22	5,33
20.0°	4,85	5,05	5,44	5,61	5,94	6,58	6,40	6,22	8,18	8,20	7,85	7,67	5,99	6,58	5,33	3,94	4,39	5,00	6,50	4,67	5,89
21.0°	5,36	5,59	5,99	6,17	6,55	7,26	7,04	6,86	9,02	9,04	8,64	8,46		7,04	5,87	4,32	4,85	5,51	7,16	5,18	6,50
22.0°	5,87	6,12	6,55	6,76	7,16	7,95	7,70	7,52	9,88	9,91	9,45	9,25		7,70	6,45	4,75	5,31	6,05	7,85	5,72	7,11
23.0°	6,40	6,68	7,14	7,37	7,82	8,66	8,41	8,18	10,77	10,80	10,31	10,11		8,41	7,04	5,18	5,79	6,58	8,56	6,22	7,75
24.0°	6,96	7,26	7,77	8,03	8,48	9,42	9,14	8,89	11,71	11,73	11,23	10,97		9,14	7,65	5,61	6,30	7,16	9,30	6,78	8,43
25.0°	7,54	7,85	8,41	8,69	9,19	10,21	9,91	9,63	12,67	12,73	12,14	12,14		9,88	8,26	6,10	6,81	7,75	10,06	7,32	9,12
26.0°																			10,87		9,86
27.0°																			11,68		10,59
28.0°																			12,55		11,38
29.0°																			13,44		12,17
30.0°																			14,33		13,00

Figure 7-30
Same Chart as in Figure 7-29 with Readings in Millimeters.
(Courtesy of Caterpillar Tractor Co.)

Index:
1—Engine front cover
2—Retaining bolts
3— Automatic timing advance unit

Figure 7-31
3406 Engine, Automatic Timing Advance Mechanism.
(Courtesy of Caterpillar Tractor Co.)

Figure 7-32
D-343 Engine, Pump Drive Shaft Clamping Arrangement.
(Courtesy of Caterpillar Tractor Co.)

start of injection in relation to the engine speed increase. The mechanical type is explained in Section 7-10.

Although the *hydraulic* variable timing unit is so called, it relies on the action of two small flyweights to actually create an advance of the injection pump camshaft. When used, this type of timing advance unit is driven from the engine gear train accessory drive gear, and is directly connected to the fuel-injection pump camshaft just as the straight mechanical one is. The timing unit is shown in Figure 7-33.

The major parts of this system are basically two L-shaped flyweights, a sliding control valve, and control valve spring. The drive shaft has the accessory drive gear bolted to its right-hand end and is therefore turning at engine speed. On the left-hand end of the shaft is connected a power piston, which in turn is splined to the fuel-injection pump camshaft. These internal camshaft splines allow axial

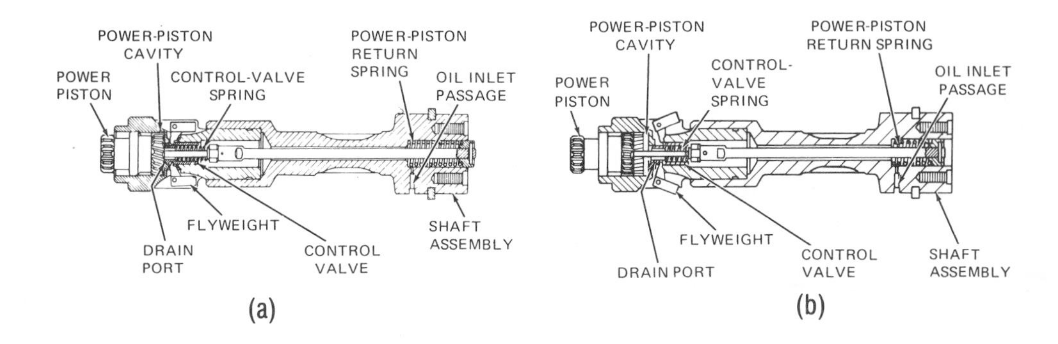

Figure 7-33
Hydraulic Variable Timing Unit: (a) Low rpm position and (b) high rpm position. (Courtesy of Caterpillar Tractor Co.)

(lengthwise) movement of the power piston without the camshaft moving itself. At the right-hand end we also have the power piston return spring. Take careful note here that the splines meshing with the camshaft are *straight* cut, whereas those at the opposite end meshing with the drive shaft assembly are *helical* cut.

Running through the hollow drive shaft is a long shaft known as the power piston control shaft; it is connected to the power piston return spring at the right-hand end and to the rod of the power piston at the other. The return spring seats within the bore of the drive shaft and is pinned to the control shaft (the spring seat is pinned to the control shaft).

When the engine is running at any rpm below approximately 1350, engine oil under lube pump pressure enters the drive shaft assembly at its right-hand end, filling the return spring cavity and working along the hollow drive shaft between the power piston control shaft to the cavity in front of the power piston, where it simply exhausts around the valve and drains back to the engine oil pan. Therefore, there is not enough oil pressure at this time to move the power piston. The centrifugal force created by the two small L-shaped flyweights is insufficient to overcome the small control valve spring.

As the engine is accelerated, the increased force developed by the revolving flyweights will reach a point where they will overcome the compression of the small control valve spring. This weight action moves the control valve to its right, throttling and then closing off the small drain hole that previously existed between the piston cavity and the valve. The buildup of oil pressure in the power piston cavity moves the piston to the left, which causes it to now overcome the force of the large spring at the opposite end.

We mentioned earlier in this discussion that the power piston was free to move axially within the mating splines of the camshaft (straight splines), but that the splines at the other end were helical cut; since a helical cut spline in moving forward causes a slight change in angle, the endwise movement of the power piston will cause the ends of the drive unit to change its angular relationship accordingly. This change will therefore advance the fuel-injection pump camshaft and consequently the start of injection.

However, any lengthwise movement of the power piston will obviously cause an increase in the force of the small control valve spring, which will start to uncover the control valve oil drain hole, allowing the power piston return spring to reestablish the power piston and control shaft back into their original position before the advance occurred.

At any speed above 1350 rpm, a state of balance can occur between the weight and spring force, just as in a governor situation. Therefore, at any time that a state of balance exists, the oil pressure will remain constant.

Remember, too, that the faster the engine rpm is, the greater the amount of advance, until the power piston will butt against the camshaft.

7-5

Rack Setting Information: 3406

Due to the broad range of Caterpillar diesel engine equipment applications, the actual rack setting can vary even with the same engine powering the same piece of equipment. The reasons for this are to allow the customer the opportunity to choose a particular rating for that specific application peculiar to its operating environment, thereby affording wide flexibility in the use of one engine. This then makes it imperative for the mechanic or serviceman to refer to the most up-to-date Caterpillar literature available prior to setting or adjusting any engine rack; otherwise, problems will occur if the setting is incorrect. Such literature is available from any authorized Caterpillar dealer for a small charge.

For instructional purposes, the 3406 engine is used to clarify this point as to how rack settings can vary with the same engine. Tables 7-2 through 7-4, show typical variations found on Caterpillar engines of the same model. Once you have studied these charts, it will become readily apparent that you cannot automatically assume that, because one engine produces a certain horsepower (kilowatt rating) at a given speed, all other engines within that family will be set the same.

Note also that, although the rack setting can be varied on the engine, in order to change the engine's output, the governor group, torque spring, and in some cases the turbocharger may have to be changed, along with the addition of an aftercooler; be aware of this at all times.

Prior to setting the fuel rack or making any changes, refer to Figures 7-27 and 7-29, which give the sequence for checking fuel injection pump timing on the engine by checking piston travel and converting it to degrees of crankshaft rotation. By following these considerations, you will be assured of obtaining the specified performance from the engine.

7-6

Torque Spring Operation

A torque spring and spacer setup is found on some models of Caterpillar diesel engines; it allows additional rack travel toward the fuel-on position within predetermined limitations once the rack stop collar makes contact with the torque spring assembly. Such a device is shown in Figure 7-34.

When the fuel rack setting is being done on an engine, if a torque spring is used, once the rack collar comes in contact with the spring, additional throttle pressure (under heavy load situations) will allow the spring to deflect until it eventually hits the stop bar, thereby preventing any additional fuel input to the engine cylinders. The point at which the rack stop collar hits the spring is established when the continuity light comes on during the rack setting check.

Torque springs are available as different groups; therefore, always check the applicable service or parts literature for a breakdown of the particular components within the engine group. For example, on 3406 engines the

Table 7-2
RACK SETTING INFORMATION: 3406 TRUCK ENGINE SETTINGS, PRE-COMBUSTION CHAMBERED, TURBOCHARGED (PC-T)

Model Description Serial Number	*Full Load RPM (±10 RPM)*	*High Idle RPM (±30 RPM)*	*Brake Horsepower Without Fan*	*Static Rack Setting*	*Governor Group Number*	*Governor Spring Number*	*Torque Spring Group*	*Turbo-charger Assembly*	*Maximum Altitude (Feet)*	*After-cooler Water System*	*Manifold Pressure Full Load (Inches of Hg)*	*Brake Specific Fuel Consumption*	*Type of Service*
3406 (PC-T)	2100	2285	325	.105″	9N72	9L6446	None	9L6309	7500-L	None	35.0 ± 3	.422	Truck Rating (A)
Truck Engine	2100	2270	280	.050″	9N72	9L6446	None	9N47	7500-L	None	35.0 ± 3	.413	Truck Rating (A)
Turbocharged													
92U1-92U2766													
92U2767-Up	2100	2300	325	.095″	9N588	9L6446	(C)	9L6309	7500-L		36.0 ± 3	.404	Truck (B)
Turbocharged	2100	2300	325	.095″	9N588	9L6446	(C)	9L6309	7500-L		36.0 ± 3	.404	Brakesaver (B)
	2100	2260	280	.045″	9N588	9L6446	(C)	9N47	7500-L		31.0 ± 3	.404	Truck (B)
	1950	2175	300	.045″	9N588	9L6446	(C)	9L6309	7500-L		30.0 ± 3	.403	Derated (B)
	1950	2125	255	.030″	9N588	9L6446	(C)	9N47	7500-L		26.0 ± 3	.400	Derated (B)

GENERAL NOTES:

Color Code of Governor Spring:
 9L6446—One Green, One Blue Stripe.

Low Idle is 600 ± 10 RPM.

RACK SETTING will be .020″ greater with engine running.

L—Lug Application.

FUEL INJECTION TIMING (STATIC) is 10° BTC; .065 inch of piston travel.

FUEL PUMP TIMING DIMENSION IS 4.3475 ± .0020 inch on engine with piston at TDC.

BHP will be 5 HP less if equipped with air compressor and alternator.

BHP will be 3 HP less if equipped with BrakeSaver.

(Courtesy of Caterpillar Tractor Company)

SPECIFIC NOTES:

(A) Air–Fuel Ratio Control Setting is −.075 with engine running.

(B) Air–Fuel Ratio Control Setting is −.030 with engine running.

(C) 92U2767-92U4736: No Torque Control Group.
 92U4737-Up: 9N2893 Torque Control Group:
  Torque Spring: 9L3625 (.022″)
  Spacer: 4N7777 (.010″)

Table 7-3
RACK SETTING INFORMATION: 3406 TRUCK ENGINE, PRE-COMBUSTION CHAMBERED, TURBOCHARGED, AFTERCOOLED (PC-TA)

Model Description Serial Number	Full Load RPM (±10 RPM)	High Idle RPM (±30 RPM)	Brake Horsepower Without Fan	Static Rack Setting	Governor Group Number	Governor Spring Number	Torque Spring Group	Turbo-charger Assembly	Maximum Altitude (Feet)	After-cooler Water System	Manifold Pressure Full Load (Inches of Hg)	Brake Specific Fuel Consumption	Type of Service
3406 (PC-TA)	2100	2300	360	.135″	9N72	9L6446	*9L5176(.024″)	9N711	7500-L	Jacket	40.0 ± 3	.404	Truck Rating (A)
Turbocharged							**8L8601(.002″)						
Aftercooled							**8L8601(.002″)						
Truck Engine							**2S9524(.010″)						
92U1-92U2766													
92U2767-Up	2100	2300	360	.120″	9N72	9L6446	*9L5176(.024″)	9N711	7500-L	Jacket	41.8 ± 3	.390	Truck Rating (B)
	2100	2300	360	.120″	9N72	9L6446	**8L8601(.002″)	9N711	7500-L	Jacket	41.8 ± 3	.390	BrakeSaver (B)
	2100	2290	340	.095″	9N72	9L6446	**8L8601(.002″)	9N711	7500-L	Jacket	36.0 ± 3	.400	Derated (B)
							**4N7777(.010″)						
	1950	2175	340	.105″	9N72	9L6446	*9N1240(.028″)	9N711	7500-L	Jacket	39.0 ± 3	.389	Economy (B)
							**9L7235(.040″)						
							**9L9347(.020″)						

*Torque Spring. **Spacers.

GENERAL NOTES:
Color Code of Governor Springs:
9L6446—One Green, One Blue Stripe.
LOW IDLE is 600 ± 10 RPM.
RACK SETTING will be .020″ greater with engine running.
L—Lug Application.
FUEL INJECTION TIMING (static) is 10° BTC, .065 inch of piston travel.
FUEL PUMP TIMING DIMENSION is 4.3475 ± .0020″ on engine, with piston at TDC.
BHP will be 5 HP less if equipped with air compressor and alternator.
BHP will be an additional 3 HP less if equipped with BrakeSaver.

SPECIFIC NOTES:
(A) Air–Fuel Ratio Control Setting is −.055, Engine Running.
(B) Air–Fuel Ratio Control Setting is −.050, Engine Running.

(Courtesy of Caterpillar Tractor Company)

Table 7-4
RACK SETTING INFORMATION: 3406 TRUCK, ENGINE, DIRECT INJECTION, TURBOCHARGED ONLY (DI-T)

Model Description Serial Number	*Full Load RPM (±10 RPM)*	*High Idle RPM (± 30 RPM)*	*Brake Horsepower Without Fan*	*Static Rack Setting*	*Governor Group Number*	*Spring Number*	*Torque Spring Group*	*Turbo-charger Assembly*	*Maximum Altitude (Feet)*	*Aftercooler Water System*	*Manifold Pressure Full Load (Inches of Hg)*	*Brake Specific Fuel Consumption*	*Type of Service*
3406 (DI-T)	2100	2360	325	.190″	9N798	9L6446	None	9L6758	7,500-L	None	34.0 ± 3	.380	(A) Truck
Truck Engine	2100	2300	280	.130″	9N49	9L6446	None	9L6308	7,500-L	None	32.5 ± 3	.392	(B) Truck
Turbocharged	2100	2300	280	.130″	9N49	9L6446	None	9L6113	7,500-L	None	33.5 ± 3	.391	(B) Slant Truck)
92U1-92U3792													
92U3793-Up	2100	2365	325	.185″	9N798	9L6446	None	9L6758	7,500-L	None	34.5 ± 3	.373	(A) Truck
	2100	2365	325	.185″	9N798	9L6446	None	9L6758	7,500-L	None	34.5 ± 3	.373	(A) BrakeSaver
	2100	2300	280	.110″	9N798	9L6446	None	9L6308	7,500-L	None	33.0 ± 3	.372	(A) Truck
	1900	2200	305	.170″	9N798	9L6446	None	9L6758	7,500-L	None	28.0 ± 3	.369	(A) Economy
	1900	2150	280	.138″	9N2626	9L6446	9N2806	9L6308	7,500-L	None	32.0 ± 3	.370	(A) Economy
	1900	2100	255	.100″	9N2626	9L6446	None	9L6308	7,500-L	None	28.0 ± 3	.371	(A) Derated
	2100	2300	280	.110″	9N2626	9L6446	9N2582	9L6308	7,500-L	None	33.0 ± 3	.372	(A) High Torque
	2100	2330	300	.150″	9N798	9L6446	—	9L6758	7,500-L	None	28.0 ± 3	.379	(A) High Torque

GENERAL NOTES:

Color Code of Governor Spring:
9L6446—One Green, One Blue Stripe.
Low idle is 600 RPM.
Rack Setting will be .020″ greater with engine running.
BHP will be 5 HP less if equipped with air compressor and alternator.
BHP will be 3 HP less if equipped with BrakeSaver.
Torque Springs and Spacers:
9N2806 Torque Spring Group:
*9L6483 (.020″)
**9L3626 (.016″)
**4N7777 (.010″)
**8L8601 (.002″)
9N2582 Torque Spring Group:
*9N1240 (.028″)
**9L9347 (.020″)
**9L7235 (.040″)

SPECIFIC NOTES:

(A) Fuel injection timing (static) is 28° BTC; .4941 inch of piston travel. Fuel pump timing dimension is 4.2003 ± .0020 inches, on engine with piston at TDC.

(B) Fuel injection timing (static) is 29° BTC; .5287 inch of piston travel. Fuel pump timing dimension is 4.2236 ± .0020 inches, on engine with piston at TDC.

(Courtesy of Caterpillar Tractor Company)

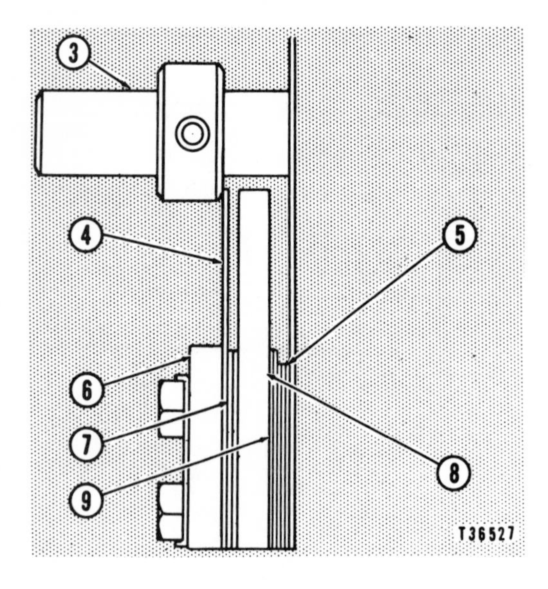

Figure 7-34
Torque Spring Assembly.
(Courtesy of Caterpillar Tractor Co.)

Index:
3—Rack
4—Torque spri
5—Shims
6—Clamp
7—Spacer
8—Stop bar
9—Spacer

spring is available in both 0.022-and 0.024-in. (0.558-mm) sizes. The stop bar is available only 0.118 in. (2.997 mm) thick; however, spacers are available in 0.002, 0.005, and 0.010 in. (0.050, 0.127, and 0.254 mm) thicknesses, with the shim being 0.020 in. (0.508 mm). Additional information can be found in Tables 7-2 through 7-4. Shims can therefore be added or removed to obtain the proper rack setting.

7-7

Adjustment of Fuel Control Rack

All Caterpillar diesel engines must have the fuel-injection pump control rack adjusted to a particular specification for the application in which it will operate. This adjustment controls the maximum amount of fuel that can be delivered to the engine cylinders and will therefore control the rated horsepower (kilowatt) produced by the engine at a particular maximum governed engine speed. Failure to properly set the rack travel can lead to complaints of a lack of power, and also can affect the performance, fuel economy, and life of the engine. Setting the rack travel too high can adversely affect the engine's life.

To properly carry out any rack adjustment, special Caterpillar tools are required. Such a tool set can be seen in Figure 7-35. This set covers all Caterpillar engines and is a master set.

The sequence of events in setting the fuel rack is basically the same regardless of the specific engine model; therefore, let us use a 3406 engine as a typical example for instructional purposes.

3406 Fuel Rack Setting

The tools required are: (1) 8S4627 circuit tester, (2) 9S8521 rod, (3) 9S8519 plug, (4) 9S7350 bracket group, (5) 9S215 dial indicator, (6) 8S4622 spring, and (7) 8S4623 dust cover.

Procedure

1. Refer to Figure 7-15(a). Remove items 1 through 3, the rack stop, spacer, and gaskets, by taking off the two bolts.
2. The governor control linkage must be dis-

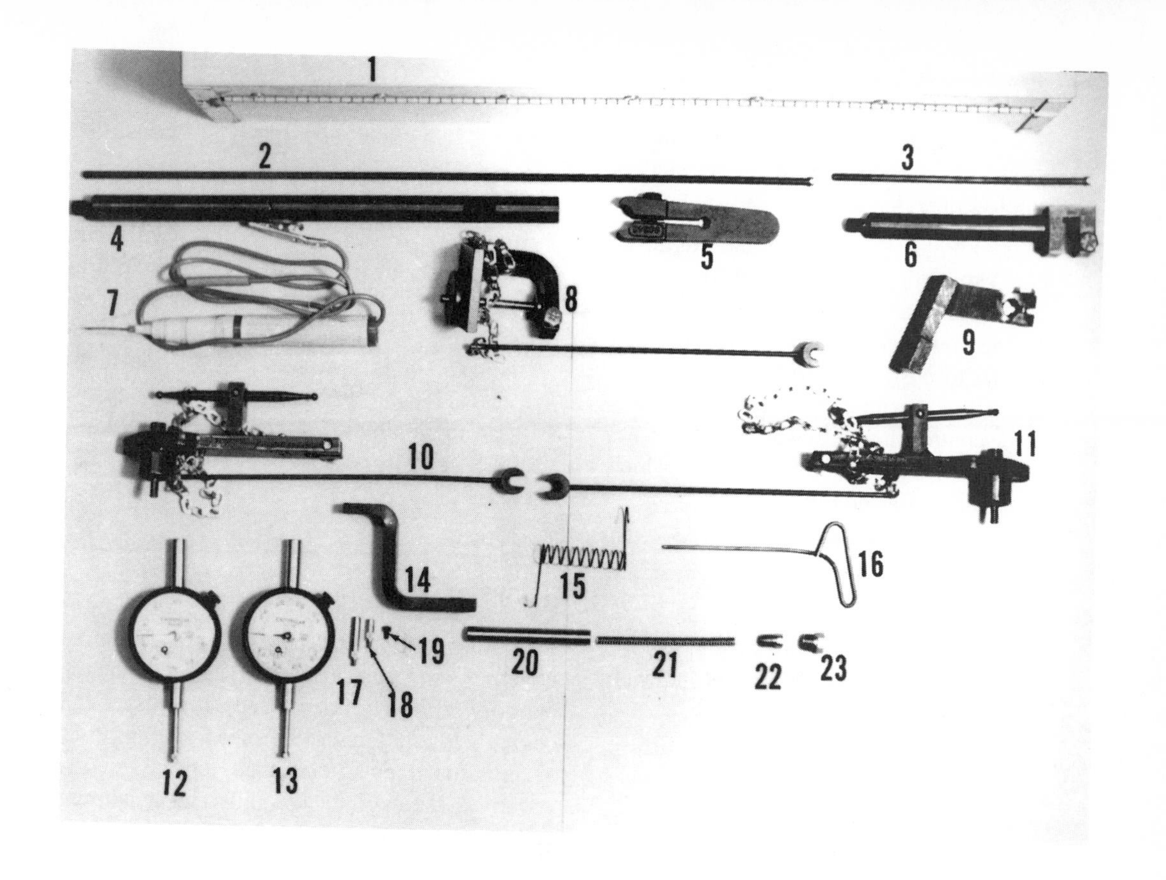

Figure 7-35
Caterpillar Service Tool Set.
(Courtesy of Caterpillar Tractor Co.)

Index:

1—956332 box assembly
8S6690 block
9S8105 liner
2—5S8088 rod assembly
3—5S8090 rod assembly
5S8086 point
4—5S8157 extension
5—9S7344 clamp assembly
S1616 bolt
6—9S239 adapter
S1616 bolt
7—8S4627 circuit tester
8—9S227 bracket group
9—9S7343 bracket assembly
S1616 bolt
10—9S7350 bracket group
11—9S238 bracket group
12—8S2283 dial indicator
8S3675 contact point (0.125 in.) (3.175 mm)
13—9S215 dial indicator
14—9S225 bracket assembly
2H191 bolt
15—9S2017 spring
16—9S8521 rod
17—3S3269 contact point (1 in.) (25.4 mm)
18—9S8883 contact point (0.50 in.) (12.7 mm)
19—3S3268 contact point (0.25 in.) (6.35 mm)
20—8S4623 dust cover
21—8S4622 spring
22—9S8518 plug
23—9S8519 plug

connected so as to allow free movement of the governor control lever throughout its complete operating range.

3. Refer to Figure 7-15(b). If the engine is equipped with a speed limiter, screw plug 4 into the underside of the governor housing and push the rod 9S8521 through the plug to push the speed limiter plunger inward to allow movement of the governor speed control lever toward the full-fuel position. Tighten the 9S8519 plug, which acts as a lock for the rod.
4. Install the 8S4623 dust cover onto the dial indicator stem after removing the standard dust cover. See Figure 7-36.
5. Refer to Figure 7-36. Install the 9S7350 bracket group (7) and 9S215 dial indicator as shown.
6. Place the governor speed control lever in the no-fuel position. Place the spacer (6) on the rod connected to the chain, over the rod that makes contact with the rack.
7. Place the dial indicator on zero, which will ensure that the fuel control rack is in the centered position. Remove the spacer (6) from the rod.
8. Attach the clip end of the 8S4627 circuit tester (11) to the brass terminal (10) on the governor housing as shown in Figure 7-37. Connect the other end of the tester to a good ground anywhere on the engine.
9. Gently move the governor speed control lever toward the fuel-on position until the tester light comes on. Pause here and very gently move the governor lever back toward the fuel shutoff position just until the light goes out. Very carefully bring the governor lever forward again until the test light just barely glimmers on. At this point, the rack stop collar (13) shown in Figure 7-38 is just contacting the torque spring or stop bar, whichever is used. Note that the air–fuel ratio control has been removed from the end of the governor housing to allow access to the fuel rack adjusting screw.
10. Note the reading on the face of the dial indicator and compare this with the specification given in Caterpillar Form No. SENR7064-03, Rack Setting Information for 3406 Diesel Truck Engines. If an adjustment is necessary, loosen the locknut (12) and turn the adjustment screw (14) to obtain the correct fuel rack setting.

7-8

Governor Adjustments

Adjustments to any Caterpillar diesel engine governor are basically the same, with the majority being confined to adjustments of the engine's low and high idle speed screws. Prior to any adjustments being made on any engine governor, the engine's low idle and maximum governed no-load speed (high idle) should be established by the use of an accurate tachometer, such as one of the readily available digital electronic magnetic tape pickup units. These readings should then be compared to the published figures given for that particular engine and application shown in the same booklet for rack setting information.

Adjustment Procedure

1. Refer to Figure 7-39. Ensure that the engine has attained its normal operating temperature prior to checking both the idle rpm and the maximum engine speed. Use an accurate tachometer.
2. If the low idle requires adjustment, turn the adjustment screw until the desired rpm is achieved. Manually disturb the engine speed by movement of the throttle and allow the rpm to drop back to idle. Check for a steady idle speed and compare to the published specification.

Index:
6—Spacer
7—9S7350 bracket group
8—9S215 dial indicator
9—8S4623 dust cover

Figure 7-36
Checking Fuel Rack Setting.
(Courtesy of Caterpillar Tractor Co.)

Index:
10—Brass terminal
11—Circuit tester

Figure 7-37
Attaching Circuit Tester to Terminal.
(Courtesy of Caterpillar Tractor Co.)

3. Adjustment of the maximum no-load (high idle) rpm is accomplished by removing the cover (3) and turning the screw (2) as necessary to attain the maximum speed desired according to published specifications. Follow the same basic procedure as for the low idle check. Note that some models of Caterpillar engines have both the low and high idle screws located in behind the cover (3).

Figure 7-38
Rack Stop Collar.
(Courtesy of Caterpillar Tractor Co.)

Index:
12—Locknut
13—Stop collar
14—Adjustment screw

7-9

Hydraulic Air–Fuel Ratio Control

Due to the strict EPA regulations regarding exhaust smoke emissions, all current high-speed-type turbocharged diesel engines employ a mechanical device whose main purpose is to ensure that at any given load and speed the amount of fuel delivered to the engine's combustion chamber will be in direct proportion to the amount of boost air pressure delivered by the engine's exhaust-driven turbocharger. To this end, Caterpillar uses a hydraulic air–fuel ratio control installed on the back side of the governor. Figures 7-40 and 7-41 show the stackup of parts used with this unit.

With reference to Figure 7-40, the activating valve and piston (4) engage the end of the fuel-injection pump rack stop collar (3). With the engine stopped, spring pressure places the rack in the full-fuel position for starting purposes. The activating valve and piston (4) have movement controlled by engine oil pressure

Index:
1—Low idle screw
2—Maximum speed adjusting screw or high idle
3—Cover
4—Engine speed limiter

Figure 7-39
Low and Maximum Speed Adjustment Location.
(Courtesy of Caterpillar Tractor Co.)

delivered from the speed limiter chamber of the governor (not shown) through passage (17) to the left-hand side of chamber (18). The action of this oil pressure against the piston (4) will therefore control the fuel rack movement. The oil in chamber (18) is itself controlled by air pressure in the engine's intake manifold delivered through tube (20), which acts against diaphragm (9), in turn moving valve (6).

Operation: Engine Stopped

With the engine stopped and, obviously, no oil pressure being delivered to passage (17), and no air pressure being delivered to tube (20), spring force (13) pushes the injection pump fuel control rack to the full-fuel position to facilitate starting, as shown in Figure 7-40. With the control valves in such a position, a drain passage (22) is open from chamber (18) so that no oil will be held in here at this time.

Operation: Engine Running (No Turbocharger Boost)

When the engine is initially started, to prevent excessive acceleration, contained within the governor assembly is a speed limiter that is dependent upon oil flow to allow placement of the throttle lever to the maximum fuel position. When the engine oil pressure is high enough, the plunger of the speed limiter is moved out of the way of the governor control mechanism, which allows oil to flow to passage (17) and chamber (18).

Since there is no load on the engine at this time, turbocharger boost pressure (air) is not high enough to cause control valve (6) to move to the left. With the activating valve (4) all the way to the left and the control valve all the way to the right, oil from passage (17) will simply drain back to the engine governor housing through the open passage (22).

Operation: Engine Running Under Load (with Turbocharger Boost)

With a load applied to a running engine, the turbocharger delivery air pressure will increase, sending this pressure into chamber (19), which will cause control valve (6) to move to its left and block off the drain holes (22). The buildup of oil pressure causes the activating valve and piston (4) to move to its right, thereby setting it in this position during engine operation. Movement of the activating valve (4) causes its left-hand end to butt up against the fuel rack stop collar, forcing it to a decreased fuel position. Therefore, the hydraulic air–fuel ratio control will automatically vary the amount of fuel to the engine in proportion to turbocharger air boost pressure, thereby preventing incomplete combustion and black exhaust smoke. These changes in turbocharger air boost pressure will vary the position of control valve 6, causing modulation of the actual oil pressure on the activating valve and piston (4).

Adjustment of Hydraulic Air–Fuel Ratio Control

Prior to making any adjustment to the hydraulic air–fuel ratio control, the fuel rack setting must be correct. See Fuel Rack Adjustment, described earlier. Any adjustment to the fuel ratio control is accomplished by removing the cover (11) shown in Figure 7-40. By using the cover over the end of the adjustment pin (16), the control valve (6) can be turned in or out, as the case may be, to effectively alter the setting.

1. When the control valve is turned in clockwise by the use of the cover, it causes the control valve (6) and activating valve and piston (4) to move farther to the left, thereby effectively changing the activating valve set position, which was

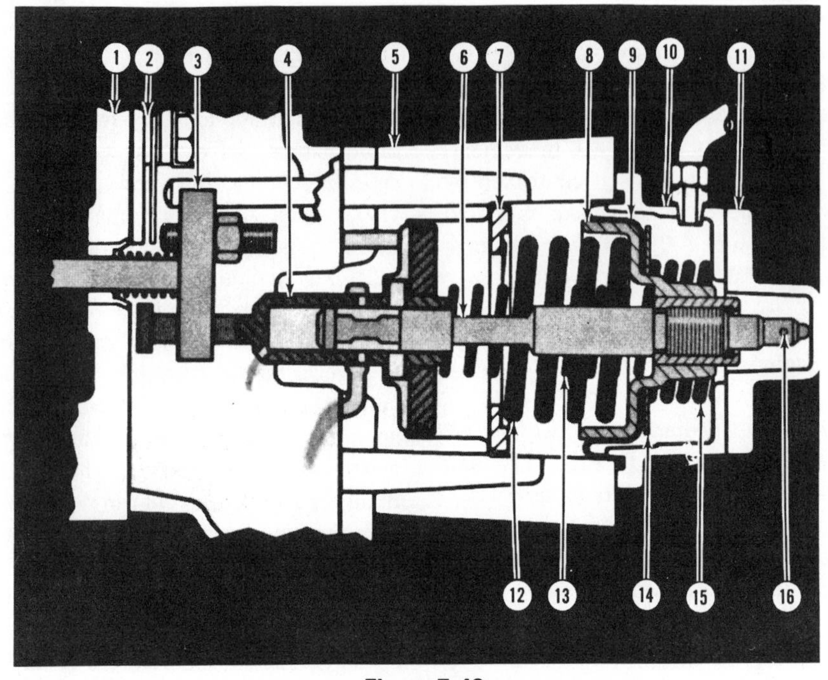

Figure 7-40
Hydraulic Air–Fuel Ratio Control: Position of Control Valve with Engine Stopped. (Courtesy of Caterpillar Tractor Co.)

Index:

1—Housing of the governor
2—Stop bar and torque spring
3—Rack stop collar
4—Activating valve and piston
5—Housing of the air–fuel ratio control
6—Control valve
7—Retainer of the spring
8—Retainer of the diaphragm
9—Diaphragm
10—Cover of the housing
11—Adjustment cover
12—Spring
13—Spring
14—Washer for the diaphragm
15—Spring
16—Adjustment pin

mentioned earlier. This allows the fuel control rack to move farther left, thereby increasing fuel to the engine.

2. Turning the control valve out or counterclockwise does exactly the opposite.

After any adjustment is made, the governor fuel control lever should be moved from the fuel-off to the fuel-on direction before any accurate reading can be made of the rack movement.

Caution: Do not start and run the engine at any time if the air–fuel ratio control is removed from the governor; otherwise, oil under pressure will leak out of the supply port to the air–fuel control.

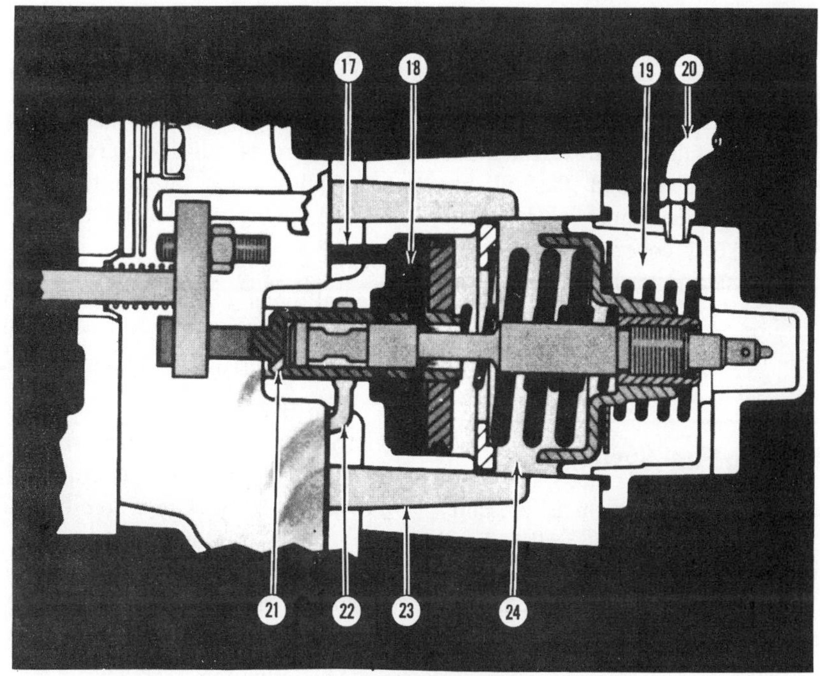

Figure 7-41
Hydraulic Air–Fuel Ratio Control: Engine Running Position.
(Courtesy of Caterpillar Tractor Co.)

Index:

17—Engine oil passage to activating valve chamber
18—Activating valve chamber
19—Air chamber, air from inlet manifold of engine
20—Tube from inlet manifold
21—Oil return (drain) passage in activating valve
22—Oil return (drain) passage in housing
23—Oil return passage to governor
24—Oil return chamber

To reduce engine exhaust smoke during acceleration, with the engine running, turn the adjustment pin out about one half-turn. If the exhaust smoke is acceptable, but acceleration is now sluggish, turn the cover pin (16) in slightly until performance and smoke are satisfactory. The EPA specifications as to smoke limitation allow some exhaust smoke under full-load acceleration conditions; they are given as so many parts per million (ppm) of soot, which is checked with a smoke opacity meter.

If, after adjusting the air–fuel ratio control, acceleration is sluggish and the engine lacks power, inspect for signs of air leaks at both the cover plate and air line. If no leaks are evident, check for a damaged diaphragm, which would prevent full fuel from being attained.

After adjustment is satisfactory, align the

cover with the pin and carefully turn the cover to align it with the closest bolt holes. If the rack was adjusted at this time, the speed limiter plunger [Figure 7-15(b)] will have to be pushed in to permit full rack travel after the engine has been stopped and the oil has drained from the air–fuel ratio control. It is advisable to use a new cover gasket and seal on the cover bolts during installation after any adjustment.

7-10

Sleeve Metering Fuel System

The sleeve metering fuel system is presently being used on smaller-bore Caterpillar diesel engines, such as the 3200 and 3300 series units. This system was designed with the following thoughts in mind.

1. To have fewer moving parts and fewer total parts.
2. Simple design with compactness.
3. It can use a simple mechanical governor. No hydraulic assist is required.
4. The injection pump housing is filled with fuel oil rather than crankcase oil for lubrication of all internal parts.
5. The same-sized plunger, barrel, and sleeve are used in all Caterpillar sleeve metering units, whether it is a V8 or in-line 4- or 6-cylinder engine.
6. The transfer pump, governor, and injection pump are mounted as one unit.
7. Employs a centrifugal timing advance for better fuel economy and easier starts.

The term *sleeve metering* is derived from the method used to actually meter the amount of fuel sent to the cylinders, a sleeve system. Figure 7-42 shows a typical layout of such a system. The basic design of the well-known Caterpillar fuel system does not change that much in this system since fuel is still injected by individual pumps. Fuel under transfer pump pressure totally fills the internal area of the fuel-injection pump and governor housing. This fuel then enters the individual pump barrels, where effective metering of the fuel is accomplished by varying the stroke of the injection plunger's own effective stroke.

Caution: Since the only lubrication for the internal parts of the injection pump is the diesel fuel, the housing must be full of fuel prior to turning over the pump camshaft.

The 3208 engine is a 4.5-in. (114-mm) bore unit with displacement of 636 in.^{3} (10.4 liters). The 3300 engine is available as a 4- or 6-cylinder engine with a bore of 4.75 in. (121 mm) and a 6-in. (152-mm) stroke, giving a displacement of 426 in.^{3} (6.98 liters) on the 4-cylinder engine and a displacement of 638 in.^{3} (10.45 liters) on the 6-cylinder engine.

Fuel from the fuel tank is drawn by the fuel transfer pump through a fuel and water separator, if so equipped, and a primary and secondary filter, if used. A priming pump typical of Caterpillar fuel systems can be used to prime the system at any time, such as after servicing of fuel filters. The fuel from the transfer pump fills the injection pump housing at a pressure of 30 ± 5 psi (205 ± 35 kPa) with the engine operating under full load. Any pressure in excess of this will be directed back to the inlet side of the transfer pump by the built-in bypass valve in the injection pump housing. A constant-bleed valve is also used to allow a continuous return of fuel to the tank so that the temperature of the fuel stays cool for lubrication purposes.

The main components of the sleeve metering fuel-injection pump are shown in Figure 7-43. The plunger is moved up and down inside the barrel by the action of the pump camshaft and roller follower assembly. The sleeve is moved up and down on the plunger by the linkage connected to the throttle and governor to effectively change the amount of fuel for injection.

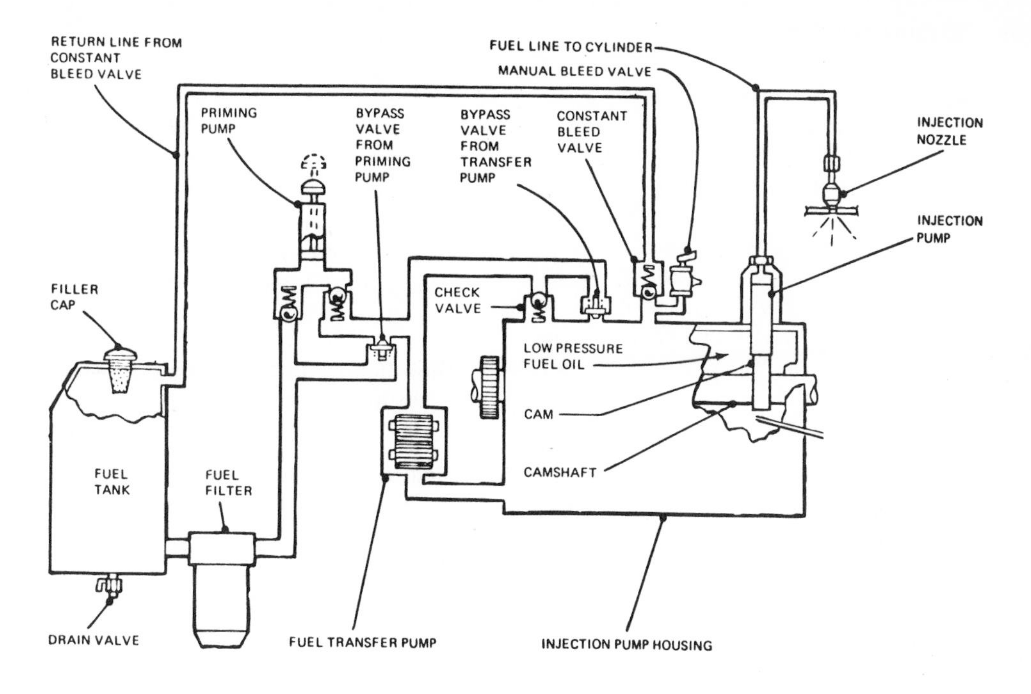

Figure 7-42
Sleeve Metering Fuel System.
(Courtesy of Caterpillar Tractor Co.)

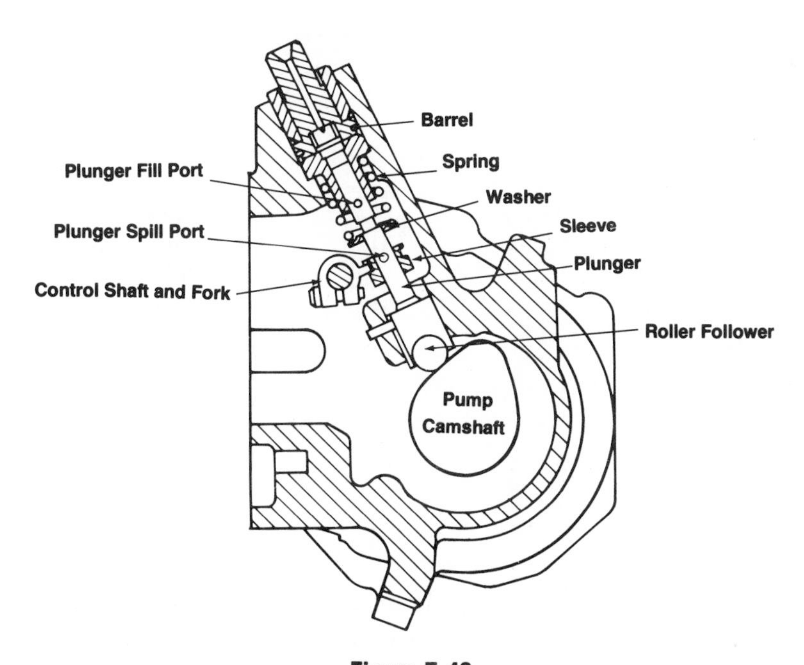

Figure 7-43
Fuel-Injection Pump and Housing.
(Courtesy of Caterpillar Tractor Co.)

Operation

1. Since the injection pump housing is constantly filled with diesel fuel from the transfer pump under pressure, anytime that the fill port is uncovered as in view 1, Figure 7-44, the internal drilling of the plunger will be primed by the inrushing fuel caused by the downward moving plunger relative to pump camshaft rotation.
2. At the correct moment, the rotation of the pump cam lobe begins to force the plunger upward until it will cause the fill port to close as it passes into the barrel. At the same time, due to the fact that the spill port is covered by the sleeve, the pump, lines, and fuel valves will be subjected to a buildup in fuel pressure, and injection will commence as shown in view 2 of Figure 7-44. As the plunger moves upward, it will cause the check valve to be lifted off its seat.
3. Injection of fuel will continue as long as both the fill and spill ports are completely closed by the barrel and sleeve.
4. Injection ends the moment that the spill port starts to edge above the sleeve, releasing the pressure in the plunger and simultaneously letting fuel escape from the pump back into the housing, as shown in view 4. Also, at the end of the stroke the check valve closes to prevent the fuel from flowing back from the injector fuel line.
5. To increase the amount of fuel injected, the sleeve is raised through the control shaft and fork so that the sleeve is effectively positioned higher up on the plunger. This means that the spill port will be closed for a longer period of time as the

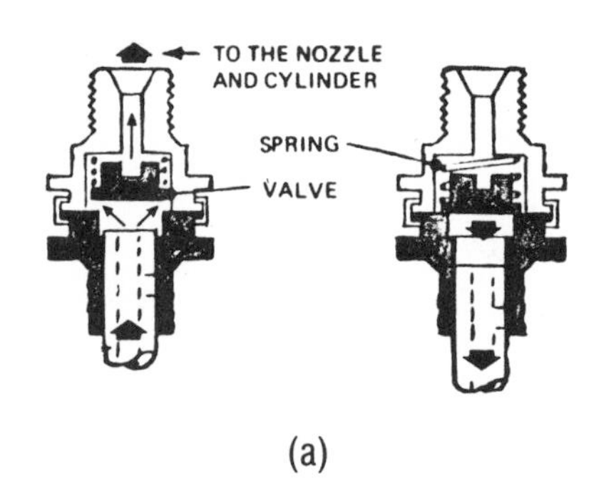

(a)

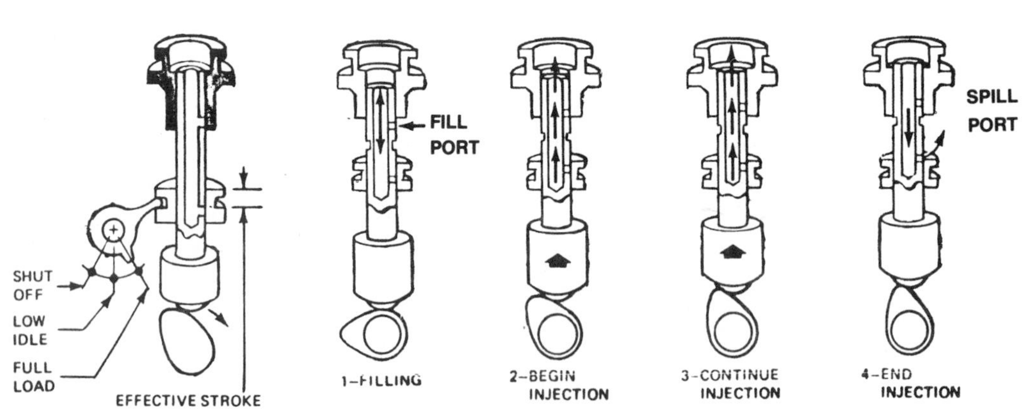

(b)

Figure 7-44
Sleeve Metering: (a) Sequence of Injection Events; (b) Left-Side View of Operation. (Courtesy of Caterpillar Tractor Co.)

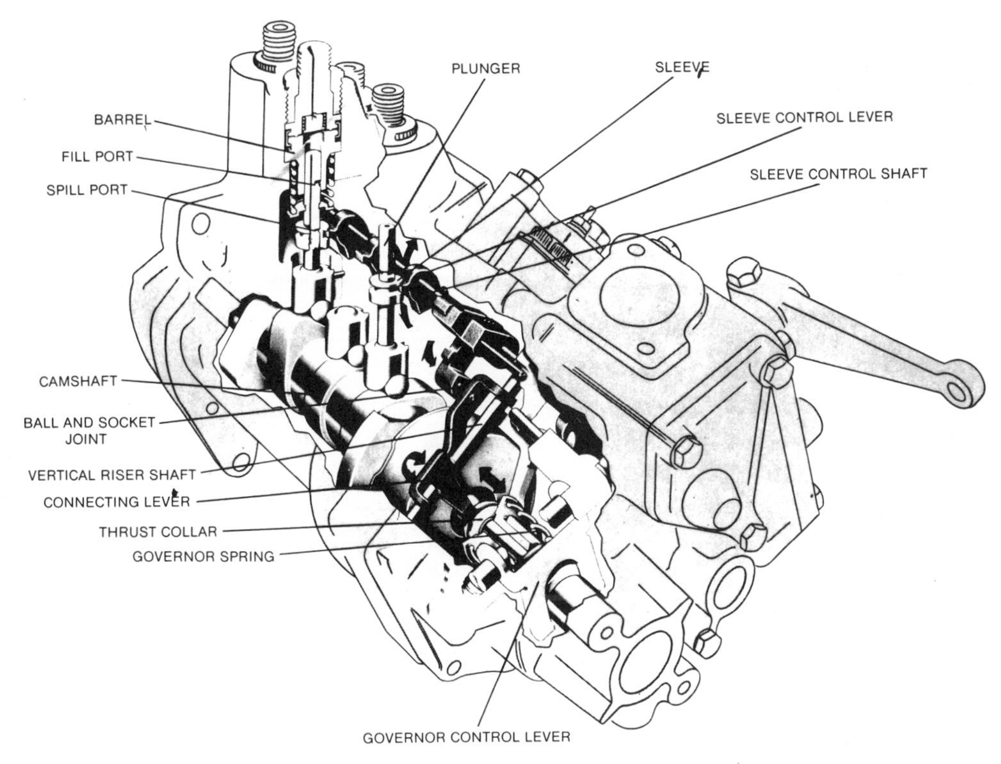

Figure 7-45
Cutaway View of Sleeve Metering Pump Assembly.
(Courtesy of Caterpillar Tractor Co.)

plunger is being raised by the lifter. Increasing the effective stroke of the plunger (time that both ports are closed) will increase the amount of fuel delivered.

7-11

Governor Action

Figure 7-45 shows a partial cutaway view of the sleeve metering pump assembly.

Basic governor components are shown in Figure 7-46.

When the operator requires more power from the engine, he either moves the hand throttle on industrial and marine applications, or steps on the throttle on a highway-truck-type unit. This causes the governor control lever to apply pressure, which compresses the governor spring, which transfers this motion to the thrust collar. Since governor action from the spring and weight motion is of the back and forth variety, the additional linkage between the injection pumps and the governor transforms this sliding horizontal governor movement from the thrust collar into a rolling, twisting motion at the sleeve control shaft. This is accomplished by a simple 90 degree lever known as a *bellcrank lever* or connecting lever. Figure 7-47 shows the connection between the bellcrank lever and the sleeve control shaft lever.

Refer to Figure 7-45. The bellcrank lever contacts the thrust collar on one end and the sleeve control shaft on the other. The bellcrank pivots on a fixed vertical bellcrank shaft to gain a mechanical advantage through the lever principle. At the sleeve shaft end, it rides in a ball and socket joint that holds it in place and minimizes linkage movement.

Therefore, any horizontal movement at the governor weight shaft and spring will cause an equally precise movement at the ball and socket joint, leading to repositioning of the sleeves. If in this case the operator has increased the throttle position, the sleeves would be lifted, thereby covering the spill ports for a longer overall effective plunger stroke, and in effect would deliver more fuel to the injectors. What is actually happening is similar to the action of all mechanical governors; that is, an increase in either the throttle or load will cause a speed change to the engine. Spring pressure is always trying to increase the fuel delivered to the engine, while the centrifugal force of the rotating governor flyweights is always trying to decrease the amount of fuel going to the engine.

Somewhere within the throttle range, however, a state of balance between these two forces will exist as long as the engine speed is capable of overcoming the load placed upon it to keep the spring and weight force in this state of balance. For more information, see the section on governors.

When the engine is stopped, the action of the governor spring force places the thrust collar to the full-fuel position; therefore, easier starting is accomplished. Once the operator cranks and starts the engine, centrifugal force will cause the governor flyweights to move outward (at approximately 400 rpm), which now opposes the spring force, and the thrust collar and spring seat will come together as they are pushed to a decreased fuel position. When the force of the weights equals the preset force of the spring established by the idle adjusting screw, these two forces will be in a state of balance, and the engine will run at a steady idle speed with the throttle at the normal idle position.

Governor action will operate from idle throughout the engine's speed range. The engine's maximum speed is controlled by a load stop pin, which is readily visible as item 17 in Figure 7-53. Very simply, rotation of the throttle lever causes the load stop lever (7, Figure 7-53) to lift the load stop pin until it comes in

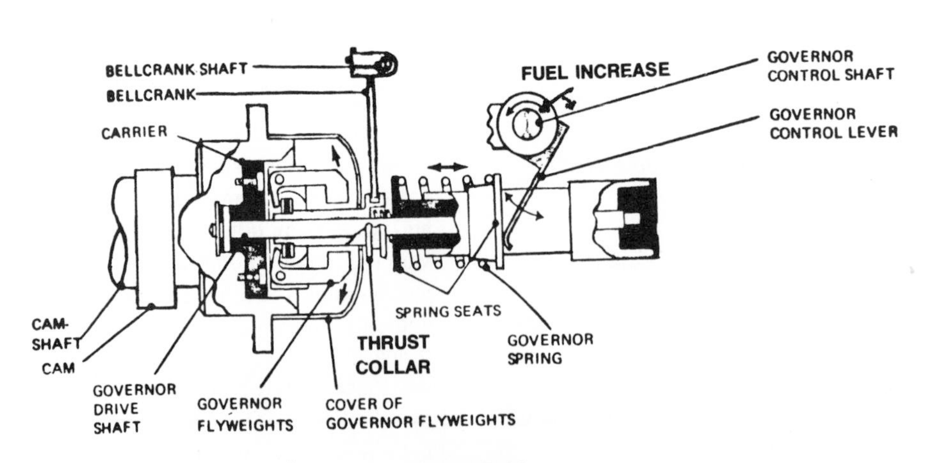

Figure 7-46
Governor Components–Sleeve Metering.
(Courtesy of Caterpillar Tractor Co.)

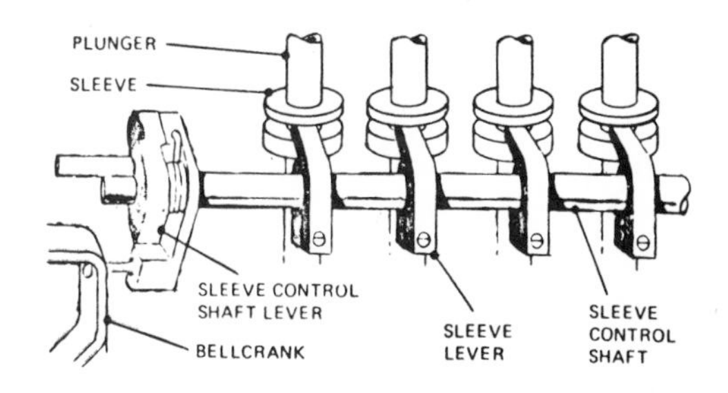

Figure 7-47
Connection Between Bellcrank Lever and Sleeve Control Shaft Lever.
(Courtesy of Caterpillar Tractor Co.)

contact with the stop bar or screw, thereby limiting any more fuel to the engine. The purpose of the dashpot piston is to prevent any surging or irregular speed regulation of the engine by the fact that the piston (1, Figure 7-53) either pulls fuel into or pushes fuel out of its cylinder through an orifice (E). The dashpot governor spring force will therefore vary with the piston movement, and as the engine load is increased or decreased, respectively, fuel will be drawn into the piston cylinder or pushed out of the piston cylinder through the orifice to give the effect of a high governor spring rate, which minimizes speed variations through oscillation during engine load changes. At any time that the ignition switch is turned off or the governor speed control lever is moved to the off position, the sleeve levers move the sleeves down, cutting off fuel to the cylinders.

7-12

Removal of Pumping Elements (SMFS)

Removal of these pumps is a relatively simple task; however, prior to undertaking such a job, ensure that the injection pump housing has been thoroughly cleaned of all external dirt. The removal procedure is the same on any model of sleeve metering pump. Also cover all injection pumps with plastic shipping-type caps to prevent the entrance of dirt.

V8 Pump

1. Remove the main cover assembly from the injection pump housing. As you take the cover off, at the drive end of the pump, remove the bypass valve spring.
2. Refer to Figure 7-48; install wrench 8S2243 over the injection pump bushing and loosen it off; then withdraw it from the bore.
3. Grasp the injection pumping assembly as shown in Figure 7-48 and withdraw it from its bore.

Caution: It is not necessary to loosen off the lever to shaft holding screws when removing, or for that matter installing, pumping units, since the sleeve at the base of the plunger will slide off the lever end as the pumping unit is withdrawn. If the lever screws are backed off, it will be necessary to readjust all the fuel pumps.

Installing Fuel-Injection Pumps

The sequence of events involved here is basically the reversal of removal; therefore, proceed as follows:

1. Carefully insert the pumping unit into its bore.
2. The sleeve at the base of the pumping unit will engage with the lever as you install it. You may have to rotate it gently.
3. Screw in the bushing that was removed in step 2 under removal, and using wrench 8S2243 tighten it to 70 ± 5 lb-ft (94.9 ± 6.77 N·m).
4. Prior to installing the injection pump main cover assembly, be sure to put the spring back on the bypass valve and that it is in

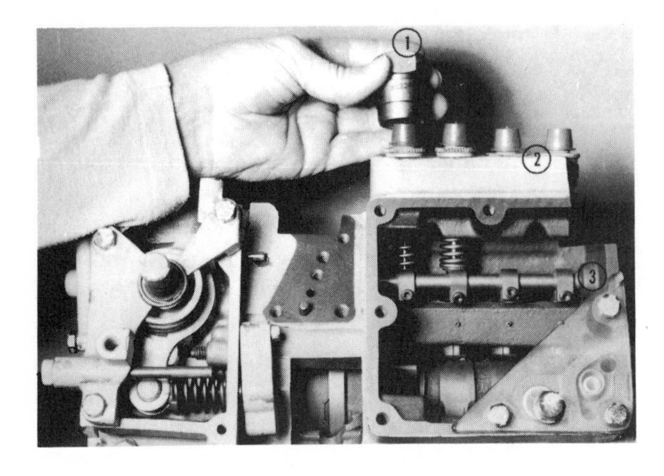

(a)

Index:
1—Tool 8S2243
2—Bushing
3—Lever screws
4—Fuel-injection pump
5—Sleeve

(b)

(4-cylinder pump model shown for demonstration purposes)

Figure 7-48
Removal of SMFS Pumping Element. (Courtesy of Caterpillar Tractor Co.)

position in the bore of the cover before tightening the hold-down bolts.

Disassembly of Sleeve Metering Fuel-Injection Pumps

As with any component of a fuel-injection system, prepare a clean, well-organized, dust-free area to lay out the parts during both teardown and assembly; otherwise, you are simply asking for possible future problems.

Figure 7-49 shows an exploded view of an injection pumping unit used with the sleeve metering fuel system. Proceed as follows:

1. Remove the injection pumping unit as described earlier.
2. With the bushing removed from the bonnet (2), take off the retaining ring (3) and remove check valve (6) and spring (4) from the bonnet.
3. Remove spring (8) and washer (5), and then plunger (9) and sleeve (10).

Note: As with all fuel-injection systems, the plunger and barrel are always matched to one another, and in this particular case the sleeve will take up a mated fit to its particular plunger; therefore, do not intermix any of these components with those from another pump.

4. Inspect all parts for scuffing, scoring, nicks, scratches, or other signs of possible damage. When reassembling the pump unit, dip all parts in clean diesel fuel.

Assembly Procedure

1. Carefully install the sleeve (10), plunger (9), spring (8), and washer (5) onto the barrel (7).
2. Take careful note that when installing the plunger into the barrel the larger hole faces up, since this is the fill hole. The spill hole is of a smaller diameter than the fill hole.
3. Carefully install the check valve and spring in the bonnet, attach the barrel and bonnet together, and install the retaining ring. Finally, install the bushing onto the bonnet.
4. Individually install all the injection pumping units back into the main injection pump housing bores.

Disassembly of the Main Injection Pump Housing

If this operation is necessary at any time, after removing the individual injection pumping units, remove the governor from the pump housing, followed by removal of the fuel transfer pump. Now proceed to remove the torque spring lever dowel from the housing, loosen the sleeve control lever screws, and slide the

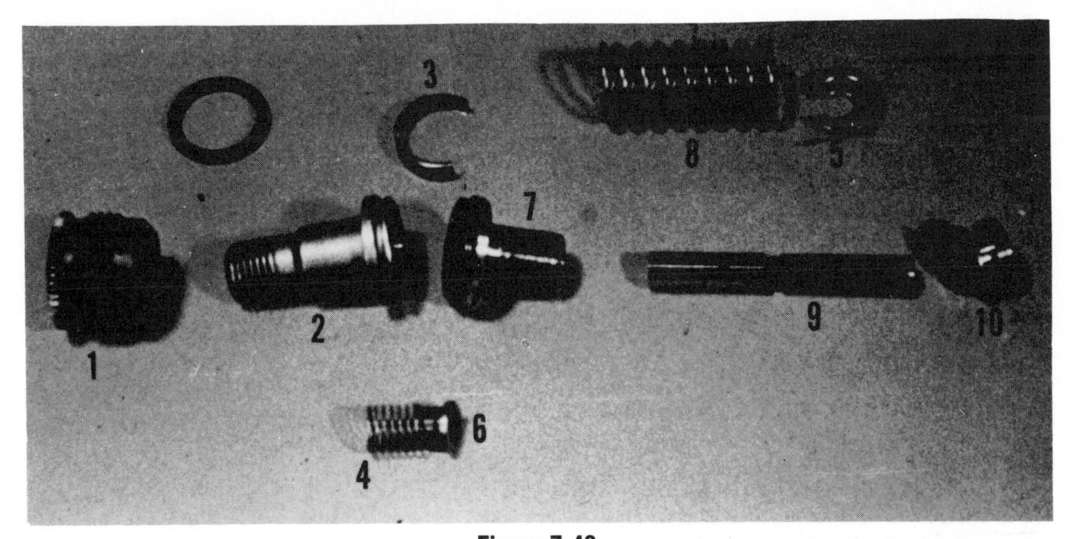

Figure 7-49
Parts Breakdown of Injection Pumping Unit Used with Sleeve Metering Fuel System. (Courtesy of Caterpillar Tractor Co.)

Index:

1—Bushing
2—Bonnet
3—Bonnet ring
4—Spring
5—Washer
6—Check valve
7—Barrel
8—Spring
9—Plunger
10—Sleeve

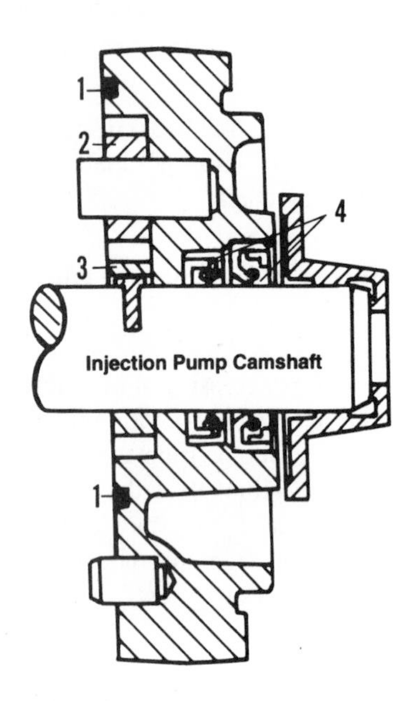

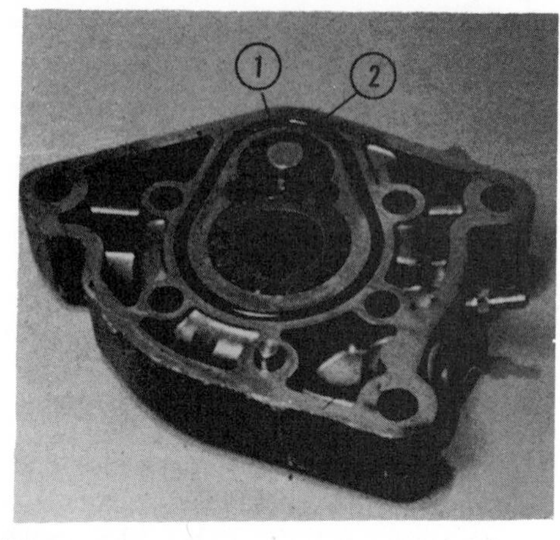

Index:
1—O ring seal
2—Pump idler or driven gear
3—Pump drive gear (mounted and keyed to injection pump camshaft)
4—Lip-type seals

Figure 7-50
Basic Transfer Pump.
(Courtesy of Caterpillar Tractor Co.)

sleeve control shaft out of the housing. Be sure to keep all the lifter follower assemblies together so that they can be reinstalled into their respective bores upon assembly. Carefully withdraw the pump camshaft from the pump housing.

When reassembling, follow the reverse procedure of teardown; however, take care when installing the lifters that their grooves are in alignment with the pins of the housing.

7-13

Fuel Transfer Pump (SMFS)

The transfer pump used with this system delivers fuel to the injection pump housing at a pressure of 30 ± 5 psi (205 ± 35 kPa) with the engine running under full-load conditions. Figure 7-50 shows the basic transfer pump, which is gear driven from the injection pump camshaft at the injection pump's drive end.

Note: When removing or assembling the fuel transfer pump, install the timing pin (item 16, Figure 7-52) in through the pump housing (injection pump), which will prevent the pump camshaft from turning. The timing pin passes in through the cover of the torque spring cover.

Fuel Pressure Check

Since the internal workings of the SMFS are lubricated by transfer pump pressure, a quick check of the pressure can be done by removing the plug on the cover of the injection pump housing. At cranking speed this should be 2 psi (13.79 kPa), at low idle 18 ± 5 psi (124.11 ± 34.48 kPa), and at high idle 30 ± 5 psi (206.85 ± 34.48 kPa).

7-14

Automatic Timing Advance Unit (SMFS)

All current Caterpillar engines employ some form of automatic timing for the fuel-injection pump. On the sleeve-metering-type engines this advance unit is mounted on the front end of the engine camshaft. The gear of the automatic advance unit mounted on the engine camshaft meshes with and drives the fuel-injection pump camshaft. Figure 7-51 shows the automatic timing advance unit.

Operation

The slides (3) are located and driven by two dowels attached to the engine camshaft gear. The slides in turn fit into notches within the weights (1), and therefore will transfer their drive from the engine camshaft gear to the weights. With the engine running, centrifugal force exerted by the rotating weight assemblies causes them to act against the force of the springs (2). Since the weights are designed with notches in them, as they move outward under centrifugal force, they cause the slides (3) to effect a change in the angle between the timing advance gear and the two drive dowels of the engine camshaft gear. This relative movement of the timing advance unit gear, which is in mesh with the fuel-injection pump drive gear, will therefore automatically advance or retard the fuel-injection pump's timing in relation to engine speed and load.

However, built into the advance unit is a maximum timing variation of 5 degrees, with the timing change starting at approximately low idle rpm and continuing on up to the engine's rated speed; therefore, you cannot adjust the automatic timing advance unit. The timing unit is lubricated by engine oil under pressure from drilled holes at the engine camshaft front bearing.

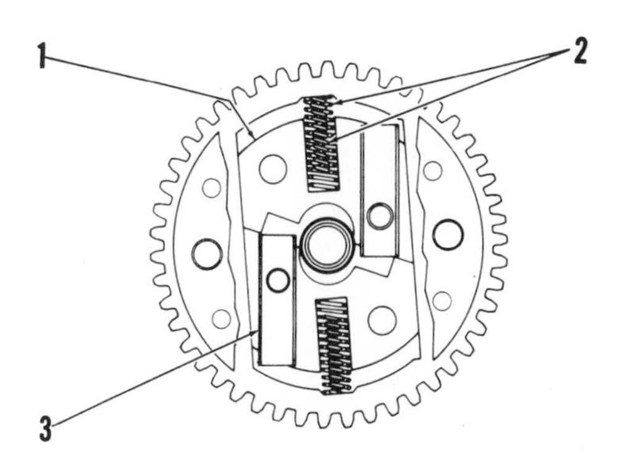

Figure 7-51
Automatic Timing Advance Unit.
(Courtesy of Caterpillar Tractor Co.)

Index:
1—Weights
2—Springs
3—Slides

7-15

Adjustment (SMFS)

To accurately adjust and tune up an engine employing a SMFS, it is necessary to have the correct Caterpillar special tools if accurate and proper adjustments are to be expected. In this instance, the 3P2200 calibration tool group is required along with some tools from the 5P4203 tool group. These tools are shown along with a listing of individual parts in Figure 7-52.

The current practice is that all dimensions and dial readings for Caterpillar engines using the SMFS are given in the metric system. The calibration adjustments for the SMFS, whether off a 4-, 6-, or 8-cylinder engine, follow the same basic pattern; therefore, for instruction purposes we shall use a V8 engine.

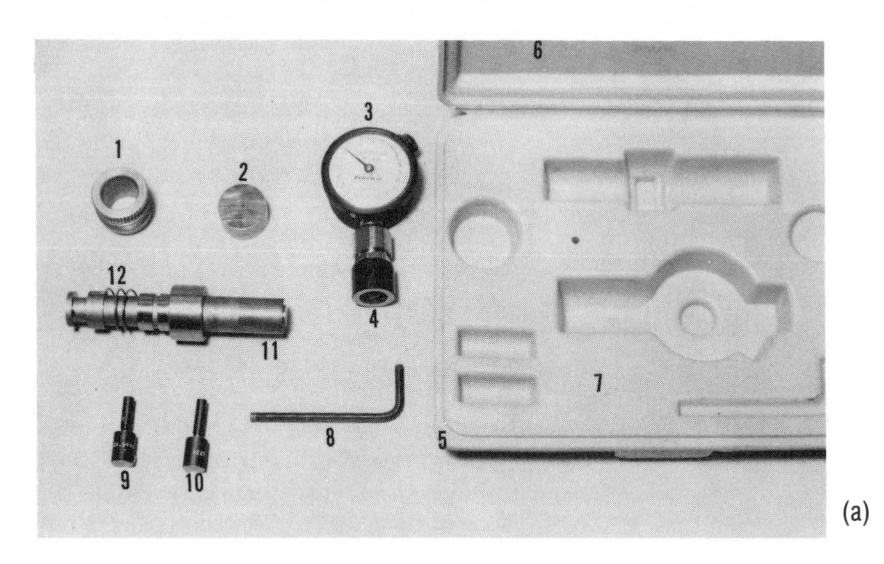

(a)

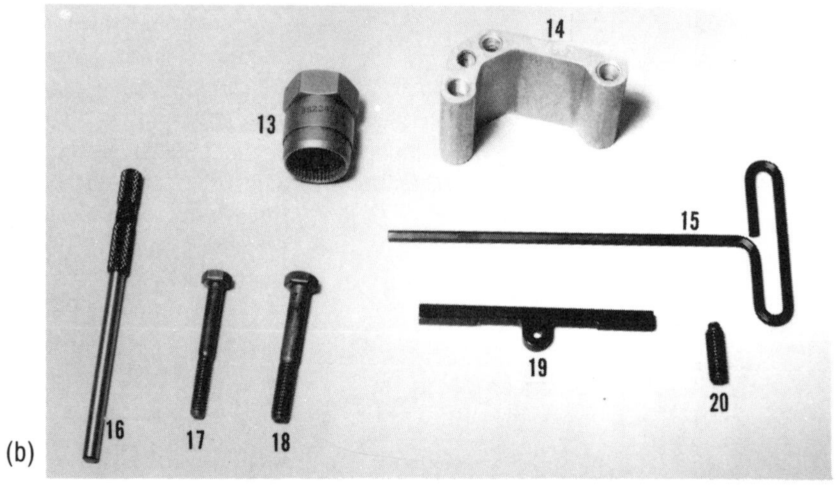

(b)

Figure 7-52
(a) Tool Group 3P2200 and (b) Tools from Group 5P4203 Required to Adjust an Engine That Uses the Sleeve Metering Fuel System. (Courtesy of Caterpillar Tractor Co.)

Index:

1—Bushing
2—Microgauge
3—Dial indicator
4—Collet assembly
5—Tool box
6—Protective box liner
7—Protective box liner
8—Allen wrench
9—Pin
10—Pin
11—Calibration pump assembly
12—Calibration pump assembly spring
13—Wrench 8S2243
14—Adapter bracket
15—T-handle-type Allen wrench
16—Timing pin
17—Bolt
18—Bolt
19—Gauge
20—Screw

Procedure

1. It may be necessary to remove the air–fuel ratio control from the fuel-injection pump housing depending on the model of engine (4- and 6-cylinder units).
2. Figure 7-53 shows a cross-sectional view of a SMFS injection pump equipped with a dashpot governor for reference purposes. However, bear in mind that the pump can be equipped with a variety of options and also installed in different ways.
3. Refer to Figure 7-53 and remove the main cover (16) and the torque control group cover.
4. Refer to Figure 7-54 and install the correct pin (9) into hole A. These pins are shown

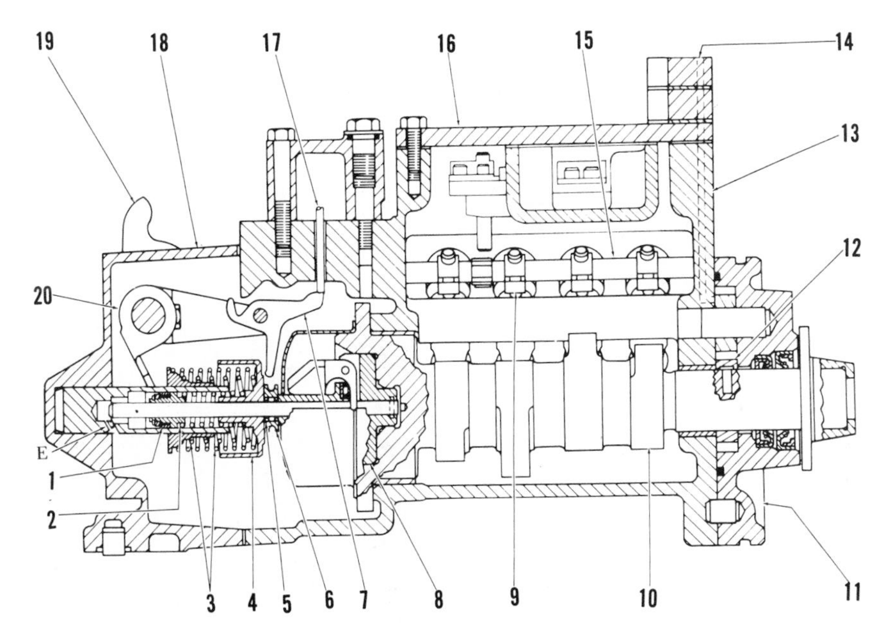

Figure 7-53
Cross Section of a Sleeve Metering Fuel System Injection Pump Equipped with a Dashpot Governor. (Courtesy of Caterpillar Tractor Co.)

Index:

1—Piston for dashpot governor
2—Spring for dashpot governor
3—Governor springs (inner: low idle; outer: high idle)
4—Spring seat
5—Overfueling spring
6—Thrust collar
7—Load stop lever
8—Carrier and governor weights
9—Sleeve levers
10—Pump camshaft
11—Fuel transfer pump
12—Fuel transfer pump drive gear
13—Injection pump housing
14—Internal fuel passage
15—Sleeve control shaft (2 on V8)
16—Main cover
17—Load stop pin
18—Governor housing
19—Speed control lever
20—Governor control shaft lever
E—Dashpot orifice

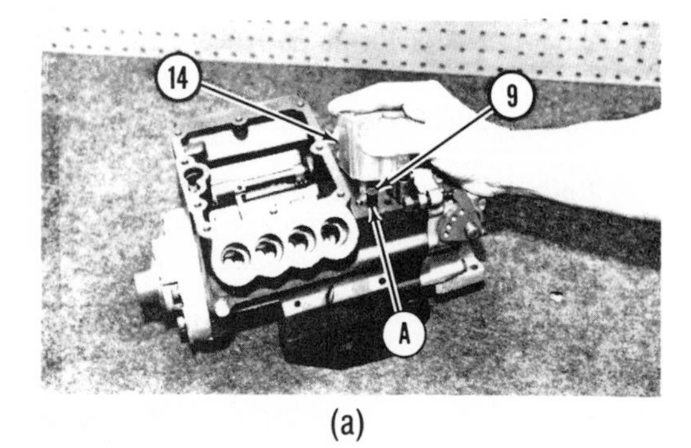

(a)

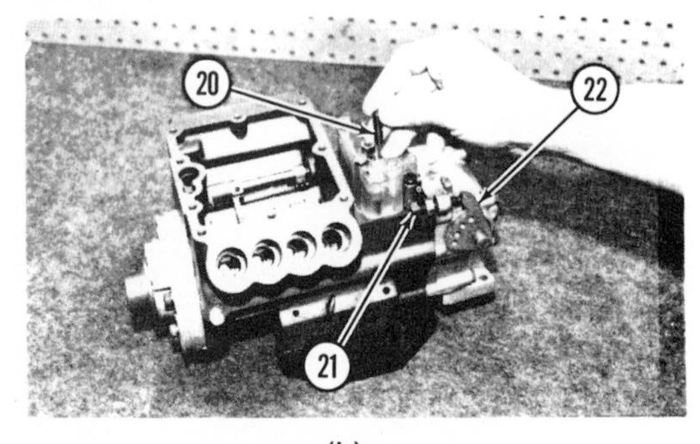

(b)

Figure 7-54
(a) Installing Adapter Bracket and Pin and
(b) Installing Screw into Adapter Bracket.
(Courtesy of Caterpillar Tractor Co.)

as items 9 and 10 in Figure 7-52; however, the pin length is stamped on the large diameter for identification purposes.

a. On 4- and 6-cylinder engine pumps, use pin (10), which is 17.3734 mm (0.684 in.) long.

b. On V8 pumps, use pin (9), which is 15.941 mm (0.6276 in.) long.

5. It is only necessary to remove those injection pumping units that are to have the calibration check; however, it is advisable to check them all if you are this far into a repair.

6. On in-line or Vee pumps, install adapter (14) to the housing as shown in Figures 7-54(a) and (b), and hold it down with bolts.

7. Install the short screw (20) and tighten it up with the Allen wrench, (8) or (15), which will push the pin (9) up against the pump housing. On Vee pumps only, remove the idle adjusting bolt (21) and install temporarily a bolt from the tool kit or any bolt that is longer than bolt (21).

8. On in-line or Vee pumps, move the speed control lever (22) to its full-load position and lock it there on in-line pumps, or lightly snug up the new longer bolt on vee pumps to hold the speed control lever in the full-fuel position.

9. *Vee pumps only:* Refer to Figure 7-55(a). Slide both sleeve levers (B) back out of the way by loosening the Allen head bolts

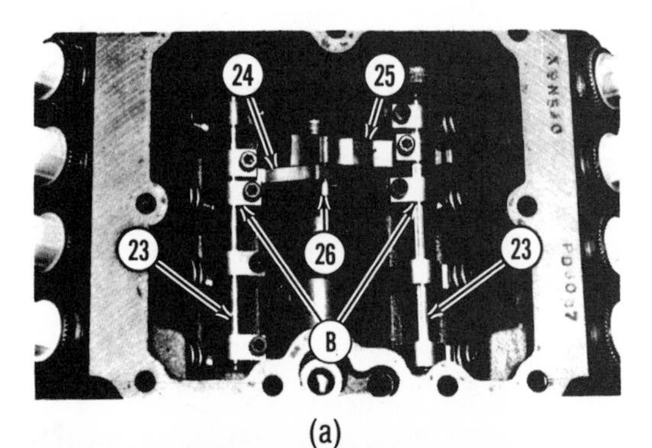

(a)

(b)

Figure 7-55
(a) Checking for Proper Crossover Lever Adjustment; (b) Adjusting Crossover Levers.
(Courtesy of Caterpillar Tractor Co.)

that hold them to their respective shafts (23). Place gauge (19) from the tool group onto the sleeve shafts as shown, and slide it forward until dowel (26) engages the hole in the center of gauge (19) without any movement of either crossover lever (24) or (25).

Note: Adjustment is correct if you can do this.

If you cannot engage the dowel (26) with the hole in the middle of gauge (19) without moving crossover levers (24) and (25), loosen Allen head bolts (27) and (28). Move lever (24) off dowel (26) and place lever (25) into a position that will allow the dowel to fit into the gauge (19) hole, and tighten bolt (28).

10. Slide the gauge (19) back and place lever (24) onto dowel (26); then tighten bolt (27). Recheck the adjustment, and if it is now correct, torque bolts (27) and (28) to 27.7 ± 2.3 cm/kg (2.7 ± 0.25 N·m) or 24 ± 2 lb-in.).
11. *All pumps:* Clean off the tool group calibration pump (11) and dip it in clean diesel fuel; also make sure that the spring (12) is properly installed, and slide the calibration pump (11) into the pump bore that is being checked so that the offset flat at its base, as shown in Figure 7-56, is toward the tang (21) of the lever (22). To ensure that the tang fits into place, gently turn the

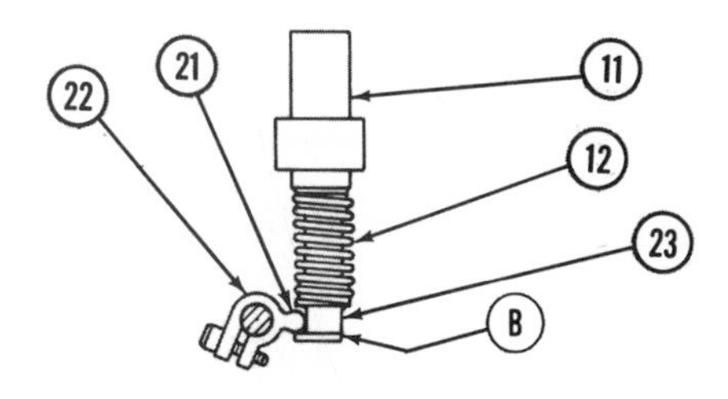

Figure 7-56
Calibration Pump Assembly Engaged with Fuel Rack.
(Courtesy of Caterpillar Tractor Co.)

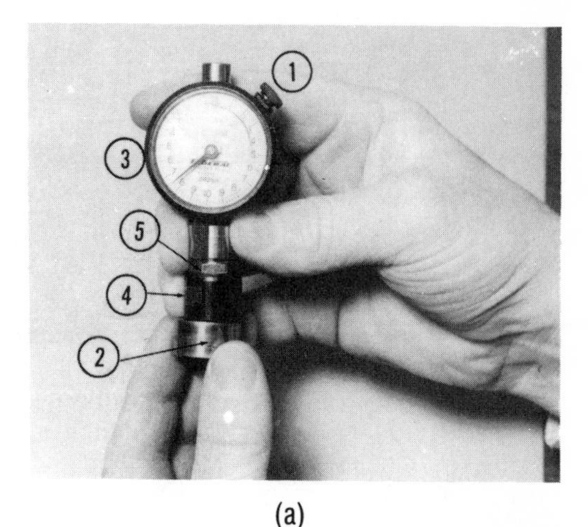

(a)

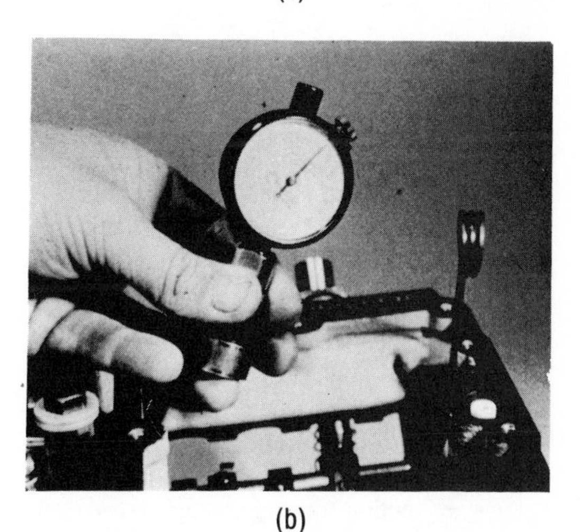

(b)

Figure 7-57
(a) Dial Indicator; (b) Zeroing-in the Dial Indicator.
(Courtesy of Caterpillar Tractor Co.)

pump (11) 45 degrees in either rotation, which will ensure that the pump is in fact all the way into its bore.

12. Install the bushing (1) from the tool group onto the calibration pump (11), and with wrench (13) tighten the bushing to 9.7 ± 0.7 mkg (95 ± 7 N·m) or (70 ± 5 lb-ft) of torque.
13. Refer to Figure 7-57; slide the collet (4)

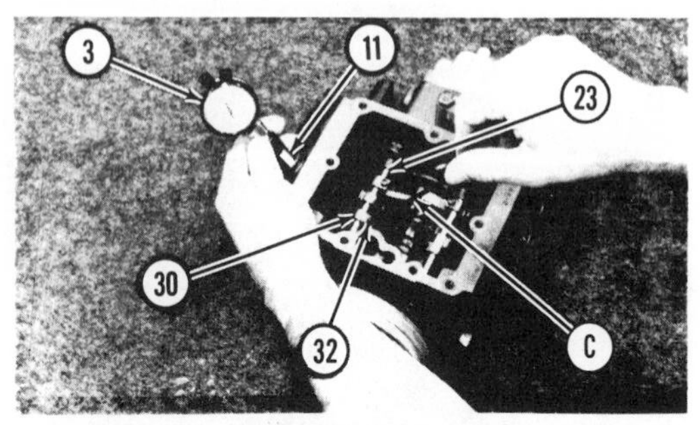

Figure 7-58
Checking Pump Plunger Travel in Relation to Fuel Rack Position.
(Courtesy of Caterpillar Tractor Co.)

onto the dial indicator very gently until the dial needle shows approximately one and one quarter turns and tighten up nut (5) just snug. Using gauge (2), hold it flat against the base of the dial stem and collet, adjust the dial gauge to a zero reading, and lock knurled screw (1); then recheck for zero.

14. Refer to Figures 7-56 and 7-58; loosen Allen head bolt and gently tighten it until lever (22) is snug against the shaft with the calibration pump (11) stationary. Push down on the Allen head wrench until it causes motion transfer through lever (22), which will raise the plunger just short of the top of the barrel as shown in Figure 7-59.

Figure 7-59
Plunger Location Relative to the Barrel.
(Courtesy of Caterpillar Tractor Co.)

15. Hold the indicator tightly in place in the center of the calibration pump (11) and push the wrench down slowly (raising the plunger) until the dial indicator reads zero. If you exceed zero, the plunger has been raised too much; therefore, repeat the adjustment and tighten the Allen head bolt on lever (22) to 27.7 ± 2.3 cm/kg (2.7 ± 0.25 N·m) or 24 ± 2 lb-in. The allowable tolerance after tightening the bolt on the dial indicator reading is 0.000 ± 0.005 on the face of the *metric* dial gauge.

Tool Group 5P4203 for Sleeve Metering Fuel Systems

Tool Group 5P4203 (Figure 7-60) must be used to check and adjust the fuel rack setting, air–fuel ratio control, and the like, on engines equipped with a SMFS.

Sleeve Metering Fuel Rack Setting

Due to the inherent design difference of the SMFS to both the flange mounted and the compact body pumps, adjustment of the fuel rack setting is accomplished by altering the torque spring setup. Two basic types of torque spring control are used on the SMFS: (1) the leaf-type torque spring (see Figure 7-65), and (2) the load-stop or coil spring type (see Figures 7-66 and 7-67).

Procedure

Prior to any adjustment of the fuel rack setting, obtain the rack setting information literature, which is readily available from your local Caterpillar dealer. Make sure that you locate the correct setting, which ties in with the engine model and serial number, and so on.

Once you have established the correct setting, proceed to check the rack setting as follows: The setting check for both the in-line and Vee engine pumps is basically the same; therefore, the sequence given will indicate a difference where one exists.

1. Obtain the complete tool group 5P4203.
2. On both in-line and Vee pumps, remove the torque control group cover. From the tool group, select either of the following pins:

   a. Pin 5P298 for Vee pumps, which is stamped with dimension 17.8507 mm (0.7591 in.) on its larger diameter.

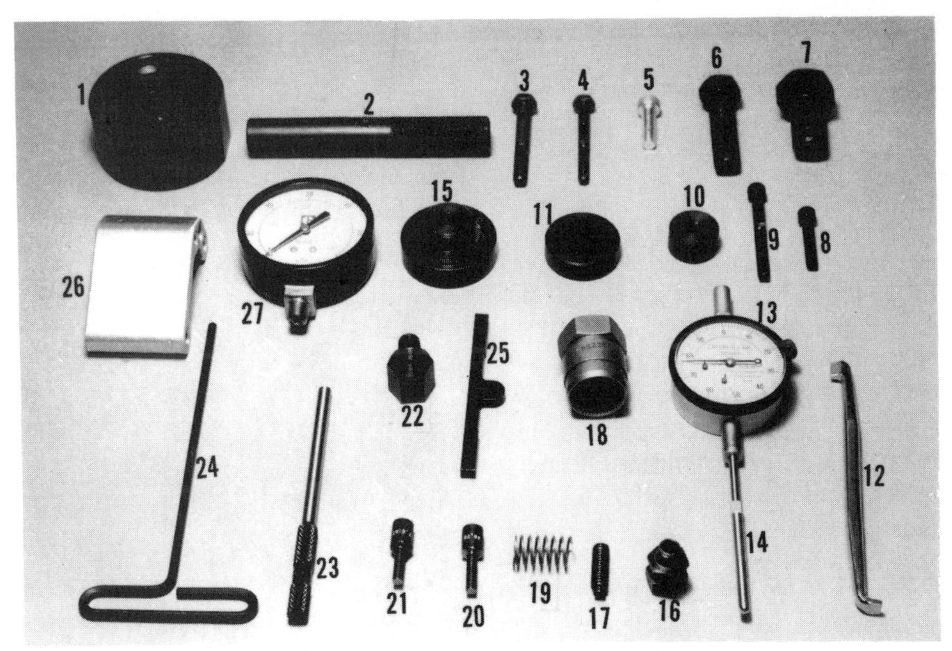

Figure 7-60
Tool Group 5P4203.
(Courtesy of Caterpillar Tractor Co.)

Index:

1—5P301 driver
2—1P529 handle
3—1D4538 bolt
4—1D4533 bolt
5—3K4910 screw
6—2H3740 bolt
7—S1603 bolt
8—4L7124 bolt
9—6L5551 bolt
10—1P463 plate
11—5P318 driver
12—5P302 bar
13—3P1567 indicator (gauge must have 5P4228 spring installed)
14—5P4809 point
15—5P319 driver
16—3P1565 collet clamp
17—8S7271 setscrew
18—8S2243 wrench
19—3J6956 spring
20—5P298 pin (17.8507 mm) (0.7028 in.)
21—5P299 pin (19.2830 mm) (0.7592 in.)
22—4H553 fitting
23—3P1544 pin
24—5P4205 wrench
25—5P4209 gauge
26—5P4226 adapter
27—3P1564 gauge

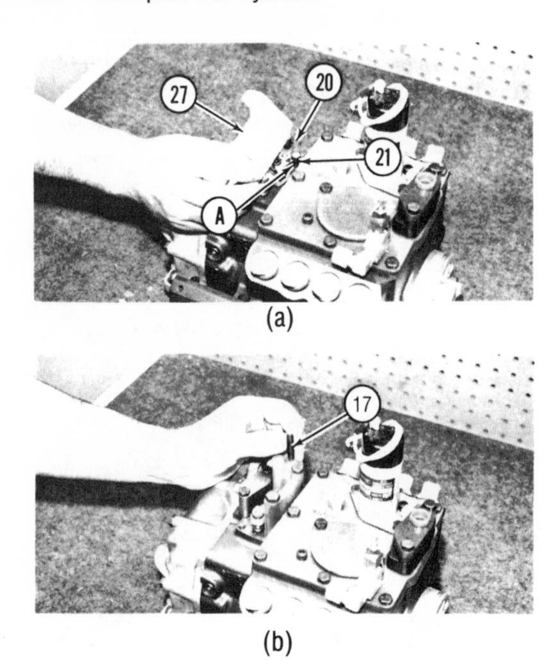

Figure 7-61
(a) Installing Pin and Adapter Bracket; (b) Installing Setscrew 8S7271 into Adapter Brackets. (Courtesy of Caterpillar Tractor Co.)

b. Pin 5P299 for in-line pumps, which is stamped with dimension 19.2831 mm (0.7591 in.) on its larger diameter.

3. Remove the air–fuel ratio control from in-line pumps.
4. Refer to Figure 7-61(a); install the correct pin into hole A and place the spring (20) over the pin (21). Install the aluminum adapter bracket (27) over the pin and spring and snug it into place with the bolts from the tool kit.
5. Refer to Figure 7-61(b); install the special setscrew 8S7271 (17) from the tool kit into the hole of the adapter bracket, and using either wrench (24) or suitable Allen wrench, lightly tighten the setscrew until the pin is up against the pump housing, which will be indicated by a step-up in effort to turn the wrench.
6. On Vee-type pumps, remove the electrically operated fuel shutoff solenoid (29 in Figure 7-62) from the top of the injection pump housing.
7. Refer to Figure 7-62; install the collet clamp (17) into the aluminum bracket adapter (27) and tie the governor speed control lever in the full-fuel position. Attach the dial indicator (14) with point (15) into the collet clamp (17), and move the indicator up or down until the large hand on the dial face is on zero and the small hand is sitting directly over a number. Tighten up nut (B) of the collet clamp to hold it in place, but be careful not to overtighten.

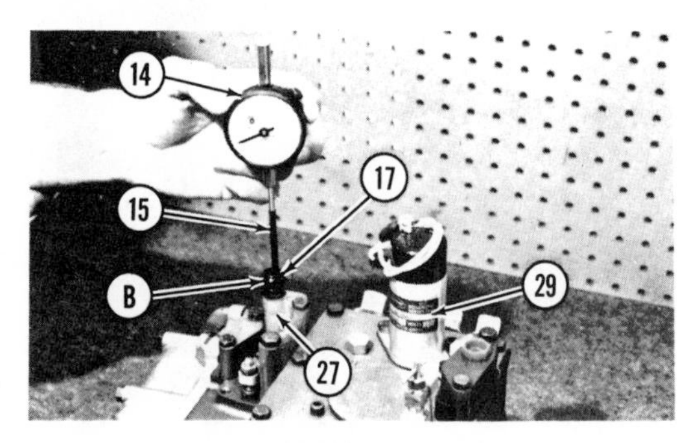

Figure 7-62
Attaching Dial Indicator onto the Collet and Adapter Bracket.
(Courtesy of Caterpillar Tractor Co.)

8. Gently turn the indicator bezel (C) to zero and carefully tighten the lock (30) as shown in Figure 7-63. Note and write down the reading on the dial indicator at this time. Connect the alligator clip end of a good circuit tester to the contact spring (D) of the torque control group, and place the other end to a clean ground.
9. *Vee-type pumps:* Refer to Figure 7-64; remove screw (17) with wrench (24) or suitable Allen wrench; then untie the governor speed control lever and return it to

its normal low idle position. Very gently move the governor speed control lever back toward the maximum fuel position just until the circuit tester light flickers on (very dim brilliance). Note and record the dimension on the dial indicator face by reading the small hand (revolution counter) plus that of the large if between numbers on the small hand. Repeat this several times to be sure that the readings are in fact consistent.

In-line pumps: Loosen (do not remove) screw (17) until the circuit tester light shows very dimly and note the dial gauge reading. Repeat several times to check for consistency.

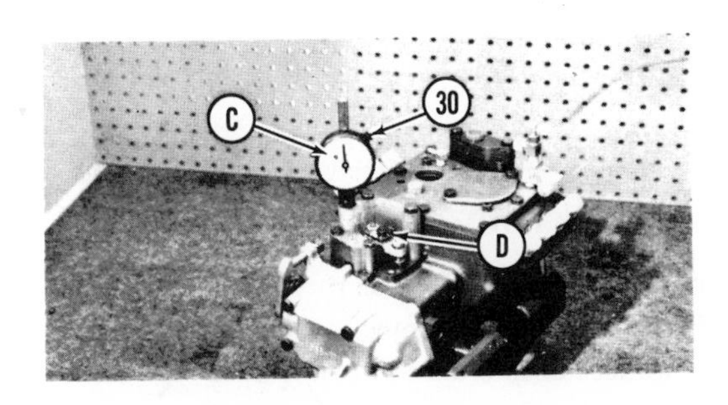

Figure 7-63
Adjusting-in the Dial Indicator.
(Courtesy of Caterpillar Tractor Co.)

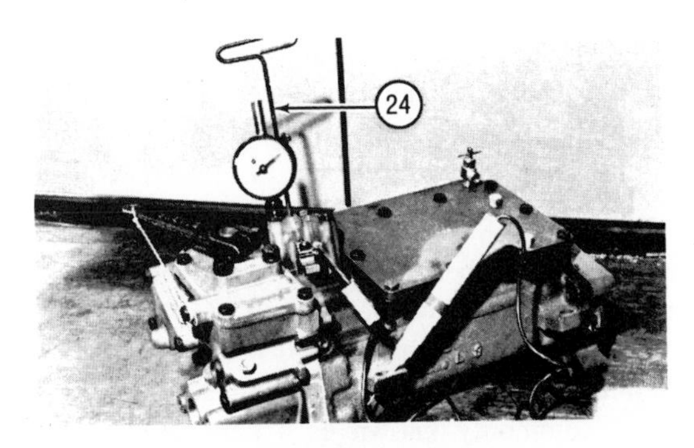

Figure 7-64
Checking Rack Movement with the Use of a Circuit Tester Light and Dial Indicator.
(Courtesy of Caterpillar Tractor Co.)

10. The dimension recorded minus the reading established earlier in step 8 will be the actual rack movement dimension. If it is found that the rack is in need of adjustment, proceed to adjust the torque spring as follows.

Adjustment of the Fuel Setting

1. For pumps equipped with a leaf-type torque spring, see Figure 7-65. To change the rack setting with this type of torque spring, remove the assembly (32) by taking off bolts (31). From the rack setting information, add or remove shims (33) until the desired setting is achieved when checked with the dial indicator, as just described.

2. For pumps equipped with the load stop or coil-spring-type torque spring, see Figures 7-66 and 7-67. Use a suitable wrench or special socket 3P2210 found in tool group 5P4203 to loosen locknut (34). With the dial indicator still set up, carefully adjust screw (35) to obtain the correct fuel rack setting; then tighten the locknut. Repeat

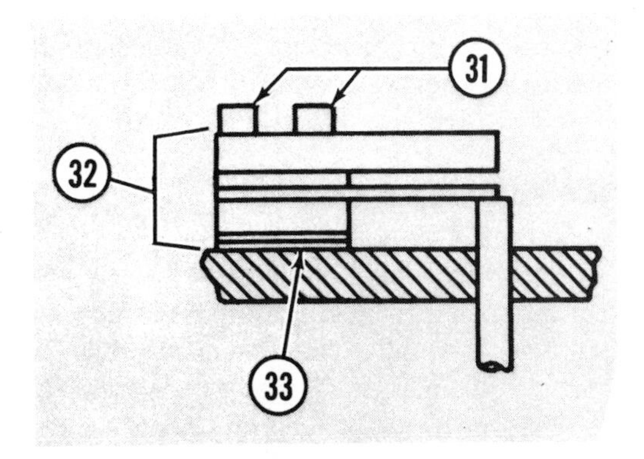

Figure 7-65
Leaf-Type Torque Spring.
(Courtesy of Caterpillar Tractor Co.)

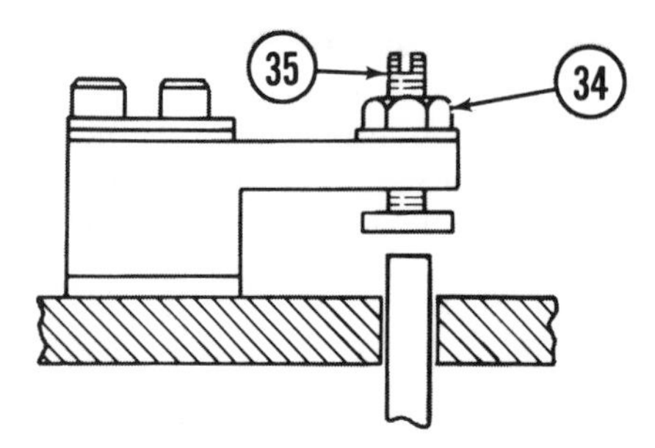

Figure 7-66
Load-Stop-Type Torque Spring.
(Courtesy of Caterpillar Tractor Co.)

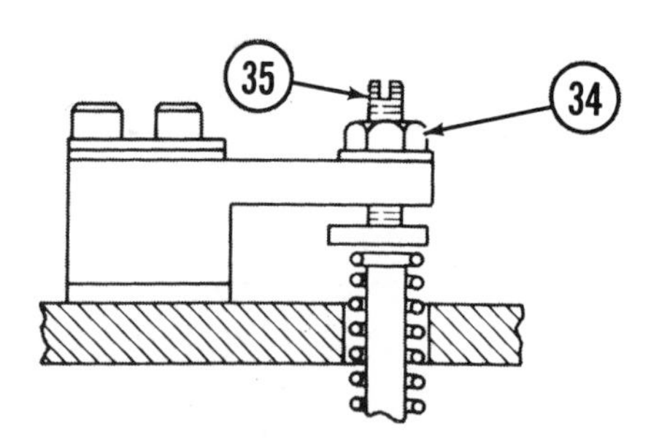

Figure 7-67
Coil-Spring-Type Torque Spring.
(Courtesy of Caterpillar Tractor Co.)

the checking procedure until you are sure that the rack setting is in fact correct.

Air–Fuel Ratio Control Adjustment

The air–fuel ratio control on the sleeve metering fuel system is for the same purpose as that used on other Caterpillar diesel engines. Anytime that the fuel rack setting is adjusted on the engine, the air–fuel ratio control should be checked as follows:

1. For air–fuel ratio controls installed on the injection pump housing on in-line engines, using pin 5P299 repeat steps 4 through 7 as described under fuel rack setting for SMFS.
2. Turn the dial indicator face bezel to zero and tighten the lock (30 in Figure 7-63). Write down the reading that is showing on the dial face at this time; then carefully loosen screw (17) with wrench (24) as shown in Figure 7-64 at least six complete turns. The reading that now exists on the dial gauge face is the air–fuel ratio control setting. The allowable tolerance from that given in the rack setting information is ±0.10 mm (0.004 in.).
3. Adjust the air–fuel ratio control by removing the cover and turning the adjustment bolt until the correct setting is shown on the dial indicator.

7-16

Fuel-Injection Nozzles

Caterpillar diesel engines can use one of three available types of fuel-injection valves or nozzles, which fall into the following categories:

1. The fuel valve that is found on their *pre-combustion* chamber engines.
2. The fuel valve found on their *direct injection* engines.
3. The *pencil nozzle* type found on their 1100 and 3200 series diesel engines manufactured by Roosa Master.

Figure 7-68 shows a typical pre-combustion chamber type of nozzle. Figure 7-69 shows a direct injection nozzle.

The nozzle valve that is used with both the pre-combustion and direct injection style engines is commonly known as a *capsule-type nozzle;* it cannot be rebuilt, but is exchanged as a unit if faulty.

Figure 7-70 shows the pencil-type nozzle. The operation of this particular nozzle is simi-

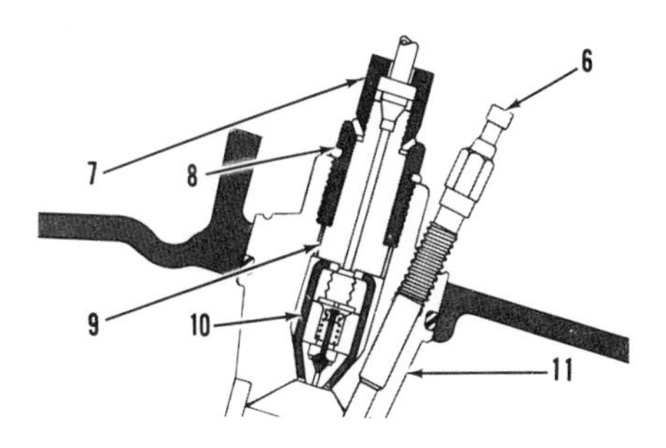

Figure 7-68
Typical Pre-Combustion Chamber Type of Nozzle.
(Courtesy of Caterpillar Tractor Co.)

Index:
6—Glow plug
7—Fuel line nut
8—Nut
9—Body
10—Nozzle assembly
11—Pre-combustion chamber

lar to other Roosa Master-type pencil nozzles; therefore, when servicing is required to this unit, refer to Chapter 8.

General Servicing: Capsule-Type Nozzles

If it is suspected that a Caterpillar diesel engine is suffering from faulty fuel-injection equipment such as nozzles, they can be removed and checked for opening pressure, spray pattern, and any other irregular signs of possible damage. However, prior to getting into nozzles and fuel-injection pumps, be sure to check for other problem areas. Check the color of the exhaust smoke and refer to Chapter 12 for interpretation and analysis of smoke color. Other items to check would include such things as (1) air starvation, (2) low compression, (3) air in the fuel, (4) water in the fuel, (5) plugged fuel filter, (6) low fuel pressure (transfer pump), and (7) burning oil.

Another simple check that can be made on the engine prior to pulling the nozzles or getting into the pump would be to run the engine at its normal low idle rpm; loosen the fuel line nut at each nozzle one at a time only. This will in effect stop the flow of fuel into the cylinder; therefore, a definite change in the sound of the engine should be noticed with some drop in engine speed if in fact the injector nozzle had been firing prior to loosening the fuel line nut. If there is no change at all in the sound of the engine when this is done, pull that particular nozzle and check it as described under checking fuel valves. If, after checking the nozzle, it is found to be satisfactory, the fuel-injection pump

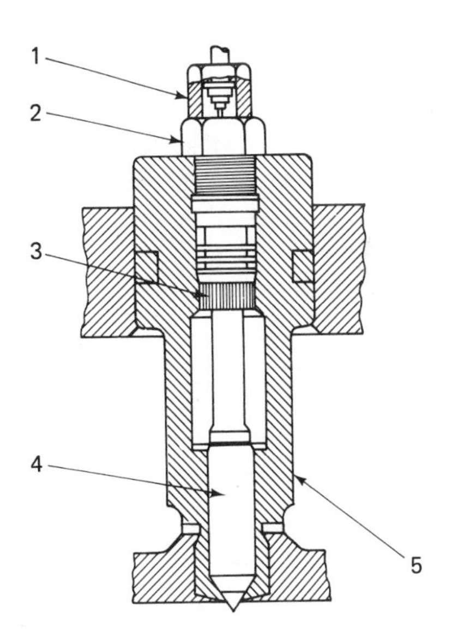

Figure 7-69
Direct Injection Nozzle.
(Courtesy of Caterpillar Tractor Co.)

Index:
1—Fuel line nut
2—Nozzle nut
3—Nozzle body
4—Nozzle valve
5—Adapter used with direct injection engine nozzle

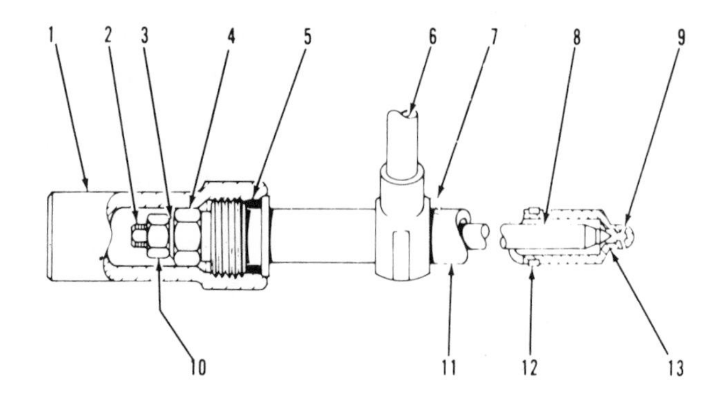

Figure 7-70
Pencil-Type Nozzle.
(Courtesy of Caterpillar Tractor Co.)

Index:

1—Cap
2—Lift adjustment screw
3—Pressure adjustment screw
4—Locknut for pressure adjustment screw
5—0 ring seal
6—Fuel inlet
7—Compression seal
8—Valve
9—Tip orifices (four); orifices in the spray tip are 0.0128 in. (0.325 mm) in diameter
10—Locknut for lift adjustment screw
11—Nozzle body
12—Carbon dam
13—Nozzle tip

feeding that nozzle would have to be checked out for performance.

Checking Fuel-Injection Nozzle Valves

This check involves basically the following:

1. Inspection of the nozzle tip or orifices for signs of carbon buildup or plugging.
2. Signs of overheating or physical damage.
3. Dirty or plugged nozzle screen in capsule-type units.
4. Opening pressure of the valve.
5. Spray pattern check.

To remove the injection nozzle from the engine, refer to Figures 7-68 and 7-69. For example, with pre-combustion chamber nozzles, the nozzle assembly and body are held in the cylinder head by a nut, and they can be removed by simply disconnecting the fuel line and removing the nut and seals. To facilitate this task, a pre-combustion chamber wrench

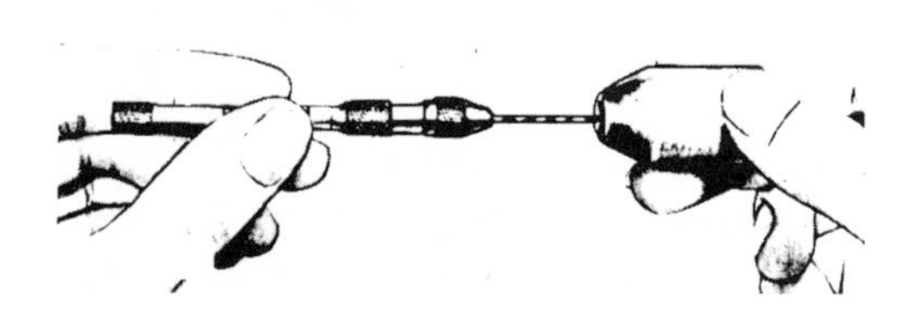

Figure 7-71
Cleaning Fuel-Injection Nozzle Assembly.
(Courtesy of Caterpillar Tractor Co.)

such as 8S225 is available. On direct injection engine nozzles, use wrench 5P961 to facilitate adapter removal.

The body and nozzle assembly can then be lifted out. Note that the nozzle assembly is assembled only finger tight on the body at all times.

If a capsule-type nozzle is found to be carboned up, remove it from the valve body and very gently clean the fuel discharge hole using the correct drill and chuck found in the available Caterpillar service tool group, such as shown in Figure 7-71.

Caution: Exercise extreme care when cleaning the nozzle. Do not attempt to remove the carbon accumulation with a wire wheel or brush; otherwise, tip damage and hole enlargement can occur.

Once the nozzle capsule has been cleaned, reassemble it to the body and mount the assembly on a suitable injector pop tester to check the following:

1. Nozzle spray pattern.
2. Nozzle valve popping or unseating pressure.
3. Rate of nozzle leakage.

Note: Always keep clear of the spray from the nozzle during pressure checking since the fuel pressure is high enough to penetrate the skin and cause serious infection.

Prior to installing the nozzle back into the cylinder head bore, check the seat of both the nozzle and the seat of the pre-combustion chamber or adapter on direct injection engines. Also ensure that you torque the nut that holds the fuel nozzle in either the pre-combustion chamber or adapter to 55 ± 5 lb-ft (7.6 ± 0.7 mkg) or (74.56 ± 6.77 N·m). If the nut is overtorqued, the nozzle can be damaged; if undertorqued, the nozzle can leak.

7-17

Use of Injection Timing Light

Caterpillar engines can be checked very quickly for proper fuel injection timing advance by the use of a timing light available from Caterpillar as tool 1P3500. The tool is manufactured by the Sun Electric Corporation, who are very well known for their automotive special tools and equipment.

Although the 1P3500 timing light is referenced here, it is because this unit will cover the complete line of Caterpillar diesel engines; however, a similar tool, 2P8280, is also available, but will handle only Caterpillar truck engine models.

In either case, both work on the same principle of fuel under pressure from the injection pump being directed through a transducer unit, which is normally a closed switch. Fuel flowing into and through the transducer provides an input signal for the timing light by opening the circuit, which will close again by spring pressure once the fuel has been delivered to the injection valve.

Caution: The small transducers have been designed for insertion into the fuel line at either the injection pump end of the line or at the injection valve end of the line. The transducers have the capability of operating only under the fuel flow rates and pressures normally encountered during *no-load* conditions; therefore, do *not* operate an engine under *full-load* situations with a transducer in the fuel line as unstable and inaccurate readings may result.

Figure 7-72(a) shows the 1P3500 injection timing light, and Figure 7-72(b) shows a larger view of the transducers themselves. The timing light is designed to operate on 115 V ac; however, a built-in inverter within the carrying case will provide 115 volts if the power source happens to be a 12- or 24-V battery. The transducers are inscribed with an arrow indicating in which direction the fuel must flow

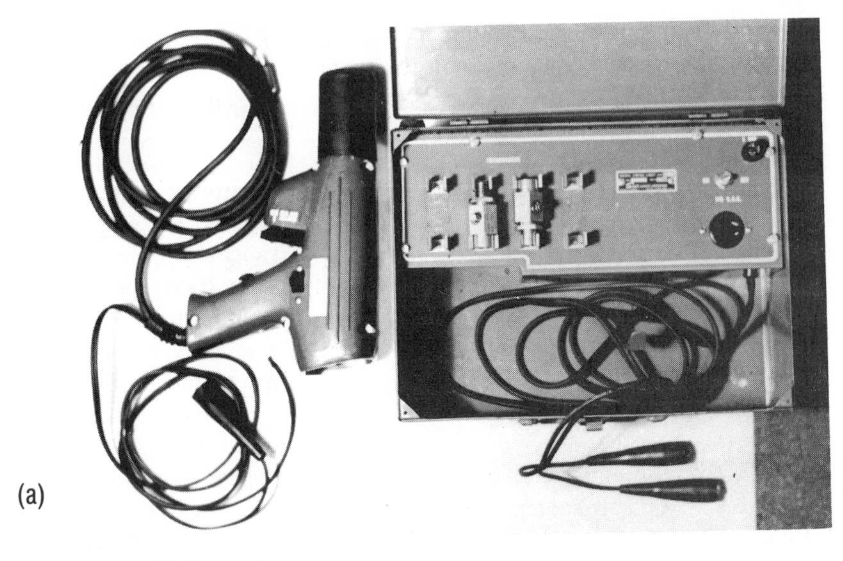

(a)

(b)

Figure 7-72
(a) Injection Timing Light 1P3500;
(b) Enlarged View of Transducers.
(Courtesy of Caterpillar Tractor Co.)

through it in order to operate. For example, on 1100 and 3100 series engines, the 1P3498 transducer can be used at the fuel-injection pump, or if more convenient the longer 1P3496 transducer can be used at the nozzle end of the fuel line. Remember that when you select a transducer for use on any engine it must be installed with the *arrow* in the direction of normal fuel flow.

In addition, the transducer can be placed on any cylinder; however, cylinder 1 is generally the first choice. The alternate cylinder would be 6 on in-line 6 engines; however, on 1100, 3100, and 3208 engines, 4 is an easier second choice.

Although the transducer can be used at both the injection pump and nozzle end of the fuel line, there is less chance for error if it is installed at the injection pump because a greater change occurs at the injection pump. This is due to the fact that, in addition to the initial timing specification plus the normal timing advance, a built-in advance of up to 5 degrees at high idle through the automatic advance unit is occurring at the pump, whereas there is a time delay involved in the fuel leaving the pump and reaching the fuel nozzle. This delay is approximately 0.0005 sec, which will consequently give a different reading at the injector than at the pump. For example, on an 1160 engine, the initial timing is plus 16 degrees, plus a built-in advance of plus 5 and an additional plus 8 degrees of timing advance, for a total of 29 degrees BTDC advance at the injection pump. Due to the delay through the fuel line, however, there is in fact a 9 degree delay through the injection line length; if the transducer is placed at the nozzle end of the fuel line, the total advance would then be only 20 degrees BTDC of the total advance.

Caterpillar special instruction form SMHS6964 lists the amount of advance for all engine models, both at the pump and the injector. Always check this Caterpillar form to ensure that the amount of advance for your particular engine model is in fact current and correct. An example of such a chart is given in Figure 7-73 for on highway truck engines *only*.

ON HIGHWAY TRUCK ENGINES
MOTEURS DE CAMIONS
LKW-MOTOREN
EN MOTORES PARA CAMIONES DE CARRETERA

MODEL Modèle Modell Modelo	SERIAL NO. N° de série Seriennummer No. de Serie	1 °ADVANCE TRANSDUCER AT PUMP ±2°	2 °ADVANCE TRANSDUCER AT INJECTOR ±2°	3 ADVANCE START ± 100 RPM	 END ±70 RPM	4 FOOTNOTES
1140	36B1-up	13	5	1200	3000	
1145	97B1-up	13	5	1200	3000	
1150	96B1-up	13	5	1200	3000	
1160	95B1-95B530	*	*	1200	3000	
1160	95B531-up	13	5	1200	3000	
1674	94B1-94B3471	12	7	1325	2200	3
1674	94B3472-94B11213	14	9	1325	2200	
1674	94B11214-up	12	7	1325	2200	3
1676	54B1-up	12	7	1320	2200	
1693	65B1-65B781	11	6	1100	2000	
1693	65B782-65B7058	11	6	1170	2015	7
1693	65B7059-up	14	9	1170	2015	7
► 3208	40S1-up, 32Y1-up	5	3	1600	2800	4
► 3208	40S27593-up	3	*	1800	2650	5
► 3208	32Y1-up	3	*	1800	2650	5
3306	76R371-up	8	3	1200	2200	
3406 PC	92U1-up	11	8	1320	2200	
3406 DI	92U1-92U3187	9	8	1320	2200	1
3406 DI	92U3188-up	6	5	1500	2040	1, 2
► 3408 PC	28V1-up	12	10	1370	2100	6
► 3408 DI	28V1-28V2047	4	3	1690	2000	1
► 3408 DI	28V2048-up	5	4	1420	2100	1

1—Not all timing lights are capable of measuring timing at low idle.
2—Engines with 6N1537 pump group (pump plate) use these figures.
3—Start of advance is 1470 rpm; end of advance is 2240 rpm on some engines.
4—Does not include 1977 model year California engines only.
5—California only 1977 model year engines.
6—Amount of advance with increase in rpm becomes less at 1650 rpm.
7—Start of advance is 1250 rpm; ends at 2080 rpm on some engines.

Figure 7-73
Example of Caterpillar Form SMHS6964
(Courtesy of Caterpillar Tractor Co.)

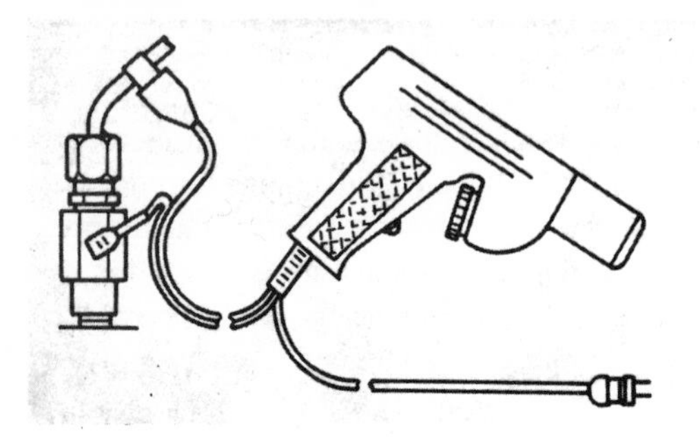

Figure 7-74
Timing Light Lead Attachment.
(Courtesy of Caterpillar Tractor Co.)

Procedure for Using Timing Light

1. Bar the engine over until the timing mark on the crankshaft pulley, damper, or flywheel inspection plate cover shows that this mark is in alignment with the stationary pointer. If the mark is incorrect, you may have to refer to the earlier section dealing with checking engine timing with the 5P6524 and 1P540 tool groups.
2. Install the correct transducer for your engine at either the injection pump or nozzle.
3. Refer to Figure 7-74; install the small brass bayonet-type clip onto the transducer pickup, the large alligator clip onto the fuel line or a suitable ground, and the large plug to the power source (115 V ac or 12- or 24-volt battery).
4. Refer to Figure 7-75; with the engine running at its normal low idle, place switch (2) in the ADVANCE position and gently press the trigger (1) and aim the flashing light at either the damper, pulley, or flywheel.
5. With the engine idling, by adjusting the knob (3) in Figure 7-75, you will be able to align the TDC scribe mark on the pulley, damper, or flywheel with the stationary pointer.
6. When you have done this, look at the number of degrees of advance recorded on the scale of the timing light face. This will in fact be the point at which fuel is injected into the cylinder combustion chamber.
7. Built into the timing light is a tachometer

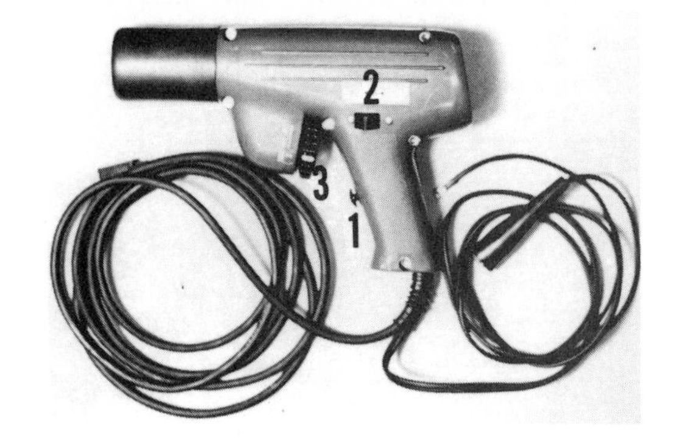

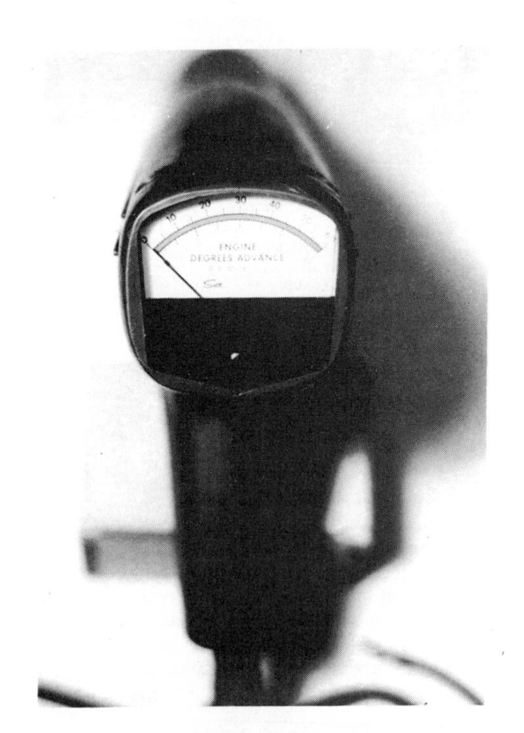

Figure 7-75
Controls and Face Readout of Timing Light.
(Courtesy of Caterpillar Tractor Co.)

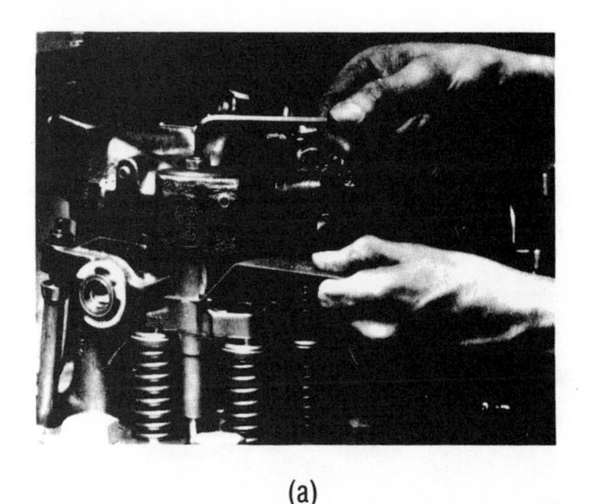

(a)

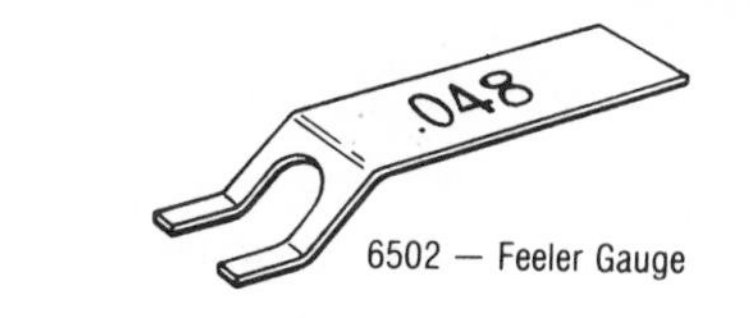

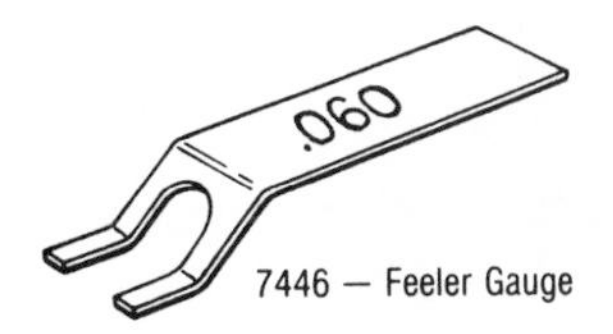

(b)

Figure 7-76
(a) Adjusting Jacobs Brake Slave Cylinder;
(b) Available Feeler Gauges.
(Courtesy of Jacobs Manufacturing Co.)

that can be used at any time by placing the switch from the ADV position to the TACH position.

8. Repeat steps 4 through 6 with the engine running at its high idle speed. Subtract the low idle reading from the high idle reading and compare the difference with those given in Caterpillar form SMHS6964. This will be the amount of timing advance.

9. Do not try to read the scale while the engine is being accelerated, but at a steady engine speed only. To observe the mark on the damper for smoothness of advance through the engine's speed range, turn knob (3) past the detent to the TIME position and place the switch in the ADV position.

7-18

Jacobs Engine Brake

The Model C-346 Jacobs engine brake is designed and approved for use on the following Caterpillar Series 3406 highway truck engines.

280 DIT	350 DITA	280 PCT	360 PCTA
290 DIT	380 DITA	325 PCT	375 PCTA
300 DIT			
305 DIT			
325 DIT			

Note: DIT—Direct injection turbocharged
DITA—Direct injection turbocharged aftercooled
PCT—Precombustion chamber turbocharged
PCTA—Precombustion chamber turbocharged aftercooled

Jacobs Slave Piston Adjustment Procedure

The engine must be shut down and cold prior to adjustment.

1. Loosen off the adjusting screw nut with a screwdriver or hex key as required; then loosen off the slave piston adjusting screw.
2. The piston must be at the top dead center position inside the engine cylinder on the compression stroke, which will ensure that

Figure 7-77
4S 6553 Engine Evaluation Tool Gourp.
(Courtesy of Caterpillar Tractor Co.)

both exhaust valves on that cylinder are closed and the valve bridges loose.

3. Insert the correct feeler gauge between the slave piston feet and bridge. Refer to Figure 7-76. Use feeler gauge 6502 on all 3406 engines. Use feeler gauge 7446 on the 3406 360 PCTA model only. Turn the adjusting screw in until a slight drag is felt on the feeler gauge. Torque the locknut to 15 to 18 lb.-ft. (20.4 to 24.4 N·m).

Figure 7-77 shows an engine evaluation test case, which can be used during troubleshooting and testing of a rebuilt engine while on a dynomometer. With this tool, engine rpm, oil pressure, turbocharger boost pressure, and full-rack travel via a built-in continuity light can all be checked. In addition, with both a mercury and water manometer gauge built into the case, crankcase pressure, air inlet restriction, and exhaust back pressure can be taken.

chapter 8
Roosa Master (Stanadyne/Hartford) Fuel Systems

8-1
Roosa Master Fuel Systems

Roosa Master diesel fuel-injection pumps are manufactured by Stanadyne's Hartford Division. These distributor-type pumps were introduced in the early 1950s and were a radical new design in comparison to the typical in-line fuel-injection pumps available at that time. These pumps have been extremely successful and today well over 11 million have been sold worldwide. The type of injection nozzle used is the revolutionary small-diameter *pencil nozzle,* which was introduced in 1965 and was the first major change in nozzle design in 30 years.

The two models of fuel-injection pumps that are presently being produced by Roosa Master are the DB2 and DM4. The description and operation of the model DB2 injection pump can be found in Section 8-2. Since the function and operation of the model DM4 is very similar to that of the DB2, we shall discuss briefly its differences and basic function of operation.

The model numbering system for the model DM4 pump is very similar to that for the DB and DB2 pump.

	a	*b*	*c*	*d*	*e*	*f*
Model Number:	D M	2	6	33	JN	2580

a: DM, M version of a D series pump

b: 2, number of plungers (2- or 4-plunger version)

c: 6, number of cylinders (2, 3, 4, 6, or 8)

d: 33, abbreviation of plunger diameter

25: 0.250 in. (6.35 mm)

27: 0.270 in. (6.86 mm)

29: 0.290 in. (7.37 mm)

31: 0.310 in. (7.87 mm)

33: 0.330 in. (8.38 mm)

35: 0.350 in. (8.89 mm)

37: 0.370 in. (9.40 mm)

39: 0.390 in. (9.91 mm)

e: JN, accessory code pertaining to combinations of special accessories such as electrical shutoff, automatic advance, variable speed droop adjustment, and the like

f: 2580, specification number, which determines the selection of parts and adjustments within a given accessory code

The model DM2 pump incorporates a single pumping chamber, whereas the DM4 has two pumping chambers; however, they are both described as *opposed plunger, inlet metering, distributor-type pumps*. The major component parts are shown in Figure 8-1. The distributor rotor located in the pump's hydraulic head is driven by the drive shaft, and the pumping plungers are actuated toward each other by an internal cam ring through rollers and shoes just as in the model DB2 pump. The

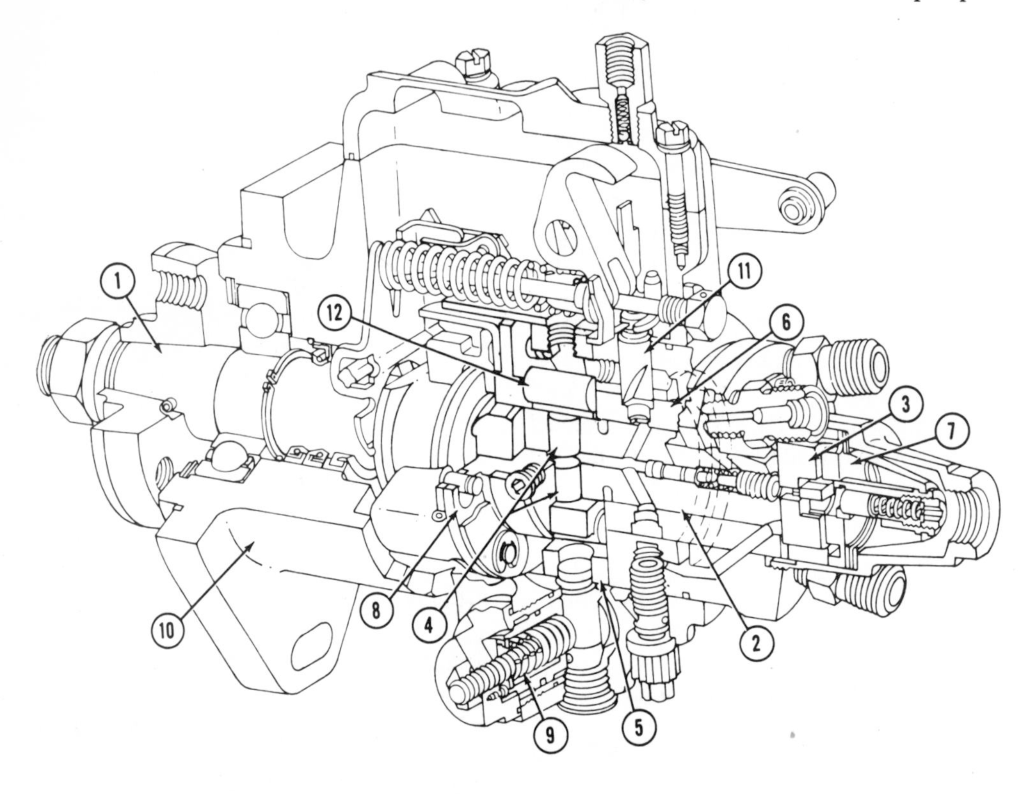

Figure 8-1
DB2 Distributor Injection Pump.
(Courtesy of Roosa Master, Stanadyne/Hartford Div.)

Index:

1—Drive shaft
2—Distributor rotor
3—Transfer pump blades
4—Pumping plungers
5—Cam ring
6—Hydraulic head
7—Regulator assembly
8—Governor
9—Automatic advance
10—Housing
11—Metering valve
12—Rollers

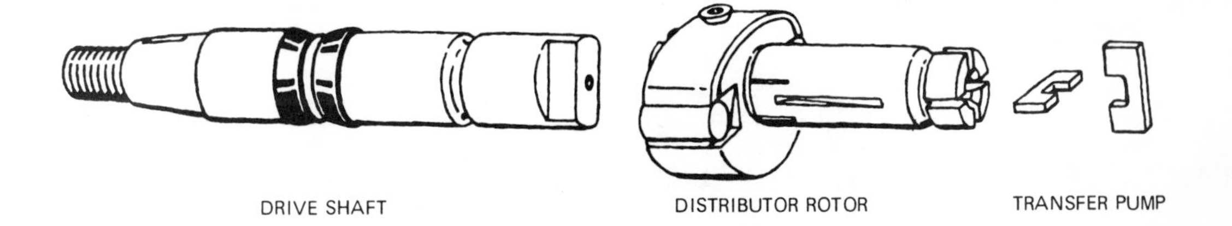

Figure 8-2
Main Rotating Parts of the DB2 Pump.
(Courtesy of Roosa Master, Stanadyne/Hartford Div.)

internal workings of all Roosa Master pumps are basically the same. Figure 8-2 shows the main rotating parts of the pump.

The fuel flow through the DM4 is the same as in the DM2, with the exception of the charging of two additional plungers. The rotor and plunger setup of both the DM2 and DM4 is shown in Figure 8-3.

Spill Port and Torque Chopper

Some DM model injection pumps are equipped with a spill port and torque chopper, which operates in conjunction with the speed advance mechanism. Figure 8-4 shows this assembly; the location of the spill port in the rotor bore is positioned between the pumping plungers and the delivery valve. If the advance mechanism has the injection pump in a retarded timing position, high-pressure fuel is able to flow through this spill passage into the charging annulus through a head charging port prior to the end of the normal injection period. This spilling of fuel will occur only in the retarded timing position, as shown in Figure 8-5. Should engine speed increase, the internal cam ring rotation will cause an advance in the start of injection, which will therefore reduce the amount of high-pressure fuel spillage from the spill port.

A point to remember here from our earlier discussions is that any reduction in engine rpm will also create a reduction in fuel transfer pump pressure; therefore, as timing is retarded, an increasing quantity of fuel will spill while injection is still occurring.

The overall effect of both the spill port and torque chopper and speed advance mechanism is to reduce exhaust smoke and engine torque with a decrease in engine speed.

Aneroid Control

At the present time, every diesel engine (4-stroke cycle) manufacturer of turbocharged engines uses some form of device to control the amount of fuel delivered to the combustion chambers in direct proportion to air flow and pressure boost from the exhaust gas driven turbocharger. Many of these engines use what is commonly called an *aneroid* to sense the intake manifold pressure during engine acceleration periods to prevent overfueling and excessive exhaust smoke. This device therefore takes into account the inherent lag factor of the exhaust gas driven turbocharger and corrects for this until it has picked up speed and can then supply enough air to fuel to avoid smoking.

Figure 8-6 shows an exploded view of the component parts of the Roosa Master aneroid. The aneroid assembly is mounted on a bracket

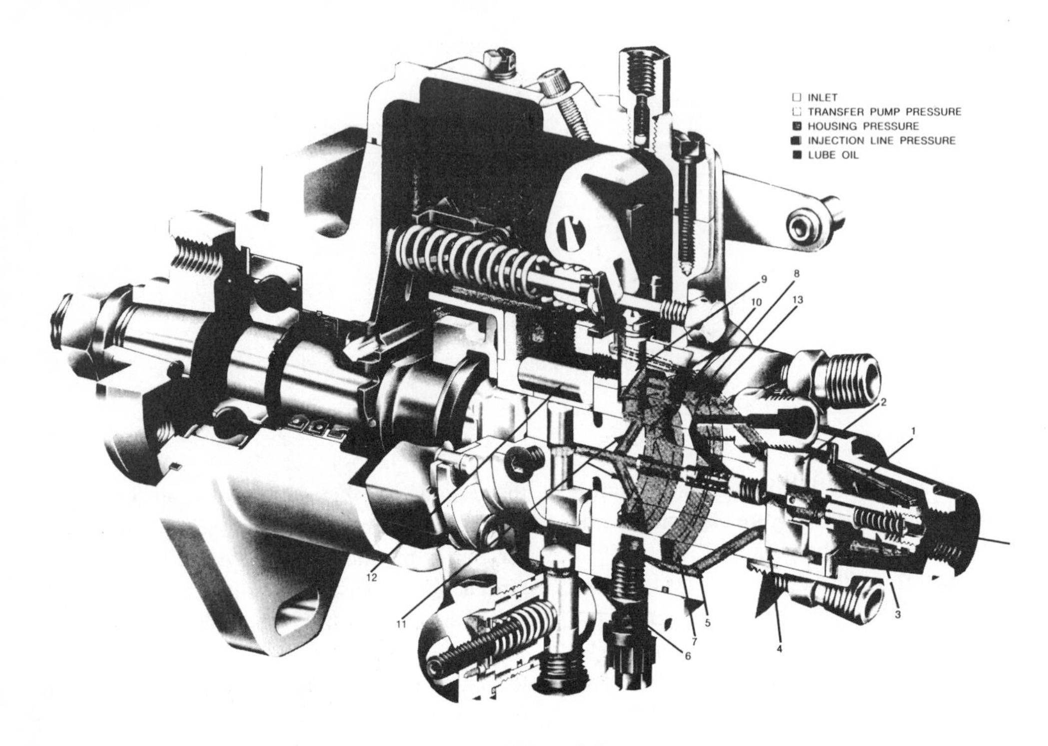

Figure 8-3
Model DM Injection Pump Fuel Flow.
(Courtesy of Roosa Master, Stanadyne/Hartford Div.)

Index:

1—Drive shaft
2—Distributor rotor
3—Transfer pump blades
4—Pumping plungers
5—Internal cam ring
6—Hydraulic head
7—Pressure regulator assembly
8—Governor
9—Automatic advance
10—Housing
11—Metering valve

attached to the top of the injection pump housing cover. Figure 8-7 shows the aneroid mounted in position.

Prior to cranking, and during starting of the engine, no turbocharger boost air pressure is being supplied to the aneroid. Fuel flow during this period past the metering valve is therefore not at its maximum. Anytime that the engine is running at its rated speed and load, turbocharger boost pressure through the intake manifold and supply line to the aneroid is at its maximum. This pressure acting on the aneroid piston will force it down against spring pressure, transmitting this movement to the operating rod, which is connected to the pump shutoff lever and placing it in the maximum *run* position.

At anytime there is a speed or load change toward the decreased range, the corresponding reduction in air pressure to the aneroid will allow the aneroid piston return spring to pull the shutoff lever in the fuel shutoff direction.

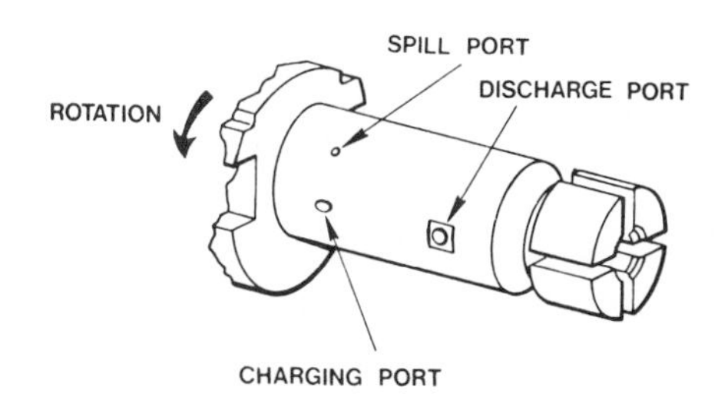

Figure 8-4
Location of Spill Port in Rotor Bore.
(Courtesy of Roosa Master, Stanadyne/Hartford Div.)

This causes the shutoff cam to push the governor linkage hook toward the rear, thereby reducing fuel flow past the pump metering valve, as shown in Figure 8-7.

The amount of fuel flow reduction is controlled by adjustment of the forward screw on the shutoff lever. This should only be done with the pump on a test stand as per specifications for the specific engine and application that the pump is scheduled for. Do not play around with this adjustment on the engine.

If you come across the older style aneroid

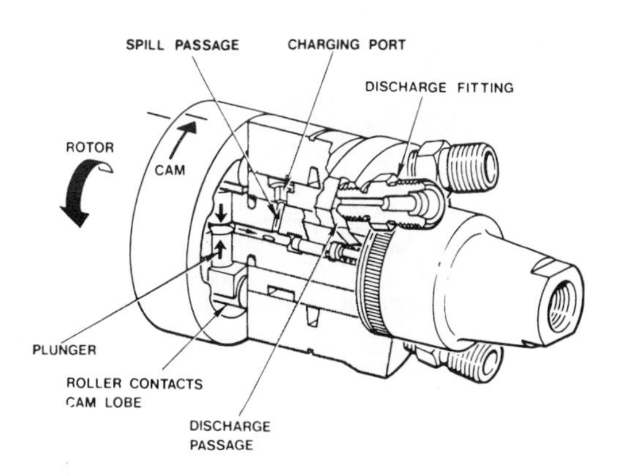

Figure 8-5
Retarded Timing Position.
(Courtesy of Roosa Master, Stanadyne/Hartford Div.)

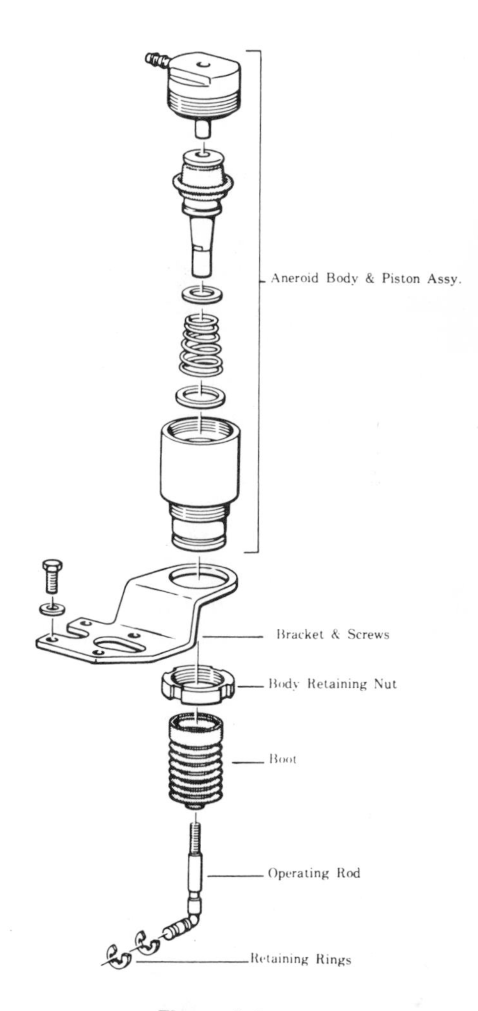

Figure 8-6
Component Parts of the Roosa Master Aneroid.
(Courtesy of Roosa Master, Stanadyne/Hartford Div.)

that incorporated an external spring and the unit is defective, replace it with the current model, as shown in Figures 8-6 and 8-7.

Checking Aneroid Operation. If it is suspected that the aneroid is faulty, with it still on the pump bracket in position, and the engine stopped, run a shop air line with a clean, dry supply and a pressure regulating valve in

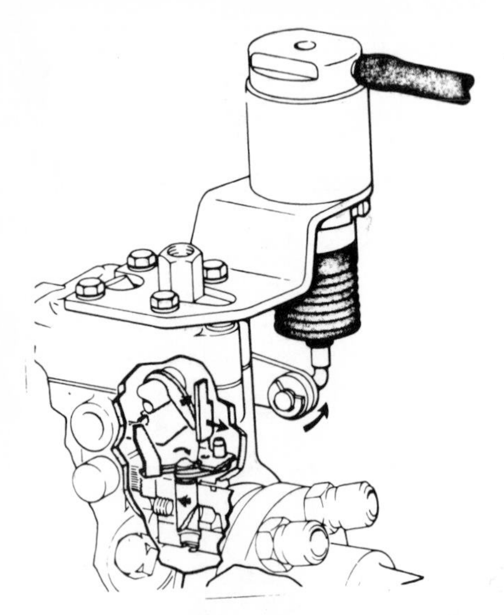

Figure 8-7
Aneroid Mounted on Injection Pump.
(Courtesy of Roosa Master, Stanadyne/Hartford Div.)

the line to the inlet of the aneroid. Tee-in a fitting to which a mercury (Hg) manometer can be attached rather than a gauge since it is much more accurate. However, if a gauge must be used, then select one with a range of 0 to 30 psi (0 to 206.85 kPa) calibrated in ½-lb (3.45-kPa) increments.

Always refer to Roosa Master specifications for the particular injection pump aneroid settings for the particular application.

Air pressure can be applied to the aneroid and its freedom of movement—signs of air leaks and any other problems determined. Do not overlook possible kinks or line leakage from the intake manifold either.

If the aneroid is faulty, remove it from its bracket, disassemble it (do not submerge the aneroid assembly prior to teardown in oil or solvents), and inspect all parts. On reassembly, a dry lubricant is used on the internal diaphragm; therefore, do not wash it off.

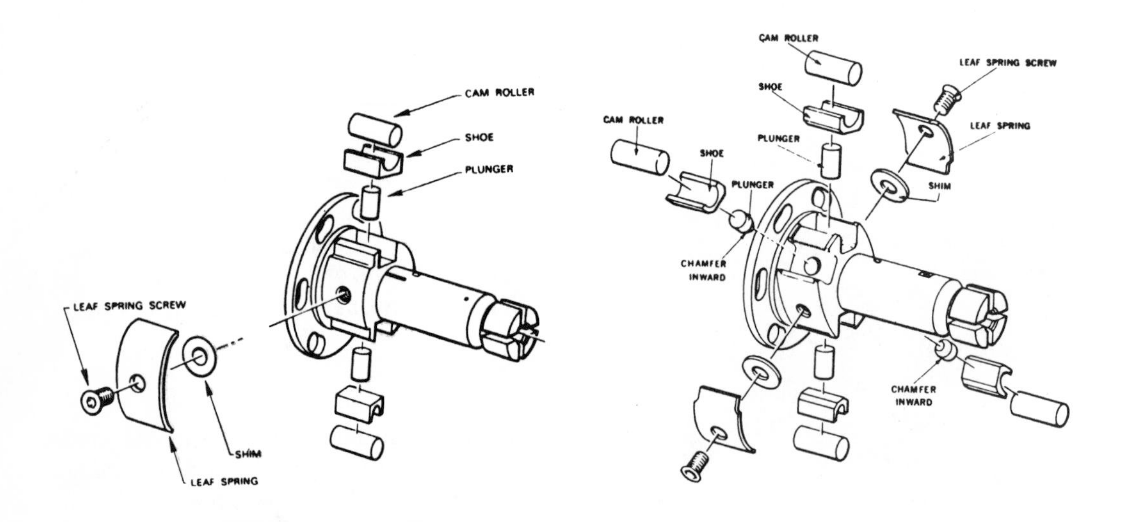

Figure 8-8
Rotor and Component Parts of the DM2 and DM4 Pumps.
(Courtesy of Roosa Master, Stanadyne/Hartford Div.)

After reassembly, and with the aneroid bracket held in a vise, tighten the retaining nut to the body of the aneroid to a torque of 80 to 90 lb-ft. (108.46 to 122 N·m) with tool 18031.

With a new dust boot over the operating rod, screw it into the piston until it bottoms, then back it out approximately two turns.

Do not attach the operating rod to the pump shutoff lever until the injection pump has been calibrated on a test stand and all preliminary adjustments have been made.

Figure 8-8 shows the component parts and relative location to that of the pump rotor.

Since the DM2 and DM4 pumps can be used on different engine applications, it would be well to point out at this time, that there are various roller shoe sizes for the DM4 pump.

These are identified by the number on their end from a minus 20 to a plus 10 in individual increments of 5.

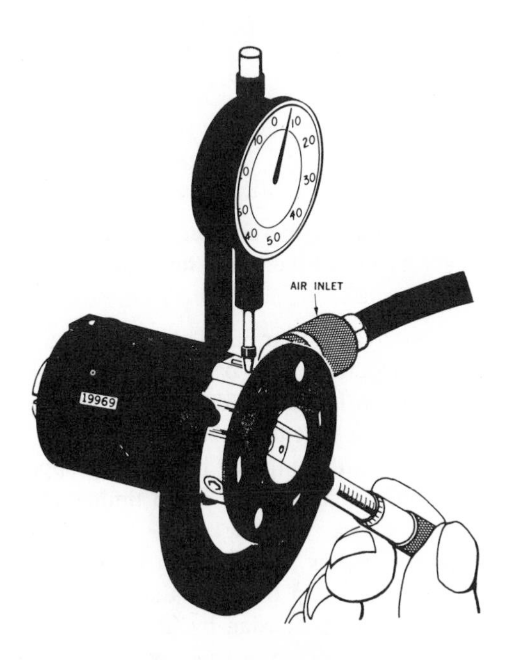

Figure 8-9
Checking the Roller-to-Roller Clearance. (Courtesy of Roosa Master, Stanadyne/Hartford Div.)

It is imperative that all four shoes are of the same dimension for the particular application. This shoe size is determined by a part number only.

Always insert a cam roller by sliding it into the shoe from the end, not the top; otherwise, damage will result.

Another important point here is that if at any time plungers are to be replaced, etched on the head of the rotor assembly is a letter indicating the specific diameter of plunger to be used with the rotor.

For example, a plunger of 0.330 in. (8.38 mm) diameter has a mated part number 11076, and the graded sizes, A through D, are identified by the part numbers 11077, 11078, 11079, and 11080, respectively. Therefore, the replacement plungers for a 0.330-in. (8.38-mm)-diameter plunger with a pump rotor etched with the letter C would be P/N 11079. However, always refer to current specifications for part numbers.

You may come across some pumps with a 2 etched on the rotor following the letter grading, which indicates that the plunger bore is 0.002 in. (0.05 mm) oversize.

Maximum Fuel Output Adjustment. Adjustment of maximum fuel delivery for all Rossa Master injection pumps must be done with the pump disassembled and mounted to the proper test fixture as shown in Figure 8-9. Of course the pump must then be reassembled and with it mounted on a fuel-injection pump test stand, its flow rate compared to specifications along with any other relevant adjustments.

For a detailed analysis and rebuild information, always refer to the manufacturer's operation and instruction manual for your particular model of injection pump.

Figure 8-9 shows a DM4 pump rotor and plunger assembly mounted to Roosa Master special test fixture no. 19969, which can be clamped into a vise on its flat sides.

The dial indicator can be mounted to its

bracket, but slide it to its outer limit of travel for now (not in contact with the plunger roller).

In order to check the distance between each set of opposed rollers, they must be forced outward by the use of clean, dry shop air supplied by a line as shown. Maintain air pressure between 40 and 100 psi (276 and 689.5 kPa). Manually rotate the rotor until the rollers attain their maximum outward movement by butting up against the leaf spring.

Using a 1- to 2-in. (25.4- to 50.8-mm) micrometer, measure the distance between the crown of the opposed cam rollers and compare with specifications for your particular pump.

If the specification is incorrect, each leaf spring must be adjusted alternately by turning its adjusting screw inward (clockwise) to increase and outward (counterclockwise) to reduce the dimension.

It may be necessary to change out or invert the leaf springs in order to obtain the correct specification on each set of rollers.

The roller settings of each pair must be within 0.003 in. (0.076 mm).

To check that the rollers are concentric to each other, i.e., to make certain that each pair will start its pumping stroke at the same time, proceed as follows:

1. Manually turn the rotor around until any roller is directly under the dial indicator stem plunger. Loosen off the clamp on the bracket holding the dial indicator in position and gently slide the indicator down until the dial gauge is pre-loaded at least 0.010 in. (0.254 mm) on its face and lock the retaining screw of the bracket. Carefully rotate the knurled bezel or outer dial to zero in the dial indicator.
2. Manually rotate the rotor in either direction until the next roller comes into contact with the dial indicator plunger and note the reading on the dial face. Do this with all 4 rollers. The allowable variation is plus or minus 0.002 in. (0.05 mm) for a total difference of 0.004 in. (0.101 mm).
3. If the reading is beyond specifications, the rollers and/or shoes can be interchanged; then recheck the centrality.

General Pump Installation

Prior to installing the injection pump to any engine, make certain that the engine has new fuel filters installed, all lines are blown clean, mounting area and timing lines in damper, pulley, or flywheel are clearly visible, and that the engine has been turned over in its normal direction of rotation to place the timing mark in position with the stationary pointer.

1. Remove the small timing cover from the injection pump to expose the timing line on the weight retainer as you manually turn over the pump drive shaft. Make certain that the timing line is in alignment with the line on the cam ring outside diameter. See Figure 8-10.
2. Slide the pump into its mated position on the engine; install the washers and nuts and torque to specifications.
3. Depending on the engine, you may now have to install the pump drive gear to the hub or shaft.
4. To remove all gear backlash in the engine to pump drive train, bar the engine backward from its normal direction of rotation at least one half-turn; then slowly bring it forward until the timing lines appear opposite one another in the pump timing window. Check the marks on the engine damper, pulley, or flywheel. If incorrect, loosen the pump flange retaining nuts and turn the pump housing by firmly grasping it and twisting to advance or retard the timing. Which way it must be turned will depend on the rotation of the pump. However, after any correction, repeat the previous procedure to ensure that the pump is

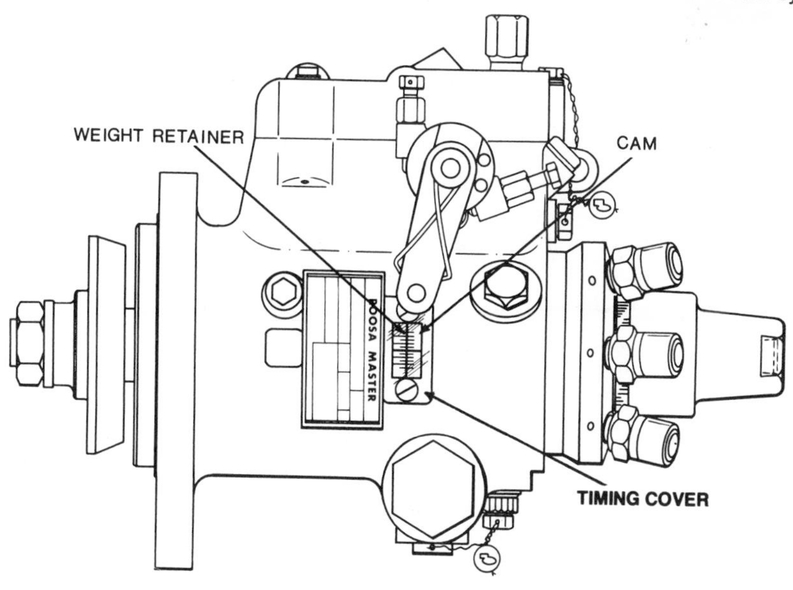

Figure 8-10
Pump Timing Lines.
(Courtesy of Roosa Master, Stanadyne/Hartford Div.)

properly timed. Install the timing cover on the pump housing.

5. Remove all protective caps on all lines and connect their nuts to the correct location and torque to specifications. Leave them loose at the individual injectors for now. Double-check that the lines are connected to the proper injector as per the engine firing order. Connect all necessary linkages and fittings.
6. Open the secondary filter bleed screw if a hand priming pump is fitted before the filter itself, or prime the filters first if no hand priming pump is fitted, until a steady flow of fuel free of air bubbles appears; then tighten the transfer pump inlet line.
7. Crank over the engine until fuel drips from each injector line and tighten the nut of each in firing order sequence, after which time the engine should fire and run. Let it idle for a few minutes while you check both lube and fuel pressures. If the engine runs rough, individually crack open the injector line nut to rebleed each one, and then tighten. If you have trouble starting the engine, you may have to rebleed the fuel system.

8-2

Oldsmobile-GMC Fuel System

The recent advent of the passenger car diesel in North America follows a pattern that has been emerging with increasing prominence in the last 5 years. The majority of diesel engine and passenger car manufacturers are now either individually or in concert with another producing diesel engines for the car and pickup truck field.

In the 1960s General Motors Corporation, through their truck and coach division, produced a 4-stroke diesel engine known as a *Toro-Flow* model. This was basically derived from a line of their existing heavy-duty gasoline engines and was reasonably successful.

The introduction of the 350-in.^{3} (5.7-liter) automotive diesel engine by the Oldsmobile Division of General Motors Corporation in 1977 was a first for a major North American passenger car company. It has met with excellent acceptance, so much that for the 1979 model year they also offered a 260-in.^{3} (4.2-liter) diesel engine in no less than 19 of their car models. The 260-in.^{3} has now been dropped from production.

Both engines are of V-8 configuration and follow the same basic design as their V-8 gasoline engine counterparts, with the exception of course that the major engine components such as the engine block, crankshaft, conrods, pistons, and valve lifters have been increased in strength to withstand the higher pressures and temperatures encountered within the diesel engine from a 22.5 : 1 compression ratio.

The diesel engine firing order is the same as that for gasoline engines: 1–8–4–3–6–5–7–2, with cylinders 1, 3, 5, and 7 being on the engine's left bank and cylinders 2, 4, 6, and 8 being on the right bank. The pre-combustion chamber design is used in these engines, since peak cylinder pressures in this style of engine are less than in a direct-injection or open chamber design. For passenger car use, this allows a slightly quieter running engine, which is desirable.

The cylinder heads contain stainless steel pre-combustion chamber inserts, which with the heads removed can be pushed out for servicing after removal of the glow plugs and injection nozzles. The glow plugs are used to facilitate starting, especially in cold weather. Run off of the engine's 12-V electrical system, they glow cherry red somewhat like the cooking element on a kitchen stove. They remain on for a very short period after the engine fires and then are cut off automatically by the electronic control module used with the car and truck applications. The glow plugs are threaded into place, whereas the injection nozzles are held in place by a bolt and spring clamp on earlier models.

Unless you are trained in the intricacies of diesel fuel-injection equipment and have access to the special equipment required to service pumps and injectors, do *not* attempt any disassembly of these components in the field.

Fuel-injection System

The fuel-injection pump and injection nozzles used are manufactured by the Roosa Master group of the Stanadyne Hartford Division. The fuel-injection pump used is the model DB2, which incorporates the same basic operating characteristics inherent in other Roosa Master pumps. The basic pump has only four main rotating parts and slightly more than 100 total component parts altogether. Prior to describing the injection pump's operation, let us take a minute to familiarize ourselves with the model numbering system used.

Model Numbering

The pump can be used with engines having either 2, 3, 4, 6, or 8 cylinders simply with a change to the hydraulic head assembly and some minor internal pump changes. For example, model number:

a	*b*	*c*	*d*	*e*
DB2	8	33	JN	3000

is interpreted as follows:

a: DB2-*D* Series Pump, *B*-Rotor, *2nd* generation

b: 8, number of cylinders

c: 33, abbreviation of plunger diameter
25, 0.250 in. (6.35 mm); 27, 0.270 in. (6.86 mm); 29, 0.290 in. (7.37 mm); 31,

0.310 in. (7.87 mm); 33, 0.330 in. (8.38 mm); 35, 0.350 in. (8.89 mm)

d: JN, accesory code that relates to a variety of special pump options such as electrical shutoff, automatic advance, and variable speed droop adjustment

e: 3000, specification number that determines the selection of parts and adjustments for a particular pump application

Fuel System Operation

The model DB2 injection pump's operation can be likened to that of an ignition distributor. However, instead of the ignition rotor distributing the high-voltage spark to each cylinder in firing order sequence through the distributor cap, the DB2 pump distributes pressurized diesel fuel as two passages align during the rotation of the pump rotor, also in firing order sequence. Figure 8-11 shows the fuel system layout as it pertains to the Oldsmobile 350 (5.7 liter) diesel engine.

The basic fuel system flow is as follows: The fuel lift pump, which is driven from the engine crankshaft, not from the camshaft as on the gasoline engine 350 (5.7 liters), draws fuel through a filter contained within the fuel tank and sends it on to an additional filter prior to entering the transfer pump. Due to the extremely fine tolerances of injection pump components, much finer fuel filtration is required than in a gasoline engine.

From the filter, the fuel is delivered into the injection pump's transfer pump at the opposite end from the pump's drive shaft. As it enters the transfer pump, the fuel passes through a cone-type filter and on into the injection pump's hydraulic head assembly.

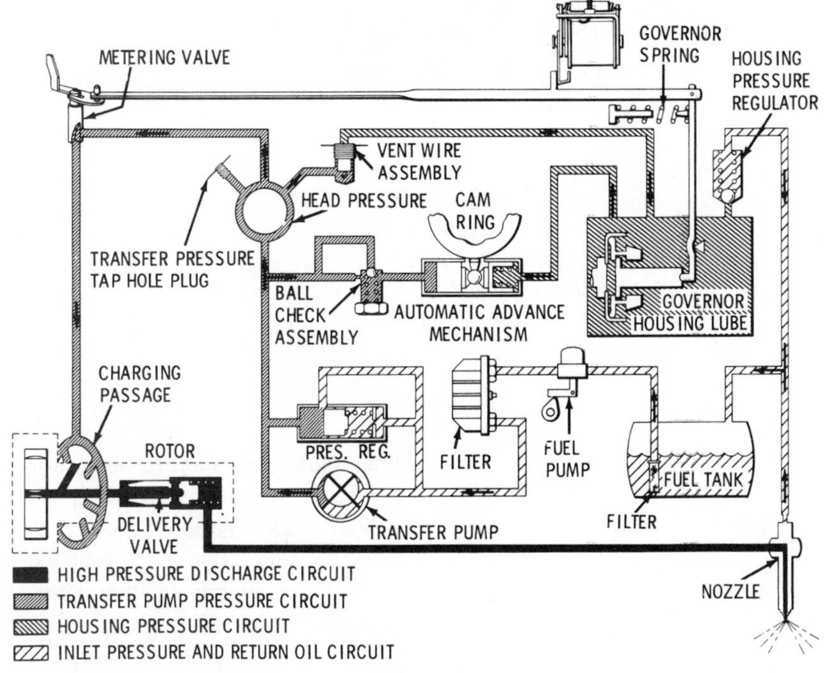

Figure 8-11
Fuel System Layout of Oldsmobile 350 Diesel Engine.
(Courtesy of Oldsmobile Div. of GMC)

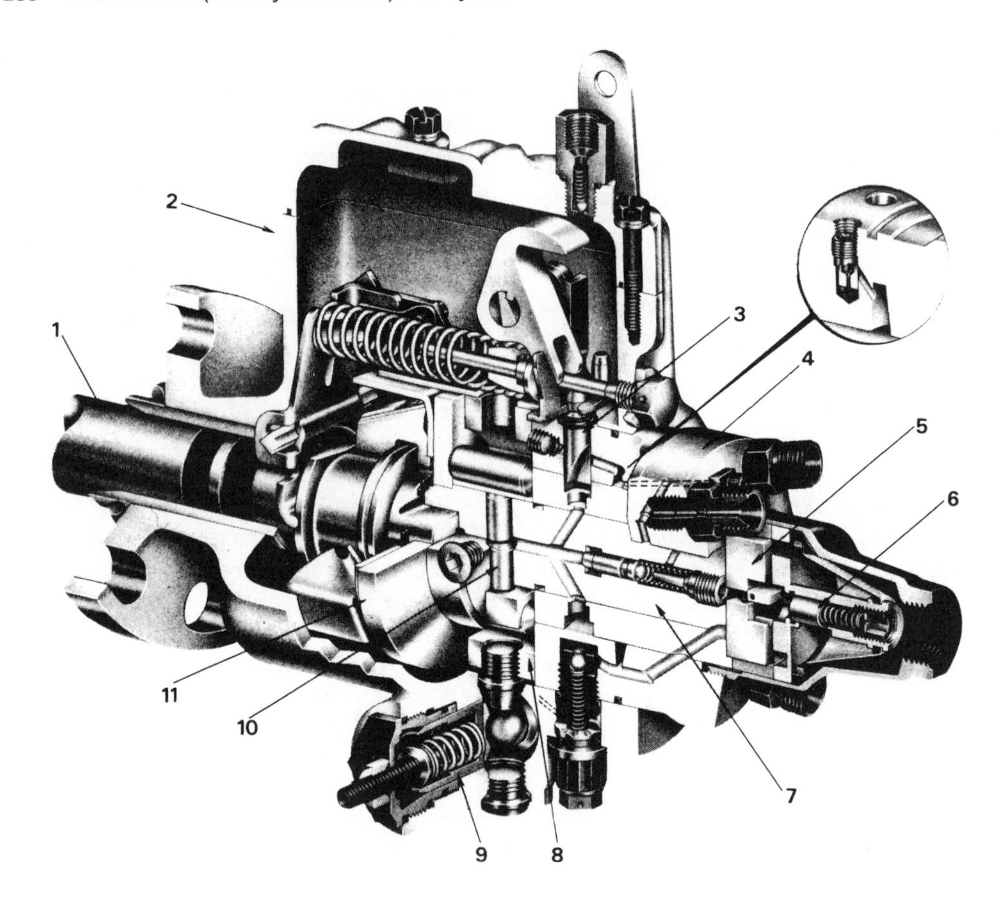

Figure 8-12
Injection Pump Assembly.
(Courtesy of Roosa Master, Stanadyne/Hartford Div.)

Index:
1—Drive shaft
2—Housing
3—Metering valve
4—Hydraulic head assembly
5—Transfer pump blades
6—Pressure regulator assembly
7—Distributor rotor
8—Internal cam ring
9—Automatic advance (optional)
10—Pumping plungers
11—Governor

Fuel under pressure is also directed against a pressure regulator assembly, where it is bypassed back to the suction side should the pressure exceed that of the regulator spring.

Fuel under transfer pump pressure is also directed to and through a ball check valve assembly and against an automatic advance mechanism piston. Pressurized fuel is also routed from the hydraulic head to a vent passage leading to the governor linkage area, which allows any air and a small quantity of fuel to return to the fuel tank through a return line, which self-bleeds any air from the system. The fuel that passes into the governor linkage compartment is sufficient to fill it, thereby lubricating the internal parts.

Most of the fuel leaving the hydraulic head is directed to a fuel metering valve, which is controlled by operator throttle position and governor action. This valve controls the amount of fuel that will be allowed to flow on into the charging ring and its ports. Rotation of the rotor by the pump's drive shaft aligns the two inlet passages of the rotor with the charging ports of the charging ring, thereby allowing fuel to flow on into the pumping chamber.

The pumping chamber consists of a circular cam ring, two rollers, and two plungers. (See Figure 8-12.) As the rotor continues to turn, the rotor's inlet passages will move away from the charging ports, allowing fuel to be discharged as the rotor registers with one of the hydraulic head outlets. With the discharge port open, both rollers come in contact with the cam ring lobes, which forces them toward one another, causing the plungers to pressurize the fuel between them and sending it on up to the injection nozzle and into the combustion chamber.

The nozzle contains basically a spring-loaded check valve that is held on its seat until fuel pressure is high enough to lift it from its seat against this spring pressure. When this occurs, fuel will be injected through holes in the end of the nozzle. When the fuel pressure drops off owing to the rotor rotation past the cam ring lobes, the rollers and plungers move outward, effectively reducing this pressure. This reduction in pressure allows the spring inside the injection nozzle to force the valve back onto its seat, effectively cutting off any further injection of fuel.

Automatic advancement is accomplished in the pump by fuel pressure acting against a piston, which causes rotation of the cam ring, thereby aligning the fuel passages in the pump sooner. Its action can be compared to the vacuum and weight advance of the breaker plate inside an ignition distributor. See Figure 8-11 for more detailed information on this aspect. Maximum engine RPM is accomplished through a mechanical governor that affects the position of the fuel metering valve; this is covered later in the chapter.

The injection pump assembly shown in Figure 8-12 depicts the general layout of the relative parts. The drive shaft engages with and drives the distributor rotor in the hydraulic head. Also contained in the hydraulic head in which the rotor revolves are the metering valve bore, the charging ports, and the head discharge fittings, which have the high-pressure steel injection lines connected to them.

Fuel Transfer Pump

The delivery capacity of the fuel transfer pump is capable of exceeding both the pressure and volume requirements of the engine, with both varying in proportion to engine speed. Excess fuel can, however, be recirculated back to the inlet or suction side of the pump by the regulator.

Figure 8-13 shows the basic component parts of the rotor and transfer pump, which is a positive displacement vane-style unit consisting of a stationary liner with spring-loaded blades that ride in slots of the rotor. The cam rollers move in or out with the rotor's rotation, causing the shoe to force the plungers inward when the lobe of the internal cam ring makes contact with them simultaneously. The leaf spring controls the maximum plunger travel and therefore the maximum fuel delivery, since it limits the travel of the roller shoes. Fuel pressure fed into the area between both plungers comes from the rotor to the charging ring, where it then enters the plunger chamber. Since fuel pressure and volume are proportional to engine speed, the plungers will reach their maximum outward travel only when the engine is running under full-load conditions.

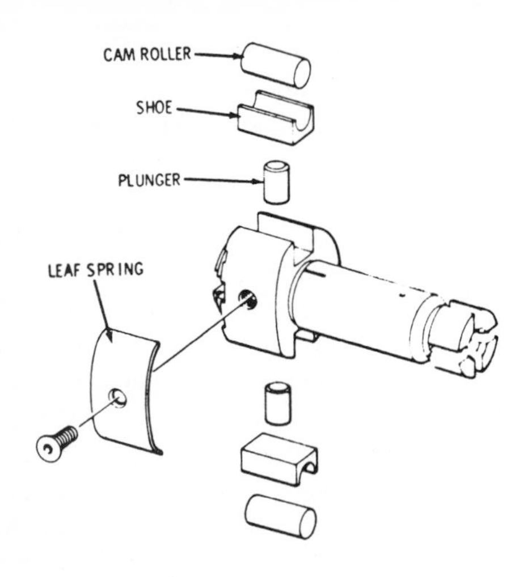

Figure 8-13
Component Parts of the Rotor and Transfer Pump.
(Courtesy of Roosa Master, Stanadyne/Hartford Div.)

Transfer pump pressure is adjusted with the pump on a test stand by increasing or decreasing the pressure regulator spring tension by clockwise or counterclockwise rotation of the spring plug. Maximum top end pressure should never exceed 130 psi (896.35 kPa).

The fuel pressure delivered to the plungers within the injection pump assembly comes from the vanes of the transfer pump riding in four slots at the opposite end from the plungers and rollers. The right-hand end of the rotor shown in Figure 8-13 shows these slots. Figure 8-14 depicts the action of the vanes as the rotor completes one full revolution of its cycle.

In Figure 8-14, as the rotor turns within the eccentric liner, the vanes riding in the ends of the rotor slots will be forced to move in and out. If we study for a moment position A of the diagram, vanes 1 and 2 are in a central position between the inlet port and are drawing fuel into the area formed between them. Moving to sequence B, we see that vanes 1 and 2 have now passed completely the inlet port of the liner, thereby trapping and carrying around with them the volume of fuel that was drawn in at position A.

As the rotor continues to turn, the vanes will be forced back into their slots because of the shape of the eccentric liner bore. Position C shows that owing to the ever-reducing space between vanes 1 and 2 the volume of fuel that was drawn in initially at point A has been forced into this smaller area, causing an increase in fuel pressure, which will leave the outlet groove in proportion to engine speed.

The pressurized fuel is directed through the groove of the pressure regulator assembly, past the rotor retainers, and into an area on the

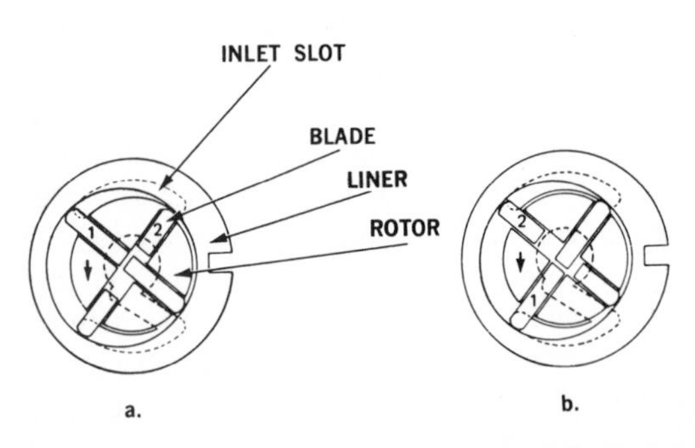

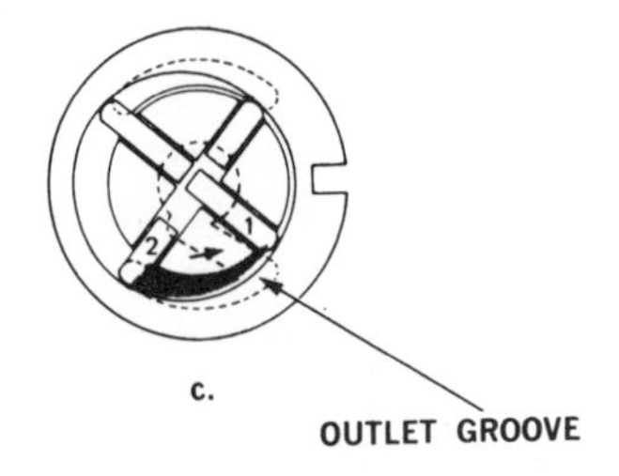

Figure 8-14
Vane Action as Rotor Completes One Full Revolution.
(Courtesy of Roosa Master, Stanadyne/Hartford Div.)

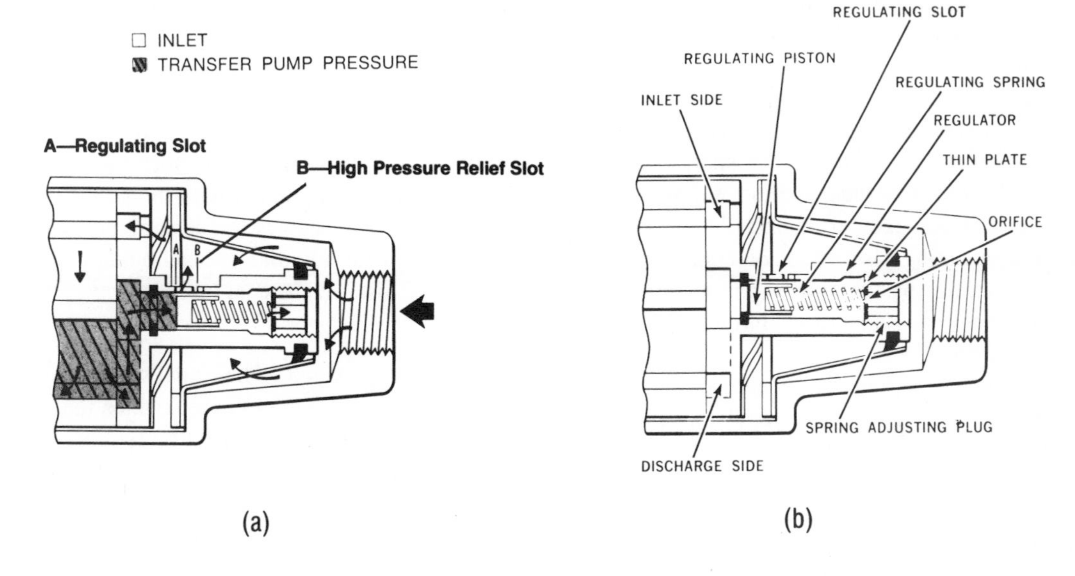

(a) (b)

Figure 8-15
Action of Regulator Assembly.
(Courtesy of Roosa Master, Stanadyne/Hartford Div.)

rotor leading to the hydraulic head passages. The fuel volume will continue to decrease at the vanes until blade 2 passes the outlet groove in the liner assembly.

Pressure Regulator Assembly

The action of the pressure regulator is the same as any pressure regulator, whether it be in a fuel, lube, or hydraulic system, that is, to limit the maximum pressure developed in the system by relieving or bypassing fuel back to the low-pressure or suction side of the system. Obviously, then, the only time that the regulating piston assembly will operate is when the engine is running. Remember from earlier discussion that the pump output will vary with engine speed.

Figure 8-15 shows the action of the regulator assembly. Fuel under pressure leaving the transfer pump is directed against the pressure regulator piston, which applies pressure to the spring in behind the piston. As the engine speed increases, so do pump volume and pressure, which will force the regulating piston back until it starts to uncover the pressure regulating slot, therefore limiting and controlling the maximum pump delivery pressure. Some regulators incorporate what is known as a high-pressure relief slot, which simply prevents excessively high transfer pump pressure should the engine be forced into an overspeed condition such as when on a vehicle.

Fuel Viscosity Compensation

Since different grades of diesel fuel may be used with the pump, and temperature variations are bound to occur in different geographical locations, the pump employs a rather unique feature within the regulating system just described that offsets pressure changes due to these two factors. A thin plate contained in the spring adjusting plug incorporates a sharp-edged orifice that allows fuel leakage that passes between the piston and its bore to return to the suction side of the transfer pump. The rate of fuel flow through any short orifice is so small with a change in fluid temperature that the pressure exerted against the back side of the regulating piston will be controlled by the fuel leakage past the piston and its bore and the resulting pressure drop through the sharp-edged orifice.

When the ambient air temperature is low, the fuel leakage past the regulating piston will be very small indeed; therefore, any pressure on the back side of the piston will be extremely low. However, if the ambient air temperature is high and the engine is running at its normal operating temperature, the viscosity of the fuel oil will be less, allowing a higher rate of fuel leakage past the piston. This will increase the fuel pressure in behind the regulating piston within the spring cavity and therefore fuel flow through the orifice.

The change in pressure behind the piston will therefore vary the piston's position as it reacts to the transfer pump's outlet pressure normally acting upon it at the front end. This variation in regulating piston position will compensate or offset the fuel leakage that would occur as the fuel thins out, and in this way the fuel transfer pump pressure can be maintained over a broad range of fuel temperature changes.

Charging and Discharging Cycle

Charging Cycle. Rotation of the rotor allows both inlet passages drilled within it to register with the circular charging passage ports. The position of the fuel metering valve connected to the governor linkage controls the flow of fuel into the pumping chamber and therefore how far apart the two plungers will be. The maximum plunger travel is controlled by the single leaf spring, which contacts the edge of the roller shoes. Maximum outward movement of the plungers will therefore only occur under full-load conditions. Figure 8-16 shows the fuel flow during the charging cycle. Anytime that the angled inlet fuel passages of the rotor are in alignment with the ports in the circular passage, the rotor discharge port is not in registry with a hydraulic head outlet and the rollers are also off the cam lobes.

Discharge Cycle. The actual start of injection will vary with engine speed since the cam ring is automatically advanced by fuel pressure acting through linkage against it. Therefore, as the rotor turns, the angled inlet passages of the rotor move away from the charging ports. As this happens, the discharge port of the rotor opens to one of the hydraulic head outlets.

Also at this time, the rollers make contact with the lobes of the cam ring, forcing the shoes and plungers inward and thus creating high fuel pressure, which flows through the axial passage of the rotor and discharge port to the injection line and injector. This fuel delivery will continue until the rollers pass the innermost point of the cam lobe, after which they start to move outward, thereby rapidly reducing the fuel pressure in the rotor's axial passage and simultaneously allowing spring pressure inside the injection nozzle to close the valve. This sequence of events is shown in Figure 8-17.

Delivery Valve Operation

To prevent after-dribble, and therefore unburnt fuel with some possible smoke at the exhaust, the end of injection, as with any high-speed diesel, must occur crisply and rapidly. To ensure that the nozzle valve does in fact return to its seat as rapidly as possible, the delivery valve within the axial passage of the pump rotor will act to reduce injection line pressure

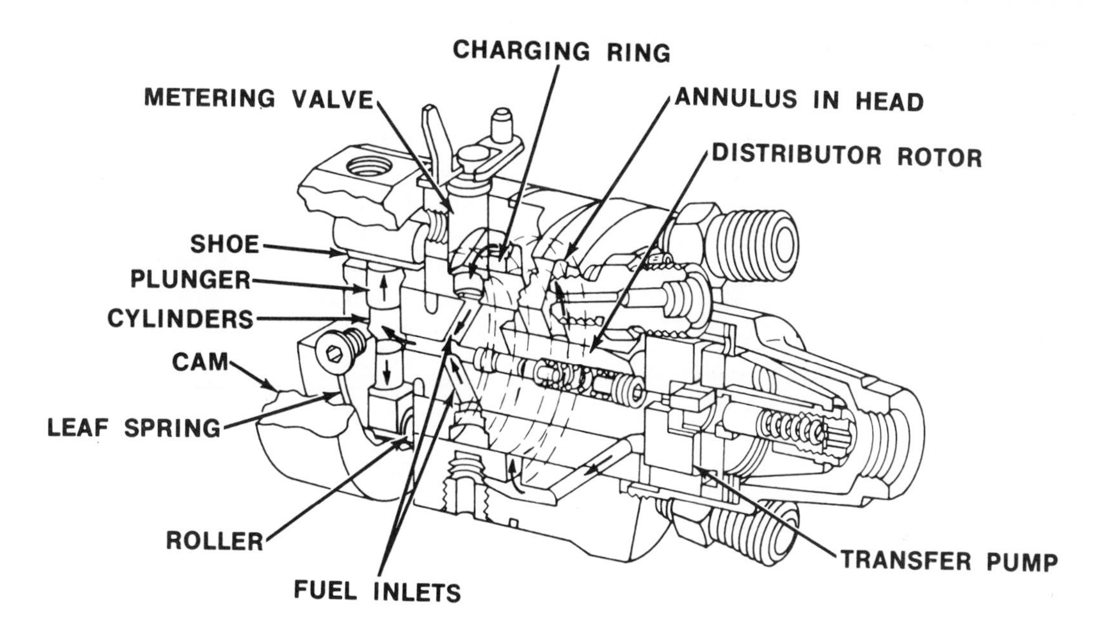

Figure 8-16
Fuel Flow During Charging Cycle.
(Courtesy of Roosa Master, Stanadyne/Hartford Div.)

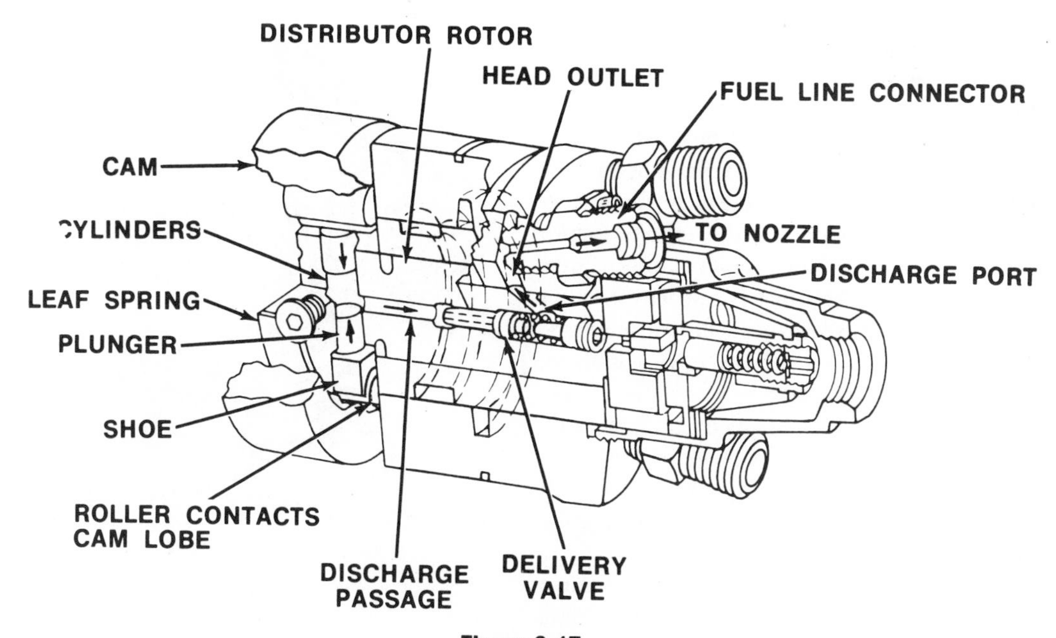

Figure 8-17
Discharge Cycle.
(Courtesy of Roosa Master, Stanadyne/Hartford Div.)

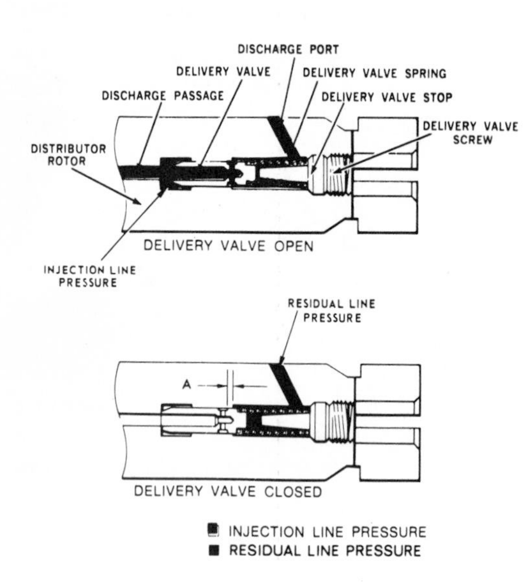

Figure 8-18
Delivery Valve Action.
(Courtesy of Roosa Master, Stanadyne/Hartford Div.)

after fuel injection to a value lower than that of the injector nozzle closing pressure.

From some of the views shown so far you will recollect that the delivery valve is located within the rotor's axial passageway. To understand its function more readily, refer to Figure 8-18. The delivery valve requires only a stop to control the amount that it can move within the rotor bore. No seals as such are required, owing to the close fit of the valve within its bore. With a distributor-type pump such as the DB2, each injector is supplied in firing order sequence from the axial passage of the rotor; therefore, the delivery valve operates for all the injectors during the period approaching the end of injection.

In Figure 8-18, pressurized fuel will move the valve gently out of its bore, thereby adding the volume of its displacement as shown (section A) to the delivery valve chamber, which is under high pressure. As the cam rollers start to run down the lobe of the cam ring, pressure on the delivery valve's plunger side is rapidly reduced, thereby ending fuel injection at that cylinder.

Immediately thereafter, the rotor discharge port closes totally, and a residual injection line pressure is maintained. In summation, the delivery valve will seal only while the discharge port is open, since the instant the port closes residual line pressures are maintained by the seal existing between the close-fitting head and rotor.

Fuel Return Circuit

We know from an earlier discussion that a small amount of fuel under pressure is vented into the governor linkage compartment. Flow into this area is controlled by a small vent wire that controls the volume of fuel returning to the fuel tank, thereby avoiding any undue fuel pressure loss. Figure 8-19 shows the location

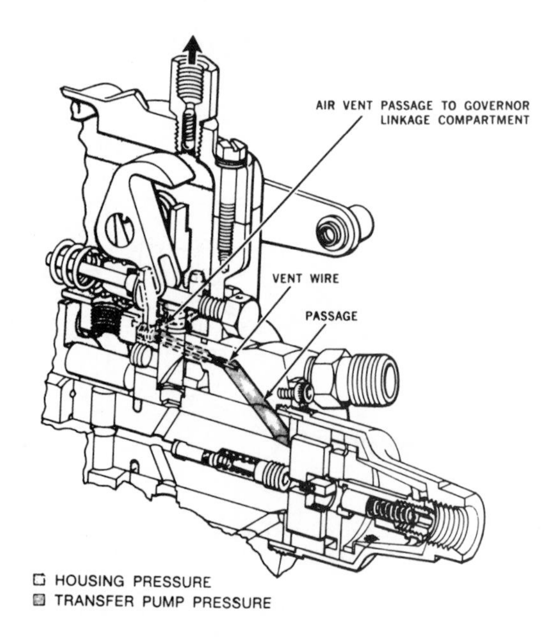

Figure 8-19
Location of Air Vent Passage.
(Courtesy of Roosa Master, Stanadyne/Hartford Div.)

of the vent passage, which is behind the metering valve bore and leads to the governor compartment via a short vertical passage. The vent wire assembly is available in several sizes to control the amount of vented fuel being returned to the tank, its size being controlled by the pump's particular application. In normal operation this vent wire should not be tampered with, as it can be altered only by removal of the governor cover. The correct wire size would be installed when the pump assembly is being flow tested on a pump calibration stand.

The vent wire passage, then, allows any air and a small amount of fuel to return to the fuel tank. Governor housing fuel pressure is maintained by a spring-loaded ballcheck return fitting in the governor cover of the pump.

Mechanical Governor

The governor function is that of controlling the engine speed under various load settings. As with any mechanical governor, it operates on the principle of spring pressure opposed by weight force, with the spring attempting to force the linkage to an increased fuel position at all times, and the centrifugal force of the rotating governor flyweights attempting to pull the linkage to a decreased fuel position. Figures 8-11 and 8-20 show the governor linkage setup.

Rotation of the governor linkage varies the valve opening, thereby limiting and controlling the quantity of fuel that can be directed to the two fuel plungers and hence that delivered to

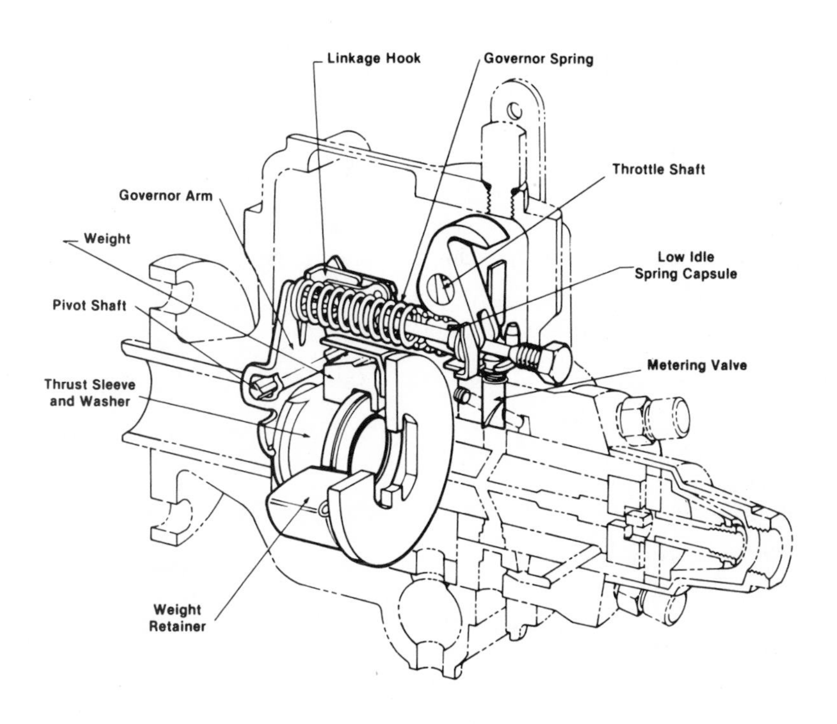

Figure 8-20
Governor Linkage Setup.
(Courtesy of Roosa Master, Stanadyne/Hartford Div.)

the injectors. The position of the throttle lever controlled by the operator's foot will vary the tension of the governor spring, and therefore the force acting through the linkage, to rotate the metering valve to an increased or decreased fuel position as the case may be.

At any given throttle position, however, the centrifugal force of the rotating governor flyweights will exert a force back through the governor linkage which is equal to that of the governor spring; in other words, a state of balance will result between these two forces. Outward movement of the weights acting through the governor thrust sleeve can turn the fuel metering valve by means of the governor linkage arm and hook, whereas throttle and governor spring position will turn the metering valve in the opposite direction.

Load Reduction and Speed Increase

If we consider a vehicle moving along the highway at a steady speed, the force of the governor weights would be in a state of balance with the governor spring pressure, which is established by throttle position. If the vehicle now starts to descend a hill, with the operator holding his or her foot in the same fixed position, the fact that the vehicle is now running downhill will reduce the load or power required by the engine to sustain the same road speed that was in effect while running along a level road. There will be a resultant increase in engine rpm and therefore a like increase in the centrifugal force of the governor flyweights, which will cause the fuel metering valve to be rotated to a decreased fuel position for a decrease in load. Therefore, the governor spring rate, throttle position, and weight force will automatically attain a state of balance even though there is a load reduction.

Load Increase and Speed Decrease

If we now consider the same vehicle moving along a level road at a given road speed, and it now encounters an incline or steep gradient, the power to move the vehicle up this hill will be greater if the same speed is desired. With the throttle held in a fixed position by the operator, engine rpm will start to drop, thereby reducing the centrifugal force of the governor

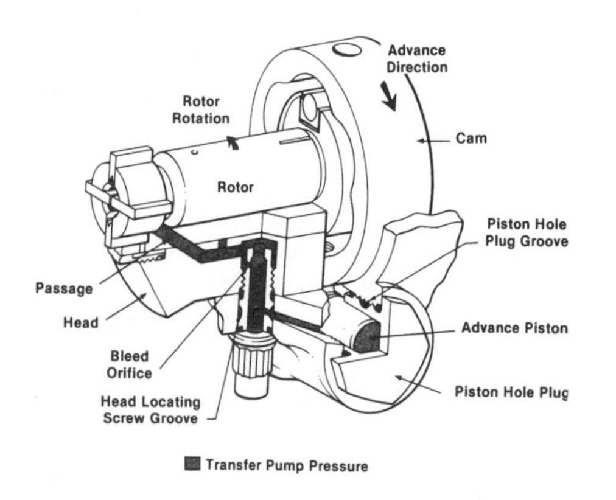

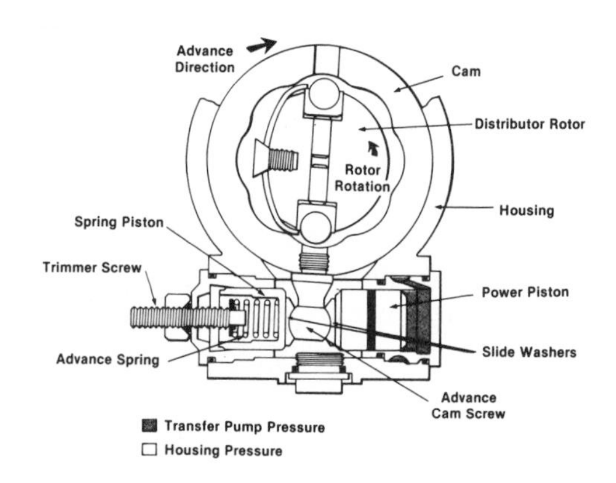

Figure 8-21
Automatic Timing Advance Mechanism. (Courtesy of Roosa Master, Stanadyne/Hartford Div.)

flyweights and upsetting the state of balance condition that had previously existed between the weights and spring. Since the change will favor the spring, it will turn the fuel metering valve in a counterclockwise direction, thereby increasing the fuel to the engine. If the vehicle is equipped with an automatic-type transmission, it will automatically downshift if the load is greater than the power ability of the engine. If, however, the vehicle is equipped with a standard transmission, then if the increased load is in excess of the engine's power ability, the operator is forced into downshifting to the next lower gear. Therefore, the engine speed throughout the operating range is dependent upon a combination of vehicle load and governor spring rate and throttle position.

At low engine speeds, such as at an idle, the centrifugal force of the governor flyweights is fairly low; therefore, a light idle spring is used for more sensitive control. The limits or arc of throttle travel is set by adjusting screws for the required engine idle speed and maximum top rpm. In addition, a light tension spring is used to remove any slack in the throttle linkage and to allow the fuel shutoff mechanism to close the fuel metering valve without having to overcome the higher force of the internal governor spring.

Automatic Timing Advance (Optional)

Similar in function to the distributor breaker plate movement in a gasoline engine, the injection pump can also be equipped with a simple direct-acting advance mechanism that is controlled by fuel pressure, which acts to rotate the cam ring and thereby vary engine pump timing. Fuel pressure from the transfer pump will vary in direct proportion to engine speed. When the advance mechanism is in effect, it will allow delivery of fuel to the injection nozzle earlier, thereby compensating for the inherent injection lag of all diesel engines. Obviously, then, the major purpose of any advance mechanism is to produce optimum power from the engine with a minimum amount of fuel usage and also minimum smoke.

Two pistons are used, which are located in a bore in the pump housing. One of these has spring pressure behind it, which will act to keep the cam ring in the retarded position at low engine speeds, while the power piston has fuel pressure directed behind it, which will force this piston and therefore the cam ring to an advanced position any time that fuel pressure is great enough to overcome the tension of the other piston's spring.

Fuel pressure fed through a drilled passage in the pump's hydraulic head registers with the bore of the head locating screw. The fuel is then directed on past a spring-loaded checkball contained within the bore of the head locating screw, after which it enters the groove on the outside diameter of the screw and passes on up to the back side of the power piston of the automatic advance mechanism. In order for any timing advance to occur, the fuel pressure must be high enough to overcome not only the opposing piston's spring pressure, but also the dynamic injection loading on the cam ring.

At low engine speeds, the start of injection timing will be controlled by the position that the injection pump was installed in and timed to the engine upon initial installation. The amount of actual advance will therefore be proportional to engine speed since fuel pressure is tied to this also. As the engine speed leaves the low range, fuel pressure will force the power piston to push against the advance pin, thereby causing relative movement of the cam ring toward the rotation of the moving rotor shaft. This simply means that injection will occur earlier in the stroke. The one-way check valve prevents the normal tendency of the cam ring to return to the retarded position during injection by trapping the fuel in the piston chamber.

Any reduction in engine speed will likewise cause a reduction in fuel pressure, allowing the other piston spring to return the

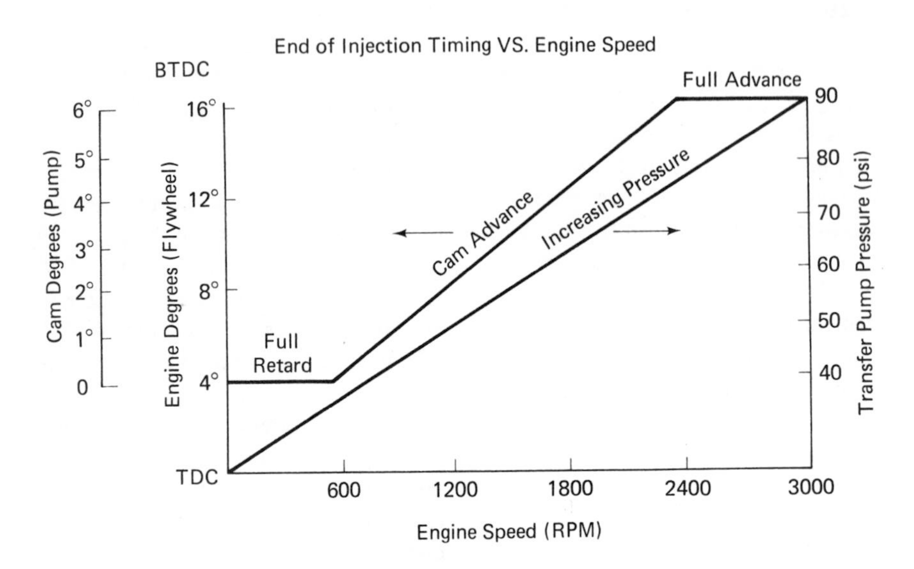

Figure 8-22
Typical Relationship Between Injection Pump and Engine Timing.
(Courtesy of Oldsmobile Div. of GMC)

cam ring to a retarded position in direct proportion to the speed and fuel reduction. Any fuel in the piston chamber can then bleed off through a control orifice in the piston hole plug (below the ballcheck valve in the head locating screw).

The actual start of cam ring movement will be determined by both piston spring pressure and the fuel pressure being high enough to overcome this spring resistance. The start of cam ring movement is therefore set by the initial adjustment of this spring, which can be done by use of the trimmer screw. For convenience, it can be found on either side of the advance mechanism; however, it should be adjusted on the fuel pump test stand. Total advancement is limited by the piston length. Figure 8-22 shows a typical relationship between injection pump and engine timing.

Any adjustment of the speed and light load advance device should be done only on a fuel-injection pump test stand, since serious engine damage could result if the engine is operated at excessive advanced timing.

Figure 8-23 shows a cutaway view of the engine at the front, which lets you see the injection pump drive mechanism. The pump drive gear is driven off the camshaft by a bevel gear drive immediately behind the chain-driven camshaft gear. Lubrication for the injection pump driven gear is by engine oil pressure directed through a passage from the top of the front camshaft bearing. Oil then passes through an angled passage in the gear shaft to the rear driven gear bearing. The fuel pump eccentric cam on the crankshaft is lubricated from the right bank main oil gallery by oil sprayed from a small orifice.

Injection Nozzle

The injection nozzles are located at each cylinder and held in place by a spring clamp and bolt screwed into the cylinder head. The nozzle is shown in Figure 8-24. Due to the compactness and thin shape of the injection nozzle, it is commonly referred to as a pencil nozzle with a closed end, since the actual nozzle valve does not project through an opening in the nozzle tip. The valve within the nozzle is held on its seat by spring pressure, and the current nozzle requires 1800 psi (12,411 kPa) to pop or open it, with an allowable tolerance of plus 100 psi (689.5 kPa) or minus 50 psi (344.75 kPa) of this figure. The nozzle lift adjusting screw is generally set for five-eighths of a turn. An edge-type filter is retained in the inlet fitting as shown. Two seals are used with the nozzle body, a nylon seal immediately beneath the inlet fitting banjo, which prevents any compression leakage, and a Teflon carbon dam seal, which prevents the buildup of carbon between the nozzle body and its bore.

Nozzle Fuel Flow

To prevent any variation in timing, the fuel lines to all nozzles are the same length. Fuel, which has been metered within the injection pump by the position of the fuel metering valve, is pressurized by the inward moving pump plungers, which send the fuel on up to the injection nozzle.

The fuel flows through the edge filter and around the valve, thereby charging or filling the nozzle body. Fuel pressure must be high enough to overcome the spring tension within the nozzle to allow fuel injection into the pre-combustion chamber. This fuel pressure acts upon a differential area at the valve tip, which is the difference between the area of the major diameter of the nozzle valve (at the valve guide) and the area at the valve seat at the tip.

When the valve opens under the action of fuel pressure, it will lift up a distance that has been set by the adjustment of the lift adjusting screw. When fuel delivery from the injection

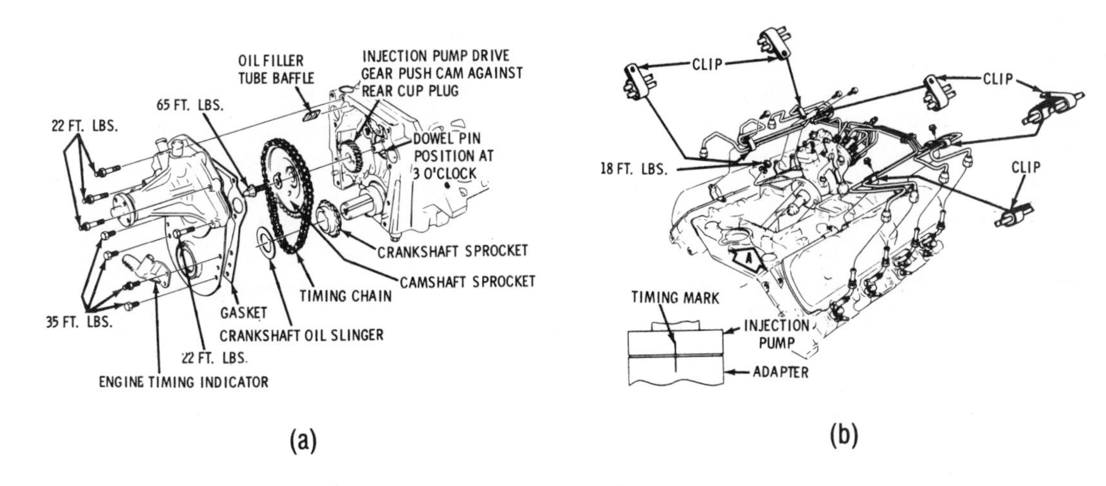

Figure 8-23
(a) Engine Front Cover Bolts; (b) Timing Marks and Injection Pump Lines.
(Courtesy of Oldsmobile Div. of GMC)

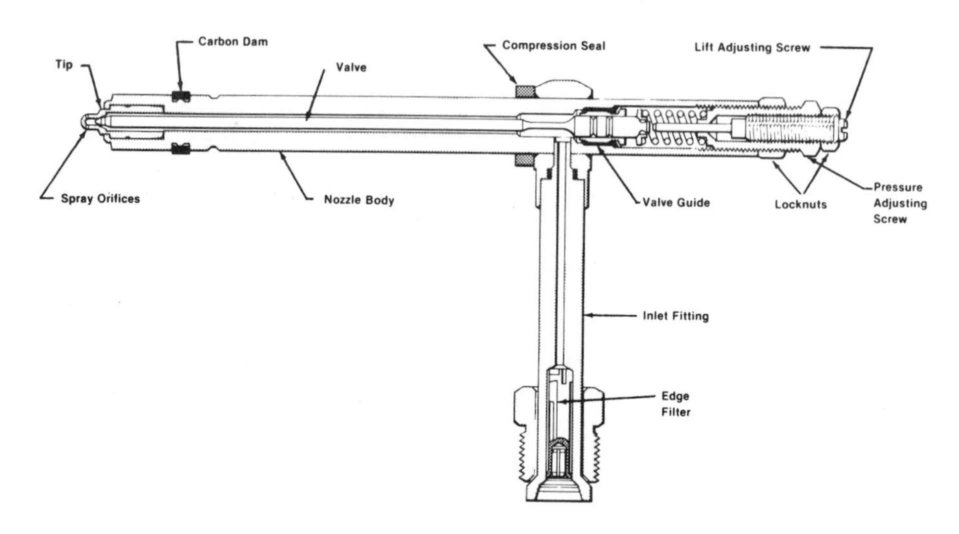

Figure 8-24
Injection Nozzle.
(Courtesy of Roosa Master, Stanadyne/Hartford Div.)

pump ends, the nozzle spring will act to rapidly seat the valve, and sealing is maintained by an interference angle between the valve and its seat, as shown in Figure 8-25.

All internal nozzle parts are lubricated by the fuel oil, a small amount of which leaks between the clearance existing at the valve and its guide during injection. This fuel then flows through a leak-off boot at the top of the nozzle and back to the fuel tank.

For nozzle testing procedures, obtain a copy of Booklet 99002 from your local fuel-injection dealer or from Roosa Master, Stanadyne/Hartford Division, Box 1440, Hartford, Connecticut, 06102, U.S.A.

Injection Nozzle Removal

At any time when nozzle removal is required, take care that you do not bend or distort fuel lines or exert undue force on component parts of the nozzle. If only one nozzle is to be removed, then only the fuel supply and return line for that one need be disturbed; however, if all nozzles are being removed for service, then all fuel supply and return lines should be removed and plastic shipping caps placed over the nozzle fuel inlet. Proceed to remove the nozzle spring clamp bolt and spacer; then to prevent any nozzle body damage, use tool number J-26952 and pull the nozzles one at a time. Exercise extreme care once the nozzles have been removed to ensure that the tips will not be damaged. If you are not immediately testing the nozzle on a pop testing machine to determine its release pressure and spray pattern, install protective shipping caps over the end of the injector.

Nozzle Installation

Prior to nozzle installation, ensure that the nozzle bore in the cylinder head is free of any carbon buildup and is not damaged or scored in any way. Once the nozzle has been removed, both a new compression and carbon stop seal,

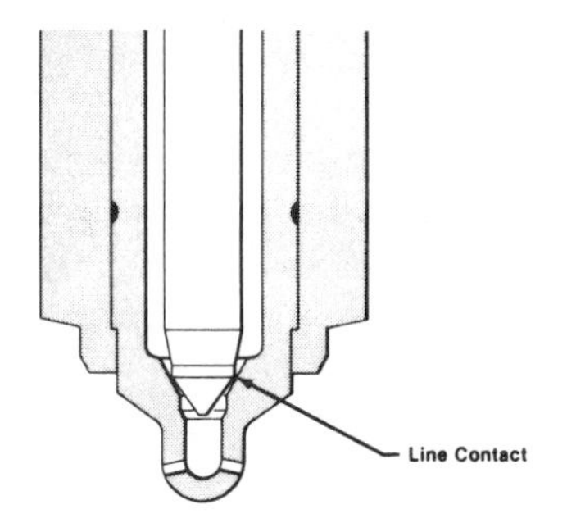

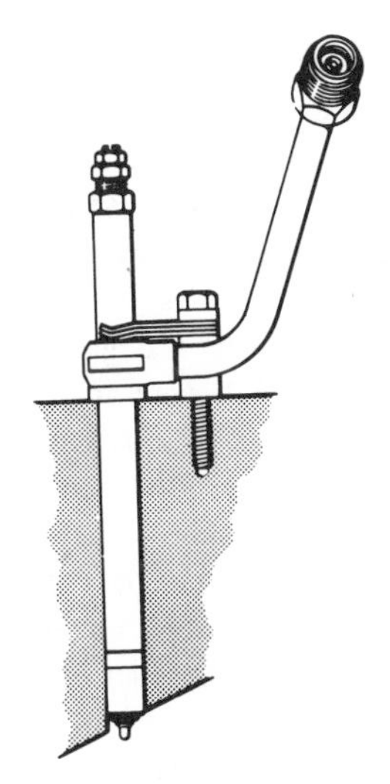

(b)

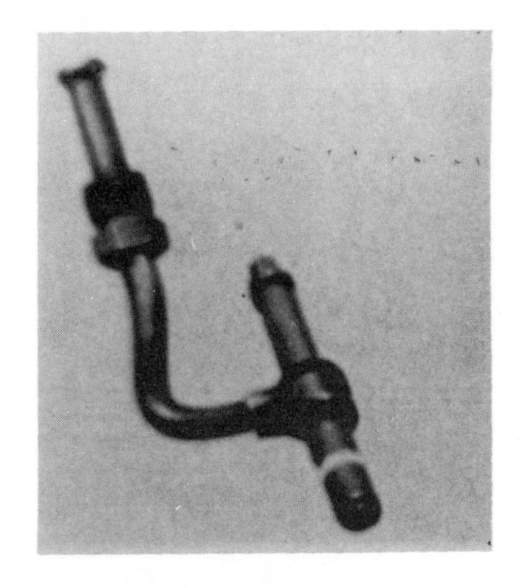

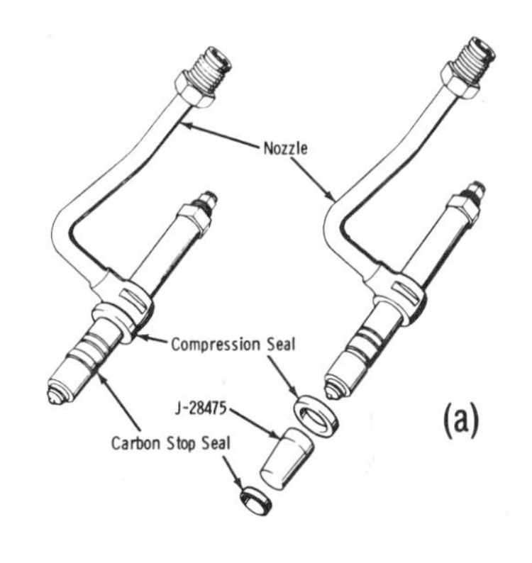

Figure 8-25
(a) Nozzle to Seat Contact; (b) Method of Holding Nozzle in Cylinder Head.
(Courtesy of Roosa Master, Stanadyne/Hartford Div.)

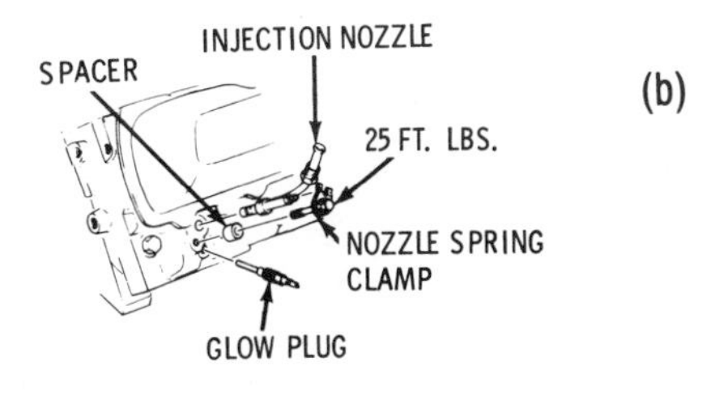

(b)

as shown in Figure 8-26, must be installed before the injector is reinstalled in the engine. With the seals installed, proceed to remove the protective shipping caps from the injector inlet and return connections, and slip the injector into position in the cylinder head as shown in Figure 8-27. Install the nozzle spring clamp and spacer, and torque the hold-down bolt to 25 lb-ft (33.89 N·m). Check all fuel lines for any damage or fretting, and blow them clean with dry compressed air; then install them into

Figure 8-26
(a) Installation of Compression Seal and Carbon Stop Seal; (b) Injection Nozzle Installation.
(Courtesy of Oldsmobile Div. of GMC)

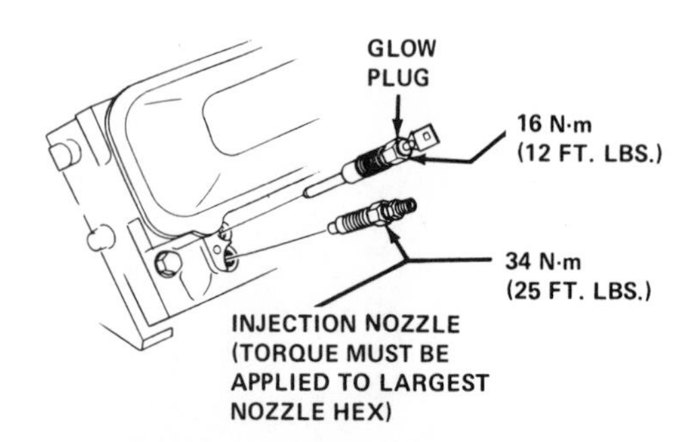

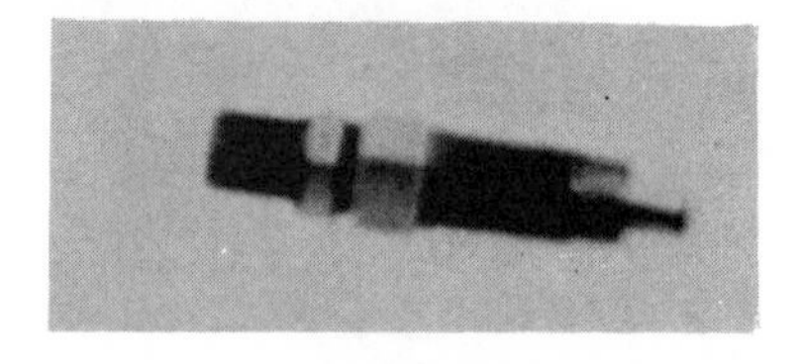

Figure 8-27
Placement of Injector into Cylinder Head. (Courtesy of Oldsmobile Div. of GMC)

position and torque the injection line nut to nozzle to 25 lb-ft (33.89 N·m). Torque the injection line nut to pump end to 35 lb-ft (47.35 N·m). If the glow plugs were removed for checking, after installation torque them to 12 lb-ft (16.26 N·m).

The Microjector

Starting with the 1980 model engines, the Oldsmobile Division of General Motors, which offers its V8 diesel engine to other car producers within the GMC line, signed a $65 million deal with Lucas CAV, whereby they will supply a ''Microjector'' which is less than half the size of the smallest pintle-type injector in use. It is only 65 mm (2.55 in.) long and weighs 52 g (0.114 lb). See Figure 8-28.

It has a unique feature in that it has a standard 14-mm spark plug thread, which allows it to be screwed directly into the cylinder head. Also no ''fuel backleak'' pipe is needed, as the nozzle valve guide is not required to seal against injection pressure. An outward opening valve creates a narrow spray evenly distributed into the pre-combustion chamber, with engine compression assisting a sharp cutoff. Both smoke and NO_x emissions plus noise have been reduced in these engine models.

The Lucas CAV microjector uses a nozzle with an outward opening, spring-loaded poppet valve, in contrast with the inward opening valve of a conventional diesel fuel injector. Since both engine compression and combustion pressure forces on an outward opening valve are additive to that exerted by the nozzle spring, opening pressure settings of the microjector are correspondingly lower than those of conventional injectors.

During injection, a degree of swirl is imparted to the fuel before it actually emerges around the head of the nozzle valve, which forms a closely controlled annular orifice with the nozzle valve seat. The resultant high-velocity atomized fuel spray forms a narrow cone suitable for efficient burning of the fuel in the pre-combustion chamber of the engine.

Servicing the Microjector. The microjector has been designed as basically a throw-away item. After a period of service, the functional performance may not meet the test specifications, as the nozzle is a matched and preset assembly. Lucas CAV recommends that after a service period of 50,000 mi (80,450 km), new microjectors be fitted automatically. The vehicle or engine manufacturers also recommend that the injectors be removed at specific service intervals for examination and testing.

A Hartridge Testmaster, or similar injector pop tester, can be used to test the microjector performance. Use only ISO 4113 test oil.

General inspection is the same as that for other nozzles described in this book. With the nozzle connected to the test equipment, tighten the high-pressure pipe nut to a torque of 14

N·m (10 lbf-ft.). After bleeding the nozzle of air, open the pressure isolating valve slightly, and operate the pump handle very slowly (between 3 and 6 seconds to complete one full stroke). Take careful note of the pressure at which the nozzle valve starts to open. Lucas CAV recommends a minimum opening pressure of 60 bars (870 lbf/in.^{2}). Otherwise, replace the injector.

Spray Pattern. Isolate the pressure gauge by closing the control valve to it; operate the pumping handle at a rate of 20 to 30 strokes per minute. Check for a uniform atomized spray in a cone shape that has a maximum inclusive angle of 30 degrees.

Unsatisfactory spray formation such as a narrow stream, excessively wide cone spray over a wide angle, or poor or no atomization is cause for replacing the microjector.

Seat Tightness. Pump up the tester handle to 20 bars (290 lbf/in.^{2}) below the previously recorded popping pressure. Wipe the nozzle face dry, and increase the line pressure to 15 bars (218 lbf/in.^{2}) below the opening pressure (popping) and maintain this value for 5 seconds. A slight sweat or dampness can be ig-

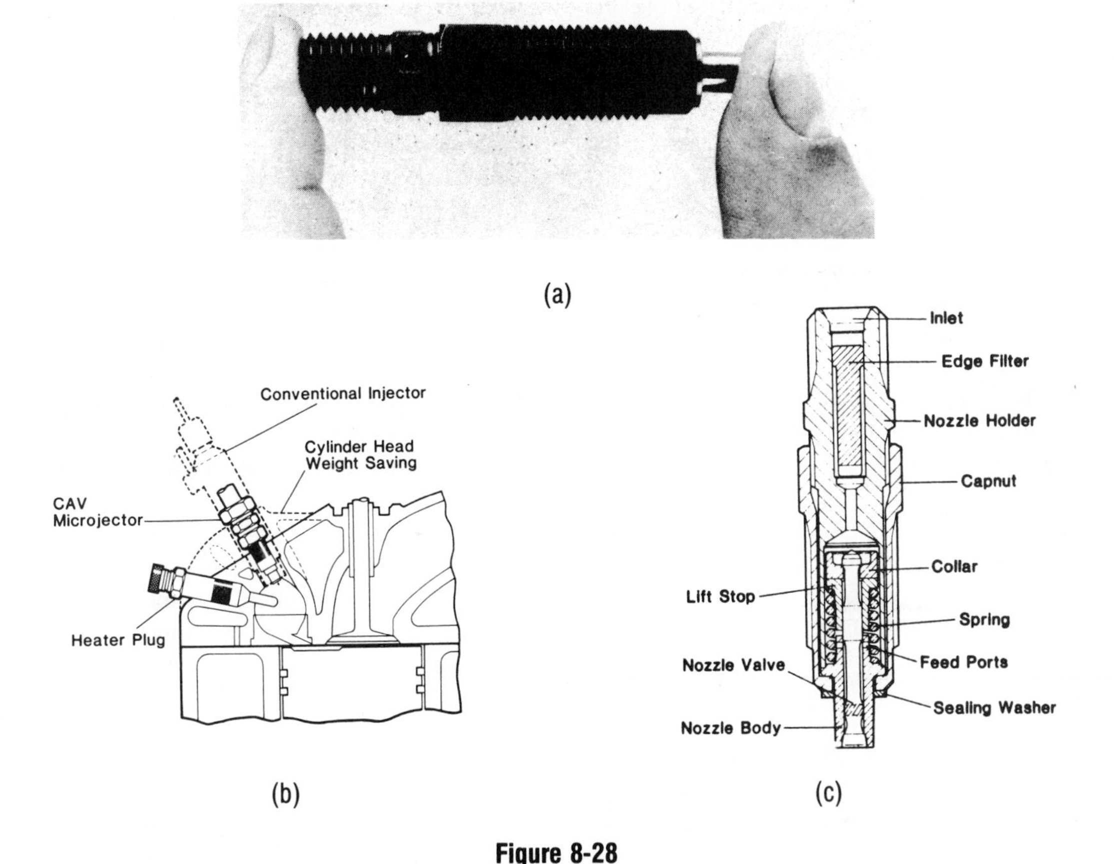

Figure 8-28
(a) Microjector Size; (b) Location of Microjector in Engine; (c) Internal Parts Location of Microjector.
(Courtesy of Lucas CAV Ltd.)

nored; however, should droplets of fuel emerge from the face of the nozzle, replace the nozzle.

Removal and Replacement:

1. Remove the high-pressure pipe nut from the top of the microjector.
2. Always fit a plastic protection cap over the fuel inlet to keep dirt out.
3. Using a 14-mm socket or spanner wrench, unscrew the microjector assembly from the cylinder head, ensuring that the copper sealing washer is still in position on the end of the nozzle; otherwise, remove the copper washer from the bore in the head.
4. Always fit a protection cap over the capnut to avoid damage to the nozzle valve face area.

To reinstall the microjector in the cylinder head, proceed as follows:

1. Using the socket (14 mm), and with a new copper sealing washer in position over the nozzle stem, install the mircojector back into the cylinder head. If the microjector is being reused, install them back into the same cylinder from which they were removed.
2. Torque the microjector capnut to 34 N·m (25 lbf-ft).
3. With the high-pressure fuel pipe reconnected, tighten it to 14 N·m (10 lbf-ft).

Caution: Should a socket not be available, and an open-ended wrench be used to tighten the pipe nut, the nozzle holder must be restrained on its hexagon with a 12-mm A/F (across the flats) spanner during the tightening operation.

Glow Plugs

The purpose of any glow plug is to heat up the air that is drawn into the pre-combustion chamber area to assist starting, especially in cold weather. Glow plugs are common on pre-combustion chamber engines, but not on direct injection diesels since they employ shaped piston crowns, which produce a very effective turbulence to the air in the cylinder. Direct injection engines also have less immediate heat loss to the surrounding cylinder area than in a pre-combustion engine and usually have a higher injector spray-in pressure.

Eight glow plugs are used, one for each cylinder; they are located just below the injection nozzles in the cylinder head. The glow plugs are controlled from an electronic module

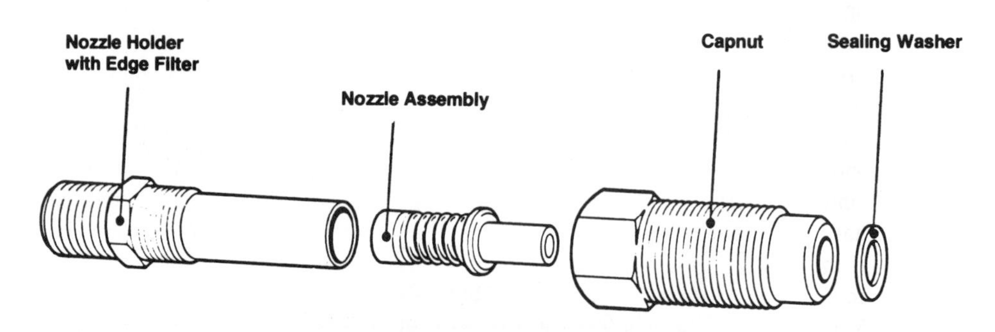

Figure 8-29
Component Parts of a Microjector.
(Courtesy of Lucas CAV Ltd.)

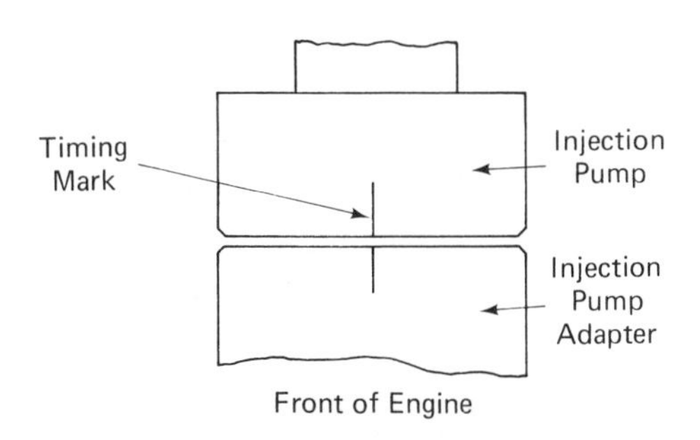

Figure 8-30
Alignment of Scribe Marks.

near the vehicle's fuse panel. When the operator turns the ignition key to the *run* position prior to starting the engine, the glow plugs are energized from the battery and will stay on for a short time after the engine is started and then automatically turn off. They can be checked for resistance once removed, and can be activated from a 12-V battery to check if they do in fact glow cherry red. Upon reinstallation, torque them to 12 lb-ft (16.26 N·m).

Two lights on the vehicle instrument panel, *wait* and *start*, signal when the engine is ready to start. When the ignition switch is turned to the *run* position, an amber *wait* light comes on, which indicates that the glow plugs are being energized. The temperature of the engine and coolant will affect just how long the *wait* light will stay on; however, as soon as the *wait* light goes out, a green *start* light will come on. If the engine has been running and is hot, the *wait* light may not come on at all, but the *start* light will, which is normal under this type of condition.

Built into the electronic module is a safety feature that will prevent discharging of the batteries if the ignition switch is left in the *run* position and not turned to *start* within approximately 5 minutes. Do not try to start the engine until the *wait* light goes out because in most cases the engine will not start. Oldsmobile cautions against using starting aids such as ether, since this can cause engine damage if used improperly and can even delay engine starting in some cases.

8-3

External Pump Adjustments

Several adjustments may be required from time to time on the injection pump, and these can be done quite readily. If minor external adjustments fail to solve the problem, do not attempt to disassemble the injection pump piece by piece on the engine. Remove the pump as an assembly and check it out on a test stand by yourself if you are experienced in fuel-injection practices and procedures, or take it to a local fuel-injection dealer for repair. Following are some of the typical checks and adjustments that you may have to do on the DB2 pump.

Checking or Adjusting Pump Timing

Visually check the scribe line marks on both the injection pump adapter and pump flange when viewing the unit from the top. They should be aligned as shown in Figure 8-30. If the timing marks are not aligned as shown, they must be set until they are. The engine would obviously be stopped while you make this adjustment. Proceed to loosen the three pump retaining nuts with tool J-26987, as this tool is necessary in order to get at all the nuts. With the nuts loose, grasp the injection pump body and rotate the pump until the marks are in alignment; then tighten the three retaining nuts to 35 lb-ft torque (47.35 N·m). If the pump is hard to rotate by hand, you can use a ¾-in. (19.05-mm) wrench on the boss at the front of the injection pump to aid in rotating the pump assembly.

If the injection pump timing is changed as just described, then throttle linkage adjustment

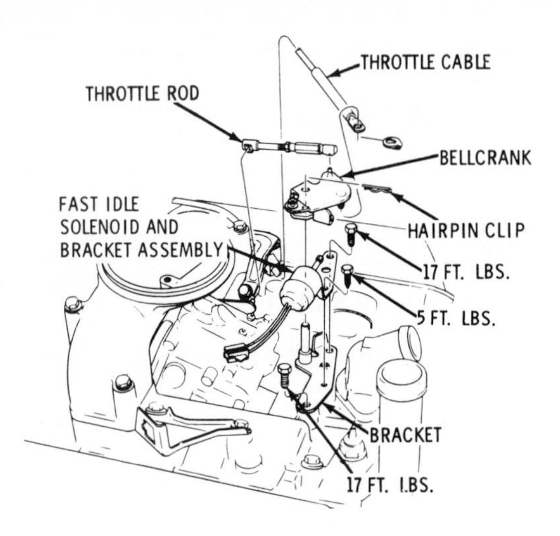

Figure 8-31
Adjusting Throttle Rod.
(Courtesy of Oldsmobile Div. of GMC)

is necessary. With the engine stopped, proceed as follows:

1. Remove the cruise control rod from the bellcrank if used and also the TV cable (200 transmission), or detent cable if a 350 transmission is used, from the bellcrank.
2. Shorten the throttle rod several turns after loosening the locknut.
3. Refer to Figure 8-31, rotate the bellcrank to the full throttle stop (hold it here), and then lengthen the throttle rod until the injection pump throttle lever contacts the injection pump's full throttle stop. Release the bellcrank and tighten the throttle rod locknut. Place the bellcrank back against the stop and double-check that the throttle rod is still correct.
4. Connect the TV or detent cable and cruise control rod if used back to the bellcrank.

Check the adjustment of the TV or detent cable as follows:

1. Refer to Figure 8-32; remove the throttle rod from the bellcrank and push the snap lock to its disengaged position.
2. Rotate and hold the bellcrank against the full throttle stop, and push in the snap lock until it is flush with the cable and fitting; then release the bellcrank and reconnect the throttle rod.

Check the adjustment of the transmission vacuum valve adjustment as follows:

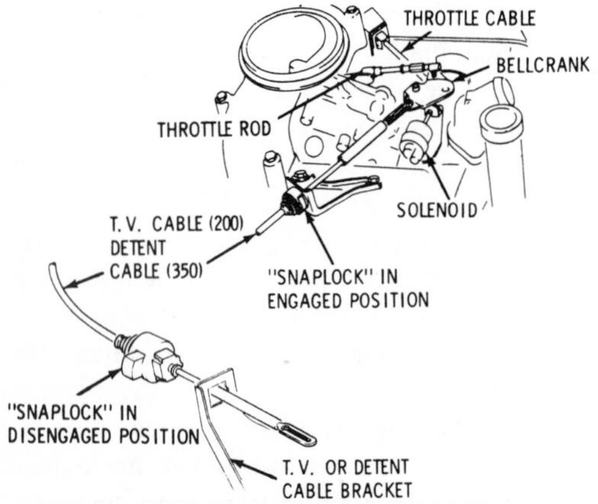

THROTTLE CABLE
BELLCRANK
THROTTLE ROD
SOLENOID
T. V. CABLE (200)
DETENT
CABLE (350)
"SNAPLOCK" IN
ENGAGED POSITION
"SNAPLOCK" IN
DISENGAGED POSITION
T. V. OR DETENT
CABLE BRACKET

Figure 8-32
Adjustment Check of TV Cable.
(Courtesy of Oldsmobile Div. of GMC)

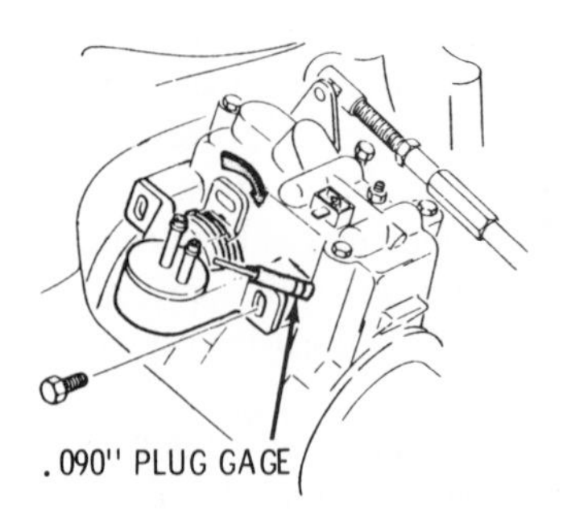

Figure 8-33
Checking Transmission Vacuum Valve Adjustment.
(Courtesy of Oldsmobile Div. of GMC)

1. Remove throttle rod from the bellcrank and loosen the transmission vacuum valve attaching bolts enough to disengage the valve from the injection pump shaft. Hold the injection pump throttle lever against the full throttle stop. See Figure 8-33.
2. Rotate the valve to the full throttle position and insert a 0.090-in. (2.286-mm) plug gauge pin, which is available from a BT-3005 carburetor kit, to hold the valve in the full throttle position. If the 0.090-in. gauge is not readily available, a similar pin can be quickly made.
3. Rotate the vacuum valve clockwise until it contacts the injection pump lever shaft, and tighten up the two bolts holding the vacuum valve to the pump.
4. Double-check that the clearance is still correct after tightening the bolts; remove the pin and release the lever, and then reconnect the throttle rod to the bellcrank.

Slow Idle Speed Adjustment

Make sure that the engine is at normal operating temperature before making this adjustment. To accurately determine engine speed during this adjustment, use one of the digital electronic magnetic tape pickup tachometers or Oldsmobile's own magnetic tachometer, tool number J-26925. This tool has a probe that is inserted into the engine timing indicator hole located at the front of the engine, as shown in Figure 8-34.

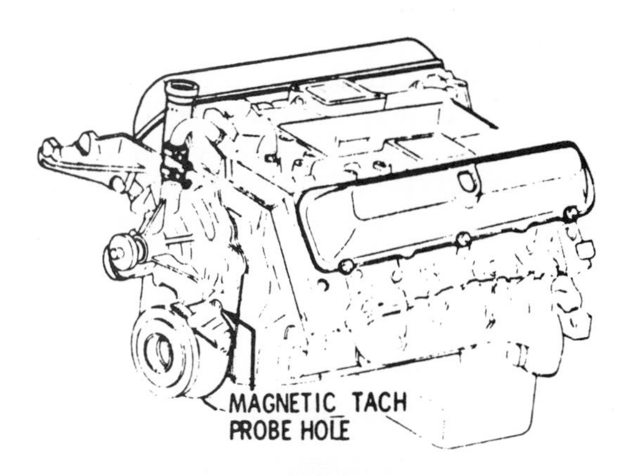

Figure 8-34
Location of Magnetic Tach Probe Hole. (Courtesy of Oldsmobile Div. of GMC)

Block the driving wheels of the vehicle and apply the parking brake. Start the engine, let it idle, and place the transmission selector lever into the D or drive range. If the vehicle is equipped with an air conditioner, make sure that it is turned off; otherwise, the air-conditioner pump will be under load and the idle speed setting will be affected. Refer to Figure 8-34; adjust the slow idle screw until the magnetic tachometer reads 575 rpm.

Fast Idle Solenoid Adjustment

As with the slow idle adjustment, the engine should be at its normal operating temperature. Adjust the fast idle solenoid plunger to 650 rpm with the engine running, wheels blocked, transmission in drive range, parking brake on, air conditioner on this time, and the compressor wires disconnected. (Disconnect the solenoid connector on vehicles that are not equipped with air conditioning.) Connect jumper wires to the solenoid terminals and ground one jumper wire. Connect the other jumper wire to 12 V to energize the solenoid while making the fast idle adjustment.

Cruise Control Servo Relay Rod Adjustment

On vehicles equipped with a cruise control feature, with the engine stopped, adjust the cruise control rod so that there is a minimum of slack or free play; then put the clip into the first free hole closest to the bellcrank and within the servo bail.

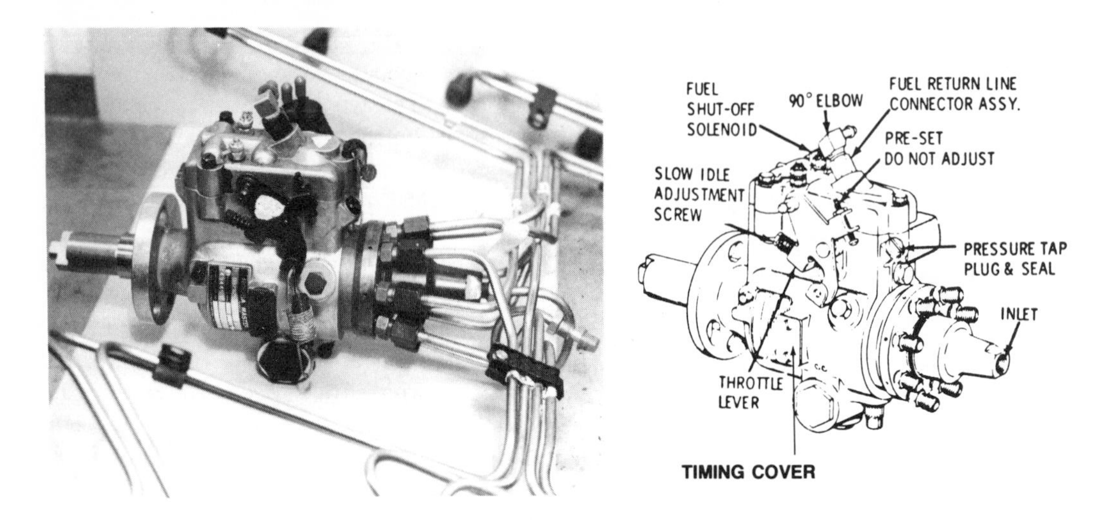

Figure 8-35
Location of Main External Components.
(Courtesy of Oldsmobile Div. of GMC)

Injection Pump Fuel Pressure Check

To do this check, the air cleaner assembly must be removed along with the air intake crossover pipe. Prior to running the engine, however, install screened covers over the intake manifolds at each cylinder head for reasons of safety when running the engine. These can be made up locally or they are available from Oldsmobile as tool number J-26996-2. Refer to Figure 8-35; remove the fuel return pressure tap plug.

Using the seal from the plug, install it onto adapter J-28526; then screw the adapter into the injection pump housing. Attach a low-pressure gauge to the adapter, and using a magnetic pickup tachometer (Figure 8-34), start and run the engine at 1000 rpm with the transmission gear selector in the *park* position. The gauge should register between 8 and 12 psi (55.16 and 82.74 kPa) with a maximum fluctuation of 2 psi (13.79 kPa).

If the pressure is low, replace the fuel return line connector assembly and repeat the procedure. If the fuel pressure is still low, the injection pump will have to be removed from the engine and repaired. Use a new seal on the pressure tap plug screw when installing.

8-4

Injection Pump Removal

Removal of the fuel pump should be done in a systematic fashion, and care should be exercised when removing all items to avoid any possible damage. Start by removing the air cleaner, and proceed to remove all items that are either in the way or are necessary in order to remove the pump. Once all these items have been removed, loosen and remove the three pump to adapter nuts and withdraw the pump assembly.

8-5

Injection Pump Installation

Prior to pump installation, you should check out the condition of the pump adapter. If there is any doubt about bushing wear or possible injection pump adapter drive shaft gear wear or

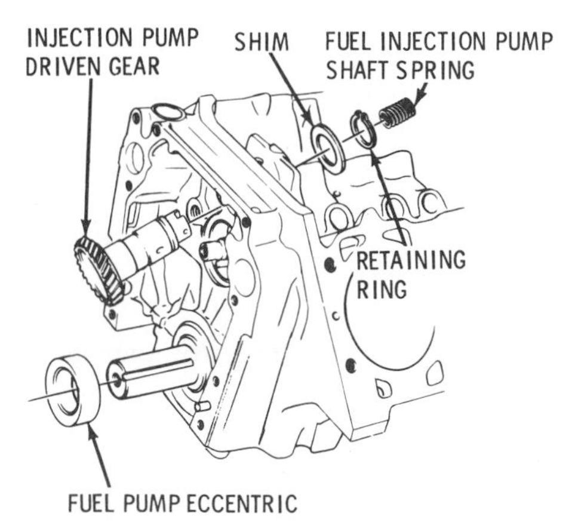

Figure 8-36
Shimming Injection Pump Gear End Play.
(Courtesy of Oldsmobile Div. of GMC)

damage, the engine front cover would have to be removed to get at the drive and driven gears. If one gear is replaced, then both must be replaced because of the matched angled drive setup used.

If the injection pump drive and driven gears are being replaced, refer to Figure 8-36 and make sure that the injection pump driven gear is within the 0.002 to 0.006 in. (0.05 to 0.152 mm) stated by Oldsmobile for gear end play. If not, shim thrust washers are available in 0.003-in. (0.076-mm) increments from 0.103 to 0.115 in. (2.616 to 2.921 mm) in order to obtain the specified clearance.

Refer to Figures 8-36 and 8-37; position the camshaft dowel pin on the front end of the camshaft where the cam gear mounts, at the 3 o'clock position. Align the 0 marks up between the injection pump drive and driven gears, and with the camshaft gear and crankshaft gear timed to one another as shown in Figure 8-38, locate and slide the injection pump drive gear onto the camshaft with care; then slide the camshaft into its bore all the way. Install the bolt into the end of the camshaft, which holds the gear and chain in position, and tighten it to 65 lb-ft (88.12 N·m) of torque.

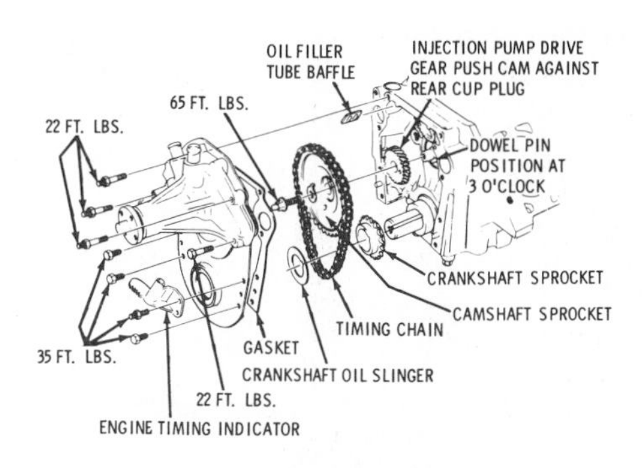

Figure 8-37
Initial Timing of Pump to Engine.
(Courtesy of Oldsmobile Div. of GMC)

Caution: Be sure to exercise care when installing the camshaft to avoid camshaft support bearing damage. Coat the camshaft and bearings with a good-quality lubricant or Oldsmobile recommended lubricant no. 1051396.

Prior to installing the injection pump adapter (Figure 8-40), bar the engine over until cylinder 1 is at TDC. Cylinder 1 is on your

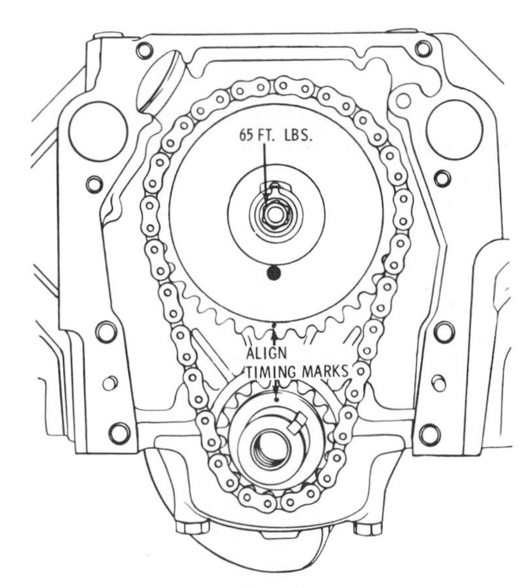

Figure 8-38
Checking Alignment of Engine Timing Marks.
(Courtesy of Oldsmobile Div. of GMC)

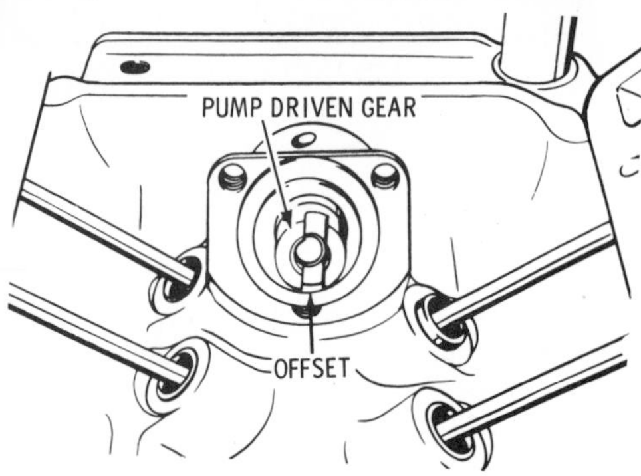

Figure 8-39
Index Offset to Right—Pump Driven Gear. (Courtesy of Oldsmobile Div. of GMC)

right when standing at the front of the engine. Make sure that the TDC mark on the crankshaft balancer is in line with the 0 or zero mark on the timing indicator, which is bolted to the engine front cover. With this mark in position, the pump driven gear shaft will be as shown in Figure 8-39 with the index *offset* to the right.

With new gears installed, or if you are installing a new injection pump adapter, a timing mark must be scribed onto the adapter to correspond with the position of piston 1 at TDC. This requires that you file the old mark off the injection pump adapter only, not off the pump. Proceed as follows:

1. Refer to Figure 8-40 (a) and (b).
2. Apply a light coating of good-quality grease to the adapter seal area and the inside bore of the manifold where the outer diameter of the seal will be positioned. Gently place the adapter into the bore area, but leave it loose.
3. To prevent seal damage, use seal installer tool J-28425. Lubricate both the inside and outside diameter of the adapter seal and tool. Place the seal onto the tool.
4. Carefully locate the seal onto the adapter and push it into position; then remove the tool and check to see that the seal is in fact properly positioned.
5. If not, repeat the procedure; then torque the adapter bolts to 25 lb-ft (33.89 N·m).

Once the adapter and seal are in position, timing tool J-26896 must be used to place a new timing scribe line onto the pump adapter, as shown in Figure 8-41. Insert the timing tool into the pump adapter. Place a torque wrench onto the tool and pull it toward cylinder 1 until a reading of 50 lb-ft (67.79 N·m) is obtained. This is to make sure that gear backlash and timing chain free play are eliminated. While holding the torque wrench in this position, strike the button on the top of the tool with a ball peen hammer just enough to place a clearly visible scribe mark on the flange.

Manually rotate the pump drive shaft so that the offset tang on the shaft is aligned with the offset on the pump driven gear shaft, as shown in Figure 8-39; then carefully guide the injection pump into position and align the scribe mark on the injection pump housing with the scribe mark on the adapter. A $\frac{3}{4}$-in.

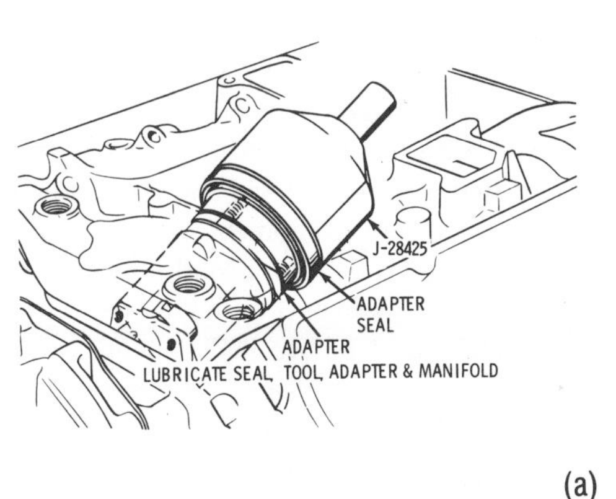

(a)

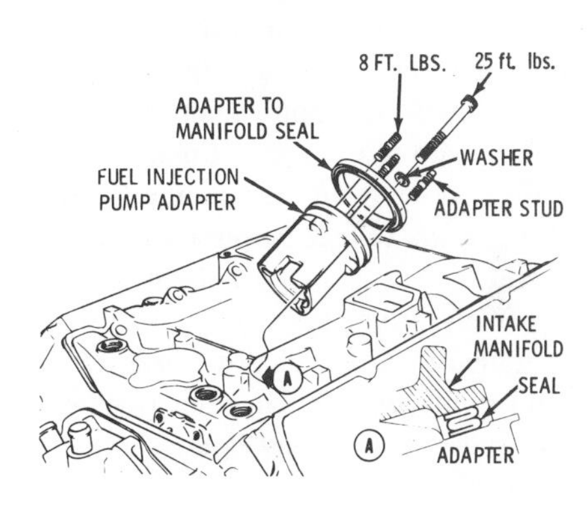

(b)

Figure 8-40
(a) Using Seal Protector to Install Adapter to Manifold Seal; (b) Exploded View of Location of Manifold Seal. (Courtesy of Oldsmobile Div. of GMC)

(19.05-mm) socket or wrench on the boss at the front end of the injection pump may be required to turn the pump so that the timing lines are aligned. Tighten the pump to adapter retaining nuts to 35 lb-ft (47.45 N·m). Remove all shipping caps from the pump and connect all fuel lines to nozzles, fuel filter bracket and filter, and all other necessary parts.

Bleeding the Fuel System

On vehicle applications, with all lines connected and the electric shutoff device wired up, place the injection pump throttle lever in its maximum speed position. All injection nozzle fuel lines should be left loose at the nozzle for now. When you turn on the ignition switch,

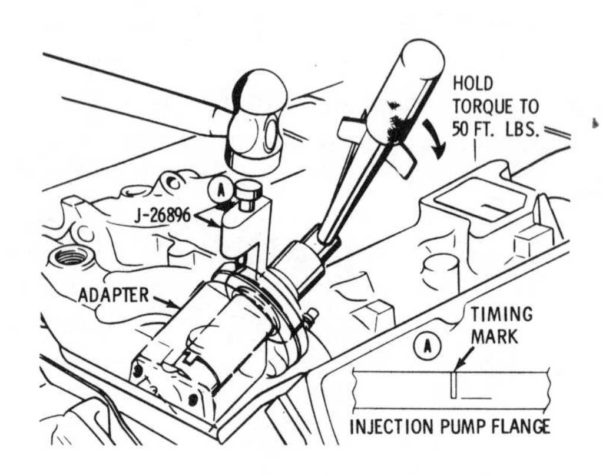

Figure 8-41
Use of Timing Tool to Scribe New Timing Line. (Courtesy of Oldsmobile Div. of GMC)

the *wait* light will come on, indicating that the glow plugs are being supplied with electricity to preheat the pre-combustion chambers. Since the nozzle lines are loose at this time, no fuel will be delivered to them. Crank the engine over until fuel starts to bleed from each nozzle fuel line, and tighten them up individually as fuel flows freely from each one. With fuel having been bled to each injection nozzle, turn the ignition key to *wait*, and once the *wait* light on the dash goes out, the *start* light will come on.

Attempt to start the engine; if it does not start, loosen the injection nozzle lines as before and rebleed the system. Repeat the starting procedure again.

Once the engine starts, run it at a fast idle and check all fuel lines for fuel leakage. Inspect the pump for any signs of leakage. Warm up the engine to its normal operating temperature and do any adjustments required.

Injection Pump Overhaul

Injection pump overhaul should only be done by those personnel experienced in the intricacies of such detailed work. Special equipment and test machines are required for this purpose; therefore, if pump repair or testing is necessary, take it to a local fuel-injection shop that is equipped to deal with such work.

If you are experienced in the overhaul and repair of fuel injection pumps and nozzles and have the necessary equipment readily available, obtain a copy of Roosa Master Operation and Instruction Manual for the Model DB2 Pump, Catalogue Number 99009, along with the specifications adjustment sheet from your local fuel-injection dealer, or write to Roosa Master, Stanadyne/Hartford Division, Box 1440, Hartford, Connecticut 06102, U.S.A.

8-6

Injection Pump Accessories

The DB2 injection pump can be used on a variety of applications; therefore, it is available with several options as required. These are (1) flexible governor drive, (2) electrical shutoff, and (3) torque screw.

Flexible Governor Drive

This is a flexible retaining ring that serves as a cushion between the governor weight retainer and the weight retainer hub. Any torsional vibrations that may be transmitted to the pump area are absorbed in the flexible ring, therefore reducing wear of pump parts and allowing more positive governor control.

Electrical Shutoff

This device is available as either an energized to run (ETR) or energized to shutoff (ETSO) model. In either case it will control the run and stop functions of the engine by positively stopping fuel flow to the pump plungers, thereby preventing fuel injection.

Torque Screw

This device is used with Roosa Master pumps to allow a tailored maximum torque curve for the particular engine application. This feature is commonly referred to as *torque backup,* since the engine torque will generally increase toward the preselected and adjusted point as the engine rpm decreases. The three factors that affect this torque are (1) the metering valve opening area, (2) the time allowed for fuel charging, and (3) the transfer pump pressure curve.

Do *not* attempt to adjust the torque curve on the engine at any time. This adjustment can be done only during a dynamometer test where fuel flow can be checked along with the measured engine torque curve or on a fuel pump test stand.

Turning in the torque adjustment screw moves the fuel metering valve toward its closed position. The amount of fuel delivered at full-load governed speed is controlled by the torque screw and not by the roller-to-roller dimension.

If additional load is applied to the engine while it is running at full-load governed speed, there will be a reduction in engine rpm; therefore, a greater quantity of fuel is allowed to pass into the rotor pumping chamber because of the increased time that the charging ports are open (slower running engine allows this to happen). Fuel delivery will continue to increase until the rpm drops to the engine manufacturer's predetermined point of maximum torque. It is at this point that the amount of fuel will be controlled by the setting of the pump roller-to-roller dimension (see page 283 on DM2 and DM4 pump fuel setting).

Torque Chopper

Another torque control system option in the pump operates in conjunction with speed light-load-type advance mechanisms. The combined effect of the lower portion of the speed advance control and spill port torque chopper is to reduce the engine's output torque at reduced engine speeds before peak torque. This feature is available through a special head and rotor assembly for the pump.

chapter 9
Cummins PT Fuel Systems

The Cummins Engine Company was founded in 1919 by Clessie Cummins and has grown to become one of the world's largest independent manufacturers of diesel engines. They have established an excellent reputation over the years of producing an engine that is used in a wide variety of applications worldwide. They presently have over 2900 distributors and dealers in 98 countries offering sales and service facilities for their product. In addition, they have parts and engine manufacturing facilities in 11 locations around the world. In the United States the main research and engineering center is located in Columbus, Indiana; an additional technical center is located in Essen, West Germany.

Over the years Cummins has produced a series of innovations such as the first automotive diesel plus the first to use supercharging and then turbocharging. During the late 1920s, Cummins tried hundreds of ideas in an effort to develop a fuel system suitable for their engines. The end result was a single-disc pump that operated somewhat like a gasoline engine distributor, except for the use of fluid instead of electricity. Until 1930, 900 rpm was considered to be about the maximum speed that a diesel engine could run due to ignition delay problems; however, tests by Clessie Cummins on a Cummins model U engine proved that in automobile tests with the distributor-type pump speeds of more than 1800 rpm were easily attainable. In 1931, the model H engine started the move to diesel-powered mobile equipment, industrial applications, and the like.

Due to the single drilling that was used for plunger pump suction and discharge, hydraulic problems at high speed created a flat torque curve and zero torque rise; therefore, the engine had little lugging ability, which slowed the acceptance of the diesel engine in highway trucks. In addition, the single-disc pump weighed a bulky 104 lb (47.17 kg).

After World War II, a double-disc pump was designed that weighed only 33 lb (14.968 kg) and was only 40% as bulky. The double-disc pump had a suction disc that indexed with the plunger pump on its own downward stroke and a distributor disc that indexed the delivery of the plunger with the injector in sequence. With this pump on an engine, a 35% torque rise was produced, and by changing the diameter and shape of the plunger cam, engines could be built with any torque rise desired; however, torque rise was limited to 10% on commercial engines produced by Cummins at that time until the effect of torque on engine life was established.

In early 1949, Clessie Cummins and Hans L. Knudsen, both retired from active duties but serving as directors of Cummins Engine Company, were still of the opinion that the original concept of using a common rail fuel system with camshaft actuated injectors would permit fuel metering to be controlled accurately by restrictions built into the injectors. Their ideas and sketches were developed and simplified mostly by Neville H. Reiners and Robert D. Schmidt of the Cummins engineering staff, and thus was born the PT fuel system, PT standing for Pressure–Time.

In early 1952 an experimental model of the PT system was used in a race car that set new qualifying records for the Indianapolis 500-mile race and proved that the system was capable of metering, injecting, igniting, and burning fuel at speeds above 4500 rpm. This knowledge led to the later development of commercial engines operating at speeds above 3000 rpm.

The PT fuel system was then introduced in 1954 and has been used on all Cummins diesels built since then. The fuel pump weighed only 13 lb (5.896 kg) and therefore dramatically simplified and cut the cost of field service.

All cylinders are commonly served through a low-pressure fuel line. The camshaft control of the mechanical injector times the ignition throughout the operating range without the timing lag problems of high-pressure, remote-mounted plunger pump systems (see Figure 9-1).

The Cummins Engine Company, like many other major diesel engine manufacturers, is presently building their engines in several different countries around the world. This company is one of the better known suppliers of high-speed diesel engines for a wide variety of applications. The engines that they build are all 4-stroke-cycle-type units of direct injection design. As with other engine manufacturers, they are constantly improving component parts to reflect a better end product.

The fuel pumps and injectors used on Cummins engines have changed over the years; therefore, the information contained herein will deal only with the Cummins PT fuel systems.

9-1

PT Fuel System

The PT fuel system is exclusive to Cummins diesel engines; it employs injectors that meter and inject the fuel, with this metering based on a pressure–time principle. Fuel pressure is naturally supplied by a gear driven positive displacement low-pressure fuel pump, and the time for metering is determined by the interval that the metering orifice in the injector remains open. This interval is established and controlled by the engine speed, which therefore determines the rate of camshaft rotation and consequently the injector plunger movement. The mechanical linkage necessary to accomplish this is shown in Figure 9-2.

Since Cummins engines are all of the 4-cycle type, the camshaft is driven from the crankshaft gear at one-half engine speed. The fuel pump turns at engine speed, with the exception of the V-555 series engine, where the fuel pump speed is 0.786 of engine speed. Because of this relationship, additional governing of fuel flow is necessary in the pump.

Although the camshaft is turning at one-half engine speed, when the engine is running at 2100 rpm, the cam is turning 17.5 times every second. As mentioned earlier, the fuel

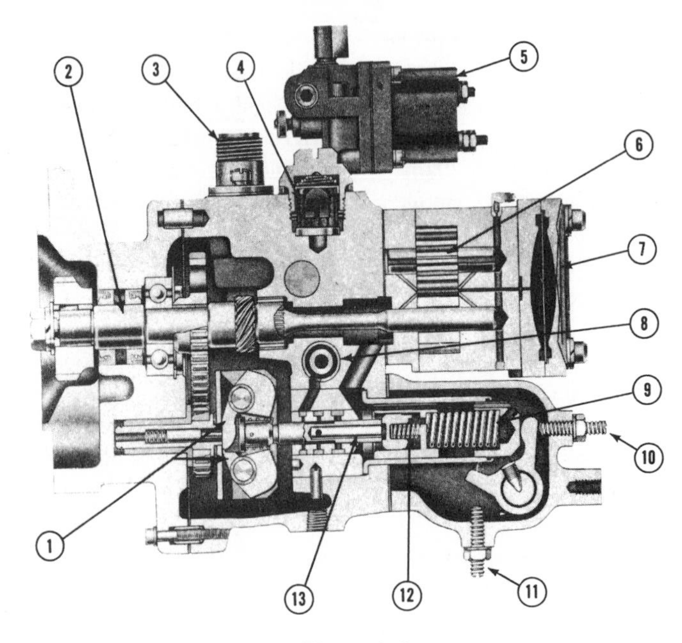

Figure 9-1
PT (Type G) Fuel Pump Cross Section.
(Courtesy of Cummins Engine Co.)

Index:

1—Governor weights
2—Main shaft
3—Tachometer connection
4—Filter screen
5—Shut-down valve
6—Gear pump
7—Pulsation damper
8—Throttle shaft
9—Governor spring
10—Idle speed adjusting screw
11—Maximum speed adjusting screw
12—Idle spring
13—Governor plunger

pump varies pressure to the injectors in proportion to engine rpm.

Downward movement of the injector plunger forces metered fuel into the cylinder as shown in Figure 9-2. This method has three important advantages:

1. The injector accomplishes all metering and injection functions.
2. The injector injects a finely atomized fuel spray for efficient combustion with a wide range of fuel from furnace oil to light jet engine fuel if necessary.
3. A low-pressure common rail system is used, which eliminates the necessity of high-pressure fuel lines running from the fuel pump to each injector.

The three components in the PT fuel pump that control the pressure at the injectors are shown in Figure 9-3. The PT fuel pump assembly is coupled to the air compressor, vacuum pump, or fuel pump drive on the engine, which is driven from the engine gear train. The fuel pump main shaft in turn drives the gear pump, governor, and tachometer shaft assemblies.

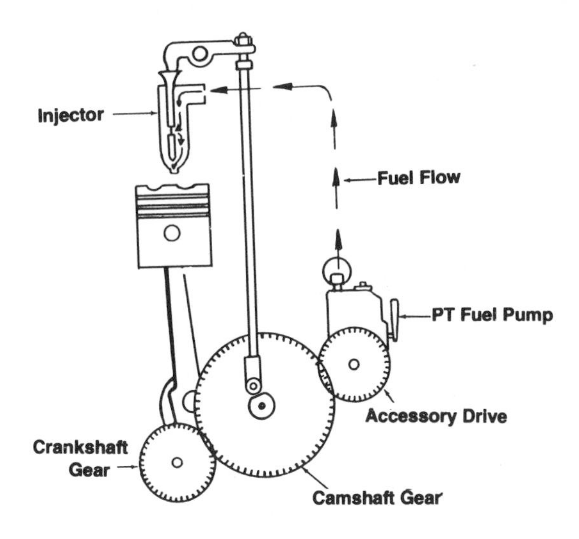

Figure 9-2
Fuel Pump and Injector Operating Mechanism
(Courtesy of Cummins Engine Co.)

The flyball type of mechanical governor controls pressure and engine torque throughout the entire operating range. It also controls the engine's idling speed and prevents engine overspeeding in the high-speed range. The throttle is simply a shaft with a hole; therefore, the alignment of this hole with the fuel passages determines pressure at the injectors.

A single low-pressure fuel line from the fuel pump serves all injectors; therefore, both the pressure and the amount of metered fuel to each cylinder are equal. Typical fuel flow diagrams for both in-line and Vee-type engines are shown, respectively, in Figures 9-4 and 9-5.

To understand the sequence of events pertaining to actual injection of fuel by the injector unit, a study of the injector operating mechanism is necessary. Since the shape of the camshaft lobes is directly related to the start

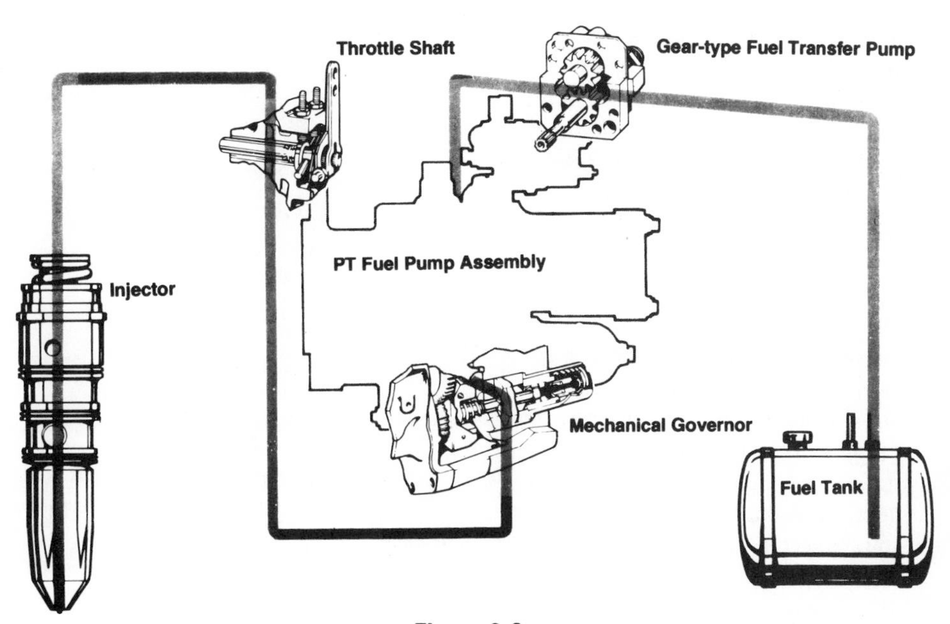

Figure 9-3
Components That Control Fuel Pressure at the Injectors.
(Courtesy of Cummins Engine Co.)

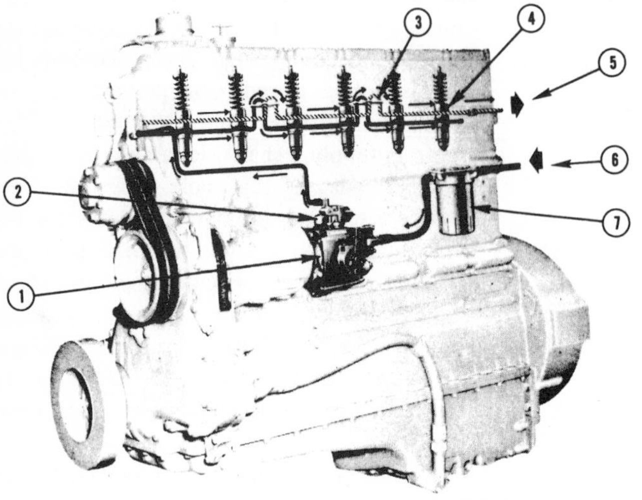

Index:
1—PT (type G) fuel pump
2—Shut-down valve
3—Fuel connection
4—Injector
5—Injector return
6—From tank
7—Fuel filter

Figure 9-4
Fuel Flow Diagram for In-Line Engine.
(Courtesy of Cummins Engine Co.)

and end of injection, let us take a look at this first. The camshaft is driven at half engine speed owing to the fact that all Cummins engines are of 4-stroke-cycle operation. The actual camshaft lobe shape is basically the same on all Cummins diesel engines. Figure 9-6 is a cross-sectional view of one of these camshafts. The cam shape is based on two circles, an inner and an outer circle.

To follow this a stage further, let's go back to basics for a minute, and using a circle to represent 720 degrees, as when using a polar valve timing diagram, the 720 degree circle can represent two rotations of the engine crankshaft. We can then place or superimpose one cam lobe shape in the center of this circle and illustrate injector push tube and injector plunger travel. Figure 9-7 shows such a setup.

Figure 9-8 shows the sequence of events described in Figure 9-7; however, Figure 9-8 actually shows the motion transfer from the camshaft lobe, to the pushtube and rocker arm, then the injector plunger.

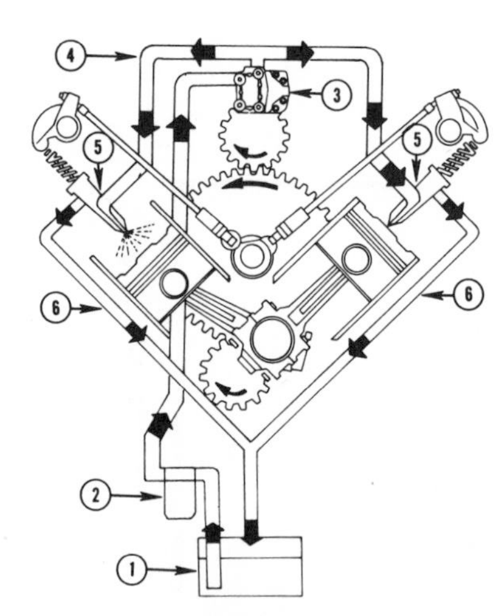

Figure 9-5
Fuel Flow Diagram for Vee-Type Engine.
(Courtesy of Cummins Engine Co.)

Index:
1—Fuel tank
2—Fuel filter
3—Fuel pump
4—Injector supply
5—Injector
6—Injector drain

Having studied Figures 9-6 and 9-7, you should now be familiar with the basic camshaft positions; therefore, Figure 9-9 shows that during the *intake* stroke, the follower roller moves from the outer cam base circle across the retraction ramp to the inner or lower base circle, which will allow the injector push tube to follow it down. Injector plunger return spring pressure lifts the plunger as the lowered push tube permits the rocker arm lever to tilt backward. As the injector plunger lifts, it allows fuel at low pressure to enter the injector at port (A) and flow through the inlet orifice (B), internal drillings, around the annular groove in the injector cup, and up passage (D) to return to the fuel tank. The amount of fuel flowing through the injector is determined by the fuel pressure before the inlet orifice (B). Fuel pressure is determined by engine speed, governor, and throttle.

As the injector plunger continues its upward movement, metering orifice (C) is uncovered and a charge of fuel is metered to the cup, the amount being controlled by fuel pressure. Passage (D) is blocked, momentarily stopping fuel circulation and isolating the metering orifice from any fuel pressure pulsations.

As the camshaft continues to rotate, and the cylinder's piston is coming up on *compression*, the follower roller crosses the inner base circle, thereby holding the plunger up for metering, and as it reaches the camshaft lobe injection ramp, the upward moving push tube working through the rocker arm assembly forces the injector plunger toward injection. Refer to Figure 9-10; you will notice that the downward moving plunger closes off the metering orifice, thereby cutting off fuel entry into the cup. At this instant, the drain outlet (D) is uncovered; fuel that was not metered to the cup can now leave the injector up passage (E), and fresh fuel enters the balance orifice. As the plunger continues down into its seating position in the cup, it forces the fuel under great hydraulic pressure through tiny holes (for example, eight holes 0.007 in. or 0.177 mm in diameter), creating a fine fuel spray for penetration of the air mass to assure complete combustion of fuel in the cylinder.

At the completion of the plunger

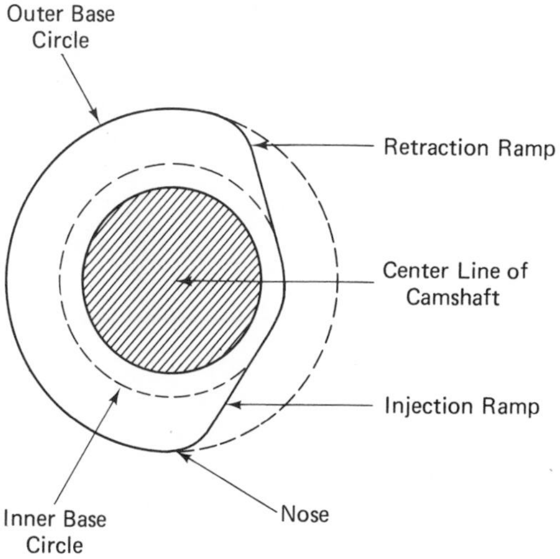

Outer Base Circle
Retraction Ramp
Center Line of Camshaft
Injection Ramp
Inner Base Circle
Nose

Figure 9-6
Cross Section of Camshaft.

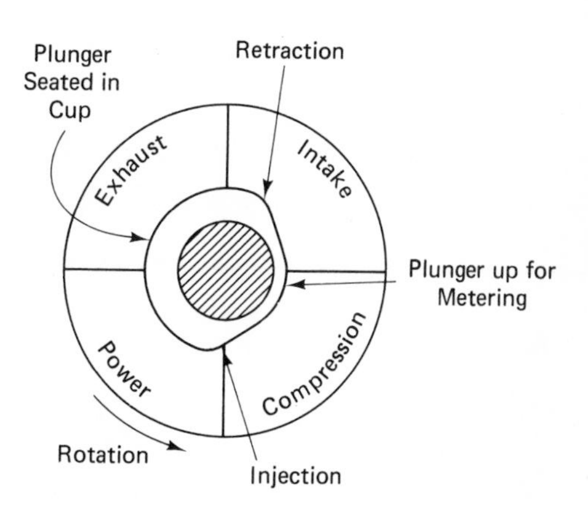

Figure 9-7
Injector Push Tube and Injector Plunger Travel.

downstroke after injection has ceased, the plunger remains seated until the next metering and injection cycle. The end of injection occurs as the roller follower reaches the nose of the cam; this ensures that the plunger remains seated in the cup because the follower is riding evenly around the concentric outer base circle of the cam lobe. During this time, however, as shown in Figure 9-11, fuel is allowed to flow freely through the injector and lubricate and cool internal parts. The fuel picks up some heat during this time, which warms the fuel in the tank, which is helpful during cold weather operation.

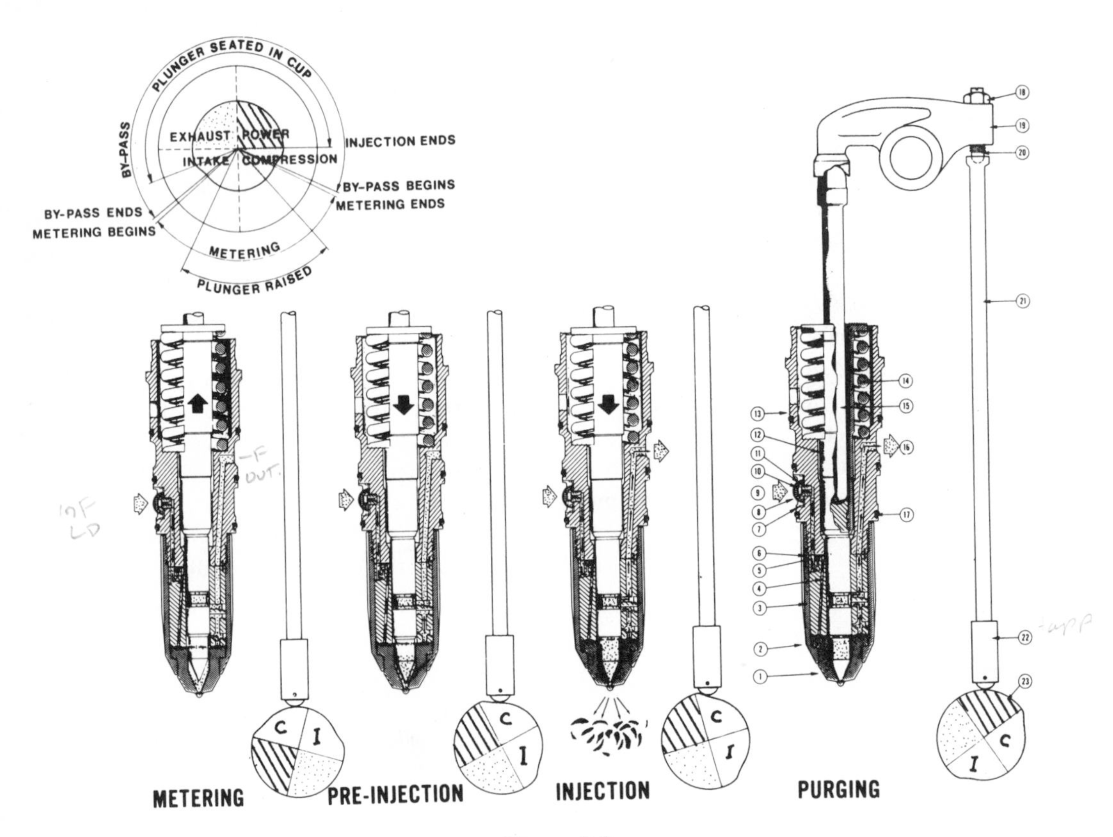

Figure 9-8
Sequence of Events in Injecting Fuel.
(Courtesy of Cummins Engine Co.)

Index:

1—Cup
2—Cup retainer
3—Barrel
4—Plunger
5—Check ball
6—Gasket
7—Clip
8—Screen
9—Fuel in
10—Orifice
11—Orifice gasket
12—Coupling
13—Adapter
14—Spring
15—Link
16—Fuel out
17—O ring
18—Nut
19—Rocker lever
20—Adjusting screw
21—Push rod
22—Tappet
23—Camshaft lobe

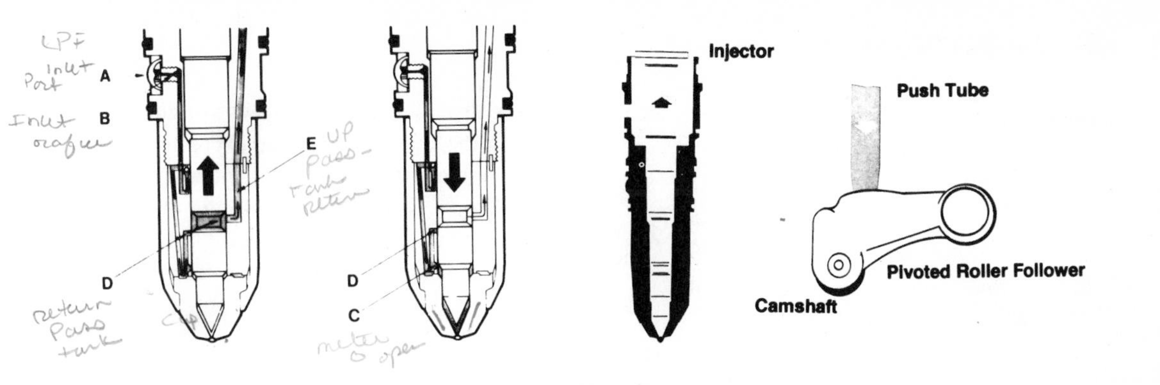

Figure 9-9
Fuel Flow into the Injector and Metering.
(Courtesy of Cummins Engine Co.)

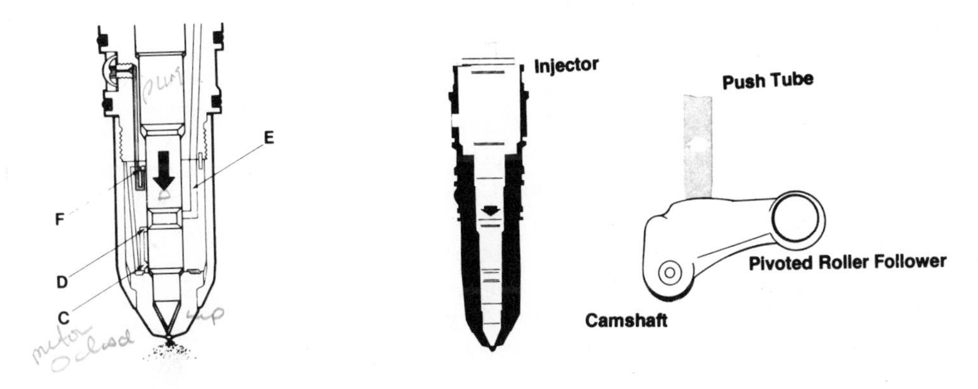

Figure 9-10
Injection of Fuel.
(Courtesy of Cummins Engine Co.)

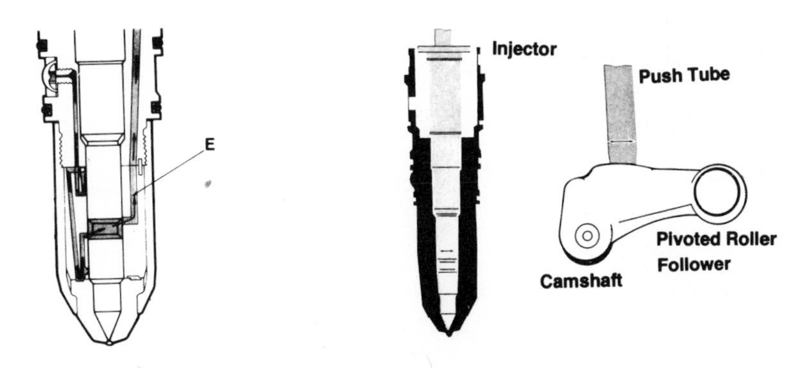

Figure 9-11
End of Injection.

9-2

Types of Injectors

Cummins injectors are of the multihole type for use in their direct injection engines. The injector spray tip contains eight holes that allow fuel to be sprayed into the combustion chamber formed by the shape of the piston crown. Over the years Cummins has updated and changed their injectors to reflect product improvement and better performance.

The injector is a simple mechanical unit that receives its fuel from the PT pump assembly under pressure and meters, injects, and atomizes this fuel through the small holes drilled within the injector cup into the combustion chamber. There are two basic types of injectors used with the PT fuel system, the flange mounted type on older engines, and the cylindrical type found on the newer models of Cummins engines.

Figure 9-12 shows a typical flanged-type injector. These injectors were only used on the older external fuel manifold engines. They contained an adjustable orifice plug in the inlet drilling that could be changed to adjust flow through the injector. The one exception to this was the matched flanged PT injector, which was delivery checked on a fuel pump test stand and then placed in sets since these did not have an adjustable orifice plug in the inlet drilling. Fuel to the flanged injector is supplied and also drained by the use of external lines. Fuel entering the injector inlet flows down and around the injector plunger between the body and cup and on up the drain passage, where it is directed back to the fuel tank.

Metering in this injector occurs as the plunger rises, letting fuel flow past the metering orifice and into the cup. While this is happening, the remainder of the fuel is free to flow out of the drain orifice. Therefore, both fuel pressure and timing control fuel volume enter the cup.

As the rocker arm forces the plunger down, the metering orifice is closed, trapping the fuel and forcing it into the combustion chamber. When the plunger bottoms in the seat, all fuel flow through the injector ceases.

Flanged injectors contain a fine mesh screen in the inlet connection that acts as a final filter before fuel enters the injector. These injectors use common inlet and return fuel lines externally on the engine.

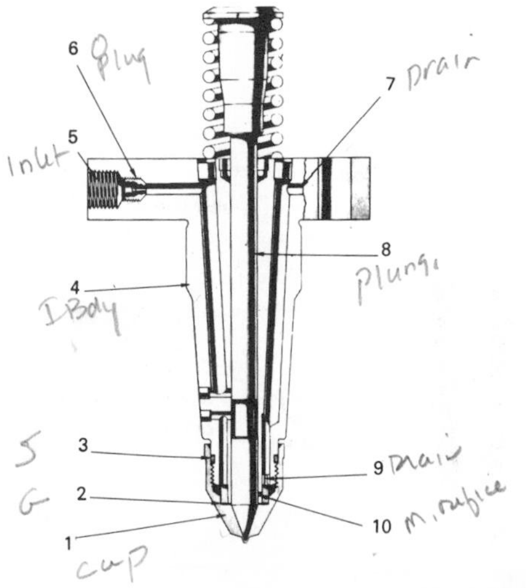

Index:
1—Cup
2—Gasket
3—Seal
4—Injector body
5—Inlet
6—Orifice plug
7—Drain
8—Plunger
9—Drain orifice
10—Metering orifice

Figure 9-12
Typical Flange-Type Injector.
(Courtesy of Cummins Engine Co.)

PT (Type D) Injectors

Cylindrical-type injectors (Figure 9-16) used on Cummins engines have both the fuel supply and return routed through internal drillings within the cylinder heads. The supply fuel entering the head surrounds a radial groove on the injector body located between two O-type seal rings. A fine mesh screen at each inlet ensures final fuel filtration. Excess fuel is drained from the injector at its upper radial groove, which aligns with a second drilling in the cylinder head.

As the injector plunger is forced down by the rocker arm, it covers the supply opening. This creates a pressure wave that causes the injector check ball to seat and simultaneously trap an amount of fuel in the injector cup. The fuel under the plunger is forced into the combustion chamber, and as the plunger continues its downward movement, it will uncover the drain opening out of the injector, which allows the check ball to rise from its seat and free fuel flows through the injector and out the drain passage for cooling purposes and purging of any gases from the cup area.

PT (Type D) Injectors, Top Stop

This injector was designed to allow better engine oil lubrication to the injector sockets, thereby reducing wear in the injector operating mechanism and permitting longer service intervals between injector adjustment. The top stop injector (Figure 9-17) operates similarly to the normal PTD unit, with the exception that the upward travel of the injector plunger is controlled or limited by an adjustable stop that is set prior to installing the injector in the engine. Proper installation and adjustment of the injector permits the plunger spring load to be carried up against the stop.

Injector Size

Located between the top and center O rings on the injector body are stamped numbers which indicate the following: 178A8717

1. 178: injector flow
2. A: 80% flow
3. 8: number of holes
4. 7: size of holes (0.007)
5. 17: degree of holes (spray angle)
6. The injector assembly number is stamped around the top periphery of the injector body.

PT Type C and D injector Cup Comparison

Although the injector cup for the type C and D units may look the same, and in fact the PT-C cup can be put into the PT-D retainer, it will not locate correctly on the injector plunger and barrel, causing improper alignment of the cup to plunger.

The AR-73309 and AR-73326 may be intermixed within the same engine.

The AR-73307 may be intermixed with the

Table 9-1
CURRENT INJECTOR NOMENCLATURE

Type	*Assembly Number*	*Cup Part Number*	*Cup Size*	*Flow (cc)*
PT (type C)	AR-73309	195339	7-.005 × 5 degrees	118
PT (type D)	AR-73326	195321	7-.0055 × 5 degrees	118
PT (type C)	AR-73307	166664	7-.006 × 3 degrees	112

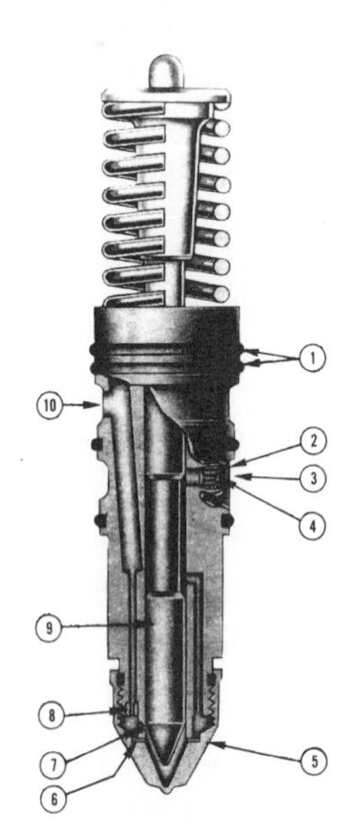

Figure 9-13
Cylindrical Injector.
(Courtesy of Cummins Engine Co.)

Index:
1—Ring seals
2—Screen
3—Fuel in
4—Delivery orifice
5—Cup
6—Gasket
7—Metering orifice
8—Drain orifice
9—Plunger
10—Fuel out

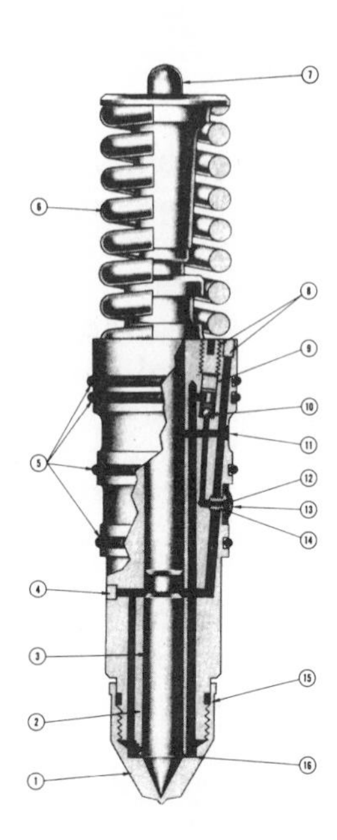

Figure 9-14
PT-B Injector.
(Courtesy of Cummins Engine Co.)

Index:
1—Cup
2—Metering orifice
3—Plunger
4—Plug
5—O ring seals
6—Injector spring
7—Injector link
8—Plugs
9—Stop
10—Check ball
11—Fuel out
12—Fuel in
13—Fuel screen
14—Orifice plug
15—O ring seal
16—Gasket

above injectors as long as it has been calibrated to provide a flow rate of 117 to 118 cc.

Note: For a detailed description, including overhaul and maintenance, of all Cummins PT injectors, obtain a copy of Cummins Bulletin No. 3379071-03 from any Cummins dealer.

The most widely used injector made by Cummins today is the PT type D unit. Preceding the use of this injector were the flanged injector, the cylindrical PT injector, and the cylindrical PT injectors types B, C, and D. The cylindrical injectors are shown in Figures 9-13 through 9-17.

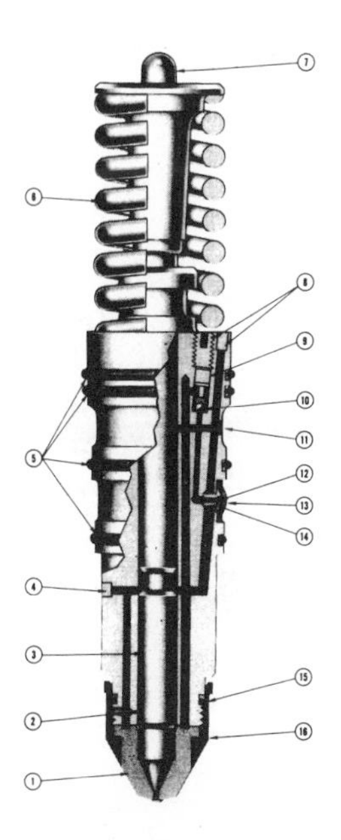

Figure 9-15
PT-C Injector.
(Courtesy of Cummins Engine Co.)

Index:
1—Cup
2—Metering orifice
3—Plunger
4—Plug
5—O ring seals
6—Injector spring
7—Injector link
8—Plugs
9—Stop
10—Check ball
11—Fuel out
12—Fuel in
13—Fuel screen
14—Orifice plug
15—O ring seal
16—Cup retainer

The injector shown in Figure 9-13 was the first cylindrical type used by Cummins in their engines; it was used in a limited number of engines with the fuel drillings cast internally in the cylinder head. They did, however, contain an adjustable orifice plug in the inlet passage and were held in the cylinder head by either a mounting yoke or mounting plate.

The PT type B injector shown in Figure 9-14 was an improvement over the first cylindrical unit in that it contained a ball-type check valve in the inlet fuel flow pasage to aid in the control of fuel flow, along with an adjustable orifice plug, which was burnished to size during flow testing of the injector on a test apparatus. Identification of the PT-B injector compared to the cylindrical injector shown in Figure 9-13 is by the socket head ball retainer plug at the top of the injector body.

The PT type C injector shown in Figure 9-15 is basically the same as the PT-B, but also contains a two-piece cup consisting of a cup and a retainer, less the gasket.

The PT type D injector shown in Figures 9-16 and 9-17 functions the same as the PT type B unit; however, the type D was designed to provide greater parts interchangeability. Also, those areas subject to wear have been localized in small components, allowing for easier and cheaper servicing. An engine presently using PT-B injectors can be changed over to use the PT-D injector as long as the total set of injectors is replaced.

Basic Injector Parts

Adjustable Orifice Plug. This plug can be readily seen in the cited injector schematics. It is used in the inlet drilling of both flanged and cylindrical injectors to allow a means of adjusting the injector's fuel delivery characteristics by either changing the orifice plug itself, or by burnishing the plug in the operating position. Some injector orifice plugs have a flange and also require the use of a gasket beneath this flange.

Drain Orifice. The injector views previously cited reflect that the drilled orifice in the

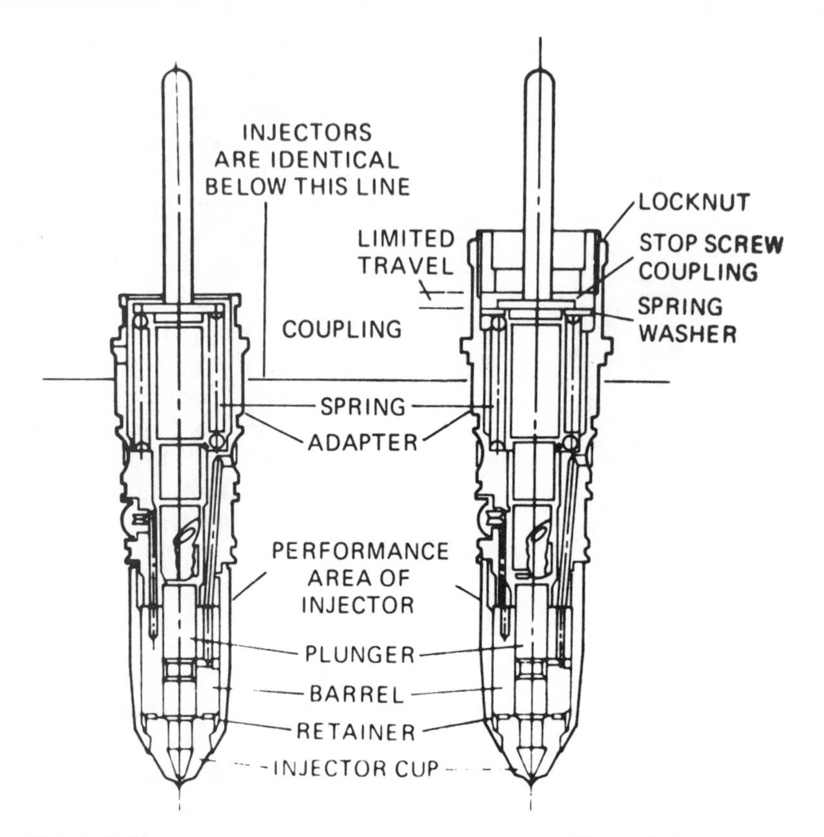

Figure 9-16
PT Type D Injector.
(Courtesy of Cummins Engine Co.)

Figure 9-17
PT Type D Top Stop Injector.
(Courtesy of Cummins Engine Co.)

cup end of the injector is used for fuel draining purposes. Do not under any circumstances attempt to alter this orifice size, which is a fixed dimension.

Metering Orifice. Fuel under pump pressure enters the injector plunger bore and cup through the metering orifice in the cup end of the injector, which is also fixed in size; therefore, do not attempt to alter this dimension in any way.

Cup Gasket. If an injector is rebuilt and fitted with oversize plungers, to provide the correct relationship between the plunger and metering orifice, a thicker cup gasket must be used. Remember, however, that PT type C injectors do not use a cup gasket.

Injector Operation

The operation of the injector has been discussed in considerable detail. You will recall that fuel injection is accomplished through a series of operations involving the rotation of the camshaft, push tube lift, and injector plunger movement. Anytime an injector is serviced, you must be certain that the correct orifices, plungers, and cups are used. You can also affect injector operation by any of the following actions:

1. Improper engine timing.
2. Mixing plungers and barrels during teardown (keep them together, since they are matched sets).
3. Incorrect injector adjustment after installation or during a tune-up adjustment.
4. Overtightening of hold-down clamps.
5. Installing an exchange set of injectors without taking time to check and correct other possible problems relating to the injector operating mechanism. This is often overlooked.

From our basic discussion thus far, you will recollect that the actual size of both the balance and metering orifices normally regulates the fuel inlet pressure to the injector, which is dependent on the fuel manifold pressure. Also, the drain orifice is a fixed size; therefore, if the injector drain line is restricted in any way the fuel will stay in the injector longer, leading to the following:

1. Possible overfueling.
2. High wear rate of injector operating mechanism.
3. Problems of slow engine deceleration.
4. High idle speeds and possible erratic fuel pump operation.

If wear occurs in the injector operating mechanism owing to either a restricted drain orifice or other reasons, such as poor adjustment or wear of push tube sockets, follower rollers, or bushings, then it is possible for the injector plunger to fail to seat correctly during the combustion and exhaust cycles, leading to the following problems:

1. Dirt or air in the fuel system.
2. Scored injector plungers.
3. Rough engine operation caused by plugging of the spray cup holes due to carbon formation.
4. Smoke out of the exhaust stack due to incomplete combustion. Caused by failure of the injection cycle to cut off quickly and sharply, allowing fuel to dribble into the cylinder at a time when it cannot burn.

Therefore, proper injector adjustment and maintenance will ensure a smooth running engine as long as the following are maintained:

1. Adequate fuel delivery pressure from the pump to the fuel manifold.
2. Selection of the proper sizes of balance and metering orifices.
3. The length of time that the metering orifice is uncovered by the upward moving injector plunger.

In summer operation, especially around extremely dusty operating environments, the entrance of dirt or dust through the fuel system can wear out fuel system components and adversely affect its operation. If dirt is allowed to enter the injector screen, it can reduce the fuel flow, leading to complaints of lack of power and rough engine operation.

In winter, when operating in cold weather, either use of a fuel with too low a cloud point or allowing equipment to stand idle for long time periods can cause wax crystals to form in the fuel, which will form on the fuel filters and injector filters and restrict fuel flow, again creating problems of failure to start or very hard starting. Also, water in the fuel in cold weather will ice up the injector screen and fuel system filters.

Orifice Sizes

A variety of orifice sizes on the balance orifice are readily available from Cummins to vary the fuel flow to the injector accordingly. The metal of the PT-D injector is soft enough that you can burnish the hole size to slightly change the injector flow rate. The metering orifice is also critical to the injector flow rate; therefore,

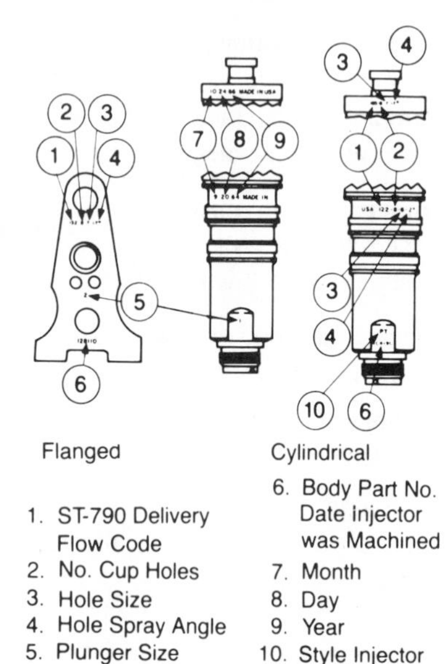

Figure 9-18
Typical Size Markings Found on Flanged and Cylindrical Injectors.
(Courtesy of Cummins Engine Co.)

always select a *balanced set of injectors* to ensure a smooth running engine.

Injector Identification

All Cummins injectors can be readily identified by size markings on the injector body, injector cup, and plunger. Figure 9-18 shows typical size markings found on both flanged and cylindrical injectors. On some of the newer PT-D injectors, a letter may be stamped after the flow code, which is indicative of a percentage of fuel flow less than 100%. For example, the letter A indicates an 80% flow.

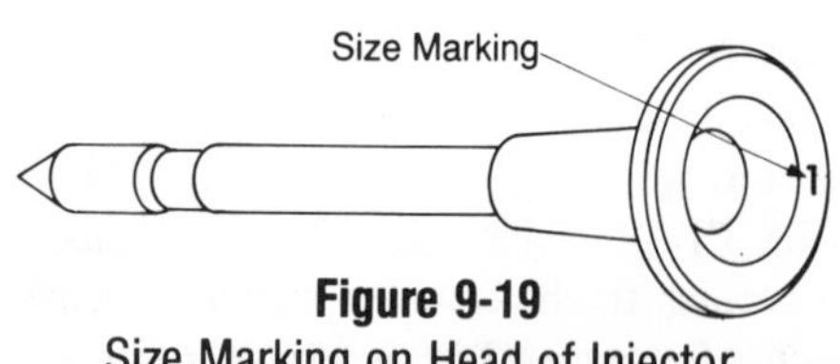

Figure 9-19
Size Marking on Head of Injector.
(Courtesy of Cummins Engine Co.)

In addition to injector body markings, each injector plunger has a size marking imprinted upon the head end, as shown in Figure 9-19. The plunger body size marking is used to establish the size of injector cup gasket used in the injector during manufacture and assists the mechanic or technician in providing the original relationship between the plunger and metering orifice at injector rebuild time. Figure 9-20 shows injector cup gasket markings that coincide with the plunger body size markings as to which cup gasket would be used with which plunger.

Figure 9-21 shows typical markings found on injector cups. Some injectors may have PT-USA stamped on them followed by several other numbers. If so, these additional numbers

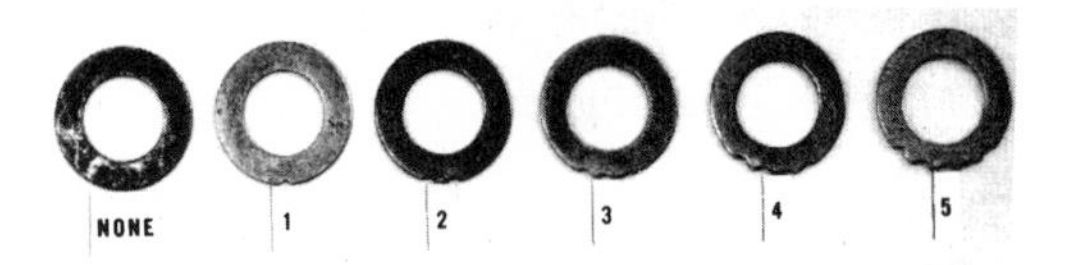

Figure 9-20
Injector Cup Gasket Markings.
(Courtesy of Cummins Engine Co.)

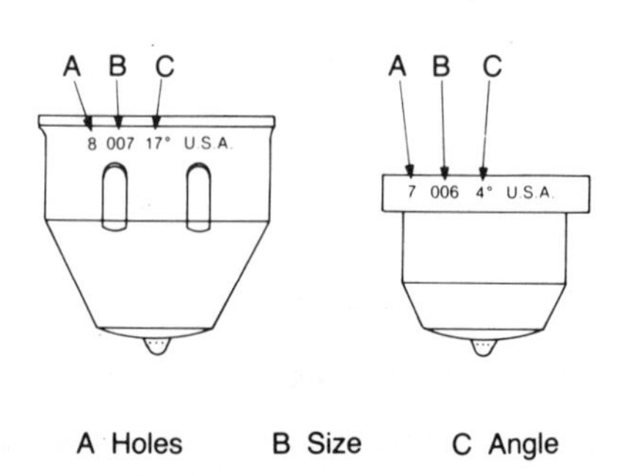

Figure 9-21
Typical Markings Found on Injector Cups.
(Courtesy of Cummins Engine Co.)

signify the month and year of cup manufacture.

9-3

Basic Injector Maintenance

The service and maintenance of fuel-injection pumps and injectors must be carried out in spotless surroundings where special tools are readily available; up-to-date service literature and equipment are also required to guarantee satisfactory results. Therefore, the following information relative to injector servicing is general in nature.

1. On both flanged- and cylindrical-type injectors, always lift out the injector plunger and spring. Remove the spring from the plunger and then replace the plunger back in the body. Injector plungers are matched to a particular injector body; therefore, *never* attempt to interchange these parts.
2. On flanged injectors, place the injector body into the ST-569 holding fixture so that the fixture spring tension is against the injector plunger as shown in Figure 9-22. Spacers listed in Cummins Injector Shop Manual are necessary to obtain the required spring tension. Proceed to loosen the injector cup with the proper Cummins injector socket wrench. Remove the wrench and very carefully unscrew the cup; then proceed to remove the injector from the holding fixture with due care and attention, and lay the part in a clean container.
3. On cylindrical injectors, remove the half-collets if used, and lift off the mounting plate (see Figure 9-23). Remove injector body O rings and throw away; then remove either the button-style ring or clamp in order to remove the injector body fuel screen. Leave the adjustable orifice alone.
4. Place the injector in the ST-569-19 adapter (Figure 9-22) and repeat step 2.

Figure 9-22
Placement of Injector Body in ST-569 Holding Fixture.
(Courtesy of Cummins Engine Co.)

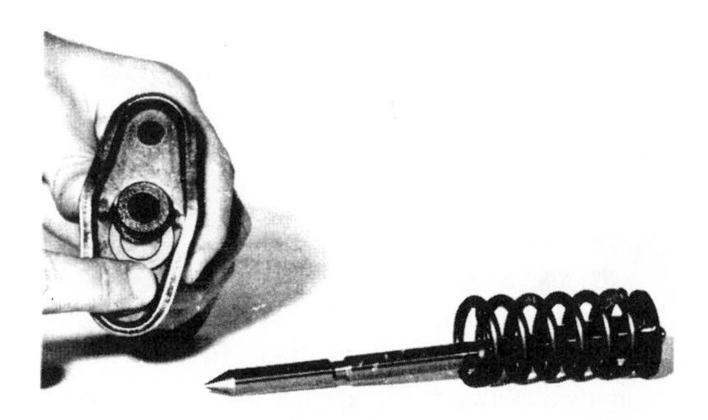

Figure 9-23
Removal of Half-Collets and Mounting Plate.
(Courtesy of Cummins Engine Co.)

5. On cylindrical PT-B, C, and D injectors, remove the ball retainer plug, gasket, and ball from the top of the injector body.

Cleaning

Injector parts must be thoroughly cleaned in special solvents such as Bendix Metal Cleane, Kelite Formula 1006, or equivalent or by an ultrasonic cleaning machine found in most fuel-injection shops. Parts should be dipped in mineral spirits in order to neutralize them after cleaning, and then dried with compressed air.

Injectors should never be buffed on a shop bench grinder-type wire wheel, because severe scratching, scoring, and spray tip hole enlargement can occur, which can drastically alter both the injector's fuel delivery and spray angle, which can cause severe engine damage.

The ultrasonic cleaning process has the advantage of removing all traces of carbon formation from injector spray tips, which stops people from using such items as drill bits and the like to try to clean the spray tip holes. It is sometimes necessary if some minute carbon particles are present in the tip holes to use special wire supplied by the manufacturer, which is slightly smaller than the hole size, for cleaning purposes.

On flat surfaces that come in contact with another within the injector, inspection with a magnifying glass is required to determine surface condition. Generally, these parts are lapped lightly at overhaul either on a surface table with crocus cloth, or on one of the automatic-type Lapmaster machines.

If inspection of the injector body plunger bore shows signs of light scoring, it may be possible to hone it and then fit it with an oversized plunger. Otherwise the injector body would have to be replaced. Also carefully check the injector body nose surface for any damage.

Carefully check the injector plunger for signs of discoloration, metal scuffing or scoring, and anything which would render it unfit for further service. Also inspect the plunger link for signs of wear. If wear is evident, then the E ring type can be removed by placing the plunger in a soft jawed fixture, using a hammer and punch, and breaking off the ring "ears," removing the ring, and replacing the link with a new retaining ring to hold the new link in place. If the sleeve type is used, a collet-type hand tap holder may be used to pull the sleeve by placing the tap holder over the link, tightening the holder, and giving it a quick pull. Both links should be pressed into place and if a sleeve-type retainer is used, press it flush to 0.010 in. (0.254 mm) below the bore surface. Make sure that you check the injector spring for such things as free and compressed length and squareness as you would with a valve spring. Check both the inlet and drain connections for any damage and repair if necessary.

Injector Cup

After cleaning, the cup spray holes should be checked with a magnifying glass for signs of hole enlargement, corrosion damage, and possible signs of overheating. High acid or sulfur content in the fuel oil or excessive engine overload can cause corrosive damage, and attempting to clean the spray hole tips with other than the correct wire size can damage them beyond use. Also inspect the cup for plunger seat pattern; however, this can be confirmed later under the cup-to-plunger leak test. PT-D cups with a black finish are completely interchangeable with older style stainless steel cups. Also check the cup retainer threads, inside and outside areas, for any damage.

Injector Reassembly

With new component parts or old parts which have been serviced, again extreme cleanliness cannot be overemphasized. Dip all parts in clean test oil or other suitable lubricant prior to assembly. New O rings should also be dipped in test oil and rolled into position on the injector body with the use of a nylon/plastic assembly tool. Take care not to cut or twist the O ring during this time.

Select a new cup gasket which matches the number stamped on the plunger and screw the cup onto the injector body finger tight, then back off about one-half turn.

Note: The two-piece PT-C injector cup is used without an injector cup gasket.

Install the plunger into the body without the spring and place the injector into the ST-569 or ST-569-19 adapter and spacer for your particular injector. Pull the fixture handle to apply tension to the holding fixture which should align the plunger to the body and torque the cup to its specified tension with the proper cup socket wrench.

Remove the injector from the fixture, take out the plunger, and install the spring. Dip the plunger in test oil and insert it into the mating body, making sure that there is no bind as it seats in the cup.

Leakage Checks

The plunger-to-body and plunger-to-cup leakage check actually gives a measurement of the amount of fuel that is leaking between the plunger and injector body plunger cup. This test will confirm whether or not the injector can be rehoned or calibrated for reuse in the engine. Refer to Cummins Injector Shop Manual for the procedure involved for the type of injector being checked.

If the injector fails to pass the leakage test, it can be disassembled and the plunger can be lapped into the cup to obtain good seat contact. Only the cup and plunger should be used when lapping is required and a lapping compound not exceeding 300 grade mixed with SAE 30 lube oil will produce the desired results.

Insert a small amount of the compound onto the end of the plunger, place it into the cup, and turn it back and forth for approximately one minute; then inspect the seating condition.

Make sure that both the cup and plunger are thoroughly cleaned of all traces of lapping compound when finished either in the ultrasonic cleaner, preferably, or suitable cleaner as mentioned earlier under Cleaning.

Cup Spray Pattern Check

This test is mandatory during injector testing to ensure that the spray pattern is crisp and even and also to determine if in fact the spray angle is correct. Damaged or enlarged spray tip holes can cause flattening out of the spray angle, leading to possible wall wash, burning of the outer periphery of the piston crown, and engine seizure. Spray test fixture ST-668 is required for this purpose. Check the Cummins Injector Shop Manual No. 3379071-03 for use of the proper cup seat spacer and test sequence for your particular injector.

ST-790 Injector Test Stand

In order to accurately flow test the injector, a test stand is required. This machine flow tests the complete injector assembly by measuring fuel delivery. In this way a matched and balanced set of injectors can be selected for installation into the engine, ensuring a smoother idling and performing unit.

The injector is actuated under conditions typical of engine operation. The test stand counts injection strokes, supplying fuel at a specified pressure and measuring it in a scaled glass container.

Since this procedure is somewhat detailed, refer to the Cummins Injector Shop Manual No. 3379071-03 for instructions.

During flow testing of injectors, adjustable delivery type units can be burnished by the use of adapter tool ST-708, which burnishes the injector inlet orifice plug instead of having to change the plug. In this way injectors can be very closely flow rated so that a matched and balanced set can be obtained. The burnishing tool has a replaceable needle point.

An alternate injector comparator, the STU-790, which is manufactured in Great Britain, can also be employed for flow checking injectors.

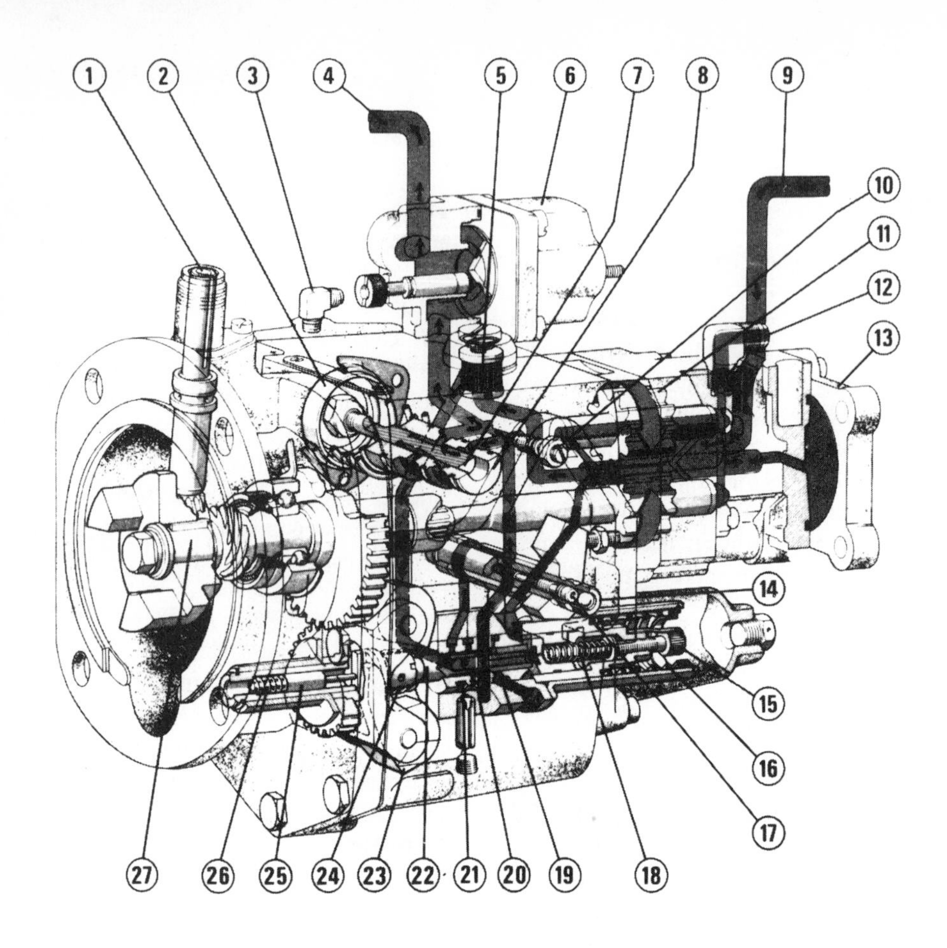

Figure 9-24
PT (Type G) AFC Fuel Pump and Pump Flow.
(Courtesy of Cummins Engine Co.)

Index:

1—Tachometer shaft
2—AFC piston
3—AFC air in
4—Fuel to injectors
5—Filter screen
6—Shut-down valve
7—AFC control plunger
8—AFC fuel barrel
9—Fuel from filter
10—AFC needle valve
11—Gear pump
12—Check valve elbow
13—Pulsation damper
14—Throttle shaft
15—Idle adjusting screw
16—High-speed spring
17—Fuel adjusting screw
18—Idle spring
19—Gear pump pressure
20—Fuel manifold pressure
21—Idle pressure
22—Governor plunger
23—Governor weights
24—Torque spring
25—Governor assist plunger
26—Governor assist spring
27—Main shaft

9-4 Cummins PT Fuel Pumps

Current production Cummins diesel engines are equipped with one of the two types of PT fuel pumps currently available; these are shown in Figures 9-24 and 9-26. The PT (type G) AFC fuel pump assembly shown in Figure 9-24 is made up of four main units: (1) gear pump, (2) standard governor, (3) throttle, and (4) air–fuel control (AFC).

The air–fuel control unit in the newer PT pumps is an acceleration exhaust smoke control device built internally into the pump body. It is designed to restrict fuel flow in direct proportion to engine air intake manifold pressure during engine acceleration conditions. It is similar in function to the aneroid control used on older model Cummins diesel engines.

Since a turbocharger is exhaust gas driven, the speed of the turbo is related to exhaust gas flow, which is in turn controlled by engine speed and load conditions. With no direct mechanical drive then to the turbocharger, during acceleration turbo speed lags behind the almost instantaneous fuel delivering capability of the pump and injectors, thereby creating a temporary air starvation situation until the turbocharger can accelerate and supply enough additional air flow. Before this can happen, however, we are supplying an overrich fuel-to-air mixture, which causes excessive exhaust smoke. Therefore, the AFC provides a more controlled combustion of fuel to air by constantly monitoring turbocharger air pressure and proportionally responding to load or acceleration changes.

AFC fuel control is monitored within the fuel pump. This necessitates running a line

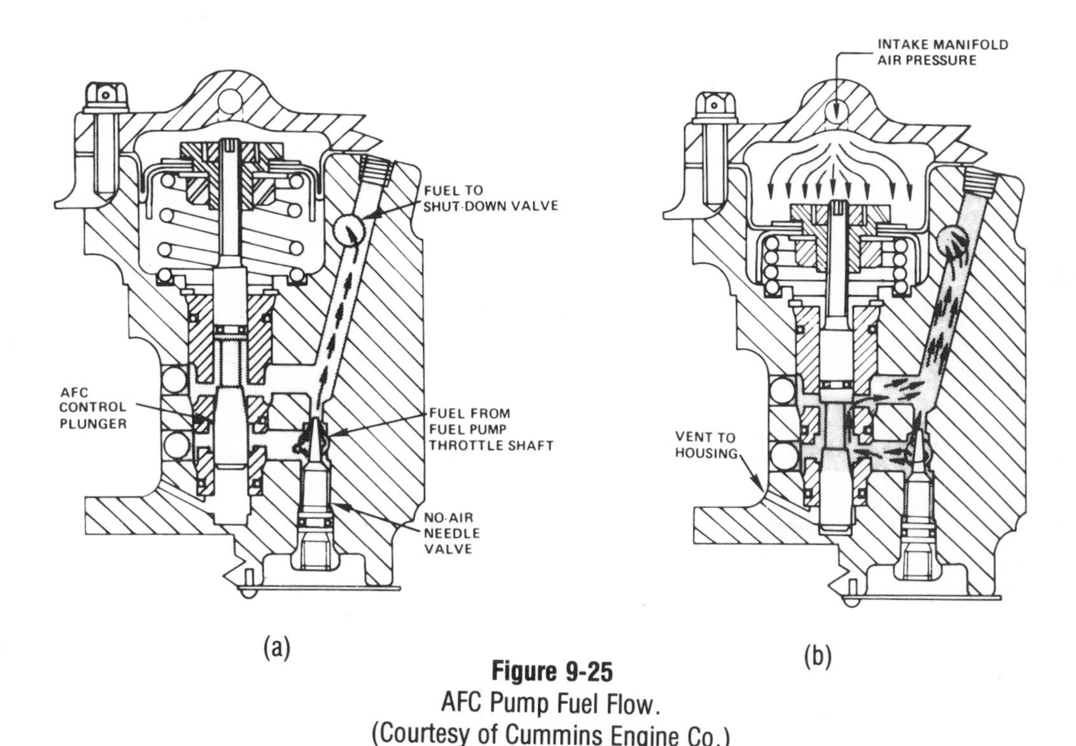

Figure 9-25
AFC Pump Fuel Flow.
(Courtesy of Cummins Engine Co.)

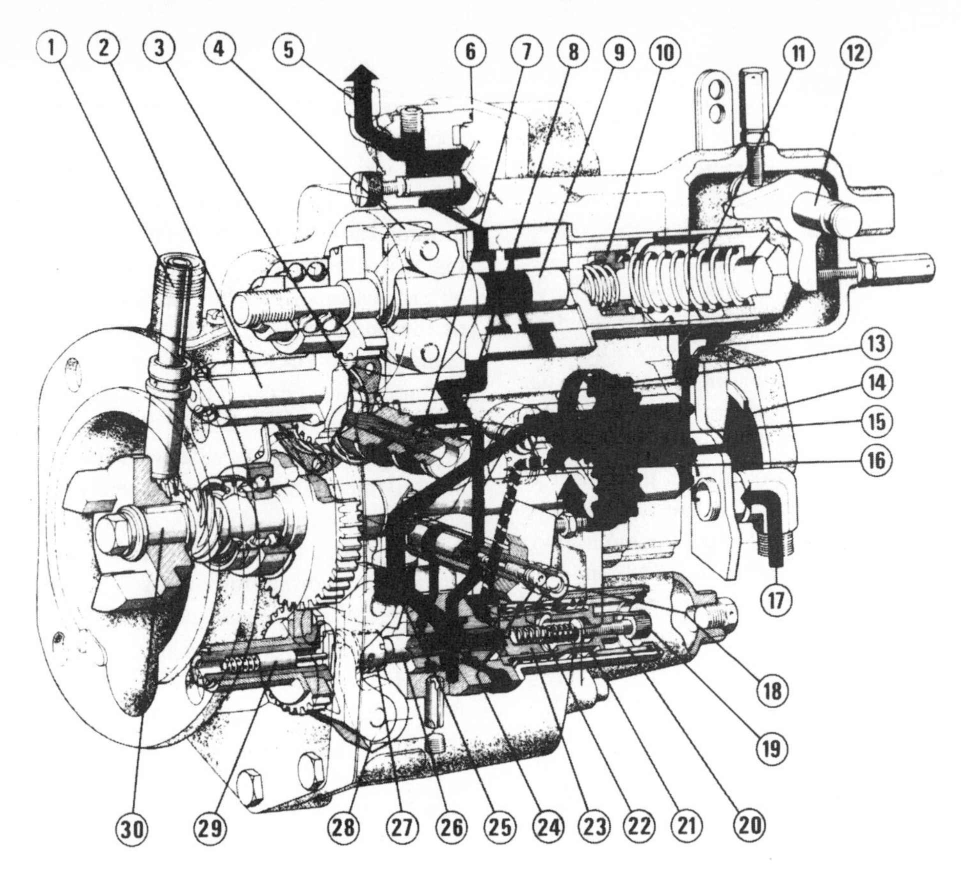

Figure 9-26
PT (Type G) Variable Speed AFC Fuel Pump and Fuel Flow.
(Courtesy of Cummins Engine Co.)

Index:

1—Tachometer shaft
2—Idler gear and shaft
3—AFC piston
4—VS governor weights
5—Fuel to injectors
6—Shut-down valve
7—AFC control plunger
8—AFC fuel barrel
9—VS governor plunger
10—VS idle spring
11—VS high-speed spring
12—VS throttle shaft
13—Gear pump
14—Pulsation damper
15—AFC needle valve
16—Pressure regulator valve
17—Fuel from filter
18—Throttle shaft
19—Idle adjusting screw
20—Spring spacer
21—High-speed spring
22—Idle spring
23—Idle spring plunger
24—Fuel adjusting screw
25—Filter screen
26—Governor plunger
27—Torque spring
28—Governor weights
29—Governor assist plunger
30—Main shaft

from the underside of the intake air manifold air pressure hole to a no. 4 fitting located in the pump AFC cover plate [see Figure 9-25(a)]. Both parts (a) and (b) of Figure 9-25 are plan views (top) of the AFC unit. View (a) is a cross section of the control plunger in the *no-air* position; view (b) shows the control plunger in the *full-air* position. Current production Cummins engines are equipped with the PT type G-AFC fuel pump. Those engines not requiring an AFC feature simply have a specially designed plug screwed into the pump housing in place of the AFC barrel assembly.

The aneroid control unit that was used on earlier Cummins engines was designed to function as an *on–off* fuel bypass device. The AFC unit acts as both a fuel pressure and flow restrictor to provide the correct air–fuel delivery rate to the engine during acceleration.

Operating Principles: AFC

The main pump operation is very similar for both the PTG and PTG-AFC; however, the major difference is that which occurs between the fuel pump throttle shaft and the shut-down valve on the AFC pump. In the straight PTG pump unit, fuel passes directly from the throttle shaft through a passage to the shut-down valve, while in the PTG-AFC pump, the fuel passes through the AFC unit after leaving the throttle shaft and before it reaches the shut-down valve on top of the pump body.

Fuel enters the AFC control after leaving the governor and passing through the throttle shaft. When no air pressure is supplied from the turbocharger, the AFC plunger closes off the primary fuel flow circuit. A secondary passage controlled by the position of the no-air needle valve supplies fuel for this condition, such as engine cranking, or at initial acceleration of the engine. The no-air needle valve is located directly above the throttle shaft under the throttle cover plate.

As intake manifold pressure increases or decreases, the AFC throttling plunger reacts to deliver a proportional increase or decrease in fuel, which prevents the fuel–air mixture from getting overrich and causing excessive exhaust smoke. The AFC plunger is positioned by action of the intake manifold air pressure acting against a piston and diaphragm opposed by a spring to a proportionate amount of travel.

Therefore, when a full-throttle lugdown condition occurs, fuel flow through the AFC unit is unrestricted. The AFC unit also controls fuel flow after deceleration in traffic, during gear shifts, running downhill with a closed throttle, or on a downgrade operation on the light-load portion of the governor droop curve.

Engine Torque Curve Adjustment

The engine torque curve can be controlled by the use of a TMD (torque modification device), TLV (torque limiting valve), and torque spring, which is located over the governor plunger at the governor weight end, as shown in Figure 9-28. This spring must be very carefully controlled as to spring rate and length, since it is the spring rate that is the factor used to determine the amount of torque, while the spring length governs the torque peak point. The engine application is the major factor that determines whether a high torque rise is required. A little torque rise is an inherent design of the system and may very well be all that is required on some applications; however, on an application such as a highway truck a high torque rise is a desirable quality.

The basic difference between the pump shown in Figure 9-26 and that shown in Figure 9-24 is the addition of an upper VS or variable speed governor. It contains the same four main units as that shown in Figure 9-24.

As fuel from the gear pump enters the drillings in the fuel pump body, as shown in Figures 9-24 and 9-26, it encounters some restriction owing to the size of the drillings. Additional pressure is placed on it by the

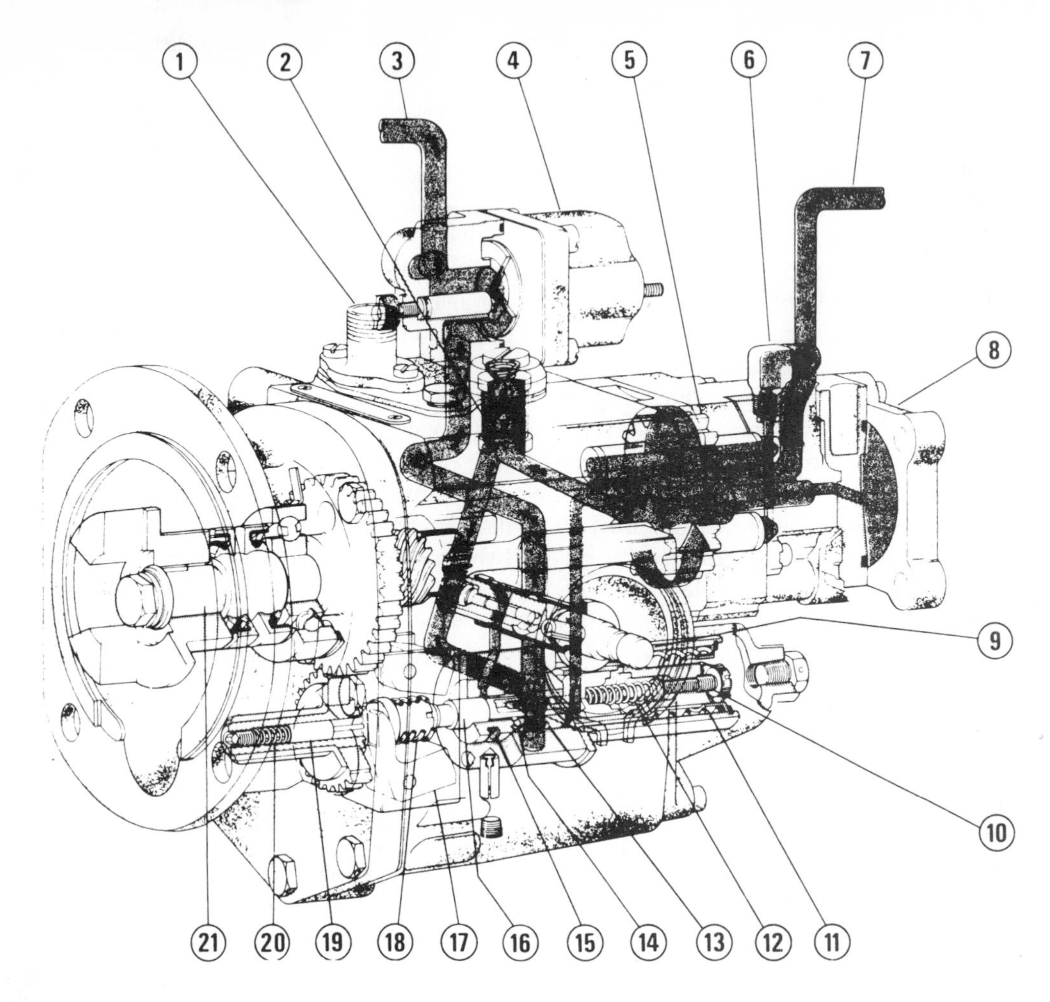

Figure 9-27
PT (Type G) Fuel Pump.
(Courtesy of Cummins Engine Co.)

Index:

1—Tachometer shaft
2—Filter screen
3—Fuel to injectors
4—Shut-down valve
5—Gear pump
6—Check valve elbow
7—Fuel from tank
8—Pulsation damper
9—Throttle shaft
10—Idle adjusting screw
11—High-speed spring
12—Idle spring
13—Gear pump pressure
14—Fuel manifold pressure
15—Idle pressure
16—Governor plunger
17—Governor weights
18—Torque spring
19—Governor assist plunger
20—Governor assist spring
21—Main shaft

presence of tighter restriction encountered in the governor assembly. Fuel under pressure flows up into the hollow center of the small filter screen located at the top of the pump.

Fuel flowing out through this screen continues through pump drillings down to the governor plunger and barrel assembly.

Note: The filter screen is located at the top of the pump assembly in the AFC pump shown in Figure 9-24 and at the base of the pump assembly in the AFC-VS unit shown in Figure 9-26.

Figure 9-27 shows the fuel flow and component parts of a standard automotive PTG pump prior to the release of the current PTG-AFC fuel pump.

Fuel entering the PT pump's governor plunger and barrel assembly flows around and into the plunger area, where the plunger rides freely in the carrier and sleeve, which is being lubricated by this fuel. Fuel flowing into the plunger travels in both directions and will therefore follow the route of least resistance. Figure 9-28 shows that spring pressure on the right-hand side holds the idle plunger against the end of the governor plunger until fuel begins to flow, at which time they are pushed apart enough for some fuel to escape.

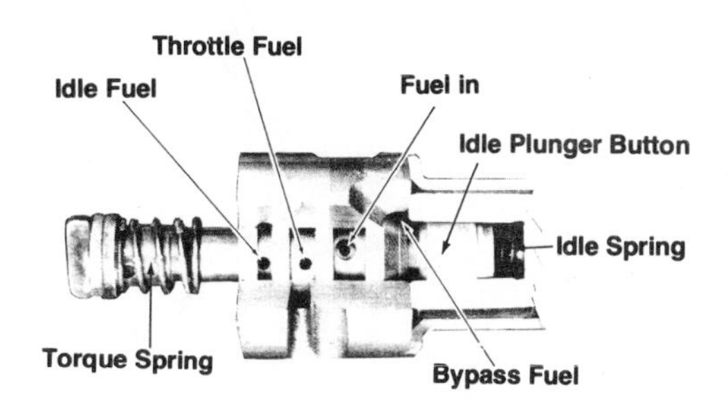

Figure 9-28
Action of Spring Pressure to Control Fuel Flow. (Courtesy of Cummins Engine Co.)

In the cutaway section shown in Figure 9-28, the amount of fuel being bypassed depends upon resistance to its flowing out in other directions through the idle and throttle openings in the plunger and barrel assembly. To simplify this action of how the fuel flow is controlled, the fuel pressure as it reaches the governor plunger is caused by the restriction to this flow by placing the surface of the idle plunger against the end of the governor plunger. Such a condition can be likened to that created when you place a thumb over the end of a garden hose minus the nozzle. Water pressure builds up in the water behind your thumb owing to restriction caused by your thumb over the end of the hose; therefore, water that does escape has an increased velocity or greater force and direction.

Under this condition, fuel is held in the governor plunger by the surface of the idle plunger, which is under spring pressure; therefore, as the volume of fuel flow increases, fuel will eventually push the idle plunger back if no other outlet is found. There are however two other outlets for governor plunger fuel, which are shown in Figures 9-29 and 9-30.

Figure 9-29 shows (1) the *idle* port (or drilling), which allows fuel to escape during *low* speeds, and (2) the *throttle* port, through which fuel escapes during times of higher speeds or loads. Whether or not fuel is routed through these two other passages is controlled by just how they are aligned with fuel from the governor plunger, and how hard it is for the fuel volume to push the idle plunger surface away from the end of the governor plunger.

The position of the throttle control shaft provides an external means of manually restricting or interrupting the fuel flow to the injectors. Figure 9-30 shows a typical fuel pump throttle setup with additional detail. The throttle system then does the following:

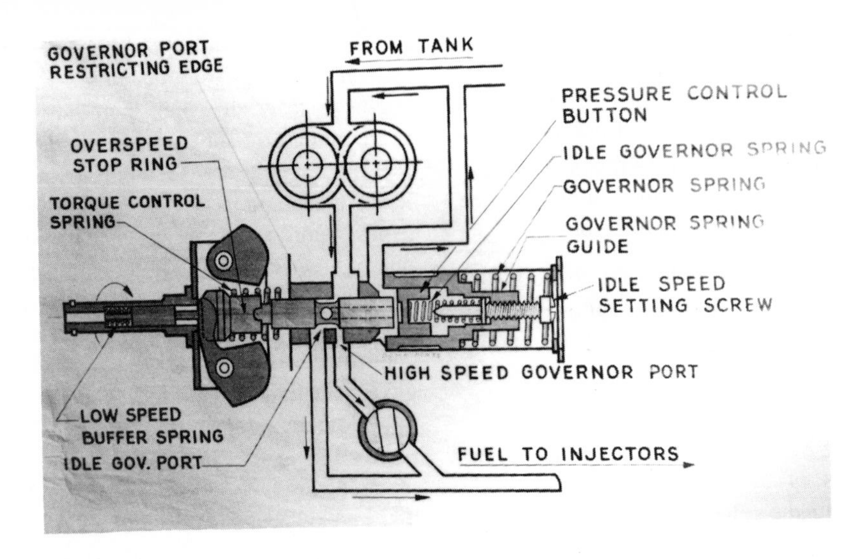

Figure 9-29
PT-Fuel Pump Control Assembly.
(Courtesy of Cummins Engine Co.)

1. Forms the idle fuel passageway.
2. Controls fuel flow for selecting the desired engine speed.
3. Controls minimum circulation to the injectors (throttle leakage).

Let's study these three functions a little more closely.

1. *Idle passage:* Idle fuel flow is controlled by the pump governor. The throttle shaft idle passage is always open to fuel pressure.
2. *Manual fuel control passage:* As mentioned earlier, fuel for engine operation must pass through the throttle shaft, which is aligned at this time with the passage in the pump body. Rotation of the throttle shaft causes misalignment of fuel passages and restricts fuel flow, thereby reducing fuel manifold pressure available to the injectors. Figure 9-30 shows that two throttle stop screws limit throttle movement. The *rear* screw allows adjustment of maximum fuel passage opening. The *forward* screw limits the closed throttle position. On AFC fuel pumps the front throttle stop screw is for throttle travel adjustments and the rear screw is for throttle leakage adjustments.

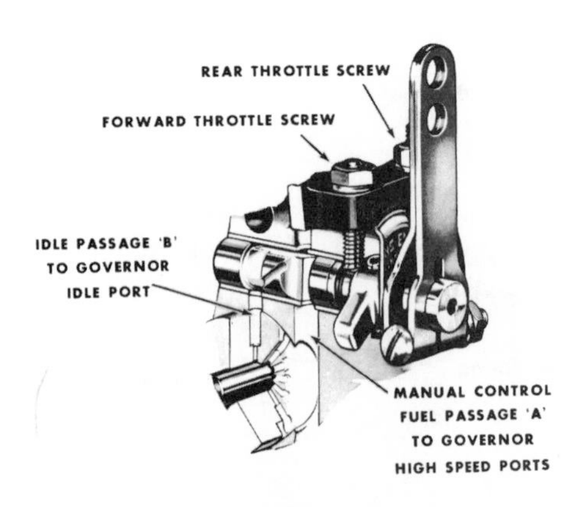

Figure 9-30
Fuel Pump Throttle Flow at Various Positions.
(Courtesy of Cummins Engine Co.)

3. *Throttle leakage:* Adjustment of the forward throttle stop screw sets the engine idle speed by controlling the amount of fuel flowing to the injectors with a closed throttle. This throttle leakage also does the following:
   a. Maintains fuel manifold pressure required at the injector metering orifice for immediate acceleration when desired.
   b. Purges air and gases from the injector.
   c. Lubricates the injector.

Figure 9-30 illustrates how fuel is directed to and through the throttle shaft with varying degrees of throttle movement.

In summation, then, fuel flow from the gear pump is as follows:

1. PTG pump: from the filter screen to the governor assembly, as shown in Figure 9-26.
2. PTG-AFC pump: from the filter screen to the throttle shaft to the no-air adjustment screw, to the air–fuel control valve, and then to the shutdown valve, as shown in Figure 9-25.
3. PTR pump: from the filter screen to the pressure regulator assembly.

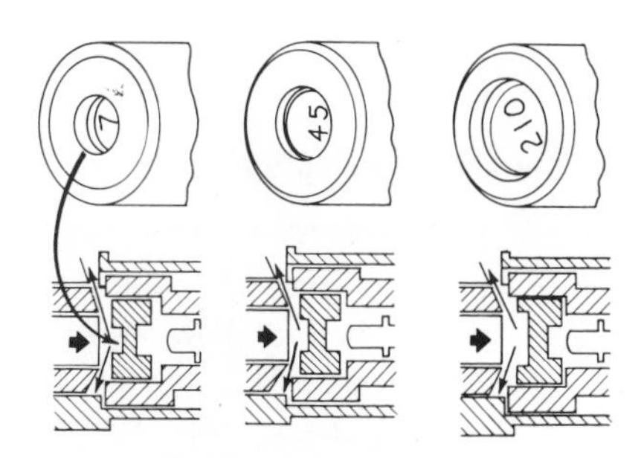

Figure 9-31
Different Sizes of Idle Spring Plungers Used to Control Amount of Fuel.
(Courtesy of Cummins Engine Co.)

In the PTG pump, to control the amount of fuel delivered to the injectors, a different size of idle spring plunger can be used, as shown in Figure 9-31. When an engine appears to be lacking power, be certain that the correct idle plunger counterbore size is being used, since either a larger or smaller counterbore may drastically change the pressure put out by the fuel pump and hence the power output of the engine.

The amount of fuel that will be delivered to the injectors at an idle speed through the idle fuel passage is controlled by the idle spring plunger resting against the governor plunger. This is in turn controlled by the idle spring and governor weight force; therefore, a plunger with a higher number allows the fuel pressure to act on a larger area, thereby allowing more fuel to escape into the pump housing.

Basic Governor Plunger Control

Without delving into the specific details of the governor assembly at this time, since what happens to the fuel flow is dependent upon the forces that change the amount of restriction to flow, we have to look at how these forces are created and consequently controlled. The governor weight assembly applies force to push the governor plunger back toward the idle plunger surface. The weight assembly is driven through gears from the engine via the engine's gear train and fuel pump mainshaft, as indicated in Figures 9-25, 9-26, and 9-29. The weights are supported and pivot on pins contained in the weight carrier assembly as shown in Figure 9-32.

Note that there are more than one set of weights available for these pumps; if the wrong weights are used, it will be impossible to properly calibrate the pump. A change of weights will entirely change the action of the governor plunger.

The weights are so positioned that when the engine is running, centrifugal force throw-

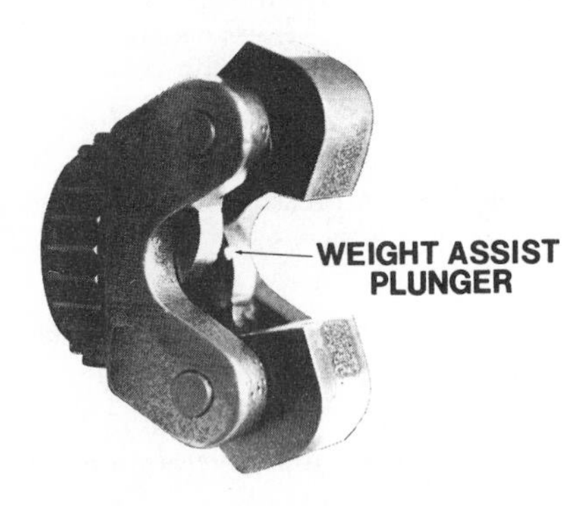

Figure 9-32
Governor Weight Assembly.
(Courtesy of Cummins Engine Co.)

ing them outward causes their feet to push against the front end of the governor plunger on each side of the drive tang, causing the plunger to spin in the barrel assembly.

Since governor weight force is proportional to engine speed, at an idle rpm plunger movement by weight force is small; therefore, a short weight assist plunger exerts pressure against the governor plunger at an idle speed to allow fuel pressure for startup and idle.

The tension of this weight assist plunger can be controlled by its spring, which can be shimmed for low-speed operation. During cold weather operation, it is advantageous to start the engine with a closed throttle in order to cut down on the amount of cold fuel being delivered to the cylinder. This is possible by the weight assist pressure holding the plunger against the idle plunger. Fuel pressure is adequate even in the reduced amount of fuel flowing to the engine through the idle port.

Figure 9-33 shows the position of the weight assist plunger. It can also be seen in Figures 9-25 and 9-26. The weight assist plunger must be installed with the smallest end toward the governor plunger.

Fuel flow through the PT-AFC (air fuel control) pump are shown in Figure 9-34(a) through (e) in the various positions of the engine stopped, starting and idling, normal driving, the beginning of high-speed governing, and complete high-speed governing.

Figure 9-35 shows the knurled thumbscrew which is located at the front of the injection pump fuel control solenoid. When a key switch for starting purposes is used, fuel flow to the injectors is controlled by energizing the electric solenoid. Stopping of the engine is accomplished by turning off the ignition key, which electrically deenergizes the power to the solenoid, allowing spring pressure to close off the internal fuel valve, thereby shutting down the engine.

Should the electric solenoid malfunction, the knurled thumbscrew can be turned in or out to allow starting and stopping of the engine by manual means.

Checking Pump Nameplate

To test a Cummins pump on a test stand or to order component parts, it is imperative that you quote the pump's serial number or know it in order that you may look up or obtain the proper test specifications. Figures 9-36 and 9-37 show the current and older methods used to identify these pumps.

Current Method of Stamping

From the nameplate you will notice that the first letter on the bottom line shows an L, indicating that the pump is in fact of left-hand rotation. However, some Vee series right-hand-rotation engines do use left-hand pumps. Right-hand rotation pumps are not stamped as are the left-hand assemblies.

If the fuel pump is fitted with a torque modification device, a second nameplate will show the regulator code number and required setting in thousandths of an inch. On Woodward governor-type pumps, the assembly number is for the fuel pump only, not the Woodward governor, since it has its own identification number.

Former Method of Stamping

1. The first item in the top line is the control parts list (0156) or designates the engine model (NHC-250).
2. The next five spaces will be the shop order number (0156-11521).
3. The following six or eight spaces give fuel pump serial number (0156-11521-846067 or 0156-11521-DP205361 possibly).
4. The L stamped on the bottom line signifies a left-hand pump.
5. The next seven spaces will be the fuel pump BM or AR numbers (0156-11521-846067 L-BM-70486).
6. The last six spaces on the second line give the calibration card number and suffix letter (0156-11521-846067 L-BM-70486-1289-C or 2868-A).

9-5

Fuel Pump Options

The basic PT fuel pump is available with special attachments for particular engine applications.

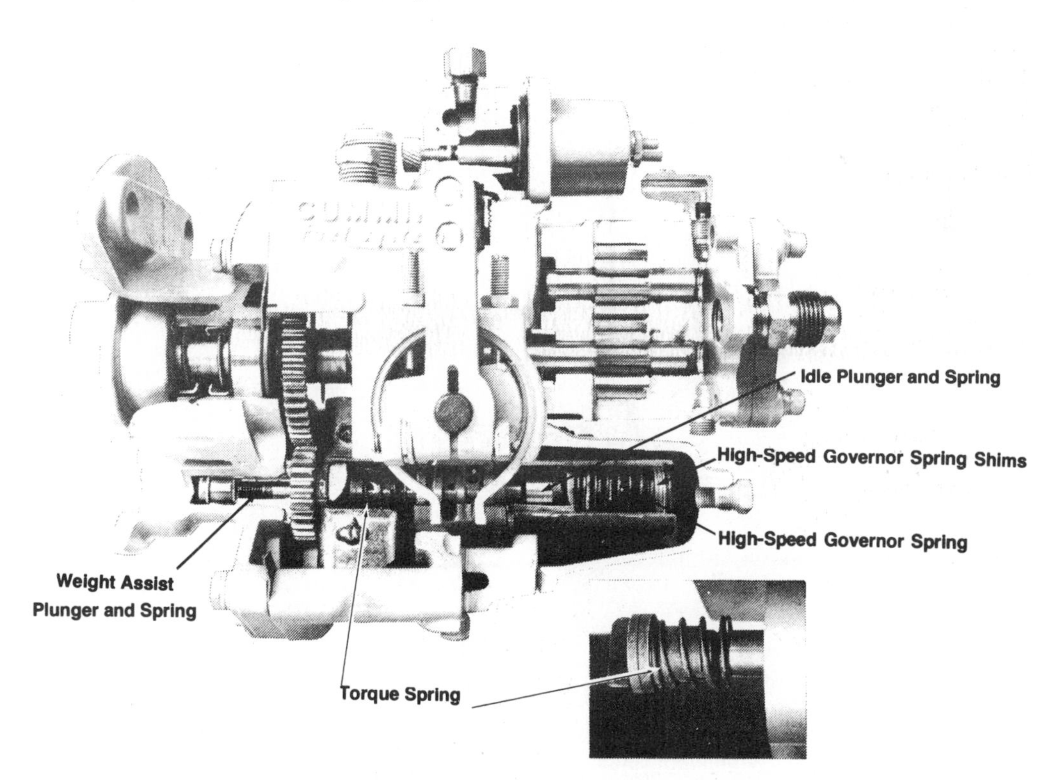

Figure 9-33
Position of Weight Assist Plunger.
(Courtesy of Cummins Engine Co.)

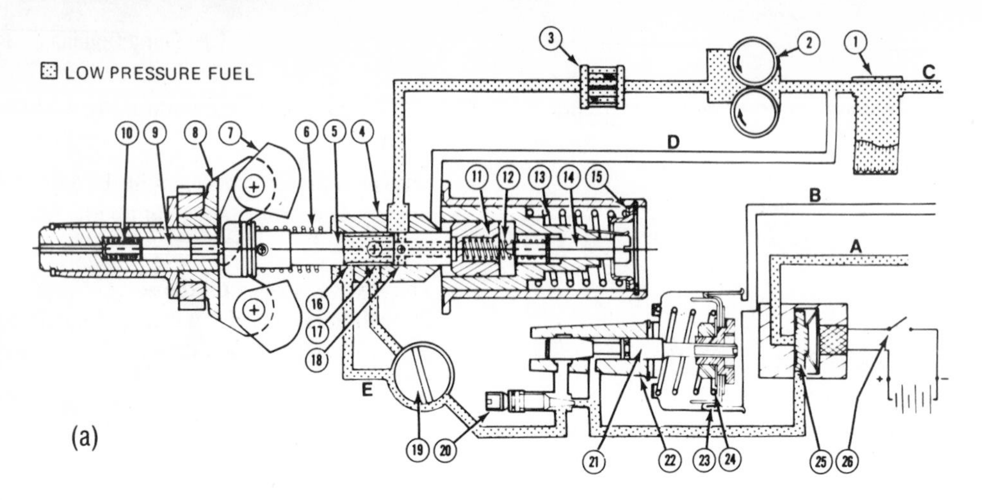

Index:

1—Primary fuel filter
2—Gear pump
3—Filter screen
4—Governor sleeve
5—Governor plunger
6—Torque control spring
7—Governor weights
8—Governor weight carrier
9—Weight assist plunger
10—Weight assist spring
11—Idle spring plunger
12—Idle speed spring
13—Maximum speed governor spring
14—Idle speed adjusting screw
15—Maximum speed governor shims
16—Idle speed governor port
17—Main governor port
18—Governor dump ports
19—Throttle
20—AFC needle valve
21—AFC control plunger
22—AFC barrel
23—Diaphragm (bellows)
24—AFC spring
25—Solenoid valve
26—Ignition switch
A—Fuel to injectors
B—Air from intake manifold
C—Fuel from tank
D—Bypassed fuel
E—Idle fuel passage

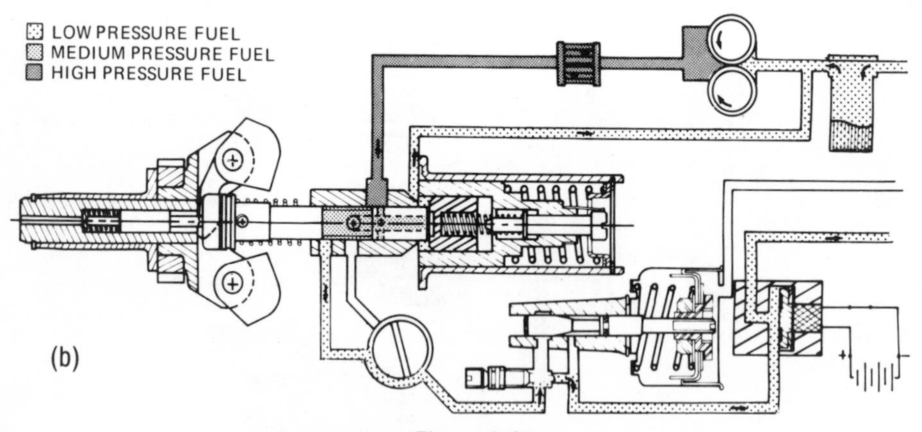

Figure 9-34
Fuel Flow Through a PT-G Fuel Pump: (a) Engine not running; (b) starting and idling; (c) normal driving; (d) beginning of high-speed governing; (e) complete high-speed governing. (Courtesy of Cummins Engine Co.)

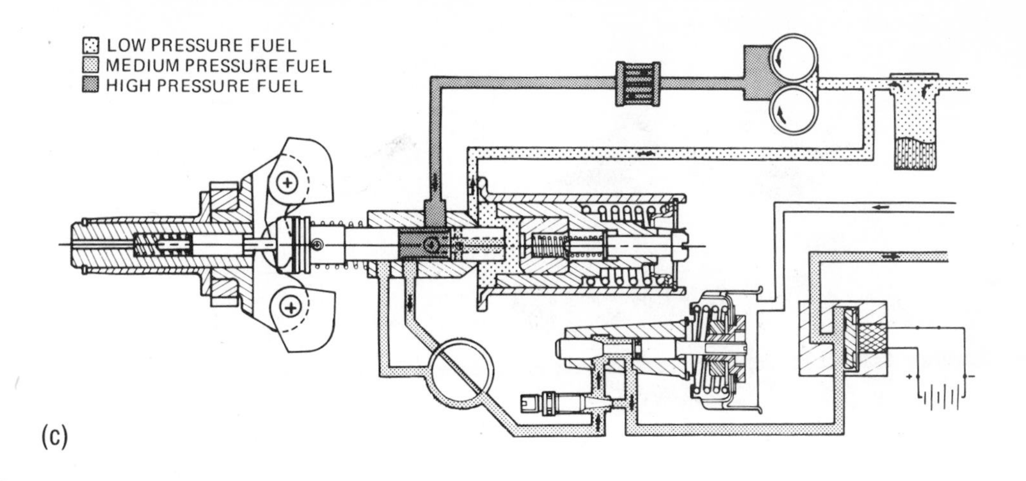
LOW PRESSURE FUEL
MEDIUM PRESSURE FUEL
HIGH PRESSURE FUEL

(c)

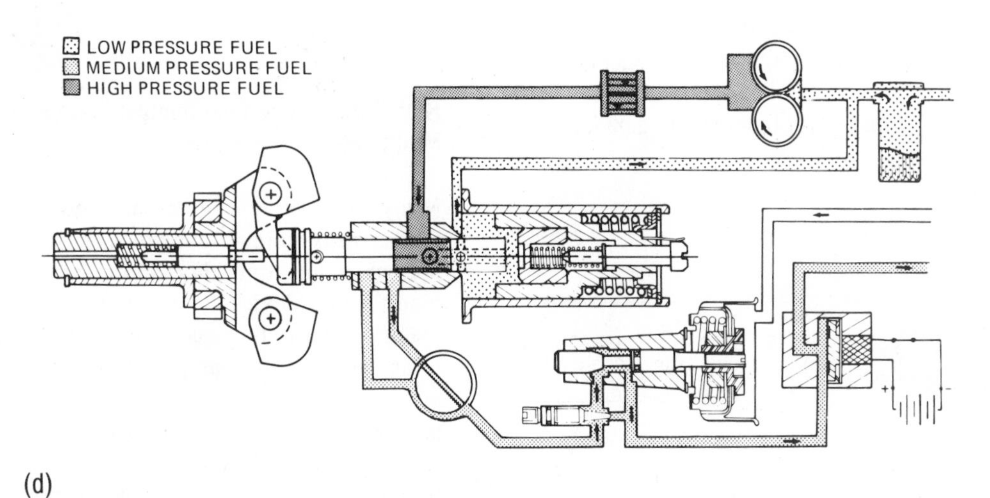
LOW PRESSURE FUEL
MEDIUM PRESSURE FUEL
HIGH PRESSURE FUEL

(d)

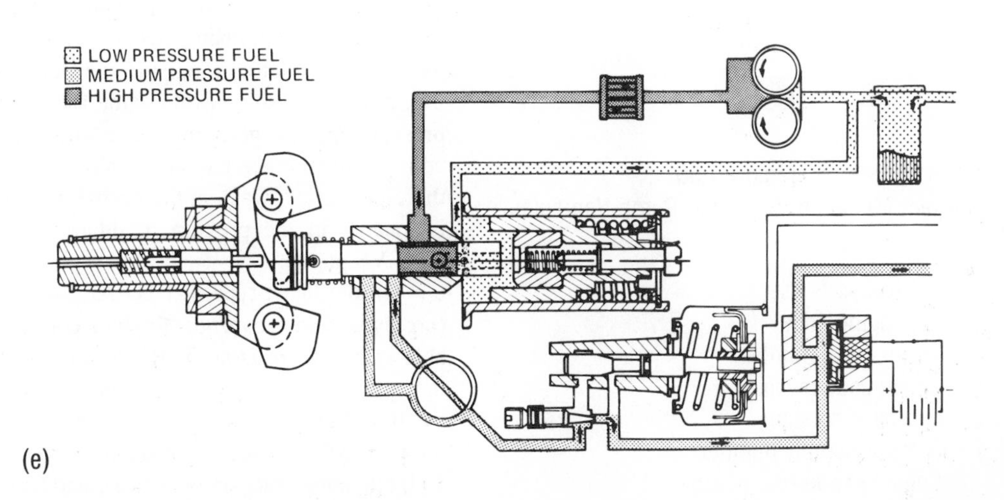
LOW PRESSURE FUEL
MEDIUM PRESSURE FUEL
HIGH PRESSURE FUEL

(e)

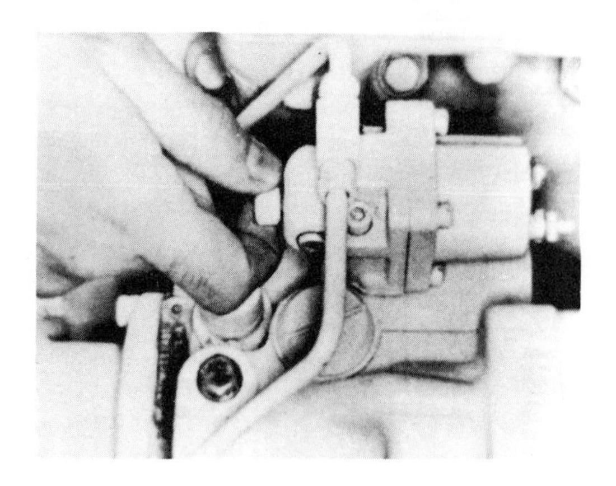

Figure 9-35
Location of Thumbscrew on Fuel Pump Housing.
(Courtesy of Cummins Engine Co.)

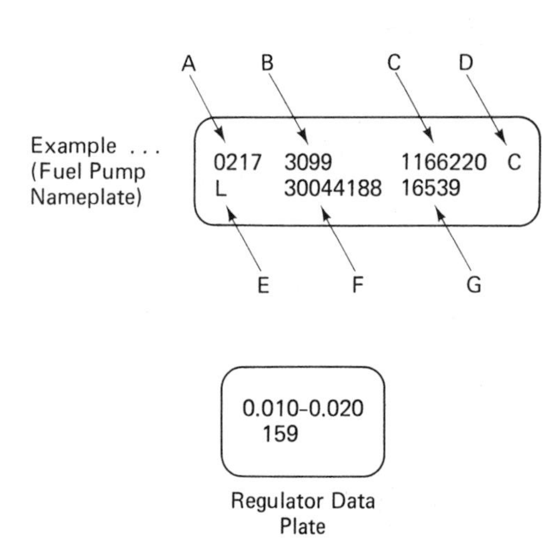

Figure 9-36
Current Method of Stamping Pump Nameplate.

Index:
A—Control parts list
B—Fuel pump code
C—Pump serial number
D—Latest code revision
E—Left-hand rotation
F—Pump assembly number
G—Engine shop order number

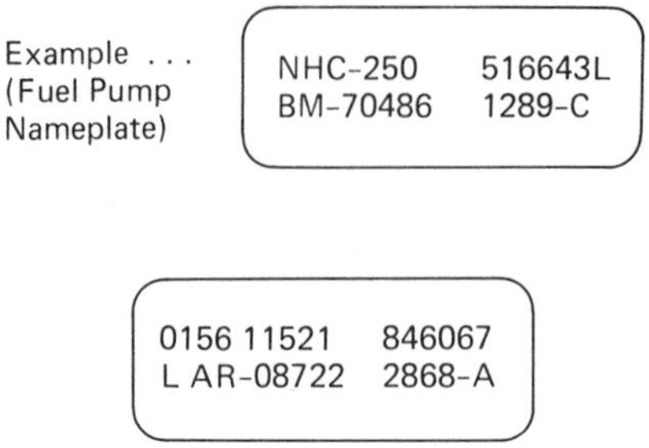

Example . . .
(Fuel Pump
Nameplate)

NHC-250 516643L
BM-70486 1289-C

0156 11521 846067
L AR-08722 2868-A

Figure 9-37
Former Method of Stamping Pump Nameplate.

High Torque Rise Fuel Pumps: Torque Modification Device (TMD)

Many automotive engines are equipped with higher peak torque pumps at a lower than normal engine rpm. For this reason a torque modification device, which is a pressure regulator similar to those used on the PTR pumps, is mounted on the pump between the housing and the shut-down valve. It is basically a bypass valve that controls the engine fuel manifold pressure by redirecting a portion of the fuel back into the top of the fuel pump housing when it exceeds the regulator spring pressure. On the PTR pump the regulator is located inside the pump housing, whereas on the PTG pump the regulator is outside the pump housing, allowing it to control engine fuel manifold pressure and not gear pump pressure. The fuel adjustment plunger has all its shims removed; therefore, it is always in the no-fuel position. It is located in the pressure regulator cavity of PTG-VS pumps.

On the latest type of PTG-AFC-VS pump, (pressure–time governor air–fuel control variable speed) there is no TMD (torque modification device) cavity since this area is taken up by the AFC assembly. However, a remote mount TMD device is available with a new TMD housing, fuel transfer tube, and fittings.

Torque Limiting Valve (TLV)

On engines used for automotive applications having gear ratios of 17 : 1 or greater in the transmission, the engine torque must be controlled when operating in this high ratio range gear only in order to protect both the transmission and drive train components from excessive torque. Therefore, a torque limiting *dual power* valve is required to control torque on these engine applications. The valve is mounted on the top of the fuel pump housing, and it is controlled by a pressure switch that is deenergized and restricts fuel pressure to the engine when the transmission is in *bull-low* or creeper gear only. The valve permits normal fuel pressure when the switch is energized in all other gear ranges. Selection of the low or creeper gear activates a mechanism to deenergize and restrict fuel pressure to the engine. The valve is located between the TMD and standard fuel shutoff valve.

Torque Limiting System (TLS)

A TLS can be used in the air line between the engine's intake manifold and the AFC or aneroid control found on any Cummins turbocharged engine. It is a nonadjustable two-way valve that protects the vehicle drive train from excessive torque when the transmission is in low gear. When low gear is selected, it activates an electric switch, closing the two-way solenoid valve, which prevents intake manifold air pressure from activating the AFC or aneroid bellows, thereby limiting the engine to the no-air fuel rate and horsepower. Once the operator shifts out of low gear the engine will operate as normal.

Air Signal Attenuator (ASA)

This unit is an auxiliary noise and smoke control nonadjustable valve used primarily for noise control, which results in a reduction in acceleration smoke on some engine models and applications. The valve is threaded into the AFC cover plate, and was only used on certain 1977 models to meet noise legislation requirements.

The valve when used basically produces a delay in the intake manifold pressure signal to the AFC mechanism in the pump body. The ASA valve looks something like an oil sending unit, and it has the air manifold to pump flex line feeding into the top of it. It has a restrictor and check valve located in a parallel circuit within the restrictor to relieve pressure in a reverse flow condition with a porous metal filter protecting the AFC unit from possible contamination.

If at any time an operator complains of very slow acceleration, you will find that it is probably due to plugging of the filter or restrictor, whereas failure of the check valve would allow the restrictor to be bypassed, which would result in faster acceleration after gear shifting. The valve can be quickly checked by blowing through each side of the valve. Air should be free to flow when blowing through the AFC cover plate end; however, the air flow should be restricted when blowing through the $\frac{1}{4}$-in. (6.35-mm) flex tube end; air should flow, but very slightly. Check the valve body itself for any signs of air leaks.

Air Actuated Road Speed Governor

On PTR-type fuel pumps an air operated road speed governor consisting of an air cylinder on the fuel pump and an air control valve on the transmission allows the air control valve to transmit air under pressure to the pump air cylinder in low and intermediate gear ranges. It also provides an air bleed passage when the transmission is in top gear.

The air cylinder actually allows air pressure to compress the governor spring, thereby affecting engine speed. In both the low and intermediate gears the engine obtains its normal full-load governed speed, whereas in top gear the governor spring tension is reduced by the exhausting of the air, allowing the engine to run at a reduced top rpm such as 1800 on a normally governed 2100 rpm engine.

Since the Cummins fuel system operates on the PT or pressure–time principle, fuel manifold pressure that supplies the injectors will obviously have a direct bearing on the engine's performance.

Several adjustments can be made to the PT fuel pump assembly while on the engine; however, others can only be done effectively on a fuel pump test stand.

9-6

PT Pump Adjustment on the Engine

Fuel Manifold Pressure

Prior to checking out the fuel manifold pressure, the governed speed of the engine should be checked and set as per Cummins specifications for the particular application. A battery operated digital tachometer used with a small strip of magnetic tape attached to the vibration damper or other suitable spot makes it relatively easy to obtain a very accurate reading of engine speed. If the engine is on a dynamometer, maximum full-load rated rpm can be easily determined. However, if the engine is in a piece of equipment, the maximum no-load engine speed should be used. A general rule of thumb on Cummins engines is to allow 10% above the rated speed as a maximum governed speed. For example, if the maximum rated full-load governed speed is given as 2100 rpm, then 10% of 2100 equals 210 rpm; therefore, the maximum no-load rpm should be 2310 (see Idle and Maximum Speed Adjustment).

Once the engine is subjected to its normal full-load condition, there may be some slight variation to its maximum full-load rpm due to such causes as fan load, engine temperature, air compressor pumping losses, generator and alternator charging rate, and other accessory loads such as a power steering pump, air conditioning compressor, and so on. These conditions are of small importance in the majority of applications.

Note: The injectors should be adjusted to the correct specifications before attempting to take a fuel manifold pressure reading.

Procedure

1. Check first that the restriction to fuel pump flow does not exceed a maximum of 8 in. of mercury. Connect a vacuum gauge ST-434 or equivalent into the gear pump as shown in Figure 9-38(a).
2. Run the engine until a stabilized water temperature of 180 to 190°F (82.2 to 87.7°C) is reached. With no load on the engine, you may be unable to obtain the normal full-load engine operating temperature; therefore, try to reach as high a temperature as possible closest to this.
3. Adjust the valves and injectors as per tune-up section.
4. Check the maximum governed engine rpm.
5. The term *snap readings* is used to express the condition of checking both the fuel flow restriction and fuel manifold pressure conditions. If either of these readings is being taken with the engine on a dynamometer, then full-load conditions can be simulated. However, if the engine is in a piece of equipment where it is not convenient or possible to apply full load to it, the snap-reading method must be used. It is not as accurate as a full-load check, but will give a reasonably good indication of the condition existing.
6. Remove the throttle linkage from the

(a)

(b)

Figure 9-38
Connecting Vacuum Gauge to Gear Pump. (Courtesy of Cummins Engine Co.)

pump throttle lever, which will allow hand operation of the throttle for snap-reading conditions.

7. For fuel manifold readings, connect gauge ST-435 or equivalent, as shown in Figure 9-38(b), by removing the $\frac{1}{8}$-in. (3.175-mm) pipe plug from the side of the fuel shutoff valve located on top of the pump and connecting the gauge line into the plug hole. If the engine is a turbocharged unit with an aneroid rather than an AFC pump, you must disconnect the line from the pump coming from the aneroid and plug off the pump hole, or simply plug off the aneroid inlet line from the inlet manifold during the test. *On AFC pumps see end of item 10.*
8. Start and run the engine for a few minutes at an idle speed. Carefully slacken off the gauge end of the fuel pressure line long enough to purge all air from the line.
9. Quickly snap the throttle to its wide-open position. You will notice that the gauge needle will rapidly advance to a high value after which it will quickly fall off. The point at which it falls off is where the fuel pump governor cuts in.
10. Carefully note the maximum gauge reading during the test. Repeat several times, average out the maximum value, and compare with Cummins published specifications for your particular engine model given in the shop manuals dealing with fuel pumps and injectors, respectively.

Note: Snap fuel pressure readings usually run 3 to 5 psi (20.68 to 34.47 kPa) higher than what would be obtained at a maximum full-load speed. On AFC pumps, apply 25 psi (172 kPa) regulated air pressure to the AFC control in order to hold the plunger in when "snap reading."

Fuel Manifold Pressure Adjustment

Fuel pressure is controlled in the PT type R fuel pump and in the PT type G high torque rise engines by a pressure regulator that is adjustable by the use of shims, which control spring pressure. This pressure regulator is readily accessible on PT-R pumps by removing the male fitting (hex head) from the side of the pump; it is found in the torque modification device on PT-G high torque rise engines.

Once the regulator assembly has been re-

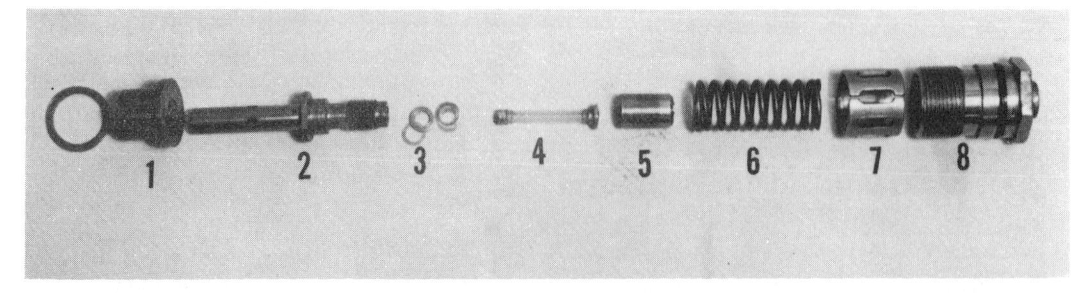

Figure 9-39
Pressure Regulator Valve Assembly.
(Courtesy of Cummins Engine Co.)

Index:
1—Valve Sleeve
2—By-Pass Valve Plunger
3—Fuel Adjusting Shim
4—Fuel Adjusting Plunger
5—Fuel Adjusting Plunger Cap
6—By-Pass Valve Spring
7—Valve Ring
8—Valve Cap

moved from the pump body, remove the fuel adjusting plunger from the bypass valve plunger by simply unscrewing the cap and removing the lock washer. See Figure 9-39. The fuel adjusting plunger is nylon with a brass head and foot that extends through the center of the bypass valve and therefore closes off a portion of the fuel adjusting bypass holes. This position is controlled by shim adjustments.

These shims are readily available as follows:

Part no. 70750: 0.010 in. (0.254 mm)

Part no. 70750-A: 0.005 in. (0.127 mm)

Part no. 70750-B: 0.002 in. (0.050 mm)

The addition of shims to the fuel adjusting plunger opens the fuel adjusting holes and thereby lowers engine fuel manifold pressure,

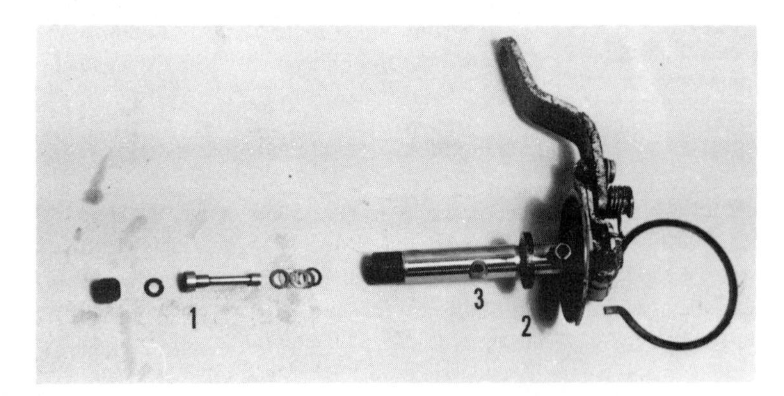

Index:
1—Throttle Leakage Shims
2—TS Seal Ring
3—TS-Hole

Figure 9-40
Throttle Shaft Assembly.
(Courtesy of Cummins Engine Co.)

whereas the removal of shims raises fuel manifold pressure.

Limited adjustment of fuel manifold pressure can also be made by adjusting the throttle shaft restriction. To do this, simply remove the large snap ring from around the throttle plate on the side of the PT pump housing and pull the throttle assembly straight out of the housing. Figure 9-40 shows the throttle shaft assembly with the restrictor shaft and shims removed.

The amount of throttle shaft restriction can be readily seen at any time by removing the throttle shaft assembly from the pump as just described. With the throttle restrictor plunger assembled, look through the hole in the throttle shaft and note the amount that the restrictor plunger closes off the hole. In most applications, especially on highway trucks, you will find that one-third to one-half is the usual amount. This of course should be determined with the fuel pump on a calibration test stand. In many cases, having set up the pump on the stand, slight changes to the throttle restrictor plunger position can cure the problem of the operator complaining of a throttle that is not crisp on initial movement.

To adjust the restrictor, simply remove the socket-head plug from the end of the throttle shaft and dump the plunger unit out of the hollow throttle shaft. Take careful note of the amount of shims on the plunger. Fuel pressure here can be raised by the addition of shims since the plunger will cover less of the hole in the hollow throttle shaft. Removal of shims will lower the fuel pressure. This is done strictly on a trial-and-error basis until the specified fuel pressure plus 5 PSI (34·47 kPa) is obtained.

While the throttle shaft is out of the pump housing, inspect the condition of the throttle shaft seal ring, since this is a possible source of an air suction leak (see Figure 9-40). If it appears to be free of any nicks, flat spots, or twists, lightly apply a very small amount of grease to the seal ring and push the throttle shaft back into place in the fuel pump housing. Run the engine or fuel pump test stand, as the case may be, in order to check the fuel manifold pressure reading. When satisfied as to its accuracy, make sure to install the throttle shaft snap ring back into place in the pump housing.

Idle and Maximum Speed Adjustment

Both of these speeds are controlled by spring pressure internally within the pump assembly by the governor action. To get at the high-speed spring assembly, remove the four bolts or socket head screws from the spring pack cover assembly at the bottom rear of the pump assembly. Figure 9-41 shows the spring pack cover assembly removed; the spring snap ring has been removed to allow withdrawal of the maximum governor speed spring and shims. Figure 9-42 shows an exploded view of the idle and high-speed spring assembly. The engine's maximum speed can be changed by adding or removing shims, which are available in the following thicknesses:

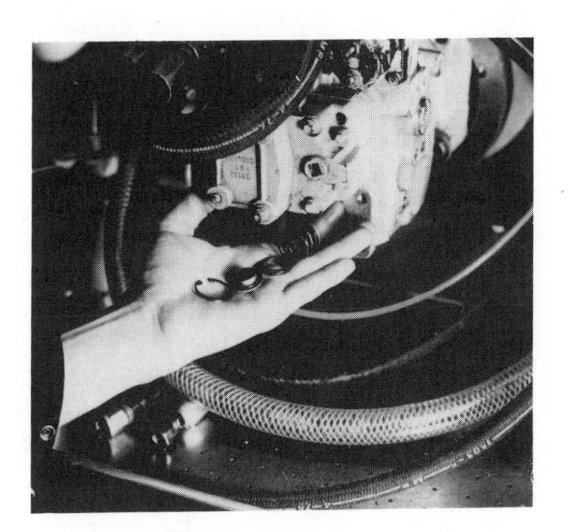

Figure 9-41
Removal of High-Speed Spring Assembly. (Courtesy of Cummins Engine Co.)

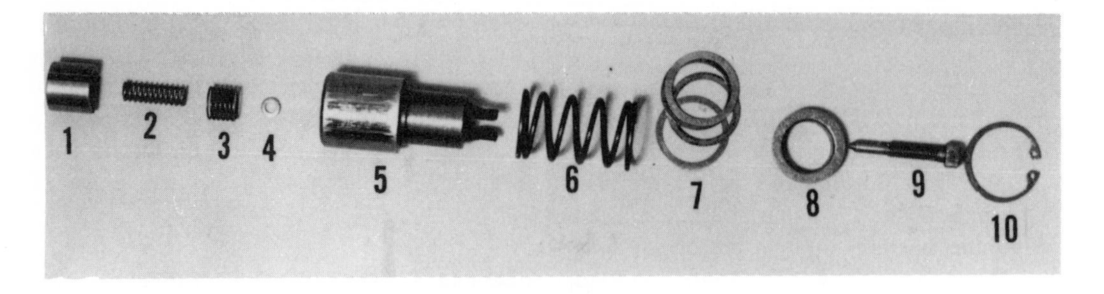

Figure 9-42
Spring Pack Cover Assembly.
(Courtesy of Cummins Engine Co.)

Index:

1—Idle Spring Plunger
2—Inner Idle Spring
3—Outer Idle Spring
4—Adjusting Screw Washer
5—Idle Spring Plunger Guide
6—Hi-Speed Spring
7—Shim
8—Hi-Speed Spring Retainer
9—Idle Adjusting Screw
10—Snap-Ring

Part no. 70717: 0.020 in. (0.508 mm)

Part no. 70717-A: 0.010 in. (0.254 mm)

Part no. 70717-B: 0.005 in. (0.127 mm)

Engine governed speed is changed 4 rpm for every 0.001 in. (0.0254 mm) of shim added or removed in the PT-R pump. Keep this in mind, because the PT-G is as follows: The addition of each 0.001 in. (0.0254 mm) will change the engine speed by approximately 2 rpm, and the removal will drop the speed the same amount on the PT-G pump.

Anytime that the maximum speed is adjusted, always bear in mind that the controlling factor as to maximum rpm is the recommended Cummins specification for the particular engine application. Therefore, do *not* under any circumstances adjust the top speed simply to please an operator.

Once you have adjusted the maximum governed speed and installed the snap ring, run the idle screw in with your fingers to the same basic position or slightly more, and install the gasket and cover.

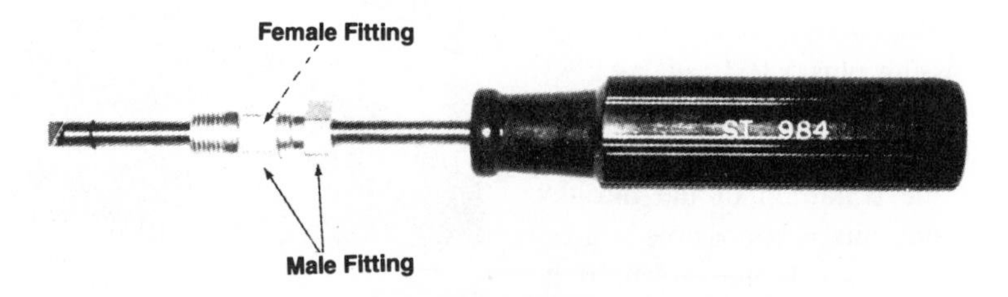

Figure 9-43
Installation of Fittings on Screwdriver ST-984.
(Courtesy of Cummins Engine Co.)

Idle Adjustment

Remove the small pipe plug at the rear of the spring pack cover. To adjust the idle speed correctly and to prevent any air being drawn into the pump, Cummins supplies a special screwdriver ST-984 for this purpose; however, one can readily be made up by selecting a small, straight-bladed screwdriver, a small O ring that fits the blade, and a male and female brass tube fitting to compress the O ring slightly and to act as a packing gland. Install the fittings as shown in Figure 9-43, install the screwdriver in through the housing until it engages the idle screw, and tighten the tube nut on the screwdriver blade snug with your fingers. Turning the screwdriver clockwise increases idle spring tension and therefore engine idle speed, and backing it out will lower the idle speed.

On AFC standard governor PT pumps, set the throttle shaft in the idle position (toward the gear pump) and hold it firmly against its stop. Check the pressure on the fuel manifold rail gauge. For example, if your pump code plate was 3055, pressure should be 33 psi (227.53 kPa) at 500 rpm. Screw the idle adjusting screw in or out to raise or lower the pressure, respectively.

Due to the variety of applications and environments that the engine will operate in, you may find it impossible to obtain the specified rail pressure by screwing the idle screw all the way in. If it bottoms out and the fuel pressure is still too low, it is because high weight assist settings are being used. Therefore, if you happen to run into this particular problem, it may be necessary to add one more spring seat washer on the spring end of the idle screw.

Throttle Travel

The throttle travel is generally set on a fuel pump test stand at the time of calibration pressure adjustment. The arc of travel of the throttle lever should be set with Cummins ST-3375355 or, if unavailable, a protractor will do. The throttle lever idle position centerline should be 27 to 29 degrees from a vertical position toward the gear pump. Figure 9-42 shows the arc travel.

With the test stand fuel flow, pressure, and pump suction restriction set as per specifications for your pump code, the throttle lever retaining capscrew and nut should be loosened to allow adjustment. When the arc of travel has

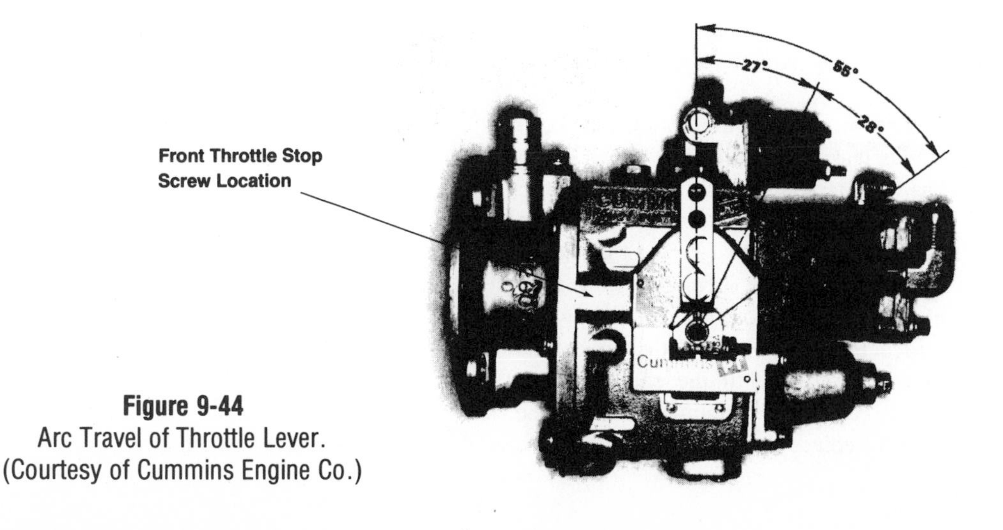

Figure 9-44
Arc Travel of Throttle Lever.
(Courtesy of Cummins Engine Co.)

been set, lock it. With the throttle lever held in a wide open position, it should be on a vertical centerline, resulting in a 28 degree arc from idle to wide-open throttle. Lock the front throttle stop screw, which you may have had to adjust with an Allen wrench to obtain the desired arc travel. Figure 9-44 shows the location of the front throttle stop screw. Engines employing hydraulic governors have their throttle lever arc travel adjusted to 55 degrees.

Note: Do not attempt this adjustment unless the pump is on a test stand.

Throttle Stop Screws

Figure 9-32 shows two throttle stop screws immediately above the throttle shaft. These screws are adjusted at the time of fuel pump calibration while on the test stand and should not require readjustment while on the vehicle. The front screw controls throttle idle position or throttle leakage; the rear screw controls maximum fuel passage opening. The exception to this is on the PT-G AFC fuel pumps, where the front screw is for throttle travel adjustments, and the rear screw is for throttle leakage adjustments.

In either case do *not* use these screws to adjust the engine idle speed since fuel flowing through the throttle at an idle speed is not under governor action control; therefore, too much fuel flowing would cause slow engine deceleration.

If, however, the throttle screw that controls throttle leakage at idle (see above statement) has been tampered with while the pump is on the engine, start and run the engine at an idle and turn in the screw very gently until there is a slight increase in engine rpm; then back out the screw two complete turns.

Remove the pipe plug on the cover and adjust the idle screw until the engine attains 490 rpm. (Use an accurate digital-type tachometer.) Replace the pipe plug and run the engine for several minutes until any air that entered the system during the idle screw adjustment is purged from the system. Very gently turn the throttle screw in until the engine's idle speed has increased to 515 rpm.

Accelerate the engine; then release the throttle and check the deceleration rate of the engine. If it is too slow, throttle leakage is excessive, and if the engine shudders or stops as the idle speed is reached, throttle leakage is insufficient; therefore, adjust the screw until the engine will decelerate and idle at an acceptable level.

Throttle Linkage Length

In many instances of a lack-of-power complaint, service personnel often overlook checking of the throttle linkage length. This can be affected by cab movement, the twists and bending of the linkage as the cab is tilted back and forward during servicing. On conventional-type cabs on highway vehicles, it is easier to check this out than on a cab-over type unit; however, with the foot pedal of the throttle hard on the cab floorboards and the throttle linkage disconnected at the fuel pump, you should be able to connect the linkage with the hole in the pump throttle bracket (also being held in full fuel) without having to back up on the throttle bracket. If you cannot, adjust the linkage as needed, but do *not* loosen the throttle lever clamp on the throttle shaft bracket.

9-7 Troubleshooting Chart

Figure 9-45 provides a troubleshooting chart for the PT (type G) fuel pump.

N, NH, NVH & VT series engines with 1/2"-13 studs.	6015-50 Adaptor
N, NH, NVH & VT series engines with 1/2"-13 tapped holes.	6015-51 Adaptor
J series engines with 3/8"-24 studs.	6015-52 Adaptor
J & C series engines with 3/8" -24 tapped holes.	6015-53 Adaptor
NH250 & NT380 series engines with 5/16"-18 tapped holes.	6015-54 Adaptor
H, NH, NVH & VT series engines with 1/2"-20 tapped holes.	No Adaptor Required

Figure 9-47
Chart for Selecting Correct Timing Adaptor.

hold-down screw threads to be screwed into the rocker housing.

5. Slide both dial indicators to the upper end of their support brackets, which will prevent any possible indicator damage when you install the timing fixture onto the cylinder head.
6. Carefully install the timing fixture over the cylinder to be checked. (The injector has of course been removed from the cylinder previously.) Position the timing tool so that the extension rod attached to the main fixture dial indicator passes through the injector tube hole on into the cylinder.

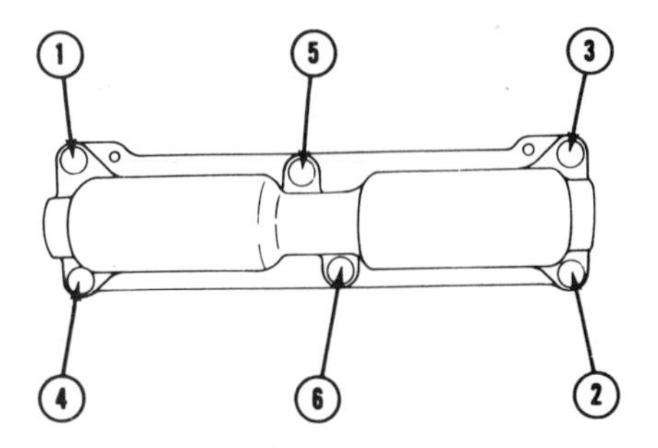

Figure 9-48
Cam Follower Torquing Sequence.
(Courtesy of Cummins Engine Co.)

7. Screw the hold-down screws into the tapped hole or studs of the cylinder head to secure it in place. (Make sure that the timing tool is straight.)
8. Loosen the capscrew E in Figure 9-46; rotate the swivel bracket so that the plunger assembly of the other dial indicator can be located into the injector push tube socket and tighten the capscrew E.
9. The timing fixture is now ready to be used.

Note: To prevent a possible false reading of the dial indicator setup, it is advisable to check the cam follower housing capscrews on the side of the engine in the sequence shown in Figure 9-48 to ensure that they are tightened to 30 to 35 lb-ft (4.1 to 4.8 kg·m).

1. With the timing fixture in position, rotate the engine crankshaft until the piston is at top dead center (TDC). Always turn the engine in its normal direction of rotation from the front, and remember that if you are using an adaptor at the rear of the engine to turn the engine over you will turn the engine in the opposite direction (only when viewed from the rear). See Figure 9-49, position (a).
2. Carefully position the timing indicator, which is in contact with the piston crown, to compress the dial indicator stem to within 0.010 in. (0.254 mm) of the inner travel stop. Gently bar the crankshaft to place the piston at exactly TDC and carefully zero in the dial indicator [Figure 9-49(a)].
3. Since all Cummins engines are of the 4-cycle type, during this check the piston must be on its compression stroke. If it is, there should be clearance at both the inlet and exhaust valve rocker arm to crosshead. If there is not, rotate the engine one

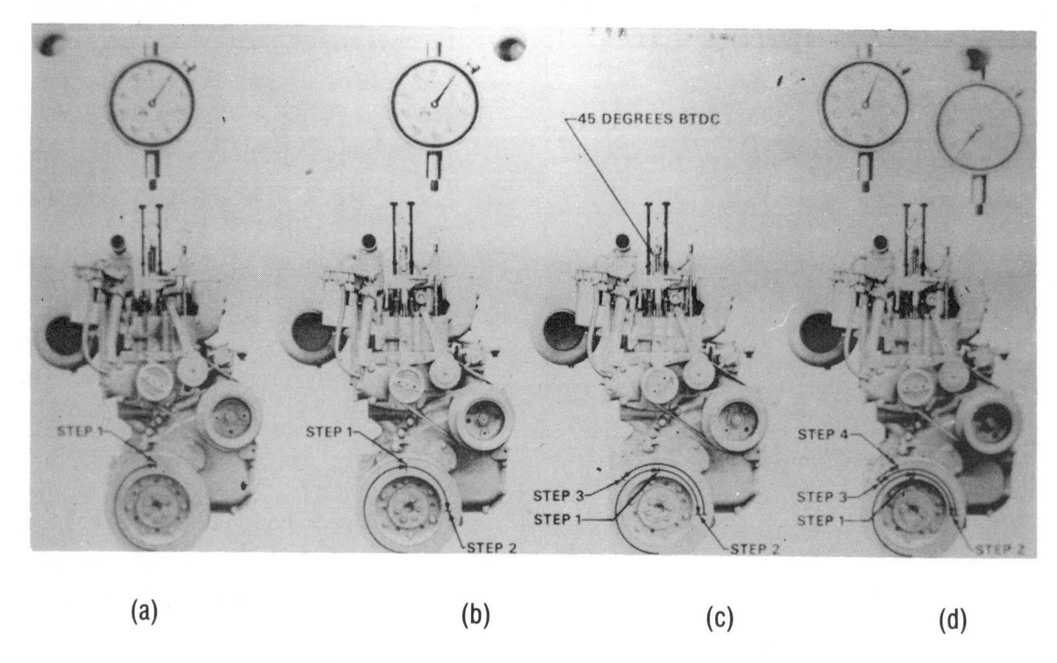

(a) (b) (c) (d)

Figure 9-49
Engine Crankshaft to Piston Position to Check Engine Timing: (a) Step 1—Top Dead Center (TDC); (b) Step 2—90° After Top Dead Center (ATDC); (c) Step 3—45° Before Top Dead Center (BTDC); (d) Step 4—19° Before Top Dead Center (BTDC) (Courtesy of Cummins Engine Co.)

full turn and again check that the dial indicator is still in a zeroed position.

4. Turn the engine over with a barring mechanism until it is at 90 degrees after TDC. Carefully position the push tube indicator rod to 0.020 in. (0.508 mm) from its inner travel stop, and secure this dial gauge in position at zero on the dial with the indicator rod secure in place [Figure 9-49(b)].
5. Slowly turn the crankshaft opposite its normal direction of rotation to effectively place the piston follower indicator rod at 45 degrees BTDC [Figure 9-49(c)].
6. Again rotate the crankshaft in its normal direction of rotation and stop 30 degrees short of TDC. Very gently turn the crankshaft until the piston indicator dial gauge reads exactly 0.0032 in. (0.0812 mm) before 0. In reality, though, this reading is 0.2032 in. (5.161 mm) before 0, because the dial indicator needle will have gone around twice as you turned the engine to 19 degrees BTDC as in Figure 9-49 (d).
7. Read the push tube dial indicator and compare its reading with that given in Table 9-2. A range of within 0.004 in. (0.101 mm) is considered to be within allowable specification.

The timing code for your engine is found on the engine dataplate located on the side of

Table 9-2
TABLE OF INJECTION TIMING

Timing Code	*Piston Travel (Inches)*	*Push Rod Travel (Inches)*		
		Nominal	*Fast*	*Slow*
A	−0.2032	−0.0415	−0.0395	−0.0435
B	−0.2032	−0.0295	−0.0275	−0.0315
C	−0.2032	−0.0335	−0.0315	−0.0355
D	−0.2032	−0.036	−0.034	−0.038
E	−0.2032	−0.029	−0.028	−0.030
Y	−0.2032	−0.039	−0.037	−0.041
Z	−0.2032	−0.026	−0.024	−0.028
AA	−0.2032	−0.031	−0.030	−0.032
AC	−0.2032	−0.028	−0.027	−0.029
AF	−0.2032	−0.045	−0.044	−0.046
AH	−0.2032	−0.035	−0.034	−0.036
AI	−0.2032	−0.034	−0.033	−0.035
AK	−0.2032	−0.041	−0.040	−0.042
AN	−0.2032	−0.046	−0.045	−0.047
AQ	−0.2032	−0.042	−0.041	−0.043
AS	−0.2032	−0.036	−0.035	−0.037
AT	−0.2032	−0.030	−0.029	−0.031
AU	−0.2032	−0.049	−0.048	−0.050
AV	−0.2032	−0.050	−0.049	−0.051
AW	−0.2032	−0.060	−0.059.	−0.061
AX	−0.2032	−0.055	−0.054	−0.056
AY	−0.2032	−0.040	−0.039	−0.041
AZ	−0.2032	−0.059	−0.058	−0.060
BA	−0.2032	−0.028	−0.027	−0.029
BB	−0.2032	−0.100	−0.099	−0.101
BC	−0.2032	−0.024	−0.023	−0.025
BD	−0.2032	−0.095	−0.094	−0.096
BH	−0.2032	−0.052	−0.051	−0.053
BI	−0.2032	−0.105	−0.104	−0.106

(Courtesy of Cummins Engine Co.)

the engine front cover: right-hand side viewed from the front.

Note: Engine timing for your engine is shown by both a code and specification stamped on the engine serial data plate located on the gear case mounting flange at the front of the engine.

8. If the readings you obtain are greater than the limits specified in the Cummins specifications, then in actuality the timing is slow. If the push tube travel is less, the timing is fast.
9. To correct either of these conditions, it is simply a case of adding or removing gaskets from the cam follower assembly to engine block (Figure 9-46). What this does is to affect the lift of the push tube, which is basically a pivoted type of roller follower. With this in mind then, the following will hold true:
   a. Removing gaskets on right-hand engines will retard the timing, and adding them will advance the timing.
   b. Adding gaskets on left-hand rotating engines will retard the timing, and removing them will advance the timing.
10. The three basic gaskets used for the NH, NT, and NTA series of engines will change the timing as follows:

	Amount of Change
Thin gasket	0.0015 in. (0.038 mm)
Medium gasket	0.003 in. (0.076 mm)
Thick gasket	0.006 in. (0.152 mm)

Although the Cummins specifications allow within 0.004 in. (0.101 mm) of the published reading as being acceptable, it is incumbent upon the service mechanic or technician to try to bring the specification to within 0.001 to 0.002 in. (0.0254 to 0.050 mm) of the published figure.

An engine timing check should include all cylinders of the engine done in the proper firing order sequence in order to reduce the number of times that you have to turn the engine over. However, since the in-line 6-cylinder 855 and 927 engines have a cylinder head covering two cylinders, only three checks are required. You may also notice that some Cummins service manuals show readings of 0.081 in. (2.057 mm) and 0.014 in. (0.355 mm) BTDC. This reading is to allow you to check camshaft lobe wear.

Timing Check: KT(A) 1150 In-Line Six

Timing of this engine and the KT(A) 2300 series V12 are similar to that of the 855 and 927 engines. The only difference is that along with

Figure 9-50
Support Block Attached to Rocker Housing K Engine.
(Courtesy of Cummins Engine Co.)

Table 9-3

Timing Code	Engine	Piston Travel Inches	Piston Travel Metric	Push Rod Travel (Inches)
AE Non Aftercooled	KT 1150	−0.2032	5.161 mm	0.108 ± 0.002 (2.743 ± 0.050 mm)
AM Aftercooled	KT 1150	−0.2032	5.161 mm	0.118 ± 0.002 (2.997 ± 0.050 mm)
KT-2300-900	2300	−0.2032	5.161 mm	0.126 ± 0.002 (3.200 ± 0.050 mm)
KTA-2300-1200	2300	−0.2032	5.161 mm	0.126 ± 0.002
KTA-2300-1050	2300	−0.2032	5.161 mm	0.126 ± 0.002

Cummins timing fixture ST-593, support block ST-593-40 is required. This support block is attached to the rocker housing as shown in Figure 9-50.

As with the timing check on the 855 and 927 series engines, the injection timing for your particular engine will be found stamped on the engine dataplate attached to the engine. Examples relating to KT(A) engines are shown in Table 9-3.

If the engine timing is not as specified, the camshaft gear key must be changed by heating the gear to between 300 and 400°F (148.88 to 204.44°C) with a heating torch and pressing the gear from the camshaft. Figure 9-51 shows the type of keys used. Key A will advance the engine timing. Key B is a straight key, and Key C will retard engine timing. The camshaft key part number is stamped on the key for identification purposes. This same type of setup is used to alter the timing on other series of Cummins Vee-type engines.

A B C

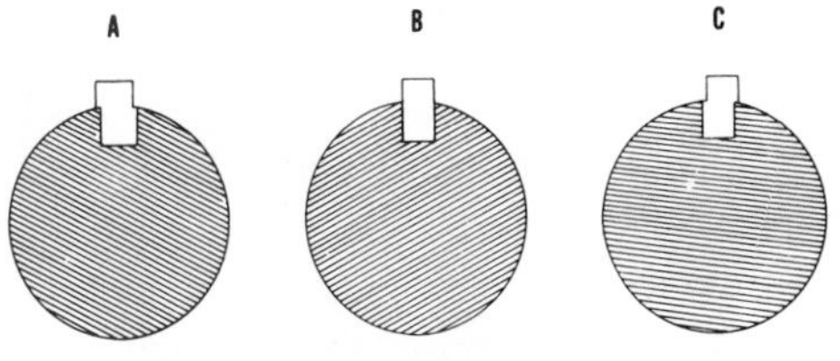

Figure 9-51
Camshaft Gear Keys K Engine.
(Courtesy of Cummins Engine Co.)

Information relating to both the camshaft key part number, the amount of offset of the key, and its relative timing effect to the engine can be found under Section 14 of the Cummins Shop Manual for your particular engine. The amount of available key offset will advance or retard timing by $\frac{1}{2}$ to 3 degrees.

Injector Plunger and Valve Adjustments

Once the engine timing is known to be correct, proper adjustment of the valves and injectors is an essential part of a tune-up on Cummins diesel engines. As with all high-speed, high-output diesel engines, both a cold set and hot set are often required, especially after a head removal or a major engine rebuild. To avoid confusion we will define these terms.

A *cold set* is done after the conditions just mentioned, or if the engine has been in operation; you must allow 4 hours minimum in ambient temperature to achieve stabilization of the engine temperature.

A *hot set* can be done after the engine has been in operation for a period of at least 10 minutes (preferably longer) with a water temperature of 180 to 190°F (82.22 to 87.77°C). Operation within this temperature range for such a period will bring the engine sump oil temperature to between 210 to 225°F (98.88 to

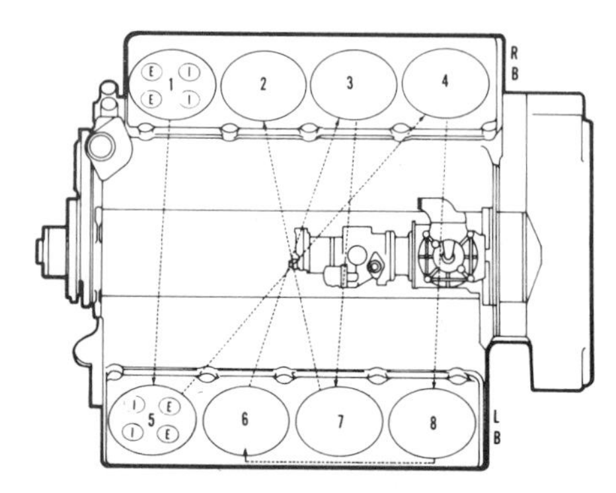

Figure 9-52
Vee Engine Cylinder Number Identification. (Courtesy of Cummins Engine Co.)

107.22°C) on the average. There is usually a temperature variation of 30 and 35°F (16.6 and 19.4°C) between the water and oil temperatures on most high-speed diesel engines.

Injector Torque

As a precaution, take a few minutes to check out the torque of the injector hold-down capscrews as per the shop manual. As with any injector, insufficient torque can cause combustion gas blowby, and overtightening can in some engines lead to cylinder head cracking. See Table 9-4.

Engine Cylinder Numbering Sequence

Since it is always easier to do both the injector and valve adjustments in the engine firing sequence, following is a brief description of the system employed by Cummins in identifying cylinder number and left or right bank on a Vee series engine.

Regardless of whether or not the engine is an in-line or Vee type, engine rotation is always established from the front. Cylinder numbers on in-line engines also start at the front as number 1, with number 6 being closest to the flywheel.

On Vee-type engines, the left and right banks of the engine are determined from the flywheel end; therefore, as you look forward you have the view shown in Figure 9-52.

Both the V6 and V8 engines use this method; however, on the V1486 and V1710 engines, which are V12 configuration units, the left and right banks are determined by standing at the front of the engine. In addition, the left and right bank cylinders are numbered 1 to 6 from the front of the engine. On the KT(A) 2300 V12 engine the left and right banks are determined in the same fashion as for the V6 and V8 engines, with the cylinder numbering sequence being 1 to 6 from the front on each bank.

Engine Firing Orders

Six-cylinder engines: Right-hand rotation, 1–5–3–6–2–4; left-hand rotation, 1–4–2–6–3–5

V8 engines: Right-hand rotation, 1–5–4–8–6–3–7–2

V6 engines: Right-hand rotation, 1–4–2–5–3–6

V-1486, V-1710: Right-hand rotation, 1L–6R–2L–5R–4L–3R–6L–1R–5L–2R–3L–4R; left-hand rotation, 1L–4R–3L–2R–5L–1R–6L–3R–4L–5R–2L–6R

KV-12, 2300 engines: 1R–6L–5R–2L–3R–4L–6R–1L–2R–5L–4R–3L

Injector and Valve Adjustment

All Cummins engines employ the same basic fuel system and therefore the same injector and valve operating mechanism setup. As mentioned earlier, they differ only in the method of altering the engine timing. The particular injec-

Table 9-4
INJECTOR CAPSCREW TORQUE

Engine Series C.I.D. (Liters)	*Type Injector*	*Type Capscrew Standard*	*Nylock*
C-464 (7.6)	Flange	10 to 12 lb-ft (14 to 16 N·m)	12 to 14 lb-ft (16 to 19 N·m)
NH-743 (12.18)	Flange	Same	Same
N-855/927) (14.02 to 15.2)	Cylindrical	Same	7 to 8 lb-ft (9 to 11 N·m)
V-1486 (24.37)	Flange	Same	12 to 14 lb-ft (16 to 19 N·m)
V-1710 (28.04)	Cylindrical		7 to 8 lb-ft (9 to 11 N·m)
V-378 (6.19)	Cylindrical	30 to 35 lb-ft (41 to 47 N·m)	
V-504 (8.26)	Same	Same	
V-555 (9.1)	Same	Same	
V-903 (14.8)	Same	Same	
KT(A) 1150 (18.86)	Same		11 to 13 lb-ft (15 to 18 N·m)
KT(A) 2300 (37.72)	Same		Same

If the injector is a PT type D or is stamped A or B after class mark, torque the Nylock capscrews to 11 to 12 lb-ft (15 to 16 N·m) in 4 lb-ft (5 N·m) increments.

tor used may also differ slightly, as was discussed under injectors. Two main methods are employed to adjust both the injectors and valves on Cummins engines:

1. The preferred method of *uniform plunger travel,* which involves adjusting the injector plunger with an injector indicator kit consisting of a dial indicator and two types of rocker arm actuators to a specific amount of travel.
2. The *torque method,* which involves the setting of the plunger adjusting screw to a specified torque reading.

In early 1972, as a result of product improvement and design changes to some of their engines, Cummins introduced the *dial indicator method* of injector adjustment. This method can be used on all engines except the V-378, V-504, V-555, V-1486, V-1710, and C-464 engines. Either the dial indicator method or torque method can be used on the NH-743, N-855, and N-927 cubic inch displacement engines. The deciding factor is that if the injector plunger free travel exceeds 0.206 in. (5.23 mm), then the torque method of adjustment must be used unless component changes such as rocker levers and/or cam followers are made that will allow 0.206 in. (5.23 mm) maximum limit of free travel to be obtained. Both the KT(A) 1150 and KT(A) 2300 series engines use the dial indicator method of adjustment.

Checking Injector Plunger Free Travel

If you are in doubt as to whether or not your engine can be set up with the dial indicator for injector adjustment, check the engine status. Those engines conforming to the U.S. Federal Clean Air Act will be adjusted by the dial indicator method. Earlier engines used the torque method.

1. Proceed as follows to check injector plunger free travel: Back off the injector adjusting screw about one and a half turns and tighten the locknut. This relieves any possibility of excessive injector actuating stresses or loading during the check.

Note: The dial indicator method of injector adjustment must *not* be used on NH-743, N-855, or N-927 series engines with the following camshafts: 1599-1, 10859-1, 10875-1, 70515, 104336, 108358, 108965, 111882, and 111884.

2. With Cummins ST-1170 dial indicator or equivalent, mount the dial indicator so that its plunger contacts the top of the injector plunger with a small amount of preload on the dial, zero in the dial gauge, and lock it in position.
3. Arrange to bar the engine over in its normal direction of rotation and record the total amount of injector plunger travel. It should not exceed 0.206 in. (5.23 mm) on any one cylinder of the engine.

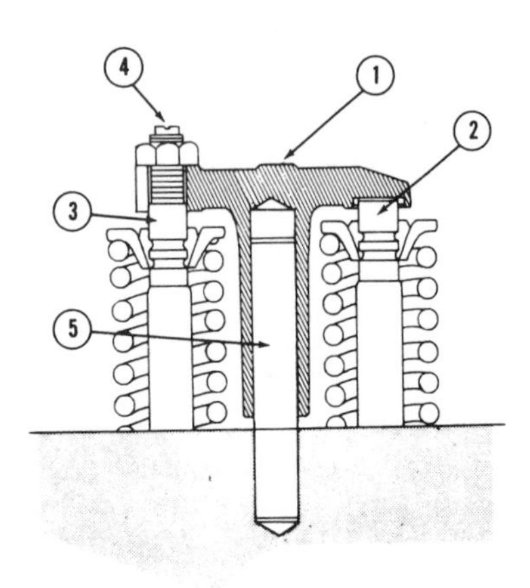

Figure 9-53
Typical Valve Crosshead.
(Courtesy of Cummins Engine Co.)

Index:
1—Crosshead
2—Valve stem
3—Valve stem
4—Adjusting screw
5—Crosshead guide

Valve Crosshead Adjustment

The adjustment for the valve crosshead or bridge mechanism is the same for all models of Cummins engines. Figure 9-53 shows a typical valve crosshead. Crosshead adjustment is not always necessary and is generally only done if the adjustment is off or during a cylinder head removal. Proceed as follows:

1. After loosening the locknut, back out the slotted adjusting screw one full turn.
2. Make sure that no bind exists between the crosshead and its guide. Apply light finger pressure on the pallet or rocker lever contact surface (item 1), and lightly run the screw down until it just touches the valve stem (item 3).
3. On a new crosshead and guide unit, to straighten the stem on its guide and to compensate for slack in the threads, turn the adjusting screw an additional one-third of a hex, which is equivalent to 20 degrees.

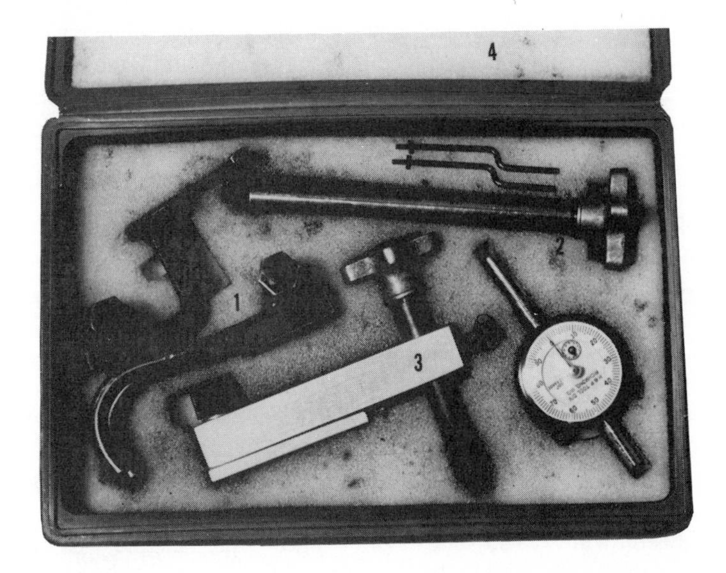

Index:
1—Barring tool in-line engines (ST-1193)
2—Barring tool V-903 (ST-1251)
3—Indicator and stand (ST-1170)
4—Instruction sheet 983004

Figure 9-54
ST-1270 Injector Adjustment Kit.
(Courtesy of Cummins Engine Co.)

4. If the crosshead and guide have been in use for some time, you may have to advance the adjusting screw as much as 30 degrees after it contacts the valve stem to straighten it on the guide.
5. If available, use torque wrench adapter ST-669 to tighten the locknut to 22 to 26 lb-ft (29.82 to 35.25 N·m); however, if this tool is not available, hold the screw with a screwdriver and tighten the locknut to 25 to 30 lb-ft (33.89 to 40.67 N·m).
6. Carefully check the clearance between the crosshead body and valve springs to ensure that at least 0.025 in. (0.64 mm) of clearance exists.

Caution: Anytime that you are adjusting valves or crossheads with an assembled setup, make sure that on those engines equipped with a compression release it is in the running position.

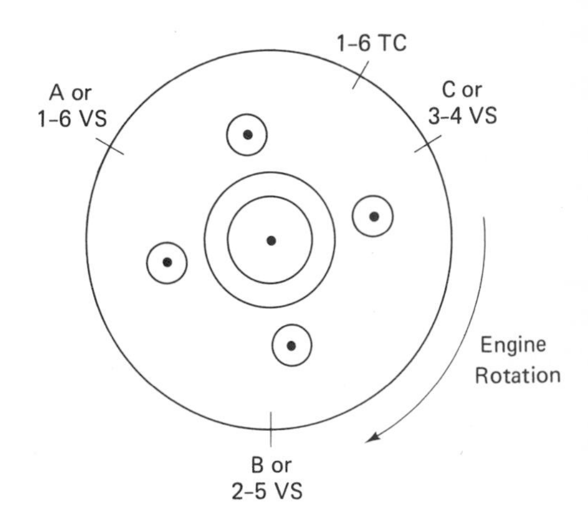

Figure 9-55
Engine Timing Marks for Setting Valves and Injectors.

Valve Set Mark Alignment

Cummins engines employ timing marks both on the engine flywheel and also on either the engine vibration damper, crankshaft pulley, or accessory drive pulley, which may be located at the front or rear of the engine depending on the particular series and model of engine. Timing marks will be found in the positions indicated in Table 9-5.

Remember, anytime you align one set of timing marks from the damper, pulley, or flywheel, two cylinders will show as being in a VS (valve set) position or TC (top dead center) position. Only *one* of these can be adjusted at this point, because the engine being of the four-cycle type, only one cylinder's piston will be on the correct stroke to allow valve or injector adjustment. The other will either be in a valve overlap condition (TC mark) or with the intake valve open on the (VS mark).

For example, if the 1–6 VS mark is aligned with the stationary pointer, check both the inlet and exhaust valve rocker arms for clearance to the crosshead. The cylinder with this clearance at the valves would be the one to be set. To adjust the other cylinder, the crankshaft would have to be turned one more complete revolution.

Dial Indicator Method of Injector and Valve Adjustment on NH-743, N-855, N-927 Cubic Inch Displacement Engines

The tools included in Figure 9-54 are required to set the injectors on these engines.
Assume that the engine is a right-hand rotating engine; therefore, it will have a firing order of 1–5–3–6–2–4. Since the engine is a 6-cylinder 4-stroke-cycle unit, it will produce one power stroke per cylinder for every two crankshaft revolutions. Therefore, 2 times 360 degrees equals 720 degrees divided by 6 cylinders, which means that each cylinder is 120 degrees apart in relative position or firing impulse.

Table 9-5

Engine Series	*Timing Mark Location*
C-464, NH-743, N-855, 927	Found on the accessory drive pulley at the front of the engine. Pulley markings align with scribe mark on the gear case cover.
V-378, V-504, V-555, V-903	Found on either the crankshaft vibration damper or pulley with the marks having to be aligned with a pointer on the engine front cover. On some earlier engines where the torque method of injector adjustment is used, the timing mark may be found 135 degrees from the existing mark location. If accessory drive pulley marks at the rear of the engine are used, these pulley marks are aligned with a cast arrow on the accessory drive support.
V-1486, V-1710	Found on the front accessory pulley to gear case pointer.
KT(A) 1150	Found on the front accessory drive pulley to gear case pointer or scribe mark.
KT(A) 2300	Found on crankshaft vibration damper and marks are aligned with a pointer on the gear case cover. Timing marks are also located on the flywheel and are aligned with a scribe mark on the flywheel housing.

You will then find that the engine timing marks will be 120 degrees apart both on the flywheel and on the accessory drive pulley. However, since there is only one flywheel, and only one accessory drive pulley, the timing marks will be stamped as indicated in Figure 9-55.

Early engines that used the torque method of injector adjustment used simply the cylinder numbers; however, the newer engines employ the letters A, B, and C as shown. When, for example, as shown in Figure 9-55, 1–6 TC is lined up with the stationary gear case timing mark or pointer, both the intake and exhaust valve could be adjusted on either cylinder 1 or 6, but not both at once. You could not, however, adjust the injector on the same cylinder that you would set the valves on because of the cam lobe position in relation to the injector plunger.

However, when the 1–6 VS (valve set) mark is lined up with the gear case timing mark, both cylinders 1 and 6 are 90 degrees ATDC (after top dead center), placing the cam lobe for the cylinder that is on the power stroke in a position that will allow you to check the actual injector plunger travel.

Figures 9-56 through 9-58 show the actual individual piston positions of such an engine with this firing order. Note that although pistons 3, 4 are in the same position, they are 360 degrees apart in the engine firing order.

Figures 9-56 and 9-57 represent two full crankshaft revolutions. Figure 9-58 is representative of the two turns of the crankshaft superimposed on top of each other. Since it is a 4-cycle engine of right-hand rotation, any piston to the right of center must be coming down the cylinder and any piston to the left of center must be going up. The piston descends on both power and intake strokes and moves up on both compression and exhaust, giving us the setup shown in Figure 9-59.

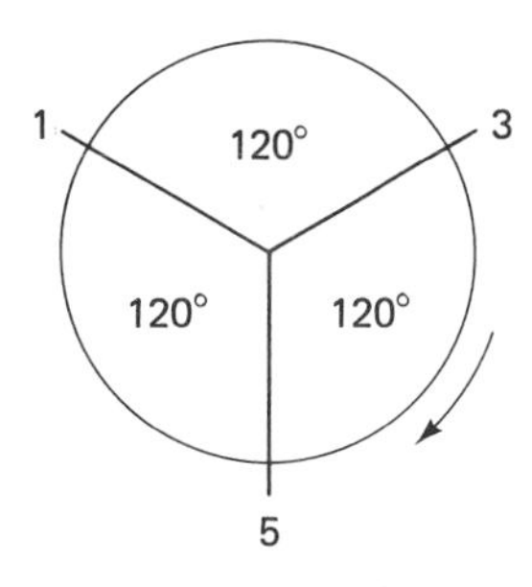

Figure 9-56
Pistons 1, 3, and 5 Are 120 Degrees Apart in Phase.

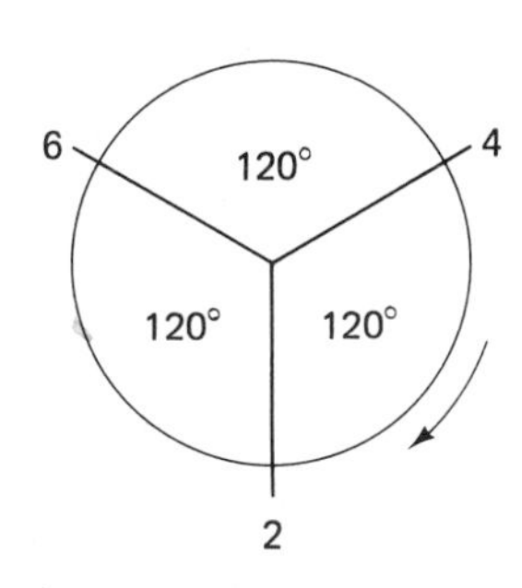

Figure 9-57
Pistons 2, 4, and 6 Are 120 Degrees Apart in Phase.

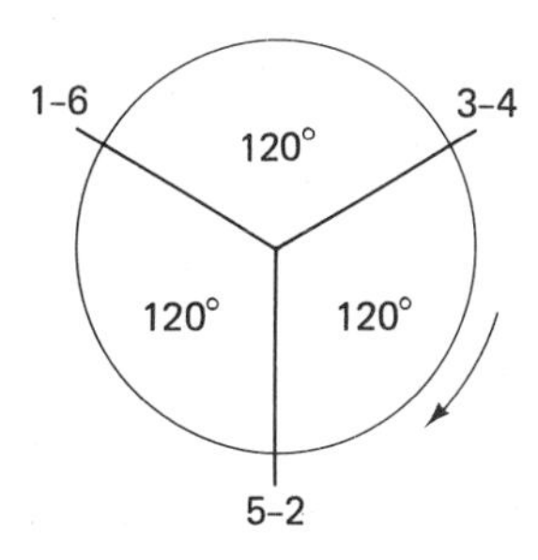

Figure 9-58
Two Turns of Crankshaft Superimposed on Each Other; All Cylinders Are Still 120 Degrees Apart in Phase.

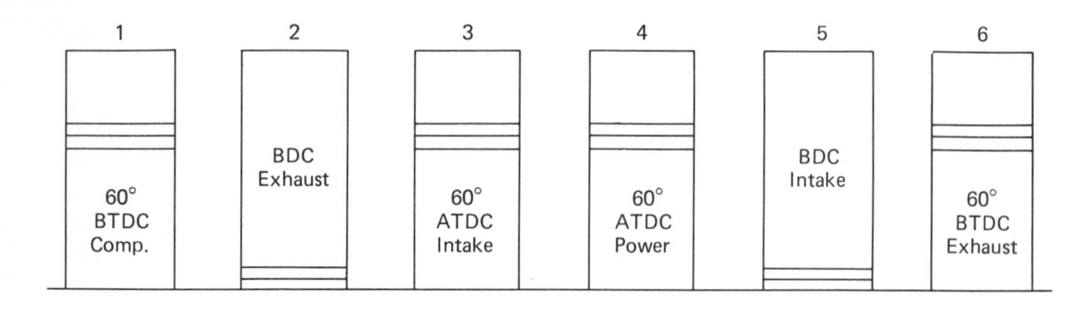

Figure 9-59
Relative Piston Positions with a 1-5-3-6-2-4 Firing Order.

Step 1. As mentioned earlier, remember to check the injector hold-down capscrew torque, and if flange-type injectors are used, tighten the inlet and drain connections to 20 to 25 lb-ft (27 to 34 N·m).

Step 2. If the engine is equipped with a Jacobs brake, you will require service tool (ST-3375096) along with ST-1270 to adjust both the injectors and valves.

Step 3. Bar the engine over until A or 1–6 VS mark on the accessory drive pulley is in line with the gear case scribe line or pointer. Check that both valves on cylinder 5 have free play between the rocker arm and crosshead. If not, start with the valves on cylinder 2, as shown in Table 9-5, since any cylinder combination may be used as a starting point. However, if we assume for instructional purposes that we have the 1–6 VS mark lined up and the valves are free on cylinder 5, then the injector plunger for cylinder 3 must be at the top of its travel.

Step 4. Check the type of injectors that are used; if they are PT-D top stop injectors, they should be cold set to zero, if not already done so, as follows:

1. The injector should be removed from the engine to properly set its plunger travel using adjusting tool ST-3375160.
2. If tool ST-3375160 is not available, a zero clearance setting can be made at the same injector adjustment position as that of the dial indicator method (inner base of the cam lobe).
3. Assuming that the engine is still at the 1–6 VS mark and injector 3 is ready to be set, tighten the adjusting screw (see Figure 9-17) until all lash is removed from the injector train. Carefully tighten the adjusting screw one more turn to seat the links and to force all oil from the socket surfaces; then very carefully back off the adjusting screw just until you feel a step up in effort, which is the point at which the spring washer contacts the stop. Using ST-3375232 torque wrench or equivalent, tighten the screw to 5 to 6 lb-in. (0.56 to 0.68 N·m). This in effect produces zero clearance by very lightly loading the link. If a torque wrench is unavailable, the zero clearance can be set approximately by turning the screw until the link is just free enough to be rotated by the thumb and forefinger. Hold the screw with a screwdriver and tighten the locknut to the correct torque.

Table 9-6
INJECTOR AND VALVE SET POSITION (DIAL METHOD)

Bar in Direction	*Pulley Position*	*Set Cylinder Injector*	*Valve*
Start	A or 1–6 VS	3	5
Advance to	B or 2–5 VS	6	3
Advance to	C or 3–4 VS	2	6
Advance to	A or 1–6 VS	4	2
Advance to	B or 2–5 VS	1	4
Advance to	C or 3–4 VS	5	1

Figure 9-60
Dial Indicator Extension Contacting Top of Injector Plunger.
(Courtesy of Cummins Engine Co.)

Step 5. Set up the ST-1270 indicator support so that the dial indicator extension rests on the injector plunger top of cylinder 3 (see Figure 9-60). Check that everything is secure and that there is no interference between the indicator and rocker lever.

Step 6. Hook the curved injector barring tool from the kit under the injector rocker arm bosses as shown in Figure 9-61. Place either a combination wrench or socket and ratchet over the hex bar of the tool, and depress the rocker lever, noting the dial indicator travel and comparing it to the specifications shown in Table 9-7.

Table 9-7
ADJUSTMENT LIMITS USING DIAL INDICATOR METHOD IN INCHES (mm)

Oil Temp.	*Injector Plunger Travel*	*Intake*	*Exhaust*
Aluminum Rocker Housing			
Cold	0.170 ± 0.001	0.011	0.023
	(4.32 ± 0.03)	(0.28)	(0.58)
Hot	0.170 ± 0.001	0.008	0.023
	(4.32 ± 0.03)	(0.20)	(0.58)
Cast-Iron Rocker Housing			
Cold	0.175 ± 0.001	0.011	0.023
	(4.45 ± 0.03)	(0.28)	(0.58)
Hot	0.175 ± 0.001	0.008	0.023
	(4.45 ± 0.03)	(0.20)	(0.58)

Step 7. If the injector plunger travel is not within the specifications listed, screw the injector lever adjusting screw down after backing off the locknut until there is a step up in effort, which indicates that the injector plunger is bottomed in the cup. Back the adjusting screw out one half-turn; then run it in again to bottom the plunger. Set the dial indicator at this time to zero.

Step 8. Carefully back out the adjusting screw until the dial indicator shows the correct travel as shown in Table 9-7. Depress the rocker lever as in step 6 several times in order to ensure that the adjustment is now correct.

Step 9. Tighten the locknut to 40 to 45 lb-ft (54 to 61 N·m) and again actuate the injector plunger with the barring mechanism to ensure that the adjustment has not moved. If Cummins tool ST-669 adapter is available, tighten the locknut to 30 to 35 lb-ft (41 to 47 N·m). The reason for the reduced torque when using this adapter is to compensate for the additional torque arm length. See Figure 9-62.

Figure 9-61
Depressing Rocker Lever Assembly.
(Courtesy of Cummins Engine Co.)

Step 10. Adjust both the intake and exhaust valves on cylinder 5 to the clearance shown in Table 9-7. If possible, use a set of *go-no go* feeler gauges when setting valves rather than a straight feeler gauge. Torque the adjusting screw locknuts to the same value as the injectors. Continue with the other cylinder adjustments as per Table 9-6.

Torque Method of Injector and Valve Adjustment on C-464, NH-743, N-855, N-927, V-1486, V-1710 Series Engines

This method of adjustment is used on those engines having an injector plunger free travel exceeding 0.206 in. (5.23 mm).

Step 1. If the engine has a compression release lever, pull it back and block it in the open position while barring the engine over; however, do *not* forget to release it when doing injector and valve adjustments.

Step 2. When using this method, both the injector and inlet and exhaust valves can be set in the VS (valve set) position. Bar the engine over until any VS mark aligns with the mark on the gear case cover. On in-line engines you will be using either A for 1–6 VS, B for 2–5 VS, or C for 3–4 VS, whereas on the V-1710 or V-1486 engines, being of V12 configuration, the accessory drive pulley would be marked, as an example, 1–6 RVS, indicating that both pistons 1 and 6 are at the same position in the cylinder on the right bank; however, only one will be in a position to be set since one will be on the power stroke and the other on the intake stroke.

Step 3. When you have established the cylinder that has free play between the rocker arm and valve crosshead, this is the one to be set. Since the injector rocker arm and push

Figure 9-62
Use of Tool ST-669 to Tighten Both the Injector and Valve Adjusting Screw Locknut.
(Courtesy of Cummins Engine Co.)

tube are heavier than both of the valves, the injector plunger is always set first; otherwise, you could change the valve setting.

Step 4: Injector Plunger Adjustment. An inch-pound (N·m) torque wrench is required for this purpose along with a screwdriver (straight blade).

1. Back off the injector plunger adjusting screw locknut at the rocker arm. Back off the adjusting screw until its ball end is free of the push tube cup; then lightly run it down until it just makes contact with the cup, indicated by a slight step up in effort. Advance the screw an additional 15 degrees to force any oil from between the push tube cup and adjusting screw ball end.
2. Loosen the adjusting screw one full turn; run it down gently until there is a step up in effort. Then using the torque wrench and screwdriver adapter, advance the adjusting screw to the specified torque value given in Table 9-8 and tighten the locknut to 40 to 45 lb-ft (54 to 61 N·m) or 30 to 35 lb-ft (41 to 47 N·m) if using Cummins tool ST-669 adapter.

Valve Adjustment

If the crosshead is in need of adjustment, proceed as described earlier in this section (Figure 9-53). The valves are set in the same VS position as the cylinders' injectors. Remember to release the compression release

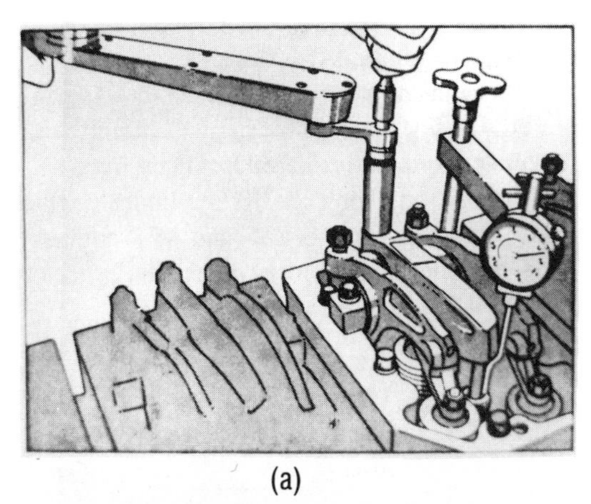

(a)

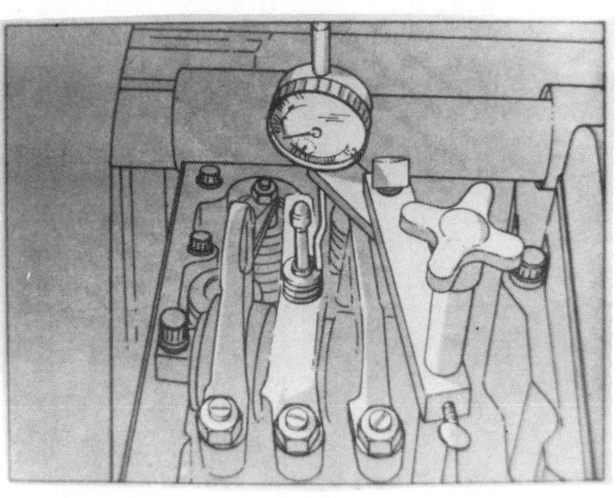

(b)

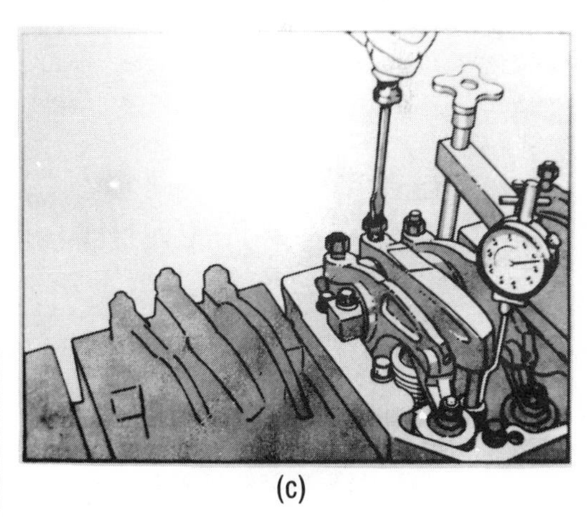

(c)

Figure 9-63
KT/KTA 1150 CID Engines: (a) Tightening Injector Plunger Rocker Lever Adjusting Screw Locknut Using Tool ST-669; (b) Dial Indicator Extension in Contact with Injector Plunger; (c) Adjustment of Injector Plunger Travel. (Courtesy of Cummins Engine Co.)

lever, if used, to the run position. Adjust the valves in the same way as in the dial indicator method and tighten the locknuts to the same torque as the injector. See Table 9-9 for the correct valve clearances.

Always make your final valve adjustment with the engine at a stabilized lube oil and water temperature.

Dial Method of Injector and Valve Adjustment on KT and KTA 1150 CID . . (37.72 L) 6-Cylinder Engine

The injector and valve adjustment for the K series engine is almost identical to that of Cummins 855 and 927 in-line six series engines. The firing order for the right-hand rotating KT/KTA 6 is the same as the 855 and 927: 1–5–3–6–2–4. The timing marks are on the backside of the crankshaft pulley around the center hub periphery where it is stamped A, B, or C, and aligned with a pointer on the gear housing cover. Also on the pulley are the letters and numbers TC 1–6; therefore, to adjust the injectors and valves, refer to Table 9-6, which gives the valve set mark position at which cylinder injectors and valves can be set. The only difference is that the K-6 engine does not have a VS (valve set) mark stamped on its pulley. (See Figure 9-63.)

Dial indicator assembly ST-3375004 is required to set up the K engine, along with rocker lever actuator 3375009. Therefore, setting up and adjusting the K 6-cylinder engine is the same as for the 855 and 927 engines, with the main difference being the injector

Table 9-8
INJECTOR PLUNGER ADJUSTMENT TORQUE METHOD IN INCH-POUNDS (N · m)

Cold Set	*Hot Set*
V-1486, V-1710 series 50 (5.7)	
NH-743, N-855, N-927 series	
Cast-iron rocker housing	
48 (5.4)	72 (8.1)
Aluminum rocker housing	
72 (8.1)	72 (8.1)
C-464 series 48 (5.4)	60 (6.8)

Table 9-9
VALVE CLEARANCE IN INCHES (mm)

Intake Valves		*Exhaust Valves*	
Cold Set	**Hot Set**	**Cold Set**	**Hot Set**
V-1486, V-1710			
0.016 (0.41 mm)		0.029 (0.74)	
NH-743, N-855, N-927			
Cast-Iron Rocker Housing			
0.016 (0.41)	0.014 (0.35)	0.029 (0.74)	0.027 (0.68)
Aluminum Rocker Housing			
0.014 (0.35)	0.014 (0.35)	0.027 (0.68)	0.027 (0.68)
C-464			
0.017 (0.43)	0.015 (0.38)	0.027 (0.68)	0.025 (0.63)

2. Bar the engine over to any valve set mark that lines up with the gear case pointer, and check that the valves are in fact loose (closed) on the cylinder to be adjusted.
3. Using either an Allen wrench or socket head wrench, back off the adjusting screw and insert a Jacobs P/N 3087 0.018-in. (0.46-mm) feeler gauge between the slave piston and crosshead, as shown in Figure 9-66.
4. Gently turn the adjusting screw in until a slight drag is felt on the feeler gauge, and tighten the locknut to 15 to 18 lb-ft (20 to 24 N·m).

Figure 9-68
Clearance of Slave Piston.
(Courtesy of Jacobs Manufacturing Co.)

Dial Method of Injector and Valve Adjustment KT and KTA-V12 Engines

Adjustment of both the injectors and valves on the V12 series K engines is not unlike putting two K 6 engines together and having a V12. The process involved is again very similar to that of the K 6 and 855 and 927 series in-line sixes.

Figure 9-67
Loosening Slave Piston Adjustment Screw
(Courtesy of Jacobs Manufacturing Co.)

The V12 K's have three locations available where both the injector and valve alignment marks can be viewed. The barring mechanism can be located on either the left or the right cylinder bank at the flywheel housing, where a cover plate on an opening designated as A or C directly above the barring mechanism must be removed when viewing the timing marks on the engine flywheel in alignment with the stationary mark on the flywheel housing.

1. If using the vibration damper or crankshaft pulley and damper combination, always match up the damper mark with the pointer mark on the gear case cover.
2. Extreme care must be exercised on the V12 engine because you can bar the engine over from either the left or right bank at the flywheel housing end of the engine. Starter motor location and available engine accessories will determine which side you can bar the engine over from.
   a. If you are barring the engine over from the right bank at the flywheel housing as shown in Figure 9-69, the A-VS timing marks on the flywheel

Table 9-11
KT AND KTA 2300 V12 ENGINE ADJUSTMENT LIMITS USING DIAL INDICATOR METHOD IN INCHES (mm)

Injector Plunger Travel	*Valve Clearance*	
	Intake	*Exhaust*
0.308 ± 0.001 (7.82 ± 0.03)	0.014 (0.36)	0.027 (0.69)

must align with the flywheel housing scribe mark when looking through the flywheel housing opening stamped A.

b. If, however, you are barring the engine over from the left bank of the engine at the flywheel housing end, the C-VS timing marks on the flywheel must be aligned with the housing scribe mark when looking through the opening stamped C on the flywheel housing (Figure 9-70).

3. V12 K series cylinder layout is shown in Figure 9-71.

KT and KTA-2300 V12 Firing Order

With a right-hand firing order, 1R–6L–5R–2L–3R–4L–6R–1L–2R–5L–4R–3L, the relative marks stamped on either the crankshaft vibration damper or flywheel would appear as shown in Figure 9-72. On any Cummins engines, when you align a valve set (VS) mark with either the crankshaft vibration damper, gear case, or flywheel timing mark, the particular cylinder that you adjust the valves and injector on has the piston at 90 degrees ATDC on its power stroke.

Injector plunger travel and both valves may be set on one cylinder at the same valve set location; however, since two cylinders will show as being in alignment at the same position with the stationary timing mark, only one will have free valves, meaning it can be set. You would have to rotate the engine one complete revolution to set up the other cylinder. For ease of explanation, cylinder 1R–6R is shown as being 90 degrees ATDC with 1R being on its power stroke.

Follow the engine firing order and adjust the injectors and valves as per Table 9-11 using the same dial indicator setup as for the

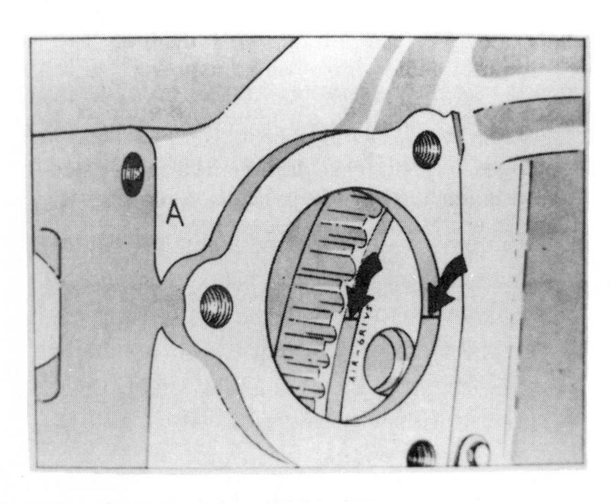

Figure 9-69
Flywheel Timing Mark A Valve Set. (Courtesy of Cummins Engine Co.)

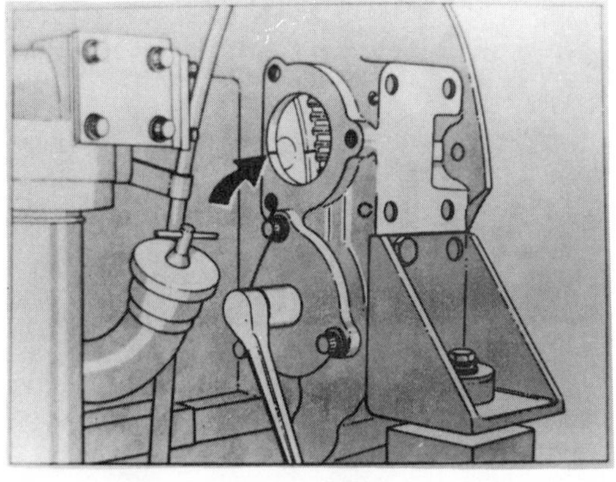

Figure 9-70
Flywheel Timing Mark C Valve Set. (Courtesy of Cummins Engine Co.)

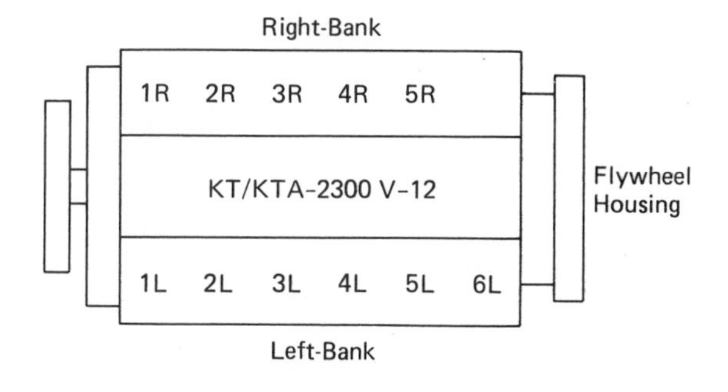

Figure 9-71
V12 K Series Cylinder Layout.

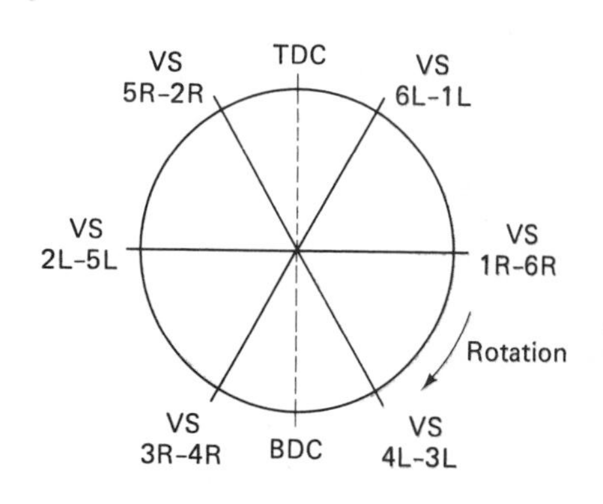

Figure 9-72
KT-2300 Flywheel Timing Marks.

KT and KTA 6 engine. Follow the step sequence that was detailed earlier under the dial indicator method for the NH-743, N-855, and 927 engines.

When adjusting the valves remember that, since this is a V12 engine, the exhaust valves are toward the front of the engine in each cylinder head on the engine's left bank and toward the rear of the engine on each cylinder head on the right-bank side. The locknut torque for the injectors and valves is the same as for other engines discussed so far.

Cummins new Vee 16 KTA-3067 has a displacement of 50 liters and uses a patented fuel control valve which shuts off 8 of the 16 cylinders during engine warm-up and light-load low-speed operating conditions to reduce white smoke.

This is one of the new high-speed, high-horsepower models of Cummins engines developing 1190 bkW (1600 bhp) with a 159-mm bore.

Figure 9-73 shows the engine firing order and cylinder location. Figure 9-74 shows the operation of the fuel control valve used to allow the engine to run on only 8 cylinders under the conditions mentioned above. The fuel pump used with this engine is designated PTH, since it features a larger capacity gear pump than previous PT pumps.

The fuel control valve remains closed at fuel rail pressures below 69 kPa (10 psi), therefore only the right bank receives fuel. When the fuel pressure increases, it causes a gradual opening of the fuel passage to the left bank, until at 276 kPa (40 psi) the left bank will receive maximum fuel flow. The patented fuel control valve doubles the amount of fuel injected into each operating cylinder of the right bank during light loads and low speed, which results in improved injection spray formation, as well as an advance in the beginning of injection. Both of these features contribute substantially to reducing causes of white exhaust smoke.

Table 9-12
INJECTOR PLUNGER ADJUSTMENT TORQUE

Oil Temperature 70°F (21.1°C)	*Oil Temperature 140°F (60°C)*
60 in.-lb (6.8 N·m)	60 in.-lb (6.8 N·m)

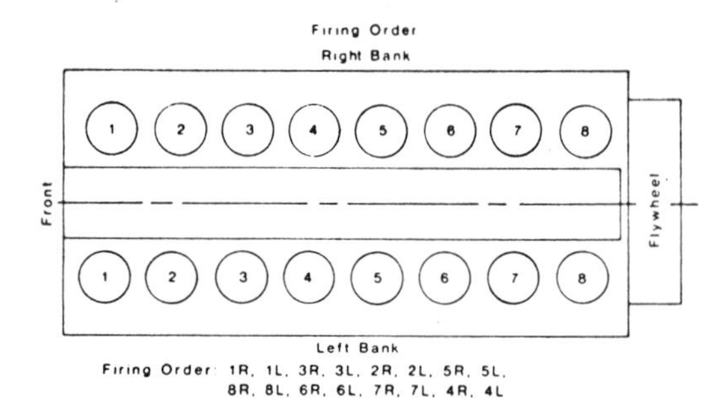

Figure 9-73
KTA-3067 (V16) Firing Order. (Courtesy of Cummins Engine Co.)

Table 9-13
ADJUSTMENT LIMITS IN INCHES (mm): V-903 USING DIAL INDICATOR METHOD

Injector Plunger Travel	*Valve Clearance Intake*	*Exhaust*
1.2 to Rocker Lever Ratio		
Injector Lever P/N 196565		
0.180 ± 0.001	0.012	0.025
(4.57 ± 0.03)	(0.30)	(0.64)
1 to 1 Rocker Lever Ratio		
Injector Lever P/N 211319		
0.187 ± 0.001	0.012	0.025
(4.75 ± 0.03)	(0.03)	(0.64)

Torque Method of Injector Plunger Adjustment for V-378, V-504, V-555, VT-555 CID Series Engines

The sequence of events for these engines is the same as that given earlier for the 855 and 927 series of engines, the only difference being in the specifications given in Table 9-12. Remember to torque the injectors prior to starting as per Table 9-4. Firing order and timing mark location can be found under the Engine Firing Orders and Valve Set Mark Alignment sections.

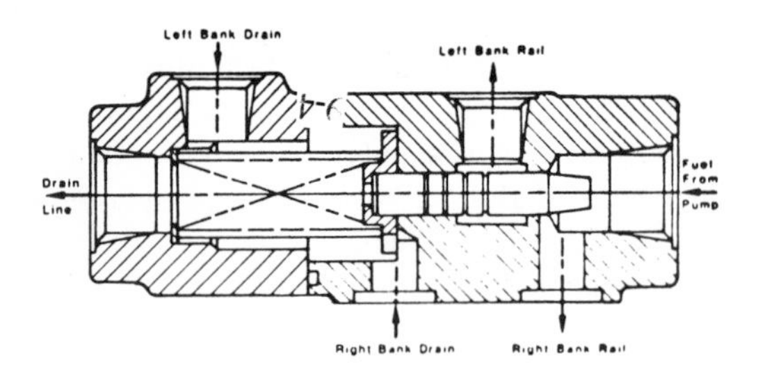

Figure 9-74
KTA-3067 Single Bank Fuel Control Valve. (Courtesy of Cummins Engine Co.)

Both the injector and valve locknut torque are the same as with other Cummins engines: 40 to 45 lb-ft (54 to 61 N·m) with a standard torque wrench and 30 to 35 lb-ft (41 to 47 N·m) when using Cummins tool ST-669.

Dial Indicator Method of Injector and Valve Adjustment for V-903 Series Engine

This method is also the same basic setup as that for any other Cummins engine employing the dial indicator method. Torque the injectors prior to starting as per Table 9-4.

Under the Engine Firing Orders and Valve Set Mark Alignment sections you will find these specifics. Since the V-903 is a V8 engine, valve set (VS) marks on the crankshaft damper will be paired in sequence: 1–6 VS, 3–5 VS, 4–7 VS, and 2–8 VS, with each pair being 90 degrees apart on the damper (4-cycle engine, 2 crank revolutions per power stroke per cylinder means 2 times 360 equals 720 degrees divided by 8 equals degrees).

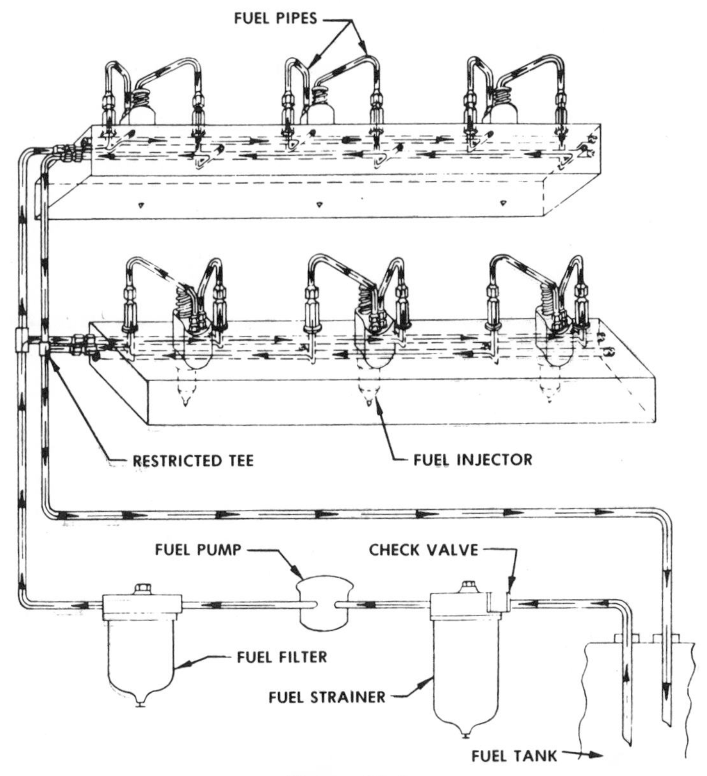

Figure 10-1
Typical VEE Fuel System on 6V Engine.
(Courtesy of Detroit Diesel Allison Div. of GMC)

5. To purge the fuel system of any air; the system is recirculatory in operation, therefore allowing any air to be returned to the fuel tank.

Purpose of Fuel System Components

Figure 10-1 shows a schematic view of a typical VEE fuel system used on a 6V engine. Since the basic fuel system employed on all Detroit Diesel engines is identical as far as components used, the description of operation for one can be readily related to any other series of DD engine. An in-line engine, for example, would only use one cylinder head, whereas a Vee engine using two would have a fuel system as shown in Figure 10-1.

The basic fuel system in Figure 10-1 consists mainly of the following:

1. Fuel injectors.
2. Fuel pipes to and from the injectors (inlet and outlet).
3. Fuel manifolds, which are cast internally within the cylinder head (older engines used external fuel manifolds running lengthwise along the head). Either way, the upper manifold is the *inlet* and the lower is the outlet or *return*. Cast in several places in the side of the head are the words *in* and *out* to prevent confusion.

4. Fuel pump (supply pump, not an injection pump).
5. Fuel strainer or primary filter.
6. Fuel filter (secondary).
7. Fuel lines.
8. One-way check valve.
9. Restricted fitting on in-line engines or a restricted TEE on Vee-type engines.

10-2

Fuel Pump

Figure 10-2 shows the typical fuel pump used on all series 53, 71, and 92 engines.

Fuel Flow

The fuel pump draws fuel from the tank past the one-way nonreturn check valve into the primary filter, where the fuel passes through a 30-micron-filtering-capacity, cotton-wound, sock-type element. From the primary filter it passes up to the suction side of the fuel pump, where the fuel is then forced out at between 65 and 75 psi (448.2 and 517.1 kPa) to the secondary filter, which is a pleated paper element of 10-micron filtering capacity. Fuel then passes up to the inlet fuel manifold of the cylinder head, where it is distributed through the fuel jumper lines into each injector. All surplus fuel (not injected) returns from the injectors through the return fuel pipes, through the restricted fitting, which maintains adequate fuel pressure in the head at all times, then back to the tank. All Detroit Diesel engines are equipped with a fuel return line restricted fitting, the actual size varying with engine injector size and application; however, every restricted fitting has the letter R followed by a number that indicates its hole size in thousandths of an inch. Therefore, a fitting with R80 stamped on it would indicate an 0.080-in. (2.032-mm)-diameter hole drilled within the fitting.

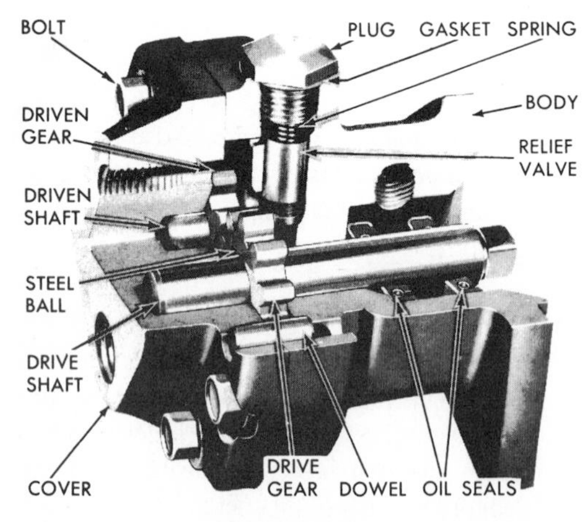

Figure 10-2
Typical Fuel Pump on Series 53, 71, and 92 Engines.
(Courtesy of Detroit Diesel Allison Div. of GMC)

These fittings may look like an ordinary brass fitting externally; therefore, care must be taken to ensure that in fact the proper restricted fitting and not just any fitting is installed into the return line. Use of too large a fitting can lead to low fuel pressure within the head manifolds and poor engine performance, whereas too small a fitting can lead to increased fuel temperatures and some restriction against the fuel flow. Refer to the particular engine service manual for any particular specifications. All engines contain one restricted fitting, with the exception of the 16V units, which use two.

The one-way check valve is used to prevent fuel from draining back to the tank from the primary filter and line when the engine is stopped.

The fuel pump is a positive displacement gear-type unit that transfers fuel from the tank to the injectors at 65 to 75 psi (448.2 to 517.1 kPa). The standard pump has the ability to deliver $1\frac{1}{2}$ gallons U.S./min (5.67 liters) or 90 gallons/hour (340.68 liters), approximately.

Since the pump constantly circulates an

excess supply of fuel to and through the injectors, the unused portion, which also cools and lubricates the injectors and purges the system of any air, returns to the fuel tank via the restricted fitting and fuel return line.

Identification of Fuel Pump Rotation

If you are in doubt as to a fuel pump's rotation, it can be identified as follows:

1. Stamped on the pump cover are the letters LH or RH, plus an arrow indicating the direction of rotation.
2. On in-line engines, the fuel pump rotation can be determined by its location on the engine. When viewed from the flywheel end: left-hand side location, LH pump rotation; right-hand side location, RH pump rotation.

   All V-71 and V-92 engines use LH rotation pumps only, whereas 149 series engines use RH rotation pumps only.
3. A similar method would be to grasp the pump in your left or right hand as it mounts on the engine. Whichever thumb covers the relief valve indicates the pump's rotation.

The letters I/L (inlet) are also stamped on the pump cover; however, if not visible, the inlet side is the hole on the pump cover closest to the relief valve plug.

The fuel pump body and cover are aligned by means of two dowels, and the body and cover are ground surfaces that contain no gasket between them, although a thin coat of sealant applied to these surfaces is recommended at installation. The relief valve bypasses fuel back to the inlet side of the pump when pressure reaches 65 to 75 psi or 448.2 to 517.1 kPa.

Fuel drawn into the suction side of the pump fills the space between the gear teeth and the pump body, where it is carried around and discharged to the outlet cavity under an average pressure of 45 to 70 psi (310.2 to 482.6 kPa). Closer study of Figure 10-3 will indicate the characteristics of the pump shown.

Figure 10-4 shows an exploded view of the pump; the stackup of the component parts is clearly visible. The standard fuel pump gears are $\frac{1}{4}$ in. wide and contain 10 teeth, whereas the high-capacity pump that is available has gear teeth $\frac{3}{8}$ in. wide (149 engine pump gear teeth are 1'' [25.4 mm]).

The drive gear is a 0.001-in. (0.0254-mm) press fit onto the shaft, and a gear retaining ball locates it on its shaft.

As shown in Figure 10-4, two oil seals are pressed into the pump bore from the flanged end for the following purposes:

1. The seal closest to the drive fork prevents lube oil from entering the fuel pump.
2. The inner seal closest to the pump gears prevents fuel oil leakage.

The installed seals do not butt up against each other, but have a small space between them. Drilled and tapped into this cavity in the fuel pump body are two small holes, one of which is usually plugged; the other is open to allow any fuel or lube oil leakage to drain, thereby indicating damaged seals. Sometimes a small fitting and tube extend from one of these holes to direct any leakage to a noticeable spot. Acceptable leakage should not exceed 1 drop/min.

A fuel pump with a star stamped on its cover indicates that the inner seal is reversed, and is used on *gravity feed* installations where the fuel tank is above the level of the fuel pump. The reversed inner seal (seal closest to the pump gears) prevents fuel seepage down the pump shaft and out the drain cavity hole, especially when the engine is shut down.

Never plug both drain holes in the pump body between the oil seals; otherwise, any fuel leakage will cause crankcase oil dilution.

Figure 10-5 shows the fuel flow through a series 12V-149 model engine from and to the tank.

With reference to Figure 10-6, the location

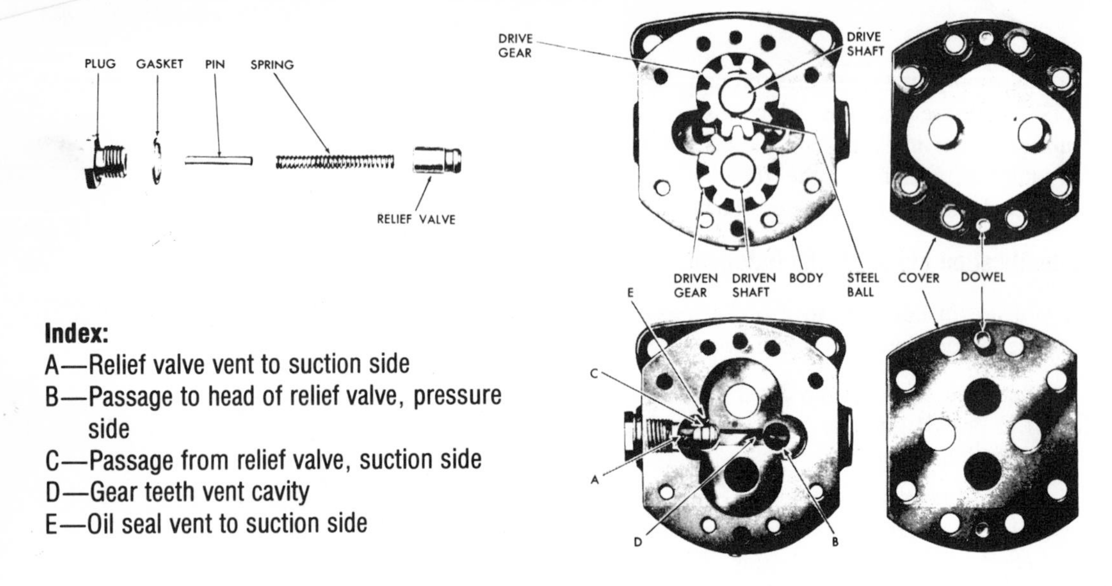

Index:

A—Relief valve vent to suction side

B—Passage to head of relief valve, pressure side

C—Passage from relief valve, suction side

D—Gear teeth vent cavity

E—Oil seal vent to suction side

Figure 10-3
Exploded View of Fuel Pump Relief Valve and Pump Gearing.
(Courtesy of Detroit Diesel Allison Div. of GMC)

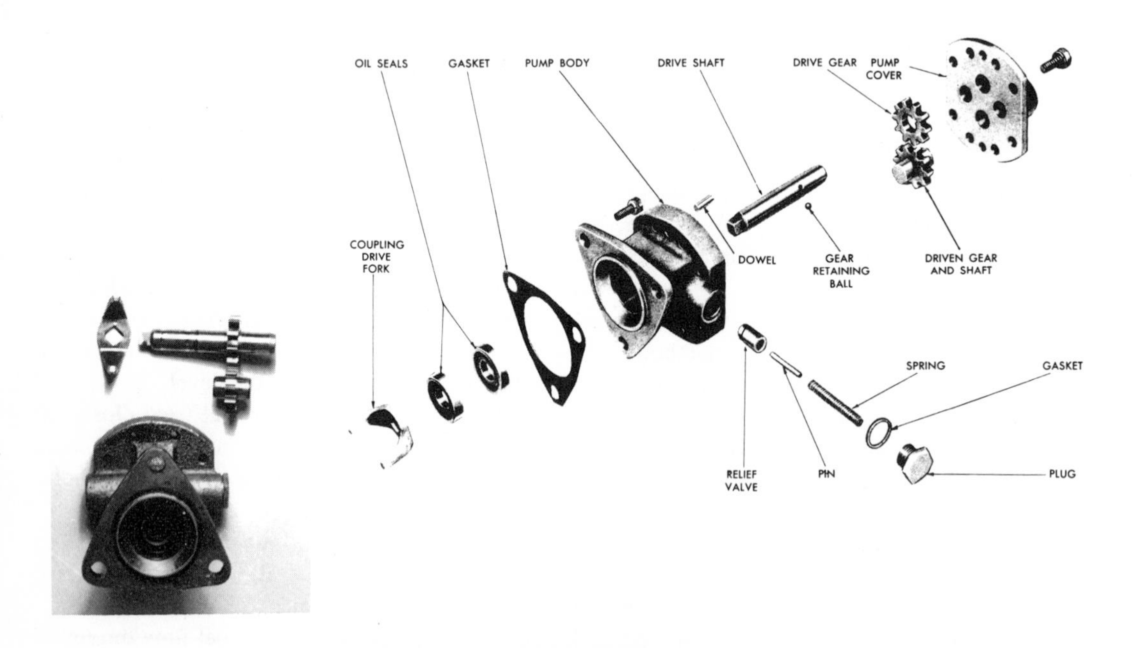

Figure 10-4
Stackup of Component Parts of Fuel Pump.
(Courtesy of Detroit Diesel Allison Div. of GMC)

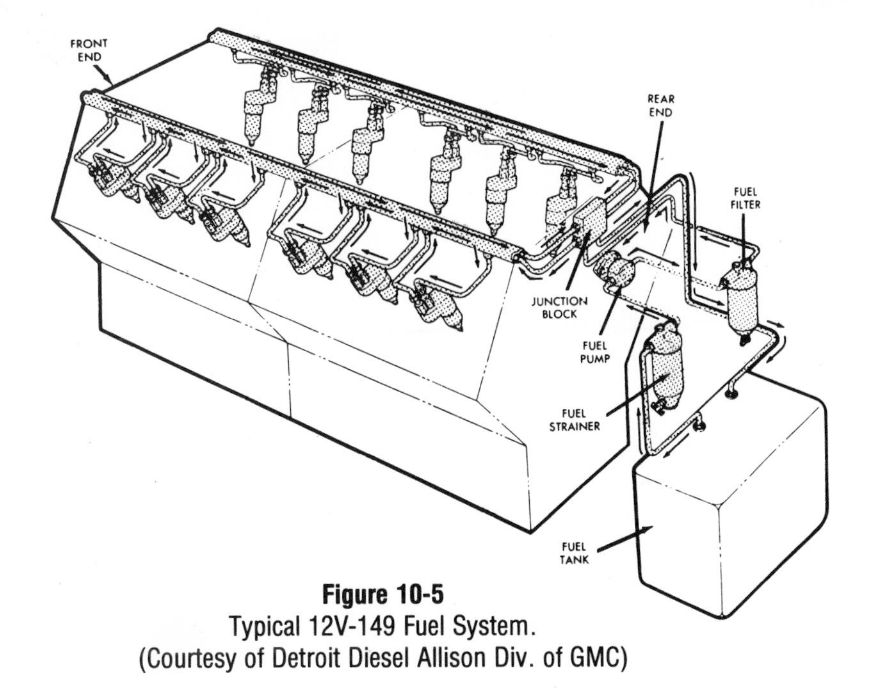

Figure 10-5
Typical 12V-149 Fuel System.
(Courtesy of Detroit Diesel Allison Div. of GMC)

of the fuel restriction orifice is shown on both the former and current engines. On current engines, the arrow must point toward the orificed plug, or fuel will not return to the tank.

Flexible fuel lines run from the fuel junction block to the fuel manifolds attached to the top of the cylinder block on each bank of the engine. Figure 10-7 depicts the location of the fuel manifolds on the engine. Also note that the fuel inlet manifold is the one facing the outside of the engine block.

10-3

149 Fuel System

The fuel pump on the 149 has the ability to deliver 350 gallons/hour (1324.9 liters) or 5.83 gallons/min (22.06 liters) at 1800 rpm engine speed (given in U.S. gallons). Mounted at the rear of the engine on an adaptor attached to the flywheel housing, it is driven off the end of the blower drive gear by a drive coupling fork attached to the end of the pump drive shaft and mates with a drive disc attached to the blower drive gear assembly (similar to other series DDA engines). The fuel pump rotation is always clockwise or RH (right hand), regardless of engine rotation.

Because of its delivery capacity, the pump is physically larger than those used on other series of engines; however, it is identical to the pump shown in Figures 10-2 through 10-4 with the following exceptions:

1. Physically larger.
2. Larger gears (1 in. wide).
3. The inlet is attached to the pump cover, and the outlet is on the side of the pump cover opposite the pressure relief valve.
4. Pumps stamped *2 or 2D* on the cover have a matched pump body and cover, and these must be kept together.
5. The current pump has a *shroud* cast onto

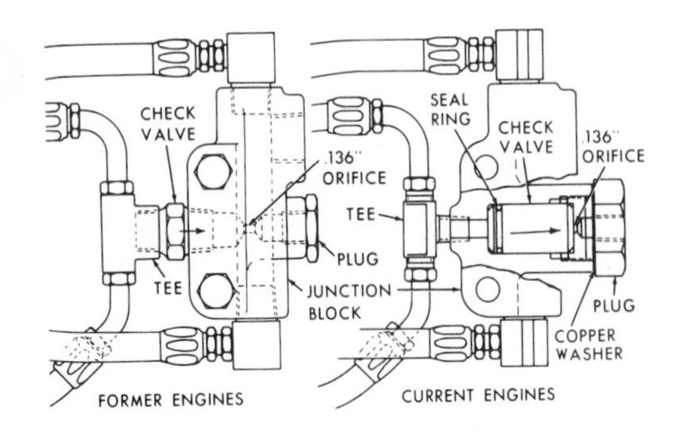

Figure 10-6
Location of Fuel Restriction Orifice.
(Courtesy of Detroit Diesel Allison Div. of GMC)

the fuel pump adaptor to retain the drive fork when removing or installing the pump; otherwise, the fork may fall into the gear train.

10-4

High-Lift Fuel Systems

There are two basic high-lift fuel systems employed by DDA; they are shown in Figure 10-8(a) and (b). A high-lift fuel system is found on certain applications of Detroit Diesel Allison engines where, due to the particular power unit installation, it is not possible to have a large enough fuel supply next to the engine. Since the standard DDA fuel system is limited to a maximum lift of 12 in. Hg (mercury), invariably the normal fuel storage is too far away for the engine fuel pump to pull. The storage tank may be located outdoors or underground; or, as in the case of some applications like marine units, the high-lift system draws fuel from the vessel's main tank, delivers it to a day tank, and from there the engine fuel pump takes over. Figure 10-8 shows a typical high-lift system that would be employed by DDA.

Fuel Flow

The normal flow in this system (Figure 10-8) would naturally be from the fuel tank through the filter (6) up into the high-lift pump (1). From here it would pass down to the fuel cup (5), where excess fuel would be allowed to return to the fuel tank. From the fuel cup (5), fuel passes to the engine primary filter (4) and on up to the engine fuel pump (3).

Fuel from the engine fuel pump is forced into the secondary fuel filter (2) and on into the inlet manifold within the cylinder head, where it is distributed to the injectors, with the return fuel passing through the restricted fitting back to the fuel cup return line to the fuel tank.

Priming the High-Lift System

If the engine loses its prime, in order to start it again only the engine fuel system requires priming. If, however, the high-lift system loses its prime, use the following procedure:

1. Open the globe valve (7).
2. Operate the manual lift pump (8).
3. When fuel is heard spilling into the fuel cup (5), close the globe valve (7).

Note: On the nonengine priming system, such as shown in Figure 10-8, priming of the high-lift system does *not* prime the engine fuel

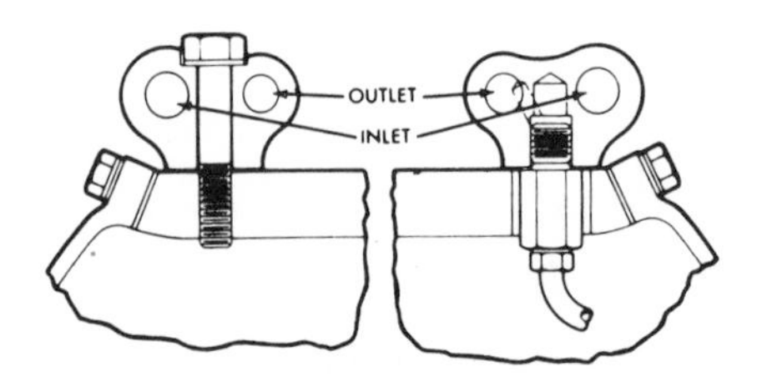

Figure 10-7
Location of Fuel Manifolds on the Engine.
(Courtesy of Detroit Diesel Allison Div. of GMC)

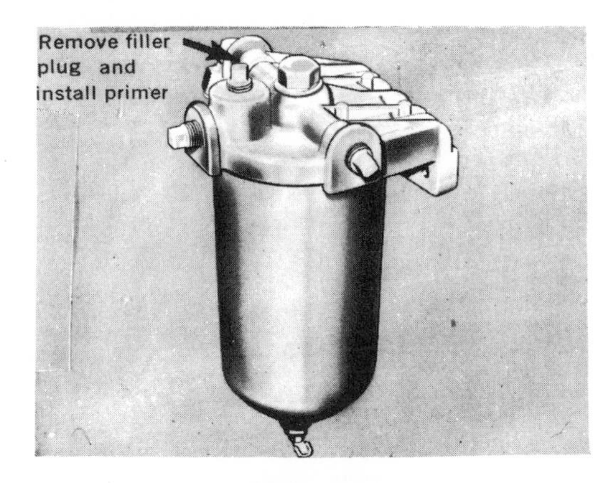

Figure 10-9
Filler Plug Location.
(Courtesy of Detroit Diesel Allison Div. of GMC)

8. The pipe plug on the top of the filter cover can be removed and using the fuel system primer J 5956 the entire fuel system can be primed.
9. If primer J 5956 is not available, a similar priming arrangement can be made that includes the following:
   a. Pump-handle-type oil squirt can.
   b. Fitting that screws into the filter cover with the other end capable of taking a piece of small-bore, rubber-type tubing.
   c. Length of rubber tubing (1 ft or 0.3 m) that will fit tightly over both the oil squirt can end and the fuel filter cover fitting.
10. A similar setup, but with the addition of an old fuel-injector fuel pipe that has the pipe cut off at one end to remove one of the fuel pipe nuts, can be used to prime the fuel injectors. By removing one injector inlet pipe and fitting the adaptor, the inlet manifold can be primed if the filters were not serviced but work has been done on the head involving a loss of prime.

Caution: If this method is used, only clean, filtered fuel should be used for priming purposes since the fuel that enters the inlet manifold does not have to pass through the secondary or primary filter unit.

11. Start the engine and check for any fuel leaks.

Figure 10-9 shows the filler plug location employed in both the primary and secondary fuel filters that can be removed for the purpose of priming the fuel system when necessary. Figure 10-10 illustrates the primer tool that can be used if available to prime the entire fuel system.

The problem in restarting an engine after it has run out of fuel stems from the fact that, after the fuel is exhausted from the fuel tank, fuel is then pumped from the primary fuel strainer and is often partially removed from the secondary fuel filter before the fuel supply becomes insufficient to sustain engine operation. Therefore, these components must be refilled with fuel and the fuel lines rid of air in order for the system to provide adequate fuel for the injectors. This situation is not only avoidable, but expensive in terms of equipment downtime.

10-6

Unit Fuel Injector Operation

The fuel injector, or what is often referred to as a *unit injector,* is used by Detroit Diesel Allison in all series of engines that they build. The Electro-Motive Division of General Motors Corporation also uses the same type of injector on their 2-cycle diesel 567 and 645 engines; therefore, the following information regarding the unit fuel injector is relevant to both these divisions of GMC. Certainly, there are some variations in basic design and in the actual testing procedures used; however, the function and operation is the same for all.

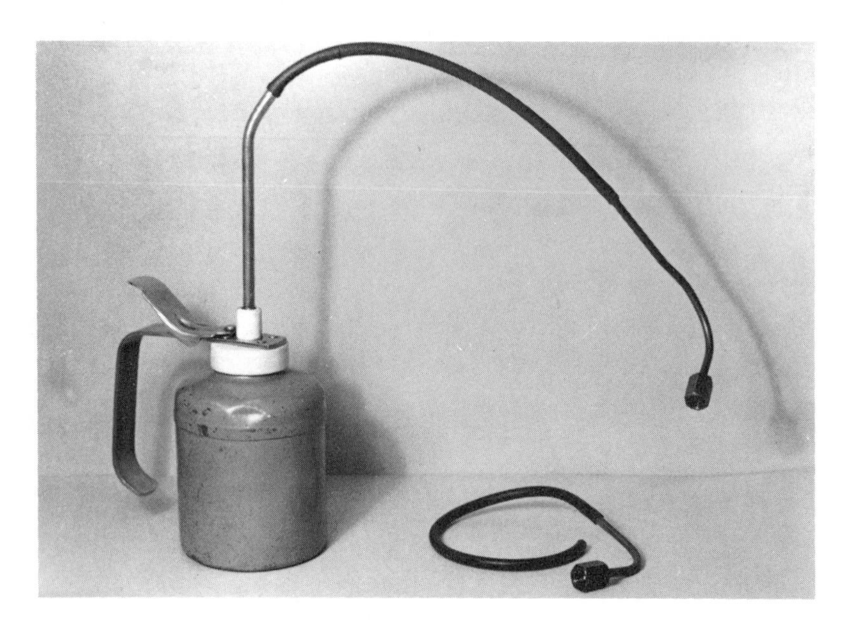

Figure 10-10
Primer Tool J 5956.
(Courtesy of Detroit Diesel Allison Div. of GMC)

These injectors were designed with simplicity in mind both from a control and adjustment outlook. They are used on direct-injection, open-type, 2-cycle combustion chamber engines manufactured by these two divisions of GMC. No high-pressure fuel lines or air–fuel mixing or vaporizing devices are required with these injectors, since the fuel from the fuel pump is delivered to the inlet fuel manifold cast internally within the cylinder head at a pressure of 45 to 70 psi or 310.3 to 482.6 kPa, and then to the injectors through fuel pipes.

Figure 10-11 shows a typical fuel injector employed by Detroit Diesel Allison in their engines. Once the fuel from the pump reaches the injector, it performs the following functions:

1. *Times injection:* Timing of the injector is accomplished by movement of the injector control rack, which causes rotation of the plunger within the injector bushing. Since the plunger is manufactured with a helical chamber area, this rotation will either advance or retard the closing of the ports in the injector bushing, and therefore the start and end of the actual injection period. Push rod adjustment establishes the height of the injector follower above the body. This in turn establishes the point or ''time'' that the descending plunger will close the bushings ports and therefore the start of injection.
2. *Meters the fuel:* The rotation of the plunger by movement of the injector control rack will advance or retard the start and end of injection. If then the length of time that the fuel can be injected is varied, so will the amount of fuel be varied.
3. *Pressurizes the fuel:* Fuel that is trapped underneath the plunger on its downward stroke will develop enough pressure to

force its way past the check valve or needle valve, as the case may be, and therefore enter the combustion chamber.

4. *Atomizes the fuel:* Fuel under pressure that forces its way past the check or needle valve must then pass through small holes or orifices in the injector spray tip. This breaks down the fuel into a finely atomized spray as it enters the combustion chamber.

Injector Assemblies

Since the injector unit is the same on all DDA engines with the exception of the plunger and spray tip, Figures 10-12 through 10-14 will allow you to compare just how similar these injectors are to one another.

Figure 10-11
Typical Fuel Injector.
(Courtesy of Detroit Diesel Allison Div. of GMC)

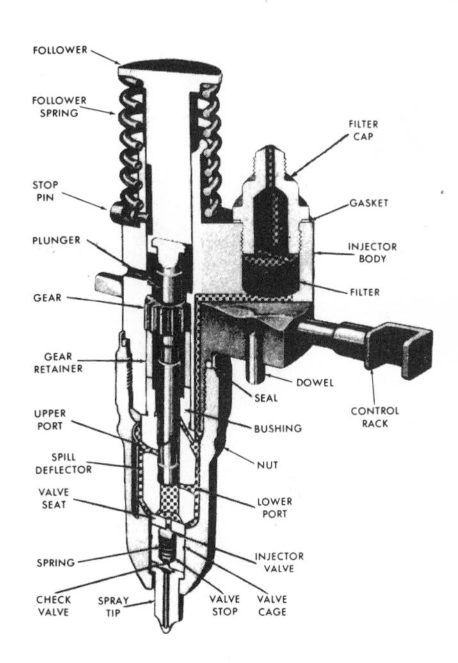

Figure 10-12
Crown Valve Injector.
(Courtesy of Detroit Diesel Allison Div. of GMC)

All current production Detroit Diesel Allison engines employ the *needle valve* type of fuel injector. There are some minor variations in the spray tip portion; however, they all operate on the same principle. The HV or high valve injector is no longer used in production engines; however, there are many DDA engines around with the HV injector, which was original equipment at the time of manufacture.

High valve injectors had a lower *popping pressure* to unseat the injector check valve, 450 to 850 psi, whereas the needle valve injectors have a popping pressure of between 2300 and 3200 psi, which is 15,858 to 22,753 kPa, required to lift the needle valve off its seat in the injector tip. The final spray-in pressure is in the region of 17,000 to 20,000 psi (137,895 kPa) with this type of injector, with the speed of penetration approaching 780 mph.

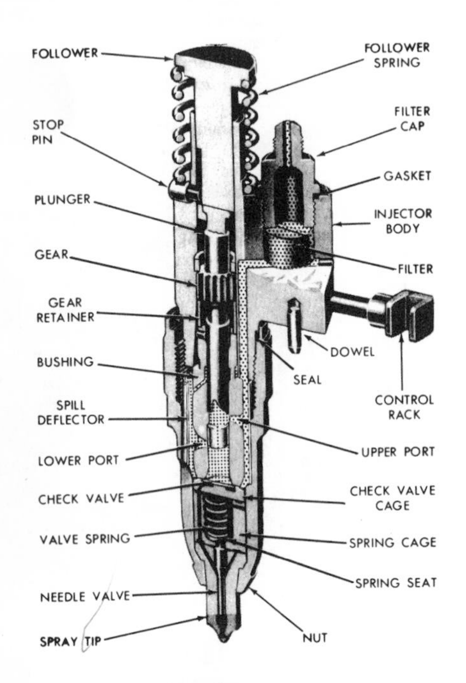

Figure 10-13
Needle Valve Injector.
(Courtesy of Detroit Diesel Allison Div. of GMC)

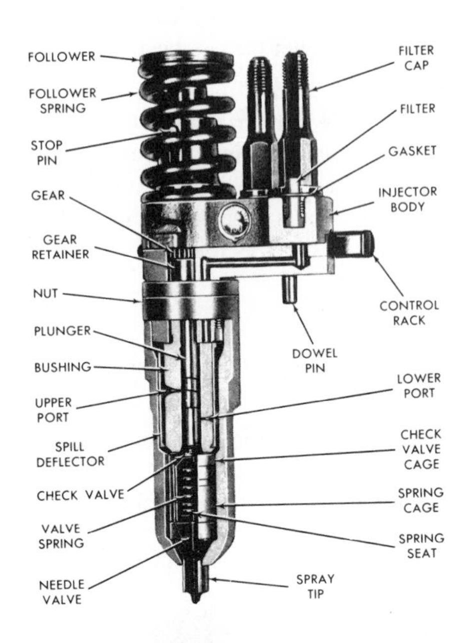

Figure 10-14
149 Injector.
(Courtesy of Detroit Diesel Allison Div. of GMC)

Due to the higher pressures encountered on injector closing pressure, there is a much cleaner and crisper cutoff to the injection period with the needle valve type of injector. There is also less fuel left beneath the valve seat to dribble out into the combustion chamber because of the smaller sac under the valve; this reduces the tendency for the engine to smoke.

Figure 10-14 depicts a series 149 engine fuel injector, which as you can see is very similar to the "N" type shown in Figure 10-13. Filter cap extensions are necessary on the 149 injector because of the location of the fuel inlet manifold running along the top of the cylinder block and the fuel supply pipe connecting it to the injector. All 149 injectors are of the needle valve type; however, the identification tag on the injector does not have the letter "N" preceding the injector displacement. Therefore, an injector with 150 stamped on the tag will deliver 150 mm^3 of fuel per stroke in the full-fuel position.

Fuel from the junction block and inlet manifold flows through the filter and drilled passages in the injector body, filling the cavity around the bushing. If the injector is not in an injection position, the fuel will flow through the upper port in the bushing and under the raised plunger. Fuel passes around the check valve, through the passages drilled in the spring cage, and into the cavity around the needle valve.

When the injector rocker arm acting on the injector follower pushes it down, the plunger will also move down, causing the upper and lower ports in the bushing to be covered, thereby trapping fuel under the descending plunger. As the trapped fuel builds up pressure, it acts upon the tapered land and small flat land of the needle valve. Between 2600 and 3200 psi (17,927 and 22,064 kPa) the valve begins to lift against the pressure of the spring, exposing the seat portion of the needle valve to this high-pressure fuel. With this pressure now working on an increased area, the needle valve will lift very rapidly, allowing the fuel to flow into the tip sac and on through the 0.010-in. holes in the spray tip into the combustion chamber. Since the fuel must force its way through these small holes, it is similar to turning on a garden hose with no nozzle on it. You will get lots of volume but little pressure; however, if you now place your thumb over the end of the hose, you will restrict the flow (less volume) but increase the pressure. Therefore, the fuel will be injected at a pressure as high as 20,000 psi (137,895 kPa), causing the fuel to enter the combustion chamber in a finely atomized state. When the downward travel of the plunger ceases, fuel pressure drops off very rapidly, allowing the needle valve spring to seat the needle valve sharply and crisply, thus preventing any further injection and minimizing after dribble and therefore smoking.

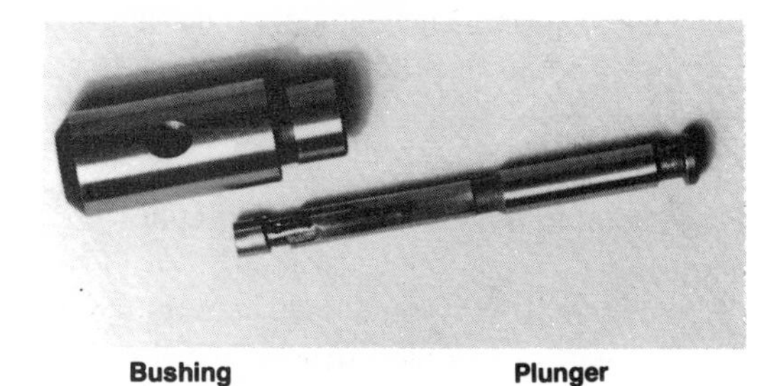

Figure 10-15
Typical New DDA Plunger and Bushing. (Courtesy of Detroit Diesel Allison Div. of GMC)

Phases of Injector Operation

The amount, rate, and timing of fuel injection is controlled by the injector plunger. Reference to Figures 10-12 and 10-13 shows that the top or neck of the plunger slides into a land area at the base of the injector follower. The follower is forced down by the action of the rocker arm; therefore, the plunger will move the same distance on each stroke. A circumferential groove cut in the plunger determines the timing and the quantity of fuel injected. The upper edge of this groove is cut in the shape of a helix, with the lower edge either being straight or also cut in a helix. A flat on the upper portion of the plunger meshes with a flat on the inside diameter of the fuel control rack gear; therefore, when the fuel control rack is moved in or out it will cause the plunger to rotate, thereby changing the position of the helix and thus the output of the injector. Figure 10-15 shows a typical new DDA plunger and bushing. At the time of manufacture by the Diesel Equipment Division of GMC, the internal diameter of the bushings is measured with an air gauge, and the bushings are segregated into groups to the nearest 12 millionths of an inch. The plungers are also measured and segregated; therefore, the plunger and bushing are mated together to maintain a specified clearance that never exceeds 60 millionths of an inch.

The bushing actually has two ports, one on each side, with one being higher than the other. Anytime that both ports are uncovered by the plunger, fuel at pump pressure (45 to 70 psi; 310.27 to 482.65 kPa) is free to flow into and out of the bushing; therefore, no pressure beyond that being delivered by the pump is possible under this condition.

Figure 10-16 depicts the phases of injector operation through the vertical or downward movement of the plunger. In Figure 10-16(a), with the plunger at the top of its stroke, fuel at pump pressure is free to flow into and out of the upper and lower ports of the bushing.

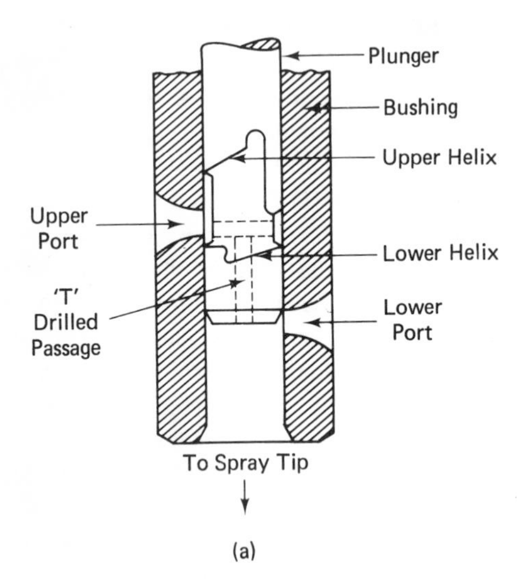

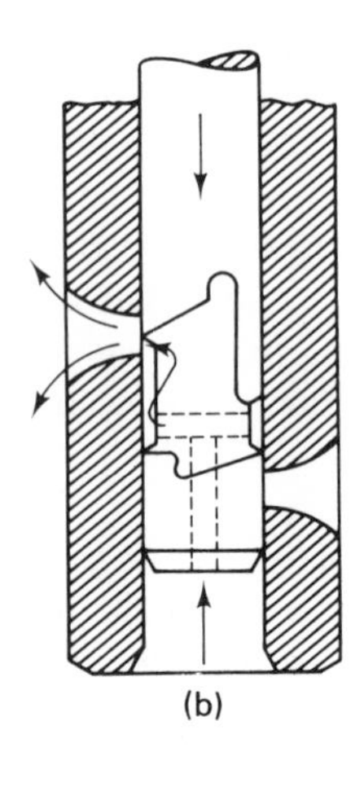

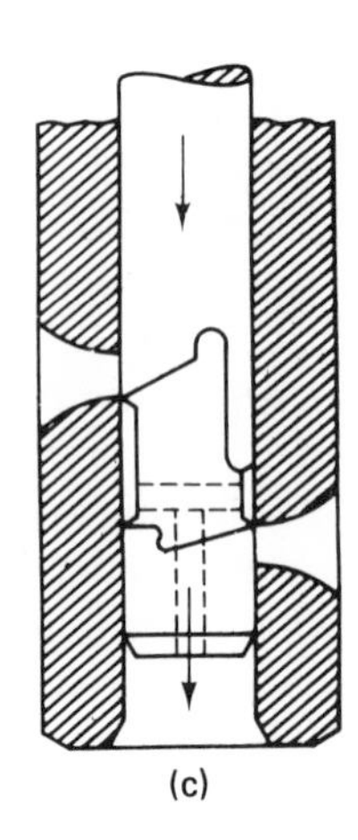

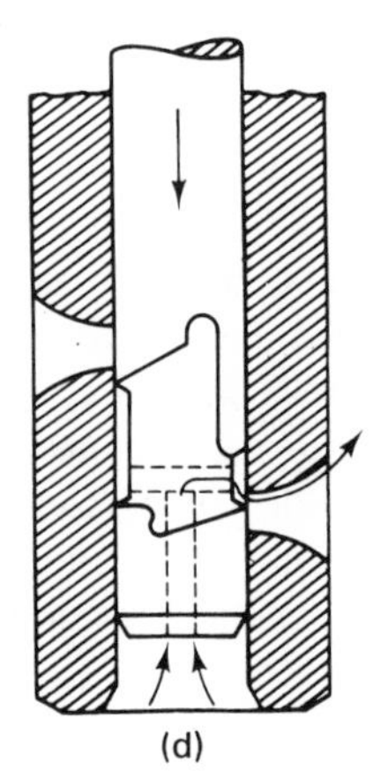

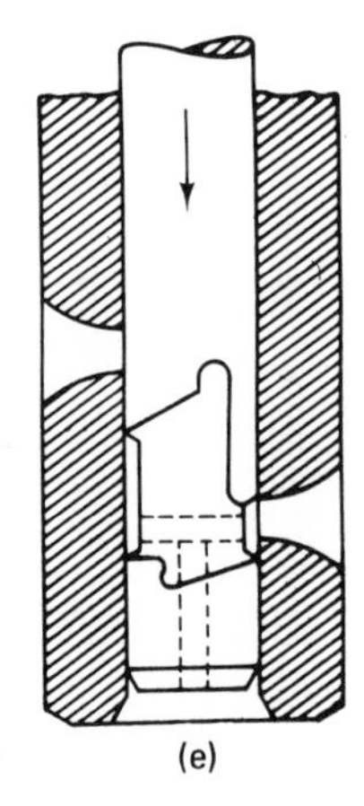

Figure 10-16
Phases of Injector Operation Through Vertical Movement of Plunger.

In Figure 10-16(b), as the plunger is being forced down by the action of the rocker arm on the injector follower, the lower port will be closed first, then the upper. Before the upper port is shut off, fuel being displaced by the descending plunger may flow up through the T drilled passage in the bottom of the plunger and escape through the upper port of the bushing.

After the upper port has been cut off [Figure 10-16(c)], fuel can no longer escape and is forced down by the plunger until it builds up sufficient pressure to pop the check valve, after which it sprays out the tip into the combustion chamber.

Injection will cease as soon as the downward moving plunger uncovers the lower port, as shown in Figure 10-16(d), because fuel under pressure will escape, striking the hardened spill deflector surrounding the bushing, up through the injector fuel outlet pipe, fuel outlet manifold, restricted fitting, and back to the fuel tank. In Figure 10-16(e), the plunger continues its downward travel, which helps to displace the hot fuel out of the lower port. The plunger is then returned by the follower spring to the top of its stroke.

Fuel Metering from No Load to Full Load

From Figure 10-16 you should now be aware that the start of injection depends on the point at which the upper land with its helix edge closes off the upper port. Therefore, the sooner the upper port is shut off, the earlier injection will commence. Basically, as the bottom port

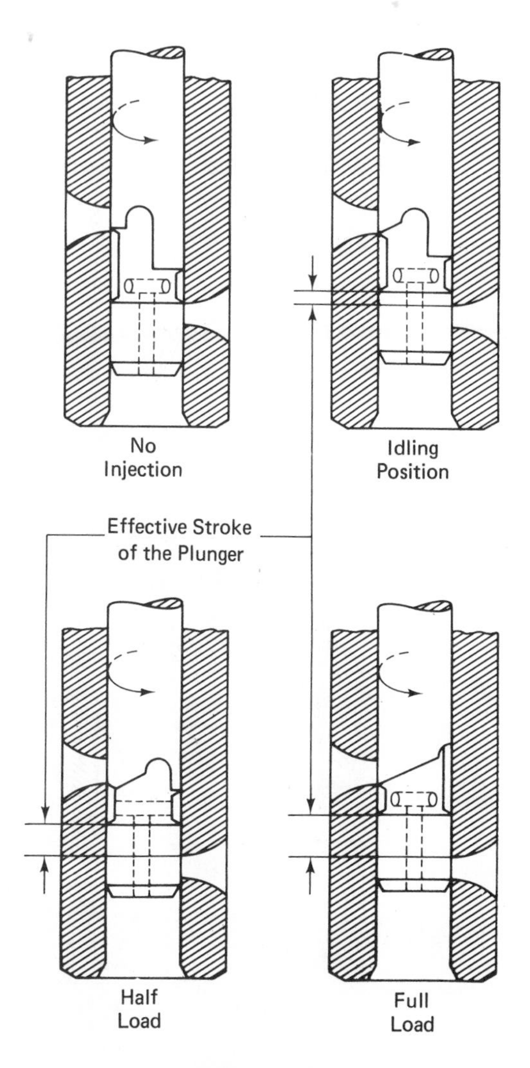

Figure 10-17
Sequence of Events from the No Injection Position to the Full-Load Position.

is closed, the upper port closes immediately behind it, and as the bottom port is uncovered, injection ceases forthwith. It may be said then that the sooner the upper port closes, the longer the time will be, until the lower port is uncovered; thus more fuel will be injected.

If the plunger is turned or rotated by the action of the fuel control rack so that the upper helix edge closes off the upper port later in its downward stroke, injection will start later and continue for a shorter period of time, meaning less fuel will be injected.

Figure 10-17 shows the sequence of events from the no-injection position to the full-load position that is controlled by the rotation of the injector plunger by the fuel control rack, which is itself actuated by either the operator on a hand or foot throttle or by engine governor action.

The no-fuel position occurs when the plunger is positioned so that the upper helix is at its extreme upper position adjacent to the upper port; by the time the upper port is closed, the lower port has already opened or been uncovered by the lower end of the plunger. The plunger is always free to slide up and down in the internal diameter of the rack control gear.

Relationship of Upper and Lower Helixes on the Injector Plunger

The angle at which the upper and lower helixes are ground on the injector plunger is controlled accurately to afford precise control of fuel injection. This relationship between the upper and lower helixes determines the amount of fuel injected for any given setting of the fuel control rack. If this relationship is altered, the performance of the injector is changed, and if this varies from one plunger to another, engine performance will be rough and erratic. As previously explained, the upper helix controls the start of injection timing. Therefore, at periods of low power, injection will start later in the piston's compression stroke, and at high power injection will have to start earlier in the compression stroke to allow adequate time for the additional fuel charge to burn. Again, in the no-fuel position, the upper helix cuts off the upper port only after the lower port has been exposed by the bottom portion of the plunger; therefore, no fuel is injected.

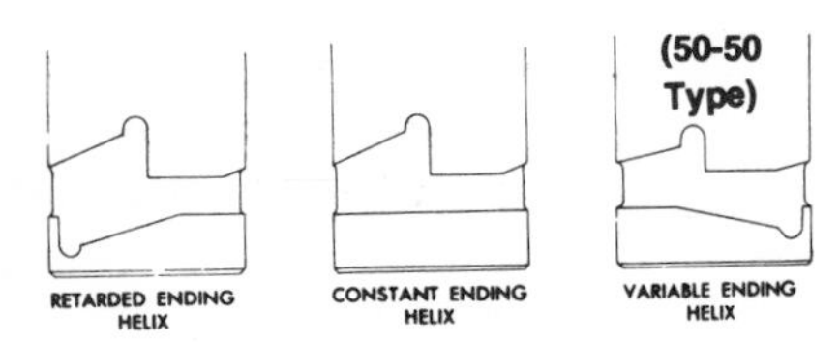

Figure 10-18
Common Types of Lower Plunger Land Machining.
(Courtesy of Detroit Diesel Allison Div. of GMC)

The lower helix is set to control the end-of-injection timing in all Detroit Diesel Allison injectors. There are three general types of lower plunger land machining employed with DDA injectors. Figure 10-18 shows the common types presently in use by DDA.

1. In the retarded type of plunger, the lower helix is a right-hand spiral (as is the upper helix), but has a much shallower angle than the upper helix. As engine speed is increased by the fuel control rack rotating the injector plunger, the upper helix advances the actual start of injection timing for greater fuel delivery, and the lower helix will also advance the shutoff timing, but to a lesser extent.
2. The lower helix controls the end of injection timing in all three types of plungers. With the constant ending type, the lower land is actually flat and is not a helix. Therefore, the lower port will always be uncovered at the same point in the stroke regardless of the fuel control position of the plunger, and thus injection is cut off or stops at the same time BTDC, no matter what speed.

10-7 Injector Removal

1. Steam clean the valve rocker cover area and adjacent surroundings to prevent the entrance of any dirt into the engine.
2. Remove the valve rocker covers; loosen and remove the injector fuel pipes. Immediately install plastic shipping caps over all injector fuel cap studs and all other fuel connectors, and open fuel lines to prevent the entrance of any dirt.
3. Crank the engine over, or bar it over, until

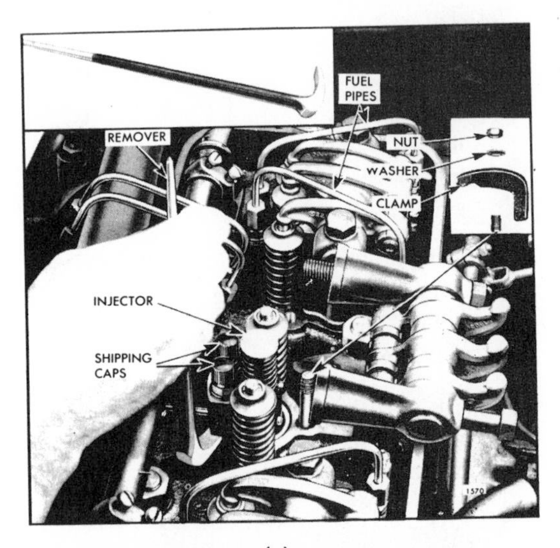

(a)

(b)

Figure 10-19
Injector Removal on: (a) 71 series, two-valve head, and (b) 71 series, four-valve head.
(Courtesy of Detroit Diesel Allison Div. of GMC)

the flats across the rocker arms (at the push rod) are all in line or horizontal (53, 71, and 92 series engines only).

4. On 149 engines, due to the fact that the fuel manifold is higher than the injector, as soon as you loosen the fuel pipe nut at the injector, slide the nut back on the pipe and push a $\frac{3}{8}$-in. or 9.525-mm-inside-diameter hose 2 to 3 ft (0.609 to 0.914 m) in length over the end of the fuel pipe. Loosen the nut on the fuel pipe at the manifold end slightly to allow the fuel pipe to swing away from the injector, and drain the fuel from the manifold into a container.

Note: Failure to do this will allow fuel oil to drain onto the cylinder head, which will in turn cause lube oil dilution leading to bearing damage of both the main and C/rods and other major problems.

5. Install plastic shipping caps over all fuel inlet and outlet holes.
6. On 53, 71, and 92 engines, loosen and remove the two rocker shaft hold-down bolts; tip back the rocker assemblies clear of the valves and injector [see Figure 10-19(a) and (b)]. On 149 engines, since these engines employ a high mounted camshaft, remove the two rocker arm shaft to camshaft bearing cap bolts; then remove the shaft and injector and valve operating mechanism [see Figure 10-20(a) and (b) opposite].
7. On four-valve cylinder heads, remove the two exhaust valve bridges by lifting them from their guides (reinstall them on the same guide).
8. Remove the injector clamp hold-down bolt and clamp; then loosen both the inner and outer adjusting screws on the injector rack control lever tube far enough to allow you to slide the lever away from the injector (some engines may only have an inner screw held by a locknut).
9. Insert a small heel bar under the injector body, taking care not to exert any pressure directly on the control rack, and gently pry the injector from the cylinder head [Figure 10-19(a) and (b)].
10. At this time, cover the injector hole in the cylinder head to prevent any entrance of foreign material. If you are removing the injector from a four-valve head 53 series

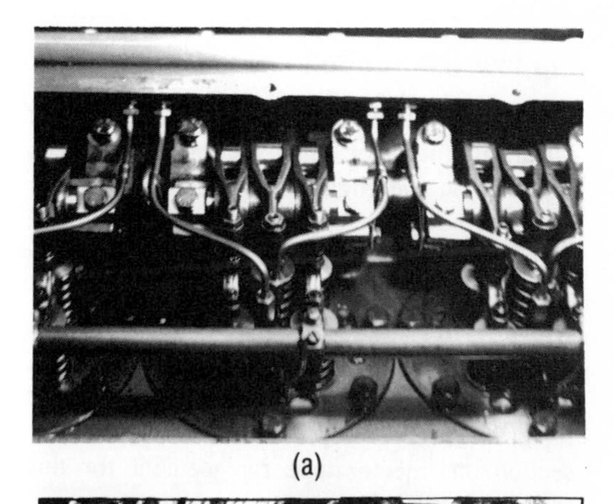

(a)

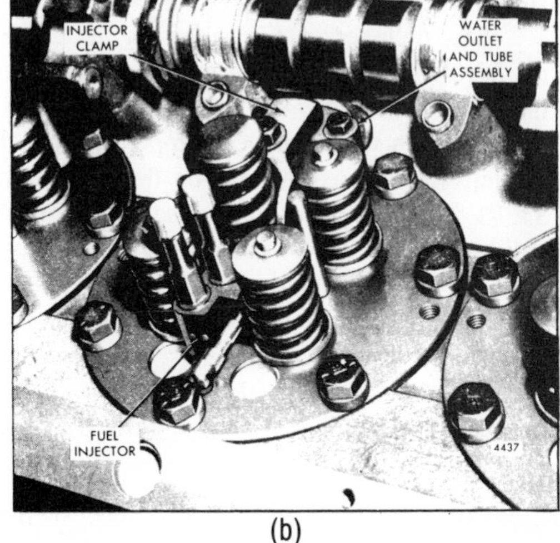

(b)

Figure 10-20
Injector Removal on: (a) 149 series, and (b) fuel control tube completely removed. (Courtesy of Detroit Allison Div. of GMC)

engine, there is no separate bridge mechanism. It is attached to the end of the rocker arm by a pin and is self-centering when in contact with the valve stems.

11. The exterior of the injector should now be cleaned with clean fuel oil and dried with compressed air prior to any additional tests.

10-8

Injector Installation

If the cylinder head is off the engine, do not install the injector until the head has been replaced on the engine; however, the injector tube can be cleaned of carbon while the head is off, which will minimize the possibility of carbon particles dropping into the cylinder. If the cylinder head has been removed for any reason other than injector replacement, the injector copper tubes in the head may be replaced if necessary; however, refer to the section on injector tube replacement for this function in all DDA service manuals.

If the cylinder head is on the engine, check the beveled seat of the injector tube where the injector nut seats for any signs of carbon deposits, which would prevent proper seating of the injector. To remove carbon deposits from 53, 71, and 92 injector tubes, use injector tube bevel reamer J 5286-9, and on 149 engines use reamer J 22342-11. When using these reamers, be very careful to remove only the carbon and not the copper from the tube itself; otherwise, the clearance between the injector and the cylinder head will be altered with possible disastrous results.

Note: It is advisable to pack the flutes of the reamer with grease to retain the carbon removed from the tube and to prevent any carbon from dropping into the cylinder.

The injector should be filled with fuel oil through the inlet filter cap until it runs out of the outlet cap prior to installation into the cylinder head.

1. Insert the injector into the injector tube, making sure that the dowel pin on the underside of the injector body fits into the mating hole in the cylinder head.
2. Slide the injector rack control lever on the control tube on the head over until it sits into the injector control rack end.
3. Install the injector clamp and special washer (with the curved side toward the injector clamp), and bolt and tighten to 20 to 25 lb-ft (27 to 34 N·m) maximum on 53, 71, and 92 injectors. On 149 engines the clamp bolt is tightened to 46 to 50 lb-ft maximum (62.36 to 67.79 N·m).

Caution: Check to make sure that the injector clamp is centered over the follower spring prior to tightening; otherwise, the spring may contact the clamp during injector operation. In addition, overtorquing of the injector clamp bolt can cause the injector control rack to stick or bind.

4. On four-valve head engines (71, 92, and 149) install the exhaust valve bridges over their guides and onto the valve stem tips (53 series four-valve head engines have the bridge mechanism attached to the end of the rocker arms).
5. Move the rocker arm assemblies into position and tighten the hold-down bolts to the following specifications:

   53 series engines: 50 to 55 lb-ft (67.79 to 74.56 N·m).

   71 and 92 series engines: 90 to 100 lb-ft (122 to 135.58 N·m).

   71 and 92 series engines (only on the two bolts attaching a load limit or power control screw if used to the rocker arm shaft brackets): 75 to 85 lb-ft (101.68 to 115.24 N·m).

   149 series engines: 90 to 100 lb-ft (122 to 135.58 N·m).

6. Remove the shipping caps from the fuel inlet and outlet studs both at the injector and cylinder head. Install the fuel pipes and tighten them as follows:

   53, 71, and 92 engines: 12 to 15 lb-ft (16 to 20 N·m).

   149 engines: 18 to 22 lb-ft (24.4 to 29.82 N·m).

   On 53, 71, and 92 engines, use of fuel line socket J 8932-01 to tighten the fuel lines is necessary, and on 149 engines socket adaptor J 21545 to tighten the fuel pipe to injector nuts and adaptor J 23385 must be used due to space limitations to tighten the fuel pipe to fuel manifold connector nuts.

Note: Extreme care must be used so as not to bend the fuel pipes during installation; also, overtightening of the fuel pipe nuts can twist or fracture the flared end of the fuel pipe, resulting in leaks that cause lube oil dilution and damage to engine bearings.

7. After all injectors have been installed, a complete engine tune-up as outlined later in this manual will be necessary. In addition, the fuel system should be primed as described earlier in this chapter with tool J 5956.
8. *Note:* If only one injector is removed from an engine, and the other injectors and the governor adjustment have not been disturbed, it is only necessary to adjust the valve clearance and time the injector for that one cylinder, and to position the injector rack control lever.

Priming New and Reconditioned Fuel Injectors

Many companies with fleets of varying sizes find it more profitable and quicker to simply purchase new or reconditioned Reliabilt exchange fuel injectors from their local DDA dealer or distributor, rather than purchase the necessary equipment to overhaul their own.

New injectors shipped from the factory are primed with a test oil and capped at the inlet and outlet fuel connections with plastic shipping caps to seal the oil inside. Distributors and dealers of DDA will also generally prime their rebuilt injectors in the same way.

The test oil inside the injectors serves two main functions:

1. It acts as a rust-preventive lubricant while the injector is in storage.
2. It ensures proper lubrication of close-fitting moving parts during initial engine start.

Plastic shipping caps normally retain this oil within the injector; however, if the injector is stored on its side or in an inverted position for a long period of time, the test oil can seep out into the container. To avoid possible scoring of the plunger and bushing in a new or rebuilt injector, prior to installing the injector into the engine, the following steps should be taken:

1. Check the injector container and Styrofoam insert for signs of test oil leakage.
2. When leakage is evident, remove the plastic caps and fill the injector with fuel oil before it is installed in the cylinder head. Add clean fuel oil at the inlet filter cap until it runs out the outlet filter cap.
3. After all injectors are installed, prime the entire fuel system as described earlier before starting the engine.

10-9

Care and Overhaul of Unit Injectors

Detroit Diesel Allison covers in great depth the maintenance and overhaul of their injectors in Section 2.1 of all the respective engine series service manuals. However, the following discusses some of the more commonly accepted tests required on the injector to determine its suitability for use, without delving into the detailed overhaul readily available within DDA service manuals.

As with all fuel injection equipment regardless of its manufacture, the amount of injector work performed will depend to a great extent on the following factors:

1. Available service facilities
2. Experience and training of maintenance personnel
3. Company policy regarding rebuild
4. Extent of service required
5. Whether company or operator owned

Whenever injection equipment is to be serviced, the need for cleanliness and organization cannot be overemphasized. Servicing must be carried out under spotless surroundings; otherwise, problems will invariably occur. Ideally, all injector repairs should be performed in a clean, well-lighted room with a dust-free atmosphere, which can be assured by means of an electric fan that draws filtered air into the room to provide slight pressure to keep any dust or dirt from entering. A suitable air outlet for fumes is also desirable.

In addition, filtered, moisture-proof compressed air will be required for drying parts.

Injector Tests

The first step here is to visually inspect the injector unit for any signs of physical damage to the spray tip, body, nut, control rack, follower and spring, and so on. The other major consideration as to whether or not the injector requires overhauling is to determine what kinds of problems existed that necessitated its removal.

For many fleets, at certain miles (kilometers) or hours, it is part of the company's preventive maintenance policy to change the injectors. This of course is totally dependent on how well the fuel system has been maintained.

The following tests to an injector are required before it can be considered satisfactory for return to service without disassembly, except for a visual check of the plunger.

1. Injector control rack and plunger movement test
2. Injector valve opening pressure test (series 53 and 71 Crown Valve S and HV only)
3. Injector valve holding pressure test
4. Injector high pressure test
5. Spray pattern test
6. Plunger inspection
7. Needle valve opening and spray pattern test
8. Injector tip pressure holding check (needle valve unit only)
9. Injector needle valve lift movement
10. Injector fuel output test (all injectors)

Injector Control Rack Freeness and Plunger Movement Test

1. Figure 10-21 shows a typical injector installed in tester J 22396.

Figure 10-21
Injector Installed in Tester J 22396.
(Courtesy of Detroit Diesel Allison Div. of GMC)

Figure 10-22
Injector Valve Opening Pressure Test.
(Courtesy of Detroit Diesel Allison Div. of GMC)

2. The adjustment screw on the tester handle must be set so that when the handle is parallel to the injector the follower is in fact fully compressed.
3. With the follower spring fully compressed, place the rack in the no-fuel position and slowly release the handle. The injector rack should fall freely under its own weight (check this several times); otherwise, it indicates internal injector damage.
4. A quick check is to hold the injector in your hand, turn it over and back, and see if the rack will free fall. If no tester is available, arrange to depress the follower spring and do the same test.

Injector Valve Opening Pressure Test (High Valve Injectors Only). With the injector mounted as shown in Figure 10-22, place the rack in the full-fuel position, and by operating the test handle, note when the injector *pops* or sprays fuel. On this type of injector, it should be between 450 and 850 psi (3103 and 5861 kPa).

Injector Valve Holding Pressure Test (All Injectors)

1. The injector should be placed in the test fixture as shown in Figure 10-22.
2. The purpose of this test is to determine whether or not the plunger to bushing fit and the various lapped surfaces in the injector are in fact leak free so that the injector will be capable of developing the necessary high pressures required for injection.
3. Bring the pressure up to around 500 psi (3448 kPa) by use of the pump handle, and immediately close the pump shutoff valve. The time taken for the fuel pressure to drop from 450 to 250 psi (3103 to 1724 kPa) should not be less than 40 seconds; otherwise, internal or external leakage is evident. Dry off the injector, repeat the test, and if there are no signs of external leakage, the internal parts are worn and require lapping or replacing.

Injector High Pressure Test (All Injectors)

1. This test will disclose fuel leaks that may not have been evident during step 3 owing to the lower pressure used at that time. Plunger and bushing clearance condition will be more apparent during this test.
2. Figure 10-23 shows the setup. With the fuel rack in full fuel, lock the test handle by means of the handle lock.
3. Operate the pump handle to raise the pressure, and vary the adjusting screw of the

Figure 10-23
Injector High Pressure Test.
(Courtesy of Detroit Diesel Allison Div. of GMC)

tester handle at the point where the injector spray will decrease appreciably and a rise in pressure will occur. At this point, both ports of the bushing are closed.

4. If you cannot obtain a pressure beyond that of the normal valve opening pressure, replacement of the plunger and bushing assembly is required.
5. If you can maintain a pressure of 1600 to 2000 psi (11,032 to 13,790 kPa), check for signs of injector leakage. A sweat or seepage of fuel at the rack is normal during this test.

Spray Pattern Test (All Injectors). With the rack in the full-fuel position, open the fuel shutoff valve, and by operating the test handle at around 40 strokes/min, closely observe the spray pattern. To determine if all the orifices in the tip are open, you can lay a piece of paper towel into the empty plastic spray deflector bowl, pump the handle as stated, and count the number of spray lines imprinted on the paper towel.

Caution: Be very careful when working with injectors to keep your hands well clear of the spray pattern; otherwise, the fuel could penetrate your skin, entering the bloodstream, and cause serious infection.

To prevent damage to the pressure gauge it is imperative that you do not exceed 100 psi (689.5 kPa) during this test. There should be no sluggishness to the start and end of injection; it should be sharp and crisp, and a finely atomized spray absent of any drops of oil at the end of the tip should be the result of a good injector. Failure to meet this criterion would necessitate injector overhaul.

Visual Inspection of the Plunger. If an injector fails to pass any of the previous tests, it would have to be disassembled and inspected. However, even though the injector may pass all these tests, it is still possible for the plunger to be worn or chipped. A chip on the bottom area of the upper or lower helix will not necessarily be indicated in any of these tests; therefore, the plunger should be removed from the injector unit as follows:

1. Support the injector as it would normally sit in the engine in holding fixture J 22396, compress the follower spring by hand, and with the aid of a small screwdriver raise the spring slightly from the injector body to decrease the tension on the small injector follower stop pin; then pull the pin out. To prevent the follower spring from flying up, gradually allow it to come up (see Figure 10-24).
2. Remove the injector from its fixture, turn

it upside down, and catch the spring and plunger as they drop out.

3. Proceed to inspect the plunger visually under a lighted type of magnifying glass, if available, for signs of wear or chips in the helix area. Refer to Figure 10-25 for conditions of unusable plungers.

Figure 10-25 shows three typical cases of unusable plungers. Case A shows an advanced state of erosion caused by poor maintenance practices; dirt in the fuel has badly damaged the plunger. Case B shows a chip at the lower helix area, and case C shows a condition that can be caused by water in the fuel or a lack of fuel under high speeds and load. If the plunger exhibits any of these characteristics, it should be replaced with a new assembly consisting of a matched plunger and bushing.

Reinstall the plunger, follower, and spring if the plunger shows no signs of wear or any of these other problems, and the injector has passed all the tests listed.

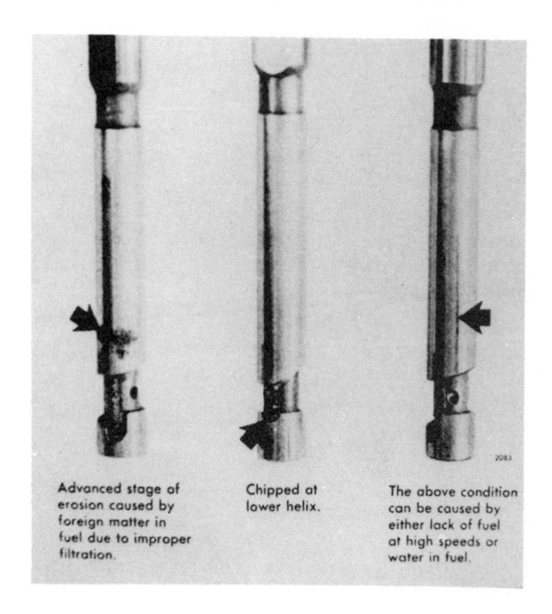

A B C

Figure 10-25
Three Unusable Plungers.
(Courtesy of Detroit Diesel Allison Div. of GMC)

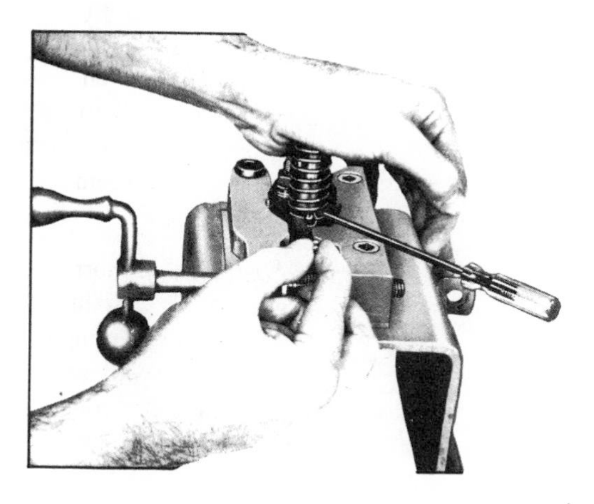

Figure 10-24
Removing Injector Plunger for Visual Inspection.
(Courtesy of Detroit Diesel Allison Div. of GMC)

Needle Valve Opening and Spray Pattern Test

1. Place the injector in the holding fixture, loosen the injector nut, and remove it; then remove all the parts above the lapped surfaces of the injector bushing (see Figures 10-13 and 10-14).

2. If any carbon is present on the spray tip seat inside of the injector nut, carefully insert carbon remover tools J 9418-1 and -5 one at a time into the nut as shown in Figure 10-26. Turn the tool in a clockwise direction to remove any carbon buildup. *Exercise extreme care* to prevent removing any metal from the spray tip seat or to avoid creating any burrs. *Be sure* to wash the nut in clean fuel oil and dry with compressed air when finished. Figure 10-26 shows a cutaway view of an injector nut with a carbon removing tool used to

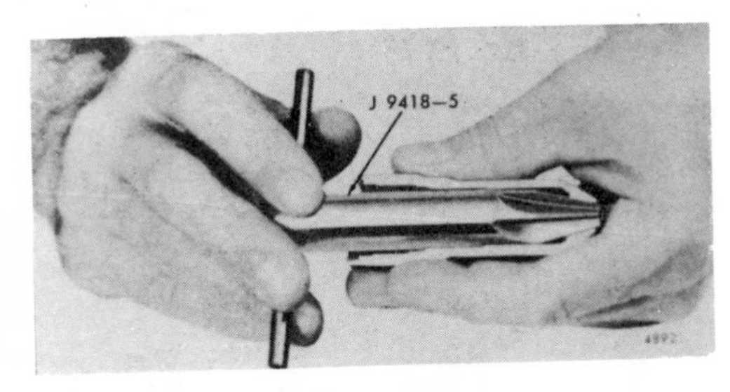

Figure 10-26
Cutaway View of Injector Nut and Carbon Removing Tool.
(Courtesy of Detroit Diesel Allison Div. of GMC)

clean the spray tip seating area, and the lower end of the nut.

3. If during the spray pattern test conducted under step 5 there were indications of carbon in the spray tip orifices, clean these and the tip cavity below the needle valve either through ultrasonic cleaning or with a brass brush, clean fuel, and pricker wire.
4. With all parts cleaned, assemble the check valve cage, spring, spring seat, spring cage, and the needle valve and tip assembly onto the auxiliary tester J 22640 with the proper dummy plunger and injector body unit in place, as shown in Figure 10-27.
5. Carefully guide the injector nut over the stacked up parts on the tester, gently thread it onto the body, and tighten it to a torque of 75 to 85 lb-ft (102 to 115 N·m) on 53, 71, and 92 series engines, or to 120 to 140 lb-ft torque (162.6 to 189.7 N·m) on 149 engines.
6. Connect the auxiliary tester J 22640 to the injector tester J 9787 as shown in Figure 10-28.
7. Make sure that you install the spray shield in the auxiliary tester prior to operating the pump handle.
8. With reference to Figure 10-28, open both shutoff valves on both testers.
9. To simulate the spray tip functioning in the engine, operate the pump handle at about 40 strokes/min. The valve should open in the spray tip at the following pressures:

   Series 53 and 71 engines: 2300 to 3300 psi (15,858.5 to 22,753.5 kPa).

   Series 92 and 149 engines: 2600 to 3200 psi (17,927 to 22,064 kPa).

Both the beginning and end of delivery should be sharp and crisp with the fuel in a finely atomized spray. If the valve opening pressure is below 2600 psi (17,927 kPa) on the 149 engine injector, or below 2300 psi on the 53, 71, or 92 injector, replace the needle valve spring. Also replace the spring if atomization is poor.

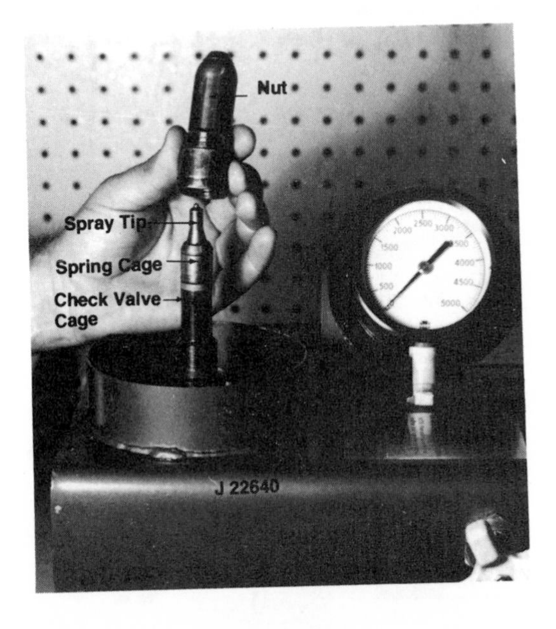

Figure 10-27
Installing Injector Components on Auxiliary Tester.
(Courtesy of Detroit Diesel Allison Div. of GMC)

On 149 engine injectors, if atomization is poor, loosen the injector nut and rotate the tip 60 degrees, tighten the nut to 120 to 140 lb-ft (162.6 to 189.7 N·m) torque, and repeat the spray pattern test. The spray tip can be rotated up to five times using 60 degree increments, after which you are back to your original position if you carry on to six. If, after rotating the tip, atomization is still poor, the needle valve and seat should be replaced.

Note: A needle valve and seat are only sold as a matched unit.

Injector Tip Pressure Holding Check. If the valve opening pressure is within the range specified, proceed to check for spray tip seat leakage by actuating the pump handle to bring the pressure up to 1500 psi (10,342.5 kPa) and hold it there for 15 seconds. Inspect the spray tip seat for leakage (you may have to lightly wipe the tip off first) to be sure that there is no more than a slight sweat. Any fuel droplets will necessitate changing the tip.

If the tip passes inspection, with the pressure at 1500 psi (10,342.5 kPa), close the shutoff valve on the auxiliary tester, and if the pressure drops more than 500 psi (3447.5 kPa) in 5 seconds (1500 to 1000 psi or 10,342.5 to 6895 kPa in less than 5 seconds), replace the needle valve and tip assembly.

Needle Valve Lift Measurement. If the needle valve assembly passes the previous needle valve tests, the needle valve lift measurement can be omitted. Otherwise, proceed as follows:

1. On 53, 71, and 92 series engines, use tool J 9462-01, and on 149 series engines, use tool J 9462-02 (see Figure 10-29).
2. Use the tool as follows; zero the dial

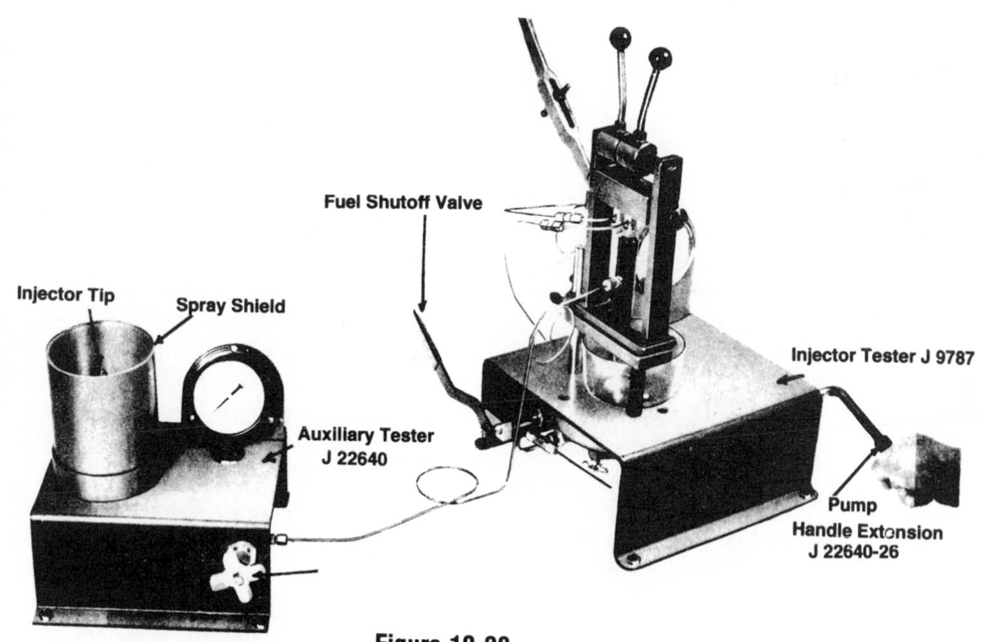

Figure 10-28
Auxiliary Tester J 22640 Connected to Injector Tester J 9787.
(Courtesy of Detroit Diesel Allison Div. of GMC)

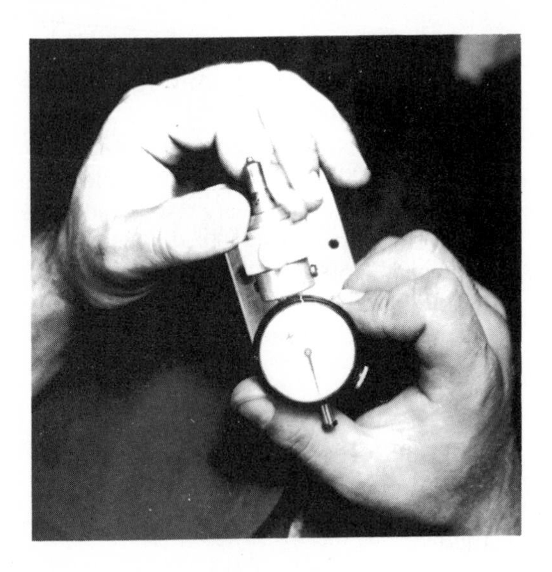

Figure 10-29
Needle Valve Lift Measurement.
(Courtesy of Detroit Diesel Allison Div. of GMC)

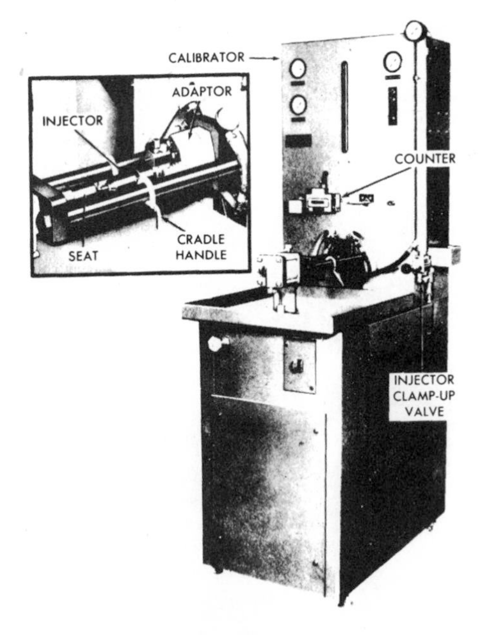

Figure 10-30
Injector Mounted in Calibration Machine for Flow Testing.
(Courtesy of Detroit Diesel Allison Div. of GMC)

indicator unit by placing it on a flat surface and turning the serrated face of the gauge until its pointer is on zero.

3. Carefully place the spray tip and needle valve assembly tight against the bottom of the gauge with the quill of the needle valve in the hole of the plunger, as shown in Figure 10-29.
4. Read the needle valve lift on the face of the dial indicator. It should be as follows:

   53, 71, and 92 injectors: 0.008 to 0.018 in. (0.2032 to 0.4572 mm).

   149 injectors: 0.012 to 0.016 in. (0.3048 to 0.4064 mm).
5. If the checking dimension exceeds the maximum figure given, the spray tip assembly should be replaced. If it is less than the minimum figure given, inspect closely for any foreign material between the needle valve and the tip seat.
6. If, however, the lift is within its specified limits, simply install a new needle valve spring and recheck the valve opening pressure and valve action. If poor atomization or low valve opening pressure exists, replace with a new spray tip assembly.

Injector Fuel Output Test (All Injectors). Testing of an injector for fuel output or delivery can only be done on a *comparator* or *calibrator*. Fuel output tests of DDA-type unit injectors have been performed for many years on comparators of Bacharach, Hartridge, or Kent-Moore manufacture. These machines activate the injector plunger stroke through an eccentric drive mechanism.

Today many companies that rebuild injectors sell them in what is commonly known as *matched and balanced sets;* each injector is flow tested and matched to the others within close limits of cubic centimeters of output per 1000 strokes on the comparator or calibrator.

Recently, improved injector calibrators have become available for injector testing pur-

Table 10-2
SUMMARY OF SERIES 53 AND 71 N INJECTORS

Engine	Inj.	Parts Catalog Type No.	Assembly Part No.	Plunger Desig-nation	P&B Part No.	Tip Part No.	Spray Tip	J 22410 Calibrator Min.–Max.	J [illegible] Comparator Min.–Max.	[illegible] Timing
53	L40	86	5228793	4L	5228679	5229026	8–.0055–165°A	41–45	4–10	1.460
53	N40	75	5228763	4N	5228691	5229034	6–.006–165°A	43–46	4–10	1.460
53	N45	76	5228773	45N	5228684	5229034	6–.006–165°A	47–51	14–20	1.460
53	N50	74	5228783	5N	5228749	5229034	6–.006–165°A	50–54	18–24	1.460
71	71N5	92	5228850	5N	5228749	5229026	8–.0055–165°A	50–54	18–24	1.460
71	N55	77	5228785	55N	5228668	5229026	8–.0055–165°A	53–57	24–30	1.460
71	N60	78	5228760	6N	5228656	5229026	8–.0055–165°A	57–61	30–36	1.460[a]
71	N65[b,c]	97	5228900	N65	5228928	5229030	8–.006–165°A	64–68	32–38	1.484[d]
71	N70[e]	80	5228770	7N	5228682	5229042[f]	7–.006–165°	71–75	38–44	1.460[g]
71	N75[e]	81	5228777	75N	5228658	5229042[f]	7–.006–165°	75–79	41–47	1.460[g]
71	N80[c,e]	82	5228780	8N	5228661	5229256	7–.0065–165°A	81–85	44–50	1.484
71	N90	100	5228790	9N	5228878	5229256	7–.0065–165°A	87–92	50–56	1.460

[a]Use 1.484 for turbocharged automotive engines.

[b]Brown tag.

[c]Except generator sets (.1460).

[d]Use 1.520 for 8V-71 TA.

[e]Multifuel use approved.

[f].030 wall spray tip.

[g]1.496 for 8V-71 TA.

Note: Comparator testing may not be accurate for needle valve injectors.

(Courtesy of Detroit Diesel Allison Division of GMC)

poses, and DDA recommends the use of a calibrator rather than a comparator, although both are capable of flow testing the injector. The calibrator activates the injector plunger stroke through a cam profile similar to the engine cam. In addition, the prescribed range of fuel output obtainable on a calibrator is not of the same number of cubic centimeters per 1000 strokes as the number obtainable on a comparator.

Note: An injector will not deliver the same amount of fuel on a comparator as on a calibrator.

Note: Current production injectors use only a filter on the inlet side of the injector body. The inlet side of the injector is always the fitting directly above the control rack, or when viewing the injector from the fuel control tube side when it is in the engine, the inlet would be on the right-hand side. Anytime that the injector is mounted on a calibrator for flow testing, keep this in mind.

Figure 10-30 shows an injector mounted into the cradle of the calibrator for flow testing. Prior to flow testing, set the counter to 1000 strokes as shown in Figure 10-31.

Figure 10-32 shows the check to determine spray tip concentricity. This check for spray tip runout is required anytime that the injector has been taken apart and then reassembled. If the total runout during one complete turn of the injector exceeds 0.008 in. (0.203 mm), remove the injector from the gauge. Loosen the injector nut, center the spray tip, and retorque the nut as per specifications. Recheck the tip runout.

Testing Reconditioned Injectors

Before the rebuilt injector can be placed into service, all the tests, with the exception of the visual inspection of the plunger, that were outlined in this section must be undertaken. If the injector fails to pass any one of these tests, it is usually an indication that defective or dirty parts have been assembled during rebuild, and would necessitate injector disassembly once again.

Figure 10-31
Setting Counter on Calibrator to 1000 Strokes. (Courtesy of Detroit Diesel Allison Div. of GMC)

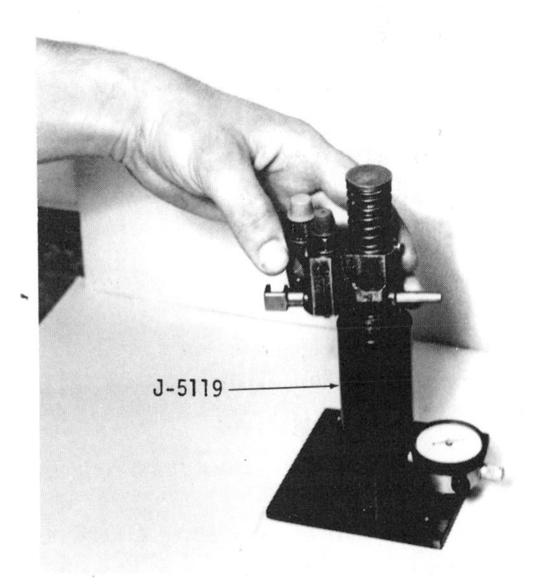

Figure 10-32
Checking Spray Tip Concentricity. (Courtesy of Detroit Diesel Allison Div. of GMC)

If the rebuilt injector is to be placed in stock after rebuild, fill it with a typical injector test oil such as J 8130. Be sure to install clean plastic shipping caps on both filter caps immediately after filling with test oil. *Do not fill the injector with fuel oil for storage purposes* since any impurities in the fuel can cause rusting of internal parts.

10-10

Injector Information Charts

Tables 10-1 through 10-7 provide specifications for Detroit Diesel Allison injectors.

10-11

Governor Operation

Detroit Diesel engines use both mechanical and hydraulic governors on their engines of the following type.

1. Mechanical limiting speed type
2. Variable mechanical speed type
3. Variable low-speed limiting speed mechanical governor
4. Mechanical constant speed governor (earlier engines)
5. Dual-range limiting speed mechanical governor
6. Woodward SG hydraulic (refer to Chapter 11 for Woodward governors)
7. Woodward PSG hydraulic
8. Woodward electric (various models)

The function of all these governors, whether mechanical or hydraulic, is to control engine speed and correct for any change in load applied or removed from the engine. They all work on the basic principle of weights against spring pressure; therefore, all governors are of the speed-sensing type. The electric governor relies on a magnetic pickup for its response.

Since the action of all the mechanical governors is the same, but with a difference only in purpose, let us study the limiting and variable speed governors. Once you understand these, you will have no problem in interpreting the action of the others.

Limiting Speed Type

This governor is found in both the single- and double-weight version, and can also be found on both in-line and Vee-type engines. Riveted on to the side of the governor housing is an identification plate, which shows the following:

1. Governor part number
2. Date of manufacture
3. Idle speed range
4. Type, such as DWLS, meaning double-weight limiting speed governor
5. Drive ratio

The governor used on the in-line engines differs from that used on the Vee in that its housing is taller because of its location on the engine, whereas the governor housing on the Vee engine is located and driven from the blower. Other than the difference in shape and mounting location, the governor operates in the same manner. Regardless of whether the limiting speed governor is of the single- or double-weight variety, the action of the governor is the same. The purpose of the limiting speed governor is as follows: (1) to control the engine idle speed, and (2) to limit the maximum speed of the engine.

The governor provides full fuel for starting purposes; the governor spring force acting through the bellcrank lever allows the differential lever to be placed into the position

Table 10-1
SUMMARY OF SERIES 53 AND 71 HIGH VALVE CONFIGURATION INJECTORS (ALL UNITS OF MEASURE IN CC)

Engine	Inj.	Parts Catalog Type No.	Assy. Standard Body	Part No. Offset Body	Plunger Desig.	P&B Part No.	Spray Tip	Tip Part No.	J 22410 Calibrator Min.–Max.	J 7041[a] Comparator Min.–Max.	Injector Timing
53[b]	35	41	—	5228535[c]	35	5228530	8–.005–165°	5228436	36–40	0– 6[d]	1.508
53	35	41	—	5228535[c]	35	5228530	8–.005–165°	5228436	36–40	0– 6[d]	1.484
53	40	47	—	5228560	40[f]	5228552	8–.005–165°	5228436	39–43	6–12[d]	1.484
53	S40[e]	51	—	5228590	40[f]	5228552	8–.0055–165°	5228473	43–47	0– 6	1.460
53	S45[e]	52	—	5228580[c]	45[f]	5228495	8–.0055–165°	5228473	48–52	6–12	[illegible]460
53	S50[e]	56	—	5228550[c]	55	5228126	8–.0055–165°	5228473	52–56	9–[illegible]	[illegible]60
71	HV55	14	5228300	—	55	5228126	6–.006–155°–H	5228296	52–56	[illegible]	[illegible]4
71	NV6	11	5228310	—	6H	5228306	6–.006–155°–H	5228296	59–[illegible]	[illegible]	[illegible]
71	HV7	12,25	5228305	5228470	7H	5228304	7–.006–165°–H	5228554	[illegible]	[illegible]	[illegible]
71	HV8	13,24	5228110	5228435	8H	5228101	7–.006–165°–H	522[illegible]	[illegible]	[illegible]	[illegible]
71	HV9	21	5228380	—	9H	5228378	7–.006–165°–H	[illegible]	[illegible]	[illegible]	[illegible]
71	S80	50	—	5228524[c]	8H	5228101	7–.006–165°–[illegible]	[illegible]	[illegible]	[illegible]	[illegible]
71	S90	48	—	5228525[c]	9X	5228549	7–.006–[illegible]	[illegible]	[illegible]	[illegible]	[illegible]

[a]Set counter at 1000 strokes except where noted.
[b]Reefer car/two speed.
[c]Low clamp bodies, standard on series 53 in-line and V's and all four-valve [illegible]
[d]Set counter at 1200 strokes.
[e]Use only with low-velocity camshafts on 3-53 (5126745-LH [illegible]
[f]Manufacturer figure as shown on actual part.
Note: Use of J 8130 injector test oil will contrib[illegible]
(Courtesy of Detroit Diesel Allison Divis[illegible]

Table 10-3
SUMMARY OF SERIES 53 AND 71 C INJECTORS

Engine	*Inj.*	*Parts Catalog Type, No.*	*Assembly Part No.*	*Plunger Designation*	*P&B Part No.*	*Tip Part No.*	*Spray Tip*	*J 22410 Calibrator Min.–Max.*	*Injector Timing*
53	C40	109	5229340	4C	5229342	5229034	6–.006–165 A	42–46	1.460[a]
53	C45	108	5229345	45C	5229346	5229034	6–.006–165 A	47–51	1.460[a]
53	C50	107	5229350	5C	5229354	5229034	6–.006–165 A	50–54	1.460[a]
71	71C5	110	5229352	5C	5229354	5229026	8–.0055–165A	50–54	1.484
71	C55	111	5229355	55C	5229356	5229026	8–.0055–165A	53–57	1.460[b]
71	C60	113	5229360	6C	5229362	5229026	8–.0055–165A	57–61	1.460[b]
71	C65	115	5229365	65C	5229366	5229192	7–.006–165 A	64–68	1.484[c]
71	C70	114	5229370	7C	5229372	5229042	7–.006–165 A	71–75	1.484

[a]Use 1.470 for 4- and 6-cylinder automotive engines.

[b]Use 1.470 for City Coach engines with throttle delay.

[c]Use 1.460 for turbocharged automotive engines.

Note: Only a calibrator reading is given for C injector output. Comparator testing may not be accurate for needle valve type injectors.

(Courtesy of Detroit Diesel Allison Division of GMC)

Table 10-4
SUMMARY OF SERIES 71 B INJECTORS

Injector	*Parts Catalog Type, No.*	*Assembly Part No.*	*Plunger Desig-nation*	*P&B Part No.*	*Tip Part No.*	*Spray Tip*	*Calibrator Min.–Max.*	*Injector Timing*
B55	128	5229555	45C	5229346	5229532	8–.006–162°[a]	53–57	1.50
B60	129	5229560	5C	5229354	5229532	8–.006–162°[a]	57–61	1.500
B65	130	5229565	65B	5229542	5229030	8–.006–165°A	64–68	1.484
71B5	131	5229550	4C	5229342	5229532	8–.006–162°[a]	50–54	1.500
B55E	132	5229575	B55E	5229559	5229553	8–.006–162°[a]	53–57	1.500
7B5E	133	5229580	7B5E	5229562	5229553	8–.006–162°[a]	50–54	1.500

[a]Designation not shown on spray tip.

(Courtesy of Detroit Diesel Allison Division of GMC)

Table 10-5
SUMMARY OF SERIES 92 INJECTORS

Injector	*Parts Catalog Type, No.*	*Assembly Part No.*	*Plunger Desig-nation*	*P&B Part No.*	*Tip Part No.*	*Spray Tip*	*Calibrator Min.–Max.*	*Injector Timing*
9270	117	5229170	F7	522	5229184	7–.007–165°A	65–70	1.460
9275	123	5229380	F75	5229282	5229184	7–.007–165°A	70–75	1.460[a]
9280	118	5229190	F8	5229191	5229184	7–.007–165°A	75–80	1.460[a]
9285	124	5229400	F85	5229278	5229184	7–.007–165°A	80–85	1.460[a]
9290	120	5229405	F9	5229292	5229184	7–.007–165°A	85–90	1.484
9295	119	5229395	F95	5229268	5229184	7–.007–165°A	90–95	1.484

[a]Use 1.484 for turbo and turbo aftercooled engines.

(Courtesy of Detroit Diesel Allison Division of GMC)

Table 10-6
SUMMARY OF MILITARY INJECTORS

Engine	*Inj.*	*Parts Catalog Type, No.*	*Assembly Part No.*	*Plunger Desig-nation*	*P&B Part No.*	*Tip Part No.*	*Spray Tip*	*Calibrator J 22410 Min.–Max.*	*Injector Timing*
53	M50	84	5228774	SN	5228749	5229048	6–.0068165°	50–54	1.460
53	M55	93	5228834	55M	5228841	5229048	6–.006–165°	55–59	1.460
53	M60	89	6228824	5N	5228749	5229042	7–.006–165°	60–64	1.460
53	M65	87	5228814	6N	5228681	5229034	6–.006–165°A	66–70	1.460
53	M70	102	5229185	7N	5228682	5229192	7–.006–165°A	73–76	1.460
71	M95	64	5228795	95M	5228804	5229256	7–.0065–165°A	97–102	1.460
71	140	127	5229495	14	5229513	5229516	7–.0075–165°A	136–144	1.484

(Courtesy of Detroit Diesel Allison Division of GMC)

Table 10-7
SUMMARY OF 149 INJECTORS

Inj.	*Parts Catalog Type No.*	*Assembly Part No.*	*Rack Part No.*	*Tip Part No.*	*Spray Tip*	*Calibrator Min.–Max.*	*Injection Timing*
120	96	5228955	5228974	5229032	7–.010–160°	118–126	2.175
130	95	522	5228975	5229032	7–.010–160°	138–146	2.175
140	99	5229020	5229019	5229032	7–.010–160°	150–158	2.175[a]
150	98	5228965	5229976	5229032	7–.010–160°	168–176	2.175[a]
165	105	5229175	5229172	5229032	7–.010–160°	184–192	2.175[a]
170	101	5229110	5229130	5229032	7–.010–160°	195–205	2.205

[a]Use 2.205 for turbocharged intercooled engines.

(Courtesy of Detroit Diesel Allison Division of GMC)

whereby, being connected to the fuel rod, it will turn the fuel control tube running along the top of the cylinder head inward. Since the control tube is connected to each individual unit injector, the engine is at full fuel for starting purposes. The exception to this is on current turbocharged engines where a starting-aid screw is used that places the fuel control racks of the unit injectors to a position corresponding to about one-half throttle for initial starting, which prevents excessive smoke, thereby allowing the turbocharger to accelerate first and therefore deliver adequate air flow. Figure 10-33 shows such a typical governor with its parts stackup.

The engine's application determines whether a single- or double-weight governor will be used. The most prominent application for the limiting speed governor is highway truck engines, since the governor has no control in the intermediate engine speed range, thereby allowing the operator to have complete control of the injector rack movement through throttle action alone. This permits very fast throttle response for engine acceleration or deceleration as the case may be.

However, look closely first at Figure 10-33 and familiarize yourself with the basic layout of the weights and how they are connected to the low and high speed spring contained within the top housing of the governor assembly. Figure 10-33 shows a cross section of a Vee type of limiting speed governor. Note that the operating shaft shown is also called a bellcrank lever in the description of operation; they are one and the same.

On in-line 71 engines, the governor is mounted and driven from the upper blower rotor shaft. On in-line 53 engines, the governor is mounted and driven from the engine rear end plate by a gear that extends through the end plate and meshes with either the camshaft gear or balance shaft gear, depending upon the particular engine model.

On V-71, V-92, and 149 engines, the governor is driven from the front end of the blower. On V-16/71 and V-16/92 engines, the governor is mounted on the front end of the rear blower.

On 8V-53 engines the governor is mounted and driven from the front of the blower (left-hand helix rotor), but on the 6V-53 engine it is mounted between the engine blower and the engine flywheel housing.

On the new 4-cycle 8.2-liter engine, the governor is mounted at the front of the engine on a mounting pad extending from the engine. Two levers are located on top of the governor cover; the small one is a stop lever, and the larger one is the throttle lever.

Operation

Engine Stopped. With the engine stopped, the governor flyweights are at rest, and the injector racks would be in the full-fuel position, other than on a current turbocharged engine.

Cranking. When the engine is cranked over, the rotational movement to the governor flyweights will cause them to move out very slightly, with this small outward action causing the weight feet to act against the hollow movable riser shaft. However, due to the slow rotational force exerted at this rpm, there is little actual linear movement of the riser shaft.

Engine Running (Idle Speed). Contained within the top of the governor housing, that is, the spring cover, are the high and low speed spring assemblies, which bear against the end of the bellcrank lever in the governor housing. The force of these springs causes the bellcrank lever to rotate the vertical operating shaft, at the bottom end of which is a fork bearing against a three-piece, self-centering thrust bearing, which is in contact with the end of the hollow riser shaft opposite the governor flyweights.

The tension of the high and low speed

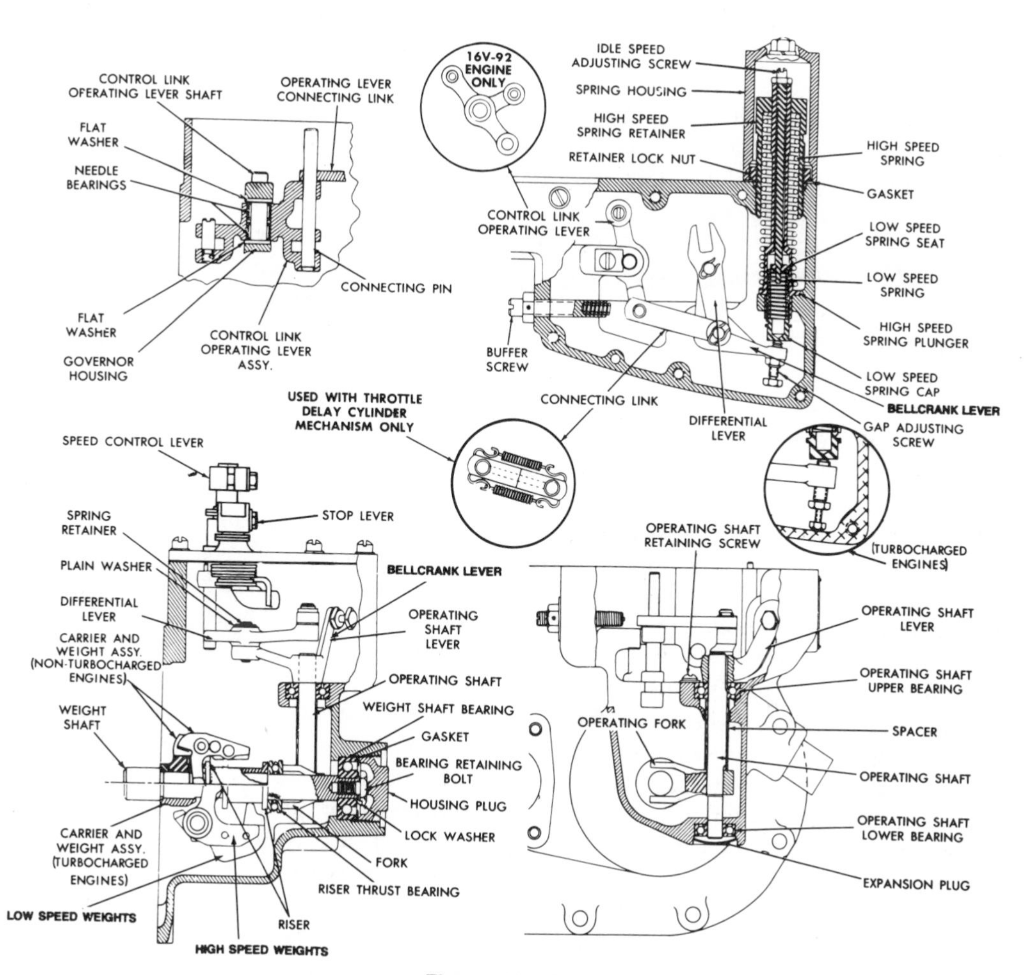

Figure 10-33
Cross Section of Vee-Type Limiting Speed Mechanical Governor.
(Courtesy of Detroit Diesel Allison Div. of GMC)

governor springs is adjustable (see the DDA tune-up section). With the throttle in the low idle position, the tension of the low speed spring is acting through the linkage against the weights. Let us consider for a moment that, if the governor is of the single-weight type, the centrifugal force of the weights will oppose the idle and the high speed spring; however, if the governor is of the double-weight type, we have two sets of weights, a low speed set (the larger weights) and a high speed set (the smaller weights).

Therefore, on the double-weight type of governor, both weight sets (low and high), which are pivoted on two pins connected to the weight carrier assembly, will oppose the tension of the low speed spring as they revolve. Actually, the smaller high speed weights ride on a shoulder of the larger low speed weights when the engine is operating in the low speed range.

As the engine fires and runs during the cranking process, the rotating weights will develop an increasing centrifugal force, which will cause them to fly outward. The outward movement will let the weight feet bearing against the hollow (horizontal) riser shaft force it lengthwise against the three-piece, self-centering thrust bearing, which will butt up against the fork connected to the base of the vertical operating shaft. The vertical operating shaft will rotate, in turn forcing the bellcrank lever connected to its top end to move in against the low speed spring cap; hence the force of the low speed (idle) spring, which is the smaller of the two springs.

Since the bell-crank lever contains a fixed pin that acts as a pivot point for the floating differential lever, any movement of the bellcrank will simultaneously cause movement of the differential lever. With the differential lever connected to the unit injectors via the control tube and fuel control rod, this movement will change the injector racks' positions.

Remember that prior to starting the engine spring force had placed the racks into the full-fuel position (other than on turbocharged engines). The amount of tension in the idle spring has been set by previous adjustment; therefore, when the centrifugal force of the rotating governor flyweights reaches a state of balance between them and the idle spring, the action of the bellcrank and differential lever will be stopped at a position that will hold the fuel control racks in a preset idle speed position.

To avoid any confusion here, it would be well to point out that, when the bellcrank lever is moved inward by the action of the weights, the differential lever moves back or away from the inward-moving bellcrank, thereby effectively pulling the fuel rod to a decreased fuel position.

Therefore, if the differential lever moves in, the fuel rod will move to an increased fuel position; however, for this to happen, the bellcrank would have to be forced out by the action of the spring or springs being greater than the force of the flyweights.

Since most engines employing limiting speed governors are set to idle at between 500 and 600 rpm, the weights will transfer their motion through the governor linkage and force the low speed spring cap in against the bore of the high speed spring plunger during this idle speed range.

Engine Running (Accelerating from Idle). If the engine is running at a steady idle speed, a state of balance exists between the centrifugal force of the revolving governor flyweights and the tension that has been preset in the idle spring. Refer to Figure 10-34. When the operator steps on the throttle, the throttle control lever mounted on top of the governor housing will transfer this motion through its pin, which engages the slotted outer end of the differential lever. This causes the differential lever to pivot around the fixed pin on the bellcrank lever, thereby moving the differential lever and fuel control rod and injector racks to an increased fuel position.

As the fuel rod moves and fuel injection increases, the engine speed will increase. On non-fuel-squeezer engines, the rotating governor low speed weights (large ones on DWLS) will reach the limit of their travel at between 800 and 1000 rpm by bottoming on a shoulder on the weight carrier assembly. However, prior to actually bottoming, their force will have caused a motion transfer through the linkage, which pushes the bellcrank lever in against the low speed spring, thereby completely compressing the spring other than the governor gap, which is 0.0015 in. (0.038 mm). On fuel-squeezer engines, a lighter set of weights is used, and the low speed weights will bottom out on the weight carrier at between 1100 and 1300 rpm with a governor gap of 0.002 in. (0.05 mm) to 0.004 in. (0.101 mm).

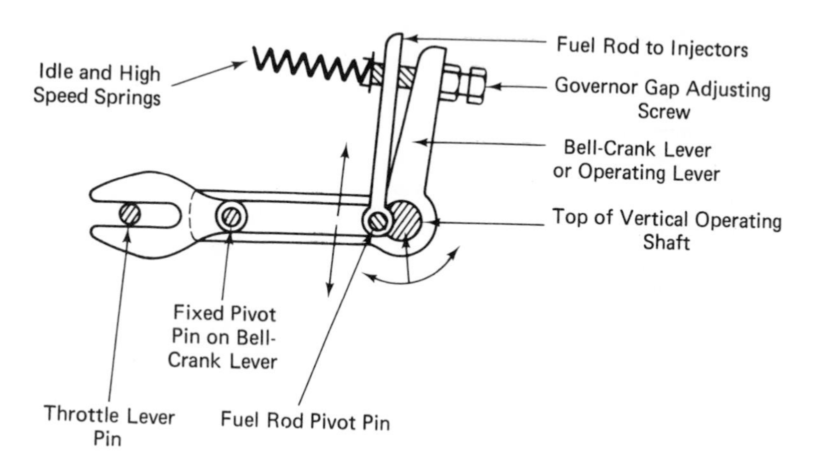

Figure 10-34
Upper Governor Housing Linkage.

When the low speed weights bottom and the low speed spring is therefore completely compressed, the governor has no control, since the operator has now entered what is commonly called the intermediate speed range. The high speed weights, which are still rotating (carried on a shoulder of the low speed weights), are not producing enough force to act upon the governor high speed spring. Within this intermediate range then, the operator has complete control of the engine by manually moving the differential lever at will.

As the engine continues to accelerate and approaches the maximum speed (dependent on the spring adjustment; see tune-up section herein), the high speed weights (smaller ones on a DWLS) will leave their seat on the shoulder of the bottomed low speed weights, thereby forcing the hollow riser shaft to move horizontally against the three-piece, self-centering bearing and operating fork connected to the vertical operating shaft. Since the bellcrank lever is connected to the top of this shaft, it will move in against the high speed spring. How far the bellcrank moves in against the high speed spring is dependent upon the throttle position and engine rpm.

If the operator pushes the throttle lever to its maximum position, the centrifugal force of the rotating high speed weights will cause the high speed spring retainer and spring to move from its seat on the governor housing, owing to weight force being greater than spring force.

Note that anytime the bellcrank lever is forced in against the springs the pivoting differential lever connected to the fuel rod is moved to a decreased fuel position. In this way, then, the amount of fuel to the engine is automatically controlled for any given condition.

Load Increase

Since there is no governor action within the intermediate speed range, a state of balance condition between the weight and spring forces can only happen in either the low or high speed range. At any time that the engine is running at a steady speed, weight and spring forces being equal will hold the rpm stable. If now a load were applied to the engine in the form of a truck climbing a hill, for a given throttle setting (fixed) the engine speed would tend to decrease, thereby upsetting the previous state-of-balance condition. This reduction in rpm would allow the stronger spring force acting against the bellcrank lever to push it away from the spring owing to the decreasing weight force. Remember, if the bellcrank lever moves out, the pivoting differential lever must move in and push the fuel control rod to an increased fuel position. As this happens, the engine speed will again start to increase until the centrifugal force of the rotating weights again equals the force of the spring, obtaining a state of balance, and maintaining a steady rpm for that throttle setting and load. However, this can only happen if the increased load applied is capable of being overcome by the available power developed by the engine; otherwise, the operator would continue to lose rpm and would have to downshift the transmission. If an automatic transmission were used, an automatic downshift would occur, but an increased throttle position may still be required to maintain sufficient road speed.

Load Decrease

As with a load increase situation, the governor can only react within the low or high speed ranges, since there is no governor action in the intermediate speed range. If the engine is running at a fixed throttle and load position, and there is a load decrease, such as a truck running downhill, the engine speed will tend to increase. As this happens, the centrifugal force of the rotating governor flyweights will also increase, transferring this motion up to the bellcrank lever, pushing it in against the spring, and allowing the pivoting differential lever to move to a decreased fuel position along with the fuel rod.

As the engine starts to reduce in speed owing to the corrective governor action, a state of balance will again be attained. If the vehicle road speed, however, were to remain unchecked, it would be possible for the road wheels to become the driving member, which would in turn increase the engine rpm to a dangerous range. Whether the throttle was depressed or in the idle position, the high rotational speed of the engine would transfer this additional rpm to the rotating governor flyweights, causing them to positively continue to push the bellcrank lever in even farther against the high speed spring, and the differential lever would eventually be forced all the way back to an equivalent idle fuel input position.

Therefore, if the engine were to overspeed and a valve were to float and hit a piston, the problem is not that the governor did not do its job properly; on the contrary, the operator or driver was at fault for not checking the vehicle's road speed by the use of the air brakes, engine or driveline brake, or hydraulic retarder.

The governor correction will be the same if the engine were in the low speed range with the low speed weights acting against the low speed spring. A load decrease could occur at idle, for example, if an accessory such as the air compressor were to go into its unloaded stage, or some similar load-carrying accessory were turned off. A load increase at idle could be caused by the air compressor cutting back in, or an air conditioning pump being turned on, or a power take-off being engaged.

Vee Governors

Figure 10-33 shows that the Vee governor parts and function are the same as that found on the in-line engines, other than the governor being somewhat more squat in shape. For information relating to these governors, refer to the section dealing with tune-up for any adjustment.

When a limiting speed governor is used on an industrial application, its maximum speed is adjusted by the use of shims behind the high speed spring at the bellcrank lever end, instead of by the use of the high speed spring adjusting nut found behind the spring housing on highway vehicle engines.

Limiting Speed Mechanical Governor (Options)

A variety of options is readily available for this governor, which are required and necessary when used on a highway truck application.

Variable Low Speed Limiting Speed (VLSLS) Governor. This governor is found on highway truck engines where the engine is used to power auxiliary equipment at an unloading area, and a high idle speed is desired. Earlier governors of this type used the two-weight setup, which gave a choice of rpm between 500 and 1200. However, the newer governors of this type are the single-weight style, which provide an idle speed range of 500 to 1800 rpm.

Figure 10-35 shows the basic layout of the parts contained within the top of the governor control housing, since the operation up to the bellcrank, which acts against the low speed spring cap, is the same as in the normal limiting speed governor.

The governor gap is 0.200 in. on the single-weight governor when used with this option on both V-71 and V-92 series engines. This is with the engine stopped, governor cover removed, fuel rods disconnected, and is taken between the low speed spring cap and high speed spring plunger with gauge J 23478, or an equivalent stackup of feeler gauges.

You will notice in Figure 10-35 that a third spring has been added to the governor. This is necessary because in the normal operation of the limiting speed governor there is no governor action within the intermediate speed range, since the low speed weights have reached the outer limit of their travel by bottoming on the weight carrier with the low speed (idle) spring completely compressed.

The high speed weights have not yet started to move from their seats on the DWLS unit, and if it is only a single-weight type, below 1800 rpm, insufficient centrifugal force is created for them to oppose the high speed spring. We therefore must have a third spring to oppose the governor weight force within this intermediate range and allow us to set the throttle to any desired position where a state-of-balance condition between the weights and springs will hold the engine speed constant for auxiliary operation.

The speed adjusting handle is generally

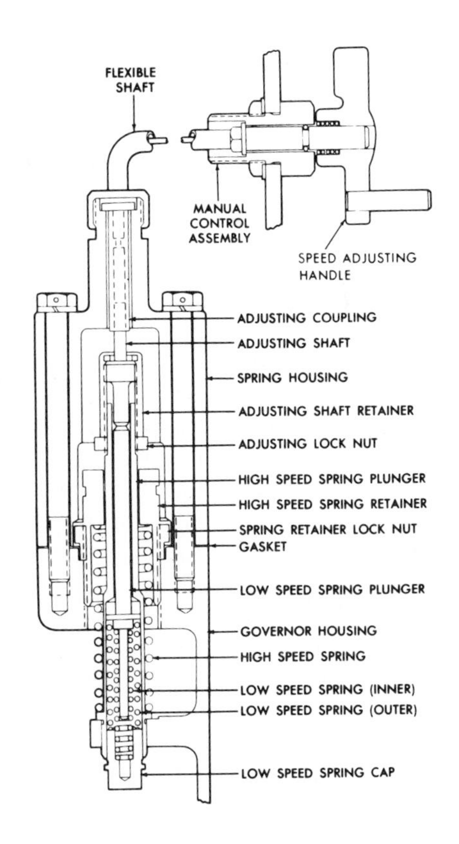

Figure 10-35
Variable Low-Speed Limiting Speed Mechanical Governor, Upper Housing Components. (Courtesy of Detroit Diesel Allison Div. of GMC)

mounted on the dash of the vehicle or in the cab. During normal highway driving, this handle would be turned all the way back to remove any tension from the third spring; therefore, the governor would operate as a normal limiting speed governor during this phase. When the vehicle is stationary (parked) at the unloading area, the throttle foot pedal would be left in the idle position, and the speed adjusting handle can be turned to the speed desired to operate the auxiliary equipment, such as a HIAB or cement drum. During this phase of its operation, the governor functions as a variable speed unit, which will maintain a fairly constant speed during the changing load situation that exists while unloading. When the unloading situation is complete, the speed adjusting handle must be turned back to the normal low idle speed, which can be felt by the fact that it will butt against its stop; it is then turned back approximately one quarter-turn.

Fast Idle Air Cylinder

This option is used with the DWLS governor and functions to allow a fast idle speed of the engine when the vehicle is parked and unloading via auxiliary equipment is desired. Figure 10-36 shows the basic setup of this unit. It consists of either two cylinders or three, depending on your choice. The three cylinders are all air operated, and are shown in Figure 10-36. They are: (1) the fast idle air cylinder, (2) the throttle locking air cylinder, and (3) an engine shutdown air cylinder if desired.

Operation. Since the limiting speed governor has no actual governor action within the intermediate speed range, it is necessary to lock the throttle lever and pin, which the slotted end of the differential lever engages into, in the idle position. If this were not done, the differential lever would float when placed into any speed range within the governor's intermediate range, especially with the changing load characteristics encountered during the unloading stage. With the throttle locked in position, air pressure is simultaneously directed to the fast idle air cylinder, which takes the place of the normal buffer screw.

The air pressure forces a piston forward within the cylinder, which in turn causes a plunger to contact the end of the differential lever, thereby moving it to an increased fuel position for a predetermined rpm. This rpm is established by changing the dual idle spring

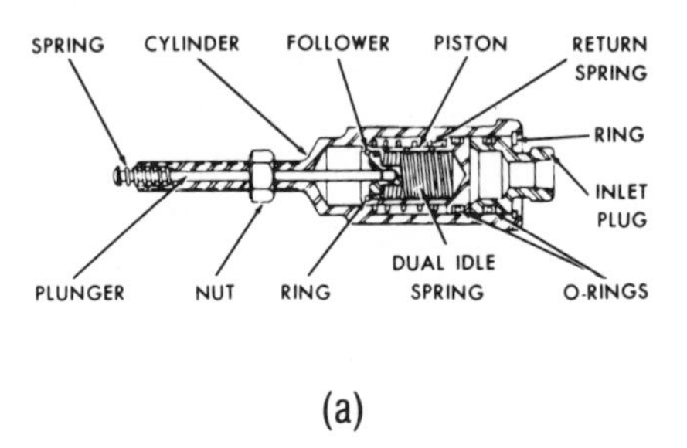

(a)

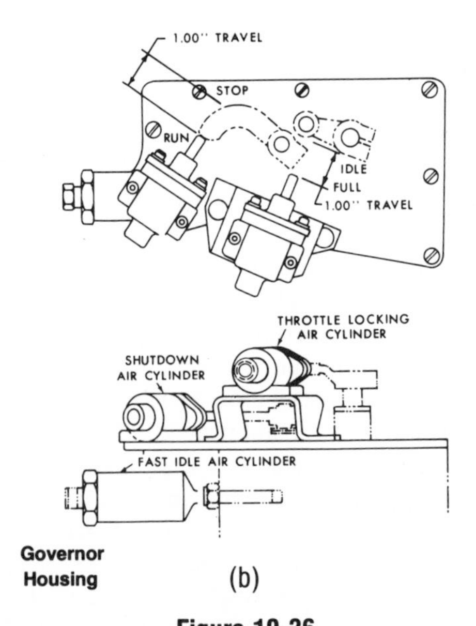

(b)

Figure 10-36
Fast Idle Air Cylinder.
(Courtesy of Detroit Diesel Allison Div. of GMC)

within the fast idle air cylinder, since the force of this spring is added to that of the governor low speed spring to establish the engine fast idle speed.

To prevent an operator from inadvertently applying air pressure to the fast idle air cylinder when the vehicle is moving, the system should be plumbed so that air is only available to the control valve of the FIAC when the vehicle's spring parking brakes are actuated. The control valve for the FIAC can be a manual type toggle on and off switch or actuated by an electric solenoid.

When load is applied to the engine during FIAC operation, the engine speed is determined by the governor droop curve. A shutdown air cylinder if used is mounted to the governor cover so that it will move the stop lever when energized to stop the engine. The governor gap with this option remains the same; however, on 53 series engines the engine should be run at between 900 and 1050 rpm because of the lighter low speed weights and heavier high speed weights.

Note: Prior to adjusting the governor gap with a FIAC option, be sure to back out the deenergized FIAC so that there is no interference against the differential lever. When setting the FIAC after an idle speed adjustment, screw it in in the same manner as you would set an engine with a buffer screw.

Mechanical Output Shaft Governor (Pierce)

Refer to Figure 10-37. On some industrial applications where a torque converter is used, in order to control the speed of the output shaft more accurately and sense its action, a governor can be mounted and driven from the torque convertor output shaft through a drive adaptor. Due to the internal slippage that exists within any torque convertor (unless a lockup clutch is used), the engine governor alone would be unable to sense a speed or load change quickly

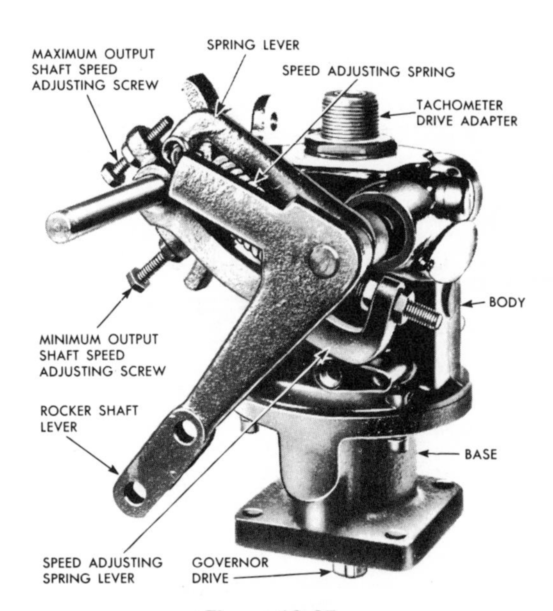

Figure 10-37
Mechanical Output Shaft Governor.
(Courtesy of Detroit Diesel Allison Div. of GMC)

enough to react. The output shaft governor, then, only senses output shaft rpm and not engine rpm.

The output shaft governor is connected to the engine governor via mechanical linkage; therefore, both the engine's idle and maximum speeds will be controlled by the engine-mounted limiting speed governor, and within the intermediate speed range of the engine, the output shaft governor connected to the limiting speed unit will sense output shaft rpm, thereby automatically varying the engine speed within this intermediate speed range.

Variable Speed Mechanical

This governor is found extensively on industrial and marine applications, since it is designed for the following functions:

1. Controls the engine idle speed.
2. Controls the maximum engine speed.

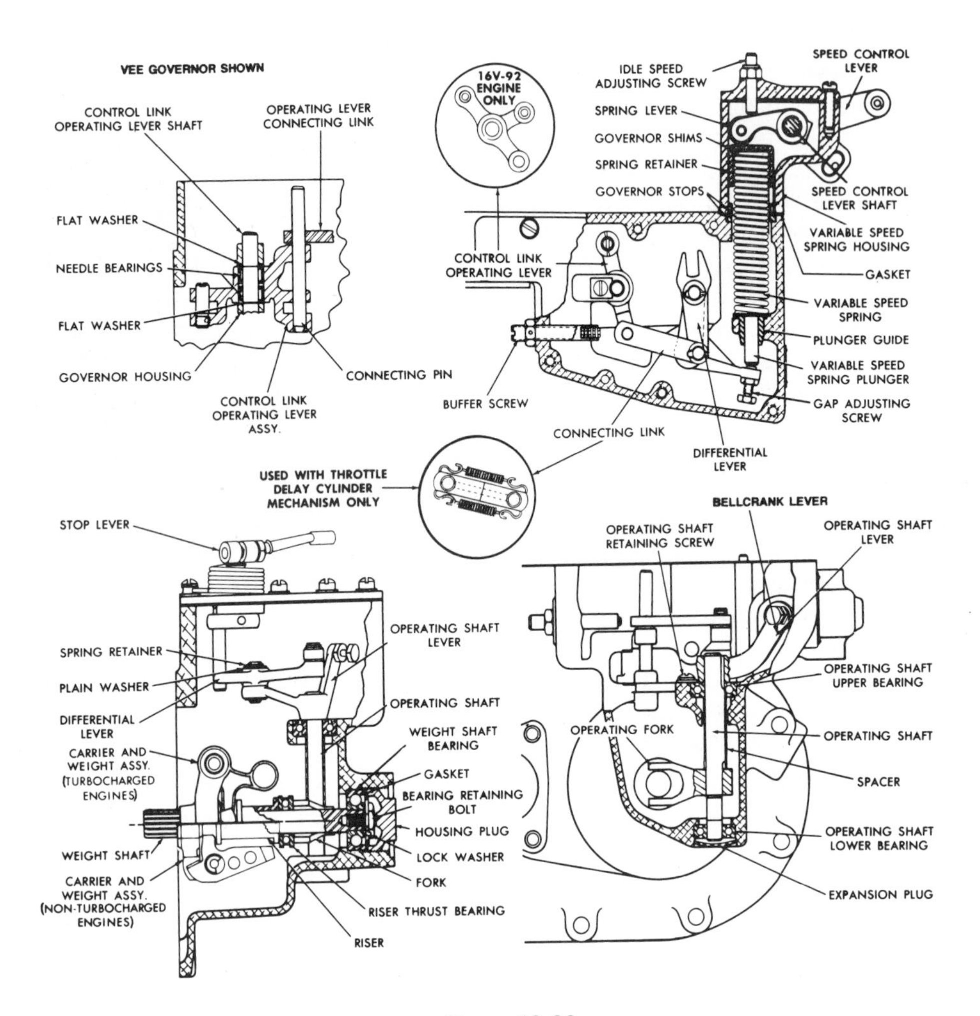

Figure 10-38
Component Parts of the Variable Speed Governor.
(Courtesy of Detroit Diesel Allison Div. of GMC)

3. Holds the engine speed at any position between idle and maximum as desired and set by the operator.

Figure 10-38 shows the component parts makeup of the variable speed governor. The response of this governor and its reaction are similar to that of the limiting speed type just discussed, with the following exceptions. Since this governor controls engine speed throughout the total rpm range, and there is no intermediate range as with the limiting speed type, only one set of weights and one spring are used and necessary.

For any given throttle setting or load from idle to maximum speed, a state of balance can exist. If, however, the load is increased or decreased, a corrective action will be initiated whereby the bellcrank lever and pivoting differential lever will be moved by the action of the governor spring or weights, respectively, to reestablish a state of balance. Again remember, the governor can only react and change the engine's rpm, especially during a load increase, if the additional load does not exceed the power capability of the engine.

This governor is readily identifiable from the limiting speed by the fact that it has only one lever on the top of the governor cover, which is the stop/run lever. The speed control lever is located vertically on the end of the governor spring housing. In addition, a large booster spring is attached between the speed control lever and a bracket on the cylinder head; it is used to assist the operator in overcoming governor resistance during throttle movement. The letters SWVS (single-weight variable speed) are stamped on the governor identification plate.

10-12

Engine Tune-up

Tune-up of a diesel engine is not based on the same criteria as for automotive-type gasoline engines. Until 1972 there was no set interval at which a full tune-up had to be performed, with the performance of the engine being the controlling factor in actually undertaking the full tune-up. However, with the 1972–1973 model year, all diesel engine manufacturers had to comply with the Environmental Protection Agency (EPA) requirements regarding exhaust emissions. This has forced engine manufacturers to check and correct as necessary at 50,000 (80,465 km) and 100,000 (160,930 km) miles (to comply with emissions warranty requirements) such items as injector timing, valve lash, idle and no-load speeds, and throttle delay or fuel modulator settings.

On DDA engines the type of governor used is dependent on the particular engine application; therefore, this setup can vary slightly between engines. All DDA mechanical governors are easily identifiable by a nameplate attached to the governor housing; the following letters are typical examples.

DWLS: double-weight limiting speed (mobile equipment).

SWLS: single-weight limiting speed (mobile equipment).

SWVS: single-weight variable speed (industrial and marine).

VLSLS: variable low-speed limiting speed (highway vehicles).

DWDRG: double-weight dual range governor (highway vehicles).

SG, PSG, SGX, UG8: Woodward hydraulic-type governors for use on industrial and generator set applications.

When performing a necessary tune-up on an engine do *not* back off all the necessary adjustments. It is only necessary to *check* these for a possible change in the settings. If, however, a cylinder head or the governor or injectors have been removed and overhauled or replaced, several initial adjustments are necessary before the engine can be started. These adjustments would consist of the first four items in the DDA tune-up sequence, with the only exception being that the valve clearance is greater on a cold engine.

Note: Normal tune-up should be performed with the engine water temperature between 160 and 185°F.

Prior to listing the tune-up sequence steps, it should be noted that, if a supplementary governing device is used, it must be disconnected before the engine tune-up; otherwise, it will interfere with the injector control rack adjustments. It should not be necessary at this time to adjust the valve bridges on each cylinder unless a cylinder head has been overhauled. The bridges should be adjusted once the heads have been installed on the engine, and unless wear occurs, or the adjustment backs off, no further adjustment is necessary. (Valve bridges are only found on four-valve cylinder heads.)

Exhaust Valve Bridge Adjustment

The following sequence should be followed for all series of DDA engines. It is only required at cylinder head overhaul, or if the adjusment screw has become loose, and is not part of normal tune-up.

1. Place the valve bridge in a *soft-jaw* vise or if available bridge holding fixture J 21772, and loosen the locknut on the bridge adjusting screw. Back out the adjusting screw several turns.

Note: Failure to follow the above sequence can result in the following damage. If the locknut is loosened or tightened with the bridge in place, the twisting action involved can result in either a bent bridge guide or bent rear valve stem.

2. Install the bridge back onto its respective bridge guide.
3. While firmly applying pressure to the bridge, as shown in Figure 10-39, turn the adjusting screw clockwise until it lightly contacts the valve stem. Carefully turn the screw an additional one-eighth to one-quarter turn and run the locknut up finger tight.
4. Install the bridge in a soft-jaw vise and, while using a screwdriver to hold the adjustment screw, tighten the locknut to 20 to 25 lb-ft torque (27.1 to 33.87 N · m).
5. Using engine oil, lubricate the bridge and guide, and reinstall it in its original position.
6. Select two 0.0015-in. (0.0381-mm) feeler gauges (pointed finger type), or cut two

Figure 10-39
Adjusting Exhaust Valve Bridge.
(Courtesy of Detroit Diesel Allison Div. of GMC)

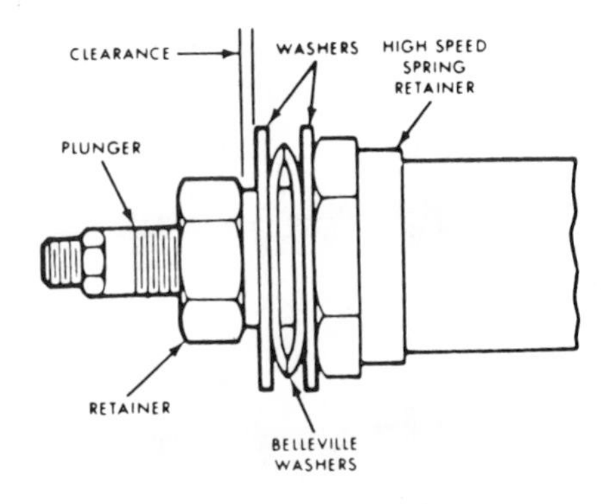

Figure 10-40
Adjusting Belleville Spring Retainer Nut. (Courtesy of Detroit Diesel Allison Div. of GMC)

thin strips that will fit under the bridge at each valve stem tip. Apply finger pressure to the pallet (center) surface of the valve bridge, and check to see that both feeler gauges are in fact tight. If they are not tight, readjust the screw as previously outlined.

7. Adjust the remaining valve bridges in the same manner.
8. Ensure that when the rocker arm assemblies are swung into position, the valve bridges are properly positioned on the rear valve stems; otherwise, damage to the valve and bridge mechanism is a possibility.

Tune-up Sequence

Ensure that the engine is at operating temperature (160 to 185°F) prior to starting the tune-up. Also, since many of the adjustments are normally made with the engine stopped, you may have to run the engine between adjustments to maintain this temperature.

Note: You should also be certain before starting the engine after any governor or speed control adjustments that when the *stop* lever is placed in this position, the injector racks will in fact move to the *no-fuel* position; otherwise, engine overspeed could result.

Note: If the engine is equipped with a *Belleville* spring retainer governor, the spring retainer nut must be backed out until there is approximately 0.060 in. (1.524 mm) of clearance between the washers and retainer nut, as shown in Figure 10-40. The Belleville washer setup is found on highway truck fuel-squeezer engines (TT; tailored torque).

1. Adjust the exhaust valve clearance.
2. Time the fuel injectors.
3. Adjust the governor gap.
4. Position the injector rack control levers.
5. Adjust the maximum no-load speed.
6. Adjust the idle speed.
7. Adjust the Belleville spring for TT (tailored-torque) horsepower.
8. Adjust the buffer screw and starting aid screw if so equipped.
9. Adjust the supplementary governing device (throttle delay).

Fuel-Squeezer Engines (71 and 92 Engines)

A fuel-squeezer engine is basically a stock engine with the addition of a turbocharger and a compression ratio of 17 : 1, compared to the normal N engine 71 with an 18.7 : 1 compression ratio and the standard 92 with a 19 : 1 compression ratio. The operating characteristics of a fuel squeezer engine are its ability to maintain reasonably constant horsepower over a wide range, and its 6% *torque rise* per 100 rpm drop from full-load rpm down to its peak torque figure. These characteristics are possible by the action of two Belleville washers adapted to the limiting speed governor found on highway vehicle engines.

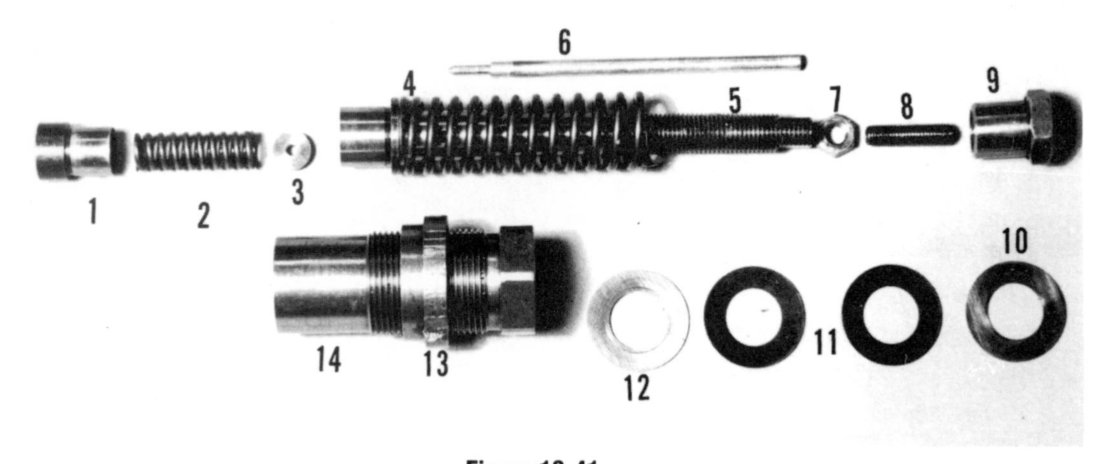

Figure 10-41
Tailored-Torque Governor Spring Pack.
(Courtesy of Detroit Diesel Allison Div. of GMC)

Index:

1—Spring cap
2—Low speed spring
3—Low speed spring seat
4—High speed spring
5—Plunger
6—Idle speed pin
7—Locknut
8—Idle speed adjusting screw
9—Retainer
10—Washer
11—Belleville spring washers
12—Washer
13—Locknut
14—Spring retainer

Published horsepower charts for fuel-squeezer engines indicate a flat horsepower (kW) performance curve; however, during dynamometer testing an engine may exhibit horsepower readings slightly above or below the flat curve. An acceptable horsepower (kW) variation from the flat published curve will be within 5%.

The spring force provided by the Belleville washers works with the governor weights to pull the injector racks out of fuel as the engine's speed is increased. Conversely, as the engine speed is reduced by increased load, the high speed spring overcomes the force of the Belleville washers and moves the injector racks to an increased fuel position. The racks move progressively into more fuel to maintain the constant horsepower until the racks are in full fuel at a speed near 1500 rpm.

Engines designated TT (tailored torque) are the latest type of 2-cycle engine produced by DDA; a TT governor (limiting speed) spring pack is shown in Figure 10-41.

Emissions Regulations for On-Highway Vehicle Engines

All on-highway vehicle engines built by Detroit Diesel Allison are certified to be in compliance with federal and California emission regulations established for each model year. To comply with these regulations, engine tune-up is dependent on the following five physical characteristics:

1. Fuel injector type.
2. Maximum full-load engine speed.
3. Engine camshaft timing (standard or advanced).
4. Fuel injector timing.
5. Throttle delay (orifice size).

Exhaust Valve Clearance Adjustment

The writer assumes that the individual involved in a typical tune-up on one of these engines has had some training in the basics of engine construction and general overhaul and, therefore, is in a position to interpret the data and sequence that follows.

On any engine, whether a gasoline or diesel engine, correct valve clearance is very important to ensure smooth efficient operation of the engine. Since DDA engines are all 2-cycle units, all valves are *exhaust* since the cylinder liner ports take the place of the intake valves (the new 8.2-liter V-8 engine is a 4-cycle type). Therefore, insufficient valve clearance on these engines can result in a loss of compression since the valves will be held off their seats. In addition, this will cause misfiring cylinders and eventually burned valve seats and seat inserts owing to the fact that the valve is unable to dissipate its heat through good seat contact. On the other hand, excessive valve clearance will result in noisy operation, especially at the lower speed ranges. In both cases there will be a noticeable loss of power.

Valve adjustment is necessary then at any time the head has been removed and replaced or the valve mechanism has been disturbed in any way. The initial setting is done with the engine cold, and must be checked and reset when the engine has reached its normal operating temperature. If the exhaust valve bridges require adjustment, see the earlier section for the procedure.

With the exception of the series 149 engines, the sequence for exhaust valve adjustment is the same for series 71 and 92 engines. The only difference on the 4-valve-head 53 series engine is that it uses a pivoted bridge attached to the end of the rocker arm assembly; therefore, the valve clearance is only checked under one valve stem to bridge since it will self-center during operation of the engine.

Series 53, 71, and 92 engines employ pushrods that are threaded into the ends of the rocker arm assembly; therefore, the exhaust valve clearance adjustment is always made at the pushrod. *Do not disturb the exhaust valve bridge adjustment screw* unless the cylinder head has been removed for servicing or the bridge adjusting screw has been loosened for any reason.

Since a 2-cycle engine only requires 360 degrees of crankshaft rotation for each power stroke per cylinder, as opposed to 720 degrees on a 4-cycle engine, all the exhaust valves can be adjusted in firing order sequence in one full turn of the crankshaft. Every DDA service manual shows the particular engine firing order under general engine specifications at the front of the manual.

On all series of DDA engines the following must be kept in mind when working with firing order sequence:

1. Engine rotation is always determined from the front of the engine as cw (clockwise) or rh (right-hand) rotation, or ccw (counterclockwise), which is lh (left-hand) rotation.
2. Also determined from the front of the engine is the cylinder numbering sequence.
3. On all Vee-type DDA engines the LB (left bank) or RB (right bank) is determined

from the rear or flywheel end of the engine.

Table 10-8 gives the firing orders for each engine within a given series.

To ensure that there is no confusion regarding identification of cylinder number and what is the LB (left bank) and RB (right bank) on the Vee-type engines, Figure 10-42 shows a typical 8V unit depicting very clearly the method used by Detroit Diesel Allison on all their engines.

Although it does not concern us at this time, determining whether a specific engine model is an A, B, C, or D unit is also done by standing at the flywheel end.

Table 10-8
FIRING ORDER SEQUENCES

	Model	*Firing Order*
53 Series	2–53	1–2 RH
	3–53	RH–1–3–2, LH–1–2–3
	4–53	RH–1–3–4–2, LH–1–2–4–3
	6V–53	RH–1L–3R–3L–2R–2L–1R
		LH–1L–1R–2L–2R–3L–3R
	8V–53	RH–3R–3L–4R–4L–2R–2L–1R
71 Series	2–71	1–2 RH
	3–71	RH–1–3–2, LH–1–2–3
	4–71	RH–1–3–4–2, LH–1–2–4–3
	6–71	RH–1–5–3–6–2–4, LH–1–4–2–6–3–5
	6V–71	RH–1L–3R–3L–2R–2L–1R, LH–1L–1R–2L–2R–3L–3R
	8V–71	RH–1L–3R–3L–4R–4L–2R–2L–1R
		LH–1L–1R–2L–2R–4L–4R–3L–3R
	12V–71	RH–1L–5L–3R–4R–3L–4L–2R–6R–2L–6L–1R–5R
		LH–1L–5R–1R–6L–2L–6R–2R–4L–3L–4R–3R–5L
	16V–71	RH–1L–2R–8L–6R–2L–4R–6L–5R–4L–3R–5L–7R–3L–1R–7L–8R
		LH–1L–8R–7L–1R–3L–7R–5L–3R–4L–5R–6L–4R–2L–6R–8L–2R
92 Series	6V–92	RH–1L–3R–3L–2R–2L–1R, LH–1L–1R–2L–2R–3L–3R
	8V–92	RH–1L–3R–3L–4R–4L–2R–2L–1R
		LH–1L–1R–2L–2R–4L–4R–3L–3R
	16V–92	RH–1L–2R–8L–6R–2L–4R–6L–5R–4L–3R–5L–7R–3L–1R–7L–8R
		LH–1L–8R–7L–1R–3L–7R–5L–3R–4L–5R–6L–4R–2L–6R–8L–2R
149 Series	12V–149	RH–1L–5L–3R–4R–3L–4L–2R–6R–2L–6L–1R–5R
		LH–1L–5R–1R–6L–2L–6R–2R–4L–3L–4R–3R–5L
	16V–149	RH–1L–2R–8L–6R–2L–4R–6L–5R–4L–3R–5L–7R–3L–1R–7L–8R
		LH–1L–8R–7L–1R–3L–7R–5L–3R–4L–5R–6L–4R–2L–6R–8L–2R

LH (left hand or ccw); RH (right hand or cw).

53, 71, 92 Exhaust Valve Clearance Adjustment

1. It is imperative that all loose dirt be either steam cleaned from around the valve rocker covers or at least washed with solvent to prevent the entrance of both dirt and foreign matter into the engine during and after their removal. The area around the governor should also be cleaned at this time.
2. Since all 53, 71, and 92 DDA engines are 2-cycle units, there is no flywheel or accessory timing marks to line up, such as on Caterpillar and Cummins diesel engines. It is only necessary to observe the position of both the injector and valve rocker arms on the cylinder that is to be adjusted.
3. The governor speed control lever should be placed in the *idle* speed position, and the *stop* lever if provided should be secured in the stop position.
4. To rotate the engine crankshaft, either the starting motor can be used or preferably an engine barring tool, such as J 22582; a $\frac{3}{4}$-in. (19-mm) square drive socket set with suitable socket to fit over the crankshaft pulley bolt will also do.

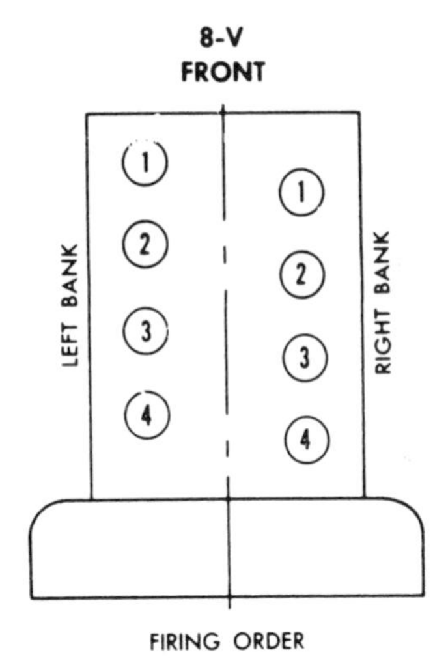

Figure 10-42
Cylinder Numbers on Typical 8V Unit. (Courtesy of Detroit Diesel Allison Div. of GMC)

Caution: When using either a barring tool or socket on the crankshaft bolt at the front of the engine, do *not* turn the crankshaft opposite its normal direction of rotation as this may loosen the bolt.

Table 10-9
VALVE CLEARANCE CHART

	Engine Cylinder Head	*Cold Setting in Inches (mm)*	*Hot Setting[a] in Inches (mm)*
Series 71	2 valve	0.012 (0.3)	0.009 (0.228)
	4 valve (bridge)	0.016 (0.4)	0.014 (0.355)
Series 53	2 valve	0.011 (0.279)	0.009 (0.228)
	4 valve (clevis)	0.026 (0.66)	0.024 (0.609)
Series 92	4 valve (bridge)	0.016 (0.4)	0.014 (0.355)
Series 110	2 valve	0.015 (0.381)	0.009 (0.228)
	4 valve (bridge)	0.016 (0.4)	0.014 (0.355)
Series 149	4 valve (bridge)	0.016 (0.4)	0.012 (0.3)

[a]Always check the hot setting with the engine between 160 and 185°F.

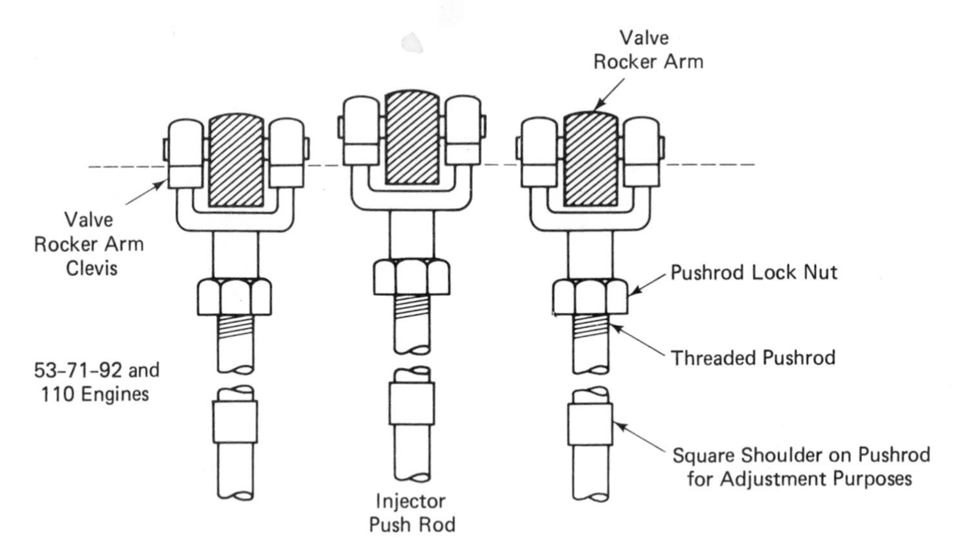

Figure 10-43
Positions of Rocker Arms When Adjusting Exhaust Valves.

5. To determine which valves or injector is in a position to be adjusted, do the following:
   a. To set the valves on any given cylinder, the center rocker arm, which is the injector arm, must be all the way up when viewing the rocker assembly from the pushrod side.
   b. When the rocker arm is in this position, you will also notice that the cast flats across the back of each rocker arm clevis will take up the position shown in Figure 10-43.
6. Loosen the exhaust valve rocker arm pushrod locknut, and from Table 10-9 select the proper feeler gauge for the particular engine that you are working on.

Note: It is advisable to use go–no go type feeler gauges for this purpose, which will ensure that all the valves are in fact set to the same clearance. See Figure 10-44.

7. Place the correct gauge between the valve bridge pallet on four-valve heads or between the valve stem and rocker arm on two-valve heads. Assume that you were setting a four-valve head 71 series engine *cold*. You would require a 0.015 to .0017 in. (0.381 to 0.431 mm) go–no go feeler gauge for this purpose. Adjust the pushrod with a $\frac{5}{16}$-in. (7.93-mm) wrench on the square shoulder until the 0.017-in. (0.431-mm) portion of the gauge can be withdrawn with a smooth pull, and tighten the locknut with a $\frac{1}{2}$-in. (12.7-mm) wrench.
8. If the adjustment is correct, you should now be able to push the 0.015-in. (0.381-mm) part of the feeler gauge through the rocker arm area freely, but the 0.017-in. (0.431-mm) portion should not pass through. You should feel the shoulder of the feeler between the two sizes actually butt up against the rocker arm pallet. If necessary, readjust the pushrod.

149 Engine Valve Adjustment. Since the 149 engines employ a high-mounted camshaft, the same first four steps just described for the 53, 71, 92, and 110 cold valve set will also apply. In addition, turn or bar the engine over until the injector follower is *fully* depressed on the particular cylinder that you are setting the valves on. Loosen the exhaust valve adjusting screw locknut on the rocker arm (see Figure 10-67), and adjust the valves as shown.

Figure 10-44
Engine Valve Adjustment.
(Courtesy of Detroit Diesel Allison Div. of GMC)

Fuel Injector Timing

Although the injector plunger is timed by the fact that it meshes with a flat on the internal rack gear inside the injector body, which is in turn timed to the fuel control rack by a dot on the gear which is centered between two dots on the injector fuel rack, the actual effective length that the plunger moves down in its bushing is controlled by the height of the injector follower above the injector body. This is adjusted by turning the injector pushrod cw or ccw as the case may be to arrive at this dimension.

This dimension is given in Section 14, Engine Tune-up, in all DDA service manuals; however, if you refer to Tables 10-1 through 10-7 you can determine the particular *timing pin* required for your particular engine injectors. Current timing pin dimensions can be found stamped on the valve rocker cover option plate.

Note: Be certain that you select the proper timing pin; otherwise, serious damage could result to the engine, not to mention poor performance.

All the injectors can be timed in firing order sequence during one full revolution of the crankshaft similar to the valves on all 2-cycle DDA engines. The new 8.2-liter 4-cycle engine would require two revolutions of the crankshaft.

The sequence for injector timing is as follows:

1. The governor speed control lever should be in the *idle* position. If a stop lever is provided, secure it in the *stop* position.
2. The crankshaft can be rotated in the same way as explained under step 4 for exhaust valve adjustment.
3. To determine which injector is in a position to be checked or adjusted, do the following:
   a. Turn the engine over until the exhaust valves are fully depressed on the cylinder that you wish to set the injector.
   b. With the rocker arms in this position, the cast flats across the back of each rocker arm clevis will take up the position shown in Figure 10-45.
4. With the injector rocker arm in the position shown in Figure 10-45, insert the small end of the timing pin (gauge) into the hole provided in the top of the injector body, with the flat portion of the gauge facing the injector follower as shown in Figure 10-46.

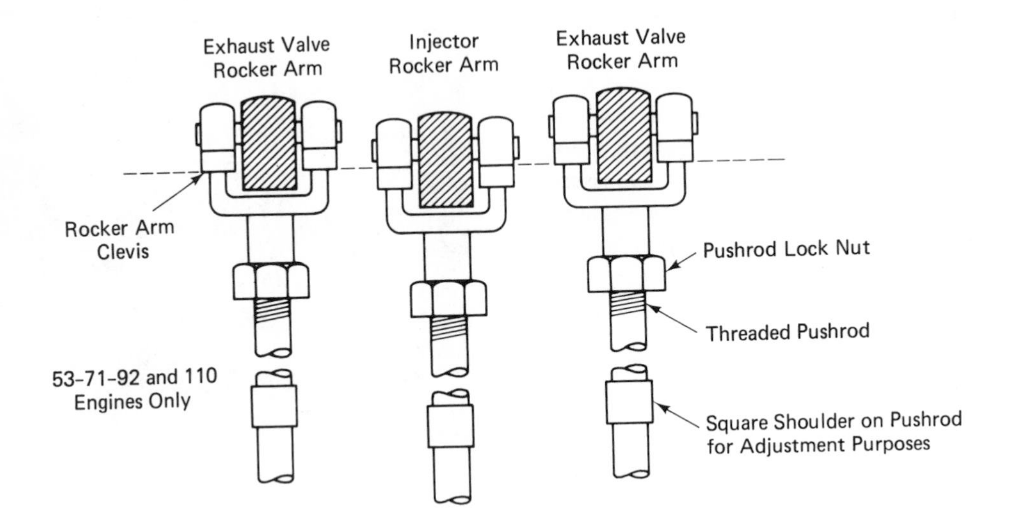

Figure 10-45
Positions of Rocker Arms When Adjusting the Injector.

5. Gently push the shoulder of the gauge by holding the knurled stem with the thumb and forefinger (see Figure 10-46) toward the follower; there should be a slight drag between the gauge and follower. You can also turn the gauge around in a circular motion to determine this same feel.
6. If this cannot be done, loosen the injector pushrod locknut and adjust it until the drag of the gauge (slight feel) has been determined; then hold the pushrod and tighten the locknut.
7. Recheck the feel, and if necessary readjust.
8. When hot setting this adjustment, wipe off the top of the injector follower and place a clean drop of oil on it. When properly adjusted, the gauge should just wipe the oil film from the follower when the slight drag is felt.
9. Time the remaining injectors in the same fashion.

Note: Adjustment of the injector timing on the 149 series engines is the same as for the valves, as shown in Figure 10-67, with the difference that both valves per bridge on the cylinder being adjusted would be fully depressed (open); then the injector would be set in the same way as just described for other series engines. On 149 engines always tighten the injector adjusting screw locknut to 30 to 35 lb-ft (40.674 to 47.453 N·m); then recheck the adjustment.

Governor Gap Adjustment

Adjustment of the governor gap is very important to the engine's performance, and if not set correctly can cause lack-of-power complaints along with additional complaints of a certain engine flat spot at a particular engine speed. This type of complaint is more prevalent on highway-truck-type applications using DWLS governors and the newer style of fuel-squeezer engines with the Belleville spring retainer setup.

The governor gap will vary on different mechanical governors; therefore, it is imperative that you establish the particular type of governor used on the engine prior to setting the governor gap. As mentioned earlier in this chapter, a plate on the side of the governor will indicate the type.

Figure 10-46
Checking Fuel Injector Timing on Model V-71 Engine.
(Courtesy of Detroit Diesel Allison Div. of GMC)

Note: On in-line 71 series engines employing a single-weight limiting speed (SWLS) governor, the gap is 0.170 in. taken with the engine stopped.

Initial Governor Gap Setting, Engine Cold

Double-Weight Limiting Speed (DWLS) Governor. To set the governor gap on this type of governor, the engine must be running. However, it is possible on the Vee-type engines, owing to the squat shape of the governor housing as opposed to that on the in-line engines, to cold set the gap prior to starting the engine. If the governor has been removed from the engine for overhaul or the governor gap has been disturbed, proceed as follows (does not apply to the 6V-53):

Procedure

1. Stop the engine and remove the governor cover. Also back out the buffer screw and starting aid screw if used.
2. Bar the engine over, look down into the governor housing, and when the weights are in a *horizontal* position, hold the engine at this point.
3. Carefully insert a large screwdriver down between the inner lobe of either one of the low-speed governor weights (the larger weights) and the weight (riser) shaft. Turn the screwdriver, which will force the weights out to their fully extended position against the stop on the weight carrier shaft. Hold the screwdriver firmly in this position.
4. Refer to Figure 10-47. With the weights held in this position, their movement through the governor linkage has forced the low speed spring cap against the high speed spring plunger, as shown.
5. Insert the feeler gauge and check for a slight drag. If the feeler gauge is tight or

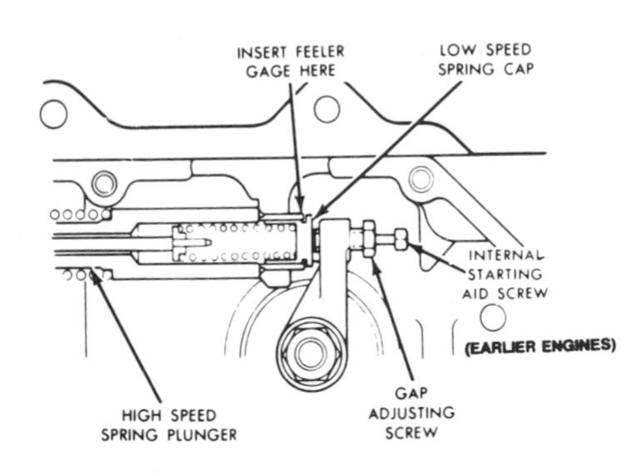

Figure 10-47
Location of Governor Gap.
(Courtesy of Detroit Diesel Allison Div. of GMC)

there is no slight drag, adjust the gap as necessary.

6. Reinstall the governor cover assembly.

Remember that, if you use this procedure to *cold* set the governor gap, it still must be set with the engine at its normal operating temperature. This sequence can be used to further check the governor gap once the engine is hot and the running check has been made.

DWLS Governor Gap (Engine Running). Although the governor gap can be adjusted on the DWLS governor (Vee engines) with the engine stopped, the proper gap can and must be done with the engine up to proper operating temperature. The DWLS governors are found on highway trucks, both on nonturbocharged engines and on the newer model of high torque rise type fuel-squeezer engines, the TT (tailored-torque) units. Prior to setting the governor gap, inspect the tag on the side of the governor housing, which will indicate whether or not the engine is equipped with a TT governor.

The DWLS governor gap is as follows:

1. Non-fuel-squeezer engines: 0.0015 in. (0.0381 mm), with the engine running between 800 and 1000 rpm.
2. Fuel-squeezer (TT) engines: 0.002 to 0.004 in. (0.050 to 0.101 mm), with the engine running between 1100 and 1300 rpm (except 16V-71 and 16V92).

Adjustment Procedure (DWLS), Vee-Type Engines. The procedure for adjusting the governor gap on a non-fuel-squeezer engine compared to that on a fuel-squeezer engine differs only slightly, with the basic differences being as follows:

1. Fuel squeezer: you must back out the Belleville spring retainer nut 0.06 in. (1.524 mm). See Figure 10-40.
2. Both types: remove the throttle delay mechanism if used.
3. Fuel squeezer: back out the starting aid screw.

With the engine stopped and at normal operating temperature of 160 to 185°F (71 to 85°C) proceed as follows:

1. Remove the high speed spring retainer cover, and back out the Belleville spring retainer nut until there is approximately 0.06 in. (1.524 mm) clearance between the washers and nut, as shown in Figure 10-40.
2. Back out the buffer screw until it extends approximately ⅝ in. (or 15.875 mm) from the locknut, all engines.
3. Check the idle speed with an accurate tachometer (digital type if possible), and if necessary adjust the idle speed screw to obtain the desired engine idle speed. EPA certified minimum idle speed is 500 rpm for trucks.
4. Stop the engine, and clean and remove the governor cover assembly and discard the gasket.
5. *Caution:* With the governor cover removed for the purpose of checking the governor gap, it is necessary to run the engine at either 800 to 1000 rpm on non-

fuel-squeezer type engines or 1100 to 1300 rpm on fuel-squeezer type engines. In *both* cases, *you are the governor* since you must manually control the engine speed by manipulation of the governor's differential lever, which is the lever that is connected to both fuel rods going to each cylinder bank of the engine. A rag placed over the governor assembly, as shown, will minimize oil throw off during a running check, unless a cutaway governor cover is being used.

6. Start and run the engine at the speed required for your particular engine, and check the gap between the low speed spring cap and the high speed spring plunger with the required feeler gauge. In both cases there should be a slight drag on the feeler gauge. If the gap is incorrect (too tight or too loose), reset the gap adjusting screw with the engine stopped, then start it and recheck. See Figure 10-48.
7. On the non-fuel-squeezer engines, where the governor gap is set at 0.0015 in. between 800 and 1000 rpm, if the gap setting is correct, the 0.0015 in. (0.0381 mm) movement can be seen by placing a few drops of clean engine oil into the governor gap at this speed range or by pressing a screwdriver against the end of the gap adjusting screw. Movement of the low speed spring cap toward the high speed spring plunger will force the oil from the governor gap in the form of a small bead.
8. Stop the engine, and using a new gasket, install the governor cover assembly and tighten the screws.

Governor; Gap on 16V-71 and 16V-92 Engines with DWLS Governor. Governor gap adjustment on these two engines is almost identical to that on the other Vee-type engines just described; however, since a 16V engine, 71 or 92, is basically two 8V's bolted together, there are several variations involved in correctly setting this adjustment. The governor on these two 16V engines is mounted on and driven from the front end of the rear blower. See Figure 10-49.

It is imperative prior to checking the governor gap on the engine that the control link levers in the governor housing and auxiliary control link housing be aligned. Refer to Figure 10-50; then position the control link levers as described next.

Positioning Control Link Levers—16V-71 and 16V-92

See Figures 10-49 and 10-50.

1. Disconnect any throttle linkage and controls from the governor speed control lever and stop lever, respectively.
2. Remove the governor and auxiliary control link housing covers.

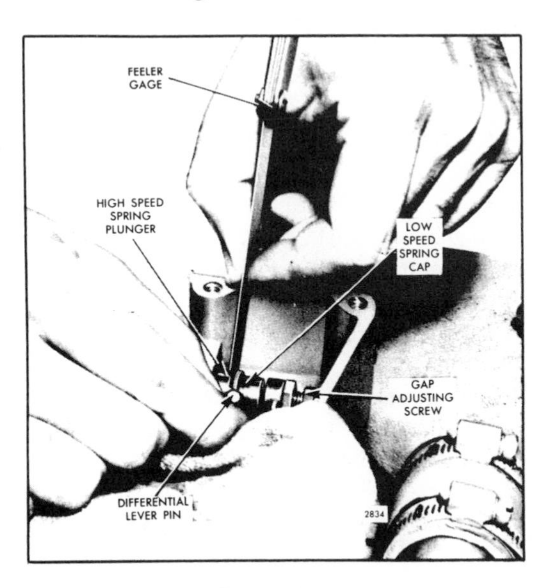

Figure 10-48
Checking Governor Gap on DWLS Governor Vee-Type Engine.
(Courtesy of Detroit Diesel Allison Div. of GMC)

3. Disconnect the adjustable link from the lever in the auxiliary control link housing.
4. Remove the small spring clip and connecting pin from the auxiliary housing governor control link lever.
5. Install gauge J 21779 (Kent Moore) so that it extends through the lever and fuel rod and into the gauge hole in the bottom of the housing. With the gauge in place, the auxiliary control link lever will be in the mid-travel position.
6. Proceed to remove the other connecting pin from the control link lever in the governor housing, and install gauge J 21780 (Kent Moore). Install the gauge so that the pin will extend through the connecting link control lever, and fuel rod, and the governor housing dowel pin extends into the small hole in the gauge. Install a governor cover bolt to lock the gauge in place, as shown in Figure 10-50. With gauge J 21780 locked in position, the governor control link lever will be in the mid-travel position and therefore parallel to the auxiliary control link lever.

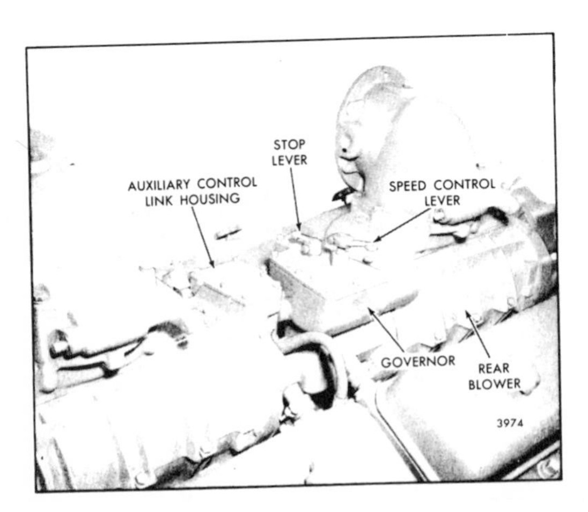

Figure 10-49
16V 71 and 92 Series Engines Showing Governor and Auxiliary Housng Location.
(Courtesy of Detroit Diesel Allison Div. of GMC)

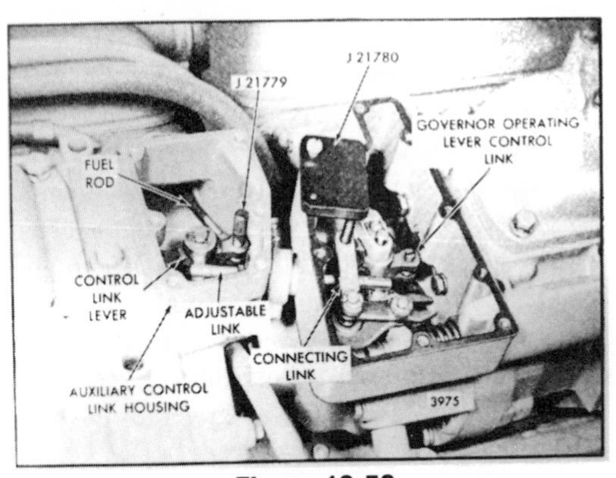

Figure 10-50
Aligning Control Link Levers.
(Courtesy of Detroit Diesel Allison Div. of GMC)

Note: This is necessary, since the 16V engine only has *one* governor. It controls all injector racks on one half of the engine through the auxiliary control link housing and the other half by direct connection to the governor. The above sequence of events will ensure that all injector control racks will respond equally to any governor movement.

7. Refer to Figure 10-50; adjust the length of the adjustable connecting link in order to retain the lever positions that were obtained in steps 5 and 6. Then install the link over the end of the auxiliary control housing link housing control link lever.
8. Proceed to remove both gauges (J 21779 and J 21780); then reinstall the control link lever connecting pins.
9. Install the governor housing and auxiliary control link housing covers. Do not use the gasket under the governor cover at this time, since the governor housing will be removed again to check the running gap.

Variable Speed Governor Gap

On engines employing SWVS governors, the gap is 0.006 in. (0.152 mm). This is taken between the spring plunger and plunger guide at the bellcrank end of the spring under the governor cover. The engine is stopped at this time; however, the governor speed control lever should be held in the full-fuel position while checking the gap.

All Series 2-Cycle Engine, Positioning the Injector Rack Control Levers. Since all the injector racks are connected to the fuel control tube and then to the governor via the fuel rod or rods, they must be set correctly to ensure that they are all equally related to the governor. Their positions determine the amount of fuel that will be injected into the individual cylinders and therefore assure equal distribution of the load. Properly adjusted injector rack control levers with the engine at full load will ensure the following:

1. The speed control lever at the maximum speed position.
2. The governor low speed gap closed.
3. The high speed spring plunger on its seat in the governor control housing.
4. The injector fuel control racks in the full-fuel position.

Failure to properly set the racks will result in poor performance and a lack-of-power complaint.

The governor's location on the engine will control which injector rack is set first. On those engines with the governor located at the front, the cylinder 1 injector rack would be set first, whereas with the governor mounted at the rear, the rear cylinder injector rack would be set first (in-line engines).

For Vee-type engines, Figure 10-42 indicated the system used in determining the right and left bank and the cylinder numbering sequence. Therefore, all Vee engines with the governor located at the front have the no. 1 left bank injector rack set first, since it is the closest rack to the governor. On those Vee engines (6V-53) with the governor located at the rear, the no. 3 left bank injector rack would be set first.

With this in mind, prior to setting the first injector rack, also do the following:

1. Disconnect any linkage attached to the governor speed control lever (hand or foot throttle cables or rods).
2. Back out the idle speed adjusting screw until there is no tension on the low speed spring (limiting speed governors only). When approximately ½ in. (12.7 mm) or 12 to 14 threads are showing beyond the locknut, when the nut is against the high speed plunger, the tension of the low speed spring will be low enough that it can be easily compressed. This allows closing of the low speed gap without possible bending of the fuel rod or rods or causing the yield link (used with throttle delay engines) spring mechanism to yield or stretch.

Note: Failure to back out the idle speed adjusting screw as stated may result in a false fuel rack setting and the problems associated with this.

3. If the engine is equipped with a throttle delay mechanism, this would have been removed prior to checking the governor gap. *Keep this unit off.*
4. Similarly, the buffer screw should *still be backed out* approximately ⅝ in. (15.875 mm) as it was prior to setting the governor gap.
5. Also, the Belleville spring retainer nut on those engines so equipped (Figure 10-40) should have the 0.06-in. (1.524-mm) clearance as shown.
6. On turbocharged or fuel-squeezer engines employing a starting aid screw, *do not touch it* at this time. Leave it backed out.

Note: When the injector racks are adjusted properly, the effort expended in moving the throttle from an idle to maximum speed position should be uniform throughout its travel. Any increase in effort while doing this could be caused by the following: (a) injector racks adjusted too tight, causing the yield link to separate; (b) binding of the fuel rods; or (c) failure to back out the idle screw.

7. Loosen all the inner and outer adjusting screws of each injector rack control lever at the control tube. The newer engines employ only one adjusting screw (the inner one) with a locknut on it. On Vee engines, loosen the screws on both banks. Be sure that all the injector rack control levers are free on the control tube. Make sure that the screws are backed off at least $\frac{1}{4}$ in.

Figure 10-51
Adjusting No. 1 Left-Bank Injector Control Rack.
(Courtesy of Detroit Diesel Allison Div. of GMC)

8. *Vee engines only:* Remove the clevis pin from the fuel rod at the right bank injector control tube lever (LB still connected to the governor).
9. On *variable speed mechanical governors,* the speed control lever must be moved to the maximum or full-fuel position and locked in this position while adjusting the injector racks.
10. On *variable speed* governors, move the stop lever on top of the governor housing to the *run* position and hold it in this position with light finger pressure. On *limiting speed* governors, move the speed control lever on top of the governor housing to the maximum speed position and hold it there with light finger pressure.
11. With either the stop lever or the speed control lever (step 10) being held lightly in position (run or full fuel), turn down the inner adjusting screw (two-screw type) or adjusting screw (one-screw type) as shown in Figure 10-51 (Vee engine example) until the no. 1 left bank injector rack control lever clevis (Figure 10-52) is observed to roll up, or an increase in effort to turn the screwdriver is noted. Tighten the screw approximately one-eighth turn more on the single-screw type; then lock it securely with the adjusting screw locknut. On the two-screw type, turn the inner adjusting screw down on the no. 1 LB of Vee engines, or the screws on the rack closest to the governor on in-line engines, until a slight movement of the control tube lever is observed or a step up in effort to turn the screwdriver is noted. Turn down the outer adjusting screw until it bottoms lightly on the injector control tube; then alternately tighten both the inner and outer adjusting screws one-eighth turn each until snug. Finally torque the screws to 24 to 36 in.-lb (3 to 4 N·m) to avoid damage to the injector control tube.

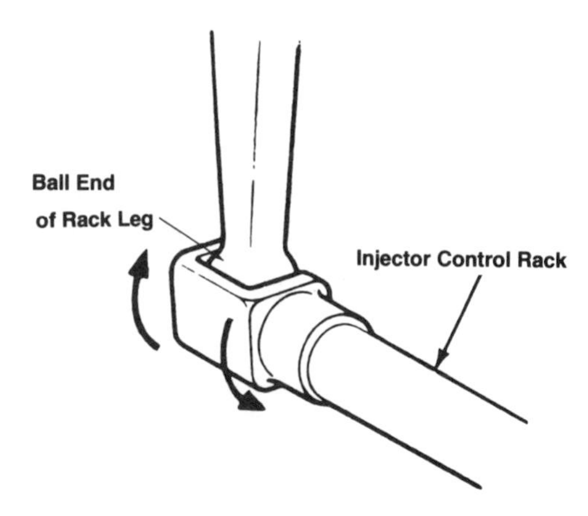

Figure 10-52
Location to Determine Injector Rack Bounce.
(Courtesy of Detroit Diesel Allison Div. of GMC)

Caution: While turning down the inner adjusting screw (one- or two-screw type), if you go too far, you will feel either the stop lever (variable speed governor) or the speed control lever (limiting speed governor) move. If this happens, you have gone too far, and the rack is being forced out of the full-fuel position. Therefore, adjust the screw until very slight movement can be felt at the stop or speed control lever; then back the screw off slightly.

At this time the no. 1 LB rack on Vee engines, or the closest rack to the governor on in-line engines, should be in the full-fuel position with the governor linkage and control tube assembly in the same position they will attain while the engine is operating at normal operating temperature under full load.

12. To be sure that you have in fact adjusted the rack correctly, hold the stop lever (variable speed governor) in the run position or the speed control lever (limiting speed governor) in the maximum fuel position. Refer to Figure 10-53, and press down on the injector rack clevis with a screwdriver blade, which should cause the rack to tilt downward; when the pressure of the screwdriver blade is released, the control rack should bounce or spring back upward. If the injector rack does not have a good bounce or spring, it is too loose; therefore, to overcome this condition, back off the outer adjusting screw very slightly and tighten the inner one an equal amount.

On single-screw-type units, loosen off the locknut, turn the adjusting screw clockwise slightly, and retighten the locknut. Recheck the rack condition for bounce.

To ensure that the rack is not set *too tight,* do the following. Move either the speed control lever (limiting speed) or the stop lever (variable speed) from the idle to the maximum speed position or from the stop to the run position, respectively. While doing this, if the injector rack becomes tight on the ball end of the rack leg (see Figure 10-52) before the end of the

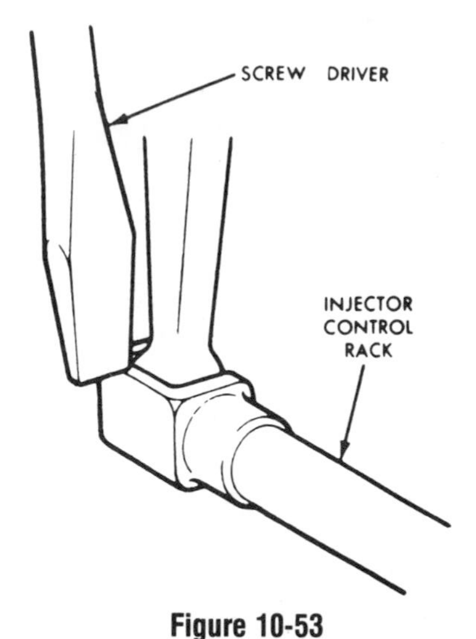

Figure 10-53
Using a Screwdriver to Check Injector Rack Bounce.
(Courtesy of Detroit Diesel Allison Div. of GMC)

lever travel, the rack also needs readjusting. To correct this condition, either loosen the screw locknut or the *inner* adjusting screw on the rack slightly and tighten the outer screw a similar distance, or back out the one-screw type and tighten the locknut. Recheck the rack bounce and movement. If an engine has been in service for a considerable period of time, the ball end of the rack leg sometimes becomes slightly scuffed. This can prevent a good bounce when setting the racks; if you encounter this problem, loosen both rack screws and slide the rack leg lever to the side of the injector, swing it upward, and lightly rub the ball end with fine emery cloth.

13. On all in-line engines the first rack that has been set is the one closest to the governor. On Vee engines the no. 1 LB is the closest to the governor, with the exception of the 6V-53 which has the no. 3 LB closest. In either case, once the first rack has been adjusted, this now becomes the *master rack,* since it has been set to the governor. To adjust the remaining injector rack control levers on the engine, proceed as follows:

    a. *In-line engines:* Remove the clevis pin from the fuel rod at the injector control tube lever; hold the injector control racks in the full-fuel position by means of the lever on the end of the control tube.

    b. *Vee engines:* Remove the clevis pin from the fuel rod at the LB injector control tube lever. Install the clevis pin in the fuel rod at the right-bank injector control tube lever and adjust the no. 1 RB rack the same way as for the no. 1 LB in step 11. To verify that both no. 1 racks are adjusted the same, insert the clevis pin at the LB fuel rod. Move the speed control lever (LSG) to the maximum speed position or the stop lever (VSG) to the run position and check the drag on the clevis pin at each bank. In addition, check the bounce on each no. 1 rack. If they are not the same, the no. 1 RB rack has to be readjusted, since the no. 1 LB was the first one set to the governor and is therefore the master rack. To increase drag or bounce on the no. 1 RB rack, turn the rack adjusting screw clockwise on the one-screw setup, or the *inner* screw clockwise on the two-screw setup after slightly loosening the *outer* screw. Turn the screws counterclockwise to decrease pin drag or bounce.

14. To adjust the remaining injector racks on each bank, remove both clevis pins from each bank and hold the LB injector control racks in the full-fuel position by means of the lever on the end of the control tube (same setup as for the in-line engines).

    a. Tighten or run down the adjusting screw (inner) of the no. 2 LB injector rack control lever until the rack clevis rolls up or a step up in effort to turn the screwdriver is noted. If you feel the control tube lever move, back off on the adjusting screw slightly and turn it clockwise gently until you are satisfied that the rack is positioned correctly. While holding the control tube in the full-fuel position, compare the bounce on the no. 2 LB rack with that of the no. 1 LB rack. They should be the same; if not, readjust no. 2 again.

Caution: Do not alter the adjustment of the no. 1 LB rack at any time. Remember that it is the master rack and has already been set to the governor.

    b. Adjust the remaining racks on the LB in the same fashion, checking the bounce of each rack with the no. 1 setting every time. They should all have the same bounce when you are finished. Repeat the same procedure for the RB injector rack adjustments, always bearing in mind that the racks on each bank are set to the no. 1 rack on that bank. Therefore, *do not alter the no. 1 LB or RB setting to suit the others*.

15. When all the injector control racks have been adjsuted, install the clevis pins in each fuel control tube to fuel rod.

    a. Move the speed control lever to the maximum fuel position on the limiting speed governor.

    b. Move the stop lever to the run position on the variable speed governor.

    Check each injector control rack for the same bounce or spring condition and also the drag on each clevis pin at each bank. If they are not the same, further checks and adjustments will be required. If one clevis is tight and the other not, one bank will invariably run hotter than the other, indicating that it is doing most of the work.

16. Once you are satisfied that you have adjusted each bank equally, secure the clevis pin with a cotter pin at each bank.

17. On limiting speed governors, turn in the idle screw adjustment until the screw projects approximately $\frac{3}{16}$ in. (4.762 mm) from the locknut, which will permit starting of the engine.

18. On in-line engines the injector racks are adjusted in the same fashion as for those on the Vee engines, the only difference being that you do not have two separate banks to adjust. Also, once the first rack has been set to the governor, do not readjust it to suit another rack's bounce.

19. Replace the valve rocker cover or covers if the engine is going to be run for any reason.

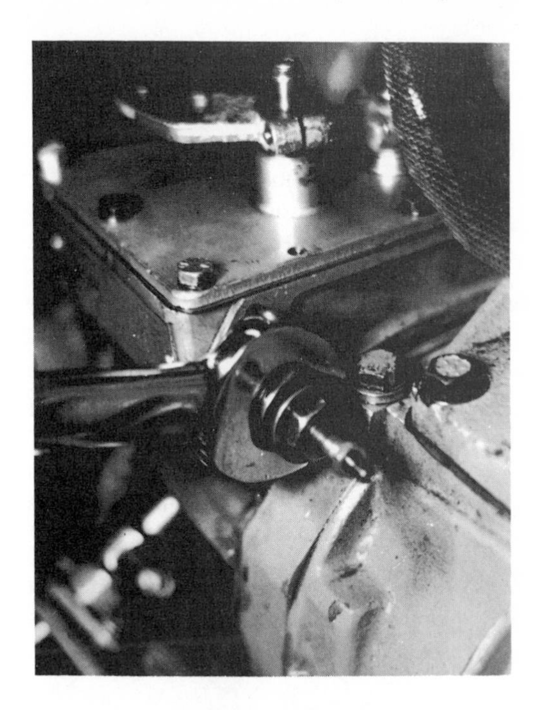

Figure 10-54
Adjusting Maximum No-Load Engine Speed on V-71 Engine.
(Courtesy of Detroit Diesel Allison Div. of GMC)

Adjust Maximum No-Load Engine Speed (Limiting Speed Governors)

The type of engine application determines the maximum governed speed of the engine, and this is set on the engine prior to leaving the factory. Due to a variety of reasons, and to ensure that the engine speed will not exceed its recommended no-load speed, which is stamped on the engine's *option plate* found on the valve rocker cover, check and set the maximum no-load engine speed as follows:

1. Make sure that the buffer screw is still backed out $\frac{5}{8}$ in. (15.875 mm) from the governor housing and locknut. If not, interference while adjusting the maximum no-load speed can occur.

2. On limiting speed governors (Figure 10-

54), loosen the spring retainer locknut and back off the high speed spring retainer nut approximately five full turns. With the engine operating at normal operating temperature of 160 to 185°F (71 to 85°C), and with no load on the engine, place the speed control lever in the full-fuel position. Turn the high speed spring retainer nut clockwise until the engine is running at the recommended no-load rpm.

Caution: On fuel-squeezer engines, the Belleville springs must be readjusted anytime that the no-load speed has been altered. The engine's no-load speed must be set 150 rpm above the rated speed prior to adjusting the Belleville springs.

Note: Hold the high speed spring retainer nut and tighten the locknut. Limiting speed governors used on industrial engines, along with some found on 53 engines, use shims at the bellcrank end of the governor spring to vary the speed, similar to a variable speed governor.

Variable Speed Governors (Maximum No-Load RPM)

The variable speed governor requires the addition or removal of *stops* or *shims,* respectively, to either increase the governed speed or decrease the governed speed of the engine. With the engine at proper operating temperature, determine the maximum no-load speed of the engine with an accurate tachometer (electronic digital preferred). If adjustment is required, proceed as follows:

1. Refer to Figure 10-55; disconnect the governor booster spring and the governor stop

Figure 10-55
Variable Speed Governor on In-Line 71 Engine.
(Courtesy of Detroit Diesel Allison Div. of GMC)

Index:

12—Governor stop/run lever shown held in the run position
21—Governor speed control lever
51—Governor variable speed spring housing
55—Idle speed adjusting screw
56—Idle screw locknut
57—Buffer screw
58—Buffer screw locknut
108—Stop/run lever retracting spring (older style)
144—Governor booster spring
145—Throttle linkage

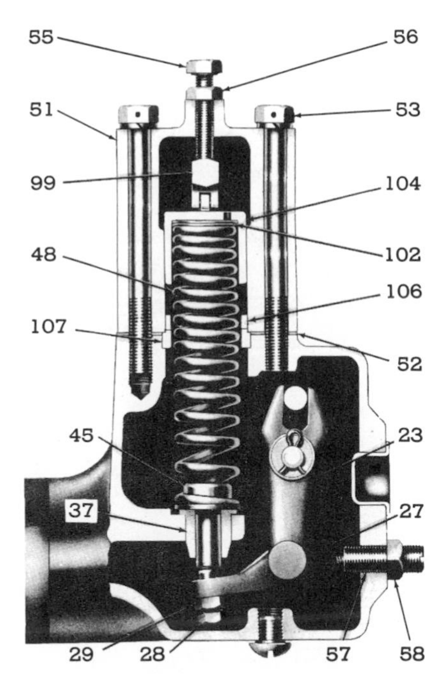

Figure 10-56
Parts Stackup of In-Line 71 Series Engine Variable Speed Governor.
(Courtesy of Detroit Diesel Allison Div. of GMC)

Index:
23—Lever-differential
27—Lever-operating shaft
28—Screw-gap adjusting
29—Locknut
37—Guide-plunger
45—Plunger-variable speed spring
48—Spring-variable speed
51—Housing-variable speed spring
52—Gasket
53—Bolt
55—Screw-idle speed adjustment
56—Locknut
57—Screw-buffer
58—Locknut
99—Lever-variable speed spring
102—Shim
104—Plunger-variable speed spring
106—Stop
107—Stop

lever spring by removing the special governor cover screw holding it in place.

2. Leave the buffer screw backed out as before.
3. Remove the two bolts and withdraw the variable speed spring housing from the governor housing.
4. From Table 10-10, determine the stops or shims required to either establish the maximum no-load rpm, or how many have to be removed to arrive at a reduced maximum no-load rpm. Also refer to Figures 10-56 and 10-57 for the location of the stops and shims.

Shims supplied for these types of governors come in both 0.078- and 0.010-in. sizes (1.981 and 0.254 mm). The addition of a 0.001-in. (0.0254-mm) shim behind the spring will alter the rpm by 1. With the use of either the split or solid stop in the governor along with a variety of shims, the desired rpm can be obtained. The maximum amount of shims that should be used should not exceed 0.325 in. (8.255 mm).

Caution: If the governor spring is removed from the housing while altering the maximum rpm, be certain when it is reinstalled that the closer wound spring coils face the idle adjustment screw; otherwise, the engine will tend to surge at the normal idle speed range and not be steady (see Figure 10-56).

If factory-supplied shims are not readily available for insertion, ordinary shim stock can be used if punched out to the desired shape, as shown in Figure 10-57. If the shims are cut or punched out as a solid circular disc with no hole in the center, when they are stacked up in the piston-type retainer, it is possible to block off the small hole in the rear of the retainer, which can lead to irregular governor operation; therefore, always punch the shims out in washer form similar to the factory-supplied units.

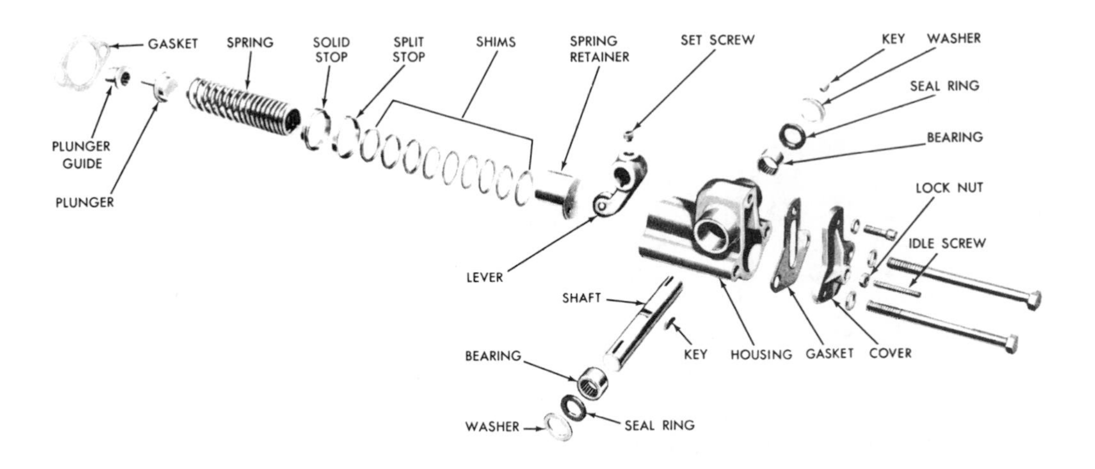

Figure 10-57
Exploded View of Variable Speed Spring Housing and Relative Location of Parts (Vee-Type Engine).
(Courtesy of Detroit Diesel Allison Div. of GMC)

Note: If the governor is adjusted whereby the engine's maximum no-load speed is raised or lowered more than 50 rpm, the governor gap should be rechecked. Similarly, if the governor gap is readjusted, the injector racks should be rechecked also.

The engine's governed no-load speed will differ from the full-load speed by as much as 140 to 220 rpm depending on the engine application and series. For a more detailed explanation of *droop,* which is basically the term used to express the difference between these two

Table 10-10
NUMBER OF STOPS TO BE USED AT VARIOUS FULL-LOAD SPEEDS

Engine Series	*Full-Load Speed (rpm)*	*Stops*	
		Solid Ring	*Split Ring*
53	2575–2800	0	0
	2101–2575	1	0
	1701–2100	1	1
	I/L 1200–1700 (*V-12-2100)	1(1*)	2(1*)
71 and 92	1200–1750	1	1
	1750–2100	1	0
	2100–2300	0	0
149			

Shims as required.

*Indicates that the V12 only uses one solid and one split stop ring when governed at 2100 rpm. Some 53 series engines use two split stops.

speeds, see the Woodward governor section.

Adjust Idle Speed

See Figure 10-58. The idle speed for an engine will vary with its particular application; therefore, always check the governor identification plate riveted onto its housing for the recommended idle range. The recommended idle speed for non-EPA-certified engines with limiting speed governors is 400 to 450 rpm on the majority of these units, but may vary with special engine applications. EPA-certified minimum idle speeds are 500 rpm for trucks and highway coaches and 400 rpm for city coaches.

With the maximum no-load speed properly adjusted, proceed to set the idle speed as follows:

1. Ensure that the engine is operating at normal operating temperature of 160 to 185°F (71 to 85°C), and that the buffer screw is still backed out to avoid contact with the differential lever.
2. On earlier engines, the idle screw had a slotted end for screwdriver adjustment; however, later engines have an Allen head screw for idle adjustment. Loosen the idle screw locknut, and turn the idle speed adjusting screw either clockwise to increase the rpm or counterclockwise to reduce the rpm until the engine operates at approximately 15 rpm below the recommended idle speed. (Some variable speed governors have a hex head adjusting screw for the idle setup.)

Note: You may find it necessary to use the buffer screw (turn it in) to eliminate engine roll or surge so that you can establish what the engine idle speed is at this time. Once this is established, *back out the buffer screw* to its previous setting, which should be $\frac{5}{8}$ in. (15.875 mm).

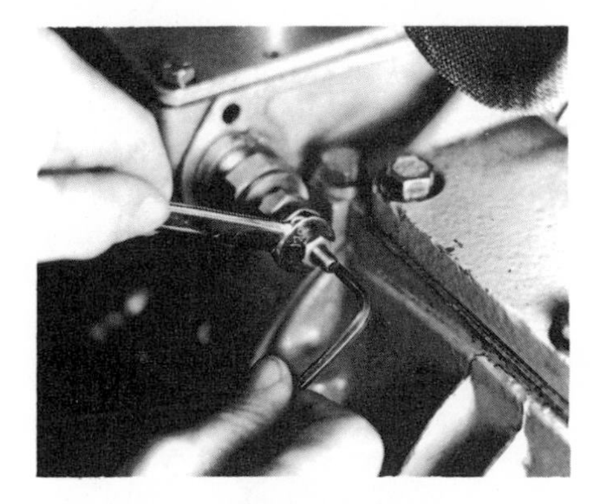

Figure 10-58
Idle Adjustment on a Vee-Type Engine. (Courtesy of Detroit Diesel Allison Div. of GMC)

3. Hold the idle screw and tighten the locknut.

Belleville Spring Adjustment (Limiting Speed Governors, Highway Vehicles)

This adjustment only applies to those engines that have the limiting speed governor; therefore, on-highway vehicles using turbocharged engines with this governor setup are classed as fuel-squeezer-type engines. The term TT for example in an engine model designation, such as an 8V-92 TT, simply means that the engine is a V8-type engine with 92 in.^{3} (1507.6 mm^3) displacement per cylinder and is equipped with a governor for TT (or tailored torque) horsepower (kW) adjustment.

TT horsepower adjustment is accomplished at the moment by two methods, depending on the equipment available, and these are: (1) by an established idle drop method and (2) by use of an established power reduction factor.

Regardless of the method used, to obtain satisfactory results the engine must be in good mechanical condition and properly tuned; therefore, do *not* attempt Belleville spring governor adjustment until an engine tune-up has been thoroughly completed.

Method 1: Idle Drop. This method is an effective, accurate means of setting TT horsepower (kW). The idle drop method employs a reduction in engine rpm in order to position the Belleville washers and both the governor low and high speed springs. The correct positioning of these components results in obtaining the desired fuel-squeezer horsepower (kW).

Note: Use of the engine's mechanically driven tachometer for this purpose is inadequate. An accurate tachometer is *mandatory* (use Kent Moore tool J 26791 or one of the latest style of electronic hand-held digital tachs when using the idle drop method). Each 1-rpm error in setting the idle drop will result in a 2- or 3-horsepower error.

Procedure (Idle Drop Method)

1. Do a complete engine tune-up. Set the no-load speed as specified according to the engine type, injector size, and governor part number.
2. Disconnect the accelerator linkage from the governor speed control lever on top of the governor housing if this has not already been done.
3. Run the engine until a stabilized engine operating temperature between 160 and 185°F (71 and 85°C) is obtained.
4. Refer to available information through a local DDA dealer or service bulletin and using the engine type, injector size, and governor part number, obtain the initial and specified idle drop numbers for the rated TT horsepower (kW) and rated engine speed at which the engine is to operate (see Tables 10-11 and 10-12).
5. Set the initial idle rpm by the use of the idle adjusting screw to that determined in step 4. For example, one particular engine would be adjusted to 915 rpm by use of the idle screw, then reduced to 848 rpm as per step 6.
6. With the governor speed control lever in the idle position, turn the Belleville spring retainer nut (Figure 10-40) clockwise on the plunger until the specified idle drop rpm is achieved. Secure the retainer nut with the locking screw. When the specified idle rpm is achieved, the engine is power controlled to the TT horsepower (kW) rating.

Note: It is imperative that the idle speeds be adjusted to the *exact* rpm and also be *steady* with no fluctuation. If they are not, check for any binding or rubbing in the fuel control system at the governor, fuel rods, injector control tubes, and racks.

7. Lower the idle speed to the normal specified operating idle rpm, using the idle adjusting screw.
8. Adjust the buffer screw and starting aid screw (see buffer screw and starting aid screw procedure later in this section).

Method 2: Power Reduction Factor. This method differs from the idle drop method in that it consists of setting the TT engine horsepower (kW) to a specific percentage below full throttle horsepower (kW) as observed on an engine, chassis, or output shaft dynamometer. The desired horsepower within a reasonable tolerance can be obtained based on the following variations:

1. Dynamometer calibration
2. Drive line efficiency
3. Fuel grade and temperature
4. Air density
5. Tire slippage

Procedure (Power Reduction Factor)

1. Perform the standard engine tune-up.

Note: The throttle delay piston must be removed and the Belleville spring retainer nut must be backed out until there is approximately

Figure 10-59
Adjustment of Belleville Washers.
(Courtesy of Detroit Diesel Allison Div. of GMC)

0.060 in. (1.524 mm) clearance between the washers and retainer nut (Figure 10-40) prior to operating the engine on the dynamometer.

2. Set the no-load speed as required by the engine type, injector size, and governor part number. This can be found stamped on the valve rocker cover option plate on the engine, and on the governor housing.
3. Run the engine until the coolant temperature is above 170°F (77°C).
4. Using an engine, chassis, or output shaft dynamometer, measure and record the *full-throttle* horsepower (kW) at 100 rpm below rated engine speed with the Belleville washers loose, as shown earlier in Figure 10-40 and in Figure 10-59.

Important: Satisfactory power adjustment can be obtained only if the full throttle horsepower (kW) and adjusted horsepower (step 4) are obtained with the engine cooling temperature in both instances being the same. (If a thermatic or air-operated fan is used, note whether it is engaged or disengaged.)

5. Select the power reduction factor shown in Table 10-13 for proper engine type, desired rated horsepower, and rated engine speed.
6. Multiply the horsepower (kW) recorded in step 4 by the factor selected in step 5. Record this value.
7. Adjust the Belleville spring retainer nut clockwise so that the observed horsepower is reduced to that recorded in step 6 at 100 rpm below the rated engine speed, with the governor speed control lever in the maximum speed position and the fan in the same mode of operation as step 4. Verify that the engine is obtaining the specified adjusted horsepower TT, within 5% at *rated* engine speed. If the adjusted TT horsepower (kW) cannot be obtained at rated engine speed, governor *droop* interference may be the cause. If necessary, to eliminate droop interference, readjust the engine no-load speed from 150 to 175 rpm above rated engine speed and repeat the power reduction factor method.
8. Check the idle speed and readjust if necessary.
9. Adjust the buffer screw and starting aid screw.

Adjust Buffer Screw

Buffer screw adjustment on DDA engines must be done carefully to avoid any unnecessary increase in the normal engine idle range and in the maximum no-load speed. Prior to buffer screw adjustment, the specified engine idling speed must be properly set to within 15 rpm of

Table 10-11
IDLE DROP SETTINGS

ENGINE TYPE	INJECTOR SIZE	CURRENT GOVERNOR PART NUMBER	THROTTLE DELAY SETTING	RATED HORSEPOWER @ RATED SPEED	INITIAL IDLE	IDLE DROP SETTING
6V-92TT	9290	5104768*	.570	270 @ 1800 260 @ 1800 250 @ 1800 240 @ 1800	880 880 880 880	810 804 798 792
6V-92TT	9290	5104768*	.570	270 @ 1850 260 @ 1850 250 @ 1850 240 @ 1850	880 880 880 880	802 802 796 790
6V-92TT	9290	5104769	.636	270 @ 1900 260 @ 1900 250 @ 1900 240 @ 1900	915 915 915 915	850 845 840 835
6V-92TT	9290	5104769	.636	270 @ 1950 260 @ 1950 250 @ 1950 240 @ 1950	915 915 915 915	848 843 838 833
6V-92TT	9290	5104768*	.636	270 @ 1900 260 @ 1900 250 @ 1900 240 @ 1900	950 950 950 950	877 873 869 865
6V-92TT	9290	5104768*	.636	270 @ 1950 260 @ 1950 250 @ 1950 240 @ 1950	950 950 950 950	875 871 867 863
6V-92TT	9290	5104769	.636	290 @ 1900	915	860
6V-92TT	9290	5104769	.636	290 @ 1950	915	858
6V-92TT	9290	5104768*	.636	270 @ 1950 260 @ 1950 250 @ 1950 240 @ 1950	950 950 950 950	875 871 867 863
6V-92TT	9290	5104769	.636	270 @ 2100	1075	995
6V-92TTA	9A90	5104768*	.636	270 @ 1900	975	900
6V-92TTA	9A90	5104768*	.636	270 @ 1950	975	899
6V-92TTA	9A90	5104769	.636	270 @ 1900 260 @ 1900 250 @ 1900 240 @ 1900	915 915 915 915	838 832 825 819
6V-92TTA	9A90	5104769	.636	270 @ 1950 260 @ 1950 250 @ 1950 240 @ 1950	915 915 915 915	837 831 824 818
6V-92TTA	9A90	5104769	.636	270 @ 2100	1075	995
8V-92TT	9A90	5104768*	.570	335 @ 1800	830	765

*Uses Belleville Spring (Orange Dye) P/N 5104535.

NOTE: Table 10-11 lists only current factory option governor assembly part numbers. See Table 10-12 for former governor assembly part numbers used on Fuel Squeezer engines.

Table 10-11 (Continued)

ENGINE TYPE	INJECTOR SIZE	CURRENT GOVERNOR PART NUMBER	THROTTLE DELAY SETTING	RATED HORSEPOWER @ RATED SPEED	INITIAL IDLE	IDLE DROP SETTING
8V-92TT	9A90	5104768*	.570	335 @ 1850	830	763
8V-92TT	9290	5104769	.636	365 @ 1900	915	850
8V-92TT	9290	5104769	.636	365 @ 1950	915	848
8V-92TT	9A90	5104768*	.570	365 @ 1900 335 @ 1900	950 950	880 870
8V-92TT	9A90	5104768*	.570	365 @ 1950 335 @ 1950	950 950	878 868
8V-92TT	9290	5104769	.636	365 @ 2100	1075	1005
8V-92TTA	9A90	5104768*	.636	335 @ 1800	880	795
8V-92TTA	9A90	5104768*	.636	335 @ 1850	880	793
8V-92TTA	9A90	5104768*	.636	365 @ 1900 335 @ 1900	950 950	877 865
8V-92TTA	9A90	5104768*	.636	365 @ 1950 335 @ 1950	950 950	875 863
8V-92TTA	9A90	5104769	.636	365 @ 1900 335 @ 1900	950 950	867 845
8V-92TTA	9A90	5104769	.636	365 @ 1950 335 @ 1950	950 950	865 843
8V-92TTA	9A90	5104769	.636	365 @ 2100	1075	1000
8V-71TT	N-75	5104767	.570	305 @ 1900	700	670
8V-71TT	N-75	5104767	.570	305 @ 1950	700	668
8V-71TT	N-75	5104767	.570	305 @ 2100	750	705
8V-71TTA	7A75	5104767	.570	350 @ 1900	700	665
8V-71TTA	7A75	5104767	.570	305 @ 1950	700	664
8V-71TTA	7A75	5104767	.570	305 @ 2100	750	715

that desired. Use an accurate electronic digital tachometer for this purpose.

With the idle speed properly set, adjust the buffer screw as follows:

1. With the engine having been adjusted to its recommended idle speed and running at normal operating temperature, refer to Figure 10-60 and turn the buffer screw in so that it lightly contacts the differential lever inside the governor housing. This is easily determined by the fact that the engine speed will pick up slightly, and the roll or surge in the engine will level out.

Note: Be very careful that you do not increase the engine idle speed more than 15 rpm with the buffer screw adjustment. This is why an accurate tachometer must be used.

Table 10-12
CURRENT AND FORMER "TT" ENGINE GOVERNOR ASSEMBLY PART NUMBERS

GOVERNOR PART NUMBER	DESCRIPTION AND STATUS		
*5104767	8V-71TT Governor Assembly Weights 1900-2100 RPM Tamper Resistant		
*5104768	6 and 8V-92TT Governor Assembly Light Weights 1800-1950 RPM Tamper Resistant (Uses Orange Dye Belleville 5104535)		
*5104769	6 and 8V-92TT Governor Assembly Light Weights 1900-2100 RPM Tamper Resistant		
*5103567	V-92TT Governor Assembly Light Weights 1900-2100 RPM Non-Tamper Resistant Released for Industrial Model 8083-7300 Only		
5102991	6 and 8V-92TT Governor Assembly Light Weights 1900-2100 RPM Tamper Resistant	Superceded by 5103620 S.I. 62-D-76	Superceded by 5104769 S.I. 27-D-77
5103620	6 and 8V-92TT Governor Assembly Light Weights 1900-2100 RPM Tamper Resistant	Superceded by 5104769 S.I. 27-D-77	
5102962	6 and 8V-92TT Governor Assembly Light Weights 1900-2100 RPM Non-Tamper Resistant	Superceded by 5103567 S.I. 62-D-76	
5103619	8V-71TT Governor Assembly Heavy Weights 1900-2100 RPM Tamper Resistant	Superceded by 5104767 S.I. 27-D-77	
5102152	8V-71TT Governor Assembly Heavy Weights 1900-2100 RPM Tamper Resistant Also Used on Early 6 and 8V-92TT Engines	Superceded by 5103619 S.I. 62-D-76	Superceded by 5104767 S.I. 27-D-76
5102455	8V-71TT Governor Assembly Heavy Weights 1900-2100 RPM Non-Tamper Resistant Also used on Early 6 and 8V-92TT Engines	Superceded by 5102962 S.I. 26-D-76 For use on 6 and 8V-92TT Engines	

*Denotes Current Governor Assembly Part Numbers.

2. Move the speed control lever to the maximum fuel position in order to check the no-load speed. If it has increased more than 25 rpm, you have gone too far on the initial adjustment; back off the buffer screw until this increase of the no-load rpm is less than 25 rpm.
3. Hold the buffer screw with a screwdriver and tighten the locknut.

Adjust Starting Aid Screw (DWLS Governors)

On those DDA engines that employ turbochargers in addition to the engine gear driven roots type blower, a starting aid screw is used that minimizes the amount of fuel that is injected during starting in order to reduce exhaust smoke. This is necessary because the turbocharger offers a resistance to the initial air

Table 10-13
POWER REDUCTION FACTOR

ENGINE TYPE	MAXIMUM RATED B.H.P.	RATED "TT" HORSEPOWER	RATED ENGINE SPEED	NO-LOAD SPEED	POWER REDUCTION FACTOR
6V-92TT – 9290 Injectors Federal Certified Throttle delay setting .636 *Throttle delay setting .570	335 @ 2100 RPM 2275 RPM Maximum No-Load Speed	270*	1800	1950	.91
		270*	1850	2000	.89
		270	1900	2050	.88
		270	1950	2100	.86
		270	2100	2250	.82
		260*	1800	1950	.88
		260*	1850	2000	.86
		260	1900	2050	.84
		260	1950	2100	.83
		250*	1800	1950	.84
		250*	1850	2000	.83
		250	1900	2050	.81
		250	1950	2100	.80
		240*	1800	1950	.81
		240*	1850	2000	.79
		240	1900	2050	.78
		240	1950	2100	.77
6V-92TT – 9290 Injectors Federal Certified Throttle delay setting .636	335 @ 2100 RPM 2275 RPM Maximum No-Load Speed	290	1900	2050	.94
		290	1950	2100	.93
6V-92TTA – 9A90 Injectors California Approved Federal Certified Throttle delay setting .636	318 @ 2100 RPM 2275 RPM Maximum No-Load Speed	270	1900	2050	.91
		270	1950	2100	.89
		270	2100	2250	.86
		260	1900	2050	.87
		260	1950	2100	.86
		250	1900	2050	.84
		250	1950	2100	.83
		240	1900	2050	.81
		240	1950	2100	.79
8V-92TT – 9290 Injectors Federal Certified Throttle delay setting .636	430 @ 2100 RPM 2275 RPM Maximum No-Load	365	1900	2050	.92
		365	1950	2100	.91
		365	2100	2250	.87
8V-92TT – 9A90 Injectors Federal Certified Throttle delay setting .570	430 @ 2100 RPM 2275 RPM Maximum No-Load	335	1800	1950	.87
		335	1850	2000	.86
		335	1900	2050	.85
		335	1950	2100	.83
8V-92TTA – 9A90 Injectors California Approved Federal Certified Throttle delay setting .636	424 @ 2100 RPM 2275 RPM Maximum No-Load	365	1900	2050	.92
		365	1950	2100	.91
		365	2100	2250	.87
		335	1800	1950	.88
		335	1850	2000	.86
		335	1900	2050	.84
		335	1950	2100	.83
8V-92TTA – 9290 Injectors Federal Certified Throttle delay setting .636	440 @ 2100 RPM 2275 RPM Maximum No-Load	365	1900	2050	.90
		365	1950	2100	.89
		365	2100	2250	.85
8V-71TT – N-75 Injectors Federal Certified Throttle delay setting .570	350 @ 2100 RPM 2275 RPM Maximum No-Load	305	1900	2050	.94
		305	1950	2100	.92
		305	2100	2250	.89

(Continued on following page.)

Table 10-13 (Continued)

ENGINE TYPE	MAXIMUM RATED B.H.P.	RATED "TT" HORSEPOWER	RATED ENGINE SPEED	NO-LOAD SPEED	POWER REDUCTION FACTOR
8V-71TTA – 7A75 Injectors California Approved Federal Certified Throttle delay .586	350 @ 2100 RPM 2275 RPM Maximum No-Load	305 305 305	1900 1950 2100	2050 2100 2250	.90 .89 .87
8V-71TTA – 7C75 Injectors Federal Certified Throttle delay setting .570	370 @ 2100 RPM 2275 RPM Maximum No-Load	305 305 305	1900 1950 2100	2050 2100 2250	.88 .87 .84
6-71TT – N-75 Injectors Federal Certified Throttle delay setting .570	272 @ 2100 RPM 2275 RPM Maximum No-Load	230 230 230 220 220 220	1900 1950 2100 1900 1950 2100	2050 2100 2250 2050 2100 2250	.91 .89 .86 .85 .86 .83

flow into the engine until the engine actually fires, and the exhaust gases are capable of accelerating the turbocharger, whereby it will supply pressurized air to the blower for cooling, scavenging, combustion, and crankcase ventilation purposes.

Adjust the external starting aid screw as follows (71 and 92 engines):

1. With the engine stopped, place the governor stop lever in the *run* position and the speed control lever in the *idle* speed position.

Figure 10-60
Buffer Screw Adjustment on a V-71 Engine. (Courtesy of Detroit Diesel Allison Div. of GMC)

2. From the following chart, determine the required setting and select the proper gauge.

Injector	*Gauge Setting in Inches (mm)*	*Tool Number*
N-75	0.385 (9.779)	J 24882
7A-75	0.385 (9.779)	J 24882
7C-75	0.385 (9.779)	J 24882
9290	0.454 (11.531)	J 23190
9A90	0.454 (11.531)	J 23190

3. Refer to Figure 10-61; measure the setting at any convenient cylinder location with the proper gauge and adjust the starting aid screw to obtain the required setting between the shoulder on the injector rack clevis and the injector body. When the starting aid screw is properly adjusted, the gauge should have an end clearance of $\frac{1}{64}$ in. (0.397 mm) in the area along the injector rack shaft between the rack clevis and the injector body.
4. Hold the starting aid screw and tighten the locknut.

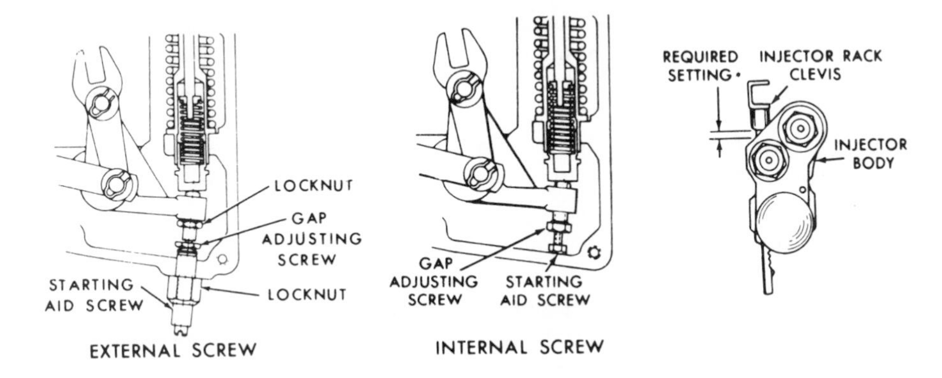

Figure 10-61
Inserting Starting Aid Gauge.
(Courtesy of Detroit Diesel Allison Div. of GMC)

5. To double-check this clearance, repeat the same procedure as outlined in step 3 after performing the following:
   a. Remove the selected gauge from the injector control rack.
   b. Position the stop lever in the *run* position.
   c. Move the speed control lever from the idle position to the maximum speed position.
   d. Return the speed control lever to the idle speed position; insert the gauge and recheck the adjustment.

Note: The reason behind the movement of the speed control lever is to help ensure proper adjustment. The injector rack clevis to body clearance can be increased by turning the starting aid screw farther in against the gap adjusting screw or reduced by backing it out.

Important: Due to the reasons mentioned for the purpose of the starting aid screw, advancing the throttle or speed control lever toward wide-open position during an engine startup will make the starting aid screw *ineffective*.

Adjust Throttle Delay Cylinder

Prior to performing a tune-up on an engine equipped with a throttle delay device, it must be removed in order to ensure that there is no interference in the governor or injector control linkage. The throttle delay mechanism (Figure 10-62) is used on many highway vehicle engines both with and without turbochargers. Its purpose is to retard the movement of the fuel control tube and injector racks toward the full-fuel injection position when the engine or vehicle is accelerating. This device aids in reducing acceleration exhaust smoke as required to conform to the U.S. Vehicle Engines Emissions Standards. Its use also helps to improve fuel economy by preventing operators from overreving the engine while upshifting the transmission. The basic throttle delay cylinder consists of a special rocker arm shaft bracket that incorporates the throttle delay cylinder, a piston, throttle delay lever, connecting linkage, oil supply orifice plug, ball check valve, U-bolt, and a yield mechanism within the governor to prevent fuel rod bending or governor damage caused by an overly zealous operator during acceleration.

A typical throttle delay (TD) unit is shown

in Figure 10-62 with the pin gauge used for setting and checking also shown in position.

Operation. Engine oil under pressure is supplied to a reservoir above the throttle delay through an orifice in the drilled oil passage in the rocker arm shaft bracket. Any time that the injector racks are moved toward the no-fuel position, free movement of the TD piston is assured by air drawn into the cylinder through the ball check valve. Sufficient movement of the piston toward the no-fuel position uncovers a port or opening that permits oil from the reservoir to enter the cylinder and displace the air. Acceleration of the engine causes the injector racks to be moved toward the full-fuel position; however, due to the fact that the TD piston must expel the oil from the cylinder through an orifice, the movement of the injector racks is momentarily retarded. On turbocharged engines, due to the slight delay in the acceleration of the turbine with an increase in engine throttle position, this will minimize incomplete combustion and therefore exhaust stack smoke. To permit full accelerator travel, regardless of the retarded rack position, a spring-loaded yield lever and spring assembly replace the standard lever on the front end of the injector control tube on in-line type engines, with a yield link as shown in Figure 10-66 used on the Vee engines.

Anytime the throttle delay piston has been removed, it should be inspected for burrs or rough edges; if any are found, remove them with crocus cloth. To determine which throttle delay gauge should be used, the engine type, injector size, and governor part number must be established. Current throttle delay gauges in use are as follows (Kent Moore tool numbers are given):

J 24882 for 0.385 in. (9.779 mm)

J 9509 for 0.404 in. (10.261 mm)

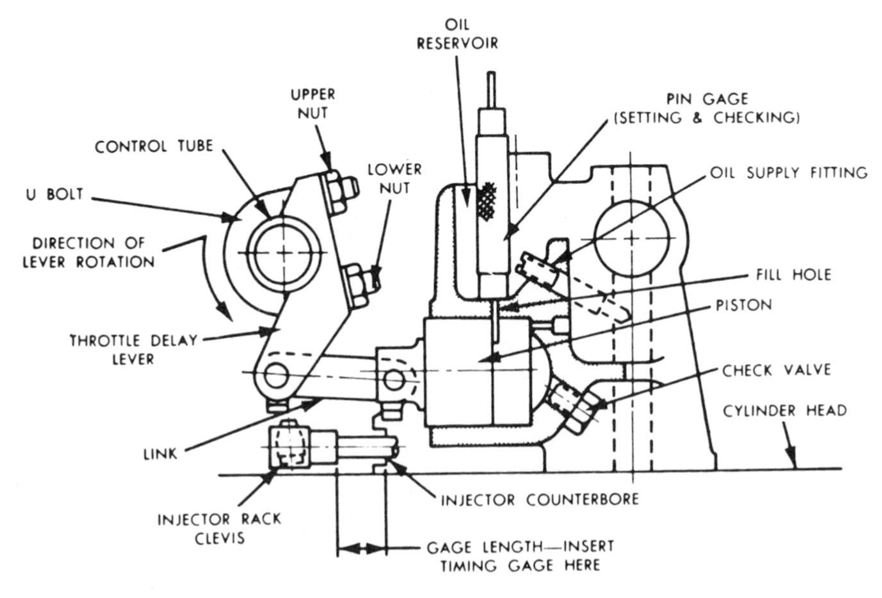

Figure 10-62
Placement of Throttle-Delay Pin Gauge Through Oil Reservoir Hole.
(Courtesy of Detroit Diesel Allison Div. of GMC)

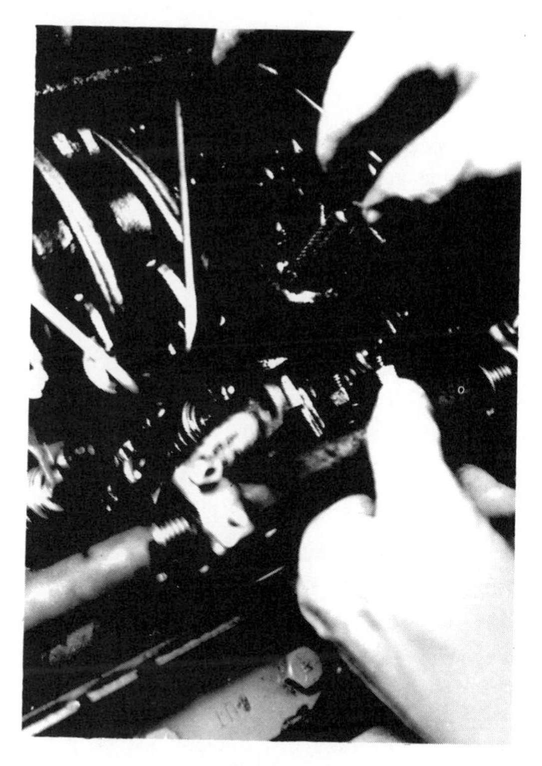

Figure 10-63
Adjusting Throttle Delay Piston with Gauge In Position on Injector Rack.
(Courtesy of Jacobs Manufacturing Co.)

J 23190 for 0.454 in. (11.531 mm)

J 25559 for 0.570 in. (14.478 mm)

J 25560 for 0.636 in. (16.154 mm)

J 25558 for 0.072 and 0.069 in. (1.828 and 1.752 mm) pin gauge

Once the proper throttle delay gauge has been selected, reinstall the throttle delay piston into its bore and adjust to that setting. With the engine stopped, proceed as follows:

1. Insert the proper gauge between the injector body and the shoulder on the injector rack clevis for the injector nearest the throttle delay cylinder (see Figure 10-63).
2. Move the governor speed control lever to the maximum speed position and hold it in that position. This will move the injector rack toward the full-fuel position and against the gauge.
3. Insert the pin gauge J 25558 (0.072-in. or 1.828-mm-diameter end) in the cylinder fill hole as shown in Figure 10-62 for 71 and 92 series engines.
4. Rotate the throttle delay lever until further movement is limited by the piston coming up against the 0.072-in. (1.828-mm) end of the pin gauge.
5. Carefully hold the throttle delay piston in this position and tighten the U-bolt nuts while exerting a slight amount of torque on the TD lever in its direction of rotation. Excessive force can cause bending of the gauge and possible piston damage.
6. Check your setting as follows:
   a. Remove the pin gauge.
   b. Carefully attempt to reinstall the pin gauge (0.072-in. or 1.828-mm end) into the cylinder fill hole. It should not be possible to reinsert the gauge at this time without moving the injector racks toward the no-fuel position.
   c. Carefully reverse the pin gauge (0.069-in. or 1.752-mm end) and insert it into the cylinder fill hole, which it should enter without resistance.

Note: If the 0.072-in. (1.828-mm) diameter end of the gauge enters the fill hole (step 6b), tighten the upper U-bolt nut. If the 0.069-in. (1.752-mm) diameter end will not enter the fill hole as per step 6c without resistance, tighten the lower U-bolt nut.

7. Release the governor speed control lever; remove both the timing gauge and pin gauge.
8. Move the injector control tube assembly

from the no-fuel to the full-fuel position to make sure that there is no binding anywhere.

After completing a tune-up on an engine, always check to see if the governor speed control lever is moved to the maximum speed position when the accelerator pedal is fully depressed.

Figure 10-64
Adjusting Slave Piston Clearance on Two-Valve Head Engine.
(Courtesy of Jacobs Manufacturing Co.)

Figure 10-65
Slave Piston Adjustment—Four-Valve Head.
(Courtesy of Detroit Diesel Allison Div. of GMC)

Caution: Prior to starting an engine after an engine speed control adjustment, or removal of the engine governor cover in order to make a running adjustment, the serviceman *must* determine that the injector racks will move to the no-fuel position when the governor stop lever is moved to the stop position. If this is not established, possible engine overspeed will result if the injector racks cannot be positioned at no-fuel with the governor stop lever. This could cause severe engine damage and also personal injury.

71 and 92 Series Jacobs Engine Brake Adjustment

Since the purpose of the engine brake units is to open the exhaust valves just prior to the piston reaching TDC, it is imperative that the Jacobs brake slave piston clearance be carefully adjusted to specifications. This clearance is 0.064 (1.625 mm) on both two- and four-valve heads. Figures 10-64 and 10-65 show the setup for the two- and four-valve heads, respectively.

Procedure. See Figure 10-64. For a two-valve head:

1. With the engine stopped and the valves in the normally closed position (injector in the delivery position), simply insert the 0.064-in. (1.625-mm) feeler gauge between the slave piston feet and valve stem cap.
2. Carefully turn the adjusting screw on top of the brake unit with an Allen wrench until a slight step up in effort is noted. Check the drag on the feeler gauge at both feet of the slave piston. Hold the adjusting screw, and torque the locknut to 15 to 18 lb-ft (20.33 to 24.4 N·m); then double-check the clearance.

For a four-valve head (Figure 10-65), the only difference is that the clearance is taken between the slave piston feet and the Jacobs exhaust valve bridge.

149 Series Engine Tune-up

The sequence of events for performing an engine tune-up on the 149 series engines is almost identical to that for other series of DDA engines. The 149 engines come equipped with one of the following types of governors:

1. Limiting speed mechanical (used on mobile equipment).
2. Variable speed mechanical (used on industrial or marine units).
3. Hydraulic (used on power generator sets mostly).

The mechanical governor tune-up sequence is as follows:

1. Adjust the exhaust valve clearance.
2. Time the fuel injectors.
3. Adjust the governor gap.
4. Position the injector rack control levers.
5. Adjust the starting aid screw if so equipped (turbocharged engines only).
6. Adjust the differential lever connecting link.
7. Adjust the maximum no-load speed.
8. Adjust the idle speed.

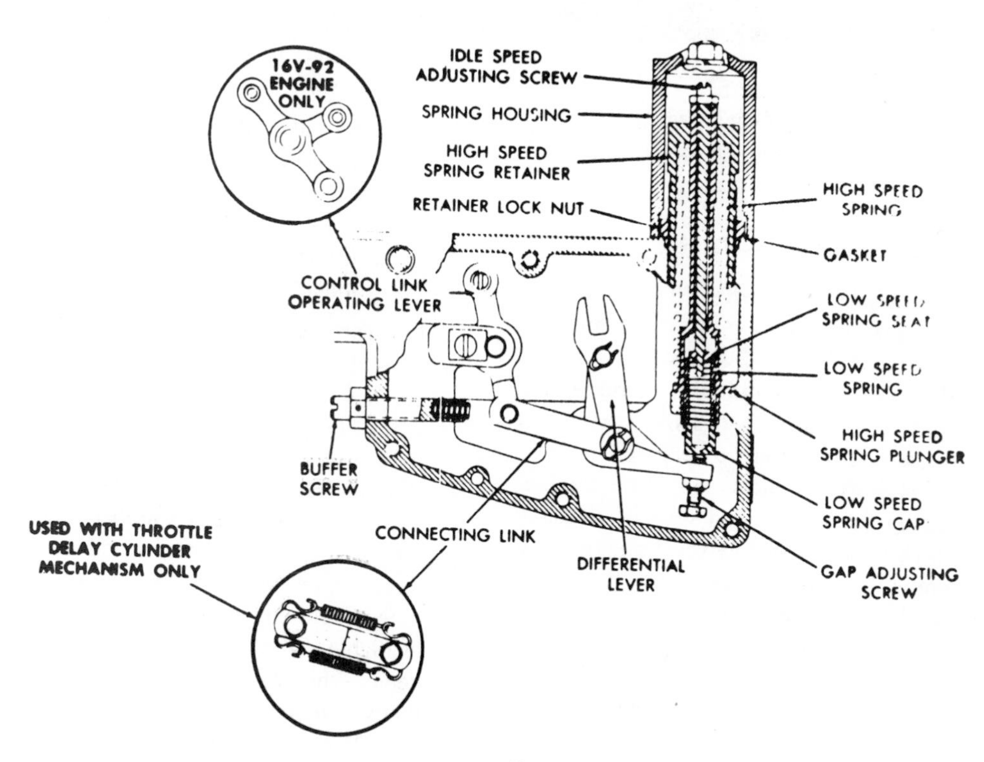

Figure 10-66
Limiting Speed Governor Yield Link.
(Courtesy of Detroit Diesel Allison Div. of GMC)

9. Adjust the buffer screw.
10. Adjust the throttle booster spring (only used on the variable speed governor).
11. Adjust stop bracket bolt (variable speed governor only).

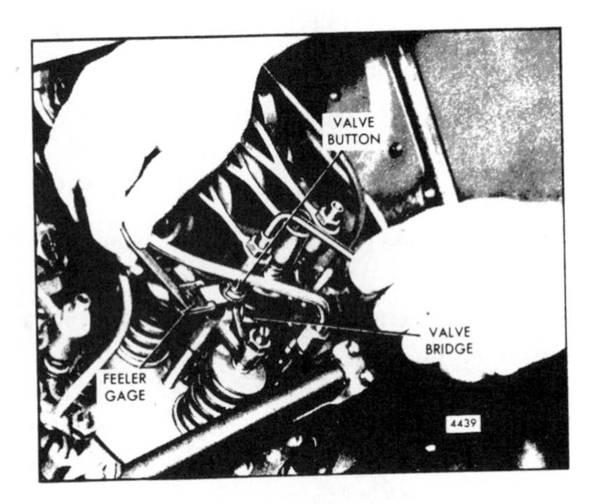

Figure 10-67
Adjusting Exhaust Valve Clearance.
(Courtesy of Detroit Diesel Allison Div. of GMC)

Mechanical Governors—Tune-Up:

1. *Adjust the exhaust valves*. This is done in the same manner as for other engines, such as the 71 and 92 series. Valve bridge adjustment when required is carried out in the same manner as described earlier in this section for 71 and 92 engines.

   However, since the 149 engines employ a high mounted style of camshaft which operates the valves through a pivoted roller type of rocker arm, the engine should be turned over with a barring mechanism until the injector follower on the cylinder to be adjusted is fully depressed by the rocker arm. Figure 10-67 shows the actual valve adjustment.

   a. Loosen off the exhaust valve adjusting screw locknut on the rocker arm as shown in Figure 10-67.
   b. Select a 0.015 to 0.017 in. (0.381 to 0.431 mm) go–no go feeler gauge and check the clearance between the end of the valve button and the pallet on the valve bridge. If necessary, using an Allen wrench, adjust the valve adjusting screw until the 0.017-in. (0.381-mm) part will withdraw with a smooth pull.
   c. Tighten the lock nut from 30 to 35 lb-ft (40.674 to 47.453 N·m) and recheck the clearance. If correct, the 0.015-in. (0.381-mm) end of the go–no go gauge should pass freely between the valve button and the bridge pallet; however, the 0.017-in. (0.431-mm) end should not. This will ensure a valve clearance of 0.016 in. (0.406 mm) on a cold setting.
   d. Readjust the setting if incorrect, then continue to set the remaining exhaust valves.
   e. As with all tune-ups, the valves should be rechecked and set with the engine at operating temperature of 160 to 185°F (71 to 85°C). It may be necessary to run the engine during adjustments to maintain this temperature.
   f. The valve clearance for a hot set is 0.012 in. (0.304 mm); therefore, an 0.011 in. to 0.013 in. go–no go feeler gauge can be used.

2. *Time the fuel injectors*. To time the injectors, first determine the proper timing pin required from the ones listed earlier in this section or from the engine service manual. If you have an injector wallet card available from any DDA distributor/dealer, this will also list the information that you require.

   There are no injector timing marks on the engine's flywheel or crankshaft pulley; therefore to set and adjust the injector with the proper pin, proceed as follows:

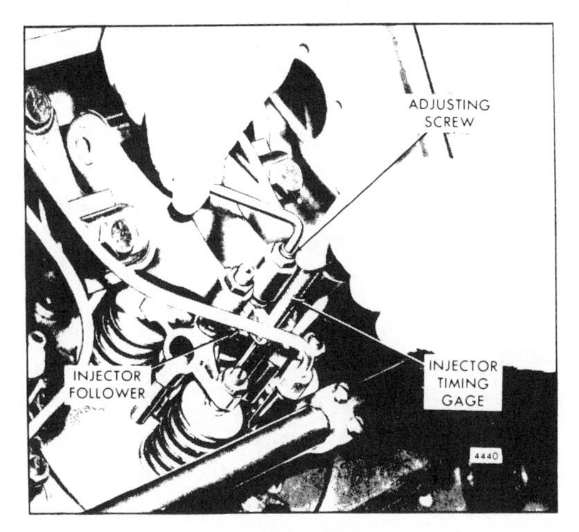

Figure 10-68
Adjusting Fuel Injector Timing.
(Courtesy of Detroit Diesel Allison Div. of GMC)

a. Turn the engine over until the exhaust valves on the particular cylinder to be set are fully open. In this position the injector is ready to be checked. See Figure 10-68.

Caution: Anytime that you plan to use the engine starting motor to turn the engine over, when adjusting either the valves or the injectors, be sure to place the throttle lever in the idle position and the stop lever in the stop position to prevent the possibility of the engine starting at this time.

b. Insert the pin end of the timing gauge into the small hole provided in the top of the injector body. Move the flat portion of the gauge toward the flat of the injector follower; you should feel a very light drag between the gauge and the follower as you do this. Another way to check this is to place a clean drop of oil onto the flat of the follower; gently rotate the timing gauge and if it wipes the oil from the flat of the follower, it is properly adjusted.

c. If the injector requires adjusting, then after backing off the injector rocker arm adjusting screw locknut, insert an Allen wrench into the adjusting screw and rotate it until the timing gauge will just pass over the flat of the follower as described in step b.

d. While holding the adjusting screw securely, tighten up the locknut to 30 to 35 lb-ft. torque. (40.674 to 47.453 N·m). Quickly recheck the adjustment to ensure that it is still all right.

e. Check the remaining injectors as just described.

3. *Adjust the governor gap.* This adjustment is as previously described for the 53, 71, and 92 series of engines earlier in this section. Just remember that on the DWLS governor as used on mobile equipment that the gap is 0.0015 in. (0.0381 mm) with the engine running between 800 and 1000 rpm.

   Whereas on an engine using a SWVS governor, the gap is 0.006 in. (0.152 mm) taken with the engine stopped and the speed control lever in the full-fuel position.

4. *Position the injector rack control levers.* The purpose of setting the racks is as described earlier for all series of DDA engines.

   a. On those engines with a DWLS governor, it is imperative that the idle speed adjusting screw be adjusted so that approximately $\frac{1}{2}$ in. (12.7 mm), equal to about 12 to 14 threads on the screw, is projecting from the lock nut when the lock nut is actually against the high speed spring plunger. Failure to do this will allow the tension of the idle spring to exert force back through the governor linkage, which could result in a false fuel rack setting. It is not necessary to do this on those engines employing a SWVS governor.

b. The buffer screw, however, on both types of mechanical governors should be backed out approximately $\frac{5}{8}$ in. (15.875 mm) from the lock nut.

c. Proceed to loosen off all of the inner and outer adjusting screws at each injector. Check to ensure that all of the racks are in fact free on the injector control tube. Regardless of the type of mechanical governor employed on the engine, the first injector rack to be adjusted is always the no. 1R (1 on the right bank) on current governor engines. Remember that the left and right bank on all DDA Vee-type engines is determined by standing at the engine flywheel end and looking forward. In addition, the cylinder number is always taken starting from the front of the engine on each individual bank.

5. It is now necessary to position the operating lever within the governor as shown in Figure 10-69. This shows the governor currently in use. Disconnect the fuel rod from the left bank injector control tube at this time at the cylinder head control tube. Reference to Figure 10-69 shows that the operating lever for the right cylinder bank should be positioned and held so that it is at a 90 degree angle to the top of the governor housing.

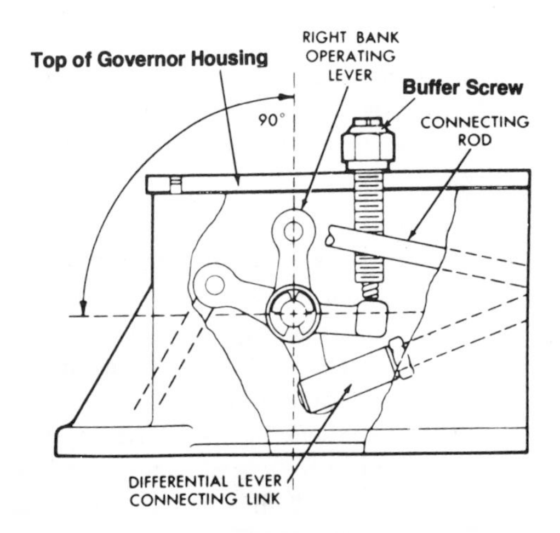

Figure 10-69
Positioning the Governor Operating Lever. (Courtesy of Detroit Diesel Allison Div. of GMC)

6. Due to the linkage setup through the governor, the no. 1 RB injector is adjusted first and with the control rack in the no-fuel position. Therefore proceed as follows (remember, only the right bank control rod to tube should be connected at this time). Turn down the no. 1 RB outer adjusting screw until the ball end of the control lever from the tube becomes snug against the injector rack with the rack in the no-fuel position. This can be sensed first by a step up in effort required to turn the screwdriver, and also by movement of the governor right bank operating lever that you are holding onto in the governor housing. Therefore, if you sense either of these conditions, back off the outer adjusting screw, and gently turn it back in until you are satisfied that the ball end of the control lever is just snug against the rack.

7. With the ball end of the no. 1 RB injector lever now snug against the injector control rack, gently turn down the no. 1 RB inner adjusting screw until it bottoms lightly on the injector control tube. Go back to the outer adjusting screw and tighten it about one-eighth turn. Do the same to the inner, and then alternate between them until they are tight. Use an 8- to 10-in. (20.32- to 25.4-cm)-long screwdriver for this purpose. Use of too long a screwdriver can result in damage to the injector control tube where the screws bottom through overtightening; therefore, after tightening both the inner and outer adjusting screws, they should be torqued to a final tightness of 24 to 36 in.-lb (2.711 to 4.067 N·m).

8. Having set the no. 1 RB injector rack, place and hold the governor differential lever in the full-fuel position after connecting the left bank injector control tube rod. Both the left and right bank fuel rods should now be connected. With the differential lever in full-fuel, proceed to turn down the no. 1 LB inner adjusting screw until the ball end of the control lever is snug, similar to that described in paragraph 6 for no. 1 RB. Turn down the outer adjusting screw until it bottoms on the control tube: then alternately tighten them as you did for no. 1 RB and finish torque them.

9. With both the no. 1 LB and RB adjusted thus far, they should now be checked to ensure that they are in balance. You may recollect that on other series of DDA engines the rack is now held in full fuel and checked for bounce with a screwdriver, as shown in Figures 10-52 and 10-53. However, the rack on the 149 series of engines differs slightly in shape from that found on other series of DDA units in that the injectors incorporate an *injector rack stop*. This is to limit the maximum travel of the injector rack. Because of this feature, check the rack as follows:

   a. Figure 10-70 shows the check to be made.

   b. Place the control racks in full-fuel when you check for balance. Grasp the no. 1 LB or no. 1 RB first, depending on which was set first. This would be determined by whether the engine uses a former or a current governor, respectively. In other words, you always check the rack that was set to the governor first, since it becomes the master rack. Referring to Figure 10-70, the object here is not to look for a bounce on the rack as you would on other series of DDA engines. Simply grasp the rack as shown, and using a push and pull rotating motion, carefully feel how much drag there is between the ball end of the control tube lever and the injector rack. If after checking both the no. 1 RB and no. 1 LB on current governor engines you feel that they are the same, then they are in a balanced state. If, however, they are not euqal, you would have to loosen either the inner or outer adjusting screws of the no. 1 RB rack on former governors (or the no. 1 LB rack on current governors) and adjust the lever until the two no. 1 racks are in balance. The key to success here is that once the no. 1 RB rack has been set on current governors, do not readjust it, and on former governors, once the no. 1 LB has been set, do not readjust it.

10. Once you are satisfied that both no. 1 racks are set the same, the remaining racks can be adjusted by holding the no. 1 RB and no. 1 LB in the full-fuel position by

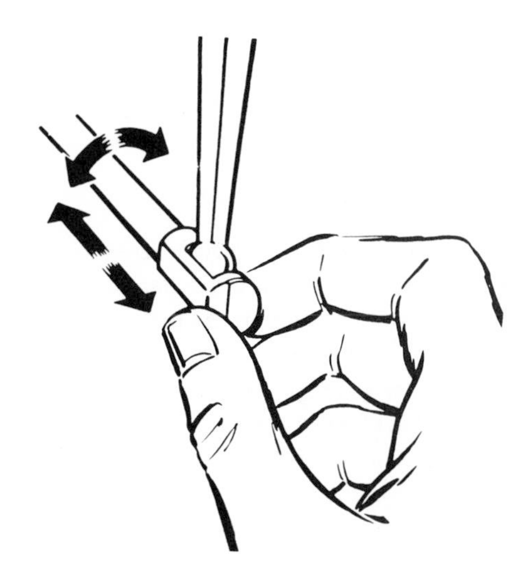

Figure 10-70
Checking Injector Rack to Clevis Drag. (Courtesy of Detroit Diesel Allison Div. of GMC)

hand or by the assistance of a spring connected to the control tube. Regardless of which method you choose, each succeeding rack that is adjusted *must* always be checked for the same drag or feel as the no. 1 rack on that bank, since it has been set to the governor and is therefore the master rack on that bank. Never attempt to readjust the no. 1 rack on either bank to compensate for differences in feel at another rack on that bank. If you disconnect the injector control tube from the fuel control rod going to the governor in order to manually hold the control tube in the full-fuel position while setting the other racks on each respective bank, any movement at the injector control tube while turning down the inner adjusting screw of a rack would indicate that you have turned the screw too far. Therefore, you should back the inner screw off and reset it. Finally, once you have adjusted *all* the injector racks on both banks, hold the speed control lever in the maximum speed position and check the drag on both no. 1 RB and no. 1 LB. If they are not the same, readjustment of the racks will be required to ensure that both banks are getting equal fuel. If this is not done, one bank will invariably be doing more work than the other, and it will show up on an exhaust pyrometer.

Adjust the Starting Aid Screw. On those engines equipped with turbochargers, the starting aid screw must be adjusted so that the injector fuel control racks on engines employing DWLS governors will be at a position less than full-fuel when the speed control lever is in the normal idle position. This has the function of reducing the amount of fuel that will be injected upon initial start-up, making starting easier.

The DWLS (double-weight limiting speed) governor employs two springs, an idle spring and a high speed spring. With the engine stopped, the tension of both springs working back through the governor linkage actually places the injector control racks in the full-fuel position for starting purposes.

On turbocharged engines, the turbo does not reach operating rpm until the engine fires; consequently, with only the gear-driven engine blower supplying air for initial starting, less air is available and therefore less fuel is needed. With a properly adjusted starting aid screw, starting is easier and smoke is reduced, giving the turbocharger time to accelerate and increase the desired air flow to the engine.

Proceed as follows:

1. Engine stopped, stop lever in the run position, speed control lever in the normal idle position.
2. Refer to Figure 10-71 and check the clearance between the injector body and the rack of the No. 3 right bank injector with the proper starting aid gauge similar to that shown in Figure 10-61. The clearance as shown in Figure 10-71 should exist when the head of the starting aid screw is up against the governor housing on current governors or against the bracket used with former governors. At the same time, if this adjustment is correct, a governor gap of 0.155 to 0.160 in. (3.937 to 4.064 mm) should exist between the high speed spring plunger and the low speed spring cap.
3. To double-check this setting, it is necessary to move the stop lever to the stop position, with the speed control lever still in the idle position; then return the stop lever to the run position. This will take up any clearance in the governor linkage. If necessary, hold the governor gap adjusting screw and turn the starting aid screw ccw to increase the clearance or cw to decrease the clearance.
4. Start the engine and recheck the governor gap of 0.0015 in. (0.038 mm) between 800 and 1000 rpm. If necessary, reset it.

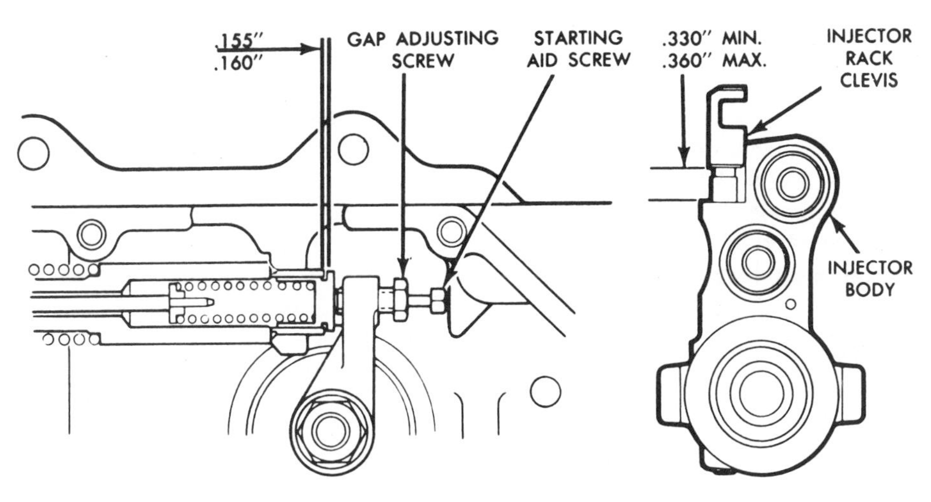

Figure 10-71
Starting Aid Screw Gap Location.
(Courtesy of Detroit Diesel Allison Div. of GMC)

Adjust the Differential Lever Connecting Link. This adjustment is necessary on engines using mechanical governors. After any disturbance to the governor linkage, in order to allow starting of the engine, it may be necessary to adjust the differential lever connecting links by loosening the locknuts and turning the ball joints on the link to establish a distance of $6\frac{13}{16}$ in. (17.303 cm) between the inner ends of the ball joints. See Figure 10-72. On both the limiting and variable speed governors, install the governor cover using a new gasket; however, the *access* cover should be removed at this time.

1. On limiting speed governors, be sure to back out the idle speed adjusting screw.
2. On variable speed governors, the stop lever spring should be removed from its screw on the governor cover so that the tension will not act through the lever. Also, to ensure full speed positioning of the speed control lever, make sure that the stop bolt is turned in far enough.

Adjust the Maximum No-Load RPM. This can be done by the use of stops and shims in behind the spring retainer housing, and follows the same format as that for any other DDA engine using a variable speed governor. See earlier description.

Adjust the Idle Speed. The idle speed is adjusted as per the idle range indicated on the governor identification plate to suit the particular application. The idle screw is located on the end of the spring retainer housing, the same as on other engine series. See earlier description.

Adjust the Buffer Screw. The buffer screw is adjusted as per that for other engines described herein. However, it is located on the top of the main governor housing, rather than on the side.

Adjust the Throttle Booster Spring. This is adjusted in the same manner as described herein for other series of DDA engines, with the exception that the booster spring is located over the top of the main governor housing, rather than on a bracket bolted to the cylinder head.

Adjust the Stop Bracket Bolt. This is a bolt threaded through the stop bracket attached to the governor top cover. It is adjusted to limit the maximum arc travel of the governor speed control lever toward full rpm. Adjust the bolt as follows:

1. Start and run the engine at an idle speed; back off the stop bolt locknut, and turn the bolt back several turns.
2. Move the speed control lever to its maximum rpm position, and note the engine rpm.
3. Turn the stop bolt in until it just makes contact with the speed control lever and decreases the engine speed by 5 to 10 rpm maximum; lock it in this position.

8.2-Liter Fuel Pincher Engine

The recent addition of the 8.2-liter fuel pincher engine to the Detroit Diesel Allison product line marks the first time that DDA has produced a diesel engine of 4-stroke cycle design. All other DDA engines are of 2-stroke cycle design.

This engine was designed specifically for the mid-range diesel truck market, and Detroit Diesel Allison undertook this engine's production based on a desire by Ford Motor Company to have a lightweight, inexpensive diesel engine for their new range of medium-duty trucks in the 1980s.

As such, 85% of all engines produced will be destined for Ford trucks in the 1980 model year. In addition, both Chevrolet and GMC desire as many of these engines as they can get. International Harvester is also in line to receive these engines for a number of their medium-duty trucks.

For those personnel who are already familiar with DDA's 2-cycle series of engines, the tune-up sequence for the 4-cycle fuel pincher does not vary that much; however, there are some differences that require additional clarification. The following pages will deal with these to enable the reader to compare these variations in the tune-up sequence with that given herein for the 2-cycle series engines.

Firing Order. The firing order for this engine follows the same pattern as that used by all gasoline engine divisions of GMC; namely that the left and right engine bank is established by standing at the rear of the engine. All odd-numbered cylinders are on the left bank, whereas all even-numbered cylinders are located on the right bank. In addition, the firing order is the same as those for the V-8 gasoline engines, namely 1–8–4–3–6–5–7–2.

Figure 10-73 shows the cylinder layout.

Relative Piston Firing Positions. Since this engine is a deviation from DDA's 2-cycle concept, it might be of value at this time to review the basic piston layout and relative piston firing positions prior to discussing valve adjustment and other stages of engine tune-up.

Figure 10-72
Variable Speed Governor with Cover Removed (Solid Connecting Link).
(Courtesy of Detroit Diesel Allison Div. of GMC)

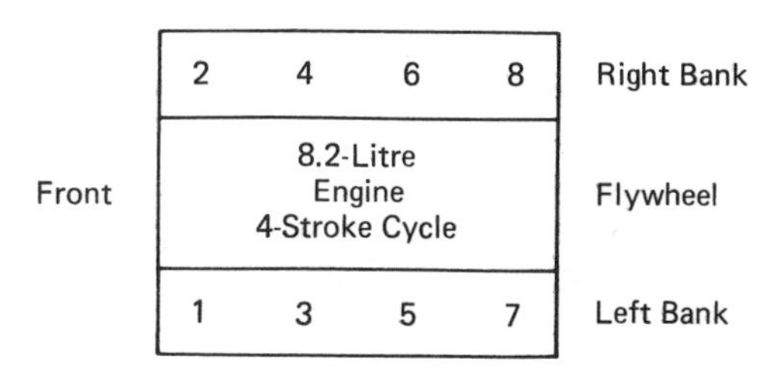

Figure 10-73
Engine Firing Order.

In all 4-stroke cycle engines, two revolutions of the crankshaft are required in order to obtain one firing or power stroke from each cylinder. Within the two revolutions of the crankshaft, we have four piston movements, namely intake, compression, power, and exhaust.

To picture this sequence of events for eight cylinders at any one time becomes rather awkward, therefore Figure 10-74 shows such a layout for your assistance. In position A, all cylinders are shown in their respective positions for a clockwise rotation engine, starting with no. 1 at TDC on power. Two turns of the crankshaft are illustrated here, or 720 degrees.

The number of degrees that the pistons are apart between firing impulses is arrived at by simply dividing 720 degrees ($2 \times 360° = 720°$) by the number of engine cylinders, which in this case is 8. Therefore, the pistons will be 90 degrees apart in phase or firing impulses.

Figure 10-75 shows the position and stroke of all pistons when the no. 1 piston is at TDC (top dead center) on its power stroke.

8.2-Liter Injector

The unit injector operates in the same way as all other DDA injectors. However, rather than the fuel being carried to and from the injector by fuel lines, the design of the 8.2 injector is such that fuel manifolds cast internally within the cylinder heads similar to Cummins fuel systems are used.

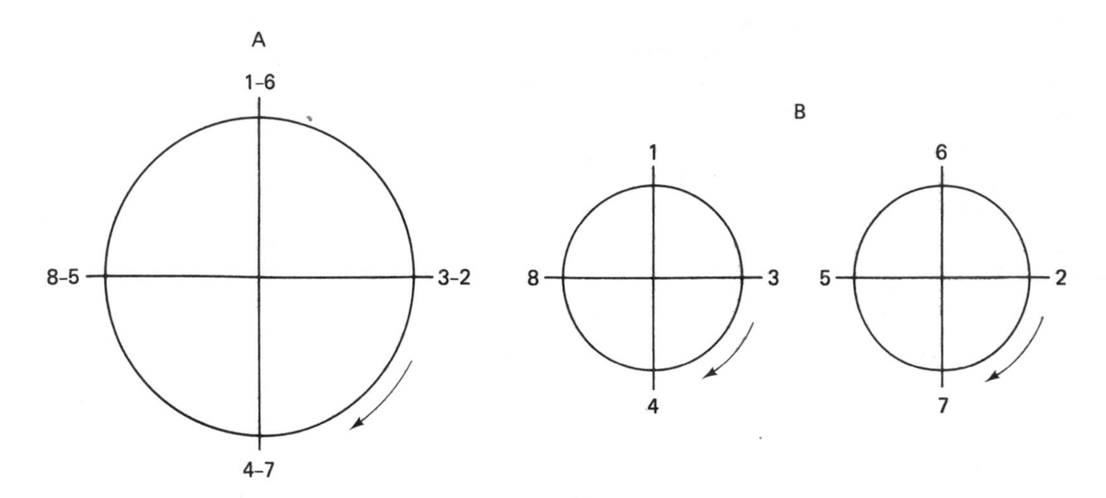

Figure 10-74
Piston Positions with No. 1 at TDC on the Power Stroke.

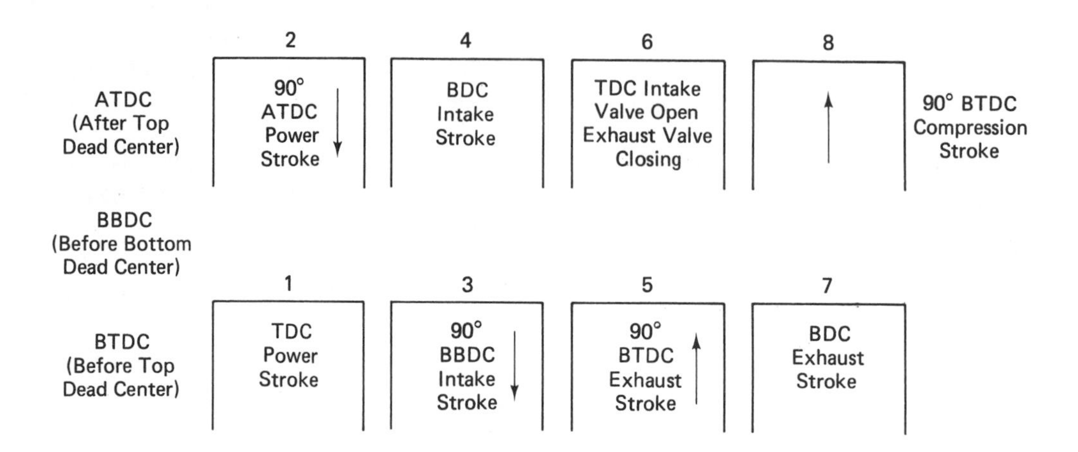

Figure 10-75
Relative Piston Position and Stroke.

Figure 10-76 shows the general layout of the 8.2 injector assembly.

Checking for a Misfiring Injector

Unlike other series of DDA engines, whereby a large screwdriver can be used to short-out the injector (see Troubleshooting, Chapter 12), under no circumstances can this particular method be employed on the 8.2-liter engine to check out the injector operation since the push-rod may leave the engine. Figure 10-77 shows a recommended method for this test.

Procedure

1. Start the engine with the rocker cover removed, and secure the throttle lever in the normal idle position.
2. Using the modified slotted screwdriver, individually push each injector into full-fuel one at a time for 3 to 6 seconds.
3. With the engine running at an idle rpm, you are effectively flooding one cylinder with maximum fuel; therefore, if the injector is firing, there will be a quick increase in engine rpm. A misfiring injector or dead cylinder will show little or no change.

Tune-up Sequence—8.2-Liter Engine

The sequence of events involved in the 8.2 tuneup differs only slightly from that of any other DDA engine series. Remember that any tuneup is only a series of checks; therefore, only when an item is found to be out of adjustment or specification should it be adjusted.

As with any tuneup, the engine must be at its normal operating temperature prior to doing any adjustments or checks.

Note: An injector timing height label assembly is attached to the front of the left-bank rocker cover, while on the front of the right-bank rocker cover is attached an emissions label.

The injector timing height label indicates the camshaft base circle injector timing height for specific cylinder positions. These timing heights are established at the factory when the engine is assembled; therefore, any time that

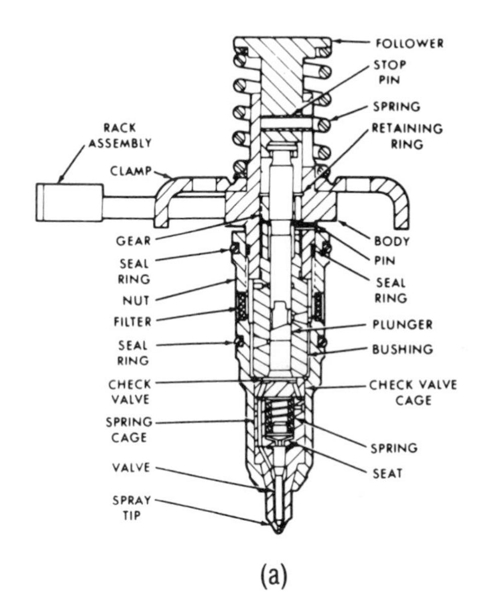

(a)

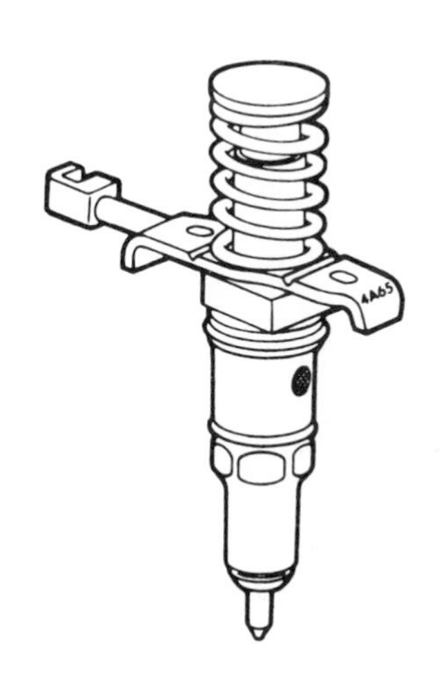

INJECTOR	*SPRAY TIP	PLUNGER
4A53	5-.20 mm(.008")-155A	53
4A65	5-.25 mm(.010")-155A	65

*First numeral indicates number of spray holes, followed by sizes of holes and angle formed by spray from holes.

(b)

Figure 10-76
Cutaway View of: (a) 8.2-liter injector assembly; (b) 8.2-liter injector identification. (Courtesy of Detroit Diesel Allison Div. of GMC)

the crankshaft, camshaft, cylinder block, or gear train are changed, a precision timing procedure similar to that done at the factory must be repeated, with a new set of heights measured and recorded in order to facilitate future injector adjustment. A new label should then be attached to the rocker cover with these new dimensions stamped on the label.

Tuneup Sequence for Mechanical Governor Engine

1. Adjust the intake and exhaust valve clearances.
2. Time the fuel injectors.
3. Position the injector rack control levers.
4. Adjust the governor gap.
5. Adjust the maximum no-load speed.
6. Adjust the idle speed.
7. Adjust the buffer screw.

The following explanations give some detail on each of the foregoing sequences. (Valve clearances: intake—0.012 in. or 0.305 mm; exhaust—0.014 in. or 0.356 mm.)

Item 1—Valve Adjustment. The intake and exhaust valves can be set by barring the engine over until the injector follower just starts to move down on that particular cylinder, then proceeding to adjust the remaining valves in firing order sequence.

Another way to adjust the valves is at the same time as the injector is being adjusted with the piston at the TDC (top dead center) position in the cylinder. This is done anytime that the base circle dimensions have been changed, such as at a major overhaul, and they are no longer the same as that stamped on the label attached to the front of the left-bank rocker cover.

This particular process is facilitated by the fact that the front side of the flywheel has four equally spaced timing pin holes, for use with a locating pin anytime that a new injector timing

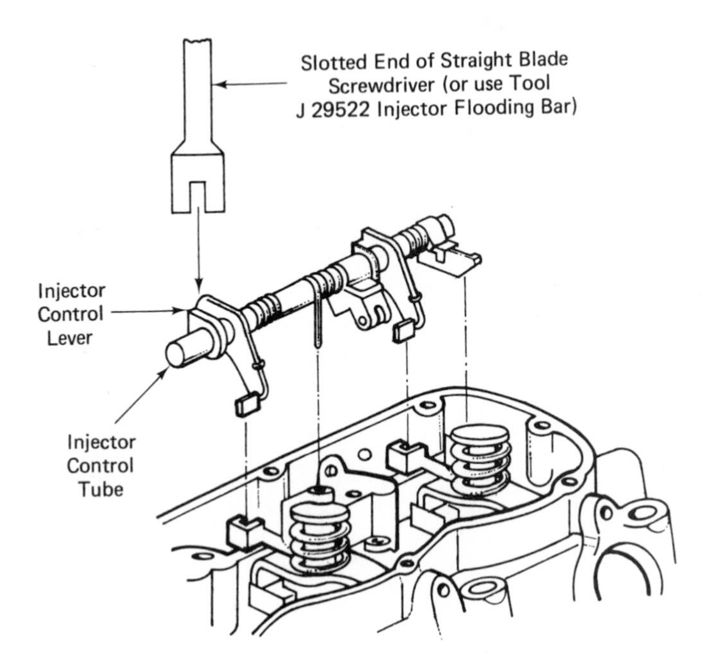

Figure 10-77
Checking for a Misfiring Cylinder.
(Courtesy of Detroit Diesel Allison Div. of GMC)

height must be established, particularly after a major engine overhaul (J 29139 flywheel timing pin and adapter).

Should this procedure be used, the following valves would be adjusted with the respective piston at TDC:

- No. 1 TDC—adjust no. 2 and no. 7 intake valves and no. 4 and no. 8 exhaust valves along with timing the no. 1 injector.
- No. 8 TDC—adjust (time) no. 8 injector only.
- No. 4 TDC—adjust injector, no. 1 and no. 8 intake valves, and no. 3 and no. 6 exhaust valves.
- No. 3 TDC—adjust (time) no. 3 injector only.
- No. 6 TDC—adjust no. 6 injector, along with no. 3 and no. 4 intake valves and no. 5 and no. 7 exhaust valves.
- No. 5 TDC—adjust no. 5 injector only (time).
- No. 7 TDC—adjust no. 7 injector, along with no. 5 and no. 6 intake valves and no. 1 and no. 2 exhaust valves.
- No. 2 TDC—adjust the no. 2 injector only (time).

Item 2—Check Injector Timing

If no changes have been made to the major components discussed earlier, then the camshaft base circle dimensions that are stamped on the label of the left-bank rocker cover at the front can be used to effectively time the injectors. Proceed as follows:

1. Manually bar the engine over in its normal direction of rotation until the no. 1 injector follower moves down approximately 0.5 mm (0.020 in.).

2. Proceed to set no. 8, no. 4, no. 3, and no. 6 injectors to the specification stamped on the rocker cover label with the use of special tool J 29014 injector timing dial indicator, as shown in Figure 10-78.
3. Once these injectors have been adjusted, manually bar the engine over approximately one full turn until the injector follower of the no. 6 cylinder moves down about 0.5 mm (0.020 in.).
4. Set no. 5, no. 7, no. 2, and no. 1 injectors with the dial indicator.

TDC Injector Timing Setting

Should the engine be overhauled and new components installed, such as a crankshaft, camshaft, gear train parts, or new engine block, then special tool J 29011-A TDC timing slug must be used to establish a precision TDC position.

Figure 10-78
Injector Timing Dial Gauge Setup. (Courtesy of Detroit Diesel Allison Div. of GMC)

When this has been established, the injector timing setting would be as follows:

Engine Family	*TDC Injector Timing Setting*
50 State Turbo Engine	44.75 mm (1.7618 in.)
California Naturally Aspirated Engine	44.45 mm (1.75 in.)
49 State Naturally Aspirated Engine	43.80 mm (1.724 in.)

Prior to using the J 29014 injector timing dial indicator shown in Figure 10-78, the gauge must be calibrated by loosening the set screw mounted in the side of the gauge, and setting the gauge on the master timing block which is part of this tool group. The indicator point of the dial gauge must make contact with the lower stepped portion of the master gauge block.

To secure the gauge on top of the master gauge block, press the two spring-loaded pins down until the roll pins are level with the slot in the block assembly.

Rotate the pins so that the groove in the top of the pin is aligned with the other pin, which will allow the dial gauge to be held to the top of the gauge block. Gently move the dial gauge up in the gauge body so that the dial lines up with the word CAL (calibrate) on the dial face.

This now sets or zeros in the dial guage, which can now be removed from the master block and installed onto the top of the injector. Gently press down on the spring-loaded pins while rotating the gauge, so that the groove points toward the other pin, when the roll pin is secured under one of the injector spring coils, which will secure the dial gauge in place.

Note:

1. For setting injector base circle timing as per the specification listed on the rocker

cover label, use the black numbers on the outer circumference of the dial gauge face.

2. For setting TDC injector timing, use the upper stepped portion of the gauge block to calibrate the dial gauge, then use the red numbers located on the inner circumference of the gauge.

Item 3—Set the Injector Rack Controls

Setting of the racks on the 8.2-liter requires the use of a dial gauge rather than the 'bounce method' common to series 53, 71, and 92 engines.

1. Loosen off the bolts on the left- and right-bank fuel rod shafts.
2. Take care to ensure at this time that both the no. 1 left-bank and no. 1 right-bank set screws are bottomed, since this lever is nonadjustable.
3. Place the fuel rack into the full-fuel position and to ensure that it will stay there, attach a light spring to it.
4. The rack design follows the same style as that for series 53, 71, and 92 engines. Arrange a dial indicator so that its stem (probe) is in contact with the center of the square end of the no. 1 left-bank injector rack. Adjust the dial gauge so that there is a small amount of pre-load registered on the face, zero in the needle, and lock the bezel.
5. Tighten the no. 2 left-bank injector screw just until the dial gauge needle starts to move (maximum allowable movement is 0.001 in. (0.0254 mm).
6. Follow steps 4 and 5 for no. 3 and no. 4 left-bank injectors, with gauge movement being kept to a maximum movement of 0.003 in. (0.076 mm) total.

Note: The left-bank injector racks must be set first, because this banks fuel rod is closer to the governor than that on the right bank.

7. Using the dial gauge assembly, repeat the same injector rack adjustments on the right bank as that for the left bank.
8. Remove the governor cover and the plug in the housing opposite the high speed spring pack (labeled *8 start*).
9. Install a long $\frac{3}{8}$ × 24 in. TPE bolt in the plug hole and screw it in until the low speed governor spring cap (same location as any other DDA limiting speed governor) gap is between 0.000 and 0.0015 in. (0.000 and 0.038 mm).
10. Install the governor cover.
11. Hold the throttle lever in the full-fuel position and lock it there with a spring, or tie it in position.
12. Manually rotate the left-bank injector control shaft firmly to full-fuel and tighten the bolt in the shaft control lever to fuel rod.
13. Repeat step 12 for the right bank.
14. Release the throttle lever from the full-fuel position and move it back to idle to reposition the racks, then back to the full-fuel position and resecure it there.
15. With your finger, push each injector rack into full-fuel and note the amount of rack movement, which should be less than 0.002 to 0.003 in. (0.051 to 0.076 mm). Should rack movement exceed this dimension, repeat steps 13 and 14.
16. Remove the long bolt that was installed through the plug hole of the governor housing in step 9.

Item 4—Governor Running Gap Check

As with any DDA engine, the governor gap should always be taken with the engine at its normal operating temperature, and the engine running at a specified rpm.

1. Remove the high speed spring cover and back out the buffer screw at least $\frac{5}{8}$ in. (15.875 mm).

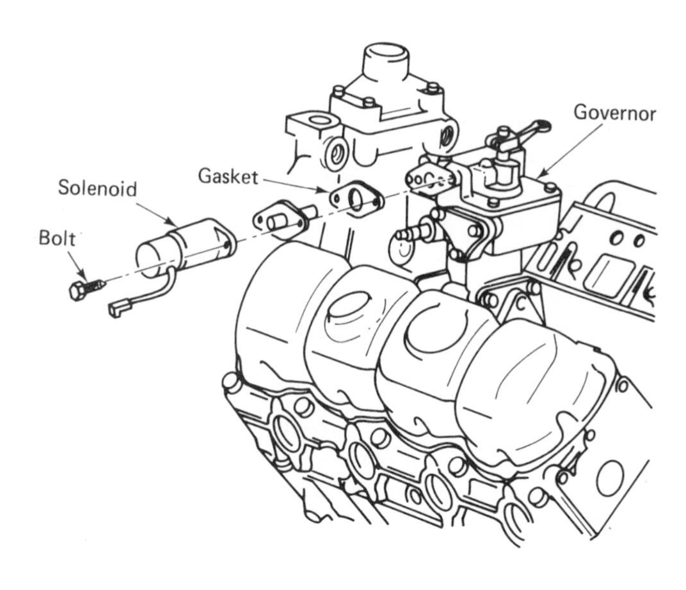

Figure 10-79
8.2-Liter Shutdown Solenoid.
(Courtesy of Detroit Diesel Allison Div. of GMC)

2. After loosening off the idle screw locknut, back out the idle screw until 0.196 in. (5 mm) protrudes beyond the locknut.
3. With the engine at operating temperature, set the idle speed to 700 rpm.
4. Stop the engine and remove the governor cover.

   Caution: Prior to starting the engine to check the governor gap, be sure that the racks can be moved to the stop position. In addition, with the top removed from the governor, do not start the engine without first having hold of the differential lever, since with no top on the governor, you are, to coin a phrase, ''the governor.''

5. Start and run the engine through use of the differential lever until the engine speed is between 950 and 1000 rpm, and check the gap between the low speed spring cap and high speed spring plunger with a 0.0015-in. (0.038-mm) feeler gauge; adjust as necessary (see governor gap adjustment earlier in this chapter).
6. Replace the governor cover, taking care to ensure that the speed control pin and fuel shutoff pins are correctly positioned.

Item 5—Adjust the Maximum No-Load Speed

The no-load rpm is established by varying the tension on the high-speed spring. (No-load rpm adjustment is also discussed under DDA Tuneup earlier in this chapter.)

No-load rpm is found on the rocker cover emissions label.

Item 6—Adjust the Idle Speed

Adjust the idle speed as mentioned earlier to 700 rpm.

Item 7—Buffer Screw Adjustment

The purpose of the buffer screw adjustment is simply to remove the roll or hunt from the engine at an idle rpm. (This is also discussed in an earlier part of this chapter in detail.)

Finally, check that full throttle travel is being obtained.

8.2-Liter Shutdown System

Unlike the other series DDA engines, there is no emergency shutdown valve similar to the air cutoff or flapper valve found on the 8.2-liter.

Figure 10-79 shows the layout of the electrically activated solenoid.

The solenoid in Figure 10-79 must be energized to the run position (ignition key on), or removed from the governor housing prior to connecting up the governor linkage to the rack control shaft, or engine overspeed may occur because the solenoid is designed to hold the injector racks in the no-fuel position until the solenoid is energized. Should the racks be held in the full-fuel position, then be connected to the governor with the shutdown in the stop position, the injectors will be locked in the maximum fuel position.

chapter 11

Woodward Governors

11-1

Function and Operation of Governors

Although governor terms are discussed in other chapters of this book, it would be helpful now if we were to again briefly review the basic operation of a governor mechanism and to also ensure that we understand the meaning of the terms used to express various reactions related to the governor. In any governor used on diesel engines the basic principle is one of *weight* force against *spring* force. In other words, the tension of the governor spring assembly is always trying to *increase* the fuel delivered to the engine cylinders, and therefore the rpm, whereas the centrifugal force of the engine-driven governor flyweights is trying to oppose and overcome the spring tension and therefore decrease the amount of fuel injected (see Figures 11-1 and 11-2).

Governor Terms

The following terms or definitions are common to all governors used with diesel engines, whether they are mechanical or hydraulic.

Speed Droop. This is the difference in engine speed (rpm) between the maximum governed no-load rpm and the maximum governed full-load rpm. The difference will vary depending on the type of governor used with the application. This difference is expressed as a percentage increase in rpm from full load to no load. The equation to determine speed droop is:

$$\text{Speed droop} = \frac{(\text{NNL} - \text{NFL}) \times 100}{\text{NFL}}$$

where NFL = normal full-load speed rpm
NNL = normal no-load speed rpm

As mentioned, the actual droop will vary with the actual engine application and is normally within the following ranges:

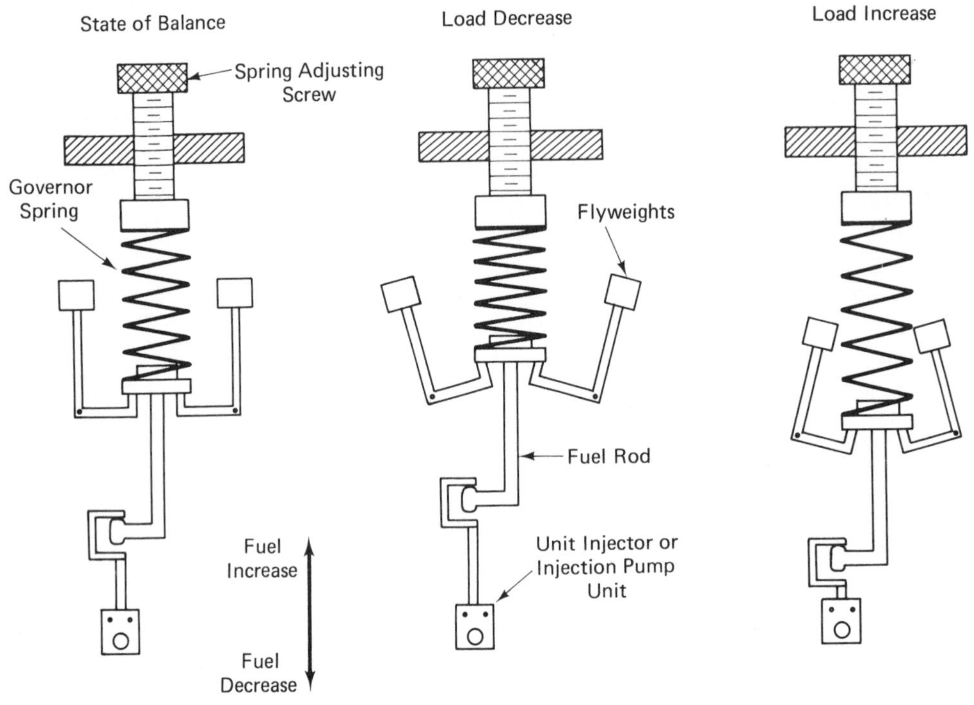

Figure 11-1
Basic Governor Mechanism.

Agricultural equipment: 8 to 13%

Industrial, automotive, marine: 5 to 10%

Diesel generator sets: ±0.5 to ±1.5%; however, 3 to 6% is required for proper paralleling, of several units

Isochronous. Hydraulic governors such as Woodward models PSG and UG8 have the ability to maintain a zero droop condition or steady-state condition of speed at any load from no load to full load.

Initial Speed Change. This is the immediate change in engine speed caused by a change in engine load. The amount of rpm increase or decrease is in direct proportion to the load change and is the main factor involved that initiates or causes the governor to react or respond.

Sensitivity. This is the amount of speed change that is necessary before the governor will sense or make a corrective change to the fuel-injection mechanism. It is generally expressed as a percentage of the governed speed of the engine. It generally varies from around 0.25% for the best mechanical governors to 0.01% with hydraulic servogovernors.

Stability. This is the ability of the governor to maintain a definite engine rpm after a speed disturbance under either constant or varying load conditions and with a minimum of false motions or overcorrections. For example, a good hydraulic servogovernor should attain a state of stability within 3 seconds or less after a sudden load change.

Deadband. This refers to that narrow speed range within the operating range where no measurable correction is made by the governor.

Hunting or Surging. This is a rhythmic variation of the engine speed that results from the governor response usually being out of phase with the feedback signal. The engine speed will continually fluctuate above and below the mean speed at which it is governed. It has various causes, and each particular engine or governor has specific items that will cause this condition. Some causes are as follows:

1. Governor moving parts gummy or sticking.
2. Worn parts in the linkage causing lost motion.
3. Low oil level in the hydraulic governor sump.
4. Dirty oil.
5. Hunts at idle: improperly adjusted buffer screw on a Detroit Diesel, for example.

Work Capacity. A governor is capable of producing a certain amount of work through its linkage and control mechanism. If this work is exceeded, the governor will be incapable of responding to the demands placed upon it. The work capacity is related to the effective centrifugal force developed at the flyweights acting on the sliding sleeve or collar assembly at the governor's mean regulating speed times the total length of the sleeve travel. Therefore, the usable work capacity is the total work capacity multiplied by the sensitivity times the required regulation.

However, in a hydraulic servogovernor the work capacity is the power cylinder displacement times the oil pressure. To reduce the

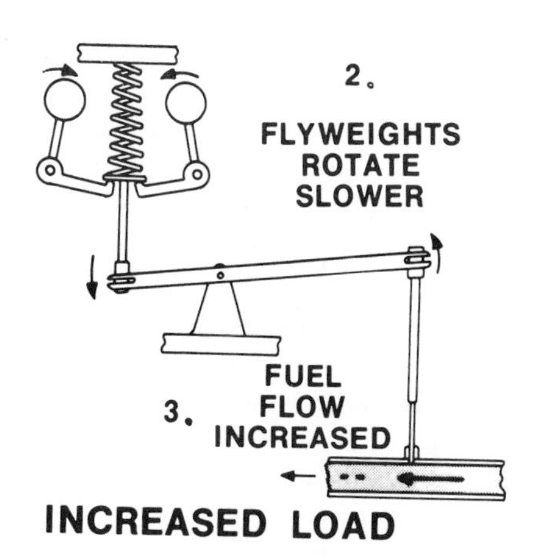
1. LOAD INCREASES, ENGINE SPEED DECREASES
2. FLYWEIGHTS ROTATE SLOWER
3. FUEL FLOW INCREASED
INCREASED LOAD

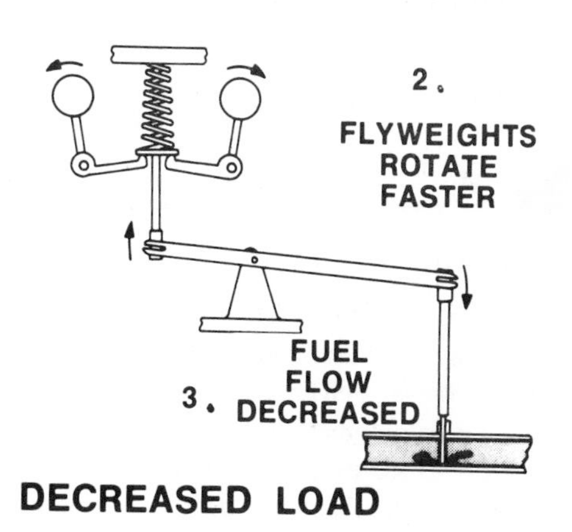
1. LOAD DECREASES, ENGINE SPEED INCREASES
2. FLYWEIGHTS ROTATE FASTER
3. FUEL FLOW DECREASED
DECREASED LOAD

Figure 11-2
Centrifugal Governor.
(Courtesy of Deere & Company)

fluctuations of momentary speed changes, the actual usable work capacity should not exceed 60% of the rated capacity. Even large diesel generators in use today seldom require a capacity in excess of 10 ft-lb (for example, a governor such as the Woodward UG8 produces a maximum of 8 ft-lb of work output over the full 42 degrees travel of its output or terminal shaft).

Torque Control. There are many additional optional features readily available for use with engine governors, and the torque control is only one of them (see governor sections in previous chapters). Basically, it is a feature used on some governors whereby the full-load stop control can be varied as a function of engine speed. Since the full-load torque curve of any engine is dependent on several factors, such as (1) combustion characteristics, and (2) fuel delivery versus speed characteristics of the injection system, the torque curve that is desired will differ for various applications. However, you will notice that regardless of the manufacture of an engine the majority of them will in fact produce their peak torque at about two thirds of their full-load speed, with 5 to 10% increases being average. However, current high-torque rise diesel engines can produce between 30% to 50% more torque at their peak torque rpm than at the rated engine speed.

11-2

Woodward Hydraulic Governors

Hydraulic governors are used extensively by all engine manufacturers worldwide when the application is such that closer speed regulation than can be obtained with a straight mechanical governor is required. The Woodward Governor Company is perhaps the best-known manufacturer of hydraulic-type governors and has also pioneered the use of electric governors and electronic control systems. Main offices are located in Rockford, Illinois (Aircraft/Hydraulic Turbine Controls Division), Fort Collins, Colorado (Engine and Turbine Controls Division), Tokyo, Japan, Sydney, Australia, Woodward Governor Nederland B.V., Hoofdorp, The Netherlands, and Woodward Governor (U.K.) Ltd., Slough, Berkshire, England.

Model SG Hydraulic Governor

This governor is the simplest form of governor presently manufactured by Woodward, and if one understands the operation of this unit, understanding of the more complex units will be much easier. This governor is a hydraulic speed droop type, which is used extensively on small, high-speed diesel, gas, or gasoline engines for which isochronous (zero droop) control is not required.

This governor can be adjusted by an internal droop screw to vary the droop required from $\frac{1}{2}$% up to approximately 7% over the full 36 degrees available travel of the governor terminal shaft connected to the engine injector. These governors are available with a useful torque output of either 12 or 24 lb-in. (1.355 or 2.711 N·m) over the 36 degrees maximum of terminal shaft rotation.

The governor is usually set to operate at either 2400 or 3600 rpm (1200 or 1800 rpm engine speed) and can control down to approximately 25% of normal rated engine speed. When the governor is used on an engine, it receives its oil supply from the engine or a separate sump since it does not have an independent sump of its own. The governor can be driven in either a clockwise or counterclockwise direction; however, the governor relief valve assembly, when viewing the governor from the nameplate end, must be on the left if the governor is to be driven clockwise (when viewed from above), or the relief valve must be on the right for counterclockwise rotation.

Figure 11-3 shows a typical SG model

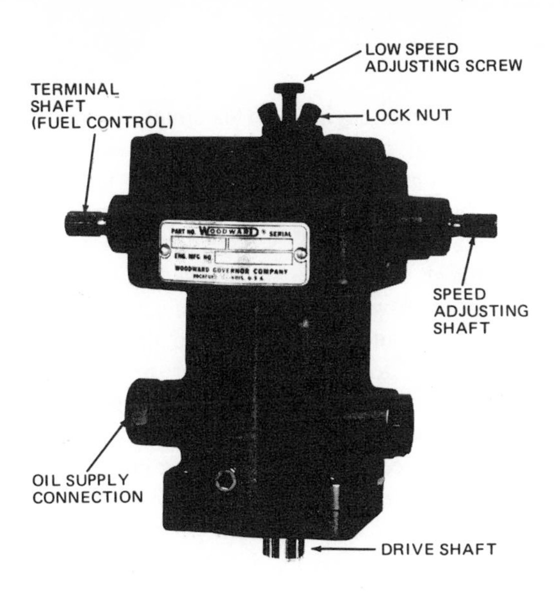

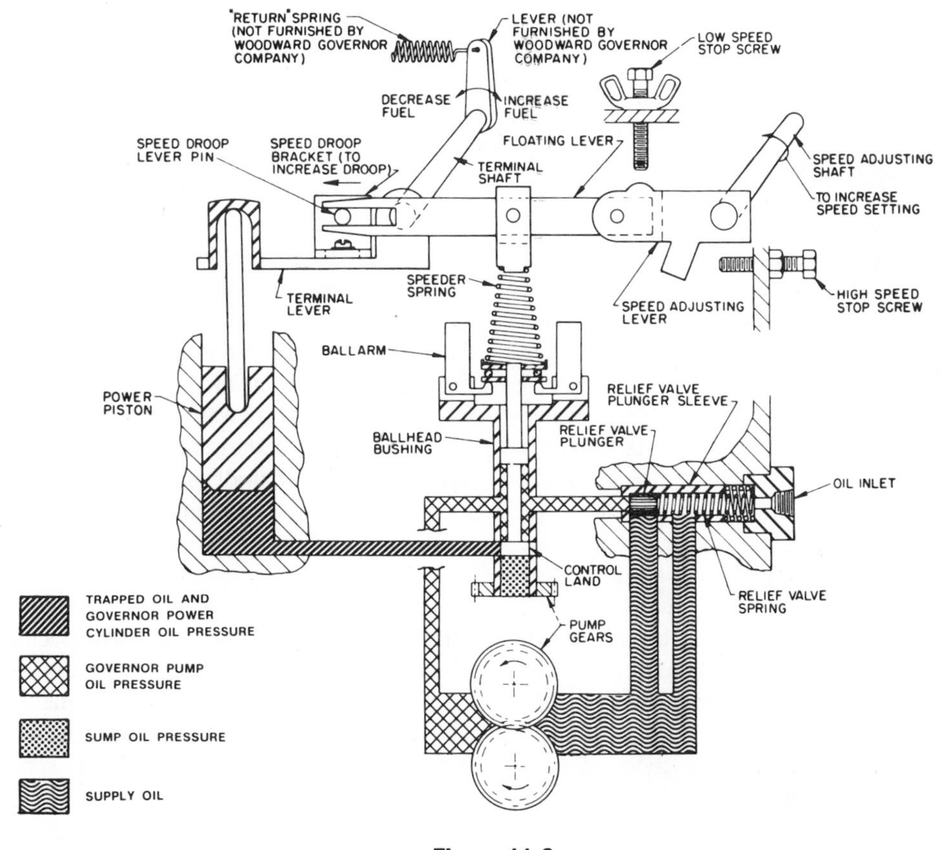

Figure 11-3
Typical SG Model Governor.
(Courtesy of Woodward Governor Co.)

governor. It can be mounted to the engine either in the vertical or horizontal position; however, if it is mounted horizontally, you must ensure that the terminal shaft is likewise horizontal and that a $\frac{3}{8}$-in. (9.525-mm) external drain line is provided to connect to a $\frac{1}{4}$-in. (6.35-mm) pipe tapped hole in the lower end of the governor cover. If this were not done, excessive oil would accumulate in the governor housing assembly, causing problems. Normally, when the governor is mounted vertically, this problem does not exist since the oil is free to drain down through the hollow governor drive shaft, which is splined into the engine drive pad.

Principles of Operation

Since this governor depends upon oil under pressure for its control, prior to starting the engine, no oil pressure will be available; therefore, the terminal shaft can be rotated to an increased fuel position to assist initial starting of the engine. On engines that require linear (straight line) motion from the governor to the fuel injectors, a special subcap assembly must be used rather than the one that is used on engines requiring rotary output. An example of this would be a Caterpillar engine, which would use an SG governor with rotary motion of the output shaft to actuate the injection pump control rack, whereas many in-line type Detroit Diesel Allison 2-cycle engines would require a linear type of motion to allow the fuel control rod to rotate the fuel control tube and activate the unit injectors. Regardless of whether rotary or linear motion is used, the governor operates and responds in the same manner for both, the only difference being in the subcap hookup for the two styles of linkage.

If a linear type of linkage is used, a fuel rod knob extending out of the governor subcap assembly can be pushed in and held there until the engine starts and oil pressure is obtained. This simply overrides the governor for starting purposes to allow placement of the fuel injector racks to the full-fuel position. When the engine is running, this can be released and the governor will assume control of the engine.

The engine can be stopped at any time by manually moving the throttle control terminal lever to the no-fuel position; however, some force is required to do this, since you are trying to force the power piston inside the governor down against lube oil pressure. The hydraulic governor will also act as an automatic shutdown device should there be insufficient oil pressure at any time.

Earlier SG governors used a solid power piston; however, the current units use the type of power piston shown in the figures herein.

Engine Stopped. Although this governor operates under the action of oil pressure, it is still a speed-sensing device just as a straight mechanical governor is. Two things, the centrifugal force of the rotating flyweights and speeder spring compression, will therefore cause a governor reaction. With the engine stopped, by moving the throttle control terminal lever to the full-fuel position, the control of the injector racks is momentarily taken away from the governor. If the throttle is left in the idle speed position during starting, the engine will take longer to start as oil pressure slowly builds up. The initial idle spring force is established by the low speed adjusting screw located on the top of the governor cover.

Engine Cranking. As the starter motor is engaged, the governor ballhead assembly, which carries the two flyweights, will rotate at cranking speed. At this low speed, the force developed by the weights is small; however, they are trying to compress the speeder spring.

Engine Running. Prior to the engine firing and running, the speeder spring force will be greater than the force being produced by the weights. During cranking, engine oil is being delivered to and through the line to the gear pump contained within the governor base at a minimum supply pressure of 5 psi (34.47 kPa). This pump then boosts the oil pressure up to 75 psi (523.8 kPa), 175 psi (1206.6 kPa), or 225 psi (1551.3 kPa), depending on the relief valve spring setting, which will unseat and bypass the oil back to the suction side of the pump.

During cranking, the pilot valve plunger within the rotating ballhead bushing is held down by speeder spring force, allowing oil under pressure to be directed to the base of the power piston. Once the engine has started, there will be a rapid oil pressure increase, plus the centrifugal force of the weights will now start to oppose the speeder spring and will therefore raise the pilot valve plunger, thereby closing off the supply port from the oil pump to the base of the power piston.

Since the upward movement of the power piston controls the fuel rack position, the instant that the weights lift the pilot valve plunger and close off the oil supply port, the racks will be held in a stable position corresponding to when a state-of-balance condition is reached between the weights and speeder spring. Engine speed then is determined by the terminal lever position, which directly affects the tension of the speeder spring. Figure 11-4 shows the position of the weights and other linkage within the governor when the engine is running at a steady speed or state-of-balance condition.

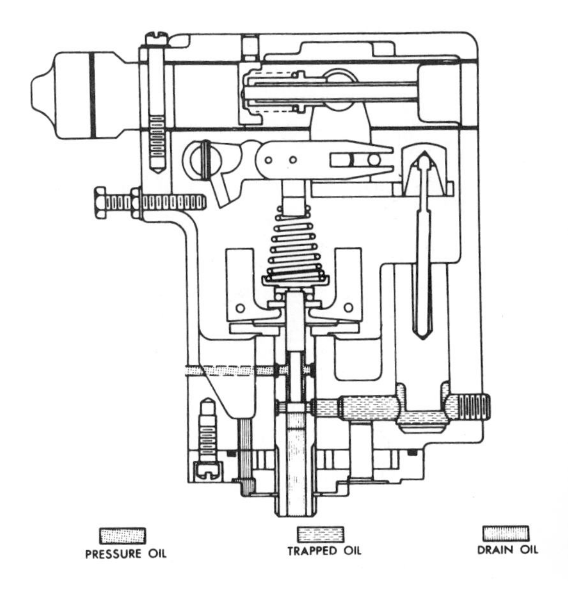

Figure 11-4
Steady Load and Speed Condition.
(Courtesy of Woodward Governor Co.)

Load Increase. Figure 11-5 shows the condition that exists within the governor during a load increase. For a given throttle setting (fixed), if a load is applied to the engine, then there will be a reduction in rpm in proportion to the governor droop setting (explained later), which will upset the previously existing state of balance in favor of the speeder spring. This will push down the pilot valve plunger and uncover the oil supply port from the oil pump. Additional oil will be sent underneath the power piston, forcing it up, which causes the

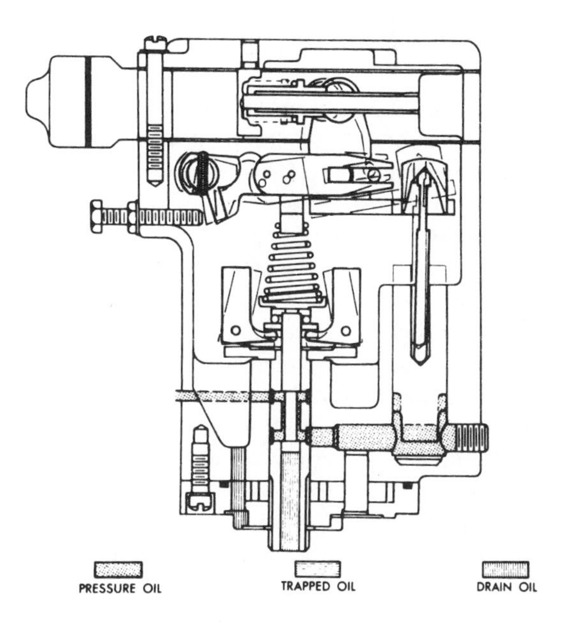

Figure 11-5
Load Increase and Speed Decrease.
(Courtesy of Woodward Governor Co.)

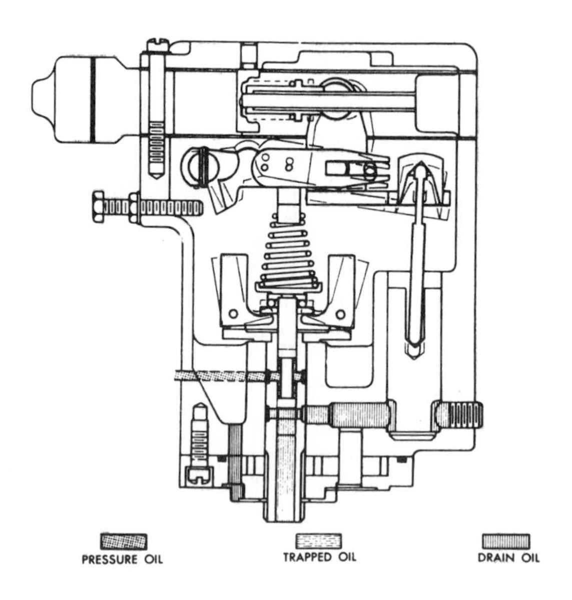

Figure 11-6
Load Decrease with Speed Increase.
(Courtesy of Woodward Governor Co.)

terminal lever to rotate the terminal shaft connected to the fuel racks in an increased fuel position.

As the engine speed starts to increase, the weight force will pull the pilot valve plunger up, thereby closing off the supply port when the state of balance again exists between the speeder spring and rotating flyweights. With the power piston now held in a steady new position, the engine will carry the increased load at a slightly reduced rpm.

How quickly the pilot valve can recenter, then, is determined by the upward moving droop pin, which will lift the slotted end of the differential lever earlier or later depending on its position within the slotted fork. If the droop adjusting screw is loosened, and the droop bracket and its pin are moved away from the pilot valve plunger, a greater amount of power piston movement is required before the speeder spring force is decreased, since the arc of travel or center distance of the droop pin to the center of the pilot valve plunger is greater. If, however, the droop bracket and pin were to be moved in toward the pilot valve plunger, this arc of travel would be less, meaning that the power piston would have to move a much shorter distance in order to cause a change in the speeder spring tension. Another way to look at this would be that if you consider a fulcrum point placed midway between the length of a seesaw, the farther out one moves from the center of the seesaw, the greater the arc of travel you will move through. This will also vary the force acting back through the fulcrum point. Therefore, moving the droop adjusting bracket in will decrease droop, and moving it out will increase droop.

Load Decrease. If the engine were running at a steady speed with a state-of-balance condition existing as shown in Figure 11-4, and the load upon the engine were decreased, we would again upset the state of balance, only this time in favor of the weights, since the engine rpm would tend to increase for a given (fixed) throttle position. As the weights fly outward, they would pull the pilot valve plunger up, allowing the trapped oil below the

Figure 11-7
Droop Adjustment.
(Courtesy of Woodward Governor Co.)

power piston to drain through the hollow drive shaft to the sump. The return spring either on the injector fuel rack or pump, or internally or externally on the governor terminal shaft, would positively force the terminal lever back. The loss of oil and terminal lever rotation would pull the fuel racks to a decreased fuel position. As the engine speed starts to decrease, the power lever lowers the speed droop pin; therefore, the floating lever lowers the spring fork to increase the speeder spring force, which will assist in recentering the pilot valve plunger faster, since the rotational speed of the weights is already decreasing along with the increased speeder spring force trying to pull them back to a balanced position. The downward movement of the pilot valve plunger closes off the drain port, stopping further power piston movement. The engine speed will stabilize, but at a slightly higher rpm than before the load decrease occurred.

Refer to Figure 11-6 for the position of the governor mechanism.

Droop Bracket. Attached to the terminal lever by an adjustable screw is a droop bracket, which also has a droop pin engaging the slotted end of the floating lever, as shown in Figure 11-7. During a load increase, the upward moving power piston will lift the outer end of the floating lever along with the terminal lever in the increased fuel direction. This action causes the speed droop pin to be raised also. At the same time, the right-hand end of the floating lever pivots around the speed adjusting lever pin as the left end (slotted) of the floating lever raises the spring fork, which will tend to decrease speeder spring force, thereby allowing the weight force to recenter the pilot valve plunger at a lower speed (with less weight action).

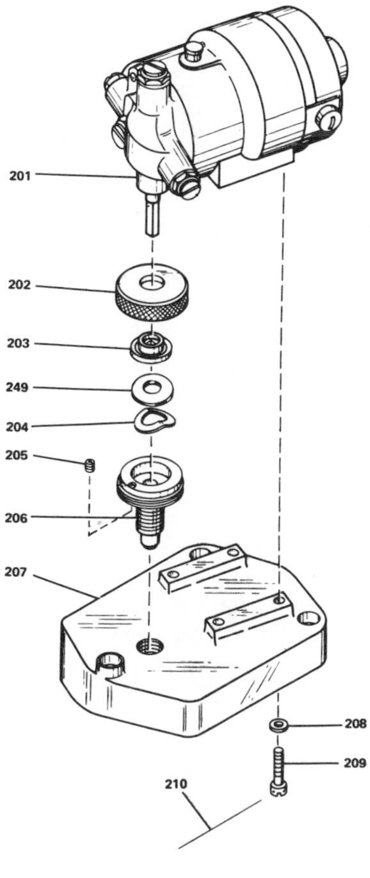

201
202
203
249
204
205
206
207
208
209
210

REF. NO.	PART NAME	NO. REQ'D.
04022-201	Motor	1
04022-202	Friction cover	1
04022-203	Friction disc	1
04022-204	Friction spring	1
04022-205	Set screw, 6-32	1
04022-206	Speed adjusting screw	1
04022-207	Cover	1
04022-208	Copper washer, 0.203 x 3/8 x 1/32	4
04022-209	Fil. hd. screw, 10-32 x 3/4	4
04022-210	Lockwire	AR
04022-249	Micarta washer	1

Figure 11-8
Split Field, Series Wound, Reversible Speed Adjusting Motor.
(Courtesy of Woodward Governor Co.)

Auxiliary Options (SG, PSG, and SGX Governors)

Figure 11-8 shows a split field, series wound, reversible speed adjusting motor, which is mounted on top of the governor cover. It is available in all standard voltages. This enables an operator on a power generator switchboard to remotely control the speed of the engine to match the frequency of an alternator with that of other units on the line before synchronizing, and to change load distribution after synchronizing. A manual speed adjusting knob with a friction clutch assembly is included on units fitted with a speed adjusting motor.

Manually turning the knob, or activating the control panel buttons, which are marked *speed increase* and *speed decrease,* causes the friction clutch within the cover, which has a small square end engaging a mating drive shaft, to drive a gear that meshes with a rack. Upward or downward movement causes the speeder spring force to be varied, thereby changing the engine speed. The amount of travel of the gear on the rack is limited; therefore, when it reaches the limit of its travel, the friction clutch will simply slip.

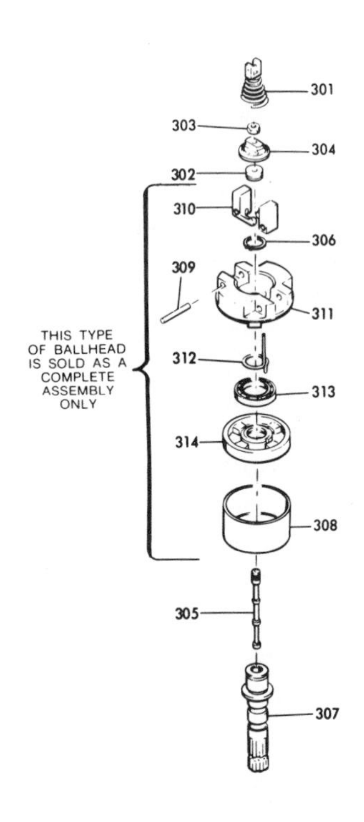

REF. NO.	PART NAME	NO. REQ'D.
04022-301	Speeder spring assembly	1
04022-302	Thrust bearing	1
04022-303	Plunger nut	1
04022-304	Spring seat	1
04022-305	Pilot valve plunger	1
04022-306	Snap ring	1
04022-307	Pilot valve bushing	1
04022-308	Ballhead cover	1
04022-309	Ballarm pin	2
04022-310	Ballarm assembly	2
04022-311	Ballhead	1
04022-312	Torsion spring	1
04022-313	Ball bearing	1
04022-314	Drive cup	1

Figure 11-9
Spring-Driven Oil-Damped Ballhead Assembly. (Courtesy of Woodward Governor Co.)

Vibration Attenuating Ballhead Assemblies

A spring-driven or a spring-driven oil-damped ballhead assembly can be used on SG and PSG governors in place of the standard solid ballhead assembly when it is necessary to attenuate undesirable torsional vibrations that could be transmitted from the engine drive. See Figures 11-9 and 11-10.

PSG Model

The PSG governor looks the same as the SG model just described; however, it differs in external appearance by the fact that it contains a compensation needle valve adjustment as shown in the following figures. The letters PSG refer to a *pressure compensated simple governor,* since it is a hydraulic governor in the same basic design as the SG, but contains buffer-type compensation.

The PSG is normally isochronous (zero droop will be maintained as long as the engine is not overloaded). On power generator applications, when ac generating sets are tied in with other units, one governor can be set to zero droop by the droop adjusting bracket, which will regulate the frequency of the entire system. If speed droop is required, however, to permit load division between two or more engines driving generators in parallel, the PSG can be adjusted between 0 and 7% droop.

The governor uses engine lube oil or an oil supply from a separate pump whose lift should not exceed 12 in. (0.3048 m), and a foot valve should be furnished. Use a 20-micron filter with a minimum capacity of 2 gallons (7.57 liters). If the governor is mounted horizontally, the needle valve must be on the bottom and a $\frac{1}{4}$-in. (6.35-mm) pipe tapped hole provided in the upper part of the governor case to drain oil away to the sump. Four check valves contained within the base plate of the governor permit rotation in either direction. Two of the passages can be plugged if rotation is only required in one direction. The oil pump within the governor is capable of producing either 75, 175, or 225 psi above inlet pressure and is controlled by the relief valve spring setting (517.12, 1206.62, or 1551.37 kPa oil pressure).

The compensation system within the governor consists of an H-shaped buffer piston with a buffer spring located on either side of it, a needle valve, and a compensating land on the pilot valve plunger. This compensation system, then, is the major difference between the PSG and the SG.

REF. NO.	*PART NAME*	*NO. REQ'D.*
04022-320	Speeder spring	1
04022-321	Pilot valve plunger	1
04022-322	Thrust bearing	1
04022-323	Retaining ring	1
04022-324	Snap ring	1
04022-325	Spring coupling assembly	1
04022-326	Ballarm pin	2
04022-327	Ballarm bearing	4
04022-328	Ballhead	1
04022-329	Ballarm	2
04022-330	Pilot valve bushing—drive gear	1

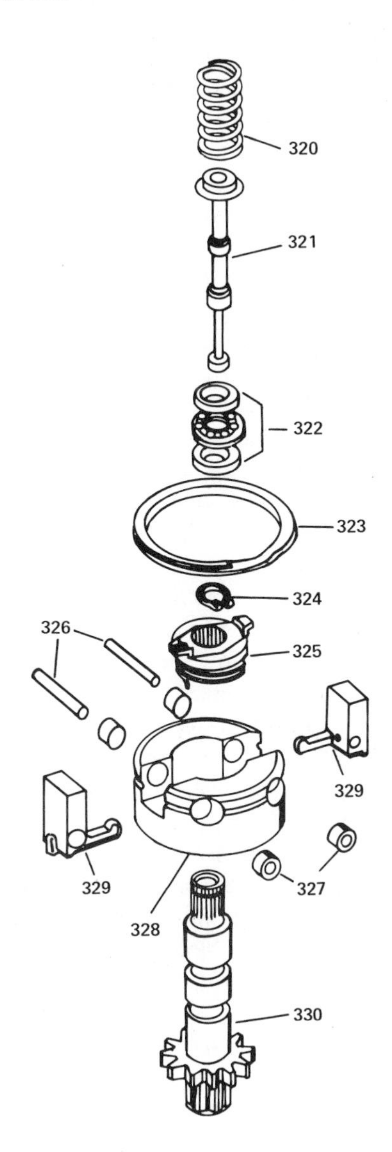

Figure 11-10
Spring-Driven Ballhead Parts.
(Courtesy of Woodward Governor Co.)

Since the speeder spring force can be adjusted, it is the initial force of this spring that will determine at what rpm the engine will attain a state of balance between the weights and speeder spring.

Engine Stopped. As with the SG governor, the PSG would have the pilot valve plunger pushed all the way down owing to the force of the speeder spring. To shorten the cranking time, place the speed control lever connected to the terminal shaft in the full-fuel position, which takes control away from the governor for initial starting purposes. Once the engine starts, move the control lever back to the desired rpm until the engine warms up.

Engine Cranking. During cranking, the centrifugal force of the flyweights will oppose the speeder spring tension, and the instant the engine starts (depending on throttle position), the weights will attain a speed proportional to the amount of force within the speeder spring. In other words, if the throttle (terminal shaft) were left in the idle position, then the rotating flyweights would only have to produce enough centrifugal force to balance out the speeder spring force at this low speed. If, however, the terminal shaft were placed in the full-fuel position, the speeder spring force, being much greater, would require a greater weight force; this would only happen at the maximum engine rpm (state of balance).

As the engine is cranking, oil pressure would flow to the base of the piston toward the underside of the pilot valve plunger compensating land, and slowly bleed past the compensating needle valve to the upper area of both the buffer piston and pilot valve plunger land. This oil pressure due to the compensating needle valve would initially be higher on the underside of both the buffer piston and pilot valve plunger land.

As the buffer piston moves up, it would compress the upper buffer piston spring, which would in turn force up the power piston. The terminal and floating levers would move to the increased fuel position, their movement being determined by the initial terminal (throttle) lever position, which would control how fast the weights would have to rotate to balance out the preset speeder spring force.

As the buffer piston is moving up, the oil pressure on the underside of the pilot valve plunger (PVP) would be pushing up the pilot valve, thereby assisting the rotating flyweights to attain their state-of-balance position. As the oil pressure on the upper area of both the buffer piston and land of the PVP attains the same pressure as that on the bottom, the buffer piston and PVP will center, which will tie in with the state of balance being reached between the weights and springs. When this occurs, the engine will run at a steady-state speed. Figure 11-11(a) shows the position of the internal governor linkage anytime that a state of balance exists.

Load Increase. How quickly the governor responds to a load change is dependent on the droop bracket adjustment, and whether or not it responds without over or under corrections is tied into the compensating needle valve adjustment. Figure 11-11(b) shows the reaction within the governor during any load increase.

Refer to Figure 11-11(b); with a load increase on the engine, the flyweights will tend to drop inward as the engine speed decreases. With the state of balance between the weights and speeder spring upset in favor of the spring, the pilot valve plunger will be forced down, which will allow pressurized oil from the pump to be directed to the underside of both the buffer piston and the receiving compensating land of the pilot valve plunger. The power piston has two diameters that are exposed to this pressurized oil from the base of the pilot valve plunger. The lower, smaller diameter is acted upon directly, and the upper annulus is connected through the bore in the power piston in which the buffer piston is carried.

The oil pressure will force the power pis-

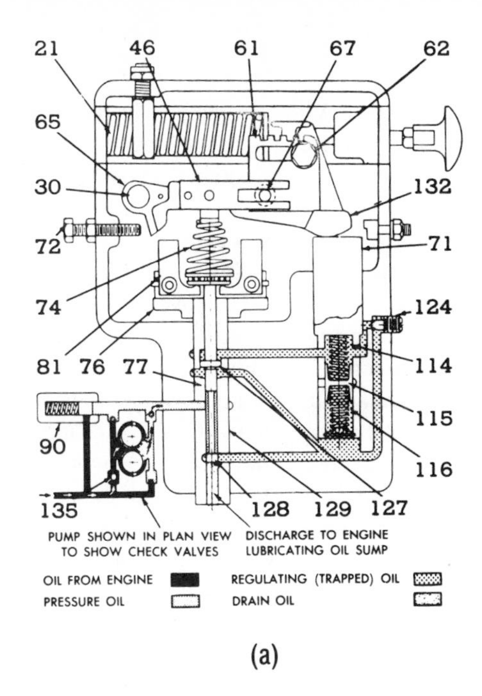

(a)

(b)

Index:

21—Spring, fuel rod
30—Shaft, speed adjusting
46—Lever, speed adjusting floating
61—Bracket, droop adjusting
62—Bolt
65—Lever, speed adjusting
67—Shaft, terminal lever
71—Piston, power
72—Screw, maximum speed adjusting
74—Spring, speeder
76—Ballhead assembly
77—Plunger, pilot valve
81—Flyweight
90—Valve assembly, relief
114—Spring, buffer (upper)
115—Piston, buffer
116—Spring, buffer (lower)
124—Valve, compensating needle
127—Land, receiving compensating
128—Land, pilot valve control
129—Bushing, pilot valve
132—Lever, terminal
135—Valve, check

(c)

Figure 11-11
(a) Position of Internal Governor Linkage When a State of Balance Exists; (b) Governor Reaction During Load Increase; (c) Position of Governor Linkage When Engine Load Is Removed. (Courtesy of Woodward Governor Co.)

ton up against the force of the terminal lever return spring, which can also be external if used with rotary motion of the terminal shaft instead of linear motion, such as would be used on some engines. The fuel racks will therefore be moved to an increased fuel position. The pressurized oil, due to the compensating needle valve, will initially be greater on the underside of the buffer piston; therefore, it forces the buffer piston up, which compresses the upper buffer spring and relieves the pressure on the lower one. Since there is a higher initial oil pressure on the underside of the compensating land of the PVP, the PVP will be pushed up, thereby recentering the flyweights and closing off the supply port. This will stop the upward movement of the power piston, which has now made the necessary fuel correction.

If the droop bracket has been set for zero droop, the engine speed will remain constant regardless of load change; however, if the droop bracket were set to its maximum of 7%, the engine speed would drop 7% when a load is applied before the governor corrected.

Load Decrease. Figure 11-11(c) shows the governor linkage position when an engine load is removed. For a given (fixed) throttle setting, if a load is removed from the engine, engine speed will increase, which causes the flyweights to fly out farther, thereby overcoming the speeder spring force. This causes the PVP to lift, which opens the control port at its base, allowing trapped oil to drain from the base of the buffer piston and PVP compensating land. Terminal shaft return spring force will push the power piston in the decreased fuel direction, therefore reducing engine rpm. This reduced oil pressure on the underside of the buffer piston and receiving compensating land of the PVP will cause the higher (temporarily) oil pressure above to recenter the PVP, followed by recentering of the buffer piston as the oil bleeds through the compensating needle valve, and both pressures above and below equalize. With a reduction in fuel input to the engine, a state-of-balance condition will again exist after the correction sequence.

PSG Adjustments. Figure 11-11 shows all the available external adjustments. To adjust the droop setting on the governor, the top cover must be removed to get at the internal adjustment screw (see Figure 11-7). By moving the bracket in toward the center of the governor, the droop pin pivot point is changed, which will decrease the droop. Moving the droop bracket away from the center of the governor will increase the droop. This is effected by the reasons explained in the description of the SG model governor.

Externally Adjusted Speed Droop. Figure 11-12 shows a PSG governor with an

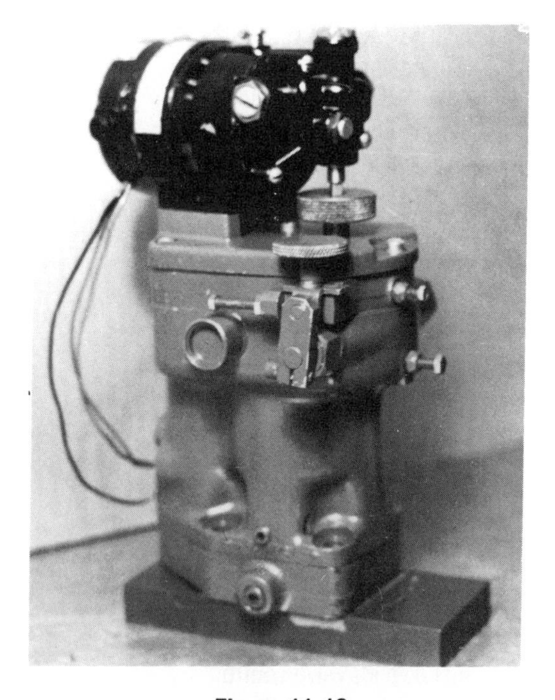

Figure 11-12
PSG Governor with External Droop Adjustment. (Courtesy of Woodward Governor Co.)

external droop adjustment. Speed droop is increased by moving the external lever forward and is reduced to zero when the lever is moved back. All droop adjustments are done on a trial-and-error basis. Make sure that the engine is at normal operating temperature prior to any final adjustments.

Compensating Needle Valve Adjustment. With the engine at operating temperature, adjust the governor for no-load rated speed by manually moving the terminal shaft to its maximum position; then adjust the high-speed stop on the side of the governor housing to obtain the speed desired. Open the compensation needle valve between two and three turns until the engine or turbine begins to hunt or surge. With a recently installed rebuilt governor, this will be more noticeable than on a unit that has been in service, since you are bleeding the system of any entrapped air. Allow the unit to surge for at least 30 seconds. Gently close the needle valve until the hunting just stops; then manually disturb the engine or turbine speed to check that the engine will return to its original steady-state speed with only a small overshoot.

Closing the needle valve farther than necessary will slow down the oil bleed back between both sides of the buffer piston and PVP compensating land, resulting in a slow return to speed following a load change, whereas overcorrection can result if it is turned out too far.

Options. The PSG is available with a temperature-compensated needle valve that adjusts the compensated oil flow with the use of bimetal strips and a spring-loaded needle valve. Adjust it in the same manner as for the non-temperature-compensated valve.

Auxiliary Equipment (PSG). In addition to those options available on the SG, such as an external electric motor for remote speed setting, the PSG can have the external droop adjustment, the temperature-compensated needle valve, spring-driven oil-damped ballhead, a torsion spring, and a pneumatic speed setting. Figure 11-13 shows such a setup, whereby remote speed adjustment is provided through a pneumatic speed setting assembly consisting of a diaphragm, housing, oil reservoir, adjusting screws, and pushrod that extends down through the governor cover and makes contact with the floating lever.

An internal return spring is also available as an option. Air signal pressure to the speed setting assembly is applied to an oil reservoir to dampen out oscillations of air compression. Oil pressure acting upon the diaphragm is transmitted to the floating lever by the pushrod, which will increase or decrease governor speeder spring force to produce a change in the speed setting. Figure 11-14 shows a schematic diagram of a PSG governor with a horizontal internal return spring, externally adjustable droop, and an electric motor speed setting.

Model SGX

Governor SGX differs from the SG model in that it is a limiting speed type, whereas the SG is a variable speed type that controls engine rpm throughout the speed range. The SGX is found on some large mobile equipment, since the flyweights are opposed by both a low idle and high idle spring setup.

At low idle the flyweights operate against the idle spring only; however, at full-load speed, the flyweights operate against both the low and high speed springs. Therefore, in the intermediate speed range, the governor operating lever is in a fixed position, which will be indicated by the pointer on the terminal lever shaft. This throttle (terminal shaft) position is controlled by the operator within the intermediate speed range since there is no governor control here. This governor operates in the same manner mechanically as the limiting speed governor described in Chapter 10. Other than having no governor control within the

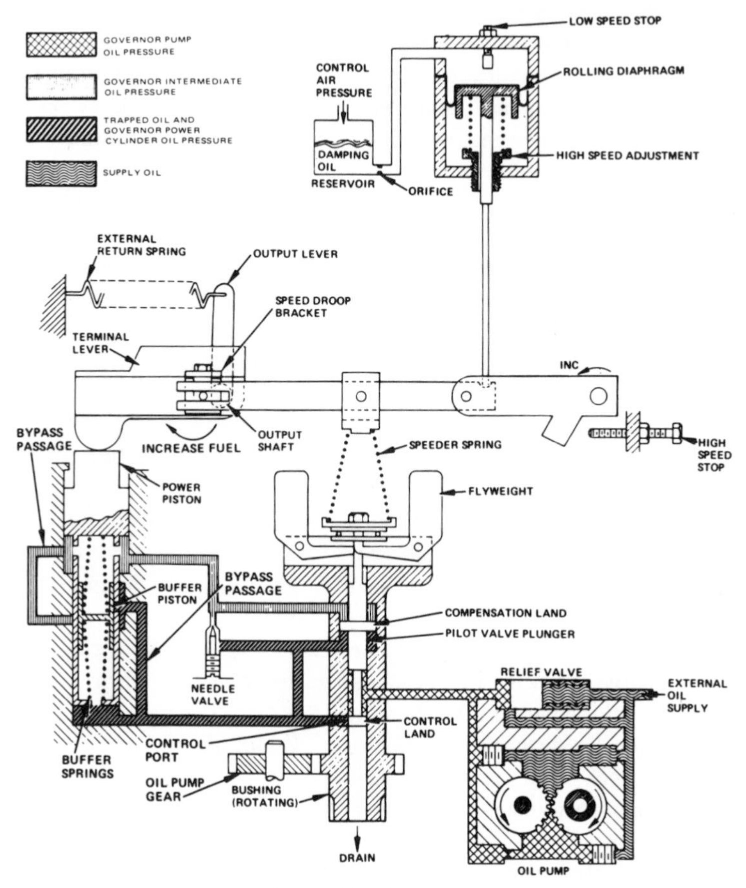

Figure 11-13
Pneumatic Remote Speed Adjustment.
(Courtesy of Woodward Governor Co.)

intermediate speed range, the SGX operates similarly to the model SG described herein.

Model UG8

The UG8 governor is available in both the *dial* and *lever* options. The lever type is common to marine applications and others, while the dial type is found extensively on power generator set applications. The basic operation of both is similar; therefore, we will concentrate on the dial model of UG8, which is very popular.

Figure 11-15 shows the basic external view of a UG8 dial governor with the individual control knobs mounted on the panel face. These are as follows:

1. *Load limit control knob:* This consists of an indicator disc that is geared to a load limit rack (see Figure 11-16); the control knob is also attached to the load limit cam. By manipulation of this knob, the load that can be applied to the engine is mechanically limited, since with the load indicator set to a preset number, the PVP (pilot

valve plunger) is lifted, which will positively stop any further increase in fuel. The engine can actually be stopped at any time by simply turning this knob to zero, which rotates the cam to force the load limit lever and shutdown strap downward; this causes the right-hand end of the load limit shutdown lever to pivot about its fulcrum, thereby lifting the PVP, which in turn releases trapped oil from under the power piston to drain.

2. *Synchronizer knob:* Turning this knob will change engine speed on a single engine; however, on engines running in parallel (power generators), turning this knob will change the engine load. The upper knob is actually the control, while the lower knob simply indicates the number of revolutions of the control knob. The lower knob cannot be turned.

3. *Speed droop knob:* The speed droop con-

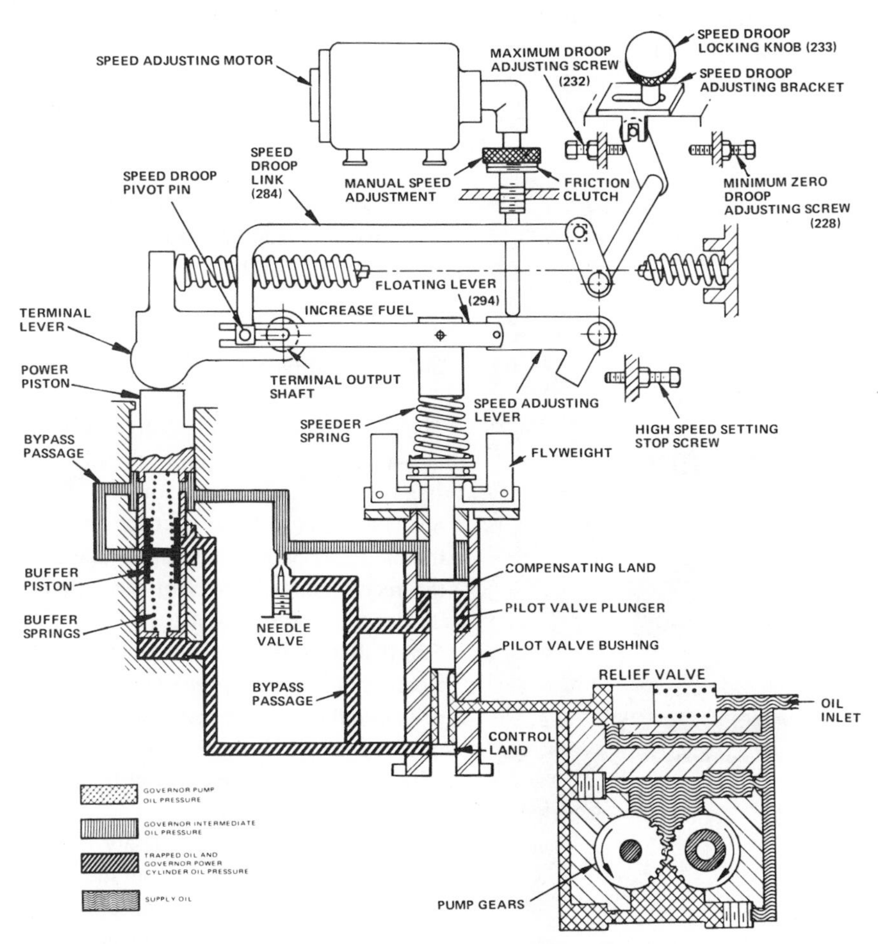

Figure 11-14
PSG Governor with Horizontal Internal Return Spring, Externally Adjustable Droop, and Electric Motor Speed Setting.
(Courtesy of Woodward Governor Co.)

trol knob shown in Figures 11-15 and 11-16 consists of the knob, a cam, and assorted linkage. Adjustment of the droop knob through its linkage will vary the compression of the governor speeder spring as the terminal shaft (throttle control) rotates. Speeder spring compression is reduced as the governor reacts to increase fuel. The droop setting would be determined by the application and whether or not it was a single-engine unit or tied in with other generator sets in parallel. A reduction in the droop control knob setting to zero allows the engine to change load without changing speed. This is generally the setting on single-engine applications; however, on interconnected engines running in parallel, the droop knobs on each engine would be adjusted for minimum droop, but still provide satisfactory load division between engines on the line. On ac generator sets running in parallel, the droop knob is usually set sufficiently high (30 to 50 on the dial) to prevent load interchange between engines. If one unit in the line has enough capacity, its governor can be set to zero droop, and it will regulate the frequency of the engine system. As long as its capacity is not overloaded, it will handle all load changes. To adjust the system's frequency, turn the synchronizer of the governor with zero droop. Operate the synchronizers of the governors that have speed droop to equally share the load between engines.

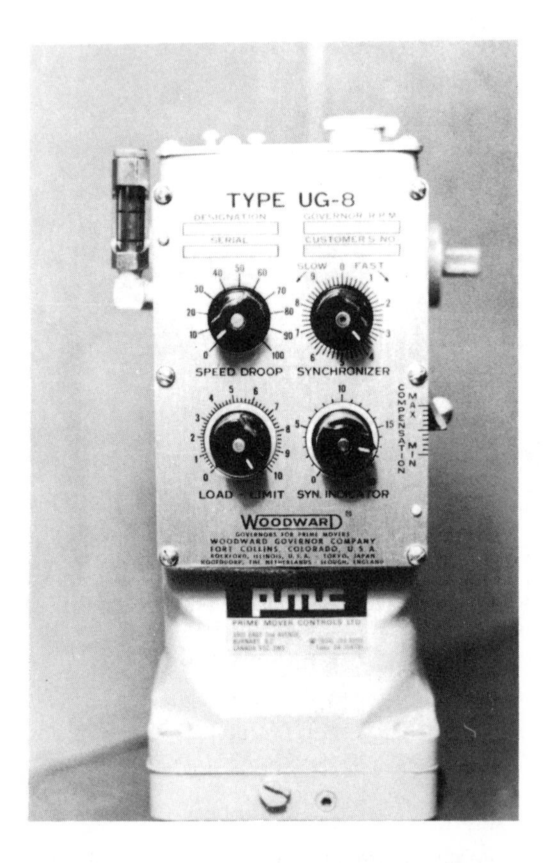

Figure 11-15
External View of UG8 Dial Governor. (Courtesy of Woodward Governor Co.)

General Information

The UG8 is a mechanical–hydraulic governor that is used on dual fuel, diesel engines, or steam turbines. The work output of the governor is 8 ft-lb (10.84 J) over the full 42 degrees travel of its output (terminal) shaft; however, the recommended travel of the output shaft is 28 degrees, providing a useful work output of 5.3 ft-lb (6.779 J). The UG8 is capable of isochronous (zero droop) control, and can be driven either cw or ccw at a recommended rated speed of the governor drive of 1000 or 1500 rpm, with a drive power requirement of $\frac{1}{3}$ hp ($\frac{1}{4}$ kW).

The governor is completely filled with oil between the lines shown on the sight gauge (approximately 1.5 quarts or 1.41 liters) with SAE 10 or 50 oil, depending on the operating temperature.

Refer to Figure 11-16 and familiarize yourself with the basic component parts layout prior to reading the explanation of the governor operation. The following explanation will also

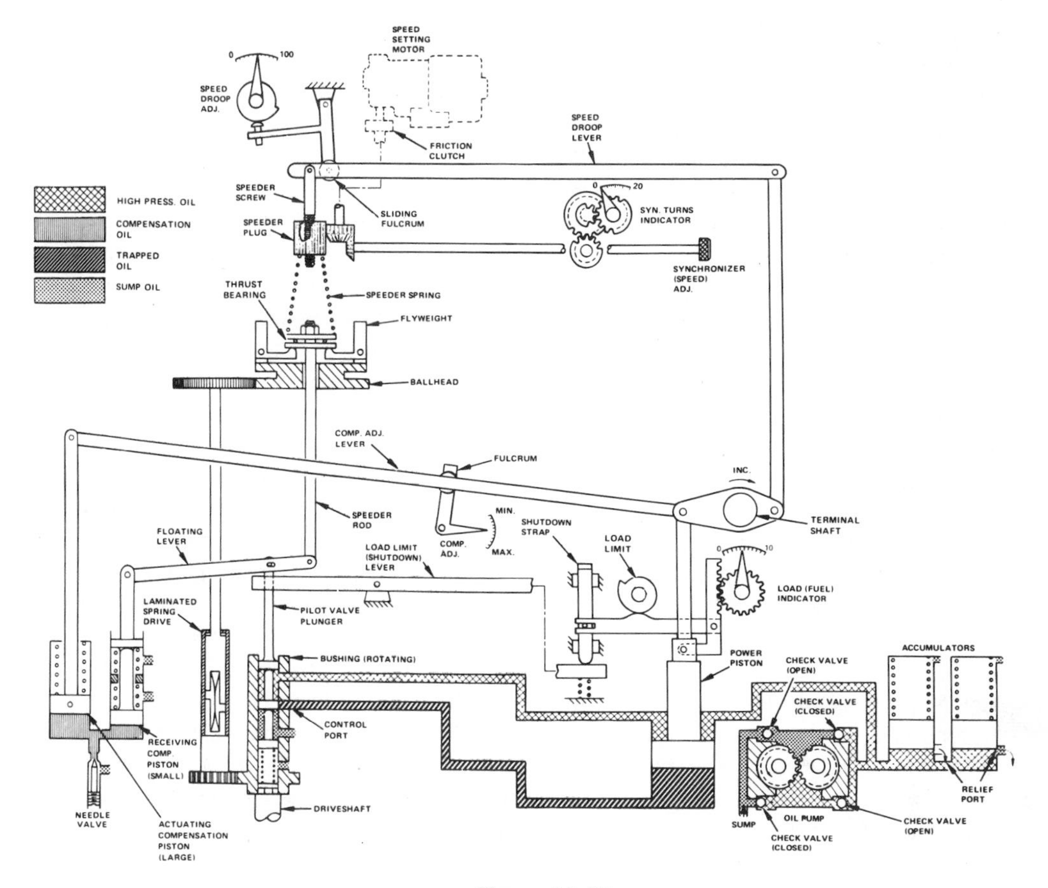

Figure 11-16
Operational Diagram of the UG8 Governor.
(Courtesy of Woodward Governor Co.)

prove useful in understanding the function of some of the major component parts within the governor.

Oil Pump. To minimize torsional vibrations, the drive from the engine to the governor is through a laminated drive, which also serves to drive the ballhead (weight) assembly at its top end. The drive shaft from the engine rotates the PVP bushing along with the oil pump drive gear contained within the controlet or base of the governor assembly. As with the PSG governor, four check valves are provided to allow cw or ccw rotation. Regardless of governor rotation, oil under pressure from the pump is directed through the check valves into the accumulator system shown in Figure 11-16.

Accumulator. Two accumulators consisting of spring-loaded pistons are housed within the controlet (base) of the governor housing assembly. Oil from the oil pump is directed into the governor system and builds to a maximum pressure of 120 psi (827.4 kPa), after which time the accumulator pistons will lift against spring pressure, dumping the oil back into the governor housing through a relief port in each cylinder. Remember that the governor is filled with oil; therefore, all the moving parts are surrounded by oil.

Power Piston. The power piston is directly connected to the engine fuel control racks through a lever and link that rotates the terminal shaft of the governor. If the power piston moves upward, fuel delivery is increased; if it moves down, fuel delivery is reduced.

The piston is shaped in such a fashion that it is stepped (differential type), whereby the bottom has a larger area than the top. Less oil pressure is therefore required on the bottom than the top to hold the piston stationary. The moment that it is the same on both sides, it must move up. The position of the PVP (pilot valve plunger) regulates oil to or from this power piston.

Pilot Valve System. The position of the PVP is controlled by the speeder spring force and the centrifugal force created by the rotating governor flyweights. When a state-of-balance condition exists, the PVP will be centered in its bore, neither supplying nor draining oil from the power piston.

Compensation System. Refer to Figure 11-16; there are two compensating pistons contained within the governor, an actuating compensating piston (larger diameter of the two) and a receiving compensating piston (the smaller one), which through linkage connected across to the power piston will move the (larger) actuating compensating piston directly opposite to the direction of movement of the power piston. Therefore, if the power piston moves upward (fuel increase), the actuating piston will be moved down, and vice versa.

Anytime that the (larger) actuating compensating piston moves, there is a flow of oil between it and the underside of the (smaller) receiving compensating piston, the flow of which can be adjusted by the compensation needle valve. When the power piston moves up, the (larger) actuating piston will move down, which forces oil under the (smaller) receiving compensating piston, pushing it upward, which lifts the PVP, thus stopping the flow of oil to the base of the power piston.

The stroke of the actuating (larger) compensating piston can be adjusted by changing the fulcrum lever pivot. This is readily available on the side of the governor housing, as shown in Figure 11-15 as the compensation pointer and in Figure 11-16.

Ballhead (Weight) Assembly. The ballhead consists of a ballhead, weights, speeder spring, thrust bearing, speeder plug, and speeder rod. The ballhead is driven by the laminated drive shaft, as shown in Figure 11-16. The speeder spring force and hence engine speed are controlled by manually turning the synchronizer knob on the governor faceplate. This connection can be seen in Figure 11-16. Turning the knob rotates a gear, which turns the speeder plug, and it will move up and down on the threaded speeder screw as desired. If, however, a split field adjusting motor is mounted on top of the governor housing, the speeder plug would be moved up and down by the small friction clutch driven from the motor for remote control applications.

Governor Operation

Refer to Figure 11-16. Prior to starting the engine, the load limit knob (fuel limit) on the faceplate should be turned to 5 on the dial, which prevents the engine from receiving excessive fuel on initial start-up and accelerating too rapidly. Once the engine has attained normal operating temperature, the load limit can be turned all the way to 10. The engine rpm can then be adjusted to its normal speed by use of the synchronizer knob.

Engine Stopped. With the engine stopped, the flyweights will be in a collapsed state. The speeder spring will push the speeder rod down, causing the inner end of the floating lever to push down the PVP, which uncovers the supply port from the oil pump. Manual rotation of the terminal shaft or load limit and synchronizer knob position will determine the initial compressive force of the speeder spring and hence fuel rack position.

Engine Cranking. When the engine is turned over, oil pump pressure will be directed to both the top and bottom areas of the power piston through the control land (center one) of the PVP. Since the bottom of the power piston is larger in diameter than the top, the piston must move up. While this is occurring, the centrifugal force of the flyweights is trying to compress the speeder spring. When the engine fires and runs, the rapid acceleration will be controlled by the position of the load limit cam, which was set to 5 on the knob prior to starting. As the flyweights move out under centrifugal force, they will lift the thrust bearing connected to the speeder rod, which will in turn lift the inner end of the floating lever and PVP. With the PVP centered, pressurized oil that was sent to the power piston will become trapped underneath it, while the oil on the top of the piston will remain at maximum oil pump pressure.

When the power piston moved up, through linkage within the governor, the actuating (larger) compensating piston moved down, displacing oil through the compensation needle valve at sump pressure only to the base of the receiving (smaller) compensating piston, which will move up, causing the outer end of the floating lever to assist in centering of the PVP.

With a state of balance existing between the weights and speeder spring, the power piston will be held in a fixed position, the PVP is centered, and the actuating and receiving compensating pistons would be level with one another in their respective bores. The engine could now be set to run at its desired rpm by manipulation of the load limit control knob and synchronizer knob.

Load Increase. As load is applied to the engine, there is a speed decrease. The flyweights move in, thereby lowering the speeder rod and inner end of the floating lever, which in turn lowers the PVP, uncovering the regulating port of the PVP. Pump oil pressure is now directed to the base of the power piston (as well as the top); however, since the bottom area is larger than the top, the power piston will start to move up.

The upward moving power piston through its linkage will cause rotation of the terminal shaft connected to the fuel racks in the increased fuel direction.

However, as the power piston moves up, the governor linkage causes the actuating (larger) piston to move down, forcing the smaller receiving compensating piston up; this compresses the compensating spring and raises the outer end of the floating lever and the PVP, which will start to close off the regulating port of the PV bushing. Immediately, power piston movement will cease, stopping the rotation of the terminal shaft at a position that will correspond to the throttle rack position for the increased load.

With increased fuel delivery to the engine, the state of balance between the weights and speeder spring will be reestablished. As they work toward this position, the speeder rod will be pulled up along with the inner end of the floating lever, which will start to level off.

The receiving (smaller) compensating piston is returned to its normal position by the action of the compensating spring, which was placed under compression when the piston moved up owing to the oil displacement from the larger actuating compensating piston when the correction was initiated. This return of the receiving piston is at the same rate as the speeder rod, which is being pulled up by the weight force, thus ensuring that the regulating port in the PV bushing is covered by the land on the PVP.

Remember, the rate at which the receiving compensating piston is returned to normal will depend upon the flow of oil through the compensating needle valve. The engine will now be running at an increased fuel delivery rate to carry the increased load.

Load Reduction. With a decrease in load, the engine speed will tend to increase for a given throttle setting. As the speed increases, the weights will fly out, thereby compressing the speeder spring and raising the speeder rod along with the inner end of the floating lever, which will in turn lift the PVP. As the PVP uncovers the regulating port, oil under pressure (trapped oil) at the base of the power piston will drain back into the sump or governor housing. The oil pressure still remaining on top of the power piston will positively force it down, thereby rotating the terminal shaft to a decreased fuel position.

Through the governor linkage, the downward moving power piston will pull the (larger) actuating compensating piston up. This draws the (smaller) receiving compensating piston down, which compresses the compensating spring as it lowers the outer end of the floating lever and PVP. As the PVP moves down, it will allow the center land of the PVP to cover the drain oil from the base of the power piston. When this happens, the power piston and terminal shaft will stop at a position corresponding to a decreased fuel input for the decrease in load. With the reduction in rpm, the weights will move in to the normal state-of-balance position, thereby lowering the speeder rod to its normal position.

While the speeder rod is returning to its normal position, the compensating spring, which was placed under compression during the governor correction process, will return the receiving compensating piston to its normal position at a rate equal to that of the speeder rod, as the oil bleeds through the compensating needle valve. The engine will now be in its corrected fuel position.

Adjustments: UG8

We will only deal with external adjustments, since any internal adjustments usually require other checks and often internal repairs to correct any problems. For a detailed analysis of governor overhaul, contact your nearest Woodward governor outlet, and obtain a copy of Bulletin 03032B, *UG8 Governor, Operating and Service Manual.*

Compensating Adjustments. If a new or rebuilt governor is installed on an engine or turbine or if the governor has simply been drained and cleaned of old oil, after refilling, the unit will have to be bled of any entrapped air. With the governor installed on the engine, be sure that prior to initial starting manual control of the fuel racks can be assumed, if necessary, in case of any problems with the unit. Use the load limit knob set to 5 on the dial to prevent overspeeding upon initial start-up. If the governor appears to be operating correctly, you can turn the synchronizer knob to control the desired engine speed, or if fitted with a drive motor, remote control can be used.

When the engine has attained its normal operating temperature, the following adjustments are required during no-load conditions to ensure that the governor is in fact capable of optimum control.

1. Loosen the self-locking nut on the compensation adjusting pointer located halfway up the side of the governor housing. See Figure 11-15 for location. Set the pointer at its maximum upward position by manually pulling it upward. This will place the fulcrum point of the compensating adjusting lever connected to the actuat-

ing compensating piston in its maximum compensation position. Retighten the nut.

2. Refer to Figure 11-15; remove the compensating valve plug from the base of the governor housing. Stop here and look into the bore where the plug was removed; you will notice a two-slot type of screw, which is actually the compensating needle valve. Since this is a self-locking screw, it is imperative that you place the screwdriver into the shallower of the two slots, and not into the deep slot located at right angles to the other. Unscrew the compensating needle valve between three and five turns with the screwdriver, and allow the engine or turbine to hunt or surge for about a half a minute to bleed entrapped air from the governor oil passages.

3. Loosen the compensation nut that was tightened in step 1, and place the pointer in its extreme downward position, which is now minimum compensation. Gradually and gently close the compensation needle valve with the screwdriver until hunting or surging just stops. Pause and make sure that the engine is running at a steady rpm. Place the screwdriver back into the needle valve and turn the screw in until it very gently bottoms. Take careful note of how many turns were required to achieve this. Slowly open the needle valve screw to the previous position where hunting or surging just stopped. Manually disturb the engine or turbine speed, and note how quickly it levels out when the throttle is released again. On governors using only one compensating spring, if the needle valve is more than one-eighth to one-quarter turn open and the engine settles out, the adjustment is satisfactory. If the governor is a two-spring unit and with the needle valve one-half to three-quarters of a turn open, the engine settles out after a manual throttle disturbance, the adjustment is satisfactory.

4. If, with the needle valve adjusted as explained in step 3, hunting still exists in the engine or turbine, carefully back off the compensation pointer locknut and raise the pointer two divisions on the scale. Turn out the compensating needle valve again, and allow the engine to hunt for 30 seconds.

5. Repeat the process described in step 3.

6. Repeat steps 3 through 5 until adjustment is acceptable.

Bear in mind when doing these adjustments that it is desirable to have as little compensation as possible. As with the PSG governor, closing of the needle valve farther than necessary will make the governor slow to return to normal speed after any load change. Also, placing the compensation adjustment pointer too far toward maximum position will cause excessive speed change when a load change occurs.

Maximum or Minimum Speed Limit Adjustment

1. With the engine running and at normal operating temperature, remove the dial faceplate from the front of the governor.

2. Remove the lower right synchronizer gear by unmeshing it and pulling it outward slightly.

3. By turning the synchronizer indicator gear, the engine speed can be increased or decreased by the fact that the high speed stop pin is either moved closer or farther around the arc of travel of the synchronizer gear, which acts as a stop to vary the speeder spring compressive force. See Figure 11-16.

4. Remesh the synchronizer indicator gear with the high speed pin engaging the intermediate gear to prevent any further in-

crease in established rpm. The high speed stop pin is closer to the gear center.

5. Run the synchronizer knob back to operate the engine at low rpm; then position the synchronizer indicator.

Friction Clutch Adjustment. Both the PSG and UG8 use a split field electric motor for remote control of the engine rpm. Remove the governor cover to get at the friction clutch located at the top of the governor linkage.

1. Remove the retaining ring with the proper type of pliers (Truarc) to avoid stretching. With the ring removed, take out the spring-loaded cover and spring from the clutch assembly. Exercise caution to prevent dropping the spring down into the governor oil sump.
2. Turning the nut will increase friction on the clutch, while ccw adjustment reduces it.
3. It is imperative that the friction clutch be adjusted with an inch-pound torque wrench to between 4.5 and 5.5 lb-in. (0.539 to 0.621 N·m) on governors equipped with speed setting motors, or to 1.5 to 2.5 lb-in. (0.169 to 0.282 N·m) on governors with a manual speed setting.

Speed droop adjustment was described earlier in this section.

11-3

Woodward Electric Governors

There has been a trend for several years to produce a range of governors with fewer moving parts and that also require a minimum of maintenance. One particular engine and turbine application that electric governors lend themselves to quite readily is the power generator set, either in single-unit form or paralleled with other units on the line.

Therefore, the main requirement for electric governors occurs when isochronous operation of paralleled generator sets is needed. The electrical load on each generator can be measured and signals sent directly into the electronic governor circuit to balance the load between the paralleled generator sets.

Electric governors also provide for greater mounting flexibility and improved transient response. If a governor drive is not available, the 1700 electric actuator can be used on diesel engines; however, in gas turbine applications, due to the additional control loop requirements, electronic governors are much better suited. Additional options with these governors can include such items as temperature control, compressor discharge pressure control, and fuel management during starting.

Let us look at some of the currently available electronic control systems and actuators presently on the market that are produced by Woodward.

8256 Series Electric Actuator Model 1700 with 8290 Series Speed Control

The 8256/8290 combination is designed for use on diesel, gas, gasoline engines, and gas turbines where a mechanical governor drive or hydraulic supply is not readily available. Even if there is a governor drive readily available, the use of this combination will greatly reduce maintenance, such as oil changes to the governor itself, normal adjustments due to wear, and adjustments due to generator set output variations. Figure 11-17 and 11-18 show a typical 8256/8290 unit.

The actuator and speed control are intended for single-unit non-parallel operation of generator sets and mechanical loads. The actuator can be mounted on the engine or in the general engine vicinity as long as it is close enough to minimize excessive linkage or reduce the work output of the actuator itself, which is 1.7 ft-lb (2.3 joules) in the fuel decrease direction, and 2.2 ft-lb (2.98 J) in the fuel increase direction.

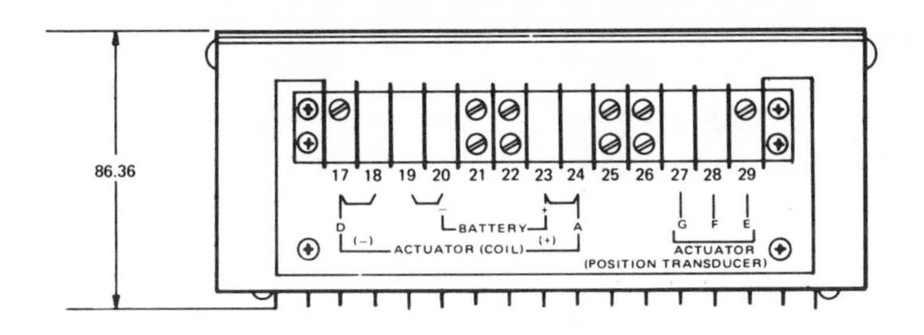

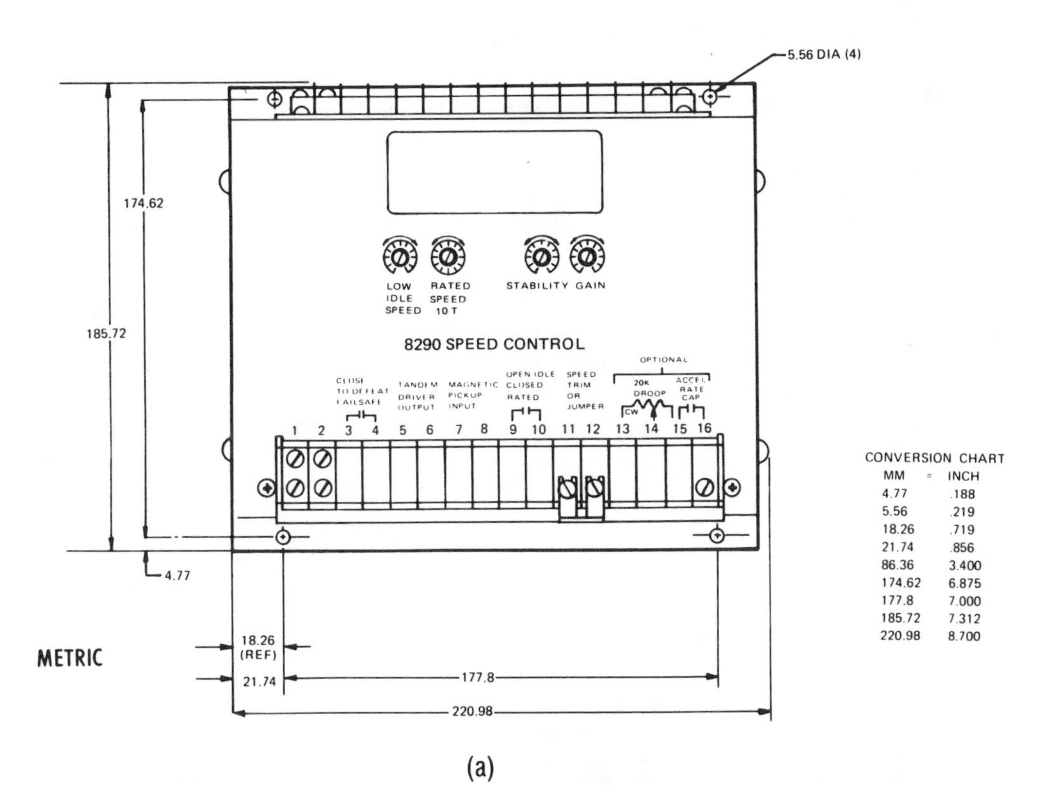

CONVERSION CHART

MM	= INCH
4.77	.188
5.56	.219
18.26	.719
21.74	.856
86.36	3.400
174.62	6.875
177.8	7.000
185.72	7.312
220.98	8.700

(a)

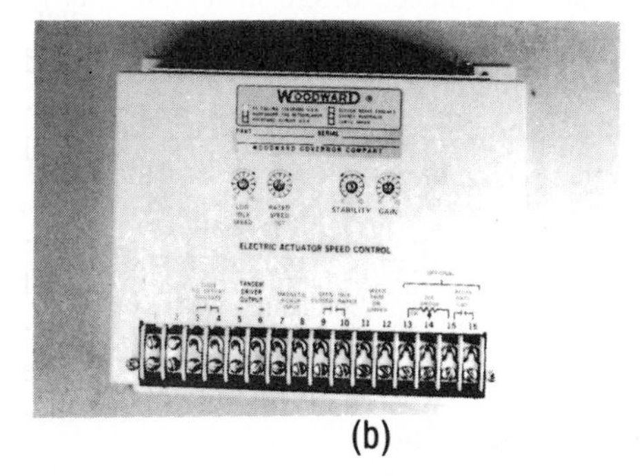

(b)

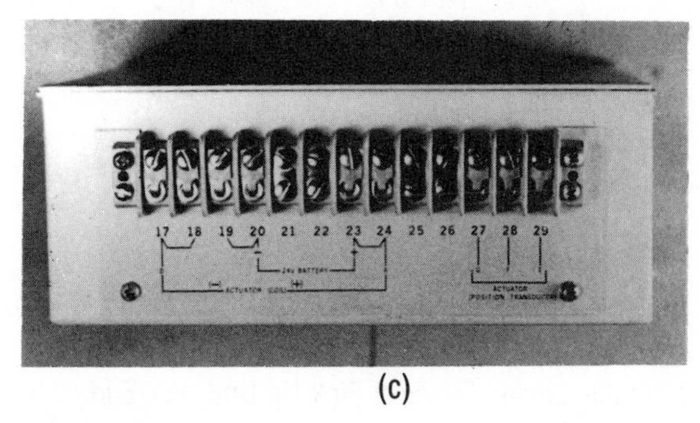

(c)

Figure 11-17
8290 Series Speed Control: (a) Diagram; (b) face view; and (c) rear view. (Courtesy of Woodward Governor Co.)

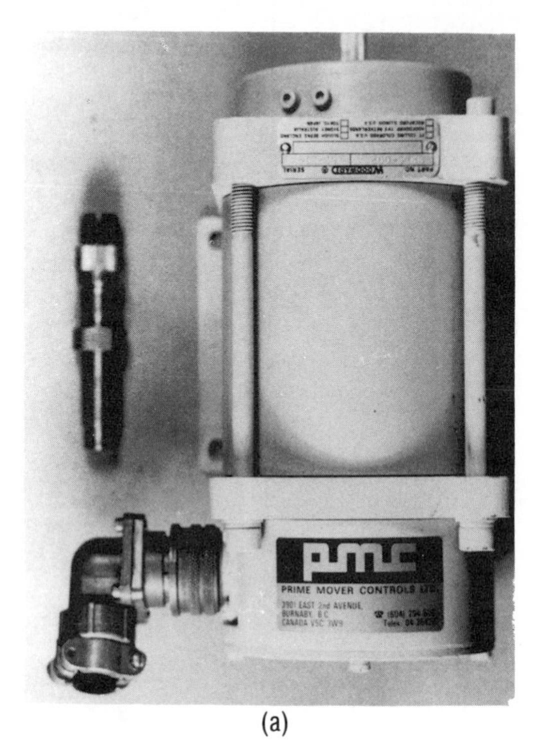

(a)

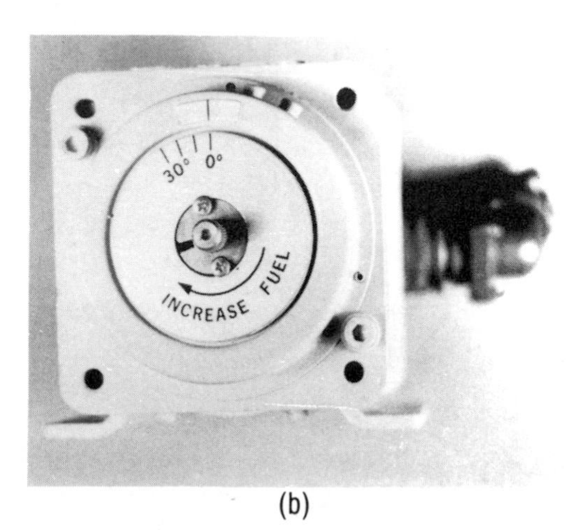

(b)

Figure 11-18
(a) 8256/1700 Electric Actuator Assembly with Magnetic Engine Flywheel Pick-Up Unit; (b) End View of 8256/1700 Electric Actuator Showing Terminal (Output) Shaft and Degrees of Rotation.
(Courtesy of Woodward Governor Co.)

Although the angular travel of the actuator shaft is 30 degrees cw to increase when viewing the end of the shaft, Woodward recommends using approximately 18 degrees between the no-load and full-load fuel positions.

A preloaded internal return spring supplies output shaft torque in the decrease fuel direction. Also, a transducer in the actuator provides a position feedback signal for accurate positioning of the fuel control to the engine or turbine. The engine speed is sensed from a magnetic pickup that is inserted into the flywheel housing, for example, on a piston engine to coincide with the number of teeth on the flywheel ring gear. All actuator adjustments are contained within the speed control module.

Magnetic Pickup

Since the magnetic pickup is designed to send a signal back to the control module, the sensed gear on the engine or turbine must be made of magnetic material. To operate properly, the number of teeth passing the pickup must be at least 300 per second, with the recommended gap between the pickup and the outside gear diameter being approximately 0.040 in. (1.016 mm) at its closest point. To make sure this is in fact correct, check that the maximum gear runout does not exceed 0.020 in. (0.508 mm). If it is not convenient to use the flywheel ring gear for magnetic pickup, and an accessory gear is selected, the spacing between the pickup and gear may have to be as close as 0.010 in. (0.254 mm).

To check the magnetic pickup installation, especially if there are problems with the governor response, the signal level from the pickup should be at least 1.0 V rms at normal engine or turbine cranking speed.

Since the gap setting may be awkward to check, if you cannot actually measure this, with the engine or turbine stopped, gently turn the pickup in until you feel a slight step up in effort (just touching gear); mark a reference point, and back out the magnetic pickup unit approximately three-quarters of a turn. Manually turn the engine or turbine over at least one full turn, and make sure that there is no bind. From the reference point marked earlier, measure the gap; then tighten the jam nut.

Magnetic pickup connectors feeding back to the control module may be ordered from Woodward if desired.

Speed Control Module Power Supply

The 8290 requires a voltage of 18 to 32 V dc, preferably supplied from the engine battery. The power leads to the control should come directly from the battery, and not through distribution points. The speed control requires separate wires to internal control and output circuits. Never disconnect power to the control circuit unless power is also disconnected from the output circuit (battery).

Number 12 AWG stranded wire should be used, with the maximum wire length being kept as short as possible; however, do not exceed 150 ft (46 m) at any time. Figure 11-19 shows the wiring circuit for the 8256 electric actuator and 8290 speed control.

The switch or circuit breaker from the positive battery terminal to contact number 23 on the module should be rated for 10 A induc-

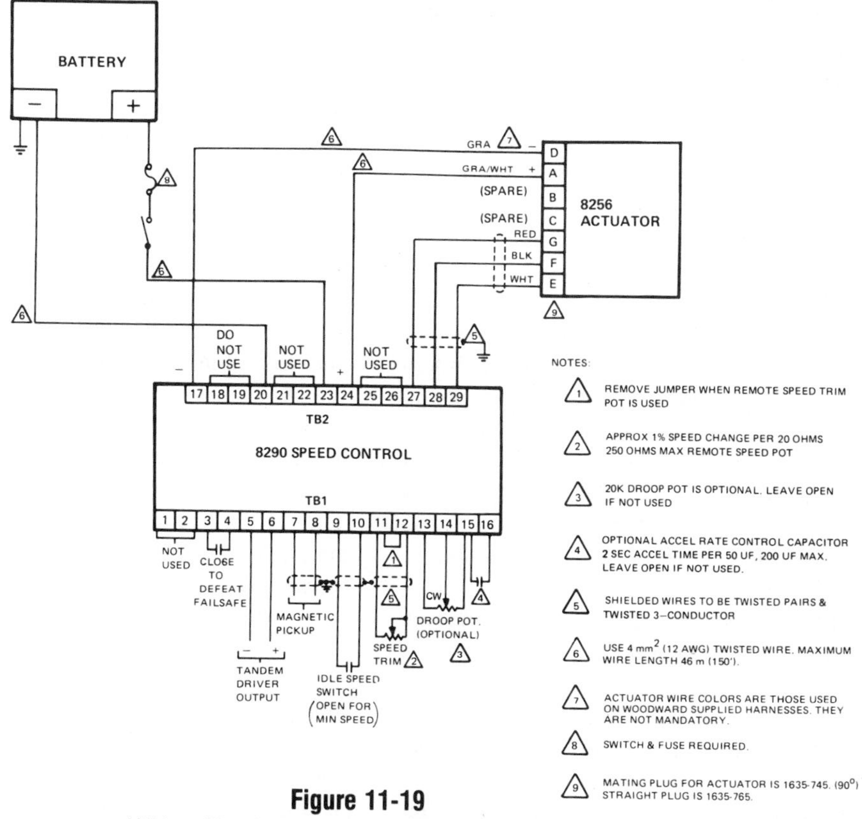

Figure 11-19
Wiring Circuit For 8256 Electric Actuator and 8290 Speed Control.
(Courtesy of Woodward Governor Co.)

tive load at 32 V dc. The 8290 speed control can be mounted in any position and operated within a temperature range of −40° to 71°C (−40° to 160°F). Ensure, however, that adequate ventilation and space are provided for cooling purposes and necessary adjustments.

Principles of Operation

Within the housing of the model 1700 electric actuator as shown in Figure 11-20 are a coil, two rotors, a rotor spacer, and two stators. The rotor and stator combination is shaped in such a way that both variable and fixed air gaps exist during rotation of the coil.

The coil receives its excitation signal from the speed loop controller within the 8290 speed control module. This coil excitation will rotate the stator so that the variable air gap poles rotate in the direction to close or reduce the air gap, while the fixed gap poles will rotate to center the rotor and stator poles. This combination produces a linear torque in proportion to the angle relationship for a given excitation signal.

The output shaft has a diameter of 0.375 in. (9.525 mm), with 36 serrations for mounting of a lever. Within the actuator is a spiral spring, which is preloaded to provide the actuator with output torque in a decrease fuel direction. This spiral spring preload therefore requires that a steady-state excitation of the actuator coil be supplied in order to position the shaft off the minimum fuel stop position.

Speed Control Operation. The magnetic pickup generates an ac voltage in direct proportion to the speed of the engine or turbine and sends it to the speed control module, where it is then converted to a dc voltage. The voltage, which is proportional to engine rpm, is monitored at the speed loop controller and compared to the speed setting voltage determined by the rated engine speed and, if used, the speed trim potentiometer. The actuator shaft is

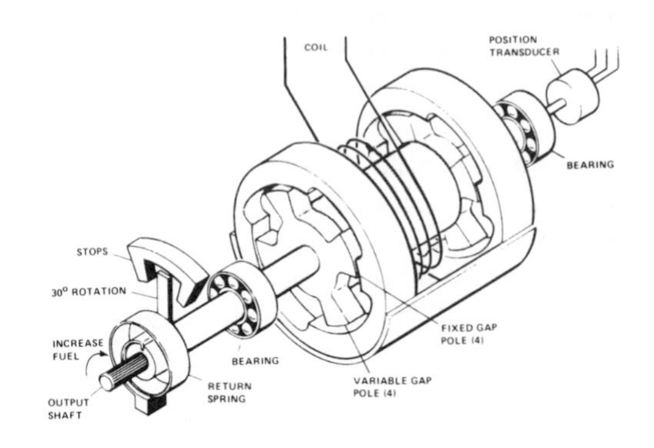

Figure 11-20
Electric Actuator Schematic.
(Courtesy of Woodward Governor Co.)

therefore controlled by a position transducer within its body, which returns a voltage proportional to its output position to the loop controller. This is then compared within the module, and an appropriate pulse width current is sent by the speed loop controller back into the actuator to establish its position.

Load Increase. If there is a load increase applied to the engine or turbine for a given throttle setting, the voltage generated at the magnetic pickup will decrease. The amplifier will increase its pulse width current to the actuator coil, therefore reducing the variable area working air gaps, which rotates the armature to the increased fuel direction in direct proportion to the electrical signal increase received from the speed control setting voltage.

Contained within the position controller is an energy limiter designed to prevent or minimize actuator overheating and battery drain should an abnormal condition occur that requires the engine fuel control to be held in the maximum fuel position for extended periods of time. Maximum actuator position at a reduced current level is then maintained. However, prior to the maximum input current being supplied to the actuator, it is designed to obtain its full angular travel to retain some reserve torque capacity if needed.

Load Decrease. If the load on the engine or turbine is decreased for a given throttle or speed setting voltage, the voltage generated by the magnetic pickup will exceed this speed setting voltage, causing the speed loop controller to reduce its signal output to the position loop controller. The reduction in the current sent to the actuator coil will allow the internal spiral return spring to rotate the output shaft of the actuator connected to the engine or turbine fuel control to a decreased fuel position in proportion to the decrease in load.

Speed Sensing Failsafe. Should there be a failure of the magnetic pickup unit, the generated voltage signal would not be received at the speed control module. If this happens, the speed control speed sensing failsafe will shut the engine down. During engine or turbine cranking, the speed sensing failsafe section will be delayed and cleared as long as the control receives a 1.0-V rms signal of more than 100 Hz. If not, shutdown will occur automatically.

Adjustments: 8256/1700/8290

Adjustments and problems with the electric governor are often a direct result of small overlooked items or an oversight on initial installation. Therefore, prior to undertaking any major repair steps, consider the following:

1. Are there any loose or corroded wires or terminals or loose or faulty plug in harnesses?
2. Is there any bind in the mechanical linkage to the fuel control racks? Is the actuator properly mounted?
3. Is adequate voltage being supplied from the battery to the output circuit?
4. Is there adequate ventilation around the speed control module?
5. Is the magnetic pickup properly gapped? Is there too much runout on the gear to the pickup?
6. Is the magnetic pickup damaged in any way? Remove and check.
7. Do you have the correct pickup for your engine or turbine?
8. Is the wire being used of the proper gauge?
9. Is the length of wire being used in excess of 150 ft (46 m)?
10. Is the actuator mounted where it receives excess heat radiation from the engine or turbine? Especially exhaust stacks?
11. Are all terminals connected to the correct terminals?
12. Are there any additional external forces acting on the actuator in the decrease fuel direction?
13. Is the amperage of the circuit breaker or relay switch properly rated?
14. Are any circuit fuses blown?

Presetting Adjustments. Refer to Figures 11-17, 11-19, and 11-21 and set the following adjustments:

1. Gently back out the screw of the rated speed control until it is all the way back to its farthest ccw position (gently now). Turn it in (cw) one complete turn to ensure adequate speed for initial startup.
2. Close the idle speed switch or jumper T9 and T10.
3. Set the external speed trim if used to its mid-travel position.
4. Ensure that the failsafe is operational.
5. Adjust the stability screw gently to its mid-position.

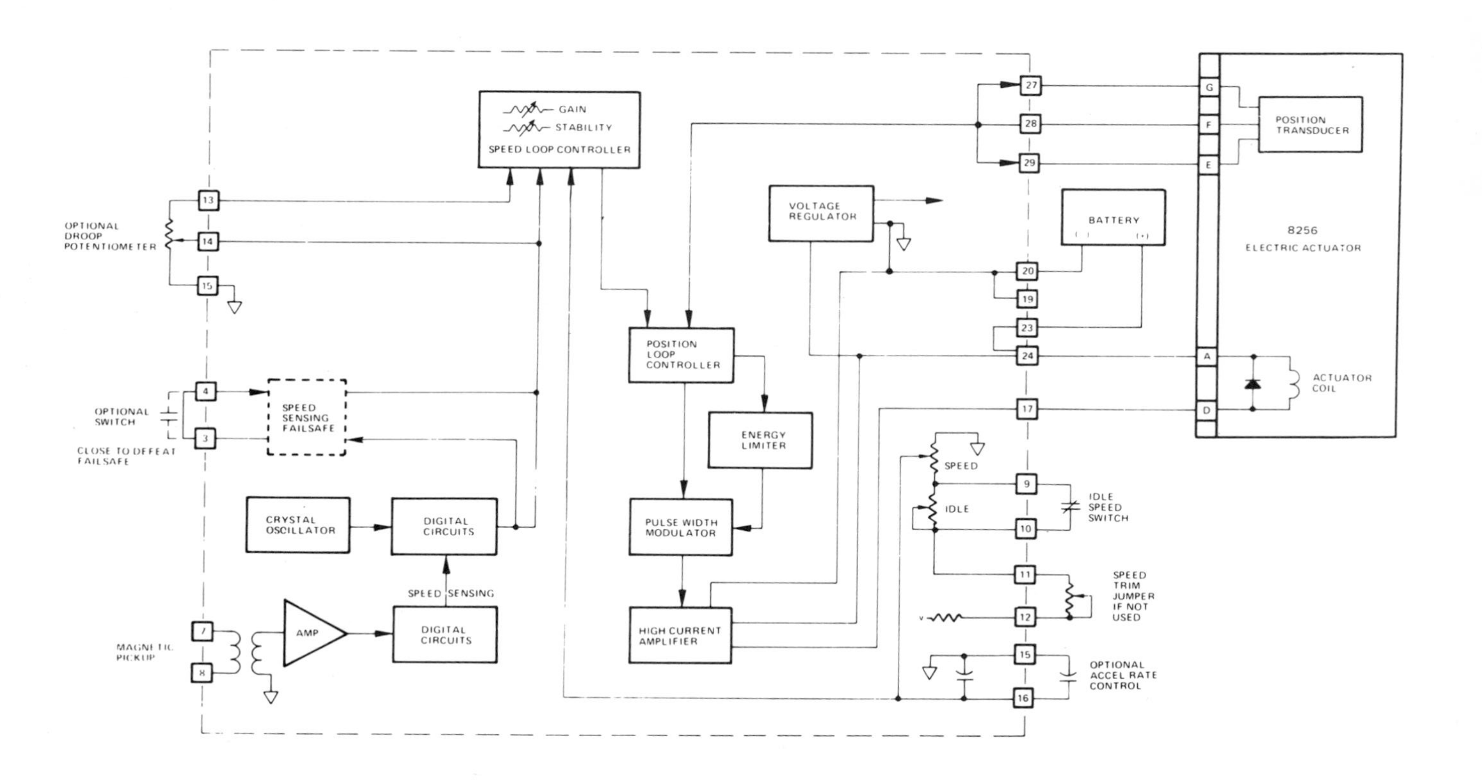

Figure 11-21
Wiring Block Schematic for 8290 Series Speed Control.
(Courtesy of Woodward Governor Co.)

6. Adjust the gain screw gently to its mid-position.
7. Gently set the low idle speed screw fully clockwise.

Checks with the Engine Stopped. With power to the governor, it may bounce quickly toward the increase fuel direction, which is normal owing to coil excitation; however, it should immediately return to zero (the actuator shaft). If the actuator shaft stays in the full-fuel position, check out the cause. However, do not leave the power turned on for any length of time, since the actuator coil will be receiving maximum current and could overheat, with subsequent damage to the internal parts.

1. With a 24-V power supply, for example, and the actuator shaft at zero degrees on its scale, the following voltages should be recorded:
   a. Terminals T20 (−) and T29 (+) should be 5 ± 1 V dc.
   b. Terminals T20 (−) and T28 (+) should be 6.25 ± 1 V dc.
   c. Terminals T20 (−) and T27 (+) should be 10 ± 1 V dc.
   d. Terminals T29 (−) and T28 (+) should be 1.25 ± 0.5 V dc.
2. With a voltmeter at terminals T29 and T28, the voltage recorded should increase and decrease when the actuator shaft is manually moved to both the increase and decrease fuel positions.
3. Voltage measured at the speed control terminals T6 (+) and T5 (−) should be 0 V dc.
4. Momentarily defeat the failsafe and ensure that the actuator shaft goes to the maximum fuel position while recording 6V dc at terminals T6 and T5.

Start-up of the Engine or Turbine. Immediately prior to starting the engine or turbine, ensure that the unit can be shut down and that all fluid levels such as fuel, water, and oil, are correct.

1. If the unit is wired with an idle speed switch, close it or install a jumper wire at terminals T9 and T10. When the unit starts, its rpm will be determined by the adjustment earlier of the rated speed potentiometer. This is only done for initial engine start-up; on any subsequent normal start-ups, the idle speed switch will be open until closed by lube oil pressure or similar means to close for acceleration to rated speed.
2. Apply power to the 8290 speed control and start the engine.

Rated Speed Setting Adjustment. With the external speed trim (if used) set at its mid-position, adjust the rated speed screw to obtain the recommended rated rpm.

Gain and Stability Adjustments. To ensure engine or turbine stability after any correction, both of these screws must be set as follows:

1. Gently turn the *gain* screw with the engine running in the increase or clockwise direction from its previously set mid-position until there is a slight unstable condition. This can be done by connecting a dc voltmeter to the speed control terminals T6 (+) and T5 (−), and slowly turning the gain screw on the speed control cover until the meter needle begins to waver or oscillate.
2. The *stability* screw can be gently adjusted either cw or ccw until stability is achieved. To check this setting, manually disturb the actuator shaft or apply varying loads to the engine. You may have to gently readjust the stability screw. If the engine still does

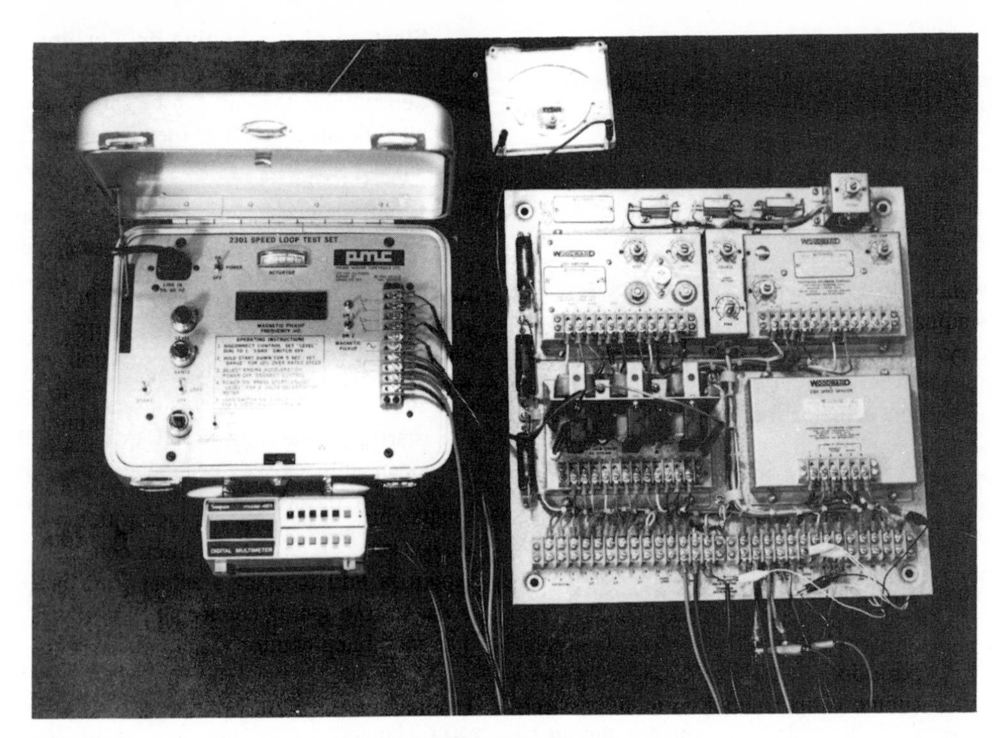

Figure 11-22
2301 Speed Loop Test Set.
(Courtesy of Woodward Governor Co.)

not settle out after a load change, turning the gain screw ccw should cure the problem. But if the voltmeter shows a slow hunt, turn the gain screw cw.

Idle Speed Adjustment. By opening the idle speed switch or removing the jumper wire from terminals T9 and T10 the engine should decelerate to a speed set by the low idle speed potentiometer. To attain the desired idle speed, gently adjust the low idle speed screw (potentiometer).

External Droop Adjustment. The 8290 speed control is set for isochronous or zero speed change regardless of engine load variation. However, if the application does require some droop, terminals T13, T14, and T15 can be wired for droop operation. The speed control will then simulate droop by decreasing the fuel input to the engine, which will cause a reduction in engine rpm. If an engine is running but not in a load-sharing mode, droop will decrease the engine speed; but if the engine is used in parallel with other units, the droop characteristic will cause a drop in load on that particular engine.

Available through Woodward is a self-contained test case for the 2301 model electronics control systems (Figure 11-22).

This 2301 speed loop test set, as it is known, can be used to do all of the adjustments necessary to the electronic control panel without it being actually connected to the prime mover (engine or turbine).

When connected, it can simulate:

(1) prime mover load, on and off, (2) speed increase and decrease, and (3) acceleration percentage characteristics.

The test case has a 115-V power supply, which can be reduced to 24 V (battery) and has generated output to replace the usual magnetic flywheel pickup signal.

As these types of governor/control modules become increasingly more prevalent, test cases like this will be used more and more.

11-4 Woodward Electronic Control Systems

Model 2301 Electronic Control

One of the most popular and widely used electronic controls produced by Woodward Governor Company is the 2301 unit shown in Figure 11-23. This unit is readily available in two versions, as a speed control or both a load-sharing and speed control, particularly on load-sharing generator sets. If the electronic control was desired on a single engine application only, where paralleling is not required, the 2301 speed control alone could be used. If, however, you desire power generator set paralleling or a tie in with local utility power, the 2301 load-sharing and speed control would be used.

Regardless of which of these 2301 units is used, an engine-mounted actuator similar in function to the 8256 described earlier is required to change the position of the engine or turbine fuel control linkage. The actuator most often used is the EG3P, which is shown in Figure 11-24(a) with a schematic of its operation shown in Figure 11-24(b).

In Figure 11-24(b), engine oil delivered to the base of the EG3P actuator is increased in pressure by the oil pump, also contained within its base, which is engine driven. This oil under pressure acts upon a power piston, which in turn causes rotation of the actuator terminal (output) shaft in relation to engine load.

The EG3P actuator pilot valve plunger is controlled by the signal received from the 2301 speed control unit, which sends out through the loop control circuit a corrected signal that is received from the magnetic pickup and constantly monitored by the preset rated speed adjustment wtihin the speed control. The oil from the pump is what actually moves the power piston, and this volume of oil is in turn controlled by the PVP and changing field strength of the solenoid coils.

Although the EG3P is extensively used along with the 2301 control module, a backup type of governor, which is a combination governor and actuator with a mechanical–hydraulic operation similar to a Woodward PSG governor described earlier, is engine mounted in the event that the electronic control unit allows the engine or turbine speed to exceed the ballhead setting; then the mechanical–hydraulic governor will take control of the terminal shaft rotation, thereby limiting the maximum engine speed. Such a

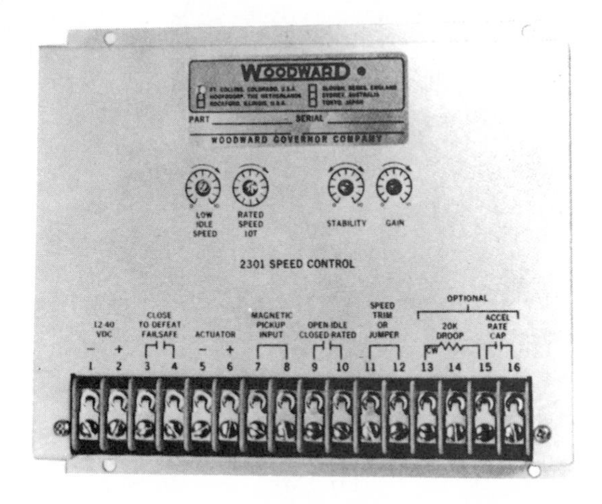

Figure 11-23
Model 2301 Electronic Control.
(Courtesy of Woodward Governor Co.)

combination governor and actuator is shown in Figure 11-25.

Governor and actuator combinations are available in work ratings to accommodate all the gas turbines and diesel engines used for power generator set service. The model EGB2P unit shown in Figure 11-25 has a work capacity of 2 ft-lb (2.711 J) and mounts on the SG, PSG, or EG3P drive pad. EGB10P and EGB35P actuators are used on UG or PG drive pads.

On diesel engines without a governor drive, the 2301 actuator can be used, which is a very economical and simple actuator that uses engine fuel pressure for an operating power input. Its work output is approximately 0.11 ft-lb (0.149 J) per 10 psi (68.95 kPa) supply pressure.

Power Generator Controls

With the increasing demand for electric utilities, many varied options are available from different manufacturers to enable automatic power plant operation. To this end, Woodward Governor Company offers combined controls and sequencers, which can handle complex sequencing and control (Figure 11-26).

While many generator sets do not require such all-encompassing sequencers, they are ideal for gas turbines, which require control of

(a)

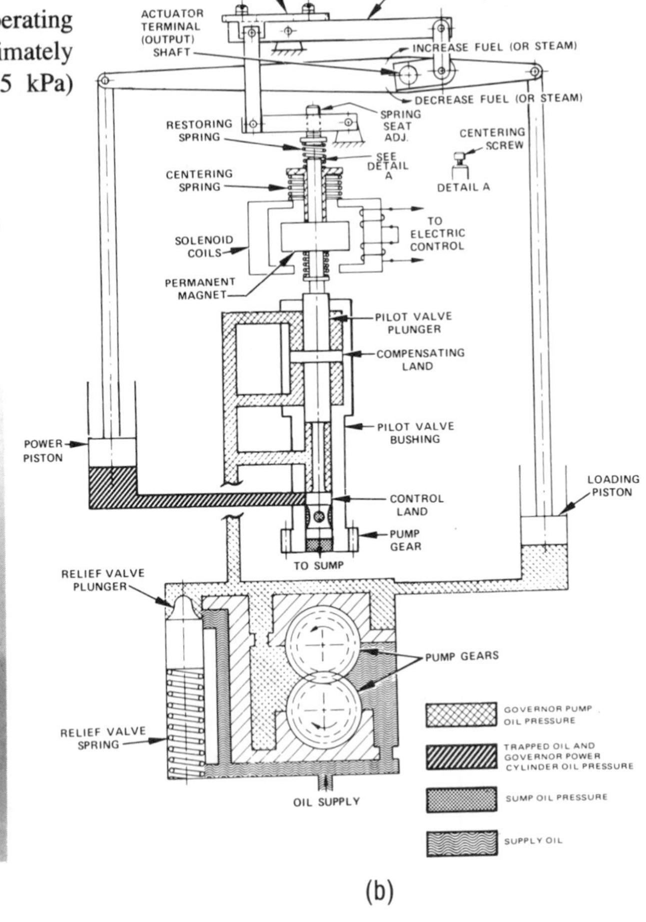

(b)

Figure 11-24
EG3P Actuator.
(Courtesy of Woodward Governor Co.)

temperature and compressor discharge pressure as well as speed and kilowatt load. The automatic synchronizing and loading of the generator requires additional control and sequencing. The model 43027 Woodward control is a modular analog system developed to meet the complex control needs of gas turbines, extraction steam turbines, and large diesel engines. Along with this unit, a microprocessed based sequencer is available to handle the logic and sequencing.

These two units, integrated into a complete control system, are an excellent team for use on complex systems. The sequencer is programmable and includes a keyboard for field changing of set points. Quick repair and system flexibility through the modular plug in design are assured.

The system can also be used on generator sets used for automatic peak shaving or cogeneration schemes, which are becoming fairly popular lately. Some of the control functions that can be assigned to such a system include load shedding, generator set call up, automatic synchronizing and loading, engine monitoring, and programmed peak shaving control.

Figure 11-25
Model EGB2P Governor and Actuator Combination.
(Courtesy of Woodward Governor Co.)

Model PGA Marine Governor

The model PGA Woodward governor is an extremely popular unit; PGA means pressure compensated air speed setting. It is available with a variety of options to suit any engine style (see Figure 11-27). The PGA marine governor can also be used on steam turbine applications. Five types of base assemblies are commonly used with the PGA governors:

1. PG standard
2. PG-UG8 standard
3. PG-UG8-90° (base rotated 90° with respect to PG-UG8 standard)
4. PG-UG40
5. PG extended square

Figure 11-26
Integrated Electronic Control and Sequencer.
(Courtesy of Woodward Governor Co.)

(a)

(b)

Figure 11-27
Model PGA Marine Governor: (a) with reciprocating power cylinder; (b) with 12-ft-lb rotary output power cylinder.
(Courtesy of Woodward Governor Co.)

The base assemblies used consist basically of the same components; however, the type of drive shaft used can be either serrated, keyed, or a special design.

If you are familiar with the operation of Woodward PSG and UG8 style governors, you should be able to readily understand the PGA operation. Basic installation, oil type, and compensation needle valve adjustments on the PGA follow the same general pattern as for other Woodward governors.

Refer to Figure 11-28 for a schematic layout of the actual governor. The governor consists of many of the same type of components as found with the UG8 and 40 governors, such as the following:

1. Two accumulators that maintain a positive working oil pressure within the main power case. If the accumulator pressure exceeds 100 psi (689.5 kPa), a pressure-reducing valve is fitted to the main power case on the basic governor. However, some particular applications can run up to 200 psi (1379 kPa).
2. Gear-type oil pump.
3. Power cylinder assembly to control the

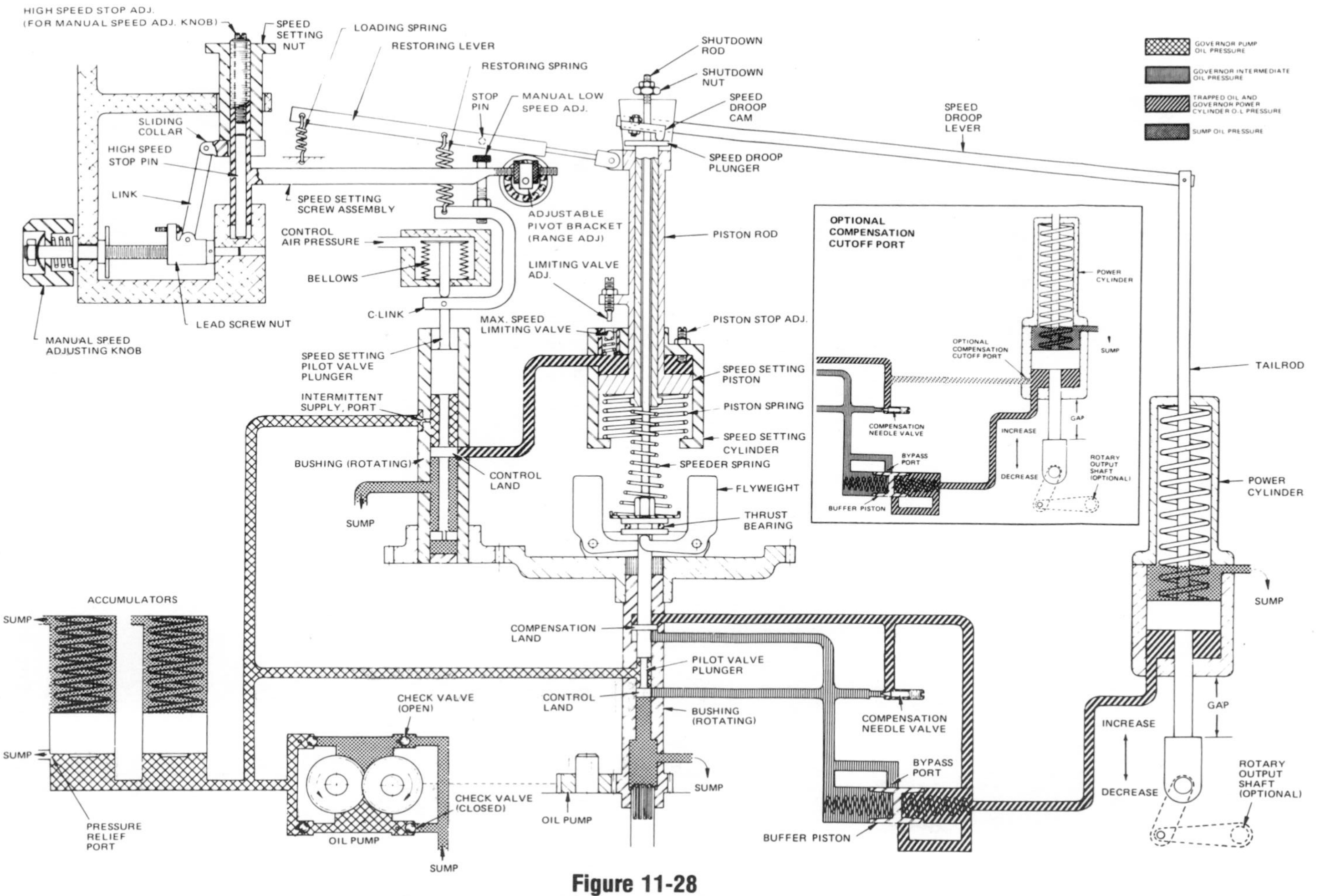

Figure 11-28
Schematic of PGA Governor with Direct Bellows.
(Courtesy of Woodward Governor Co.)

fuel racks, fuel valve, or steam valve of the engine or turbine.

4. Compensation needle valve.
5. Rotating pilot valve plunger bushing.
6. Rotating speed setting PVP bushing.
7. Buffer piston.
8. Set of flyweights and a speeder spring.
9. Pneumatic speed setting mechanism.

You will recognize many of these features from the PSG and UG8 type of governors described within this book.

The speed setting PVP is acted upon and centered by control air pressure and a restoring spring, which when equal will center the speed setting PVP. This air pressure originates from a pneumatic air transmitter or controller against a bellows, with the most commonly used air pressure range for the governor being 7 to 71 psi (48.26 to 489.54 kPa). However, the minimum control pressure should not drop below 3 psi (20.68 kPa), with the maximum being 100 psi (689.5 kPa). Woodward recommends a governor operating range of between 250 and 1000 rpm.

The PGA is capable of isochronous control or a preset droop if desired. As with the PSG and UG8 herein, duplicate suction and discharge valves at the pump allow either cw or ccw rotation; however, some governors are plugged to permit rotation in one direction only. Either a spring-driven, oil-damped, ballhead assembly is available to attenuate torsional vibrations to the governor, or a solid head can be used if the drive is relatively free of vibration.

Governor Operation

Refer to Figure 11-28; the manual speed setting knob and friction clutch can be used to adjust the engine speed setting within the normal rpm range when control air pressure is not available or not desired. When the knob is turned clockwise (speed increase), the lead screw moves out, lowering the sliding collar and allowing the loading spring to move the speed setting screw and ball bearing fulcrum down until the high speed adjusting set screw comes into contact with the high speed stop pin. When this happens, no further movement is possible, and the small friction clutch will simply slip if the knob is turned any further.

The downward movement of the speed setting screw will cause the left end of the restoring lever to be pulled down by the action of its loading spring. This lever will in turn butt up against the pneumatic low speed adjusting screw and *C-link,* which will force the speed setting PVP down, thereby uncovering the intermittent supply port of the rotating bushing to the pressurized oil flow supply from the governor oil pump.

Starting. With the engine stopped, the flyweights within the governor are tipped inward. The speed setting piston would be forced back against the piston stop adjusting screw by its return spring. The position of this screw will establish the initial tension of the governor speeder spring, which will in turn force the PVP underneath it down, thereby opening its control land to whatever oil pressure is generated by the oil pump during the cranking period.

If the engine or turbine is not set up for control air pressure to the bellows, the manual knob can be set to approximately half-load position for starting purposes. However, since air boost to the bellows can take some time, starting may be delayed until the bellows moves down, opening up the speed setting PVP.

Rotate the manual knob as stated until the engine has started, at which time it can be backed out as desired. Some engines may employ a manual override for starting purposes only. As the engine is cranked over, oil pressure will be directed to both the speed setting inlet port as well as the PVP inlet area below the weights, which are starting to move out under the influence of centrifugal force. Immediately, when the engine starts, there is a very rapid increase in the oil pump pressure and flyweight force. The oil pressure to the top of the speed setting piston will push it down, thereby increasing the speeder spring force. This downward movement of the speed setting piston also pulls the piston rod with it, which will in turn cause a cw rotation of the restoring lever. With the pneumatic low speed adjusting screw held against the stop pin of the restoring lever by the restoring spring, the speed setting PVP is lifted upward as the lever rotates, until the plunger is again centered, which will correspond to the speed position attained by the piston.

While all this is happening, oil pressure is also being directed through the PVP under the weights, which will cause the power piston within the power cylinder to be raised, thereby increasing the terminal (output) shaft position toward an increase fuel input to the engine. When the centrifugal force of the rotating governor flyweights equals the speeder spring force, the PVP, which was gradually being pulled up, will recenter, closing off any further oil flow to the power piston. Both the oil to the speed setting piston and the power piston are now trapped, which will maintain a set fuel rack position and therefore a steady engine rpm as determined by the initial setting of the speeder spring force.

Engine Running. With the engine running, the force on the speeder spring will establish the maximum rpm of the engine or turbine. The speed setting can be changed either manually by turning the speed adjusting knob or pneumatically by control air pressure acting on the bellows of the speed setting assembly, as shown in Figure 11-28, which will move the PVP downward to increase the speed.

Load Increase. With the engine running at a steady-state rpm, and either a load applied to the engine for this fixed throttle setting or the speed setting manually or pneumatically changed, the result is that the rotating governor flyweights will tend to move inward, caused by either the increased speeder spring compression or by the decrease in engine rpm.

The pilot valve plunger (PVP), being forced down, uncovers the supply port from the oil pump, which directs oil into and against the left-hand side of the buffer piston, thereby displacing it to the opposite side. The volume of oil displaced by the buffer piston will act to move the power piston up, thereby increasing the output shaft position toward the increase fuel direction. Simultaneously, the pressurized oil passing through the PVP land is directed to the bottom of the PVP compensation land. Due to the position and setting of the compensation needle valve, oil pressure is initially greater on both the left-hand side of the buffer piston and lower side of the PVP compensating land. This pressure will therefore tend to push the PVP back up, thereby assisting the rotating flyweights to attain a state-of-balance condition with the speeder spring, which recenters the PVP just before the engine or turbine has accelerated to its new position caused by the load increase. Power piston movement will stop the instant that the PVP is recentered.

The rate of oil pressure dissipation through the compensation needle valve will depend upon its adjustment; however, if properly set, this rate of dissipation will be the same or proportional to the increase in flyweight force, so that the pressure on each side of the buffer piston and PVP compensating land will be the same. When the compensation needle valve is correctly adjusted, this pressure differential on both sides will be reduced to an equal amount at the same instant that the weight and spring forces balance. If the compensation needle valve is not properly adjusted, speed overshoot or undershoot can occur, and the engine will not attain a stable balance quick enough.

If at any time a large load or speed change is made, a bypass port in the buffer piston will be uncovered, allowing pressurized oil to flow directly to the power piston. During such a condition, since pressure compensation is nonexistent, the engine may tend to overshoot or undershoot slightly prior to attaining a steady-state speed again.

With the increased air pressure to the bellows the C-link moving down will stretch the restoring spring, until its tension is then great enough to pull back the PVP and recenter it.

Load Decrease. Decreasing the load or moving the speed control to a lower position will do the same basic thing to the governor reaction. The flyweights will overcome the speeder spring force, lifting the PVP and letting trapped oil drain from the buffer piston's left-hand side first, as well as from the underside of the compensation land on the PVP. The higher initial oil pressure on the right-hand side of the buffer piston will tend to push it to the left. The resultant volume increase on the right side of the buffer piston, combined with the bleed back through the needle valve, will allow the power piston to be forced down by its spring to an engine fuel decrease situation. This higher initial oil pressure above the PVP compensating land tends to push the PVP down, which will assist the speeder spring in recentering the PVP, again, just before the engine or turbine has decelerated to its new setting.

Compensation dissipation through the needle valve follows the same pattern as described under a load increase situation. An optional compensation cutoff port located in the power cylinder wall blocks the compensation oil passage to the needle valve, which prevents normal equalization of compensation pressures. This occurs as a result of large changes in the speed setting or large load decreases, which will move the power piston initially to a no-fuel position.

With the buffer piston forced off center, pressure transmitted to the upper side of the PVP compensation land will increase, which, added to the speeder spring force, tends to temporarily increase the governor speed setting. The instant, however, that the engine drops below the temporary speed setting the governor initiates a corrective reaction in order to stop a large transient underspeed.

Compensation will resume, and engine speed will stabilize when the power piston moves up enough to open up the compensation passage. When a compensation cutoff port is used, adjust the governor-to-fuel rack linkage so that the power piston gap does not exceed $\frac{1}{32}$ in. (0.793 mm) when running at an idle rpm no load or less than 4 degrees from minimum fuel.

Speed Setting Piston Movement. The rate of oil flow into the speed setting piston cylinder (SSPC) is controlled by the diameter of the orifice within the rotating bushing, which aligns with the oil supply port only once every revolution. Varying the size of the orifice can alter the rate of speed setting piston movement from idle to maximum speed, anywhere from 1 to 50 seconds.

Turbosupercharged engines require some time to allow their supercharger to accelerate with the rpm increase of the engine, and conversely require slower engine deceleration to minimize and prevent compressor surge. This timing can be within a nominal range of 1 to 15 seconds; however, the SSPVP has an additional land that covers the drain port of the bushing. A vertical slot in the drain land registers with a second orifice of the rotating bushing once per revolution, thereby also limiting its rate of flow to drain. The slot width of the drain land controls the opening time of the drain port and therefore the deceleration time involved.

Turning the manual speed adjusting knob ccw will decrease the speed setting of the engine by allowing the lead screw to move in, raising the sliding collar. This lifts the speed setting screw and ball bearing fulcrum, the left end of the restoring lever, and the SSPVP, draining oil from the SS cylinder and allowing it to move up under its spring force, which will decrease the speeder spring force. Counterclockwise restoring lever movement will recenter the plunger once the piston attains its new lower speed.

The distance between the ball bearing fulcrum and the point at which the restoring lever is attached to the piston rod will vary the ratio of change of the restoring spring force for a given movement of the speed setting piston. If this distance is shortened, then for a given control air pressure, it will decrease the governor speed range, whereas lengthening this distance will increase the governor speed range for a given air pressure.

Pneumatic Low Speed Adjusting Screw. This screw can be set to come into contact with the stop pin of the restoring lever or adjusted with a definite clearance. Setting with a clearance would be found on applications where, if the governor is adjusted to stop the engine upon loss of control air pressure, the PVP would be raised, draining oil from the SSPC. As the piston moves up, the restoring lever does not tend to recenter the SSPVP.

On other applications, as the speed setting piston moves upward due to control air pressure dropping below the minimum, or because intentional or accidental interruption of control air pressure occurs, the stop pin in the restoring lever simultaneously will push down on the pneumatic low speed adjusting screw, which will recenter the SSPVP when the piston reaches its low speed position. These engines are generally equipped with an auxiliary shutdown device.

Normal Engine Shutdown. A shutdown rod running through the center of the speed setting piston rod is also attached to the top of the main PVP. When control air pressure is turned off, the state of balance between the bellows and restoring spring is upset in favor of the spring, which raises the speed setting PVP, allowing oil to dump from the SSPC. The return spring under the SSP forces it back past its normal low speed position. After an additional $\frac{1}{16}$-in. (1.587-mm) movement, the fulcrum block at the top end of the piston rod makes contact with the lower shutdown nut, which lifts the rod and PVP. Oil will now drain from the base of the power cylinder, allowing movement of the fuel control linkage to the zero position.

On certain applications, an external method of shutdown may be required, since the speed setting piston stop is used as a positive low speed stop; therefore, the rod shutdown nuts are generally omitted.

Maximum Speed Limiting Valve. Located in the top of the speed setting cylinder is the maximum speed limiting valve, which can be adjusted by a screw mounted on the piston rod of the speed setting piston. When the speed piston is at its maximum position, approximately 5 rpm later, the piston will unseat the valve and drain oil from the speed piston cylinder to the oil sump. This condition occurs whether the engine speed setting is changed pneumatically or manually.

Piston Stop Set Screw. This screw passes through the speed setting cylinder and therefore limits the upward travel of the piston to $\frac{3}{32}$ in. (2.38 mm) above the normal low speed setting position of the piston. This has the advantage of reducing starting time when cranking, since a lower volume of oil is needed to force the piston down to its low speed position.

Temperature Compensation. As with some other models of Woodward governors, a bimetal strip in the restoring lever of some earlier model PGAs compensated for changes in oil temperatures. However, in current governors a temperature-compensated speeder spring is used instead, since the governor speed settings are more stable with less tendency to drift, either by operating temperature or ambient changes.

Speed Droop Control Linkage. Speed droop is normally expressed as the difference in engine rpm from the no-load to full-load speeds, and is given as a percentage of the maximum rated speed.

Attached to the upper end of the speed setting piston rod is a fulcrum block, with a lever and fulcrum pin assembly connected between the block and power piston tail rod. There is also an adjustable cam attached to this fulcrum pin and a movable plunger within the SSP rod. Any movement of the power piston causes a reaction of the speed droop lever, which causes the cam to rotate and contact the top of the plunger; this will in turn move the plunger resting on top of the speeder spring up or down. The amount of plunger movement is proportional to the actual cam lobe location with respect to the fulcrum pin centerline.

For a zero droop setting, both the pin and cam lobe must be in the same centerline. If, however, a certain droop is required, the cam lobe must be located at a predetermined distance from the fulcrum pin centerline.

Note: The cam should never be positioned on the opposte side of the fulcrum pin centerline (toward the pneumatic receiver); otherwise, negative droop will occur, resulting in unstable engine or turbine operation.

PGA Adjustments

If any air is suspected of being trapped within the housing at any time, especially after changing the oil or repairing the governor, loosen the air vent plug on the side of the governor base as shown in Figure 11-26. The compensating needle valve should be all the way back to about two turns open, which will cause the engine to hunt while running. After no more bubbles are evident, tighten the vent plug and gently close the compensating needle valve until the engine runs without any hunt or surge.

The needle valve should be kept open as far as possible to prevent sluggish governor response. This setting will of course vary with each particular engine; however, it can range from one-sixteenth of a turn to as much as two full turns open. If, however, the governor is equipped with optional buffer piston springs, the needle valve will not require more than one-sixteenth of a turn open to ensure smooth operation of the engine. As with all Woodward hydraulic governors employing needle compensation adjustment, do not close the needle valve fully; otherwise, poor governor action will result.

Due to the wide range of options available with the PGA governor, if when opening the needle valve the engine does not hunt, leave the air vent plug open, and manually increase the speed setting to cause the power piston to move to full fuel several times and all the way back to idle speed; then close the vent screw.

It may be necessary on buffer-spring-type engines to replace the springs with the next higher ones available, if with the needle valve almost closed, the engine just will not stay at a stable position when running.

Speed Setting Adjustment. Refer to Figure 11-28. Many available options are used on the PGA governor; it is not intended to duplicate Woodward Governor Company Service Bulletin Number 36604, which deals in detail with the overhaul, maintenance, and all the adjustments to the governor and its accessories. However, to set the governor's maximum and minimum operating speeds, proceed as follows:

1. If the governor can be removed and placed on a test stand, this is the preferred method.
2. Follow the sequence given; otherwise, satisfactory results will not be obtained.
3. Regardless of whether the adjustments are done on the engine or on a test stand, the governor must be at its normal operating temperature to achieve proper settings.
4. If a shutdown solenoid is fitted, it must be energized, if it is set to shut down the engine when de-energized.
5. If a pressure-type shutdown unit is used, pressurize it above the shutdown point if it is adjusted to shut down the engine upon low oil pressure.
6. Manually turn the speed adjusting knob ccw until you feel the small friction clutch slip (minimum speed position).
7. Refer to Figure 11-28 and adjust the high speed set screw until its upper end is flush with the top of the T-shaped speed setting screw.
8. Adjust the speed piston stop screw $\frac{1}{2}$ in. (12.7 mm) above the top of the speed setting cylinder.
9. Proceed to set the governor low speed setting as follows:
   a. Direct control air pressure corresponding to the desired idle speed to the governor.
   b. Turn the speed setting nut on the speed setting screw ccw until the desired low speed is reached with the minimum control air pressure. Be very sure that the pneumatic speed adjusting screw is not contacting the stop pin of the restoring lever, and that the piston stop set screw is not interfering in any way with the speed setting piston's upward movement.

To calibrate the governor's speed range to the control air pressure range, proceed as follows:

1. Be careful not to overspeed the engine or turbine during this time.
2. Gradually increase the control air pressure to the required maximum value.
3. Take very careful note now that the limiting check valve in the top of the speed setting cylinder is not unseating owing to the maximum speed limiting valve adjusting screw contacting it.
4. The ball bearing pivot bracket can be adjusted by loosening the top screw in the pivot bracket on the arm of the speed setting screw. By now loosening the knurled nut on the side of the bracket and tightening the other nut, the following results can be obtained:
   a. If the engine attains its maximum desired rpm before the control air pressure reaches its maximum value, adjust the ball bearing pivot bracket fulcrum by moving it toward the speed setting cylinder. This will proportionately decrease the engine speed in relation to air pressure.
   b. If, however, the maximum air pressure is attained prior to the engine reaching its desired maximum rpm, move the ball bearing pivot away from the speed setting cylinder.
5. The low speed setting always requires readjustment after the ball bearing pivot bracket has been moved as above. Therefore, repeat step 9 under the low speed setting and steps 2 through 4 above until the required low speed corresponds with the required minimum control air pressure, and both high speed and maximum control pressure are attained simultaneously.
6. With the engine running (or governor on a test stand) at its maximum rpm/air control pressure, allow the engine speed to

stabilize (Figure 11-28). Gently turn the limiting valve adjusting screw cw until the engine rpm just starts to drop. Take careful note of the screw location here; then back it out one-quarter to one-half turn and lock it. This adjustment controls the speed setting piston stroke within its cylinder, and therefore prevents possible accidental engine overspeed.

7. Reduce the control air pressure to its minimum value.
8. Turn the piston stop setscrew cw until a step up in effort is felt as it just makes contact with the top of the speed setting piston in its bore; then back it out three full turns, which should be approximately $\frac{3}{32}$ in. (2.38) mm), and lock it in this position. (This adjustment opens the fuel control during cranking to minimize the time required to start the engine.) Some applications, however, require an external means of shutting down the engine, where the piston stop setscrew is used to limit the upward movement of the piston at the low or minimum speed point.

Figure 11-29
Left-Side View of Governor with Cover Removed.
(Courtesy of Woodward Governor Co.)

9. The shutdown nuts located at the top of the shutdown rod must be properly adjusted with the engine running or the governor on a test stand by very carefully lifting the shutdown rod up far enough just to remove the end play (you will feel a step up in effort when this is reached) (Figure 11-29). If the engine speed drops below its low speed setting, you have gone too far; therefore, repeat the process. With the rod held up, adjust the shutdown nut (Figure 11-29) to obtain 0.050 in. (1.27 mm) clearance above the fulcrum block upper end on the speed setting rod; then lock it in position. If shutdown capability is not required through the shutdown nut, make certain that the nuts are adjusted right to the top of the rod.
10. Adjustment of the pneumatic low speed stop screw to place it 0.040 to 0.050 in. (1.016 to 1.27 mm) below the restoring lever stop pin at low speed is required if the governor is used to stop the engine when the control air pressure is turned off or accidentally interrupted. Turn off the control air to stop the engine, and readjust the stop screw for 0.002 to 0.005 in. (0.0508 to 0.127 mm) clearance between the head of the screw and restoring lever stop pin.
11. If it is desired simply to have the engine return to its low idle rpm when the control air is turned off, adjust the pneumatic low speed stop screw until the desired rpm is attained with no control air pressure. This is generally about 20 rpm below idle.
12. Adjust the maximum speed setting for the manual speed setting knob as follows:
    a. Turn off control air to the governor.
    b. If, however, the governor is adjusted to stop the engine with interruption or air loss, turn the speed adjusting knob cw until the engine or turbine rpm increases slightly before turning off the air supply.

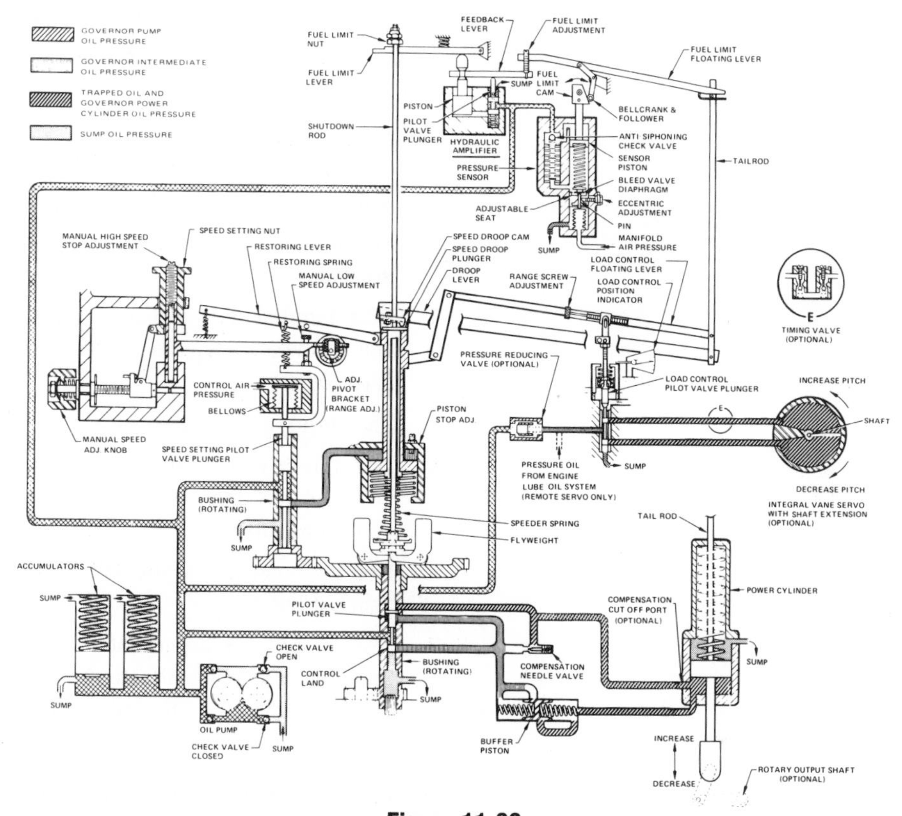

Figure 11-30
Schematic of Fuel Limiter, Optional Load Control Override Linkage, and Vane Servo. (Courtesy of Woodward Governor Co.)

c. Turn the speed knob cw to place the engine while running at its high speed position.

d. Adjust the high speed adjusting set screw in the speed adjusting screw cw until it just makes contact with the high speed stop pin. Be careful that you do not run it in too far; otherwise, engine rpm will decrease.

13. Do not forget to turn the manual speed adjusting knob fully ccw to the minimum speed position before resuming normal engine operation by pneumatic control again.

Load Pitch Control and Manifold Pressure Fuel Limiter

The load pitch control system operates through a vane servo that is connected to a transmitter, which in turn receives a signal from the vane servo, and relays it to the pitch control unit to change propellor pitch and therefore the engine load. The load control pilot valve and governor linkage controls the vane servo. (See Figure 11-30.) The function of the fuel limiter is to prevent overfueling of the engine due to turbo-supercharger lag during acceleration due to load increase. It functions similar to an *aneroid* control used on smaller high speed diesels.

chapter 12 General Troubleshooting

The ability to be able to effectively and quickly diagnose an engine problem is basically related to the following:

1. A thorough understanding of the basic fundamentals of what actually goes on within an internal combustion engine.
2. The amount of experience of the mechanic involved.
3. How familiar the mechanic is with a particular make of engine; also, how up to date you are.
4. The ability to be analytical.
5. The ability to control one's temper when an irrate customer or operator is pushing you for an answer.
6. Don't second guess yourself; if in doubt, check it out.
7. If necessary, do not hesitate to refer to the manufacturer's specifications or troubleshooting charts in the respective engine service manual.

People often make reference to someone as being a really good mechanic. How do you think that person was able to achieve such respect? Obviously, experience is in many instances a series of a great many mistakes that were learned on the way up through the apprenticeship stage and early journeyman years. But one must have a genuine desire to want to succeed and be nothing less than the best in the field of diesel mechanics. Certainly, in this ever-changing technological era, and especially with high labor costs and overhead, it is easy to become simply the proverbial parts replacer instead of a highly skilled and dedicated craftsman. Company policy can, of course, dictate to what degree they desire their maintenance personnel to replace component parts.

In many instances a new part may be required; however, there are many, many instances when a new component part is installed with the assumption that this will in fact cure the problem at hand. Within a short time, the same problem exists, leading you to scratch your head and ask why.

Don't accept at face value, unless a part shows particular excessive wear or damage, that it is nonserviceable. Learn to accept where possible nothing less than the best; place yourself in the position on every job that you do that the engine or equipment is your own. It is hard work to stay abreast of the many changes that are constantly going on in this business, and that is one of the reasons that it is such a challenge to you as an individual and as a skilled craftsman.

People will remember your abilities as a first-class mechanic only as long as you continue to produce first-class work. Foul up once, and that is the last job that stays in their minds, regardless of how many jobs you completed successfully for them previously.

With these few considerations and thoughts, tackle a troubleshooting problem with an open and keen mind. Don't panic, take it easy, and eventually you will find that the majority of problems are in many instances of a minor nature. Walk before you run.

The problems that can relate to the fuel system are obviously somewhat diversified in nature. The method that you choose to effectively pinpoint the particular problem will be dependent upon how familiar you are with the make of engine. However, if you systematically collect all the information as to what led up to the problem at hand, you should be able to analyze on a step-by-step basis the reason for the existing problem. Remember, satisfactory operation of the engine is primarily dependent upon the following items:

1. An adequate supply of clean, relatively cool air, which once in the cylinder can be compressed to a high enough pressure to effect proper combustion.
2. The injection of the correct amount of fuel at the proper time during the compression stroke.
3. Use of the proper grade of fuel for the environment in which the engine operates.
4. The ability to maintain the fuel oil if possible at an optimum temperature range of 90 to 95°F (32 to 36°C) for high speed diesel operation (maximum allowable of 150°F (65°C).
5. Clean, filtered water and sediment-free fuel.
6. Maintenance of the proper engine water temperature. Most high speed diesel engines operate between 165 and 190°F or 73 and 88°C. Satisfactory water treatment.
7. Maintenance of exhaust back pressure within specifications.
8. Use of the proper grade of oil with proper service intervals.
9. Proper selection and application of the engine for what it was intended.

When collecting information prior to analyzing a problem, keep an open mind. There are always those who are more than ready to tell you what the problem is. Listen to their suggestions by all means, but remember that it is you who are the trained and skilled mechanic; therefore, proceed with this in mind. It is very easy to become side-tracked and led into believing that what an operator happens to say is in fact the cure for the problem at hand. Maybe it is! However, think for a minute before jumping in.

In these days of high labor costs, it is worth every penny that the customer pays if, after 5 or 10 minutes of basic checks and collecting your thoughts, you are able to effectively arrive at a solution to the problem, rather than going off on a haphazard guess and idea that you may just be lucky enough to stumble upon the problem. With the high costs involved in purchasing equipment, most companies have a reasonably good maintenance program that in most instances reflects itself in minimum engine failures and downtime. When a problem occurs then, you will find that it is of a minor nature in very many instances. Therefore, don't automatically suspect and look for a major reason for failure prior to considering some of the foregoing thoughts.

There are many types and shapes of internal combustion engines on the world market today found in a wide variety of applications. Obviously, the more familiar you are with a particular make of engine, the more confident you will be when faced with a problem on that engine. However, if you take into account the seven items listed earlier, you should have little trouble in effectively finding the problem. After all, every engine has pistons, crankshafts, and pumps and injectors, which are all very similar in basic operation.

With this in mind, let us consider some of the readily available aids that we can use to assist us in troubleshooting an engine. Whether the engine is a 2- or 4-cycle matters little, since the majority of problems that you will come across relate to both types. Of course there are certain problems that can be peculiar to one type, but let's leave that until later.

Most manufacturers supply special tools that makes working on their engine that much easier; however, you can in most troubleshooting situations resort to the use of a few locally and readily available tools to reach the same conclusions.

12-1

Manometers

One of the most effective troubleshooting tools that you can use is a set of manometers, one a water type and the other of mercury. These can be bought either in a solid tube fashion mounted on a stand or cabinet fixture, which is usually more common in a shop setup. However, many mechanics prefer to use what is known as a *slack tube* type of manometer, which is a clear heavy plastic type of tube. It is less susceptible to breakage and is easily packed around in a tool box or service truck. Regardless of the type used, they will both perform the same function. Both are known as U-tube manometers because of their shape and are available in sizes of 12 and 18 in. (30.48 and 45.72 cm), with longer units available if required.

Purpose of Manometers

Manometers measure either a pressure or vacuum reading on the engine. This is done through a scale connected to the manometer, as

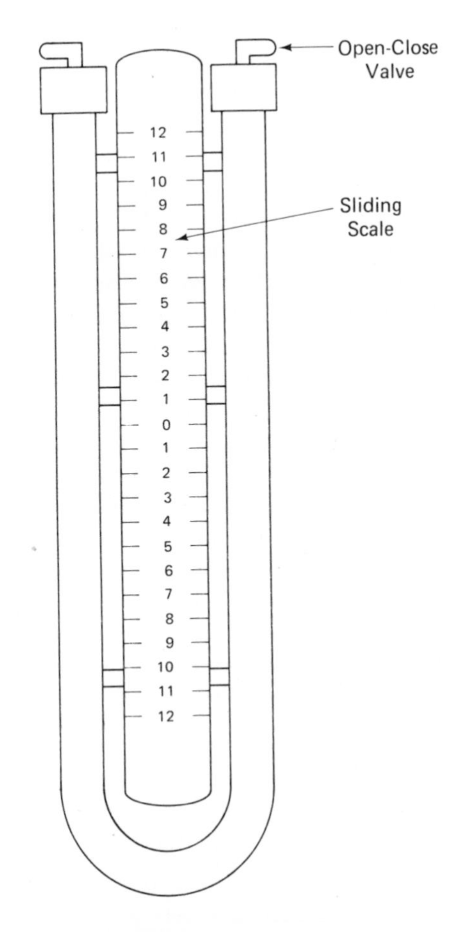

Figure 12-1
Manometer with Sliding Scale.

PRESSURE CONVERSION CHART

1 in. of water (2.54 cm)	= 0.0735 in. of mercury (1.866 mm)
1 in. of water (2.54 cm)	= 0.0361 psi (0.248 kPa)
1 in. of mercury (2.54 cm)	= 0.491 psi (3.385 kPa)
1 in. of mercury (2.54 cm)	= 13.6 in. of water (34.54 cm)
1 psi (6.895 kPa)	= 27.7 in. of water (70.358 cm)
1 psi (6.895 kPa)	= 2.036 in. of mercury (5.17 cm)

shown in Figure 12-1, which can be adjusted prior to its use to a zero position. The scale is calibrated in either English or metric or a combination of both. The scale reflects water or mercury displacement within the U-shaped tube in either inches or millimeters. The majority of engine manufacturers list relative specifications in their respective engine service manuals for the particular test that you wish to take. A typical pressure conversion chart is given in this section for converting from inches or millimeters to either pounds per square inch (kilopascals) or back and forth between water and mercury.

Manometers can be used for the following purposes on a diesel engine:

1. To measure air inlet restriction (AIR).
2. To measure air box pressure (ABP) on 2-cycle engines or turbocharger discharge pressure on 2- or 4-cycle engines.
3. To measure crankcase pressure.
4. To measure exhaust back pressure (EBP).

Prior to looking at each one of these possible tests, let us study exactly how to use the manometer.

Proper Use of a Manometer

Water manometers usually have a small amount of colored dye added to the water prior to filling the tube, which makes it easy to read the level of fluid displacement within the tube. On mercury manometers it is sometimes advantageous to hold a bright light behind the tube, which makes reading easier, since mercury can tend to discolor the tube after long use and by some contamination during exhaust back pressure checks. Some manometers are calibrated in such a manner that their sliding scale reads on only one side, whereas others are divided equally on both sides for the purpose of reading the displacement per side. Usually, the slack tube type has the scale calibrated on both sides; therefore, when a reading is taken, *both sides* must be added together to obtain the pressure or vacuum reading. The reason for this is that, with the slack tube type being of pliable heavy plastic, the internal bore may vary slightly. This can cause the column of fluid to move farther on one side than the other; therefore, total displacement is very important.

The fixed type of solid tube manometer

PRESSURE OR STRESS CONVERSION (ENGLISH TO METRIC)

1 inch of mercury	= 3.377 kilopascals (kPa)
1 inch of water	= 0.2491 kPa
1 pound/square inch	= 6.895 kPa

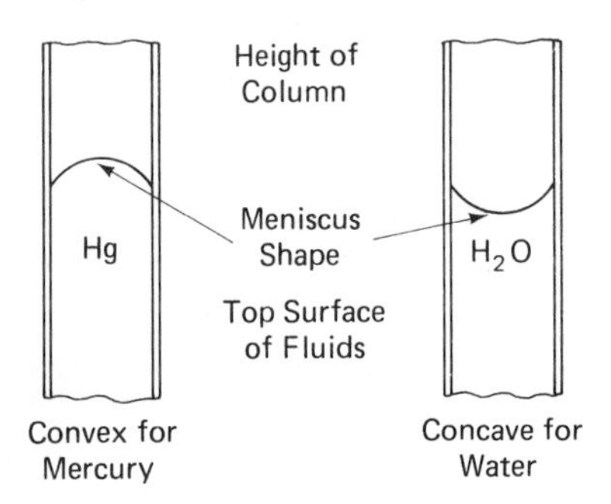

Figure 12-2
Comparison of Column Heights for Mercury and Water Manometers.

sometimes has the scale graduated so that ½ in. (12.7 mm) on the scale is actually representative of 1 in. on a slack tube unit. Therefore, if the solid tube is scaled on only one side in this fashion, only one side need be read.

Figure 12-2 shows that water will wet the inside of the manometer tube; therefore, when reading this manometer, sight horizontally between the bottom of the concave water surface and the scale. However, mercury, being 13.6 times heavier than water, does not wet the inside of the tube; therefore, when reading this manometer, sight horizontally between the top of the convex mercury surface and the scale.

Use of the Manometer

1. Open both valves at the top approximately one turn.
2. Allow the internal fluid to settle; possibly shake the manometer slightly. Sighting the fluid level horizontally, move the sliding scale until the zero on the scale is level with the miniscus shape as in Figure 12-2. This is known as zeroing in the manometer.
3. Connect a rubber tube or hose that will fit fairly tightly over both the manometer valve and engine measurement point.
4. On a manometer scaled on both sides up and down the scale, it does not matter to which side you connect the hose. However, if the scale is marked up only one side, connect the tube or hose to the valve on that side.
5. Run the engine at the specified rpm and note the displacement of the fluid as per specifications given by the engine manufacturer.

Measuring Air Inlet Restriction

On any engine the purpose of this check is to establish how much actual restriction there is to air flow. This can establish just how efficiently the air intake system design is. The freer breathing the engine is, the lower will be the measured restriction to air flow, meaning longer intervals between air cleaner servicing and overall higher thermal efficiency of the engine if it can operate at this low restriction range the majority of the time. The air flow requirements for a particular type and size of engine can be readily determined from the charts shown in Figures 12-3 and 12-4.

Many complaints of a lack of horsepower, smoking exhaust, increased exhaust temperatures, and poor fuel economy can be directly attributed to nothing more than high air inlet restriction. Therefore, the size and type of air cleaner unit, duct work size diameter, length of duct work, and the number of bends and elbows all add to the overall restriction of an air intake system. The shortest, most direct setup is the best; however, due to the many varied types of diesel engine applications, this is not always possible, but you should try to minimize this restriction.

On a nonturbocharged or naturally aspirated engine, the manometer is measuring the air pressure within the intake manifold in relation to atmospheric pressure. On turbocharged engines it measures the same thing, but on the suction side of the turbocharger only.

The term *volumetric efficiency* (VE) is used to specify just how efficiently the engine consumes or takes in air. It is the difference in the weight or volume of air contained in the cylinder with the piston at the bottom of its stroke with the engine stopped, which is atmospheric pressure of 14.7 psi (101.356 kPa), compared to the weight or volume of air in the cylinder with the piston at the bottom of its stroke with the engine running. This usually runs between 85 and 90% on naturally aspirated engines and is as much as 1.3 to 2 times atmospheric on turbocharged and supercharged engines. Thus, the higher the volumetric efficiency, the better the engine thermal efficiency.

On small engines using direct-mounted air cleaners, restriction is usually only caused by plugging of the air filter. However, on larger engines, especially the high horsepower, high speed engines, air cleaner design and the like can all come into play, including location of the filter itself.

The tables in Figures 12-3 and 12-4 are

4-CYCLE ENGINES

CUBIC INCH DISPLACEMENT	ENGINE RPM							
	1250	1500	1750	2000	2250	2500	2750	3000
100	28	35	41	46	52	58	64	69
120	35	42	49	56	63	69	75	84
140	41	49	57	65	73	81	89	97
160	46	56	65	74	83	93	102	111
180	52	63	73	83	94	104	114	125
200	58	69	81	93	104	116	127	139
220	64	76	89	102	114	127	140	153
240	70	84	98	111	125	139	153	167
260	75	90	105	120	135	150	165	180
280	82	98	114	130	146	162	176	194
300	87	104	121	139	156	173	191	208
320	92	111	130	148	167	185	204	222
340	98	118	138	158	177	197	217	236
360	104	125	146	167	187	208	229	250
380	110	132	154	176	198	220	242	264
400	115	139	162	185	208	232	255	278
420	121	146	170	194	218	244	268	292
440	127	152	178	202	229	255	280	305
460	133	160	186	213	240	266	293	320
480	138	166	194	221	250	278	306	334
500	144	173	202	231	260	289	318	347
520	150	180	210	240	270	300	330	360
540	156	187	218	250	281	312	344	375
560	162	194	226	259	292	324	356	389
580	168	201	235	269	302	335	369	401
600	174	208	244	277	312	346	382	416
620	179	215	251	287	323	359	395	431
640	185	222	259	296	334	371	408	435
660	191	229	267	305	343	382	420	458
680	197	237	276	315	354	394	434	473
700	201	243	284	325	365	406	445	486
720	209	251	292	334	375	417	459	500
740	214	257	300	343	385	429	472	514
760	219	264	308	352	396	440	484	528
780	226	272	316	361	408	453	497	542
800	232	278	324	371	418	484	510	556
850	246	296	345	384	444	493	542	591
900	260	312	365	417	469	521	574	625
950	275	330	385	440	495	550	605	660
1000	280	348	406	463	521	580	638	695
1050	302	362	425	487	547	608	669	730
1100	319	382	446	510	574	638	702	765
1150	333	400	467	534	600	667	734	800
1200	348	418	487	557	626	696	765	835
1250	362	485	587	579	632	724	796	868
1300	376	453	528	609	678	754	829	904
1350	390	469	548	625	704	782	860	938

Figure 12-3

Air Flow Requirements for Four-Cycle Engines.
(Courtesy of Farr Air Cleaner Co.)

used for approximate sizing only and are based on a volumetric efficiency of 0.80; therefore, engines operating at higher VE will require different air flows. If the engine uses a supercharger, multiply the cubic feet per minute (cfm) requirements as shown for your engine displacement by 1.6; if it is turbocharged, multiply the cfm requirements as indicated in the tables by 2.

If the 2-cycle engine is also turbocharged, as are many of the newer 53, 71, 92, and 149 Detroit Diesel Allison engines, multiply the cfm requirements for your engine by 1.3. If your engine is of larger displacement than shown in these tables, double or triple the reading given. Always choose the next larger displacement shown if your size is in between those shown.

An air inlet restriction check should take into account the following two major considerations:

1. The restriction imposed on the system by the air cleaner and element.
2. The restriction imposed on the system by the piping or associated ductwork to the engine itself.

To check the first condition, you will find that the majority of heavy-duty diesel air cleaners incorporate a small plug screwed into the outlet to the engine. This plug can be removed and a suitable small fitting installed in its place to

2-CYCLE ENGINES

CUBIC INCH DISPLACEMENT	ENGINE RPM							
	1250	1500	1750	2000	2250	2500	2750	3000
100	94	112	132	150	170	188	206	226
120	112	136	158	180	203	225	228	270
140	132	158	184	211	238	264	290	316
160	150	180	210	240	270	300	330	360
180	170	203	237	270	304	348	372	406
200	188	226	264	301	339	376	414	452
220	208	248	289	330	372	414	456	496
240	226	270	316	361	406	451	496	540
260	244	293	342	391	440	489	538	586
280	264	316	368	421	474	527	580	632
300	282	338	395	451	508	564	620	678
320	300	360	420	481	542	602	660	720
340	320	384	448	512	576	640	704	768
360	338	406	474	542	610	678	746	812
380	358	428	500	572	644	716	788	858
400	376	452	526	604	678	753	828	904
420	394	473	553	632	710	790	868	946
440	414	496	579	663	745	827	910	992
460	432	520	606	691	779	865	952	1038
480	450	542	632	722	813	903	994	1083
500	466	564	658	752	846	940	1034	1128
520	484	586	685	782	880	978	1076	1174
540	508	608	711	812	913	1016	1117	1219
560	526	632	737	842	948	1053	1158	1264
580	546	654	764	874	981	1090	1200	1309
600	564	677	789	903	1016	1128	1241	1354
620	682	700	816	933	1049	1166	1282	1399
640	602	722	842	964	1083	1204	1324	1444
660	620	744	869	992	1117	1241	1365	1490
680	638	768	895	1023	1151	1279	1407	1535
700	658	790	923	1053	1185	1316	1448	1580
720	678	814	948	1083	1219	1354	1489	1625
740	696	836	974	1113	1253	1392	1531	1670
760	714	856	1001	1143	1286	1429	1572	1715
780	734	880	1027	1174	1320	1467	1614	1760
800	752	904	1053	1204	1354	1505	1655	1806

Figure 12-4
Air Flow Requirements for Two-Cycle Engines.
(Courtesy of Farr Air Cleaner Co.)

connect to the manometer (water). Often you may find a small bore tube running from this up to an air cleaner restriction service indicator on the dash of heavy equipment. Or you may find a direct-mounted restriction indicator on the air cleaner. These indicators are set to show a red area in a small clear window when the system restriction reaches 25 in. of water. Although a higher restriction is available in these gauges, 25 in. (5.08 cm) is the acceptable maximum restriction. However, many of the current turbocharged and after-cooled constant horsepower or high-torque rise diesel engines are limited to 20 in. Hg restriction or 50.8 cm.

To check total system restriction, a small plug can be removed from the air inlet manifold on nonturbocharged engines several inches from the air inlet housing or manifold. However, on turbocharged engines, a plug is usually located 4 to 6 in. (10.16 to 15.24 cm) away from the suction side of the turbocharger; otherwise, you would be obtaining a pressure reading if it was taken on the turbocharger outlet side.

Open the valves on top of the manometer, zero in the scale, and connect a tube or hose between the fitting in the air inlet and the manometer valve. Start and run the engine at the recommended engine manufacturer's rpm, and carefully read and record the water displacement within the manometer. Remember that you have to add both sides of the scale together, unless it is graduated on only one side for a total reading. Compare your readings to specifications to determine the condition of the air system.

Causes of High Air Inlet Restriction

1. Plugged air filter.
2. Too small an air filter.
3. Crushed inlet piping or ductwork.
4. Inlet piping diameter too small.
5. Inlet piping too long.
6. Too many bends or elbows in the piping to the engine.
7. Plugged precleaner on heavy-duty units.
8. Plugged turbocharger inlet screen if used.

An additional check that should be done on the air intake system is to start and run the engine at an idle rpm. Take a pressurized can of ether starting fluid and spray around each individual joint of the inlet piping. If there is a slight surge or increase in the engine idle speed, it signifies that the engine is sucking air at this point, which means that unfiltered air is entering the system.

Nothing will wear piston rings faster than unfiltered air since it mixes with the cylinder lube oil and forms an excellent lapping compound. If you do not have a can of starting fluid, seal off the inlet piping at both ends and apply about 2 psi (13.79 kPa) of pressure to it while coating the outside joints with a solution of soapsuds.

Some air filter manufacturers can supply a small dust sight gauge, which can be inserted into the inlet piping. It has a small tab that projects into the air stream of the piping visible through a clear plastic inspection window. If any unfiltered air enters the system, dust or dirt will be deflected by the tab against the clear inspection window, causing it to cloud over. You can also shine a flashlight into the window to inspect the internal cleanliness of the system at any time.

As little as two tablespoons of fine dust can reduce the compression in an engine to the point that it requires new piston rings. Thus, it is easy to see why the air inlet system is as important as the actual fuel system to efficient operation of the engine.

Air Box Pressure and/or Turbocharger Boost Pressure

Many engines today employ exhaust-driven turbochargers that draw air into the cold end (air inlet), compress it to a point in excess of atmospheric, and force it on into the engine. Depending on just how much increase there is in pressure, many of these engines first send the air through what are known as *aftercoolers* or *intercoolers*. These devices can be either water or air cooled, which reduces the temperature of the air being supplied to the engine to keep it as dense as possible in order to extract as much energy as possible from it.

Many 2-cycle engines employ both a gear-driven engine blower and an exhaust-gas-driven turbocharger unit. This supplies a tremendous amount of air to the engine for the following purposes:

1. Combustion within the engine cylinder.
2. Scavenging of the burnt gases from the cylinder.
3. Cooling of the cylinder components.
4. Crankcase ventilation.

Regardless of whether the engine is a 2- or 4-cycle unit, the turbocharger's air flow is directly related to engine load and speed, since there is no direct mechanical link between the engine and the turbocharger. As a consequence, the turbocharger will accelerate and decelerate with the engine operation.

On 4-cycle engines, the turbocharger boost pressure is taken by removing a small pipe plug from the intake manifold. Since this pressure boost can extend up to 14 to 20 psi (96.53 to 137.9 kPa), a mercury manometer or suitable type of gauge is required. After removing the plug and zeroing in the manometer, install a fitting into the manifold, couple up a hose or tube between them, and start and run the engine. This must be done while the engine is under load, as this is when the turbocharger will receive its greatest flow of exhaust gases and therefore produce its highest delivery pressure. Note and compare your readings with those given in the engine manufacturer's service manual.

On 2-cycle engines the turbocharger delivers its air to an engine-driven blower and then into the engine; therefore, both the individual turbocharger pressure and overall air box pressure can be checked if desired. However, most manufacturers simply list a minimum acceptable air box pressure. On Detroit Diesel type 2-cycle engines, this is taken by attaching a rubber hose or tube over one of the air box drain tubes (see this chapter).

On other 2-cycle engines there is a fitting that can be removed from the air manifold area for the same basic purpose. On the outlet side of the turbocharger, a plug can be removed, a fitting installed, and a manometer reading taken. As with the 4-cycle engine, the 2 cycle must also be under load during this test.

Causes of Low Air Box Pressure or Turbocharger Boost

1. Anything that creates a high air inlet restriction condition, because if the turbo or blower cannot receive an adequate air supply, it cannot deliver its rated pressure.
2. Leaking O ring seals between the inlet manifold seals from the turbocharger on sectional manifolds if used.
3. Leaking inlet manifold-to-cylinder head gasket.
4. Leaking hand hole cover plate gaskets on Detroit Diesel 2-cycle engines and others.
5. A clogged or plugged blower inlet screen (found on nonturbocharged-type Detroit Diesel 2-cycle engines).
6. Leaking blower-to-engine block gasket.
7. Leaking end plate-to-block gasket on DDA-type 2-cycle engines (would also create a high crankcase pressure condition).
8. Possible damaged blower rotors.

Usually, if there is an air leak on the turbocharger outlet side, some form of a whistling sound is evident due to the pressure loss. Although a low air box or turbocharger boost is much more prevalent, you may occasionally run into problems of too much or a high air box or turbo pressure.

Causes of a High Air Box or Turbocharger Pressure

1. Anything that creates a high exhaust back pressure (see EBP causes).
2. Excessive carbon formation in the cylinder liner port area of 2-cycle engines, generally caused by excessive periods of idling and light-load operation.

High air box pressure on a 2-cycle engine will work back against the blower, thereby reducing its efficiency. This causes reduced scavenging, poor combustion, and higher exhaust temperatures.

Exhaust Back Pressure

Due to noise considerations especially, all internal combustion engines employ a system to quiet or muffle the sound of the exhaust gases as they leave the engine. Since these gases are under pressure, the use of the exhaust system will induce a certain amount of back pressure to the flow of these gases out of the engine. Even allowing for noise considerations, a slight pressure in the exhaust stream is needed to drive turbochargers, and in applications of excessive idling or light-load operation or cold weather operation, a built-in exhaust back pressure within limits is desirable.

The exhaust back pressure is measured in inches of mercury with either a manometer or pressure gauge. On nonturbocharged engines a small pipe plug is located about 1 to 2 in. (2.54 to 5.08 cm) toward the muffler or silencer just where the exhaust manifold companion flange bolts to the exhaust piping. However, on turbocharged engines the pipe plug will be found after the turbocharger, since if you were to measure the exhaust pressure on the manifold side a higher than normal reading would be obtained. If no plug is available, a small hole can be drilled and tapped to accept a small plug in the positions so indicated. Make sure that you always use a brass plug since a steel one will tend to freeze after some usage.

On Vee-type engines, especially with double exhaust stacks, it is advisable to check both for back pressure. If it is a Vee engine with only one exhaust stack, the reading is generally taken about 4 to 6 in. (10.16 to 15.24 cm) from the Y-connection toward the muffler or silencer. If, however, you have a high exhaust back pressure on this type of setup, check both banks of the engine. With the mercury manometer connected up and zeroed in, start the engine, and note and record the readings at the specified engine speed both at no load and full load.

Causes of High Exhaust Back Pressure

1. Stuck rain cap at the end of a vertical exhaust stack.
2. Crushed exhaust piping or muffler.
3. Excessive carbon buildup within the exhaust system.
4. Too small a muffler.
5. Exhaust piping too long.
6. Exhaust pipe diameter too small.
7. Too many bends or elbows in the exhaust piping.
8. Obstruction lodged in the exhaust piping.

Excessive exhaust back pressure can increase the temperature considerably, leading to lack of power, smoking exhaust, burned-out exhaust valves, poor fuel economy, and shortened engine life.

Crankcase Pressure

Crankcase pressure in an internal combustion engine within certain limits is a highly desirable feature. This small amount of pressure tends to keep any dust or light dirt within suspension, preventing it from being deposited on internal engine parts. Different methods are employed by various engine manufacturers to obtain this; however, a continued loss of engine oil through the engine breather tube, crankcase ventilator, or dipstick tube hole in the block is indicative of high crankcase pressure. The specific causes will vary between engines; however, most of them are common in nature.

Since crankcase pressure is very low, a water manometer is used for this purpose. A hose or tube can be readily adapted to the dipstick tube or shroud and connected at the manometer on the other end after zeroing it in. If it is a larger-bore-type engine, a plug is usually readily accessible in the engine base or inspection doors for this purpose. With the manometer connected as stated, start and run the engine at the recommended speeds, note and record your readings, and compare them to service manual specifications.

Causes of High Crankcase Pressure

1. Too much oil in the crankcase.
2. Plugged crankcase breather tube or ventilator.
3. High exhaust back pressure.
4. Leaking cylinder head gasket.
5. Worn or broken piston rings.
6. Loose piston pin retainers (solid type) found on some 2-cycle engines.
7. Worn blower oil seals (mainly 2-cycle engines).
8. Leaking cylinder block-to-end plate gasket (DDA 2 cycle).
9. High air box pressure (2 cycle).
10. Blower or turbocharger air pressure leakage into an oil drain passage (blower-to-block gasket on DDA 2-cycle engines).
11. Damaged piston or liner.

Having taken these four types of readings and compared them to manufacturer's specifications, you are now in a position to quickly establish the condition of the air system, exhaust, and crankcase breathing system.

12-2

Checking the Fuel System

If the engine is misfiring, running rough, and lacking power, you can start and run it at an idle rpm. If it is a 4-cycle engine with a high pressure fuel system, you can loosen the fuel line retaining nut at the injector one half-turn and note if there is any change in the sound of the engine. If the injector is firing properly, there should be a positive change as you loosen the line; otherwise, the injector is not firing properly. On 2-cycle high speed engines such as Detroit Diesel, simply short out the injector as described under DDA troubleshooting. An exhaust smoke analysis is very helpful in determining the possible causes of poor performance. See Table 12-1.

If the engine fails to reach its maximum governed speed and generally seems to be starving for fuel, install a fuel pressure gauge into the secondary filter, run the engine, and check the fuel pressure with the engine manufacturer's specifications. On Detroit Diesel engines, perform a fuel spill back check (see this chapter). Some engines employ a small filter screen located just under the cover of the fuel transfer pump; check that this is not plugged.

If a fuel strainer or fuel water separator is used, check them for plugging and excessive amounts of water. Check that all fuel lines are free of sharp bends and kinks. Check the tightness of all fittings and connections from the suction side of the transfer pump back to the fuel tank. Install a clear portion into the suction line to check for air bubbles.

You may have to undertake a restriction

Table 12-1
EXHAUST SMOKE ANALYSIS CHART

Smoke Color	*Probable Causes*
Black or gray	1. Incompletely burned fuel caused by a high exhaust back pressure or high air inlet restriction condition. 2. Excessive or irregular fuel distribution caused by improperly timed injection pump or elements. Improperly timed unit-type injectors or improperly adjusted fuel injector or pump controls. May require injector removal for testing and checking. Possibly due to engine lugging condition. 3. Improper grade of fuel; check manufacturer's specifications
White	1. Low compression. 2. Use of a low cetane fuel. 3. Faulty injectors. 4. Water leakage into the combustion chamber.
Blue	1. Worn intake valve guides or seals. 2. Worn piston rings, broken oil control rings or wrong expanders behind oil rings causing reduced radial pressure, or improperly installed oil scraping rings. 3. Loose, solid-type piston pin retainers. 4. High crankcase pressure. 5. Excessive oil in the intake manifold or air box (2 cycle) caused by leaking turbocharger or blower oil seals. Oil leakage into the air box from a damaged blower-to-block gasket. 6. Glazed cylinder liners through use of the wrong type of oil, improper run-in procedures, or excessive periods of idling and/or light load.

check to the fuel flow as discussed in Chapters 9 and 12 under troubleshooting.

Check the fuel pump drive for security and proper engagement. Ensure that there are no external fuel leaks, especially at the pump or injectors. Also, if more than one fuel tank is employed, check to see that the balance line valve is open between them; if a three-way valve is employed, check that it is in the correct position.

In certain instances you may also find that there is a restriction to fuel flow from inside the fuel tank caused by sediment or some foreign object that has dropped into the tank either during filling or maintenance checks.

One complaint that you may occasionally come across is that the engine runs well in the early part of a shift, but stalls and lacks power as the day wears on. This could be caused by debris, such as a piece of wood or bark, especially around logging equipment, which only creates a restriction to fuel flow as the level in the fuel tank drops and the debris is drawn over the suction line.

If the engine has recently been overhauled or the injection pump or injectors serviced, double-check the injection pump timing, injector release pressure, or injector timing.

If the engine has a considerable amount of hours or miles on it, it very well may be in need of a tune-up; however, this alone may not be the cause of the problem. Too many people immediately assume that if an engine is lacking power the answer is to tune it up. Although many large companies have developed a sequence of checks to be carried out at certain intervals of time, a tune-up should be done only if other checks show that everything else is according to specifications.

When doing a tune-up, do not back off all adjustments and start from scratch. Check each adjustment first and if necessary readjust. One of the first checks that should be carried out is to disconnect the throttle linkage and manually hold the speed control lever on the governor to the full-fuel position and accurately record the maximum governed engine rpm. Reconnect the throttle linkage, place it in the full-fuel position, and compare the readings. If not the same, adjust the linkage to correctly obtain the maximum engine rpm. Similarly, the maximum governor speed adjustment may require adjustment. Ensure that there is no binding anywhere in the fuel control linkage.

Fuel Temperature

On high speed diesel engines, fuel temperature can adversely affect the horsepower output of the engine. The optimum fuel temperature should be kept between 90 and 95°F (32 and 35°C). With each 10°F temperature rise beyond this figure, there is approximately a 1% loss in horsepower due to expansion in the fuel. Therefore, if you were running at a fuel temperature of 135 to 140°F (57 to 60°C), theoretically your engine would be producing approximately 4% less horsepower. On a 350-hp unit, this would amount to about 14 hp (10.4 kW). Maximum temperature is 150°F (65°C).

On larger, slow speed engines using Bunker C-type fuel, this can be forgotten since this fuel is preheated to 150°F prior to entering the injection system for better flow properties.

Compression Checks

A compression check may be necessary to determine the condition of the valves and rings. On many engines this check is taken with the use of a dummy injector and the engine running. Such a check can be found in Chapter 12 under troubleshooting. Each make of engine will have some variation in the sequence of events required to do this; therefore, check the respective engine manufacturer's service manual for the routine and specifications.

A crankcase pressure check taken with the use of a water manometer could tip you off to worn rings, as can an exhaust smoke analysis, hard starting, and low power.

Engine Will Not Start

Although there can be a variety of reasons for not starting, these generally fall into one of the following areas:

1. Cranking problem: a low battery or low air pressure, high circuit resistance in the electrical system caused by corroded terminals, loose terminals, faulty wiring, and so on. It could also be a solenoid failing to engage or starter drive or internal starting problems, to name but a few.
2. Lack of air: see air inlet restriction checks.
3. Lack of fuel: see fuel system checks and the troubleshooting charts, Tables 12-2 and 12-3.

Table 12-2
TROUBLESHOOTING INJECTION PUMP AND FUEL SYSTEM (HIGH PRESSURE TYPE)

Fault	*Possible Cause*	*Remedy*
Fuel not reaching injection pump. Engine fails to start.	Tank empty or tank valve closed.	Fill tank; open valve.
	Primary or secondary filter plugged.	Replace filters.
	Defective in-tank pump.	Repair or replace electric fuel pump.
	Fuel lines clogged or damaged.	Clean or replace fuel lines.
	Defective damper valve.	Replace damper valve.
	Defective transfer pump.	Replace transfer pump.
Fuel reaches injection pump but does not inject to nozzle.	One or more injection lines plugged or damaged.	Replace lines.
	Plunger seized or spring broken.	Repair or replace injection pump.
Fuel reaches nozzles but engine fails to start or starts hard.	Cranking speed too slow.	Recharge battery or correct faulty cranking motor system. Lube oil may be too heavy at low temperature.
	Intake air temperature low.	Use starting aids. Check ether device for proper operation. (Check operator's manual.)
	Swirl destroyer stuck open.	Check through opening in air intake manifold.
	Accelerator not in full-fuel position.	Correct accelerator linkage.
	Improper fuel or water in fuel.	Drain fuel system and pump housing. Use specified fuel and prime system.
	Poor compression.	Correct engine as required.
	Pump out of time with engine.	Correct timing.
	Worn or scored plungers.	Repair or replace injection pump.
Engine starts and stops.	Water in fuel or contamination.	Drain system and pump housing; use new fuel and prime system.
	Air in system.	Prime system.
	Fuel filter clogged.	Replace filters.
	Fuel tank vent plugged.	Clean vent.
	Waxed fuel filters.	Replace filters; use proper fuel.

(Continued on following page.)

Table 12-2 (Continued)

Fault	*Possible Cause*	*Remedy*
	Fuel lines clogged or damaged.	Clean or replace fuel lines.
Erratic engine operation (surging or misfiring).	Improper or contaminated fuel.	Drain system; use new fuel and prime system.
	Low transfer pressure.	Change filters, check for fuel line restriction, replace damper valve or transfer pump.
	Fuel filters clogged.	Replace filters.
	Fuel lines clogged or damaged.	Clean or replace fuel lines.
	Injection pipes leaking.	Correct leaks.
	Pump out of time.	Correct timing.
	Nozzles faulty.	Repair or replace nozzles.
	One or more injection lines plugged or damaged.	Replace lines.
	Poor compression.	Correct engine as required.
Engine idles poorly.	Improper low idle adjustment.	Make low idle adjustment.
	Water in fuel.	Drain system and pump housing; use new fuel and prime.
	Air in fuel.	Bleed system.
	Accelerator linkage incorrect.	Correct linkage as required.
	Incorrect timing.	Check and correct timing.
	Damaged swirl destroyer.	Correct swirl destroyer.
	One or more injection lines damaged.	Replace lines.
	Poor compression.	Correct engine as required.
	Defective low speed flyweight spring.	Calibrate injection pump.
	Plungers worn or scored.	Repair or replace injection pump.
Engine has low power but shows no increase in smoke.	Accelerator linkage restricted.	Adjust accelerator linkage.
	Governor high idle adjustment.	Make high idle adjustment.
	Low transfer pressure.	Change filters, check for fuel line restriction, replace damper valve or transfer pump.
	Loss of one cylinder.	Injection line plugged.
	Faulty maximum fuel setting.	Calibrate injection pump.
	Plungers worn or scored.	Repair or replace injection pump.
	Exhaust system restricted.	Check pipe diameter or muffler for damage.

Table 12-2 (Continued)

Fault	*Possible Cause*	*Remedy*
	Swirl destroyer in *on* position.	Check linkage for bind to prevent jamming.
	Air cleaner restricted.	Check manifold vacuum. Service air cleaner if vacuum exceeds 20 in. of water or 1.47 in. of mercury.
	Nozzles faulty.	Repair or replace nozzles.
	Pump out of time.	Correct timing.
High smoke, with no loss or with increase in power.	Increase in fuel delivery.	Check nozzles.
General.	General mechanical condition of engine poor.	Correct as required before extensive repair to the fuel system.

(Courtesy of International Harvester Co.)

Table 12-3
TROUBLESHOOTING FAULTY NOZZLES

Fault	*Possible Cause*	*Remedy*
Excessive leak-off.	Dirt between pressure face of nozzle, spring retainer, or plate and nozzle holder.	Clean nozzle.
	Loose nozzle retainer nut.	Inspect lapped faces and tighten retainer nut.
	Defective nozzle.	Replace nozzle.
Nozzle blueing.	Faulty installing or tightening.	Replace nozzle.
	Insufficient cooling.	Correct cooling system.
Nozzle opening pressure too high.	Incorrect shim adjustment.	Replace nozzle.
	Nozzle valve dirty or sticky or opening is clogged.	Clean nozzle.
	Seized nozzle.	Replace nozzle.
Nozzle opening pressure too low.	Incorrect shim adjustment.	Readjust nozzle.
	Nozzle valve spring broken.	Replace spring and readjust pressure.
	Nozzle seat worn.	Install new or reconditioned nozzle.

(Continued on following page.)

Table 12-3 (Continued)

Fault	*Possible Cause*	*Remedy*
Nozzle drip.	Nozzle leaks because of carbon deposit or sticking nozzle valve.	Clean nozzle.
	Defective nozzle.	Replace nozzle.
Spray pattern distorted.	Carbon deposit on tip of nozzle valve.	Clean nozzle.
	Nozzle hole partially blocked.	Clean nozzle.
	Defective nozzle.	Replace nozzle.

(Courtesy of International Harvester Co.)

Note: Use new gasket each time nozzle is installed.

High Horsepower Complaint

On several of the new-style high speed, high torque rise engines used in highway truck applications, the pump and/or governor must be adjusted within very close tolerances to allow the engine to produce an increase in torque with a decrease in rpm. It is possible on some of these engines, such as DDA fuel-squeezer units, to obtain a higher horsepower than specified; therefore, if you run into this complaint on some of these newer engines, consult the engine manufacturer's service manual or refer to the fuel section found herein for your type of engine.

Crankcase Oil Dilution

This complaint is sometimes referred to as the engine making oil, since the oil level becomes increasingly higher during operation. Fuel oil leakage from an overtightened injector fuel line is one of the common causes; therefore, always use the recommended torque figure when tightening up fuel lines. On Cummins engines, check the injector O rings for damage, the same applies to Detroit Diesel Allison's new 8.2 liter ''fuel pincher'' 4-stroke cycle engine.

Piston Scuffing, Scoring, and Possible Seizure

This problem can often be caused by injectors either dribbling raw fuel into the combustion chamber, owing to a faulty check valve, or by a combination of water and dirt entering the injector. On multihole-type fuel injectors, water can blow the tip off the end of the injector. With dirt passing through the small spray tip holes, they can become enlarged, leading to a flattening out of the fuel spray-in angle. This can create what is commonly called *wall wash,* since the fuel will tend to penetrate the outer periphery of the piston crown, causing burning of the outer circumference of the piston and leading to increased piston temperatures and fire ring seizure or breakup. If the fuel sprays onto the cylinder wall, this will create wall wash and lube oil dilution, leading to eventual scuffing and scoring of the cylinder and piston.

Engine Timing

Improper engine timing can lead to actual physical mechanical damage, such as valves hitting pistons. However, this will depend on just how far out the timing is and is more acute on some engines than on others.

If the injection pump timing or injector timing is off, problems of smoking exhaust, low power, high fuel consumption, and internal engine damage can result; therefore, always ensure that the engine is timed as per manufacturer's specifications and that injection pump and injectors are timed for the particular application for which the engine is being employed.

There is also a noticeable change in the engine's exhaust temperature with a timing change. The degree of change will vary; however, most heavy-duty on- and off-highway trucks and equipment employ dash-mounted pyrometers, which constantly monitor the exhaust gas temperatures and are readily visible at a glance. One pyrometer is employed per engine bank on Vee engines. If these are not installed, the individual cylinder temperature can be read by turning a numbered switch that is connected through thermocouples at each cylinder. This type is more common to larger engines used in stationary industrial and marine applications. If the engine does not have either of these, a hand-held contact-type pyrometer can be used to individually check cylinder exhaust temperatures.

Exhaust temperatures will vary between engines; therefore, always check the manufacturer's specifications regarding normal and maximum allowable temperatures. The size of the injectors, fuel pump delivery rate, and engine speed and load will all contribute to exhaust temperature. On turbocharged engines the temperature is usually taken after the turbocharger. Two-cycle engines will generally run 150 to 200°F (65 to 93°C) cooler than a 4-cycle engine.

An engine timing light that operates off fuel pressure through a transducer pickup can be used on engines that employ a high pressure fuel system. Caterpillar uses such a device to advantage on their engines (see Chapter 7, fuel systems).

Engine Vibration

Misfiring cylinders due to low compression or faulty injectors, improper timing of individual pumping units or injectors, valves set too tight, improperly balanced cylinder banks or individual injector racks, water in the fuel, and plugged fuel filters are some of the typical causes of engine vibration. However, it may be caused by accessory items on the engine, necessitating a more thorough analysis with a vibration meter.

Exhaust Smoke Analysis

Always make exhaust smoke checks with a minimum water outlet temperature as specified by the engine manufacturer, usually around 160°F (70°C); it is also advisable if possible to check the opacity of the exhaust smoke at both no load and full load.

Although there are smoke meters readily available on the market, not everyone has such a device; therefore, a Ringelmann-type smoke chart can be used to approximate the density of the exhaust smoke emanating from the stack (see Figure 12-5).

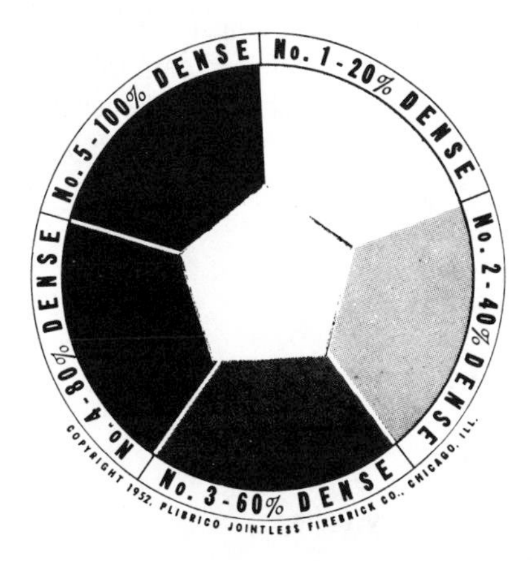

Figure 12-5
Ringelmann-Type Smoke Chart.
(Courtesy of Detroit Diesel Allison Div. of GMC)

Less than 5% opacity is hardly visible to the naked eye. The acceptable standards for exhaust smoke opacity are controlled by the EPA. Each engine produced by a manufacturer must meet certain limits of maximum allowable exhaust smoke opacity under a variety of conditions, including acceleration and lug down conditions.

Many states in the United States are very strict on exhaust gas emissions; therefore, it is encumbent upon the maintenance personnel to ensure that all engines meet these specifications; otherwise, costly fines can result. Also, if the engine is producing abnormal amounts of smoke, it is an indication of poor performance and is reflected as a direct economic loss to the user.

A Ringelmann smoke scale enables the user to observe conveniently the approximate density of the smoke coming out of the engine's exhaust stack. The scale should be held at arm's length, at which distance the shaded areas on the chart can be compared to the shade or density of the smoke coming from the exhaust stack. The observer's line of observation should be at right angles to the direction of smoke travel and not be less than 100 ft (30.48 m) or more than a $\frac{1}{4}$ mile (0.4 km) from the stack. The background directly beyond the top of the exhaust stack should be free of buildings or other dark objects and direct sunlight. By recording the changes in smoke density, the average *percentage of smoke density* for any period of time can be determined.

Hard Starting

Engine Will Not Rotate

1. Low battery voltage.
2. High circuit resistance.
3. Loose battery or starter connections.
4. Faulty starter solenoid.
5. Faulty starter.
6. Insufficient air tank pressure (air starter).
7. Faulty starting valve or switch.
8. Safety valve or switch locked out.
9. Excessive air leaks to air starter.
10. Jammed starter mechanism.
11. Mechanical engine damage.

Low Cranking Speed

1. Same as under engine will not rotate, items 1, 2, 3, 5, 6, 7, 9.
2. Improper lube oil viscosity, especially in cold weather.
3. Tight component parts after initial rebuild.
4. Excessive heat buildup in starter (after a maximum of 30 seconds of usage, allow electrical starters to cool for at least 2 minutes).

Low Compression

1. Sticking or burned valves.
2. Worn or broken piston rings.
3. Leaking cylinder head gasket.
4. Improper valve clearance adjustment.
5. Blower or turbocharger not functioning.
6. Cracked or holed piston.

No Fuel or Low Delivery

1. Air leaks on suction side of system.
2. Fuel flow obstruction (plugged filters).
3. Wax crystals formed in fuel (low ambients).
4. High restriction to pump flow caused by fuel line diameter being too small or fuel line being too long or having too many bends. May also be trying to lift fuel too great a height.
5. Fuel valve partially or fully closed.

6. Faulty fuel pump.
7. Bind in governor or throttle linkage, preventing injection pump or injector racks from attaining run position or full-fuel position.
8. Fuel tank to line blockage.

Faulty Starting Aid

1. Check glow plug operation.
2. Check fluid starting aid operation.
3. Fuel heater faulty.

No Fuel or Insufficient Amount

Sucking Air

1. Low fuel supply.
2. Loose connections from the transfer pump suction side back to the fuel tank.
3. Leaking strainer or primary filter gasket.
4. Possibility of cylinder pressure blowing back through a faulty injector tip assembly.

Fuel Flow Restricted

1. Plugged fuel filters.
2. Fuel lines crushed, collapsed, kinked.
3. Stuck, closed one-way fuel line check valve.
4. Wax crystals in fuel due to cloud point of fuel temperature being too high.
5. Fuel suction line diameter too small.
6. Reducing fitting installed at suction side or discharge side of pump.
7. Restricted fitting installed at fuel inlet manifold, such as on Detroit Diesel engines.

Faulty Fuel Pump

1. Relief valve not seating.
2. Worn pump gears or body.
3. Sheared pump drive or not engaged.
4. Ruptured pump diaphragm.

Other Areas

1. Fuel lines excessively long.
2. Fuel lift exceeds pump's capability.
3. Restricted fitting missing from fuel outlet and return manifold (Detroit Diesel).
4. Excessive fuel leakage on pressure side of system.
5. High fuel temperature.

The preceding outline covers the items that you will most often encounter when faced with typical fuel system problems. It is not meant to be an all-encompassing chart; therefore, some things may apply to one make of engine and others to some other make of engine. However, you will find it most helpful in many everyday situations. The engine service manual for the specific engine that you are dealing with should be referred to at any time that you are in doubt as to a particular test, check, or specification.

12-3 Abnormal Engine Operation

Engine Idles Normally But Misfires Above Idle

1. Plugged fuel filter.
2. Faulty pump timing advance mechanism.
3. Low fuel pressure above idle (early opening bypass valve).

Engine Fails to Return to Idle

1. Binding linkage or misadjusted.
2. Internal injection pump or unit injector malfunction.

Engine Speed Continues to Climb Beyond Maximum Governed Speed

1. Check and adjust maximum speed control linkage adjustment.
2. Engine is receiving an external source of fuel (excessive lube oil leakage past rings, blower, or turbocharger seals).
3. Internal governor problem.

Fuel Leakage, Engine Runs Normally

1. Loose or damaged fuel lines.
2. Injection pump seal leakage.

Engine Noise Similar to a Bearing Knock

1. Air in fuel system.
2. Sticking nozzle in the open position.
3. Low nozzle or injector opening pressure.

Excessive Combustion Noise

1. Improper timing.
2. Lube oil leakage into the cylinder.
3. Internal injection pump or injector problem.

Engine Fails to Shut Down

1. Binding control linkage.
2. Improperly adjusted linkage.
3. Key switch solenoid valve faulty; does not return fuel valve to *off* position.

Engine Runs Rough or Tends to Stall Frequently

1. Restricted fuel return system.
2. Low engine coolant temperature.
3. Insufficient fuel.
4. Low cylinder compression.
5. Misfiring injectors (air in system).
6. Governor linkage wear.
7. Improper governor adjustments (idle screw).
8. Wrong grade of fuel.
9. Water in fuel.
10. Faulty valve adjustments.
11. Excessive crankcase pressures.
12. Pump or injector timing wrong.

Engine Lacks Power

1. Engine requires a tune-up.
2. High air inlet restriction.
3. High exhaust back pressure.
4. Low coolant temperature.
5. Engine or injection pump, unit injector timing off.
6. Restricted fuel return to tank.
7. Low compression.
8. Insufficient fuel or irregular fuel distribution.
9. High fuel temperature.
10. High air inlet temperature.
11. High altitude operation.
12. Throttle linkage or governor linkage binding preventing full-fuel situation.
13. Problems associated with low compression.
14. Engine overworked; not enough power for particular application.

An increase in fuel temperature can cause a reduction in engine horsepower. Similarly, an increase in air temperature (ambient) causes the air to expand and therefore become less dense. On a turbocharged engine, this is offset by the increase in air flow and pressure increase, plus the use of an aftercooler or intercooler.

On most high speed engines, a power decrease can be expected of between 0.15 and 0.5 hp per cylinder (0.11 and 0.373 kW) depending on the fuel injector's or pump's delivery capability for each 10°F (5°C) air temperature rise above 90°F (32°C). Therefore, on complaints of low horsepower, always check to ensure that these two temperatures are within specifications; otherwise, you could spend a lot of time trying to solve the complaint.

Engine Overspeed

1. Maximum governed rpm adjustment improperly set.
2. Internal governor problem.
3. Oil pullover from an oil bath air cleaner or other external fuel source such as blower or turbocharger seals.
4. Running the engine with the governor linkage disconnected.
5. Operator problem: this particular problem is not unusual on mobile equipment and highway truck operation. If an operator allows the engine rpm to climb beyond the maximum safe road speed for a particular gear, what in effect happens is that the vehicle's road wheels become the driving member. As there is a direct mechanical link from the road wheels to the differential and the drive line, this increased road wheel speed will work through the transmission, causing the engine to be the driven member instead of the driving member. During this time it matters not that the operator has his foot on the throttle, since the governor will react to pull the engine to a decreased fuel situation. Even if the operator has the throttle in the idle speed position, the road wheels being the driving member can still spin the engine to a point that the valves will strike the piston crown, leading to mechanical failure of the engine. Therefore, caution drivers and operators about excessive road speed when going down long inclines and steep hills.

Detonation

Do not confuse the normal combustion sound within the engine for this complaint. Some engines do run louder than others, and many of them have a peculiar sound common to that particular engine or application. Pressure pulsations within the engine cylinder create the condition often referred to as *diesel* knock; it is an inherent characteristic of all diesel engines.

Experience will tune your ear to pick up sounds other than the normal combustion pressure sounds. However, it is often helpful even to an experienced mechanic to isolate any irregular noises with the use of an engine *stethoscope,* which amplifies any sounds remarkably well. Often a piece of welding rod or even a lead pencil placed on the engine with the other end at your ear can magnify sounds reasonably well.

If detonation occurs, check for the following:

1. Lube oil picked up by the air intake stream to the engine, which cannot only cause detonation but engine overspeed.
2. Low coolant temperature caused by excessive periods of idling and light-load operation. Or cold weather operation without proper attention to maintaining operating temperatures.
3. Faulty injectors: leaking fuel, fuel spray-in pressure low.

Filter Cartidge Service Indicator

An air filter service gauge (Figure 12-6) is widely used by major engine manufacturers to indicate the condition of the filter element. It works on a pressure drop principle caused by the increase in air inlet restriction, which gradually rises as dirt trapped by the filter accumulates. This causes a red indicator flag to show in the window area, which locks in place

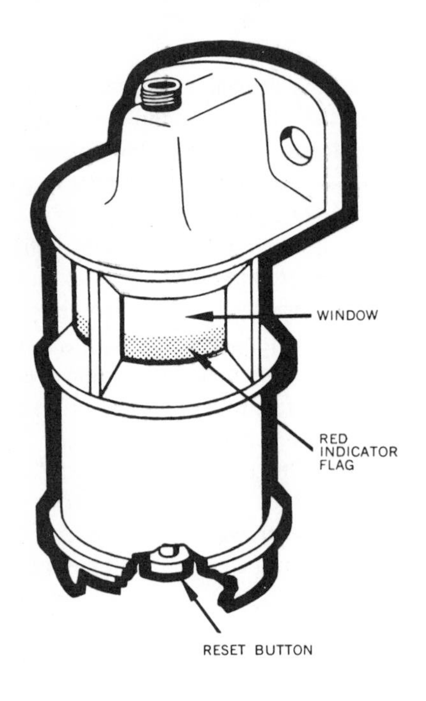

(a)

when it reaches the top position, signifying that the filter cartridge should be changed.

The indicator is factory preset to signal when normal maintenance is required. The restriction at which the red flag will appear can be specified; however, most are 20 or 25 in. of water.

The gauge can be either remote or direct mounted to the air filter. No electrical connections are required; to reset the gauge after servicing, simply push the reset button.

Dust Sight Gauge

This gauge should be mounted as close as possible to the air intake manifold in a position where the tab is easily seen. Do *not,* however, locate it downstream of a turbocharger. Any dust entering the air system upstream from the dust sight will immediately cloud the glass as it is deflected by the metal tab. By shining a flashlight into the window, the internal condition of the air system can be monitored (Figure 12-7).

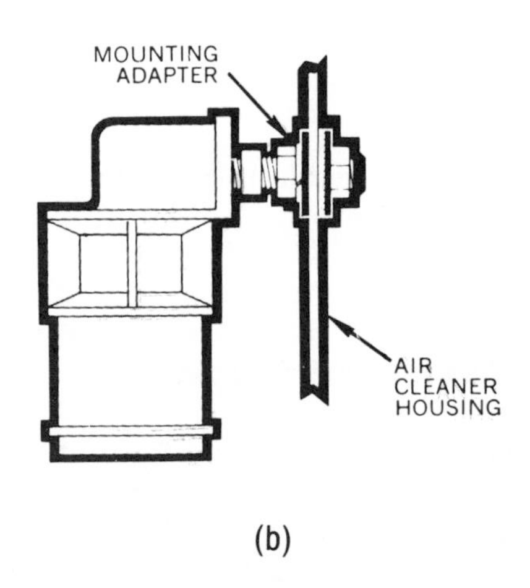

(b)

Figure 12-6
Air Filter Service Gauge.
(Courtesy of Farr Air Cleaner Co.)

Figure 12-7
Dust Sight Gauge.
(Courtesy of Farr Air Cleaner Co.)

Pyrometers

Exhaust temperature gauges, or what are more commonly called *pyrometers,* are extremely helpful when checking an engine for a lack of power complaint. Most recent highway trucks with diesel engines are equipped with dash-mounted pyrometers, which can readily assist you in determining if both engine banks are running at the same temperature on Vee-type engines. On in-line engines the pyrometer can establish whether or not the engine is operating within the range specified by the engine manufacturer.

Larger diesel engines use thermocouples installed at each cylinder; however, they are seldom found on individual cylinders on high speed diesels applied to mobile or off-highway equipment. Stationary and marine applications, however, are using them much more regularly now than they have in the past.

The most common form of pyrometer uses a pickup or thermocouple consisting of two wires of different metals welded together at their ends, which is known as a *hot junction*. The metals used in these wires are selected for their response to temperature and ability to withstand high heat. As the hot junction is exposed to a heat source, a small electric current is generated at the junction; it flows through the wires to the measuring instrument, which is a *millivoltmeter*. The amount of current flow is proportional to the heat created at the hot junction.

Many companies offer pyrometers that can be readily used by one person during troubleshooting. These are of the hand-held type; they have a heat probe that will register temperature upon contact with the surface to be checked. Newer pyrometers are offered with a digital readout and are therefore very helpful.

Many major engine manufacturers offer special instrument groups that are handily packaged into portable tool boxes and are excellent for troubleshooting their engines as well as others.

Dynamometers

The quickest and most effective method of determining whether or not an engine is producing its rated horsepower is through the use of a dynamometer. At the present time, a variety of load-testing machines are available for any purpose and application. Basically, they are as follows:

1. A truck chassis dynamometer; the rear driving wheels of the vehicle are forced to drive against either a single or double set of rollers connected to the dynamometer. This allows road wheel horsepower to be read directly from the instrument cluster.
2. A stationary dynamometer that can have an engine coupled to either end of it for convenience, but only one engine can be tested at a time.
3. A portable, compact, relatively lightweight dynamometer that can be bolted to the engine flywheel housing and driven from the engine flywheel. This type can also be readily adapted to truck applications simply by disconnecting the drive line and coupling up the dynamometer unit.

There are a variety of dynamometer manufacturers; some of the more common are Go-Power, Kahn, Clayton, Taylor, and Froude.

12-4

Troubleshooting DDA Engines

With the exception of the recent 8.2-liter 4-cycle engine V8, the series 53, 71, 92, and 149 DDA engines are all 2-cycle engines, and there are problems that will occur with these particular units that will be peculiar to them alone, and may not occur on a 4-stroke cycle engine. However, with their simplicity of design and operation, if one is familiar with one model or series of DDA engine, you will find it relatively simple to effectively troubleshoot a problem on any of their engines if you systematically follow a given procedure.

The following problems and their cures will deal strictly with the fuel-injection section of the engine. There may be some necessity to include cross-related problems within the engine; however, these shall be kept to a minimum where possible.

One of the quickest and most accurate methods of effectively troubleshooting any engine is through the use of both a water (H_2O) and mercury (Hg) manometer. The design of all DDA engines is such that many apparent problems can be quickly diagnosed in as little as 15 to 20 minutes or less through the use of manometers.

Let us then consider the function and operation of these particularly useful tools. (Refer to the section on general troubleshooting for a more detailed description of manometers.)

Both water and mercury manometers are used on all DDA engines for the following purposes:

1. To measure air inlet restriction, commonly referred to in manufacturer's service manuals simply as AIR.
2. To measure air box pressure (ABP) or turbocharger boost pressure.
3. To measure crankcase pressure.
4. To measure exhaust back pressure (EBP).

The water (H_2O) manometer is used for very low pressure or even vacuum readings if desired; therefore, both the AIR and crankcase pressure would be checked with this manometer.

The mercury (Hg) manometer is used for pressures in excess of 1 psi (kPa); therefore, it is used for both ABP and EBP or turbocharger boost pressure.

Prior to using any manometer, you must first obtain from the engine service manual the manufacturer's specifications relating to the particular test that you are undertaking. This can be found under Section 13, Engine Operating Conditions, of all DDA manuals. You will find listed therein both a specification and a specific engine rpm at which the check should be made.

One of the most common complaints that comes up is that the engine lacks power or is sluggish to throttle response. Additional checks can be undertaken after you have done a manometer check.

It may well be, after installing a new injector, that there is still a problem. If this is the case, you should proceed to do a compression check on the suspected cylinder. If the engine was emanating blue smoke from the exhaust stack when you first checked out the complaint, this would indicate that the engine is burning oil. Since all DDA engines are of the 2-stroke-cycle type, this is usually an indication of piston ring problems if the engine is at its normal operating temperature. If white smoke is visible at the exhaust stack, this indicates a misfiring cylinder probably due to low compression, especially if you have shorted out the injector as stated. Black or gray smoke can be caused by excessive or irregular fuel distribution. For all the time it takes, on lack of power complaints, removal of the hand hole cover inspection plates will allow you to quickly determine the condition of the rings in each cylinder (see under compression check following).

Locating a Misfiring Cylinder

With the engine at its normal operating temperature, remove the valve rocker covers.

1. Start and run the engine at its normal idle speed and listen to the sound of the exhaust gases escaping from the exhaust pipe outlet. If there is a popping sound, it may very well be a burnt-out exhaust valve. If there is no unusual sound, you

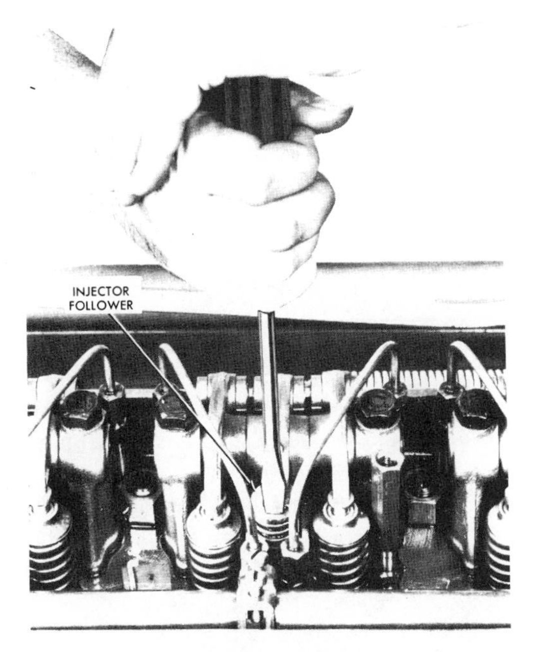

Figure 12-8
Shorting Out an Injector by Depressing the Follower.
(Courtesy of Detroit Diesel Allison Div. of GMC)

can be reasonably assured that the valves are normal.

2. Stop the engine and individually check the exhaust valve clearances. (Remember, this is a 2-cycle engine; therefore, there are no intake valves.) If necessary, adjust them to specifications.

3. Start the engine and let it idle. Refer to Figure 12-8. Take a screwdriver as shown and manually depress the injector as far as it will go and hold it there. This effectively bypasses fuel internally within the injector, preventing it from firing; therefore, when you hold the injector follower down, there should be a positive change in the engine's sound and rpm. If there is no change when you do this, that particular injector is faulty. This sequence of events is similar to shorting out a spark plug on a gasoline engine.

Caution: Do not use this procedure on the new 8.2 liter 4-cycle engine. Refer to Chapter 10.

4. Repeat this test on each additional cylinder.

5. If it is necessary to change an injector, ensure that there is no damage to the particular cylinder's injector operating mechanism. Then remove the old injector and install the new one as described under injector removal earlier in this chapter.

Checking Cylinder Compression Pressure

Since a certain amount of time is required to do a compression check on the engine, you should first analyze the color of the smoke coming out of the exhaust stack; then proceed as follows:

1. Refer to Figure 12-9. Remove the hand hole cover inspection plate from the side of the cylinder block. This allows free access into the air box area and the cylinder liner port area. Select a blunt (nonpointed) tool and push against the compression rings to check for free spring or tension. If there is no sign of this, the piston ring is stuck in its groove. An additional check would be to carefully note whether or not the compression rings have a visible groove all the way around the center circumference. This groove is placed there at the time of manufacture; if it is not visible, the rings are very badly worn. If it is visible in some spots, but not others, irregular ring wear is evident. You can also check for damage to the piston ring lands and skirt area at this time.

2. If piston rings are badly worn in one or more cylinders, this would be noticeable as high crankcase pressure when using a water manometer. All rings badly worn

would be reflected by blue exhaust smoke, lack of power, hard starting, and rough running.

3. Prior to taking a compression check, it is imperative that the engine be at normal operating temperature. If only one cylinder is suspected of having low compression, start with it. If doing them all, start with cylinder 1.
4. A cylinder compression check is taken on all DDA 2-cycle engines at a speed of 600 rpm; while on the 4-cycle 8.2-liter fuel pincher model, it is taken at 700 rpm; therefore, you cannot hope to use a typical hand-held automotive-type gauge. Several suitable test gauges are readily available from well-known tool suppliers, or an old injector nut and body can be readily adapted for this purpose. A good machinist can easily make up a dummy-type injector for this also. Figure 12-10 shows the gauge installed ready for the compression check.
5. To install the dummy injector and pressure gauge unit, it is first necessary to remove the fuel jumper pipes from the inlet and outlet of the injector. Place plastic shipping caps over the injector fuel holes. On 149 engines, place plastic shipping caps temporarily over the connections at the inlet and return fuel manifold connections to minimize lube oil dilution from fuel oil spillage. Remove the rocker arm hold-down bolts, and tip the assembly back. Loosen the injector clamp bolt and remove the injector. On 149 engines only, remove the injector rocker arm and install Kent-Moore spacer J 22503-7 in place of it.
6. Install the proper adapter (dummy injector), and clamp it in place with the hose and gauge attached. Using an old fuel

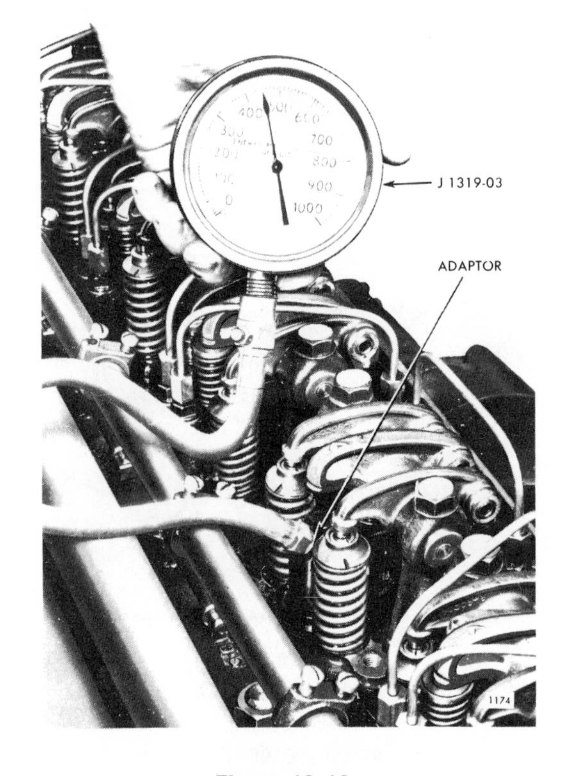

Figure 12-10
Compression Checking Gauge Installed Ready for Testing.
(Courtesy of Detroit Diesel Allison Div. of GMC)

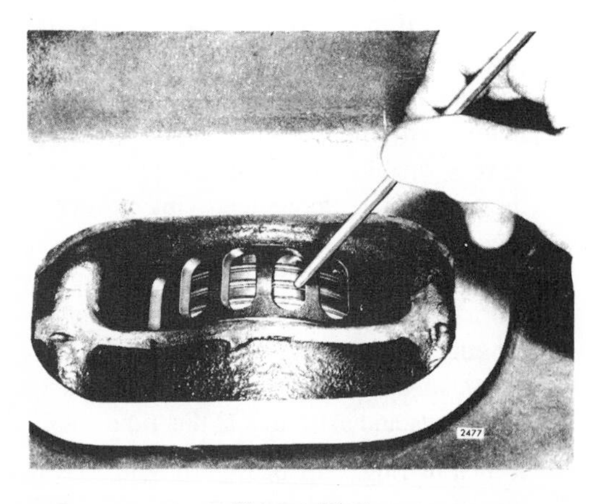

Figure 12-9
Inspecting Condition of Piston Rings.
(Courtesy of Detroit Diesel Allison Div. of GMC)

pipe, connect it between the fuel inlet and return manifold connections (all engines).

7. It is advisable, if at all possible, especially on the larger model engines to use an old rocker cover that has suitable sections cut out of it to facilitate running the engine during the compression test. This will minimize oil throw off.
8. Start the engine and run it at 600 rpm until the pressure on the test gauge reaches its maximum point. Note and record the cylinder pressure. The pressure variation should not exceed 25 psi (172.37 kPa) between cylinders. To determine what the minimum acceptable pressure is for your engine, check the DDA service manual for your particular engine under Section 13, Operating Conditions. There are quite a variety of minimum acceptable standards, which vary with altitude.
9. In addition to stuck or broken rings, compression leakage can also occur at the cylinder head gasket, valve seats, injector tube, and in extreme cases through a cracked or holed piston.

Fuel Flow or Fuel Spill-Back Check

If a lack of power, stalling, or rough running condition complaint exists, an exhaust smoke analysis check can be helpful in determining the possible cause. Black smoke usually indicates air starvation, whereas gray smoke can be either due to excessive or irregular fuel distribution or the improper grade of fuel. White smoke can be caused by a misfiring cylinder or water in the combustion chamber, and blue smoke is usually burning oil.

The injectors can be quickly shorted out as described earlier, and manometer checks can be taken to establish that there is not more than one problem. You may want to check the fuel pump drive quickly by inserting the end of a small wire through one of the pump drain holes as you crank the engine over. Vibration or wire movement will indicate that the shaft is rotating. Sticking of the fuel pump relief valve toward the open or fuel bypass position can create low pump delivery pressure.

You may recollect from earlier discussions on DDA fuel systems that the system is termed a recirculatory one because of the high return of fuel back to the tank. This fuel must first pass through a *restricted fitting,* the size of which will vary among engines and injector size. Each fuel line restricted fitting is stamped with a number that signifies the actual hole size inside it. For example, an R80 fitting is a restricted fitting with an 0.080-in. (2.032-mm) hole size. The purpose and function of this fitting is to restrict the flow of fuel returning from the cylinder head fuel return manifold to ensure a minimum fuel pressure within the cylinder head at the injectors of 30 to 35 psi (206.85 to 241.32 kPa). Normal fuel pressure is between 45 to 70 psi (310 to 483 kPa) from the pump.

To conduct a fuel flow test or what is more commonly referred to as a *fuel spill-back check,* proceed as follows:

1. Check first that you have the correct size of restricted fitting for your model and engine series. This can be found listed in all DDA service manuals, section 13.2.
2. The amount of fuel spill back varies with the restricted fitting size. A general rule of thumb average for fuel spill back on engines employing a standard fuel pump is approximately related to its size. For example, an engine using a 0.055-in. (1.397-mm) restricted fitting should return 0.5 U.S. gallon (1.892 liters) per minute. An 0.080-in. (2.032-mm) fitting should return 0.8 U.S. gallon (3.028 liters) per minute at 1200 rpm or 0.9 U.S. gallon (3.406 liters) per minute at 1800 to 2300 rpm. In other words, if for some reason you did not have specifications readily at hand, by using the basic rules stated you will be able to establish whether or not

you have adequate fuel flow. Remember, however, that 16V series 71 and 92 engines employ two restricted fittings per engine; therefore, you have to double the amount of fuel return as per the fittings size. For example, two 0.070-in. (1.778-mm) fittings would return 1.4 U.S. gallons (5.299 liters) per minute at 1800 to 2100 rpm.

3. Disconnect the fuel return line at a convenient place that will allow you to readily run the fuel into a clean, adequately sized container. See Figure 12-11.

4. You will need a watch with a second hand on it, or if in a shop, a large wall clock with a second hand will do. On nonturbocharged engines a fuel spill-back check is normally taken at an engine speed of 1200 rpm, although it can be checked out at 1800 rpm and beyond to ensure continuity of flow as per specifications. On turbocharged engines, the fuel spill back is normally taken at 1800 rpm, but it can also be taken at the higher rpm ranges as specified under fuel spill back to ensure continuity of fuel flow.

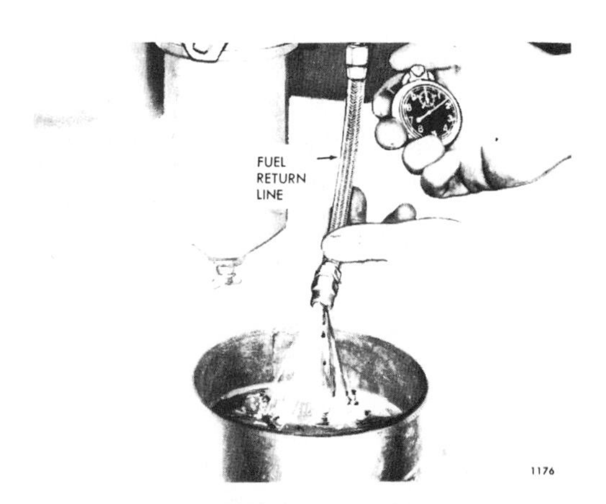

Figure 12-11
Checking Amount of Fuel Return per Minute on a Fuel Spill-Back Check.
(Courtesy of Detroit Diesel Allison Div. of GMC)

5. Start and run the engine at the specified speed for 1 minute, after which you can determine whether or not the system is receiving an adequate supply of fuel. While you are doing this check, immerse the fuel return line into the container to check for any sign of air bubbles rising to the surface. This would indicate that air is being drawn into the fuel system on the suction side of the fuel pump. Check all fuel line connections from the suction side of the pump back to the fuel tank, including the seal ring at the primary filter and at the strainer or fuel water separator if used. Remember, from the outlet or discharge side of the pump the fuel is under pressure; therefore, a fuel leak would occur from here on up to the cylinder head fuel manifold rather than sucking in air. When checking for air bubbles at the container during a fuel spill-back test, ensure that the fuel line is in fact submerged totally. Otherwise, agitation and aeration on the surface of the fuel may lead you to believe that the system is sucking air.

6. If the amount of fuel returned is less than specified in DDA manuals, replace the primary fuel filter, remove the pipe plug from the top of the secondary filter, and install a fuel pressure gauge. Start and run the engine again at 1200 or 1800 rpm as the case may be and measure the amount of fuel returned to the container. Also note what fuel pressure registers on the gauge at the secondary filter. Normal fuel pressure should be between 45 and 70 psi (310.27 and 482.65 kPa).

7. If the fuel return and pressure are still low, replace the secondary filter element and repeat the previous procedure.

8. If low fuel pressure and return still exist, tee-in to the fuel pump outlet; start and run the engine to establish what pressure the pump is producing. If the relief valve within the pump is not stuck in a partially

open position, the pump requires replacing due to wear.

9. If, after replacing the fuel pump, low fuel flow persists, then tee-in a vacuum gauge at the primary filter inlet line. Start and run the engine and note what the maximum restriction to fuel flow is. The maximum allowable on a system with new filters is 6 in. of mercury (mercury) or 12 in. of mercury on a dirty system. Check that the fuel line size is as recommended for your engine as stated by DDA. In addition, if the fuel tank is in excess of 20 ft (6.096 m) away from the pump, the next size of line should be used. Also, if you are lifting the fuel vertically more than 4 ft (1.219 m), you will have to go to a high lift fuel system, as discussed earlier under DDA fuel systems.

Although not a common problem, do *not* neglect checking out the fuel tank for any foreign objects that may very well be blocking the fuel flow. There have been several instances in my own experience, especially around logging equipment, where an engine will run fine until the level in the tank drops low enough to allow a piece of wood chip or bark to be held against the fuel suction line and suddenly create a lack-of-power complaint, rough running, and even stalling.

Another possible problem area can be plugged injector filters. This is not common since most equipment owners usually change their fuel filters on a reasonably steady basis. (See the earlier section on fuel injector filters.) If it is found that the injector filters are in fact plugged, it is advisable to remove them for service and replace them with a matched rebuilt set.

A quick check for plugged filters is to remove the fuel return jumper line from the injector. Install an old line onto the injector, which is bent to take fuel away from the head area. Crank the engine with the starter and note if a steady gush of fuel emanates from the fuel line. If not, the injector filter is plugged.

Engine Runs Out of Fuel

If an engine runs out of fuel, it is strictly due to carelessness on either the equipment operator's part or the maintenance personnel. Downtime caused by this situation can be expensive, especially if it happens on the road or in a remote off-highway location. If you have to restart an engine due to this condition, check the sequence given earlier in this chapter, priming the fuel system.

Matched Injector Sets

When changing a set of injectors, always ensure that you ask for a matched and balanced set. This will give you a set of injectors that produces the same flow rate within close tolerances. The engine will idle smoother and also perform better throughout the speed range.

Appendix

Glossary*

Acceleration smoke limiter. A device which limits the smoke of a diesel engine during acceleration by temporarily limiting the amount of fuel injected into the engine cylinders during speed and/or load transients below the steady-state limit.

Calibration. (1) *Balancing:* The setting of the delivery of an injection system or the setting of the rack pointer on a single unit pump in relation to predetermined positions of a quantity control member. (2) *Adjustment:* Fixing fuel delivery and speed adjustments to specified engine requirements.

Camshaft pump. An injection pump containing a camshaft to operate the pumping element or elements. It can be classified as "in-line," "distributor," "submerged," etc.

Camshaft pump mountings. (1) *Base mounted:* A pump mounted on a surface of the engine which is parallel to the axis of the pump camshaft. (2) *Cradle mounted:* A special form of a "base mount" in which the base is contoured to permit rotation of the pump around the axis of the pump camshaft. (3) *Flange mounted:* A pump mounted on a surface of the engine which is at a right angle to the axis of the pump camshaft.

Closed nozzle. A nozzle incorporating either a poppet valve or a needle valve, loaded in order to open at some predetermined pressure. (1) *Poppet nozzle:* A closed nozzle provided with an outward opening, spring-loaded poppet valve. (2) *Differential nozzle:* A closed nozzle provided with a spring-loaded needle valve. (3) *Pintle nozzle:* A closed nozzle provided with a spring-loaded needle valve. The body of the nozzle has a single large orifice into which enters a projection from the lower end of the needle, this projection being so formed as to influence the rate and shape of the fuel spray. (4) *Hole-type nozzle:* A closed nozzle provided with one or more orifices through which the fuel issues. Nozzles with more than one orifice are known as multihole nozzles.

Control pinion (control sleeve). A collar engaging the plunger and having a segment of gear teeth, integral or attached, which mesh with the control rack. By this means, linear motion of the control rack is transformed into rotary movement of the plunger to regulate the amount of fuel delivered by the pump.

Control rack (control rod). The rack or rod by means of which the fuel delivery is regulated.

Delivery valve assembly. A valve installed in a pump, interposed between the pumping chamber and outlet, to control residual line pressures and which may or may not have an unloading or retraction function.

Delivery valve holder. A device which retains the delivery valve assembly within the pump.

Differential angle. The difference between the angles of the seat face of the valve and that of the seat in the body provided to insure its effective sealing.

Differential ratio. The ratio between the guide diameter of the needle valve and the effective diameter of the needle valve seat.

Distributor pump. An injection pump where each metered delivery is directed to the appropriate engine cylinder by a distributing device.

Dribble. Insufficiently atomized fuel issuing from the nozzle at or immediately following the end of main injection.

Excess fuel device. Any device provided for giving an increased fuel setting for starting only, generally designed to restore automati-

---

*Courtesy of The Society of Automotive Engineers.

cally action of the normal full load stop after starting.

Fuel injection tubing. The tube connecting the injection pump to the nozzle holder assembly.

Fuel pump housing. The main casing into or to which are assembled all the components of the injection pump, and it may accommodate the camshaft in the case of camshaft pumps; or the camshaft, or driveshaft in the case of distributor type pumps.

Full load stop. A device which limits the maximum amount of fuel injected into the engine cylinders at the rated load and speed specified by the engine manufacturer.

Helix hand. The hand of the helix in plungers is designated right or left, the same as a thread.

Helix lead. The axial advance of the helix edge in one revolution.

Helix (scroll). A term used to describe the control edge of a spill groove provided on the plunger, usually of helical form. The helices may be upper or lower or both and may be the same hand or opposite. They can also be duplicated on both sides of the plunger.

Hydraulic governor. A mechanical governor having a hydraulic servo-booster to increase output force.

Hydraulic head assembly. The assembly containing the pumping, metering, and distributing elements (and may include the delivery valve) for distributor-type pumps.

Ignition injection. A small charge of fuel used to ignite the main gas charge in dual fuel engines.

Injection lag. The time interval (usually expressed in degrees of crank angle) between the nominal start of injection pump delivery and the actual start of injection at the nozzle.

Injection pump. The device which meters the fuel and delivers it under pressure to the nozzle and holder assembly.

Injection pump assembly. A complete assembly consisting of the fuel pump proper, together with additional units such as governor, fuel supply pump, and additional optional devices, when these are assembled with the fuel injection pump to form a unit. (1) *Right-hand mounted:* When the pump is mounted on the right-hand side of the engine commonly viewed from the engine flywheel end. (2) *Left-hand mounted:* When the pump is mounted on the left-hand side of the engine commonly viewed from the engine flywheel end.

Injection timing. The matching of the pump timing mark, or the injector timing mechanism, to some index mark on an engine component, such that injection will occur at the proper time with reference to the engine cycle. Injection advance or retard is respectively an earlier, or later, injection pump delivery cycle in reference to the injection cycle.

Inlet metering. A system of metering fuel delivery by controlling the amount of fuel entering the pumping chamber during the filling or charging portion of the pump's cycle.

Inlet valve. A valve used to admit fuel to the pump barrel.

In-line pump. An injection pump with two or more pumping elements arranged in line, each pumping element serving one engine cylinder only. A pump which has the elements arranged in line and in more than one bank, for instance, in two banks forming a "V," is a specific case of an in-line pump.

Leak-off. Fuel which escapes between the nozzle valve and its guide. (This term is also used to describe the leakage past the plunger of a fuel pump.)

Load-sensing governor. An engine speed control device for use on engine-generator sets to control engine fuel settings as a function of electrical load to anticipate resulting changes in engine speed. It may or may not incorporate a mechanical speed-sensing device as well.

Maximum-minimum governor. Any one of the above varieties which exerts control only at the upper and lower limits of the designed engine speed range, intermediate speeds being controlled by the operator setting the fuel delivery directly by throttle action.

Mechanical governor. A speed sensitive device of the centrifugal type, which controls the injection pump delivery solely by mechanical means.

Needle valve (in a closed nozzle). A needle valve has two diameters, the smaller at the valve seat. The fuel injection pressure acting on a portion of the total valve area lifts the valve at the predetermined pressure, then acts on the total area. The end opposite the valve seat is never subjected to injection pressure.

Nozzle. The assembly of parts employed to atomize and deliver fuel to the engine.

Nozzle body. That part of the nozzle which serves as a guide for the valve and in which the actual spray openings may be formed. These two parts, the body and the valve, are considered as a unit for replacement purposes.

Nozzle and holder assembly. The complete apparatus which injects the pressurized fuel into the combustion chamber.

Nozzle holder assembly. The assembly of all parts of the nozzle and holder assembly other than those comprised in the nozzle.

Nozzle holder cap. A cap nut or other type of closure which covers the outer end of the nozzle holder.

Nozzle holder shank length. The distance from the top of the cylindrical shank to the seating face of the nozzle holder.

Nozzle opening pressure. The pressure needed to unseat the nozzle valve.

Nozzle retaining nut. The nozzle holder part which secures the nozzle or nozzle tip to the other nozzle holder parts.

Nozzle tip. The extreme end of the nozzle body containing the spray holes (may be a separate part).

Open nozzle. A nozzle incorporating no valve.

Overspeed governor. A mechanical speed-sensitive device that, through mechanical or electrical action (operation of a switch), acts to shut down the engine and limit the speed by cutting off fuel and/or air supply should the engine speed exceed a preset maximum.

Peak injection pressure. The maximum fuel pressure attained during the injection period (not to be confused with opening pressure).

Pilot injection. A small initial charge of fuel delivered to the engine cylinder in advance of the main delivery of fuel.

Pintle valve (in a closed nozzle). A special type of a ''needle valve'' wherein an integral projection from the lower end of the needle is so formed as to influence the rate and/or shape of the fuel spray during operation.

Plunger and barrel assembly (or plunger and bushing assembly). The combination of a pump plunger and its barrel constituting a pumping element. The plunger and barrel assembly may also perform the additional functions of timing and metering.

Plunger control arm. A lever attached to a collar or sleeve engaging the plunger, or attached directly to the plunger, its other end engaging possibly adjustable fittings on the control rod. This transforms linear motion of the control rod to rotary motion of the plunger to regulate the amount of fuel delivered by the pump.

Pneumatic governor. (1) *Vacuum or suction governor:* One operated by a change in pressure created by the air actually consumed by the engine. (2) *Air governor:* One operated by air displaced by a device provided for this particular purpose and driven by the engine.

Poppet valve. An outwardly opening valve used with certain forms of closed nozzles.

Port closing. A term referring to the fuel injection pump of the port and helix or sleeve metering type in which timing is determined by the point of the closing of the port by the metering member, corresponding to the nominal start of pump delivery.

Port and helix metering. A system of metering fuel delivery by means of one or more helical cuts in the plunger and one or more ports in the barrel. Axial rotation of the plunger alters the effective portion of the stroke by changing the points at which the helices close and/or open the port or ports.

Port opening. A term referring to a fuel injection pump of the port and helix or sleeve metering type in which timing is determined by the point of the opening of the port by the metering member, corresponding to the nominal end of pump delivery.

Pressure adjusting screw (shims). The screw (shims) by means of which the spring load on the nozzle valve is adjusted to obtain the prescribed opening pressures.

Pump rotation. (1) *Clockwise:* The rotation of the pump camshaft or driveshaft is clockwise when viewed from the pump drive end. (2) *Counterclockwise:* The rotation of the pump camshaft or driveshaft is counterclockwise when viewed from the pump drive end.

Retraction volume. The volume of fuel retracted from the high-pressure delivery line by action of the delivery valve's retraction piston in the process of the delivery valve returning to its seat following the end of injection.

Sac hole. The recess immediately within the nozzle tip and acting as a feeder to the spray hole(s) of a hole-type nozzle.

Seating face. The face upon which the nozzle and holder assembly seats to make a gastight seal with the cylinder head. Commonly, this face is on the nozzle retaining nut.

Secondary injection. The fuel discharged from the nozzle as a result of a reopening of the nozzle valve after the main discharge.

Sleeve metering. A system of metering fuel delivery by incorporating a movable sleeve with which port opening and/or port closing is controlled.

Spill valve. A valve used to terminate injection at a controllable point on the pumping stoke by allowing fuel to escape from the pumping chamber.

Spindle. A spindle transmits the load from the spring to the valve.

Spray angle. The included angle of the cone embracing the axis of the several spray holes of a multihole nozzle. In the case of nozzles for large engines, more than one spray angle may be needed to embrace all the sprays; for example, an inner and an outer spray angle.

Spray dispersal angle. The included angle of the cone of fuel leaving any single orifice in the nozzle or tip including pintle type.

Spray inclination angle. The angle which the axis of a cone of spray holes makes with the axis of the nozzle holder.

Spray orifice/orifices. The opening or openings in the end of the nozzle or up through which the fuel is sprayed into the cylinder.

Spring retainer. The spring retainer encloses the spring and carries the adjusting screw or shims.

Submerged pump. A pump with the mounting flange raised to limit pump projection above the mounting face.

Supply pump. A pump for transferring the fuel from the tank and delivering it to the injection pump.

Tailshaft governor. A mechanical speed-sensitive device commonly mounted on an engine driven torque convertor to monitor its tailshaft speed. It is mechanically connected to the normal engine governor such that engine output will be governed to maintain a constant tailshaft speed regardless of torque load.

Timing device. A device responsive to engine speed and/or load to control the timed relationship between injection cycle and engine cycle.

Torque control. A device which modifies the maximum amount of fuel injected into the engine cylinders at speeds below rated speed to obtain the desired torque output.

Unit fuel injector. An assembly which receives fuel under supply pressure and is then actuated by an engine mechanism to meter and inject the charge of fuel to the combustion chamber at high pressure and at the proper time.

Unit pump. An injection pump containing no actuating mechanism to operate the pumping element or elements. It can be classified as "in-line," "distributor," "submerged," etc.

Index

A

B

C

D

E

F

G

H

I

J

K

L

M

N

O

P

R

S

T

U

V

W

Y

Z